A Commemorative Issue in Honor of Centennial of the Discovery of Vitamin D-The Central Role of Vitamin D in Physiology

A Commemorative Issue in Honor of Centennial of the Discovery of Vitamin D-The Central Role of Vitamin D in Physiology

Editor

Carsten Carlberg

MDPI • Basel • Beijing • Wuhan • Barcelona • Belgrade • Manchester • Tokyo • Cluj • Tianjin

Editor
Carsten Carlberg
University of Eastern Finland
Finland

Editorial Office
MDPI
St. Alban-Anlage 66
4052 Basel, Switzerland

This is a reprint of articles from the Special Issue published online in the open access journal *Nutrients* (ISSN 2072-6643) (available at: https://www.mdpi.com/journal/nutrients/special_issues/commemorative_vitamin_Dr).

For citation purposes, cite each article independently as indicated on the article page online and as indicated below:

LastName, A.A.; LastName, B.B.; LastName, C.C. Article Title. *Journal Name* **Year**, *Volume Number*, Page Range.

ISBN 978-3-0365-6918-5 (Hbk)
ISBN 978-3-0365-6919-2 (PDF)

Contents

About the Editor . . . . **vii**

Preface to "A Commemorative Issue in Honor of Centennial of the Discovery of Vitamin D-The Central Role of Vitamin D in Physiology" . . . . **ix**

Carsten Carlberg
A Pleiotropic Nuclear Hormone Labelled Hundred Years Ago Vitamin D
Reprinted from: *Nutrients* **2023**, *15*, 171, doi:10.3390/nu15010171 . . . . **1**

Michael F. Holick
The One-Hundred-Year Anniversary of the Discovery of the Sunshine Vitamin D_3: Historical, Personal Experience and Evidence-Based Perspectives
Reprinted from: *Nutrients* **2023**, *15*, 593, doi:10.3390/nu15030593 . . . . **5**

Carsten Carlberg
Vitamin D in the Context of Evolution
Reprinted from: *Nutrients* **2022**, *14*, 3018, doi:10.3390/nu14153018 . . . . **27**

Shahid Hussain, Clayton Yates and Moray J. Campbell
Vitamin D and Systems Biology
Reprinted from: *Nutrients* **2022**, *14*, 5197, doi:10.3390/nu14245197 . . . . **39**

Carsten Carlberg
Vitamin D and Its Target Genes
Reprinted from: *Nutrients* **2022**, *14*, 1354, doi:10.3390/nu14071354 . . . . **55**

Natacha Rochel
Vitamin D and Its Receptor from a Structural Perspective
Reprinted from: *Nutrients* **2022**, , 2847, doi:10.3390/nu14142847 . . . . **69**

Miguel A. Maestro and Samuel Seoane
The Centennial Collection of VDR Ligands: Metabolites, Analogs, Hybrids and Non-Secosteroidal Ligands
Reprinted from: *Nutrients* **2022**, *14*, 4927, doi:10.3390/nu14224927 . . . . **83**

Michał A. Żmijewski
Nongenomic Activities of Vitamin D
Reprinted from: *Nutrients* **2022**, *14*, 5104, doi:10.3390/nu14235104 . . . . **137**

James C. Fleet
Vitamin D-Mediated Regulation of Intestinal Calcium Absorption
Reprinted from: *Nutrients* **2022**, *14*, 3351, doi:10.3390/nu14163351 . . . . **157**

René St-Arnaud, Alice Arabian, Dila Kavame, Martin Kaufmann and Glenville Jones
Vitamin D and Diseases of Mineral Homeostasis: A *Cyp24a1* R396W Humanized Preclinical Model of Infantile Hypercalcemia Type 1
Reprinted from: *Nutrients* **2022**, *14*, 3221, doi:10.3390/nu14153221 . . . . **171**

Marjolein van Driel and Johannes P. T. M. van Leeuwen
Vitamin D and Bone: A Story of Endocrine and Auto/Paracrine Action in Osteoblasts
Reprinted from: *Nutrients* **2023**, *15*, 480, doi:10.3390/nu15030480 . . . . **183**

Nejla Latic and Reinhold G. Erben
Interaction of Vitamin D with Peptide Hormones with Emphasis on Parathyroid Hormone, FGF23, and the Renin-Angiotensin-Aldosterone System
Reprinted from: *Nutrients* **2022**, *14*, 5186, doi:10.3390/nu14235186 . . . . **201**

Guillaume T. Duval, Anne-Marie Schott, Dolores Sánchez-Rodríguez, François R. Herrmann and Cédric Annweiler
Month-of-Birth Effect on Muscle Mass and Strength in Community-Dwelling Older Women: The French EPIDOS Cohort
Reprinted from: *Nutrients* **2022**, *14*, 4874, doi:10.3390/nu14224874 . . . . **219**

Shariq Qayyum, Radomir M. Slominski, Chander Raman and Andrzej T. Slominski
Novel CYP11A1-Derived Vitamin D and Lumisterol Biometabolites for the Management of COVID-19
Reprinted from: *Nutrients* **2022**, *14*, 4779, doi:10.3390/nu1422477 . . . . **231**

Juhi Arora, Devanshi R. Patel, McKayla J. Nicol, Cassandra J. Field, Katherine H. Restori, Jinpeng Wang, Nicole E. Froelich, et al.
Vitamin D and the Ability to Produce $1,25(OH)_2D$ Are Critical for Protection from Viral Infection of the Lungs
Reprinted from: *Nutrients* **2022**, *14*, 3061, doi:10.3390/nu14153061 . . . . **245**

Hadeil M. Alsufiani, Shareefa A. AlGhamdi, Huda F. AlShaibi, Sawsan O. Khoja, Safa F. Saif and Carsten Carlberg
A Single Vitamin D_3 Bolus Supplementation Improves Vitamin D Status and Reduces Proinflammatory Cytokines in Healthy Females
Reprinted from: *Nutrients* **2022**, *14*, 3963, doi:10.3390/nu14193963 . . . . **261**

Alberto Muñoz and William B. Grant
Vitamin D and Cancer: An Historical Overview of the Epidemiology and Mechanisms
Reprinted from: *Nutrients* **2022**, *14*, 1448, doi:10.3390/nu14071448 . . . . **271**

Karina Piatek, Martin Schepelmann and Enikö Kallay
The Effect of Vitamin D and Its Analogs in Ovarian Cancer
Reprinted from: *Nutrients* **2022**, *14*, 3867, doi:10.3390/nu14183867 . . . . **313**

Ewa Marcinkowska
Vitamin D Derivatives in Acute Myeloid Leukemia: The Matter of Selecting the Right Targets
Reprinted from: *Nutrients* **2022**, *14*, 2851, doi:10.3390/nu14142851 . . . . **323**

Xiaoying Cui and Darryl W. Eyles
Vitamin D and the Central Nervous System: Causative and Preventative Mechanisms in Brain Disorders
Reprinted from: *Nutrients* **2022**, *14*, 4353, doi:10.3390/nu14204353 . . . . **337**

William B. Grant, Barbara J. Boucher, Fatme Al Anouti and Stefan Pilz
Comparing the Evidence from Observational Studies and Randomized Controlled Trials for Nonskeletal Health Effects of Vitamin D
Reprinted from: *Nutrients* **2022**, *14*, 3811, doi:10.3390/nu1418381 . . . . **357**

Elina Hyppönen, Karani S. Vimaleswaran and Ang Zhou
Genetic Determinants of 25-Hydroxyvitamin D Concentrations and Their Relevance to Public Health
Reprinted from: *Nutrients* **2022**, *14*, 4408, doi:10.3390/nu14204408 . . . . **389**

About the Editor

Carsten Carlberg

Carsten Carlberg graduated in 1989 with a PhD in biochemistry from the Free University Berlin (Germany). After holding positions as a postdoc at Roche (Basel, Switzerland), a group leader at the University of Geneva (Switzerland), and a docent at the University of Düsseldorf (Germany) he is, since 2000, a full professor of biochemistry at the University of Eastern Finland in Kuopio (Finland) and, since 2022, ERA chair for nutrigenomics at the Polish Academy of Sciences in Olsztyn (Poland). His work focuses on mechanisms of gene regulation by nuclear hormones, in particular on vitamin D. At present, Prof. Carlberg focuses projects on the epigenome-wide effects of vitamin D on the human immune system in the context of cancer.

Preface to "A Commemorative Issue in Honor of Centennial of the Discovery of Vitamin D-The Central Role of Vitamin D in Physiology"

In 2022, we celebrated the 100 year anniversary of the naming of vitamin D by Dr. McCollum. At the beginning of the last century, small molecules were found to cure several diseases caused by nutritional deficiencies and were, therefore, termed vitamins. Vitamin D received the index "D" because it was the fourth vitamin named after vitamins called A, B, and C. The specific purpose of vitamin D was described as the protection of bone growth and the prevention of rickets. In the last 100 years, tremendous advances in the understanding of the physiological role and mechanisms of action have been achieved. Vitamin D2 was identified in 1932 and vitamin D3 was identified in 1937, but it took until 1978 to prove that vitamin D3 is synthesized in human skin after UV-B irradiation. At the beginning of the 1960s, 1,25(OH)2D3 (calcitriol) was found to be a biologically active form of vitamin D. The vitamin D receptor (VDR) was isolated in 1969, and its gene was cloned in 1987. This was the basis for understanding vitamin D as a direct regulator of hundreds of genes in various tissues. VDR's ligand-binding domain was crystalized in 2000, and since 2010, we know VDR's genome-wide binding pattern in various cellular models. In parallel, 1,25(OH)2D3 and its synthetic analogues have been known since the early 1980s as regulators of monocyte differentiation during hematopoiesis as well as inhibitors of cellular proliferation in various in vitro and in vivo models. Today, it is obvious that an appropriate vitamin D status is essential for maintaining the health and well-being of everyone. Thus, vitamin D3 does not only regulate our calcium level in order to allow proper bone formation, but it also modulates our immune system and, in this way, indirectly protects us from cancer.

This vitamin D centenary issue provides a fantastic opportunity to present recent developments and the latest research in the fascinating broad range of today's vitamin D biology, from evolution to systems biology. In 21 articles, leading authors in the field presented original work and review articles to the whole biomedical community.

Carsten Carlberg
Editor

Editorial

A Pleiotropic Nuclear Hormone Labelled Hundred Years Ago Vitamin D

Carsten Carlberg [1,2]

[1] Institute of Animal Reproduction and Food Research, Polish Academy of Sciences, PL-10-748 Olsztyn, Poland; c.carlberg@pan.olsztyn.pl

[2] School of Medicine, Institute of Biomedicine, University of Eastern Finland, FI-70211 Kuopio, Finland

This year we are celebrating 100 years of the naming of vitamin D, but the molecule is, in fact, more than one billion years old [1]. *Nutrition Bulletin* and *Endocrine Connections* are also honoring the centenary of vitamin D's discovery, but with 21 original publications and review articles written by experts in the field, *Nutrients* represents the largest collection [1–21].

At the beginning of the last century, small molecules were found to cure several diseases caused by nutritional deficiencies and were, therefore, termed vitamins. These diseases are xerophthalmia (a clinical spectrum ranging from night blindness to complete blindness), anemias, and neurological disorders, such as beriberi and scurvy (a disability affecting the repair of bone, skin, and connective tissue), which can be healed and prevented using the supplementation of vitamins A, B and C, respectively. Thus, when McCollum and colleagues demonstrated in 1922 that experimentally induced rickets in rats could be cured by a factor isolated from cod liver oil, they followed the nomenclature termed and named it vitamin D [22].

The term "vitamin" implies that these molecules should be regularly supplied as part of our diet or as pills. However, some vitamins can be endogenously produced by our bodies, such as vitamin D_3 in UV-B-exposed skin [23]. Thus, changes in our lifestyle, such as spending more time indoors, wearing clothes outdoors, and living at latitudes with seasonal variations of UV-B intensity, made vitamin D a vitamin. The insufficient production or supplementation of vitamin D_3 causes a low vitamin D status, which is determined by the serum concentration of the most stable vitamin D_3 metabolite, 25-hydroxyvitamin D_3 (25(OH)D_3). Severe vitamin D deficiency, defined as 25(OH)D_3 serum levels below 30 nM (12 ng/mL), can lead to bone malformations, such as rickets in children and osteomalacia in adults [24], and at all ages, lead to a malfunctioning immune system [25]. The latter may increase the risk for severe consequences of infectious diseases, such as COVID-19 (coronavirus disease) [26] or tuberculosis [27], as well as for the onset and progression of autoimmune diseases, such as type 1 diabetes [28] and multiple sclerosis [29].

Vitamin D and vitamin A differ from other vitamins due to the interesting property that a few metabolic steps can convert them into nuclear hormones that bind with high affinity to members of the nuclear receptor superfamily [17]. In the case of vitamin D_3, this is the metabolite 1,25-dihydroxyvitamin D_3 (1,25(OH)$_2D_3$) activating the transcription factor VDR (vitamin D receptor) [4]. In fact, 1,25(OH)$_2D_3$ binds to VDR with a K_D of 0.1 nM, which is a significantly higher affinity than that of other nuclear receptors for their specific ligands. This suggests that the genomic signaling of 1,25(OH)$_2D_3$ via VDR is the predominant means of vitamin D's mechanism of action [30]. However, there are also indications for non-genomic vitamin D signaling [19] as well as the biological activity of vitamin D metabolites other than 1,25(OH)$_2D_3$ [16].

In genomic signaling, in all VDR-expressing tissues and cell types, vitamin D controls the transcription of hundreds of target genes [4]. The Genotype-Tissue Expression (GTEx) project provides gene expression data from 54 tissues obtained from 948 post-mortem

Citation: Carlberg, C. A Pleiotropic Nuclear Hormone Labelled Hundred Years Ago Vitamin D. *Nutrients* **2023**, *15*, 171. https://doi.org/10.3390/nu15010171

Received: 15 December 2022
Accepted: 21 December 2022
Published: 30 December 2022

donors [31]. The selected tissues are representative of more than 400 tissues and cell types that constitute our body. At present, the publicly available dataset (https://gtexportal.org) (accessed on 12 December 2022) is the gold standard for comparing tissue-specific gene expression. The expression of the *VDR* gene is highest in the skin, intestines, and colon and lowest in different regions of the brain (Figure 1). In other tissues, such as the blood and kidneys, the *VDR* gene shows intermediate levels of expression. The shortcut interpretation of this gene expression panel is that vitamin D, via the activation of VDR, primarily impacts the tissue of its synthesis and that its main action is in the gut, while in the brain, it may have no direct function. However, one has to distinguish the role of vitamin D as a controller of calcium transport in the gut [7] from its regulatory function, e.g., in the immune system [3]. Therefore, immune cells may not need as many VDR proteins as intestinal cells.

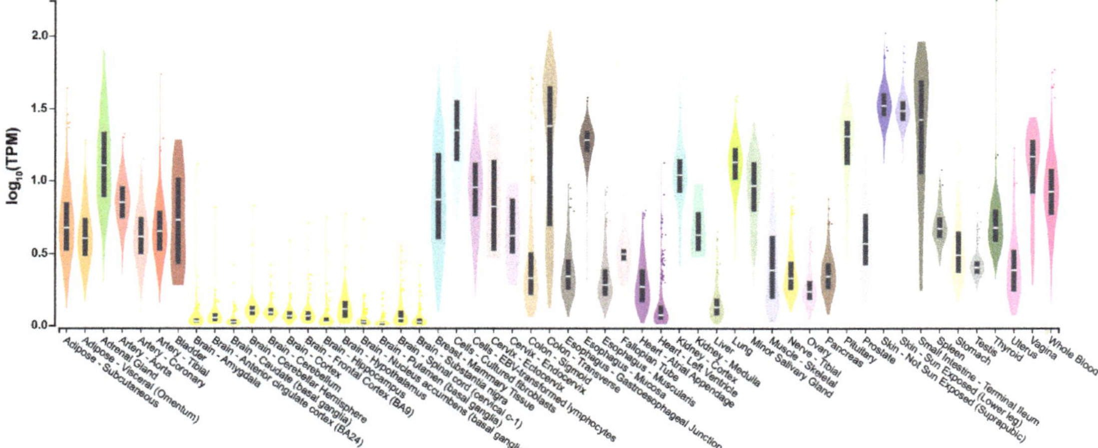

Figure 1. Expression of the *VDR* gene in 54 different human tissues. Normalized RNA sequencing (RNA-seq) data are shown in TPM (transcripts per million), where all isoforms were collapsed into a single gene. Box plots display the median as well as 25th and 75th percentiles. Points indicate outliers that are 1.5 times above or below interquartile range. Data are based on GTEx analysis release V8 (dbGaP Accession phs000424.v8.p2) [31].

From an evolutionary perspective, the calcium-absorbing and bone-remodeling function of vitamin D (Figure 2) was obtained less than 400 million years ago, when species left the ocean and needed a stable skeleton [1]. Thus, the best-known role of vitamin D was not why evolution created vitamin D endocrinology. However, regarding this physiological function, VDR and its ligand became dominant regulators [7,20]. In this context, vitamin D learned to control the expression of parathyroid hormone (PTH) and fibroblast growth factor 23 (FGF23) [11]. The peptide hormones are expressed in the parathyroid gland and osteocytes, respectively, and up- and down-regulate the production of $1,25(OH)_2D_3$ (Figure 2).

Most likely, the first function of VDR was the regulation of energy metabolism [32]. Energy is essential for basically all biological processes, but particularly for the function of innate and adaptive immunity [33]. Vitamin D and its receptor obtained via the control of immunometabolism a modulatory role on immunity [34] (Figure 2).

Taken together, despite the story of its naming, vitamin D is not only a vitamin that prevents bone malformations. In contrast, via $1,25(OH)_2D_3$, vitamin D is a direct regulator of gene expression, resulting in pleiotropic physiological functions.

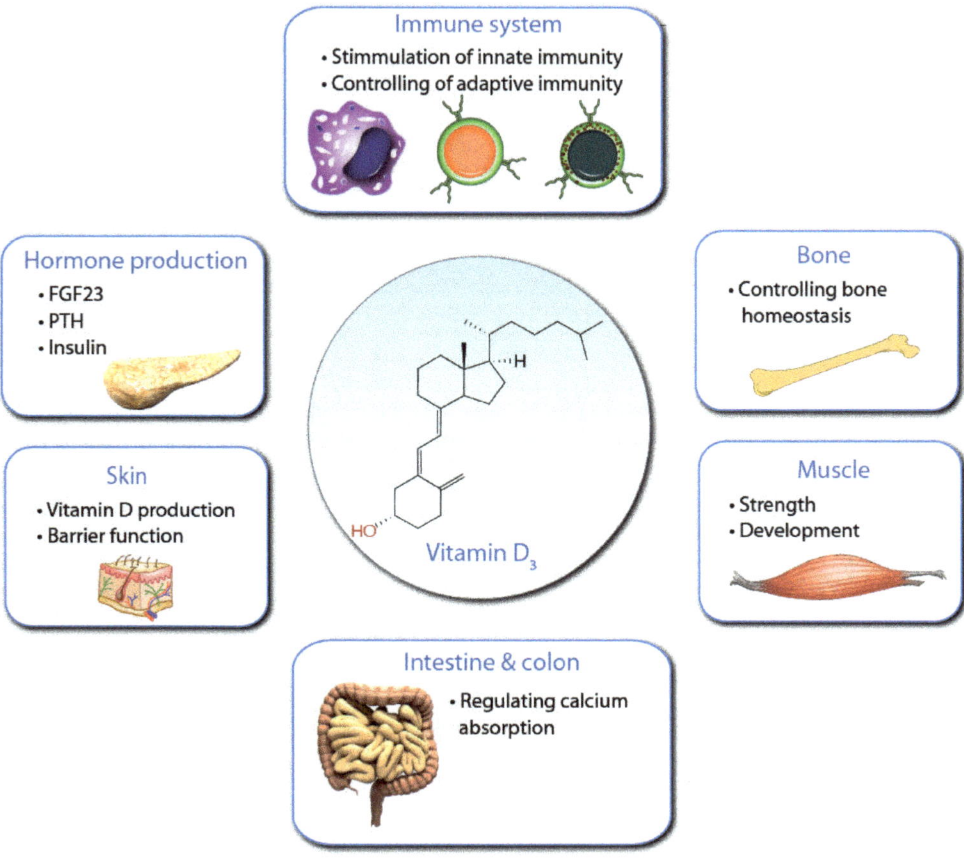

Figure 2. Pleiotropic physiological functions of vitamin D. Vitamin D_3 (center) can be produced endogenously in UV-B exposed skin. Via its metabolite $1,25(OH)_2D_3$, vitamin D_3 activates VDR in various tissues and cell types, where it regulates the indicated major physiological functions.

Funding: This publication is part of a project that has received funding from the European Union's Horizon2020 research and innovation program under grant agreement no. 952601.

Conflicts of Interest: The author declares no conflict of interest.

References

1. Carlberg, C. Vitamin D in the context of evolution. *Nutrients* **2022**, *14*, 3018. [CrossRef]
2. Alsufiani, H.M.; AlGhamdi, S.A.; AlShaibi, H.F.; Khoja, S.O.; Saif, S.F.; Carlberg, C. A single vitamin D_3 bolus supplementation improves vitamin D status and reduces proinflammatory cytokines in healthy females. *Nutrients* **2022**, *14*, 3963. [CrossRef]
3. Arora, J.; Patel, D.R.; Nicol, M.J.; Field, C.J.; Restori, K.H.; Wang, J.; Froelich, N.E.; Katkere, B.; Terwilliger, J.A.; Weaver, V.; et al. Vitamin D and the ability to produce $1,25(OH)_2D$ are critical for protection from vral infection of the lungs. *Nutrients* **2022**, *14*, 3061. [CrossRef]
4. Carlberg, C. Vitamin D and its target genes. *Nutrients* **2022**, *14*, 1354. [CrossRef]
5. Cui, X.; Eyles, D.W. Vitamin D and the central nervous system: Causative and preventative mechanisms in brain disorders. *Nutrients* **2022**, *14*, 4353. [CrossRef]
6. Duval, G.T.; Schott, A.M.; Sanchez-Rodriguez, D.; Herrmann, F.R.; Annweiler, C. Month-of-birth effect on muscle mass and srength in community-dwelling older women: The French EPIDOS cohort. *Nutrients* **2022**, *14*, 4874. [CrossRef]
7. Fleet, J.C. Vitamin D-mediated regulation of intestinal calcium absorption. *Nutrients* **2022**, *14*, 3351. [CrossRef]
8. Grant, W.B.; Boucher, B.J.; Al Anouti, F.; Pilz, S. Comparing the evidence from observational studies and randomized controlled trials for nonskeletal health effects of vitamin D. *Nutrients* **2022**, *14*, 3811. [CrossRef]

9. Hussain, S.; Yates, C.; Campbell, M.J. Vitamin D and systems biology. *Nutrients* **2022**, *14*, 5197. [CrossRef]
10. Hypponen, E.; Vimaleswaran, K.S.; Zhou, A. Genetic determinants of 25-hydroxyvitamin D concentrations and their relevance to public health. *Nutrients* **2022**, *14*, 4408. [CrossRef]
11. Latic, N.; Erben, R.G. Interaction of vitamin D with peptide hormones with emphasis on parathyroid hormone, FGF23, and the renin-angiotensin-aldosterone system. *Nutrients* **2022**, *14*, 5186. [CrossRef] [PubMed]
12. Maestro, M.A.; Seoane, S. The centennial collection of VDR ligands: Metabolites, analogs, hybrids and non-secosteroidal ligands. *Nutrients* **2022**, *14*, 4927. [CrossRef] [PubMed]
13. Marcinkowska, E. Vitamin D derivatives in acute myeloid leukemia: The matter of selecting the right targets. *Nutrients* **2022**, *14*, 2851. [CrossRef] [PubMed]
14. Munoz, A.; Grant, W.B. Vitamin D and cancer: An historical overview of the epidemiology and mechanisms. *Nutrients* **2022**, *14*, 1448. [CrossRef] [PubMed]
15. Piatek, K.; Schepelmann, M.; Kallay, E. The effect of vitamin D and its analogs in ovarian cancer. *Nutrients* **2022**, *14*, 3867. [CrossRef] [PubMed]
16. Qayyum, S.; Slominski, R.M.; Raman, C.; Slominski, A.T. Novel CYP11A1-derived vitamin D and lumisterol biometabolites for the management of COVID-19. *Nutrients* **2022**, *14*, 4779. [CrossRef] [PubMed]
17. Rochel, N. Vitamin D and its receptor from a structural perspective. *Nutrients* **2022**, *14*, 2847. [CrossRef]
18. St-Arnaud, R.; Arabian, A.; Kavame, D.; Kaufmann, M.; Jones, G. Vitamin D and diseases of mineral homeostasis: A Cyp24a1 R396W humanized preclinical model of infantile hypercalcemia type 1. *Nutrients* **2022**, *14*, 3221. [CrossRef]
19. Żmijewski, M.A. Nongenomic activities of vitamin D. *Nutrients* **2022**, *14*, 5104. [CrossRef]
20. van Driel, M.; van Leeuwen, J.P.T.M. Vitamin D and osteoblasts: An endo/auto/paracrine story. *Nutrients* **2022**, *14*. under review.
21. Holick, M. The one hundred year anniversary of the discovery of the sunshine vitamin D: Historical, personal and evidence based perspectives. *Nutrients* **2023**, *15*. under review. [CrossRef]
22. McMollum, E.V.; Simmonds, N.; Becker, J.E.; Shipley, P.G. Studies on experimental rickets: An experimental demonstration of the existence of a vitamin which promotes calcium deposition. *J. Biol. Chem.* **1922**, *52*, 293–298. [CrossRef]
23. Holick, M.F. Photobiology of vitamin D. In *Vitamin D*, 3rd ed.; Elsevier: Amsterdam, The Netherlands, 2011; pp. 13–22. [CrossRef]
24. Holick, M.F. Resurrection of vitamin D deficiency and rickets. *J. Clin. Investig.* **2006**, *116*, 2062–2072. [CrossRef] [PubMed]
25. Malmberg, H.R.; Hanel, A.; Taipale, M.; Heikkinen, S.; Carlberg, C. Vitamin D treatment sequence is critical for transcriptome modulation of immune challenged primary human cells. *Front. Immunol.* **2021**, *12*, 754056. [CrossRef] [PubMed]
26. Bilezikian, J.P.; Bikle, D.; Hewison, M.; Lazaretti-Castro, M.; Formenti, A.M.; Gupta, A.; Madhavan, M.V.; Nair, N.; Babalyan, V.; Hutchings, N.; et al. Vitamin D and COVID-19. *Eur. J. Endocrinol.* **2020**, *183*, R133–R147. [CrossRef]
27. Sita-Lumsden, A.; Lapthorn, G.; Swaminathan, R.; Milburn, H.J. Reactivation of tuberculosis and vitamin D deficiency: The contribution of diet and exposure to sunlight. *Thorax* **2007**, *62*, 1003–1007. [CrossRef]
28. Hypponen, E.; Laara, E.; Reunanen, A.; Jarvelin, M.R.; Virtanen, S.M. Intake of vitamin D and risk of type 1 diabetes: A birth-cohort study. *Lancet* **2001**, *358*, 1500–1503. [CrossRef]
29. Munger, K.L.; Levin, L.I.; Hollis, B.W.; Howard, N.S.; Ascherio, A. Serum 25-hydroxyvitamin D levels and risk of multiple sclerosis. *JAMA* **2006**, *296*, 2832–2838. [CrossRef]
30. Carlberg, C. Genome-wide (over)view on the actions of vitamin D. *Front. Physiol.* **2014**, *5*, 167. [CrossRef]
31. Kim-Hellmuth, S.; Aguet, F.; Oliva, M.; Munoz-Aguirre, M.; Kasela, S.; Wucher, V.; Castel, S.E.; Hamel, A.R.; Vinuela, A.; Roberts, A.L.; et al. Cell type-specific genetic regulation of gene expression across human tissues. *Science* **2020**, *369*, 6589. [CrossRef]
32. Hanel, A.; Carlberg, C. Vitamin D and evolution: Pharmacologic implications. *Biochem. Pharmacol.* **2020**, *173*, 113595. [CrossRef]
33. Muller, V.; de Boer, R.J.; Bonhoeffer, S.; Szathmary, E. An evolutionary perspective on the systems of adaptive immunity. *Biol. Rev. Camb. Philos. Soc.* **2018**, *93*, 505–528. [CrossRef] [PubMed]
34. Vanherwegen, A.S.; Gysemans, C.; Mathieu, C. Vitamin D endocrinology on the cross-road between immunity and metabolism. *Mol. Cell Endocrinol.* **2017**, *453*, 52–67. [CrossRef] [PubMed]

MDPI

Review

The One-Hundred-Year Anniversary of the Discovery of the Sunshine Vitamin D_3: Historical, Personal Experience and Evidence-Based Perspectives

Michael F. Holick

Department of Medicine, Boston University Chobanian & Avedisian School of Medicine, Boston, MA 02118, USA; mfholick@bu.edu; Tel.: +1-617-358-6139

Abstract: The discovery of a fat-soluble nutrient that had antirachitic activity and no vitamin A activity by McCollum has had far reaching health benefits for children and adults. He named this nutrient vitamin D. The goal of this review and personal experiences is to give the reader a broad perspective almost from the beginning of time for how vitamin D evolved to became intimately involved in the evolution of land vertebrates. It was the deficiency of sunlight causing the devastating skeletal disease known as English disease and rickets that provided the first insight as to the relationship of sunlight and the cutaneous production of vitamin D_3. The initial appreciation that vitamin D could be obtained from ultraviolet exposure of ergosterol in yeast to produce vitamin D_2 resulted in the fortification of foods with vitamin D_2 and the eradication of rickets. Vitamin D_3 and vitamin D_2 (represented as D) are equally effective in humans. They undergo sequential metabolism to produce the active form of vitamin D, 1,25-dihydroxyvitamin D. It is now also recognized that essentially every tissue and cell in the body not only has a vitamin D receptor but can produce 1,25-dihydroxyvitamin D. This could explain why vitamin D deficiency has now been related to many acute and chronic illnesses, including COVID-19.

Keywords: vitamin D; McCollum; sunlight; rickets; 25-hydroxyvitamin D; 1,25-dihydroxyvitamin D; food fortification; vitamin D receptor; vitamin D_2; vitamin D_3; COVID-19

Citation: Holick, M.F. The One-Hundred-Year Anniversary of the Discovery of the Sunshine Vitamin D_3: Historical, Personal Experience and Evidence-Based Perspectives. *Nutrients* **2023**, *15*, 593. https://doi.org/10.3390/nu15030593

Academic Editor: Carsten Carlberg

Received: 25 December 2022
Revised: 18 January 2023
Accepted: 19 January 2023
Published: 23 January 2023

1. Prequel

The sunshine vitamin D is likely to be the oldest hormone to be produced almost from the beginning of time when eukaryotes began to evolve in the oceans more than 1–1/2 billion years ago. The first evidence for this was the observation that a phytoplankton species *Emiliania huxlei*, which has existed for more than 750 million years unchanged in the Sargasso Sea, contained ergosterol (provitamin D_2; 1 mcg/g dry weight), which was converted to previtamin D_2 when exposed to simulated sunlight. Why these early eukaryotic organisms would contain such a large amount of ergosterol and produce vitamin D_2 is unknown. It has been speculated that early photosynthetic eukaryotes that required sun exposure for their energy source were also at risk of being exposed to ultraviolet radiation (UV) that could damage their precious DNA, RNA and proteins. The UV absorption spectra for provitamin D_2, previtamin D_2 and vitamin D_2 have high-extinction coefficients that completely overlap the UV-absorption spectra for these UV-sensitive macromolecules, thereby protecting them from solar ultraviolet radiation damage. It was also suggested that ergosterol in the plasma membrane not only served as an ideal sunscreen but also, upon exposure to ultraviolet radiation, was converted to previtamin D_2. Previtamin D_2, which is a planar structure embedded into the membrane, is thermally unstable and isomerizes into a non-planar vitamin D_2 structure. This process would have caused a significant structural change in the bilipid layer of the plasma membrane, resulting in vitamin D_2 being released into the environment. This process would have opened a tiny pore in the membrane that could have permitted the entrance of calcium and other cations into the cell. This could help explain why sunlight, vitamin D and calcium are so intimately related. When vertebrate life forms left the oceans for terra firma, they were leaving a calcium-rich environment into a poor calcium environment. It is possible that the association of the production of vitamin D_2 in the plasma membrane of a phytoplankton, permitting the entrance of calcium into the cell, may have been taken advantage of in land vertebrates. During exposure to sunlight, land vertebrates would have produced vitamin D_3 in their skin, which would have entered their circulation and increased the efficiency of calcium absorption. Although it is unknown when vitamin D_3 was first metabolized and activated, it is likely that this occurred as marine vertebrates began to leave the oceans and explore and ultimately dominate terra firma. Most vertebrates require sunlight for their vitamin D requirement to maintain a healthy skeleton. Some species such as cats, other heavily furred vertebrates, and mole rats obtain their vitamin D from dietary sources. Fish-eating bats and vampire bats easily obtain an adequate amount of vitamin D from their diet, and recently, it was reported that insects, including mosquitoes exposed to sunlight, produce vitamin D_3 and are excellent dietary sources for insect-eating bats [1–4].

2. The Discovery of the Sunshine Vitamin D

By the 16th century, there was a migration of the population into crowded and air-polluted cities (Figure 1). Simultaneous with this migration was the appearance of a devastating crippling bone-deforming disorder that affected most of the children.

Figure 1. Photograph from Glasgow, Great Britain, in about 1870 showing that the buildings are built near each other. Ref. [5] Holick, copyright 1994. Reproduced with permission.

In 1650, Glisson, DeBoot and Whistler published their observations and concluded that this was a distinct disease that caused deformities of the skeleton, including enlargement of the joints of the long bones and rib cage, curvature of the spine and thighs, short stature, generalized muscle weakness and enlargement of the head (Figure 2). This disease, commonly known as rickets or English disease, continued to spread throughout European cities, and when the industrial revolution arrived in the 19th century, rickets became endemic in the northeastern United States. By the 20th century, it was estimated based on autopsy studies performed in Boston and Leiden that between 80–90% of children living in industrialized cities in Europe and the United States were suffering from this catastrophic bone-softening skeletal disorder. Besides causing bone deformities and growth retardation, vitamin D deficiency also gave rise to a flattened deformed pelvis with a small pelvic outlet in females of childbearing age. As a result, young women had a difficult time with child birthing, often experiencing serious complications including death of both the mom and newborn. In 1889, Murdock Cameron, a Scottish surgeon, began routinely performing C-sections to save both the mother and infant from a prolonged, difficult, and life-threatening childbirth [6–8].

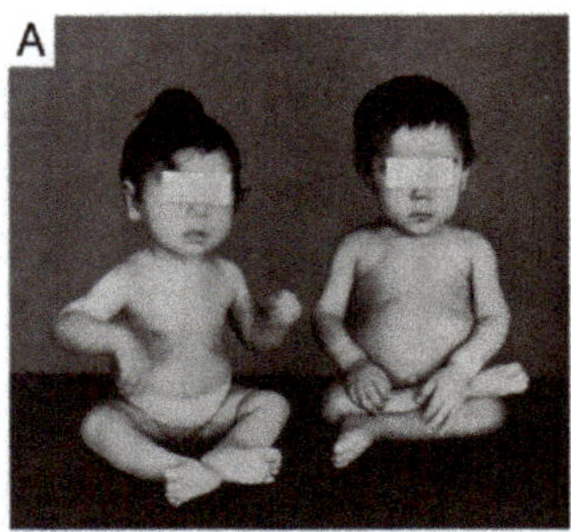

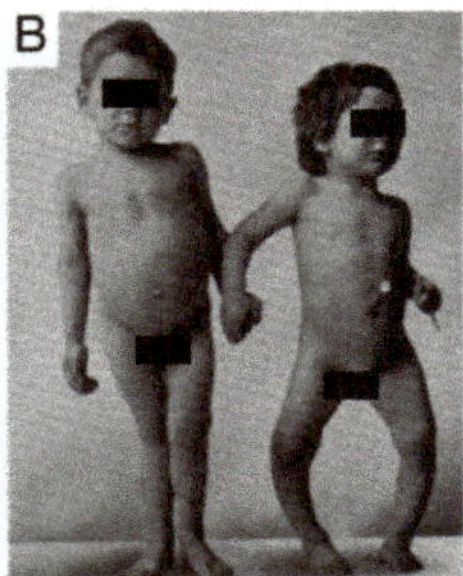

Figure 2. Skeletal deformities observed in rickets. (**A**) Photograph from the 1930s of a sister (left) and brother (right), aged 10 mo and 2.5 y, respectively, showing enlargement of the ends of the bones at the wrist, carpopedal spasm, and a typical "Taylorwise" posture of rickets. (**B**) The same brother and sister 4 y later, with classic knock knees (genu valgum) and bowed legs (genu varum), growth retardation, and other skeletal deformities. Ref. [7] Holick, copyright 2006. Reproduced with permission.

The first association relating inadequate sunlight exposure as a cause for rickets was made by the Polish physician scientist Sniadecki in 1822 [9]. He connected the dots and realized that children living in industrialized Warsaw who were not exposed to direct sunlight were plagued with rickets, whereas children who lived in rural farm areas and were outdoors and exposed to sunlight had no evidence of this bone-deforming disease. This was followed by Palm who in 1890 recognized that the lack of sunlight was a common denominator that could be associated with the high incidence of rickets in children living in the inner cities in Great Britain when compared to children living in underdeveloped countries [10]. He encouraged systematic sunbathing as a means for preventing and curing rickets, as did Sniadecki [9].

It was common folklore in the 19th century in fishing communities on the coast of England to give their children cod liver oil as a way of preventing rickets. Bretonneau in 1827 treated a 15-month-old child with severe rickets with cod liver oil and noted the incredible speed at which the patient was cured. His student Trousseau further demonstrated that oils from aquatic mammals including seals and whales as well as oily fish including herring were also effective in treating rickets [7,8]. These observations suggested that rickets was caused by a nutritional deficiency. Finally, in 1918, Mellanby reported that he could produce rickets in beagles by placing them on an oatmeal diet and then reversing the disease by giving them cod liver oil [11]. This observation convinced the scientific community that rickets was caused by a nutritional deficiency and the hunt was on to determine what nutrient was deficient. It was originally concluded that vitamin A in cod liver oil was responsible for its antirachitic activity. However, Elmer McCollum, who had discovered vitamin B, was not convinced that the antirachitic factor was vitamin A. It was known, based on the observation by Hopkins, that vitamin A was very sensitive to heat and could easily be destroyed by heating and aerating it [8]. McCollum used this information and conducted a study where he aerated and heated cod liver oil and demonstrated that it no longer had vitamin A activity but was still capable of healing rickets. This eureka discovery resulted in McCollum naming this new nutrient vitamin D [12].

At the same time, sunlight was beginning to be appreciated for its health-promoting properties. For his observations that direct sunlight exposure was effective in treating cutaneous tuberculosis and other skin disorders, Finsen received the Nobel Prize in 1903 [8]. In 1919, Huldschinski demonstrated that exposing children to radiation from a sun quartz lamp (mercury arc lamp) or carbon arc lamp resulted in marked increases in the mineralization of the long bones in the children's X-rays. He evaluated a similar group of children not exposed to his lamps and reported no significant change in their X-rays. (Figure 3) He concluded that exposure to ultraviolet (UV) radiation was an infallible remedy against all forms of rickets in children. He also exposed one arm of some children to one of his lamps and demonstrated that the improvements in the mineralization in the X-rays of the children were identical in both arms. He concluded that the UV exposure to the skin was causing some type of photochemical reaction, producing a substance that entered the circulation and had a systemic effect on the skeleton [13,14]. These observations prompted Alfred Hess to expose seven rachitic children in New York City to varying periods of sunshine, and he reported marked improvement in rickets of each child as evidenced by calcification of their epiphyses (growth plates at the end of the long bones). He also concluded that children of color were at higher risk for developing rickets and required longer exposure times to sunlight to have the same effect that was observed in white children [15] (Figure 4).

Figure 3. Photographs of researchers who made crucial contributions to vitamin D and rickets research. (**A**) Jędrzej Sniadecki, (**B**) Kurt Huldschinsky, (**C**) Alfred Hess, (**D**) Harry Steenbock, and (**E**) Elmer McCullum. Ref. [8] Holick, copyright 2013. Reproduced with permission.

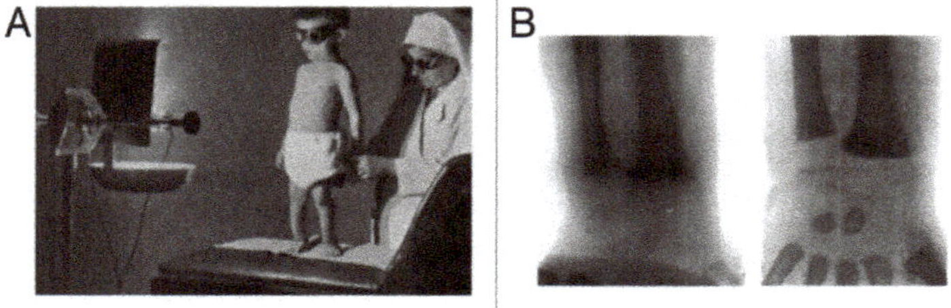

Figure 4. UV radiation therapy for rickets. (**A**) Photograph from the 1920s of a child with rickets being exposed to UV radiation. (**B**) Radiographs demonstrating florid rickets of the hand and wrist (**left**) and the same wrist and hand taken after treatment with 1 h UV radiation 2 times per week for 8 weeks. Note mineralization of the carpal bones and epiphyseal plates (**right**). Ref. [7] Holick, copyright 2006. Reproduced with permission.

These two disparate observations for the cure of rickets caused confusion. In 1921, Powers et al. fed a rachitic diet to rats and demonstrated that exposure to ultraviolet radiation was effective in preventing rickets, similar to dietary vitamin D [16]. However, the true appreciation for the health benefits of preventing rickets by exposing foods to UV radiation did not occur until 1924. Steenbock and Hess independently reported that exposure of foods, cotton seed and corn oil and especially the sterol fraction of foods with UV radiation produced antirachitic activity in rodents and chickens. Whereas Hess commented that "the question of the therapeutic value of this procedure is of secondary importance", Steenbock patented the process for the UV irradiation of foods and other substances as a means of increasing their antirachitic properties. These included yeasts, cereals, grain, oils, fats, butter and milk. The dairy industry appreciated the importance of providing milk to children with antirachitic activity and initiated the addition of ergosterol, obtained from yeast, to milk, followed by UV irradiation. Later, UV-irradiated ergosterol was added to milk to provide antirachitic activity [17,18]. By the early 1930s, essentially all milk in the United States and most European countries including Great Britain encouraged the fortification of milk with vitamin D_2. At the same time, the United States Department of Labor sent out a brochure encouraging parents to give their children a coat of tan from sun exposure to prevent rickets and to improve bone health (Figure 5). Within a few years, these interventions resulted in the essential eradication of rickets in the industrialized countries. Vitamin D fortification was so popular worldwide that beer, soda pop, hot dogs, custard and even soap and shaving cream were fortified with vitamin D (Figure 6). Parents were also able to purchase a mercury arc lamp from their local pharmacy and expose their children at home to ultraviolet radiation to prevent rickets (Figure 5).

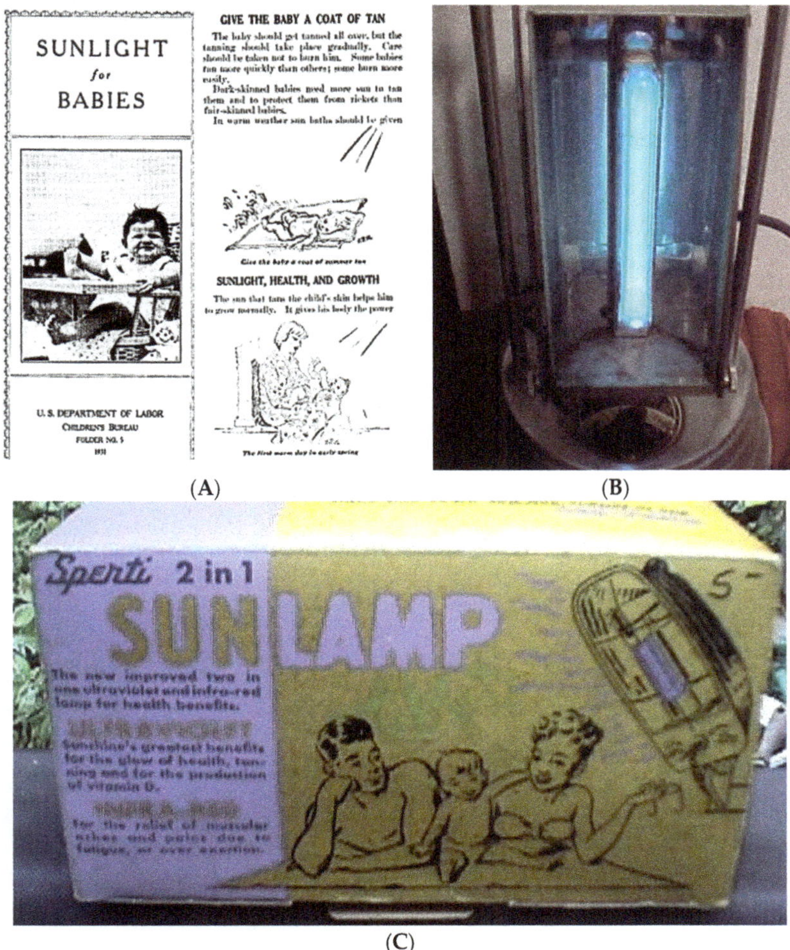

Figure 5. (**A**). Brochure from the US Department of Labor promoting sensible sun exposure for children in 1931. (**B**). Mercury arc lamp turned on for the use of preventing rickets in children in the 1930s. (**C**). The Sperti lamp available in the 1940s was used to expose children to ultraviolet radiation to promote bone health and to prevent rickets. Holick copyright 2023. Reproduced with permission.

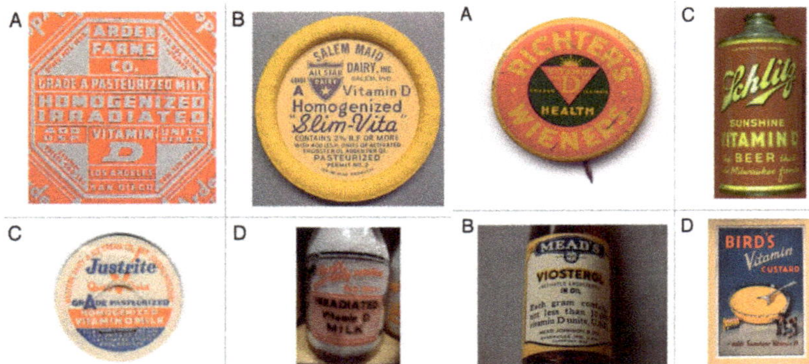

Figure 6. (**Left panel**). (**A**) Seal of a milk bottle that denoted that the milk was irradiated with UV radiation and contained vitamin D. (**B**) Cap of a milk bottle stating that activated ergosterol had been added to the milk. (**C**) Cap of milk bottle stating that the milk had been fortified with vitamin D. (**D**) Seal of a bottle of milk that denoted that the milk had been irradiated and contained vitamin D. Holick, copyright 2013. Reproduced with permission. (**Right panel**) (**A**) Seal denoting that this product was fortified with vitamin D. (**B**) Bottle of oil denoting that it contained irradiated ergosterol. (**C**) Beer can denoting that it was fortified with vitamin D. (**D**) Advertisement denoting that Bird's custard contained vitamin D. Ref. [8] Holick, copyright 2013. Reproduced with permission.

The vitamin D craze abruptly came to a halt when, in the early 1950s, an outbreak of infants with elfin facies, mild mental retardation, cardiac problems, and hypercalcemia was reported. The Royal College of physicians initiated an investigation and concluded that this outbreak in children was due to ingestion of excessive amounts of vitamin D in milk. This was based on a published observation that pregnant rats that received high doses of vitamin D delivered pups with some of the same clinical presentations including altered facial appearance, supravalvular aortic stenosis and hypercalcemia. Since it was difficult to monitor the content of vitamin D in milk, as a precautionary measure and based on the recommendation from the Royal College of Physicians, legislation quickly followed banning the fortification of any food or personal use product with vitamin D [19,20]. This ban quickly spread across Europe, Asia, India, and South America. Most of these countries still ban the fortification of foods with vitamin D with a few exceptions. Finland and Sweden now permit milk, margarine, and some cereals to be fortified with vitamin D. India recently introduced in 2017 a vitamin D_2 fortification program for milk and cooking oil. In retrospect, it is more likely that the children in Great Britain were suffering from the rare genetic disorder, Williams syndrome, which is associated with elfin facies, supravalvular aortic stenosis, mild mental retardation, and a hypersensitivity to vitamin D that causes hypercalcemia [21]. Unfortunately, this ban on vitamin D fortification persists in many countries throughout the world. It is responsible for the increased risk of rickets in children and vitamin D deficiency and insufficiency in children and adults.

3. Structural Identification of Vitamin D_2 and Vitamin D_3

When the oily sterol extract from yeast, containing ergosterol, was exposed to UV radiation, it was assumed that this resulted in the production of the antirachitic factor, vitamin D. However, it was quickly realized that there were other photoproducts present in the irradiated oil, and therefore, the UV-exposed sterol oil was called visterol. (Figure 6, right panel B). When animal oily sterols that mainly contained cholesterol were exposed to UV radiation, it resulted in the oils processing antirachitic properties. It was presumed that there was the contaminant, ergosterol, in the cholesterol oil that was transformed after UV irradiation into vitamin D. In 1932, Windaus in Germany and Bourdillon in England isolated pure crystalline vitamin D from irradiated ergosterol. Windaus and his colleagues named the compound vitamin D_2, whereas the British investigators called their crystalline compound ergocalciferol [22,23]. Originally irradiated ergosterol was given the name vitamin D_1. Once it was realized that the irradiated ergosterol contained vitamin D and lumisterol, the term was no longer used.

It was originally believed that vertebrate animals such as pigs and humans produced vitamin D_2 in their skin when exposed to sunlight. Waddell, in 1934, found that irradiated ergosterol was less effective as an antirachitic factor in chickens when compared to the same amount given to rachitic rodents. He concluded that there was a contaminant in cholesterol, and it was not ergosterol, that was responsible for the antirachitic activity in cholesterol oil exposed to UV radiation [24]. This conundrum was finally solved when Windaus and colleagues synthesized 7-dehydrocholesterol, and after its irradiation, the isolated vitamin D had a high antirachitic activity in chickens, whereas vitamin D_2 had only 1–3% the activity in chickens that it had in rats. Further investigations revealed that the vitamin D isolated from halibut and blue fin tuna oil was identical to the vitamin D generated from 7-dehydrocholesterol. Windaus named his vitamin D as vitamin D_3. He received the Nobel Prize in 1928 for his work on the constitution of sterols, including the identification of vitamins associated with them. Schenck, in 1937, finally determined that Windaus's vitamin D in his irradiated sterols was vitamin D_3 based on its successful crystallization [25]. The structures for vitamin D_2 and vitamin D_3 can be found in Figure 7.

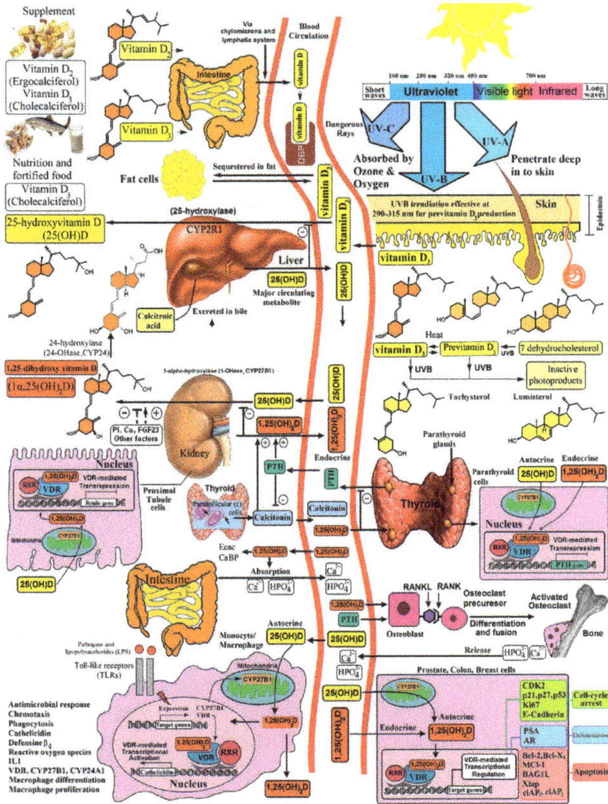

Figure 7. Schematic representation of the synthesis and metabolism of vitamin D for skeletal and extra-skeletal function. During exposure to sunlight, 7-dehydrocholesterol in the skin is converted to previtamin D_3. Previtamin D_3 immediately converts by a heat-dependent process to vitamin D_3. Excessive exposure to sunlight degrades previtamin D_3 and vitamin D_3 into inactive photoproducts.

Vitamin D_2 and vitamin D_3 from dietary sources are incorporated into chylomicrons and are transported by the lymphatic system into the venous circulation. Vitamin D (D represents D_2 or D_3) made in the skin or ingested in the diet can be stored in and then released from fat cells. Vitamin D in the circulation is bound to the vitamin D-binding protein (DBP), which transports it to the liver, where vitamin D is converted by the vitamin D-25-hydroxylase to 25-hydroxyvitamin D [25(OH)D]. This is the major circulating form of vitamin D that is used by clinicians to measure vitamin D status (although most reference laboratories report the normal range to be 20–100 ng/mL, the preferred healthful range is 30–60 ng/mL). It is biologically inactive and must be converted in the kidneys by the 25-hydroxyvitamin D-1a-hydroxylase (1-OHase) to its biologically active form 1,25-dihydroxyvitamin D [1,25$(OH)_2$D]. 1,25$(OH)_2D_3$ is then taken up by target cells and targeted to intracellular D-binding proteins (IDBP) to mitochondrial 24-hydroxylase or to the vitamin D receptor (VDR). The 1,25$(OH)_2D_3$-VDR complex heterodimerizes with the retinoic acid receptor (RXR) and binds to specific sequences in the promoter regions of the target gene. The DNA-bound heterodimer attracts components of the RNA polymerase II complex and nuclear transcription regulators. Serum phosphorus, calcium fibroblast growth factors (FGF-23), and other factors can either increase or decrease the renal production of 1,25(OH)2D. 1,25$(OH)_2$D feedback regulates its own synthesis and decreases the synthesis and secretion of parathyroid hormone (PTH) in the parathyroid glands. 1,25$(OH)_2$D increases the expression of the 25-hydroxyvitamin D-24-hydroxylase (24-OHase) to catabolize 1,25$(OH)_2$D to the water-soluble, biologically inactive calcitroic acid, which is excreted in the bile. 1,25$(OH)_2$D enhances intestinal calcium absorption in the small intestine by stimulating the expression of the epithelial calcium channel (ECaC) and calbindin 9K (calcium-binding protein, CaBP). 1,25$(OH)_2$D is recognized by its receptor in osteoblasts, causing an increase in the expression of the receptor activator of the NF-kB ligand (RANKL). Its receptor RANK on the preosteoclast binds RANKL, which induces the preosteoclast to become a mature osteoclast. The mature osteoclast removes calcium and phosphorus from the bone to maintain blood calcium and phosphorus levels. Adequate calcium and phosphorus levels promote mineralization of the skeleton. Autocrine/paracrine metabolism of 25(OH)D occurs when a macrophage or monocyte is stimulated through its toll-like receptor 2/1 (TLR2/1) by an infectious agent such as Mycobacterium tuberculosis or its lipopolysaccharide the signal upregulates the expression of VDR and 1-OHase. A 25(OH)D level of 30 ng/mL or higher provides adequate substrate levels for 1-OHase to convert 25(OH)D to 1,25$(OH)_2$D in mitochondria. 1,25$(OH)_2$D travels to the nucleus, where it increases the expression of cathelicidin, a peptide capable of promoting innate immunity and inducing the destruction of infectious agents such as M. tuberculosis. It is also likely that the 1,25$(OH)_2$D produced in monocytes or macrophages is released to act locally on activated T lymphocytes, which regulate cytokine synthesis, and activated B lymphocytes, which regulate immunoglobulin synthesis. When the 25(OH)D level is approximately 30 ng/mL, the risk of many common cancers is reduced. It is believed that the local production of 1,25$(OH)_2$D in the breast, colon, prostate, and other tissues regulates a variety of genes that control proliferation, including p21. Ref. [26] Holick, copyright 2013. Reproduced with permission.

4. Understanding How Vitamin D Works to Promote Bone Health

By 1940, it was acknowledged that rickets could be prevented and cured in two ways: by irradiation of the skin to solar or artificial UV radiation or by the ingestion of vitamin D (D represents D_2 or D_3). It was recognized that rachitic children had low blood concentrations of phosphate and calcium and that treatment with vitamin D corrected them to normal. It was unclear however how vitamin D affected this. In 1923, Orr et al. observed decreased fecal excretion of calcium in rats exposed to UV radiation. They concluded that this was likely due to vitamin D increasing intestinal calcium absorption [27]. This was confirmed in 1953 by Nicolaysen et al. who established that the important function of vitamin D was for stimulating intestinal calcium absorption [28]. It was also generally believed that vitamin D was playing a fundamental role in bone mineralization. However, studies by Carlsson demonstrated that vitamin D, rather than stimulating the direct deposition of calcium into bone, stimulated the mobilization of calcium from the skeleton [29]. We now recognized that the fundamental function of vitamin D is to maintain the serum calcium and phosphate in the normal range for maintaining most metabolic functions, including neuromuscular and cardiac activity [7,30]. Vitamin D also maintains an adequate circulating and extracellular calcium–phosphate product, resulting in the passive deposition of calcium hydroxyapatite, i.e., mineral into the skeleton [7,30]. There is also evidence that vitamin D enhances intestinal phosphate absorption, which is also critically important for energy and muscle function through the action of ATP and is a major component of calcium hydroxyapatite in the skeleton (Figure 7) [30].

5. The Saga for the Quest to Identify the Active Form of Vitamin D_3

In the early 1950s, using radiolabeled calcium, it was determined that when a vitamin-D-deficient rat or chicken received vitamin D, intestinal calcium absorption began to increase, with maximum absorption occurring at around 24 h. Kodicek and coworkers reasoned that to understand the mode and sites of action of vitamin D, investigations were needed to understand its storage, distribution and metabolism. They produced ^{14}C-beled vitamin D_2 with a low specific activity of 0.46 mCi/mmol and administered pharmacologic doses to rats. They concluded, based on determination of radioactivity in the tissues and excreta by reverse chromatography, that vitamin D_2 was likely the biologically active form of the vitamin [31]. In the early 1960s, Norman and DeLuca prepared randomly labeled ^{3}H-vitamin D_3 and ^{3}H-vitamin D_2 with much higher specific activities of 7.3 and 3.25 mCi/mmol, respectively, and gave them to vitamin-D-deficient rats. The chromatography results from their study revealed that there was accumulation of radioactivity mainly in the liver and to a lesser degree in the ileum and kidneys and much less in the adrenal glands. This resulted in the investigation of the metabolism of tritiated vitamin D_3 in rat kidney and intestinal mucosa. At least three lipid-soluble radioactive compounds other than ^{3}H-vitamin D_3 were detected, and all showed partial vitamin D activity [32]. The DeLuca group then made [1, 2-^{3}H] vitamin D_3 with very high specific activity and found by silicic acid chromatography a polar metabolite labeled as metabolite IV circulating in rats. They found that this metabolite had at least the same antirachitic potency as vitamin D_3. Furthermore, they discovered that the amount of this metabolite continued to increase with increasing doses of vitamin D_3. Utilizing this fact, Blunt et al. gave four hogs 6.25 mg of vitamin D_3 daily for 26 days and then collected their blood. They separately gave a hog [1, 2-^{3}H] vitamin D_3 and collected its blood. Organic extractions were made of both blood pools and were then combined. The lipid extract underwent a series of straight-phase and reverse-phase chromatographies, resulting in the recovery of 1.2 mg of metabolite labeled as peak IV. The vitamin D_3 metabolite was identified as 25-hydroxyvitamin D_3 [33]. Suda et al. performed a similar procedure, only this time using pharmacologic amounts of vitamin D_2 and tritiated vitamin D_2 and identified the polar metabolite as 25-hydroxyvitamin D_2 (Figure 8) [34]. These observations in pigs were interesting, but it was unknown whether this metabolism also occurred in humans.

(A)

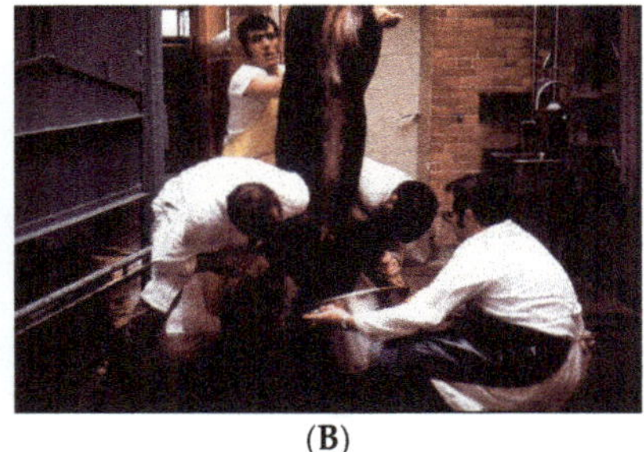

(B)

Figure 8. (**A**). This hog was treated with 6.25 mg vitamin D_2 by Dr. Suda; (**B**) the blood was then collected. After purification of the blood lipid extract, the vitamin D_2 metabolite was identified as 25-hydroxyvitamin D_2. Holick copyright 2023. Reproduced with permission.

6. A Personal Experience and Perspective for the Identification of the Major Circulating Form of Vitamin D_3 in Human Blood

Clearly, a different strategy would be needed to determine if 25-hydroxylated vitamin D was present in human circulation. I arrived at the University of Wisconsin Madison in 1968 in the Department of Medical Microbiology. This was a time when there was great excitement about the discovery of DNA and RNA. Molecular biology was in its infancy, and my goal was to achieve a Ph.D. degree by investigating the molecular biology of microorganisms. However, no one in the department had any interest in this area, and I, after completing the year, went to the Biochemistry Department and explained to them that I had been accepted into the University's graduate program and that they should accept me into the graduate program in the Biochemistry Department. Much to my amazement, I was accepted. I quickly realized, however, that I had a big problem to overcome. I was considered to have completed my first year of graduate work, and to proceed into the second year, I needed to take the biochemistry preliminary exam. Although I had a chemistry degree with a special interest in organic chemistry and had biochemistry laboratory experience, I had never taken a biochemistry course. As a result, I barely passed the exam. I wanted to find a research mentor who had a special interest in molecular biology. However, the professors, after seeing my poor performance on the preliminary exam, informed me that there were no openings but that I should see Professor DeLuca, who was making important observations about vitamin D. I explained to various professors that I had no interest in vitamin D since I considered it a boring subject. I was informed that this was my only option. When I met Dr. DeLuca in June 1969, he was pleased that I had significant biochemistry research experience and a publication while I was in the chemistry program at Seton Hall University. However, after seeing my preliminary exam results, he informed me that I needed to go into a softer science program such as physiology or pharmacology and that biochemistry would be too difficult for me. I reassured him that it was my goal to receive a Ph.D. degree in biochemistry and that I would do whatever was necessary to accomplish this. He said that he did not want to discourage me but to be realistic, I needed to first receive a master's degree that would likely take 2 years followed by being quizzed by the biochemistry faculty to determine if I could proceed into the Ph.D. program. He said that the Ph.D. program would take at least 4–5 years. Thus, he said it could take up to 7 years before I would receive a Ph.D. degree. I accepted the challenge, and he accepted me into his research program in July 1969. As the newest graduate student, I was given a desk against the wall where the entrance was located with no laboratory space. In July of 1969, Dr. DeLuca received a phone call from Dr. Avioli, at Washington University in St. Louis, informing him that he has been treating hypoparathyroid patients with pharmacologic doses of vitamin D_3 (1.5 mg daily) and had been collecting their heparinized blood plasma. He asked Hector if he would like the blood plasma to see if his laboratory could identify 25-hydroxyvitamin D_3 in the blood plasma. Since all his graduate students and postdoctoral fellows had ongoing research projects and I had none, I was informed by Dr. DeLuca that this would be my project. He instructed me that all I needed to do was to follow the exact chromatography procedures that were successfully used to identify 25-hydroxyvitamin D_2 and 25-hydroxyvitamin D_3 in hog blood. With great enthusiasm, I began by obtaining ^{3}H-tracer from the blood of vitamin-D-deficient rats that were injected with [1, 2-^{3}H] vitamin D_3. I performed a lipid extraction on 600 mL of human blood plasma and rat serum and combined the lipid extracts. I proceeded to follow the exact chromatography methods used to successfully obtain purified 25-hydroxyvitamin D from hog blood. I conducted all the chromatographies that were previously reported. After exhausting all chromatographic procedures known at the time, I realized at the end of November 1969 that there was a lipid contaminant in human blood that was not present in hog blood and that the contaminant could not be separated from the vitamin D metabolite for its identification. This was Thanksgiving weekend. Since I could not afford to travel home to New Jersey to be with my family, I went to the laboratory. I unlocked the door early in the morning and walked in distressed because of my predicament of failing to purify the vitamin D metabolite. As I was walking to my desk, I noticed on one of the shelves on the research bench that was next to my desk a bottle containing Sephadex LH-20. I remembered from my college days that Sephadex, which is a glucose polymer, can separate proteins based on size. I reasoned that maybe I could use Sephadex LH-20 to separate the vitamin D metabolite from other contaminants based on size. I made a slurry of Sephadex LH-20 with methanol and poured the slurry into a column. After the Sephadex LH-20 settled, I placed my human plasma blood lipid extract, which had undergone multiple chromatographies, on the column and collected fractions. The vitamin D metabolite peak was identified by determining the location of the tritiated peak IV. The peak was recovered and the metabolite was now pure. I recovered 25 mcg that was identified as 25-hydroxyvitamin D_3 (25(OH)D_3). This metabolite is the major circulating form of vitamin D_3 that is clinically measured to determine a person's vitamin D status. What I did not realize at the time was that I had developed a new chromatography method known as liquid gel chromatography that became the standard chromatography technique in most vitamin D laboratories and was instrumental for purifying the active form of vitamin D_3 [35,36]. I had completed my master's research in 3 months, and for these discoveries, I was awarded a master's degree in January 1970.

7. The Hunt to Identity the Active Form of Vitamin D_3

It was found that 25(OH)D_3 was biologically more potent than vitamin D_3, and it was concluded that it was the active form [33]. However, it soon became evident that ^{3}H-25(OH)D_3 was rapidly metabolized to a more polar metabolite that quickly appeared in the intestine. Haussler et al. [37] reported a more polar metabolite of 25(OH)D_3 in nuclear fractions of the intestine in chickens that received ^{3}H-vitamin D_3. At the same time, Lawson et.al. [38] reported that during the formation of this more polar metabolite, the 1^{-3}H on their [4-^{14}C,1α-^{3}H] vitamin D_3 was lost, suggesting that this polar metabolite had some alteration on carbon 1. The realization that this more polar metabolite not only had all the biologic actions that vitamin D_3 and 25(OH)D_3 possessed but that it acted more quickly in stimulating intestinal calcium absorption. It became clear that 25(OH)D_3 was not the biologically active form, but rather, it was a more polar metabolite that needed to be identified.

This realization initiated a frenzy of activity in the Kodicek, Norman and DeLuca laboratories. Various strategies were undertaken, the first being in the DeLuca lab where they gave hogs pharmacologic doses of vitamin D_3 and then collected their intestines. Dr. Robert Cousins, who was a postdoctoral fellow in DeLuca's laboratory, was in charge. It was quickly realized that it made little sense to be giving pharmacologic doses of vitamin D_3 and expecting marked increases in the concentrations of the active form in the intestine. As a result, in the summer of 1970, it was decided that what was needed was many intestines from chickens that were receiving presumably physiologic amounts of vitamin D_3 in their diet. Dr. Cousins, Dr. Suda and I traveled an hour north in the department of biochemistry's pickup truck to the BrakeBush facility where they processed 20,000 chickens per day (Figure 9A). We collected approximately 400 pounds of chicken intestines and returned to the University of Wisconsin with the intestines. We quickly realized that the intestines had begun to putrefy since we had not brought ice with us, and it was a hot steamy day. It was decided that the entire laboratory would go back to the BrakeBush facility to collect, clean, and process approximately 400 pounds of chicken intestines and freeze them immediately on dry ice (Figure 9B,C). The realization quickly set in upon arriving back at the University that we had no ability to process such a large amount of intestinal material. The biochemistry department had large fermentation stainless steel containers in the basement. Dr. Cousins, Dr. Suda and I began to process the intestines in a sausage meat grinder. The mashed intestines were then added to a large fermentation tank (Figure 9D). Several hundred gallons of chloroform, methanol and water were added and stirred. An additional amount of chloroform was then added, which caused a separation of the organic phase and which contained the lipids that were present in the intestines, from the water phase, which contained the water-soluble compounds. The several hundred gallons of the chloroform phase was taken to dryness and then chromatographed on a huge Shepadex LH-20 column that I designed. It became obvious to me that it was going to be futile to chromatograph and purify the minute quantity of the active form of vitamin D_3 that was present in several hundred grams of intestinal lipid extract.

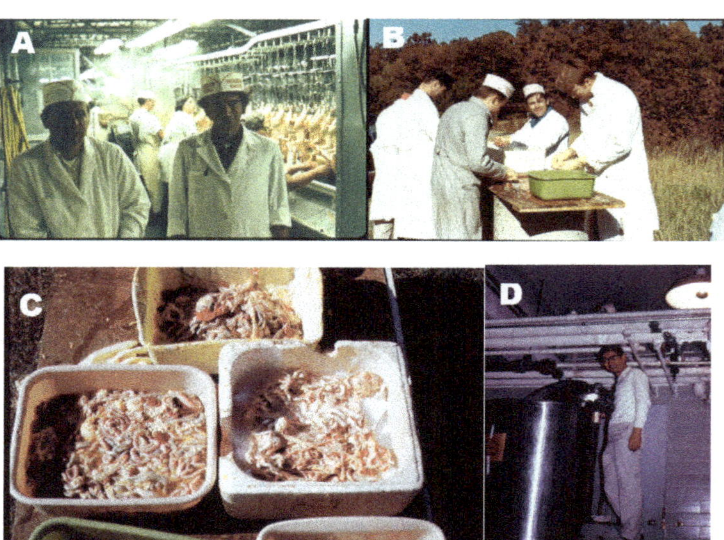

Figure 9. (**A**) The first attempt. Dr. Robert Cousins on the left and me collecting intestines from the trough behind us at the BrakeBush facility. (**B**) The second collection of intestines with several members of the DeLuca laboratory helping out. From left to right is Dr. Ian Boyle, Dr. Chuck Frolich, Dr. Jack Omdahl and Dr. Hector DeLuca. (**C**) Cleaned and processed intestines placed on dry ice. (**D**) Dr. Tatsuo Suda inspecting the huge stainless steel fermentation tank that contained approximately 400 pounds of mashed intestines and several hundred gallons of chloroform and methanol for the lipid extraction. Holick copyright 2023. Reproduced with permission.

Thus, while this process was underway, I went to Dr. DeLuca and suggested to him that we know based on multiple studies that there was approximately 1 ng of the active metabolite in 1 g of intestine. I suggested to him that a better way of isolating this intestinal metabolite was to give vitamin-D-deficient chickens intravenously 100 IUs of vitamin D_3 that contained [1, 2-^{3}H] vitamin D_3. Based on my calculation, I explained to him that I would need 1500 chicken intestines to obtain approximately 10 mcg of the active metabolite that could then be extracted, chromatographed, and ultimately purified for identification. He gave me the green light. I asked Dr. Jack Omdahl, who was a postdoctoral fellow in Dr. Deluca's laboratory, with extensive experience in working with vitamin-D-deficient chickens, for his assistance. Due to the capacity of the department of agriculture to house chickens, I was able to only order 500 chicks. I placed them on a vitamin-D-deficient diet that we produced in the biochemistry department. After 6 weeks on the vitamin-D-deficient diet, Jack gave an oral dose of 100 IUs (2.5 mcg) of vitamin D_3 containing trace amounts of [1, 2-^{3}H] vitamin D_3 in vegetable oil to all 500 chickens. The chickens were sacrificed 24 h later, and Jack and I removed and cleaned the small intestines and placed them in a freezer. At the same time, another 500 chickens were ordered and placed on the vitamin-D-deficient diet and underwent the same procedures. By November 1970, I had 1000 intestines stored in the freezer and was waiting for my last 450 chickens that were scheduled to be sacrificed in early December. However, at the end of November, late at night, we heard a news flash on BBC radio stating that the Kodicek laboratory had made a major discovery related to vitamin D. We did not know what that discovery was. Dr. Deluca became extremely concerned that this meant that the Kodicek group had successfully identified the active form of vitamin D. I was asked to proceed immediately to process the 1000 chicken intestines I had in the freezer, with the intent of identifying the active form of vitamin D_3 as quickly as possible. My response was that I needed 1500 chicken intestines to be successful and that I did not want to proceed. Challenging my mentor was not easy but I initially stood my ground. I then relented, reiterating my concern that I would not be successful but that I would do as directed and begin the processing of the thousand chicken intestines the next morning. As I was beginning to collect the frozen intestines, early the next morning, Dr. DeLuca appeared and informed me that I had been right in the past with my research activities and that he would permit me to wait until I collected the additional 500 intestines. By the middle of December 1970, I had completed the lipid extraction of all 1450 chicken intestines. Based on the known specific activity of the [1, 2-^{3}H] vitamin D_3 that was given to the chickens, I estimated that I had approximately 11 mcg of the active metabolite in 22 g of oily lipid. I immediately began multiple chromatographies on my Sephadex columns (Figure 10A,B). It was now coming close to Christmas, and I realized that we were in stiff competition with the Kodicek and Norman laboratories. I flew home to New Jersey on Christmas Eve and wished my parents a Merry Christmas and flew back to Madison Wisconsin on Christmas day to continue to purify the active vitamin D_3 metabolite that was called peak V. When I returned, I learned that the Kodicek group published its newsworthy finding in the journal Nature. At this time, it was assumed that because this active metabolite appeared so quickly in the intestine that it was the intestine that was producing it. Therefore, it was obvious that to generate the active form of vitamin D_3 in sufficient quantities for its identification, all you would need to do is incubate intestinal homogenates with 25(OH)D_3. After numerous attempts by several laboratories including the DeLuca laboratory, incubation of intestinal homogenates with [^{3}H]-25(OH)D_3 did not yield any of the desired polar metabolite. The Kodicek group not only made homogenates of chicken intestines, they made homogenates of many of the other organs. They then incubated each of the homogenates with [4-^{14}C,1-^{3}H] 25(OH)D_3. What they observed was truly remarkable. They found that only kidney homogenates could produce the active metabolite [39]. For me, this revelation was very disconcerting, since it meant that an unlimited amount of active vitamin D_3 could be easily produced by incubating kidney homogenates with 25(OH)D_3. It was becoming increasingly more likely that the Kodicek group was close to purifying the active metabolite for its identification. By early January 1971, I realized after 17 chromatographies, I had approximately 8 mcg of active metabolite that was associated with a large amount, that I estimated to be about 2000 mcg, of lipid contamination (Figure 10C). There were no more chromatographies that I could do that would make any difference, and I concluded that I had failed. However, within a few hours, I reasoned that we knew that the active form of vitamin D had a secondary hydroxyl group on carbon 3 and a tertiary hydroxyl group on carbon 25. It was presumed that there was an additional secondary hydroxyl group on carbon 1 based on the observation by the Kodicek group and that this metabolite had lost its $1\alpha{-}^3$H on carbon 1. I considered that it was unlikely that the lipid contaminants in my preparation had the same number and type of hydroxyl groups. I reasoned that if I trimethylsilylated the active metabolite, and then selectively, by acid hydrolysis, removed the trimetylsilyl ether groups from the secondary hydroxyl groups, this would result in the 25-hydroxyl being protected with a trimethylsilyl ether. This modification would decrease the polarity of the vitamin D metabolite, thereby altering its chromatographic properties. I proceeded with this strategy and chromatographed the dihydroxyated active form on a Sephadex LH-20 column and recovered it in pure form. The purified monosilyl ether derivative was then hydrolyzed, and after purification, 2 micrograms of pure metabolite was recovered. The ultraviolet absorption spectrum showed a lambda max 265 nm and a lambda min 228 nm that was identical to the absorption spectrum for the 5,6-triene system for vitamin D_3. A small amount was immediately placed in the mass spectrometer, and it revealed a mass spectrum with a molecular ion of 416 *m/e*. This was consistent with an additional hydroxyl group being present in the metabolite. The fragmentation pattern was consistent with the metabolite having the additional hydroxyl group in the A ring (Figure 11). Further chemistry on nanogram quantities of the purified active form of vitamin D_3, followed by the skillful mass spectroscopy by Dr. Heinrich Schnoes (Figure 10D), finally revealed its structure as 1,25-dihydroxyvitamin D_3 ($1,25(OH)_2D_3$) (Figures 7 and 11) [40].

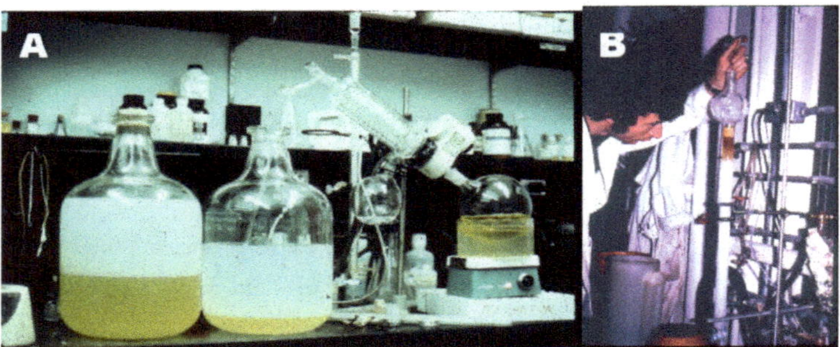

Figure 10. (**A**) Lipid extraction of 1450 chicken intestines. The yellow chloroform organic phase that contained the active metabolite was being dried down under vacuum in preparation for its first chromatography on a Shephadex LH-20 column (**B**) The dried down lipid extract from the intestines was first chromatographed on a Shephadex LH-20 column and eluted with a mixture of 65% chloroform in hexane. (**C**) In one of the final chromatographies, the semi-purified active vitamin D_3 metabolite was dissolved in methanol and chromatographed on a Shephadex LH-20 column prepared in methanol. (**D**) Dr. Heinrich Schnoes in his office. Holick copyright 2023. Reproduced with permission.

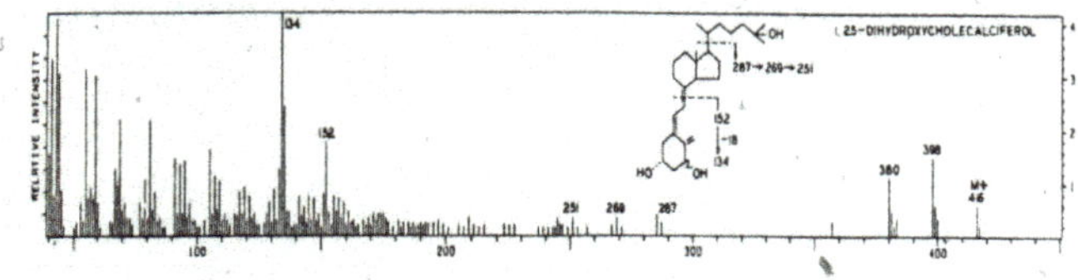

Figure 11. The mass spectrum of the purified metabolite demonstrated a molecular ion of 416 *m/e* consistent with an additional hydroxyl group on the metabolite. The fragments of 152 and 134 revealed that the hydroxyl group was somewhere in the A ring, most likely on carbon 1. Several chemical reactions were performed on nanogram quantities of the pure metabolite that provided evidence that the extra hydroxyl group was on carbon 1. Holick copyright 2023. Reproduced with permission.

During that same time, the Kodicek group isolated 60 mcg of the metabolite with approximately 30% purity, using Sephadex LH-20 chromatography, from chicken kidney homogenates. They reported that the isolated metabolite had a UV absorption lambda maximum of 269 nm and a lambda minimum at 232 nm. The mass spectrum showed a molecular weight of 416, identical to what we had observed. Because they only had 30% purity, the only way they were able to conclude that the additional hydroxyl group was on carbon 1 was based on their observation that the 1-^{3}H on their [4-^{14}C,1α-^{3}H] vitamin D_3 was lost while it was being metabolized. They concluded that their metabolite was also 1,25-dihydroxyvitamin D_3 [41]. The Norman group also used the Kodicek kidney homogenate strategy and confirmed several months later that the structure of the active form of vitamin D was 1,25-dihydroxyvitamin D_3. [42] These three observations were submitted for publication on 12 February to PNAS, 19 February to Nature and 10 May 1971 to Science, respectively [40–42]. We had won the race in just one week. For my research accomplishments, I received my Ph.D. Degree in May 1971, a little less than 2 years after I joined the DeLuca laboratory.

The final evidence that the one hydroxyl group was on C-1 was confirmed by the 21- step synthesis of 1,25-dihydroxyvitamin D_3 by my roommate Eric Semmler and me [43]. We demonstrated that the synthetic metabolite migrated identically with the biologically produced active metabolite and had the same biologic activity on intestinal calcium absorption and bone calcium mobilization [43]. This was further confirmed several years later when my group made 1β, 25-dihydroxyvitamin D_3 and demonstrated that it was biologically inactive [44]. Soon after these discoveries, Dr. Glennville Jones, while a postdoctoral fellow in Dr. DeLuca's laboratory, began using the newly introduced high-performance liquid chromatography and identified 1,25-dihydroxyvitamin D_2 as the biologically active form of vitamin D_2. He and other investigators went on to identify numerous other vitamin D metabolites [45]. However, to date, it remains that 1α,25-dihydroxyvitamin D_3 is the biologically active form of vitamin D_3.

8. Clinical Uses for 1α,25-Dihydroxyvitamin D_3 and 1α-Hydroxyvitamin D_3

Once 1,25-dihydroxyvitamin D_3 was identified, there was a great interest in producing it chemically. It was decided that it would be worthwhile to see if the 1α-hydroxyl group could be introduced into the less expensive starting material, cholesterol. The synthesis of 1α-hydroxyvitamin D_3 was accomplished. The same methodology was then used to successfully produce 1,25-dihydroxyvitamin D_3 [46]. 1α-hydroxyvitamin D_3 was shown to be biologically active after it was metabolized in the liver on carbon 25 to form 1,25-dihydroxyvitamin D_3 [47].

Several milligrams of both compounds were produced. They were immediately sent out to clinicians around the world who found that these active vitamin D compounds were extremely effective in treating metabolic calcium and bone diseases in children and adults with kidney failure [48,49]. It was known that kidney-failure patients had a resistance to the biologic actions of vitamin D that was difficult to understand. The revelation that the kidneys were responsible for metabolizing 25(OH)D_3 to 1,25-dihydroxyvitamin D_3 revealed the reason why. It was also observed that a rare form of hereditary rickets, known as vitamin-D-dependent rickets type I or pseudo-vitamin D deficiency rickets, was effectively treated with physiologic doses of 1,25-dihydroxyvitamin D_3 [50,51]. Both active vitamin D compounds were also effectively used to treat hypocalcemia in patients with hypoparathyroidism and pseudohypoparathyroidism [52,53]. These clinical successes prompted Dr. Milan Uskokovic at Hoffmann LaRoche to develop a streamlined method for producing the active metabolite, and it became commercially available as an FDA-approved pharmaceutical in the early 1970s.

9. Enter the Era of Noncalcemic Genomic Health Benefits of Vitamin D

In 1979, Stumpf et al. reported that ^{3}H-1,25(OH)$_2$$D_3$ was concentrated in the nuclei of most tissues in the body of a vitamin-D-deficient rat [54]. This provided evidence that the vitamin D receptor (VDR) not only existed in calcium-regulating tissues, but also was present in tissues that were not related to calcium and bone metabolism. The physiologic significance of this revelation was not appreciated until it was observed that 1,25(OH)$_2$D$_3$ inhibited the proliferation and induced terminal differentiation of HL-60 human myeloid leukemia cells [55]. Several laboratories in the early 1980s began to report that some cultured cancer cells possessed a VDR and that incubation of these cells with 1,25(OH)$_2$D$_3$ resulted in decreased proliferative activity [56]. The observation that mice with M1 leukemia had prolonged survival when they were treated with 1α(OH)D$_3$ prompted great interest in seeing whether the same could be true for patients with preleukemia [57]. Eighteen patients with preleukemia were treated with 1,25(OH)$_2$D$_3$. Although the patients performed quite well early in the treatment, they ultimately developed hypercalcemia, and all died in blastic phase [58]. As a result of these initial studies, a huge effort was made to develop 1,25(OH)$_2$D$_3$ and its active analogs as a potential therapy for treating a variety of cancers, including prostate cancer. Unfortunately, these therapies were not found to be successful [59]. It appeared that the cancer cells cleverly designed mechanisms to bypass the antiproliferative and pro-differentiating properties of 1,25-dihydroxyvitamin D$_3$ and its active analogs. As a result, 40+ years later, there still has never been a vitamin D analog that has proven to be successful in treating any form of cancer.

At the same time, Bikle et al. [60] and our group [61] discovered that keratinocytes had a VDR, and when these cells were incubated with 1,25(OH)$_2$D$_3$, they demonstrated a dramatic response. The keratinocytes decreased their proliferative activity and became fully differentiated. It had been reported that one osteoporotic patient being treated with 1α-hydroxyvitamin D$_3$ had improvement in their psoriasis. We had at the same-time initiated psoriasis clinical trials with both topical and oral 1,25(OH)$_2$D$_3$, which demonstrated remarkable improvement in disease activity (Figure 12). Pharmaceutical companies began making analogs to see if they could reduce the calcemic action of 1,25(OH)$_2$D$_3$ while maintaining the same or more potent anti-proliferative and pro-differentiating activity. Several analogs were developed, and they and 1,25(OH)$_2$D$_3$ remain a first-line treatment for patients with minimal psoriasis [62–65].

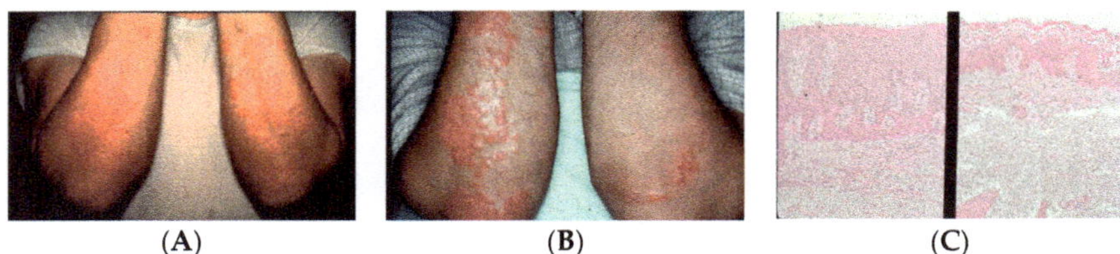

(A) **(B)** **(C)**

Figure 12. (**A**) A young man with chronic psoriasis with no effective treatment. He was a participant in our clinical research trial evaluating the effectiveness of topical-applied 1,25-dihydroxyvitamin D3 that I formulated in Vaseline with a concentration of 50 mcg/gram. This was a double-blind placebo-controlled trial with one side receiving topical Vaseline and the other side receiving topical Vaseline containing 1,25-dihydroxyvitamin D3. (**B**) After 2 months, there was dramatic improvement on the right forearm that had received the active metabolite. (**C**) A skin biopsy was taken from each forearm and the histology of the skin biopsy from the right forearm shown on the right side of figure (**C**) revealed a dramatic improvement by marked decreases in the keratinocyte proliferation with induction of terminal differentiation to normalize the skin. Holick copyright 2023. Reproduced with permission.

Although it was dogma that only the kidneys could produce 1,25(OH)$_2$D$_3$, many laboratories began reporting that a wide variety of tissues and cells not only had a vitamin D receptor but also expressed the 25-hydroxyvitamin D-1α-hydroxylase. Of particular interest was the observation that when macrophages were activated, toll-like receptors were turned on to induce the macrophage to express the 25-hydroxyvitamin D-1α -hydroxylase. This resulted in the ability of the macrophage to metabolize 25(OH)D$_3$ to 1,25(OH)$_2$D$_3$ (Figure 7). The reason for this activation was that once produced, 1,25(OH)$_2$D$_3$ interacted with the macrophage VDR, resulting in expression and production of cathelicidin, a defensen protein responsible for killing and lysing infectious agents, including bacteria and viruses [66]. It is also now recognized that $1,25(OH)_2D_3$ produced by immune cells is a major regulator of both innate and acquired immunity, as illustrated in Figure 13 [67]. When healthy vitamin-D-deficient/insufficient adults were given 600 IUs or 10,000 IUs daily for 6 months, an evaluation of their circulating immune cells before and after revealed that even on 600 IUs daily, 128 genes were being influenced, whereas 1289 genes were being influenced in those taking 10,000 IUs daily [68]. What was also observed was although all the healthy vitamin-D-deficient/insufficient participants substantially raised their circulating concentrations of 25(OH)D in the range of 60–90 ng/mL, only 50% had a robust response. This observation demonstrates that there were other factors such as VDR polymorphisms that could influence the responsiveness of vitamin D, promoting genomic effects on the immune system (Figure 14)

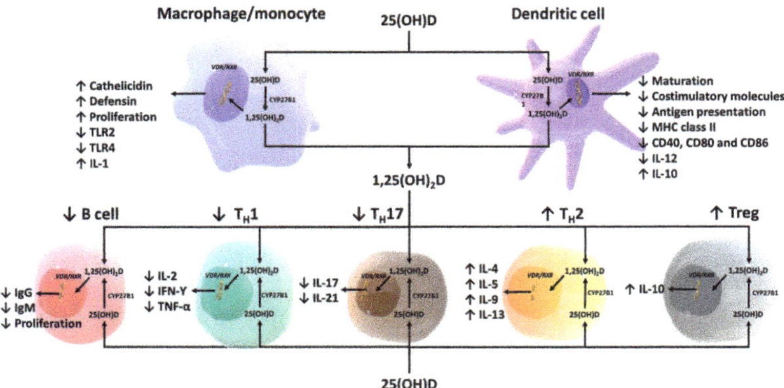

Figure 13. Schematic representation of paracrine and intracrine function of vitamin D and its metabolites and actions of 1,25-dihydroxyvitamin D on the innate and adaptive immune systems. Abbreviations: 1,25(OH)2D: 1,25-dihydroxyvitamin D; 25(OH)D: 25-hydroxyvitamin D, IFN-Y: interferon-Y; IL: interleukin; MHC: membrane histocompatibility complex, TH1: T helper 1; TH2: T helper 2; TH17: T helper 17; Treg: regulatory T cell, TNF-α: tumor necrosis factor-α; TLR2: toll-like receptor 2; TLR4: toll-like receptor 4. Ref. [67] Holick MF, copyright 2020. Reproduced with permission.

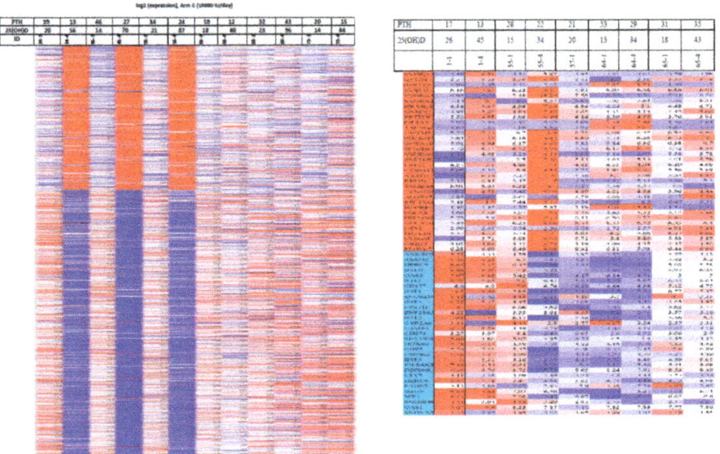

Figure 14. (**Left**) Heatmap of 1289 vitamin D-responsive genes whose expression response variation in 6 vitamin-D-deficient subjects taking 10,000 IUs per day of vitamin D_3 for 6 months, showing that

3 subjects had a robust response in gene expression compared to the other 3 subjects who had minimum to modest responses even though these subjects raised their blood levels of 25(OH)D in the same range of ~60–90 ng/mL. (**Rgiht**) Heatmap of only 128 vitamin D-responsive genes in 5 vitamin D deficient subjects taking 600 IUs per day for 6 months. Abbreviation: 0 m: 0 month; 6 m: 6 months; 25(OH)D: 25-hydroxyvitamin D; PTH: parathyroid hormone. Ref. [68] Holick MF, copyright 2019. Reproduced with permission.

10. Associating Vitamin D Deficiency with Acute and Chronic Illnesses

Most tissues and cells in the body have a VDR and can produce $1,25(OH)_2D$ in a wide variety of cells, including prostate, breast, colon, skin and the brain. The reason that the local production of $1,25(OH)_2D_3$ does not potentially cause hypercalcemia is that once it is produced and enters the nucleus, $1,25(OH)_2D$ immediately induces the 25-hydroxyvitamin D-24-hydroxylase. This enzyme begins to rapidly metabolize $1,25(OH)_2D_3$ to an inactive water-soluble carboxylic acid derivative (calcitroic acid). Simultaneously, $1,25(OH)_2D_3$ up- and downregulates more than 1000 genes responsible for cellular proliferation, differentiation, a variety of cellular metabolic activities, antiangiogenesis and apoptosis (Figure 7) [30,68].

Epidemiology studies have related vitamin D deficiency with a multitude of chronic illnesses, including the autoimmune disorders multiple sclerosis, type 1 diabetes and rheumatoid arthritis, cardiovascular disease, type 2 diabetes, neurocognitive dysfunction and infectious diseases, including COVID-19 (Figure 15) [69–71]. These observations will be discussed in more detail by other authors participating in this special edition.

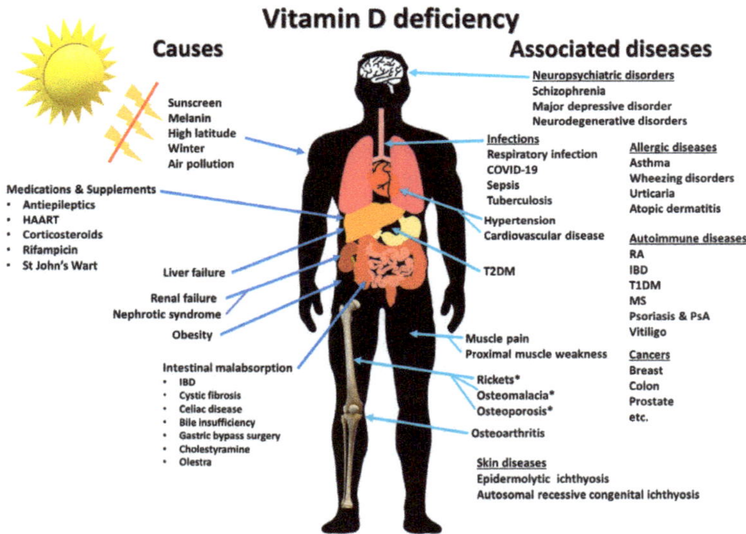

Figure 15. Summary of causes of vitamin D deficiency and diseases and disorders associated with vitamin D deficiency. Abbreviations: HARRT: highly active antiretroviral therapy; IBD: inflammatory bowel diseases; MS: multiple sclerosis; PsA: psoriatic arthritis; T1DM: type 1 diabetes mellitus; T2DM: type 2 diabetes mellitus; RA: rheumatoid arthritis. * denotes diseases that are direct consequences of vitamin D deficiency. Ref. [67] Holick MF, copyright 2020. Reproduced with permission.

11. The Future of Vitamin D for the Next 100 Years

The discovery of the antirachitic nutrient/hormone, vitamin D, by McCollum 100 years ago has had and continues to have far-ranging health benefits. This is due to the discovery that vitamin D must undergo sequential metabolism in the liver to 25(OH)D, which is then converted to its active form, $1,25(OH)_2D$, in the kidneys for regulating calcium and phosphate metabolism. The new appreciation that essentially all organs and cells have a VDR, and that many cells can produce 1,25(OH)$_2$D, opens a new chapter for vitamin D playing an important role in promoting good health and well-being by reducing the risk for chronic illnesses (Figures 7 and 15).

Vitamin D has by no means revealed all its biologic functions and clinical potential. It is hoped that there will be a resolution for recommendations for how much vitamin D and sensible sun exposure is necessary for maximum health. There continues to be debate as to what the definition of vitamin D deficiency is based on a circulating blood concentration of 25(OH)D. There is consensus that to achieve and maintain a healthy skeleton, the circulating concentration of 25(OH)D should be at least 20 ng/mL. To maintain maximum bone health and to prevent any evidence of vitamin D deficiency osteomalacia, the circulating concentration should be at least 30 ng/mL [72]. There is mounting evidence that when a blood concentration of 25(OH)D is at least 30 ng/mL, the noncalcemic health benefits of vitamin D become more apparent, based on association studies and randomized controlled trials. It is documented that Maasai herders and the Hazda maintain a circulating concentration of 25(OH)D in the range of 40–60 ng/mL. This concentration range has been associated with many of the noncalcemic health benefits including reduced risk for cardiovascular disease, neurocognitive dysfunction, several cancers and infectious diseases [73]. To achieve a circulating concentration of 25(OH)D in the range of 40–60 ng/mL requires a daily average intake of 2000–5000 IUs or the same amount of vitamin D$_3$ produced in the skin from sun exposure, as demonstrated in the Maasai and Hazda. An adult in a bathing suit exposed to enough sunlight to cause a minimal erythemal response results in the production of an amount of vitamin D equivalent to ingesting 15,000–20,000 IUs of vitamin D [72].

There is some evidence to suggest that vitamin D itself has its own health benefits. When screening for compounds that were most effective in stabilizing endothelial membranes, vitamin D$_3$ was found to be much more effective than its metabolites as well as a multitude of other compounds [74]. Extremely high doses, in the range of 20,000–50,000 IUs daily, have been effective in treating autoimmune disorders including psoriasis, multiple sclerosis, rheumatoid arthritis and vitiligo. The toxicity that is associated with ingesting these high doses of vitamin D is mitigated by having the patients on an extremely low calcium diet and maintaining good hydration [67,75].

It is likely in the future with the advent of the UVB-emitting LEDs that a variety of devices will become available for personalized in-home use for generating adequate amounts of vitamin D in the skin [76]. At this time, it is generally accepted that obtaining vitamin D from supplements and diet is the same as obtaining vitamin D$_3$ from sun exposure. There is evidence that during sun exposure, not only previtamin D$_3$ is produced in the skin, but there are other photoproducts of both previtamin D$_3$ and vitamin D$_3$ that have unique biologic activities not related to the classical nuclear function through the VDR [77] (Figure 7).

The reintroduction of 25(OH)D$_3$ (calcifediol) for the treatment of vitamin D deficiency offers the advantage of its hydrophilic properties by being more bioavailable in patients with fat malabsorption syndromes and obesity. In the future, we may see a combination supplement containing 25(OH)D$_3$ and vitamin D$_3$ [78].

The enthusiasm for the health benefits of vitamin D, which was initiated 100 years ago with its identification by McCollum, has not abated and continues to prosper.

Funding: This research received no external funding.

Acknowledgments: I am grateful for the editorial assistant of Nazli Ucar, Nipith Charoenngam and Arash Shirvani.

Conflicts of Interest: Grants from Carbogen-Amcis, Solius Inc., consultant for Solius Inc. speaker's Bureau Pulse LTD. and Menarini Inc.

References

1. Holick, M. Phylogenetic and evolutionary aspects of vitamin D from phytoplankton to humans. In *Verebrate Endocrinology;* Pang, P.K.T., Schreibman, M.P., Eds.; Fundamentals and Biomedical Implications Academic Press, Inc. (Harcourt Brace Jovanovich): Orlando, FL, USA, 1989; Volume 3, pp. 7–43.
2. Holick, M.F. Vitamin D: A millenium perspective. *J. Cell. Biochem.* **2003**, *88*, 296–307. [CrossRef] [PubMed]
3. Pasanen, A.L.; Yli-Pietila, K.; Pasanen, P.; Kalliokoski, P.; Tarhanen, J. Ergosterol content in various fungal species and biocontaminated building materials. *Appl. Environ. Microbiol.* **1999**, *65*, 138–142. [CrossRef] [PubMed]
4. Holick, M.F.; Stoddard, P.K. Mosquitoes Exposed to Sunlight in Florida Are Capable Making Vitamin D3. *Anticancer. Res.* **2022**, *42*, 5091–5094. [CrossRef]
5. Holick, M.F. McCollum Award Lecture, 1994: Vitamin D-new horizons for the 21st century. *Am. J. Clin. Nutr.* **1994**, *60*, 619–630. [CrossRef] [PubMed]
6. Smerdon, G.T. Daniel Whistler and the English disease; a translation and biographical note. *J. Hist Med. Allied. Sci.* **1950**, *5*, 397–415. [CrossRef] [PubMed]
7. Holick, M.F. Resurrection of vitamin D deficiency and rickets. *J. Clin. Investig.* **2006**, *116*, 2062–2072. [CrossRef] [PubMed]
8. Wacker, M.; Holick, M.F. Sunlight and Vitamin D: A global perspective for health. *Derm. Endocrinol.* **2013**, *5*, 51–108. [CrossRef]
9. Mozołowski, W. Jędrzej Sniadecki (1768–1838) on the Cure of Rickets. *Nature* **1939**, *143*, 121. [CrossRef]
10. Palm, T.A. The geographical distribution and etiology of rickets. *Practitioner* **1890**, *45*, 270–342.
11. Mellanby, E. Nutrition Classics. The Lancet 1:407-12, 1919. An experimental investigation of rickets. *Nutr. Rev.* **1976**, *34*, 338–340. [CrossRef]
12. McCollum, E.V.; Simmonds, N.; Becker, J.E.; Shipley, P.G. Studies on experimental rickets: XXI. An experimental demonstration of the existence of a vitamin which promotes calcium deposition. *J. Biol. Chem.* **1922**, *53*, 293–312. [CrossRef]
13. Huldschinsky, K. Heilung von Rachitis durch künstliche Höhensonne. *Dtsch. Med. Wochenschr.* **1919**, *45*, 712–713. [CrossRef]
14. Huldschinsky, K. *The Ultra-Violet Light Treatment of Rickets;* Alpine Press: Trenton, NJ, USA, 1928; pp. 3–19.
15. Hess, A.F.; Unger, L.J. The cure of infantile rickets by sunlight. *J. Am. Med. Assoc.* **1921**, *77*, 39.
16. Powers, G.F.; Park, E.A.; Shipley, P.G.; McCollum, E.V.; Simmonds, N. The prevention of rickets in the rat by means of radiation with the mercury vapor quartz lamp. *Proc. Soc. Exp. Biol. Med.* **1921**, *19*, 120–121. [CrossRef]
17. Steenbock, H.; Black, A. The reduction of growth-promoting and calcifying properties in a ration by exposure to ultraviolet light. *J. Biol. Chem.* **1924**, *61*, 408–422.
18. Hess, A.F.; Weinstock, M. Antirachitic properties imparted to inert fluids and to green vegetables by ultraviolet irradiation. *J. Biol. Chem.* **1924**, *62*, 301–313. [CrossRef]
19. Palermo, N.E.; Holick, M.F. Vitamin D, bone health, and other health benefits in pediatric patients. *J. Pediatr. Rehabil. Med.* **2014**, *7*, 179–192. [CrossRef]
20. Vitamin D toxicity A British Paediatric Association Report. Infantile hypercalcaemia, nutritional rickets, and infantile scurvy in Great Britain. *Br. Med. J.* **1964**, *1*, 1659–1661. [CrossRef]
21. Pober, B.R. Williams-Beuren syndrome. *N. Engl. J. Med.* **2010**, *362*, 239–252. [CrossRef] [PubMed]
22. Windaus, A.; Linsert, O.; Lüttringhaus, A.; Weidlich, G. Crystalline Vitamin D2. *Ann. N. Y. Acad. Sci.* **1932**, *492*, 226.
23. Askew, F.A.; Bourdillon, R.B.; Bruce, H.M.; Jenkins, R.G.C.; Webster, T.A. The distillation of Vitamin D. *Proc. Roy. Soc.* **1931**, *B107*, 76.
24. Waddell, J. The provitamin D of cholesterol. I. The antirachitic efficacy of irradiated cholesterol. *J. Biol. Chem.* **1934**, *105*, 711. [CrossRef]
25. Schenck, F. Über das kristallisierte Vitamin D3. *Naturwissenschaftem* **1937**, *25*, 159. [CrossRef]
26. Hossein-nezhad, A.; Holick, M.F. Vitamin D for Health: A Global Perspective. *Mayo Clin. Proc.* **2013**, *88*, 720–755. [CrossRef]
27. Orr, W.; Holt, L.E., Jr.; Wilkins, L.; Boone, F.H. The Calcium And Phosphorus Metabolism In Rickets With Special Reference to Ultraviolet Ray Therapy. *Am. J. Dis. Child* **1923**, *26*, 362. [CrossRef]
28. Nicolaysen, R.; Eeg- Larsen, N. The Mode of action of vitamin D. In *Ciba Foundayion Symposium on Bone Structure and Metabolism;* Wolstenholme, G.W., O'Connor, C.M., Eds.; Little Brown: Boston, MA, USA, 1956; pp. 175–186.
29. Carlsson, A. Tracer experiments on the effect of vitamin D on the skeletal metabolism of calcium and phosphorus. *Acta. Physiol. Scand* **1952**, *26*, 212. [CrossRef]
30. Holick, M.F. Vitamin D Deficiency. *N. Engl. J. Med.* **2007**, *357*, 266–281. [CrossRef]
31. Kodicek, E. Metabolic Studies on Vitamin D. In *Ciba Foundation Symposium on Bone Structure and Metabolism;* Wolstenholme, G.W., O'Connor, C.M., Eds.; Little Brown: Boston, MA, USA, 1956; pp. 161–174.
32. Norman, A.W.; Deluca, H.F. The preparation of super prescription of H3 vitamins D2 and D3—Their localization in the rat. *Biochemistry* **1963**, *2*, 1160–1168. [CrossRef]
33. Blunt, J.W.; DeLuca, H.F.; Schnoes, H.K. 25-hydroxycholecalciferol. A biologically active metabolite of vitamin D3. *Biochemistry* **1968**, *7*, 3317–3322. [CrossRef]
34. Suda, T.; DeLuca, H.F.; Schnoes, H.K.; Blunt, J.W. 25-hydroxyergocalciferol. A biologically active metabolite of vitamin D2. *Biochemistry* **1969**, *8*, 3515. [CrossRef]
35. Holick, M.F.; DeLuca, H.F. A new chromatographic system for vitamin D_3 and its metabolites: Resolution of a new vitamin D_3 metabolite. *J. Lipid Res.* **1971**, *12*, 460–465. [CrossRef]

36. Holick, M.F.; DeLuca, H.F.; Avioli, L.V. Isolation and identification of 25-hydroxycholecalciferol from human plasma. *Arch. Intern. Med.* **1972**, *129*, 56–61. [CrossRef]
37. Haussler, M.R.; Myrtle, J.F.; Norman, A.W. The association of a metabolite of vitamin D_3 with intestinal mucosa chromatin in vivo. *J. Biol. Chem.* **1968**, *243*, 4055–4064. [CrossRef]
38. Lawson, D.E.M.; Wilson, P.W.; Kodicek, E. Metabolism of vitamin D. A new cholecalciferol metabolite, involving loss of hydrogen at C-1, in chick intestinal nuclei. *Biochem. J.* **1969**, *115*, 269–277. [CrossRef]
39. Fraser, D.R.; Kodicek, E. Unique biosynthesis by kidney of a biological active vitamin D metabolite. *Nature* **1970**, *228*, 764–766. [CrossRef]
40. Holick, M.F.; Schnoes, H.K.; DeLuca, H.F. Identification of 1, 25-dihydroxycholecalciferol, a form of vitamin D_3 metabolically active in the intestine. *Proc. Natl. Acad. Sci. USA* **1971**, *68*, 803–804. [CrossRef]
41. Lawson, D.E.M.; Fraser, D.R.; Kodicek, E.; Morris, H.R.; Williams, D.H. Identification of 1,25- dihydroxycholecalciferol, a new kidney hormone controlling calcium metabolism. *Nature* **1971**, *230*, 228–230. [CrossRef]
42. Norman, A.W.; Myrtle, J.F.; Midgett, R.J.; Nowicki, H.G.; Williams, V.; Popjak, G. 1, 25-dihydroxycholecalciferol: Identification of the proposed active form of vitamin D in the intestine. *Science* **1971**, *173*, 51–53. [CrossRef]
43. Semmler, E.J.; Holick, M.F.; Schnoes, H.K.; DeLuca, H.F. The synthesis of 1α, 25-dihydroxycholecalciferol—A metabolically active form of vitamin D_3. *Tetrahedron Lett.* **1972**, *40*, 4147–4150. [CrossRef]
44. Holick, S.A.; Holick, M.F.; McLaughlin, J.A. Chemical synthesis of [1β-^{3}H] 1α, 25-dihydroxyvitamin D_3: Biological activity of 1β, 25-dihydroxyvitamin D_3. *Biochem. Biophys. Res. Commun.* **1980**, *97*, 1031–1037. [CrossRef]
45. Jones, G.; Strugnell, S.; DeLuca, H.F. Current understanding of the molecular actions of vitamin D. *Physiol. Rev.* **1998**, *78*, 1193–1231. [CrossRef]
46. Holick, M.F.; Semmler, E.J.; Schnoes, H.K.; DeLuca, H.F. 1α -Hydroxy derivative of vitamin D_3: A highly potent analog of 1α, 25-dihydroxyvitamin D_3. *Science* **1973**, *180*, 190–191. [CrossRef]
47. Holick, M.F.; Holick, S.A.; Tavela, T.; Gallagher, B.; Schnoes, H.K.; DeLuca, H.F. Synthesis of [6-^{3}H]-1α-hydroxyvitamin D_3 and its metabolism in vivo to [6-^{3}H]-1α, 25-dihydroxyvitamin D_3. *Science* **1975**, *190*, 576–578. [CrossRef]
48. Chan, J.C.M.; Oldham, S.; Holick, M.F.; DeLuca, H.F. The use of 1α-hydroxyvitamin D_3, in chronic renal failure. *J. Am. Med. Assoc.* **1975**, *234*, 47–52. [CrossRef]
49. Silverberg, D.S.; Bettcher, K.B.; Dossetor, J.B.; Overton, T.R.; Holick, M.F.; DeLuca, H.F. Effect of 1, 25-dihydroxycholecalciferol in renal osteodystrophy. *Can. Med. Assoc. J.* **1975**, *112*, 190–195.
50. Balsan, S.; Garabedian, M.; Sorgniard, R.; Holick, M.F.; DeLuca, H.F. 1,25-Dihydroxyvitamin D_3 and 1α-hydroxyvitamin D_3 in children: Biologic and therapeutic effects in nutritional rickets and different types of vitamin D resistance. *Pediat Res.* **1975**, *9*, 586–593. [CrossRef]
51. Fraser, D.; Kooh, S.W.; Kind, J.P.; Holick, M.F.; Tanaka, Y.; DeLuca, H.F. Pathogenesis of hereditary vitamin D-dependent rickets. An inborn error of vitamin D metabolism involving defective conversion of 25-hydroxyvitamin D to 1α, 25-dihydroxyvitamin D_3. *N. Engl. J. Med.* **1973**, *289*, 817–822. [CrossRef]
52. Kooh, S.W.; Fraser, D.; DeLuca, H.F.; Holick, M.F.; Belsey, R.E.; Clark, M.B.; Murray, T.M. Treatment of hypoparathyroidism and pseudo hypoparathyroidism with metabolites of vitamin D: Evidence for impaired conversion of 25-hydroxyvitamin D_3 to 1α, 25-dihydroxyvitamin D_3. *N. Engl. J. Med.* **1975**, *293*, 840–844. [CrossRef]
53. Neer, R.M.; Holick, M.F.; DeLuca, H.F.; Potts, J.T., Jr. Effects of 1α-hydroxyvitamin D_3 and 1, 25-dihydroxyvitamin D_3 on calcium and phosphorus metabolism in hypoparathyroidism. *Metabolism* **1975**, *24*, 1403–1413. [CrossRef]
54. Stumpf, W.E.; Clark, S.A.; Sar, M.; DeLuca, H.F. Topographical and developmental studies on target sites of 1,25$(OH)_2$-vitamin D_3 in skin. *Cell Tissue Res.* **1984**, *238*, 489–496. [CrossRef]
55. Tanaka, H.; Abe, E.; Miyaura, C.; Kuribayashi, T.; Konno, K.; Nishi, Y.; Suda, T. 1,25-Dihydroxycholeciferol and human myeloid leukemia cell line (HL-60): The presence of cytosol receptor and induction of differentiation. *Biochem. J.* **1982**, *204*, 713–719. [CrossRef]
56. Colston, K.; Colston, M.J.; Feldman, D. 1,25-Dihydroxyvitamin D_3 and malignant melanoma: The presence of receptors and inhibition of cell growth in culture. *Endocrinology* **1981**, *108*, 1083–1086. [CrossRef]
57. Honma, Y.; Hozumi, M.; Abe, E.; Konno, K.; Fukushima, M.; Hata, S.; Nishii, Y.; DeLuca, H.F. 1γ,25-Dihydroxyvitamin D_3 and 1γ-hydroxyvitamin D_3 prolong survival time of mice inoculated with myeloid leukemia cells. *Proc. Natl. Acad. Sci. USA* **1982**, *80*, 201–204. [CrossRef]
58. Koeffler, H.P.; Hirjik, J.; Itri, L. 1,25-Dihydroxyvitamin D_3: In vivo and in vitro effects on human preleukemic and leukemic cells. *Cancer Treat Rep.* **1985**, *69*, 1399–1407. [PubMed]
59. Jones, G.; Calverley, M.J. A dialogue on analogs: Newer Vitamin D drugs for use in bone disease, psoriasis and cancer. *TEM* **1993**, *4*, 297–303.
60. Bikle, D.D.; Nemanic, M.D.; Whitney, J.O.; Elias, P.O. Neonatal human foreskin keratinocytes produce 1,25-dihydroxyvitamin D3. *Biochemistry* **1986**, *25*, 1545–1548. [CrossRef]
61. Smith, E.; Walworth, N.C.; Holick, M.F. Effect of 1α, 25- dihydroxyvitamin D_3 on the morphological and biochemical differentiation of cultured human epidermal keratinocytes grown under serum-free conditions. *J. Investig. Dermatol.* **1986**, *86*, 709–714. [CrossRef] [PubMed]

62. Perez, A.; Chen, T.C.; Turner, A.; Raab, R.; Bhawan, J.; Poche, P.; Holick, M.F. Efficacy and safety of topical calcitriol (1,25-dihydroxyvitamin D_3) for the treatment of psoriasis. *Brit. J. Dermatol.* **1996**, *134*, 238–246. [CrossRef]
63. Perez, A.; Raab, R.; Chen, T.C.; Turner, A.; Holick, M.F. Safety and efficacy of oral calcitriol (1,25-dihydroxyvitamin D_3) for the treatment of psoriasis. *Brit J. Dermatol.* **1996**, *134*, 1070–1078.
64. Kragballe, K.; Gjertsen, B.; DeHoop, D.; Karlsmark, T.; van de Kerkhof, P.; Larko, O.; Nieboer, C.; Roed-Petersen, J.; Strand, A.; Tikjob, G. Double-blind, right/left comparison of calcipotriol and betamethasone valerate in treatment of psoriasis vulgaris. *Lancet* **1991**, *337*, 193–196. [CrossRef]
65. Holick, M.F. Will 1, 25-Dihydroxyvitamin D_3, MC 903, and their analogues herald a new pharmacologic era for the treatment of psoriasis? *Arch Dermatol.* **1989**, *125*, 1692–1697. [CrossRef] [PubMed]
66. Adams, J.S.; Ren, S.; Liu, P.T.; Chun, R.F.; Lagishetty, V.; Gombart, A.F.; Borregaard, N.; Modlin, R.L.; Hewison, M. Vitamin d-directed rheostatic regulation of monocyte antibacterial responses. *J. Immunol.* **2009**, *182*, 4289–4295. [CrossRef] [PubMed]
67. Charoenngam, N.; Holick, M.F. Immunologic Effects of Vitamin D on Human Health and Disease. *Nutrients* **2020**, *12*, 2097. [CrossRef] [PubMed]
68. Shirvani, A.; Kalajian, T.A.; Song, A.; Holick, M.F. Disassociation of Vitamin D's Calcemic Activity and Non-calcemic Genomic Activity and Individual Responsiveness: A Randomized Controlled Double-Blind Clinical Trial. *Sci. Rep.* **2019**, *9*, 17685. [CrossRef]
69. Garland, C.F.; Garland, F.C.; Gorham, E.D.; Lipkin, M.; Newmark, H.; Mohr, S.B.; Holick, M.F. The role of vitamin D in cancer prevention. *Am. J. Public Health* **2006**, *96*, 252–261. [CrossRef]
70. Grant, W.B. An estimate of the global reduction in mortality rates through doubling vitamin D levels. *Eur. J. Clin. Nutr.* **2011**, *65*, 1016–1026. [CrossRef]
71. Bae, J.H.; Choe, H.J.; Holick, M.F.; Lim, S. Association of Vitamin D status with COVID-19 and its severity. *Rev. Endocr. Metab. Disord.* **2022**, *23*, 579–599. [CrossRef]
72. Holick, M.F.; Binkley, N.C.; Bischoff-Ferrari, H.A.; Gordon, C.M.; Hanley, D.A.; Heaney, R.P.; Murad, M.H.; Weaver, C.M. Evaluation, Treatment & Prevention of Vitamin D Deficiency: An Endocrine Society Clinical Practice Guideline. *J. Clin. Endocrinol. Metab.* **2011**, *96*, 1911–1930.
73. Luxwolda, M.F.; Kuipers, R.S.; Kema, I.P.; Dijck-Brouwer, D.A.; Muskiet, F.A. Traditionally living populations in East Africa have a mean serum 25-hydroxyvitamin D concentration of 115 nmol/l. *Br. J. Nutr.* **2012**, *108*, 1557–1561. [CrossRef]
74. Gibson, C.C.; Davis, C.T.; Zhu, W.; Bowman- Kirigin, J.A.; Walker, A.E.; Tai, Z.; Thomas, K.R.; Donato, A.J.; Lesniewski, L.A.; Li, D.Y. Dietary vitamin D and its metabolites non-genomically stabilize the endothelium. *PLoS ONE* **2015**, *10*, e0140370. [CrossRef]
75. Mahtani, R.; Nair, P.M.K. Daily oral vitamin D3 without concomitant therapy in the management of psoriasis: A case series. *Clin. Immunol. Commun.* **2022**, *2*, 17–22. [CrossRef]
76. Kalajian, T.A.; Aldoukhi, A.; Veronikis, A.J.; Persons, K.; Holick, M.F. Ultraviolet B Light Emitting Diodes (LEDs) Are More Efficient and Effective in Producing Vitamin D3 in Human Skin Compared to Natural Sunlight. *Sci. Rep.* **2017**, *7*, 11489. [CrossRef] [PubMed]
77. Slominski, A.T.; Chaiprasongsuk, A.; Janjetovic, Z.; Kim, T.K.; Stefan, J.; Slominski, R.M.; Hanumanthu, V.S.; Raman, C.; Qayyum, S.; Song, Y.; et al. Photoprotective Properties of Vitamin D and Lumisterol Hydroxyderivatives. *Cell Biochem. Biophys.* **2020**, *78*, 165–180. [CrossRef]
78. Charoenngam, N.; Mueller, P.M.; Holick, M.F. Evaluation of 14-day Concentration-time Curves of Vitamin D3 and 25-Hydroxyvitamin D3 in Healthy Adults with Varying Body Mass Index. *Anticancer Res.* **2022**, *42*, 5095–5100. [CrossRef] [PubMed]

MDPI

Review

Vitamin D in the Context of Evolution

Carsten Carlberg [1,2]

[1] Institute of Animal Reproduction and Food Research, Polish Academy of Sciences, 10-748 Olsztyn, Poland; c.carlberg@pan.olsztyn.pl
[2] School of Medicine, Institute of Biomedicine, University of Eastern Finland, 70211 Kuopio, Finland

Abstract: For at least 1.2 billion years, eukaryotes have been able to synthesize sterols and, therefore, can produce vitamin D when exposed to UV-B. Vitamin D endocrinology was established some 550 million years ago in animals, when the high-affinity nuclear receptor VDR (vitamin D receptor), transport proteins and enzymes for vitamin D metabolism evolved. This enabled vitamin D to regulate, via its target genes, physiological process, the first of which were detoxification and energy metabolism. In this way, vitamin D was enabled to modulate the energy-consuming processes of the innate immune system in its fight against microbes. In the evolving adaptive immune system, vitamin D started to act as a negative regulator of growth, which prevents overboarding reactions of T cells in the context of autoimmune diseases. When, some 400 million years ago, species left the ocean and were exposed to gravitation, vitamin D endocrinology took over the additional role as a major regulator of calcium homeostasis, being important for a stable skeleton. *Homo sapiens* evolved approximately 300,000 years ago in East Africa and had adapted vitamin D endocrinology to the intensive exposure of the equatorial sun. However, when some 75,000 years ago, when anatomically modern humans started to populate all continents, they also reached regions with seasonally low or no UV-B, i.e., and under these conditions vitamin D became a vitamin.

Keywords: vitamin D; evolution; energy metabolism; immune system; calcium homeostasis; migration of *Homo sapiens*

Citation: Carlberg, C. Vitamin D in the Context of Evolution. *Nutrients* **2022**, *14*, 3018. https://doi.org/10.3390/nu14153018

Academic Editor: Bruce W. Hollis

Received: 4 July 2022
Accepted: 21 July 2022
Published: 22 July 2022

1. Introduction

Evolution is a dominant driver of the development of biological processes and their adaption to changes in the environment. The statement "Nothing in biology makes sense except in the light of evolution" [1] underlines that evolution is an essential component for understanding the mechanisms of these processes. Accordingly, this review will discuss vitamin D, its nuclear receptor (NR) VDR and their molecular action in the light of evolution.

It was exactly 100 years ago that vitamin D was named a vitamin, because it is able to cure experimentally induced rickets in dogs and rats [2]. Since rickets is a bone malformation disorder in children, this and many other studies linked vitamin D to calcium homeostasis and bone remodeling [3]. However, calcium homeostasis is only one of multiple biological processes being regulated by vitamin D, such as detoxification, energy metabolism as well as innate and adaptive immunity [4]. In fact, the relationship between vitamin D and bone remodeling developed as one of the most recent evolutionary functions of vitamin D.

Furthermore, the consequences of the migration of modern humans from equatorial East Africa to regions of higher latitude will be reflected in relation to vitamin D's possible role in skin lightening, particularly in European populations [5].

2. Sterols and Vitamin D Synthesis

Sterols are lipophilic molecules carrying a four-ring skeleton, a hydroxyl group at carbon (C) 3 and a flexible side chain at C17 (Figure 1). The use of sterols in cell membranes is a characteristic difference between eu- and prokaryotes [6]. Some 2.45 billion years ago, atmospheric molecular oxygen (O_2) concentrations drastically rose in the so-called great oxidation event [7] (Figure 2). This stimulated the development of complex eukaryotes with the help of new enzymes and biochemical pathways [8]. A key example is the biosynthesis of sterols, where four different types of enzymes require O_2. Moreover, aerobic metabolism, such as oxidative phosphorylation, significantly increases the generation of energy from nutrients. In parallel, some of these new enzymes and pathways had the primary role of protecting from oxygen toxicity, which also may have been the primary role of sterols [9]. Thus, the occurrence of oxygen and sterols are tangled; without oxygen there is no sterol synthesis and many sterols protect from damage created by O_2 and reactive oxygen species. Furthermore, some sterols can be considered as oxygen sensors [10].

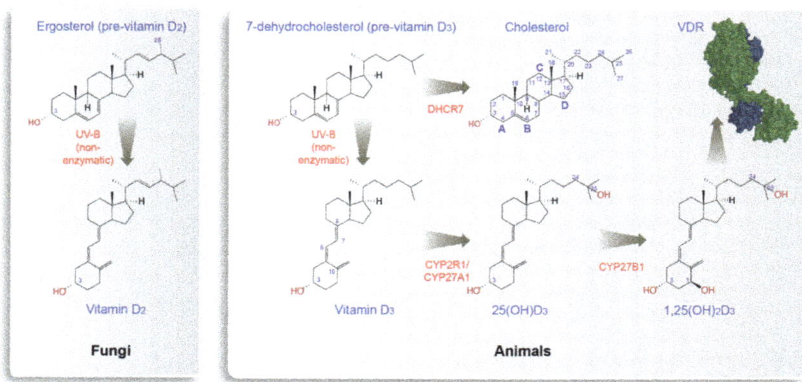

Figure 1. Principles of endogenous vitamin D_2 and vitamin D_3 production. In fungi, vitamin D_2 is produced non-enzymatically when the sterol ergosterol is exposed to UV-B radiation. In cholesterol synthesizing animals (as well as in some plants, such as phytoplankton), 7-dehydrocholesterol reacts to vitamin D_3 using the energy of UV-B. Animals express enzymes that convert vitamin D_3 first to 25(OH)D_3 and then to 1,25(OH)$_2$$D_3$. In animals (but not in fungi), vitamin D endocrinology developed and 1,25(OH)$_2$$D_3$ acts as a hormone binding with high affinity to the nuclear receptor VDR (green, the co-receptor RXR is shown in blue). In the example of cholesterol, the numbering of rings (A-D) and carbons (1-27) is indicated, while only key carbons are marked in the other molecules.

There is a large number of naturally occurring sterols and species can be phylogenetically distinguished by their sterol profile. Sterols are primarily distinguished by the modification at C24 in their side chain. In animals, 24-desmethylsterols such as cholesterol (no additional group, 27 carbon atoms in total) are typical, while fungi have 24-demethylsterols, such as ergosterol (28 carbon atoms) (Figure 1). In contrast, plants produce a wide range of more than 250 different sterols, the most common of which are sitosterol, campesterol and stigmasterol [11]. Interestingly, cholesterol represents 1–2% of the plant sterol content, i.e., cholesterol is not unique to animals but can also be produced at least by some plants, such as algae [12]. Cholesterol is not only critical for membrane fluidity but also an important precursor for bile acids and steroid hormones [13]. Most eukaryotic species, including humans, can synthesize sterols de novo, but some others, such as insects, depend on a supply of sterols via their diet [14].

When sterols like ergosterol (pre-vitamin D_2) in fungi or the direct cholesterol precursor 7-dehydrocholesterol (pre-vitamin D_3) in animals and phytoplankton are exposed to UV-B radiation of 280–315 nm, they transform non-enzymatically to vitamin D_2 and vitamin D_3, respectively (Figure 1). First, the double bond between C7 and C8 of both types of sterols absorbs the energy of the radiation, which creates thermodynamically unstable pre-vitamin D molecules, in which their B ring is opened between C9 and C10, creating secosteroids [15]. Then elevated temperature catalyzes the isomerization of pre-vitamin D into vitamin D. Since both reactions do not require any enzyme, it is likely that vitamin D_2 and vitamin D_3 are evolutionary very old molecules that occurred as early as the ergosterol and cholesterol biosynthesis pathways evolved some 1.2 billion years ago (Figure 2). For example, for at least 750 million years, phytoplankton has produced vitamin D_3 [16]. However, in early evolution, vitamins D_2 and vitamin D_3 may have been primarily side products of sterol biosynthesis in UV-B exposure, i.e., they had no signaling function since a respective endocrine system (see Section 3) had not evolved. Interestingly, continuous UV-B exposure can convert pre-vitamin D_3 into lumisterol, tachysterol and other photoproducts [17], i.e., vitamin D_3 precursors and metabolites are able to perform UV scavenging by rearranging double bonds within the molecule [18]. Thus, the pre-endocrine function of vitamin D was and still is the protection of DNA and proteins from mutagenesis and degradation, respectively.

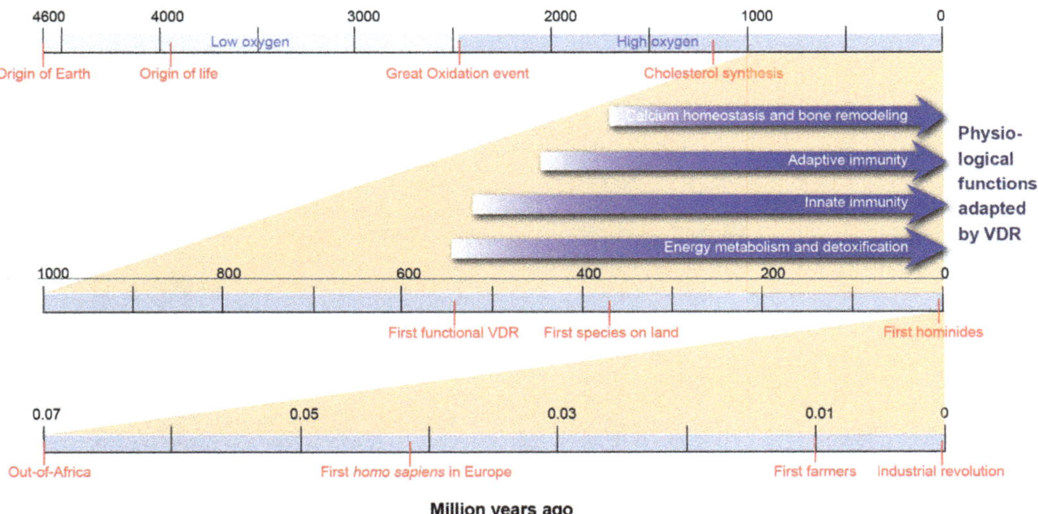

Figure 2. Timeline of evolution. The evolutionary history of the past 4.6 billion years (**top**), 1 billion years (**center**) and 70,000 years (**bottom**) are depicted. Important events discussed in this review are indicated and the time of their approximate occurrence is marked.

Interestingly, the accumulation of vitamin D_3 in the marine food chain [12,19] explains why salmon, as well as the liver of cod ("cod liver oil"), have a high content of the vitamins [20,21].

3. Evolution of Vitamin D Endocrinology

The core protein of an endocrine system is its receptor. A high-affinity receptor for vitamin D, the transcription factor VDR evolved some 550 million years ago [22]. However, contrary to its name, VDR is activated neither by vitamin D_2 nor by vitamin D_3 [23]. Carrying only one hydroxy group, both secosteroids are not polar enough to bind VDR. In fact, two hydroxylation reactions are required to form with 1α,25-dihydroxyvitamin D_3 (1,25(OH)$_2$D$_3$) a vitamin D metabolite that offers three hydroxy groups for specific high-affinity binding to the ligand-binding domain (LBD) of VDR (Figure 1). This implies that the 25-hydroxylases cytochrome P450 (CYP) 2R1 and CYP27A1, as well as the 1α-hydroxylase CYP27B1, are key components of vitamin D endocrinology. They transform vitamin D_3 into 25-hydroxyvitamin D_3 (25(OH)D$_3$) and 25(OH)D$_3$ into 1,25(OH)$_2$D$_3$, respectively. Furthermore, as described for other hormones, the levels of 1,25(OH)$_2$D$_3$ need to be tightly regulated. This happens via the 24-hydroxylase CYP24A1 that converts 1,25(OH)$_2$D$_3$ to 1,24,25(OH)$_3$D$_3$ and inactivates the VDR ligand in this way [24]. Despite their hydroxyl groups, all vitamin D metabolites are lipophilic and need to be carried in hydrophilic serum and cellular liquids by transport proteins. Thus, for a functional endocrine system, specific receptor(s), metabolizing enzymes and transport proteins need to evolve [25].

VDR belongs to the transcription factor family of NRs, which in humans is formed by 48 genes [26]. Comparative genomics demonstrates that the closest relatives to VDR are the NR1I subfamily members pregnane X receptor (PXR) and constitutive androstane receptor (CAR), and the NR1H subfamily members liver X receptor (LXR) α and b, as well as the farnesoid X receptor (FXR) [27]. This indicates that VDR and its five relatives have a common ancestor and that the individual receptor genes developed by whole genome duplications in early vertebrate evolution [28]. Interestingly, the six NRs function as sensors for cholesterol derivatives, such as $1,25(OH)_2D_3$, oxysterols and bile acids [29]. Moreover, FXR, VDR, CAR and PXR detect toxic secondary bile acids, such as lithocholic acid, and get activated by them [30–33]. This suggests that the prime function of the common ancestor of NR1H- and NR1I-type NRs was to act as a bile acid sensor. Accordingly, one of the first functions of VDR and its relatives was the regulation of genes encoding for enzymes of marine biotoxin degradation [4,34].

Detoxification reactions represent a specialized form of metabolism that allows a response to environmental conditions, such as the rise in toxic compounds. However, the most dominant environmental challenge of species is their diet, which is primarily composed of macro- and micronutrients. This created an evolutionary pressure, with the push of which the sensing of the levels of nutritional molecules like fatty acids, cholesterol and vitamins became the main function of NRs, such as peroxisome proliferator-activated receptors (PPARs), LXRs, retinoid acid receptors (RARs) and VDR [35,36]. This function is closely linked to the control of energy metabolism, which was and still is a prime task of many NRs, including VDR [37]. Accordingly, a significant proportion of the hundreds of VDR targets are metabolic genes [38–41].

Archetypical NRs were orphan receptors, as some members of the NR superfamily still are [42]. Comparative genomics suggests that in a stepwise evolutionary adaption, orphan, NRs changed critical amino acids within their LBD, so that a ligand-binding pocket got accessible to potential small lipophilic ligands. The 40 or more amino acids forming this pocket are specifically adapted to the shape and polarity of the ligand. Some 550 million years ago, this evolutionary adaptation process resulted in the first known VDR that binds $1,25(OH)_2D_3$ at sub-nanomolar concentrations was found in the early jawless vertebrate sea lamprey (*Petromyzon marinus*) [22], meaning that VDR had evolved into a classical endocrine receptor, such as those for the steroid hormones estrogen, testosterone and progesterone. Crystal structure analysis of lamprey's VDR ligand-binding domain [43] confirmed similar binding of $1,25(OH)_2D_3$ as identified for human VDR [44]. In vertebrate evolution, amphibians, reptiles, bony fish, birds and mammals also learned to express functional VDR proteins [45]. Most species have only one *VDR* gene, but the genome of teleost fishes underwent a third whole genome duplication and contains even two *VDR* genes [46].

Since the levels of $1,25(OH)_2D_3$ in lamprey are similar to that in higher vertebrates, respective enzymes, such as CYP2R1 and CYP27B1, must have co-evolved with VDR [22]. Similar co-evolution also happened for the vitamin D transport protein vitamin D binding protein (encoded by the *GC* gene) [25]. This indicates that some 150 million years before the first species left the ocean and had the need for a stable skeleton, vitamin D endocrinology was already established. Thus, from an evolutionary perspective, the control of calcium homeostasis was rather a secondary than a primary goal for establishing the vitamin D endocrine system.

4. Evolution of the Physiological Functions of Vitamin D

Possible harming invaders created since the early times of life on Earth a strong evolutionary pressure for developing defense mechanisms, such as an immune system. The innate immune system is evolutionarily older and found already in many non-vertebrate species, such as insects. It involves a number of barriers, such as skin and mucosa, and uses a limited set of pattern recognition receptors that detect only general features of possible pathogens. In contrast, the adaptive immune system developed some 500 million years ago in ectothermic cartilage fishes and uses antigen receptors, such as B and T cell receptors, that have a very high affinity and specificity to their antigens [47,48].

The growth of immune cells and their function in defense and tissue repair takes significant amounts of energy [49]. Key vitamin D target genes in this context are *PFKFB4* (6-phosphofructo-2-kinase/fructose-2,6-biphosphatase 4) in dendritic cells [50] and *FBP1* (fructose-bisphosphatase 1) in monocytes [51]. Therefore, the regulatory function of vitamin D and its receptor on energy metabolism were essential during the development of the immune system (Figure 2). Moreover, vitamin D modulates innate immunity through further target genes, such as those encoding for the antimicrobial peptide CAMP (cathelicidin) [52] and the toll-like receptor 4 co-receptor CD14 [51] in monocytes. Furthermore, in dendritic cells, which present antigens to T cells of the adaptive immune system, many genes respond to vitamin D [53]. In this way, vitamin D was and still is involved in efficient responses to pathogens, such as the intracellular bacterium *Mycobacterium tuberculosis* [54]. Moreover, the cluster of *HLA* (human leukocyte antigen) genes on human chromosome 6, many of which are vitamin D targets [41], is a "hotspot" of vitamin D-induced chromatin accessibility [55]. Thus, most non-skeletal functions of vitamin D, like the modulation of the immune system, developed before its regulation of calcium homeostasis and bone remodeling had been established (Figure 2).

Some 385 million years ago, the next important step in vertebrate evolution happened: some species moved from the ocean onto land and had to develop a skeleton supporting locomotion under gravitational forces [25] (Figure 2). At earlier times, calcified cartilage and dermal bone had already been developed by cartilage fishes like sharks. Bone fishes even had replaced this cartilage with bone [56]. In the calcium-rich environment of water (approximately 10 mM), this transformation was not limited by calcium abundance. However, the calcium-poor conditions on land created an evolutionary pressure to tightly regulate the concentration of calcium in intra- and extracellular compartments of the body. Since the largest amounts of calcium are stored in bones, they serve as reservoirs to balance variations in the supply of the mineral by diet. In this process, vitamin D, as well as the peptide hormone PTH (parathyroid hormone), took the lead role. For example, the calcium channel TRPV6 (transient receptor potential cation channel subfamily V member 6), as well as the calcium-binding proteins CALB1 (calbindin 1) and CALB2 are encoded by vitamin D target genes [57].

Vitamin D regulates the activity of bone-resorbing osteoclasts by the cytokine RANKL, which is encoded by the vitamin D target gene *TNFSF11* (TNF superfamily member 11) [58]. Moreover, also bone mineralization is controlled by proteins encoded by vitamin D target genes, such as SPP1 (osteopontin) and BGLAP (bone gamma-carboxyglutamate protein, also called osteocalcin). Bone remodeling, i.e., the resorption of extracellular matrix by osteoclasts as well as bone formation by osteoblasts, requires, like immune functions, also a lot of energy. [59]. Thus, bone remodeling and immunity are connected via their dependency on energy metabolism [60]. Moreover, hematopoietic stem cells find in the interior of large bones, the bone marrow, a niche, i.e., a place where proliferating immune cells are effectively shielded from radiation and, in parallel, supported by calcium. Interestingly, already in bone fishes like zebrafish (*Danio rerio*) vitamin D regulates hematopoietic cell growth during embryogenesis [61]. Thus, the close connection of calcium homeostasis and bone remodeling to immunity, i.e., the co-evolution of both systems, illustrates why vitamin D shifted into this additional task.

Taken together, vitamin D evolved from one of the multiple factors controlling (energy) metabolism and immunity to a dominant regulator of calcium homeostasis and bone remodeling. This explains why bone malformations were observed as the first symptom of vitamin D deficiency.

5. How Does the Evolution of *Homo sapiens* Relate to Vitamin D?

The anatomically modern human (*Homo sapiens*) evolved just some 300,000 years ago in East Africa [62] and spread then over the whole continent. In order to protect from sunburn and skin cancer induced by the intensive equatorial sun, the skin of these humans was profoundly pigmented [63,64]. Despite dark pigmentation, there was and still is sufficient vitamin D_3 synthesis [65]. Just some 75,000 years ago, modern humans started to migrate to Asia and from there to Oceania, Europe and the Americas [66,67] (Figure 2). In Europe and in northern parts of Asia they experienced cold winter climates that let them cover their skin by clothes. In addition, the intensity of UV-B is at higher northern latitudes far lower, and in winter, for a few months, the radiation does not reach the surface [68]. Both clothing and northern latitude reduced the amount of vitamin D_3 produced in the skin and could cause vitamin D deficiency. The medical consequences of vitamin D deficiency, bone malformations and reduced potency of the immune system, may have created an evolutionary pressure that could have pushed for a reduced skin pigmentation [69], in order to explain today's North–South gradient in skin color [70].

Skin pigmentation depends on the load of keratinocytes with melanosomes [71], which are melanin-loaded organelles that origin from melanocytes [72] (Figure 3). The UV absorbing pigment melanin is produced via oxidation and polymerization of the aromatic amino acid tyrosine. The brown/black eumelanin is the most common form of melanin, while pheomelanin is yellow/red [73]. The difference in skin pigmentation of human populations and individuals (as well as that of their eyes and hair) primarily depends on SNPs (single nucleotide polymorphisms) in genes encoding for key proteins in melanogenesis [72]. The most relevant SNPs are those related to the genes *SLC24A5* (solute carrier family 24 member 5), *SLC45A2* and *OCA2* (OCA2 melanosomal transmembrane protein) [67,74] that encode for a potassium-dependent sodium/calcium exchanger, an ion transporter and a pH regulator in melanosomes, respectively [72]. Thus, the loss of function of these key proteins in melanin production leads to reduced skin pigmentation.

Modern humans arrived in Europe some 42,000 years ago [75,76] with dark skin like their African ancestors, but many of them had blue eyes due to variations of their *OCA2* gene [75,77] (Figure 3). By interbreeding, they outnumbered the ancestral Neanderthal hominins, which had lived in Europe already for some 400,000 years [77–79]. In net effect, today's Europeans have, on average, 2.3% Neanderthal DNA in their genomes [80]. These hunter–gatherers lived first in ice-free southwestern Europe [81] and started some 11–12,000 years ago to colonize also northern Europe [75]. Based on archeogenomic data the evolution and timing of trait changes within European populations had been discovered [82] (Figure 3). First, some 8400 to 6000 years ago, people from northwestern Anatolia spread over southern Europe. These Anatolian farmers started the Neolithic revolution in Europe by introducing the concept of agriculture, i.e., the domestication of animal and plant species, to the hunter-gatherers. Moreover, by interbreeding with the indigenous European population, the Anatolian farmers also brought them their *SLC24A5* gene variant for lighter skin. In a second wave, some 5000 years ago, Yamnaya pastoralists from the Eurasian steppe arrived in Europe and settled preferentially in the North (Figure 3). They introduced the horse, the wheel, their Indo-European languages as well as lighter skin due to SNPs in their *SLC45A2* and *SLC24A5* genes to the preexisting European populations [83–85]. Thus, the relative admixture of the hunter-gatherers, Anatolian farmers and Yamnaya pastoralists explains the variation in skin color (as well as many other traits) of present Europeans.

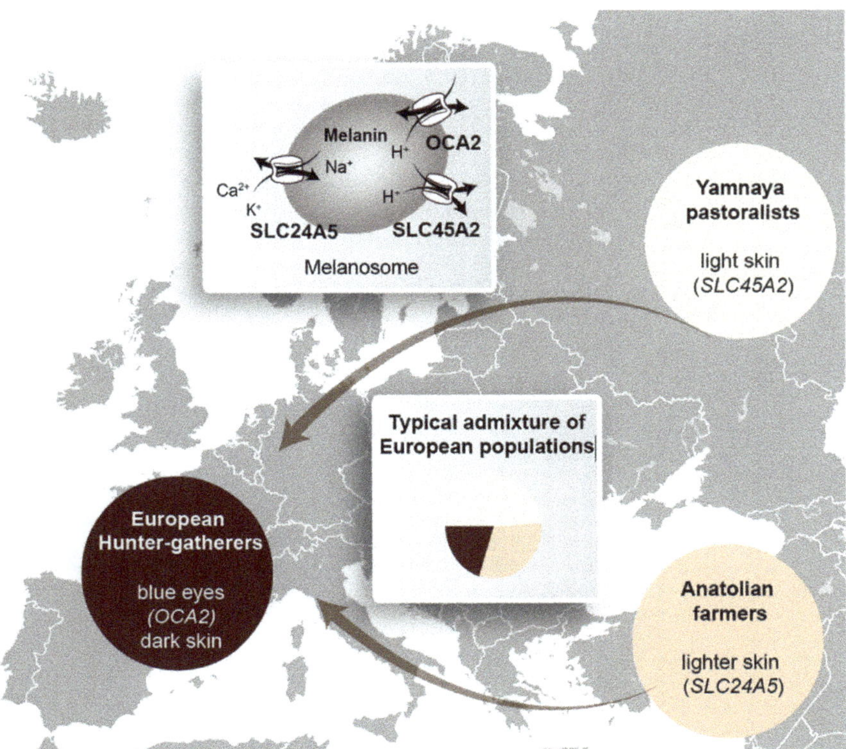

Figure 3. Schematic representation of the admixture of the European population. All European populations derived from European hunter–gatherers, Anatolian farmers and Yamnaya pastoralists [84]. A pie chart indicates the typical admixture of the founding populations (**center**). Melanin is produced in melanosomes under the control of the proteins OCA2, SLC24A5 and SLC45A2 (**top**).

Archeogenomic data demonstrated that *Homo sapiens* hunter-gatherer populations lived in Europe with dark skin for more than 30,000 years. Did they suffer from consequences of vitamin D deficiency? With the exception of people living an urban lifestyle already some 2000 years ago in the Roman empire [86], older bone samples do not show signs of malformation. Explanations could be a dominant outdoor lifestyle in southern Europe (the rest of Europe was covered by ice) or, in part, vitamin D_3 supplementation via a marine-based diet for populations living close to the coast. However, the most dominant effects were SNPs in the regulatory regions of the *DHCR7* (7-dehydrocholesterol reductase) gene, which reduced its expression, and by this, its enzymatic activity [82,87]. The resulting increased concentrations of 7-dehydrocholesterol in the skin led then to a more efficient synthesis of vitamin D_3 (Figure 1). Genome-wide association studies (GWAS) confirmed that the vitamin D status (as measured by 25(OH)D_3 serum levels) significantly depends on SNPs of the *DHCR7* gene [88]. Furthermore, other GWAS demonstrated the dependence of the vitamin D status on genes related to vitamin D endocrinology, such as *CYP2R1*, *CYP24A1* and *GC*, but not to skin color [89,90]. This suggests that at least in (western) Europe, skin lightening did not happen due to an evolutionary pressure caused by vitamin D deficiency but by interbreeding with populations from northwestern Anatolia and the northern Caucasus. Thus, the light skin color of today's Europeans is primarily based on the migration of populations from western Asia and the Near East to Europe [75]. Nevertheless, concerning their vitamin D status and their ability to populate also northern regions, the European populations benefitted from skin lightening.

6. How Vitamin D Became a Vitamin?

With the exception of highly developed societies, such as in the Roman empire, in the past, humans exposed larger percentages of their skin to the sun for a far longer part of the day than nowadays. In their evolutionary origin in East Africa, humans were every day around the year exposed to extensive UV-B radiation, which induced sufficient vitamin D_3 synthesis. Therefore, over a period of more than 200,000 years, they got used to a constantly high vitamin D status of 100 nM 25(OH)D_3 or more [65]. However, within the last 50–75,000 years, the migration toward regions with a latitude above 37 °N let them experience seasonal changes in sun exposure and periods of the year when vitamin D_3 cannot be produced endogenously. Furthermore, as a result of the industrial revolution, humans adapted to an urban lifestyle with predominant indoor work and activity. Both conditions, vitamin D winters and indoor preferences, often led to vitamin D deficiency in industrialized countries. For example, in England in the 19th century, rickets, also called the "English disease", was a very common disorder in children [91,92]. Moreover, also the severity of tuberculosis was and still is significantly increased in vitamin D deficient individuals [93]. Thus, not evolution but human migration and lifestyle changes made vitamin D_3 a vitamin.

7. Conclusions

Vitamin D_2 and vitamin D_3 started their "career" more than a billion years ago as side products of sterol biosynthesis, in fungi, some plants and animals that scavenge UV-B radiation. Just half a billion years later, the endocrinology of vitamin D evolved and the biologically active form of vitamin D_3, 1,25$(OH)_2D_3$, became a hormone. Via its high-affinity receptor VDR 1,25$(OH)_2D_3$ regulates genes that are involved in detoxification, energy metabolism, immunity and calcium homeostasis. With this pleiotropic functional profile vitamin D is an important contributor to organismal homeostasis and health. This explains why rather recently (from an evolutionary perspective) changes in human lifestyle caused a reduced endogenous vitamin D_3 production. Since, in parallel, the majority of human populations are not adapted to a marine-based diet [94], i.e., their average dietary intake of vitamin D is low, vitamin D deficiency became a common problem. Worldwide, vitamin D deficiency affects more than a billion people [95] and causes health problems, such as bone malformations and a decreased potency of the immune system.

Funding: This publication is part of a project that has received funding from the European Union's Horizon2020 research and innovation program under grant agreement no. 952601.

Conflicts of Interest: The author declares no conflict of interest.

References

1. Dobzhansky, T. Nothing in biology makes sense except in the light of evolution. *Am. Biol. Teach.* **1973**, *35*, 125–129. [CrossRef]
2. McMollum, E.V.; Simmonds, N.; Becker, J.E.; Shipley, P.G. Studies on experimental rickets: An experimental demonstration of the existence of a vitamin which promotes calcium deposition. *J. Biol. Chem.* **1922**, *52*, 293–298. [CrossRef]
3. Levine, M.A. Diagnosis and management of vitamin D dependent rickets. *Front. Pediatr.* **2020**, *8*, 315. [CrossRef] [PubMed]
4. Bouillon, R.; Carmeliet, G.; Lieben, L.; Watanabe, M.; Perino, A.; Auwerx, J.; Schoonjans, K.; Verstuyf, A. Vitamin D and energy homeostasis-of mice and men. *Nat. Rev. Endocrinol.* **2014**, *10*, 79–87. [CrossRef] [PubMed]
5. Hanel, A.; Carlberg, C. Skin color and vitamin D: An update. *Exp. Dermatol.* **2020**, *29*, 864–875. [CrossRef] [PubMed]
6. Summons, R.E.; Bradley, A.S.; Jahnke, L.L.; Waldbauer, J.R. Steroids, triterpenoids and molecular oxygen. *Philos. Trans. R. Soc. Lond. B Biol. Sci.* **2006**, *361*, 951–968. [CrossRef]
7. Sessions, A.L.; Doughty, D.M.; Welander, P.V.; Summons, R.E.; Newman, D.K. The continuing puzzle of the Great Oxidation Event. *Curr. Biol.* **2009**, *19*, R567–R574. [CrossRef]
8. Raymond, J.; Segre, D. The effect of oxygen on biochemical networks and the evolution of complex life. *Science* **2006**, *311*, 1764–1767. [CrossRef]
9. Desmond, E.; Gribaldo, S. Phylogenomics of sterol synthesis: Insights into the origin, evolution, and diversity of a key eukaryotic feature. *Genome Biol. Evol.* **2009**, *1*, 364–381. [CrossRef]
10. DeBose-Boyd, R.A. Feedback regulation of cholesterol synthesis: Sterol-accelerated ubiquitination and degradation of HMG CoA reductase. *Cell Res.* **2008**, *18*, 609–621. [CrossRef]

11. Japelt, R.B.; Jakobsen, J. Vitamin D in plants: A review of occurrence, analysis, and biosynthesis. *Front. Plant Sci.* **2013**, *4*, 136. [CrossRef] [PubMed]
12. Martin-Creuzburg, D.; Merkel, P. Sterols of freshwater microalgae: Potential implications for zooplankton nutrition. *J. Plankton Res.* **2016**, *38*, 865–877. [CrossRef]
13. Head, B.P.; Patel, H.H.; Insel, P.A. Interaction of membrane/lipid rafts with the cytoskeleton: Impact on signaling and function: Membrane/lipid rafts, mediators of cytoskeletal arrangement and cell signaling. *Biochim. Biophys. Acta* **2014**, *1838*, 532–545. [CrossRef]
14. Capell-Hattam, I.M.; Brown, A.J. Sterol evolution: Cholesterol synthesis in animals is less a required trait than an acquired taste. *Curr. Biol.* **2020**, *30*, R886–R888. [CrossRef]
15. Holick, M.F. The cutaneous photosynthesis of previtamin D_3: A unique photoendocrine system. *J. Investig. Dermatol.* **1981**, *77*, 51–58. [CrossRef] [PubMed]
16. Holick, M.F. Vitamin D: A millenium perspective. *J. Cell Biochem.* **2003**, *88*, 296–307. [CrossRef] [PubMed]
17. Holick, M.F.; MacLaughlin, J.A.; Doppelt, S.H. Regulation of cutaneous previtamin D_3 photosynthesis in man: Skin pigment is not an essential regulator. *Science* **1981**, *211*, 590–593. [CrossRef]
18. Wacker, M.; Holick, M.F. Sunlight and vitamin D: A global perspective for health. *Dermatoendocrinol* **2013**, *5*, 51–108. [CrossRef]
19. Boyce, D.G.; Lewis, M.R.; Worm, B. Global phytoplankton decline over the past century. *Nature* **2010**, *466*, 591–596. [CrossRef]
20. Byrdwell, W.C.; Horst, R.L.; Phillips, K.M.; Holden, J.M.; Patterson, K.Y.; Harnly, J.M.; Exler, J. Vitamin D levels in fish and shellfish determined by liquid chromatography with ultraviolet detection and mass spectrometry. *J. Food Compos. Anal.* **2013**, *30*, 109–119. [CrossRef]
21. Rajakumar, K. Vitamin D, cod-liver oil, sunlight, and rickets: A historical perspective. *Pediatrics* **2003**, *112*, e132–e135. [CrossRef] [PubMed]
22. Whitfield, G.K.; Dang, H.T.; Schluter, S.F.; Bernstein, R.M.; Bunag, T.; Manzon, L.A.; Hsieh, G.; Dominguez, C.E.; Youson, J.H.; Haussler, M.R.; et al. Cloning of a functional vitamin D receptor from the lamprey (Petromyzon marinus), an ancient vertebrate lacking a calcified skeleton and teeth. *Endocrinology* **2003**, *144*, 2704–2716. [CrossRef] [PubMed]
23. Hanel, A.; Bendik, I.; Carlberg, C. Transcriptome-wide profile of 25-hydroxyvitamin D_3 in pimary immune cells from human peripheral blood. *Nutrients* **2021**, *13*, 4100. [CrossRef] [PubMed]
24. Jones, G.; Prosser, D.E.; Kaufmann, M. Cytochrome P450-mediated metabolism of vitamin D. *J. Lipid. Res.* **2014**, *55*, 13–31. [CrossRef]
25. Bouillon, R.; Suda, T. Vitamin D: Calcium and bone homeostasis during evolution. *BoneKEy Rep.* **2014**, *3*, 480. [CrossRef]
26. Escriva, H.; Bertrand, S.; Laudet, V. The evolution of the nuclear receptor superfamily. *Essays Biochem.* **2004**, *40*, 11–26. [CrossRef]
27. Krasowski, M.D.; Ni, A.; Hagey, L.R.; Ekins, S. Evolution of promiscuous nuclear hormone receptors: LXR, FXR, VDR, PXR, and CAR. *Mol. Cell Endocrinol.* **2011**, *334*, 39–48. [CrossRef]
28. Reschly, E.J.; Bainy, A.C.; Mattos, J.J.; Hagey, L.R.; Bahary, N.; Mada, S.R.; Ou, J.; Venkataramanan, R.; Krasowski, M.D. Functional evolution of the vitamin D and pregnane X receptors. *BMC Evol. Biol.* **2007**, *7*, 222. [CrossRef]
29. Hanel, A.; Carlberg, C. Vitamin D and evolution: Pharmacologic implications. *Biochem. Pharmacol.* **2020**, *173*, 113595. [CrossRef]
30. Makishima, M.; Lu, T.T.; Xie, W.; Whitfield, G.K.; Domoto, H.; Evans, R.M.; Haussler, M.R.; Mangelsdorf, D.J. Vitamin D receptor as an intestinal bile acid sensor. *Science* **2002**, *296*, 1313–1316. [CrossRef]
31. Makishima, M.; Okamoto, A.Y.; Repa, J.J.; Tu, H.; Learned, R.M.; Luk, A.; Hull, M.V.; Lustig, K.D.; Mangelsdorf, D.J.; Shan, B. Identification of a nuclear receptor for bile acids. *Science* **1999**, *284*, 1362–1365. [CrossRef] [PubMed]
32. Staudinger, J.L.; Goodwin, B.; Jones, S.A.; Hawkins-Brown, D.; MacKenzie, K.I.; LaTour, A.; Liu, Y.; Klaassen, C.D.; Brown, K.K.; Reinhard, J.; et al. The nuclear receptor PXR is a lithocholic acid sensor that protects against liver toxicity. *Proc. Natl. Acad. Sci. USA* **2001**, *98*, 3369–3374. [CrossRef] [PubMed]
33. Guo, G.L.; Lambert, G.; Negishi, M.; Ward, J.M.; Brewer, H.B., Jr.; Kliewer, S.A.; Gonzalez, F.J.; Sinal, C.J. Complementary roles of farnesoid X receptor, pregnane X receptor, and constitutive androstane receptor in protection against bile acid toxicity. *J. Biol. Chem.* **2003**, *278*, 45062–45071. [CrossRef] [PubMed]
34. Reschly, E.J.; Krasowski, M.D. Evolution and function of the NR1I nuclear hormone receptor subfamily (VDR, PXR, and CAR) with respect to metabolism of xenobiotics and endogenous compounds. *Curr. Drug Metab.* **2006**, *7*, 349–365. [CrossRef] [PubMed]
35. Carlberg, C. Nutrigenomics of vitamin D. *Nutrients* **2019**, *11*, 676. [CrossRef]
36. Müller, M.; Kersten, S. Nutrigenomics: Goals and strategies. *Nat. Rev. Genet.* **2003**, *4*, 315–322. [CrossRef]
37. Muller, M.; Mentel, M.; van Hellemond, J.J.; Henze, K.; Woehle, C.; Gould, S.B.; Yu, R.Y.; van der Giezen, M.; Tielens, A.G.; Martin, W.F. Biochemistry and evolution of anaerobic energy metabolism in eukaryotes. *Microbiol. Mol. Biol. Rev.* **2012**, *76*, 444–495. [CrossRef]
38. Seuter, S.; Neme, A.; Carlberg, C. Epigenome-wide effects of vitamin D and their impact on the transcriptome of human monocytes involve CTCF. *Nucleic Acids Res.* **2016**, *44*, 4090–4104. [CrossRef]
39. Pasing, Y.; Fenton, C.G.; Jorde, R.; Paulssen, R.H. Changes in the human transcriptome upon vitamin D supplementation. *J. Steroid Biochem. Mol. Biol.* **2017**, *173*, 93–99. [CrossRef]
40. Ferreira, G.B.; Vanherwegen, A.S.; Eelen, G.; Gutierrez, A.C.; Van Lommel, L.; Marchal, K.; Verlinden, L.; Verstuyf, A.; Nogueira, T.; Georgiadou, M.; et al. Vitamin D_3 induces tolerance in human dendritic cells by activation of intracellular metabolic pathways. *Cell Rep.* **2015**, *10*, 711–725. [CrossRef]

41. Hanel, A.; Carlberg, C. Time-resolved gene expression analysis monitors the regulation of inflammatory mediators and attenuation of adaptive immune response by vitamin D. *Int. J. Mol. Sci.* **2022**, *23*, 911. [CrossRef] [PubMed]
42. Mullican, S.E.; Dispirito, J.R.; Lazar, M.A. The orphan nuclear receptors at their 25-year reunion. *J. Mol. Endocrinol.* **2013**, *51*, T115–T140. [CrossRef]
43. Sigueiro, R.; Bianchetti, L.; Peluso-Iltis, C.; Chalhoub, S.; Dejaegere, A.; Osz, J.; Rochel, N. Advances in vitamin D receptor function and evolution based on the 3D structure of the Lamprey ligand-binding domain. *J. Med. Chem.* **2022**, *65*, 5821–5829. [CrossRef]
44. Rochel, N.; Wurtz, J.M.; Mitschler, A.; Klaholz, B.; Moras, D. Crystal structure of the nuclear receptor for vitamin D bound to its natural ligand. *Mol. Cell* **2000**, *5*, 173–179. [CrossRef]
45. Kollitz, E.M.; Zhang, G.; Hawkins, M.B.; Whitfield, G.K.; Reif, D.M.; Kullman, S.W. Molecular cloning, functional characterization, and evolutionary analysis of vitamin D receptors isolated from basal vertebrates. *PLoS ONE* **2015**, *10*, e0122853. [CrossRef] [PubMed]
46. Kollitz, E.M.; Hawkins, M.B.; Whitfield, G.K.; Kullman, S.W. Functional diversification of vitamin D receptor paralogs in teleost fish after a whole genome duplication event. *Endocrinology* **2014**, *155*, 4641–4654. [CrossRef] [PubMed]
47. Flajnik, M.F. A cold-blooded view of adaptive immunity. *Nat. Rev. Immunol.* **2018**, *18*, 438–453. [CrossRef]
48. Cooper, M.D.; Alder, M.N. The evolution of adaptive immune systems. *Cell* **2006**, *124*, 815–822. [CrossRef]
49. Ganeshan, K.; Nikkanen, J.; Man, K.; Leong, Y.A.; Sogawa, Y.; Maschek, J.A.; Van Ry, T.; Chagwedera, D.N.; Cox, J.E.; Chawla, A. Energetic trade-offs and hypometabolic states promote disease tolerance. *Cell* **2019**, *177*, 399–413.e12. [CrossRef]
50. Vanherwegen, A.S.; Eelen, G.; Ferreira, G.B.; Ghesquiere, B.; Cook, D.P.; Nikolic, T.; Roep, B.; Carmeliet, P.; Telang, S.; Mathieu, C.; et al. Vitamin D controls the capacity of human dendritic cells to induce functional regulatory T cells by regulation of glucose metabolism. *J. Steroid Biochem. Mol. Biol.* **2019**, *187*, 134–145. [CrossRef]
51. Heikkinen, S.; Väisänen, S.; Pehkonen, P.; Seuter, S.; Benes, V.; Carlberg, C. Nuclear hormone 1α,25-dihydroxyvitamin D_3 elicits a genome-wide shift in the locations of VDR chromatin occupancy. *Nucleic Acids Res.* **2011**, *39*, 9181–9193. [CrossRef] [PubMed]
52. Gombart, A.F.; Borregaard, N.; Koeffler, H.P. Human cathelicidin antimicrobial peptide (CAMP) gene is a direct target of the vitamin D receptor and is strongly up-regulated in myeloid cells by 1,25-dihydroxyvitamin D_3. *FASEB J.* **2005**, *19*, 1067–1077. [CrossRef] [PubMed]
53. Bscheider, M.; Butcher, E.C. Vitamin D immunoregulation through dendritic cells. *Immunology* **2016**, *148*, 227–236. [CrossRef] [PubMed]
54. Coussens, A.K.; Wilkinson, R.J.; Hanifa, Y.; Nikolayevskyy, V.; Elkington, P.T.; Islam, K.; Timms, P.M.; Venton, T.R.; Bothamley, G.H.; Packe, G.E.; et al. Vitamin D accelerates resolution of inflammatory responses during tuberculosis treatment. *Proc. Natl. Acad. Sci. USA* **2012**, *109*, 15449–15454. [CrossRef] [PubMed]
55. Carlberg, C.; Seuter, S.; Nurmi, T.; Tuomainen, T.P.; Virtanen, J.K.; Neme, A. In vivo response of the human epigenome to vitamin D: A proof-of-principle study. *J. Steroid Biochem. Mol. Biol.* **2018**, *180*, 142–148. [CrossRef]
56. Venkatesh, B.; Lee, A.P.; Ravi, V.; Maurya, A.K.; Lian, M.M.; Swann, J.B.; Ohta, Y.; Flajnik, M.F.; Sutoh, Y.; Kasahara, M.; et al. Elephant shark genome provides unique insights into gnathostome evolution. *Nature* **2014**, *505*, 174–179. [CrossRef]
57. Van de Peppel, J.; van Leeuwen, J.P. Vitamin D and gene networks in human osteoblasts. *Front. Physiol.* **2014**, *5*, 137. [CrossRef]
58. Kim, S.; Yamazaki, M.; Zella, L.A.; Meyer, M.B.; Fretz, J.A.; Shevde, N.K.; Pike, J.W. Multiple enhancer regions located at significant distances upstream of the transcriptional start site mediate RANKL gene expression in response to 1,25-dihydroxyvitamin D_3. *J. Steroid Biochem. Mol. Biol.* **2007**, *103*, 430–434. [CrossRef]
59. Karsenty, G.; Khosla, S. The crosstalk between bone remodeling and energy metabolism: A translational perspective. *Cell Metab.* **2022**, *34*, 805–817. [CrossRef]
60. Tsukasaki, M.; Takayanagi, H. Osteoimmunology: Evolving concepts in bone-immune interactions in health and disease. *Nat. Rev. Immunol.* **2019**, *19*, 626–642. [CrossRef]
61. Cortes, M.; Chen, M.J.; Stachura, D.L.; Liu, S.Y.; Kwan, W.; Wright, F.; Vo, L.T.; Theodore, L.N.; Esain, V.; Frost, I.M.; et al. Developmental vitamin D availability impacts hematopoietic stem cell production. *Cell Rep.* **2016**, *17*, 458–468. [CrossRef] [PubMed]
62. Hublin, J.J.; Ben-Ncer, A.; Bailey, S.E.; Freidline, S.E.; Neubauer, S.; Skinner, M.M.; Bergmann, I.; Le Cabec, A.; Benazzi, S.; Harvati, K.; et al. New fossils from Jebel Irhoud, Morocco and the pan-African origin of *Homo sapiens*. *Nature* **2017**, *546*, 289–292. [CrossRef] [PubMed]
63. Greaves, M. Was skin cancer a selective force for black pigmentation in early hominin evolution? *Proc. Biol. Sci.* **2014**, *281*, 20132955. [CrossRef]
64. Fajuyigbe, D.; Lwin, S.M.; Diffey, B.L.; Baker, R.; Tobin, D.J.; Sarkany, R.P.E.; Young, A.R. Melanin distribution in human epidermis affords localized protection against DNA photodamage and concurs with skin cancer incidence difference in extreme phototypes. *FASEB J.* **2018**, *32*, 3700–3706. [CrossRef] [PubMed]
65. Luxwolda, M.F.; Kuipers, R.S.; Kema, I.P.; Dijck-Brouwer, D.A.; Muskiet, F.A. Traditionally living populations in East Africa have a mean serum 25-hydroxyvitamin D concentration of 115 nmol/l. *Br. J. Nutr.* **2012**, *108*, 1557–1561. [CrossRef] [PubMed]
66. Pagani, L.; Lawson, D.J.; Jagoda, E.; Morseburg, A.; Eriksson, A.; Mitt, M.; Clemente, F.; Hudjashov, G.; DeGiorgio, M.; Saag, L.; et al. Genomic analyses inform on migration events during the peopling of Eurasia. *Nature* **2016**, *538*, 238–242. [CrossRef] [PubMed]

67. Skoglund, P.; Mathieson, I. Ancient genomics of modern humans: The first decade. *Annu. Rev. Genomics Hum. Genet.* **2018**, *19*, 381–404. [CrossRef]
68. O'Neill, C.M.; Kazantzidis, A.; Ryan, M.J.; Barber, N.; Sempos, C.T.; Durazo-Arvizu, R.A.; Jorde, R.; Grimnes, G.; Eiriksdottir, G.; Gudnason, V.; et al. Seasonal changes in vitamin D-effective UVB availability in Europe and associations with population serum 25-hydroxyvitamin D. *Nutrients* **2016**, *8*, 533. [CrossRef]
69. Jablonski, N.G.; Chaplin, G. The evolution of human skin coloration. *J. Hum. Evol.* **2000**, *39*, 57–106. [CrossRef]
70. Deng, L.; Xu, S. Adaptation of human skin color in various populations. *Hereditas* **2018**, *155*, 1–12. [CrossRef]
71. Cichorek, M.; Wachulska, M.; Stasiewicz, A.; Tyminska, A. Skin melanocytes: Biology and development. *Postepy Dermatol. Alergol.* **2013**, *30*, 30–41. [CrossRef] [PubMed]
72. Pavan, W.J.; Sturm, R.A. The genetics of human skin and hair pigmentation. *Annu. Rev. Genomics Hum. Genet.* **2019**, *20*, 41–72. [CrossRef] [PubMed]
73. Lindgren, J.; Moyer, A.; Schweitzer, M.H.; Sjovall, P.; Uvdal, P.; Nilsson, D.E.; Heimdal, J.; Engdahl, A.; Gren, J.A.; Schultz, B.P.; et al. Interpreting melanin-based coloration through deep time: A critical review. *Proc. Biol. Sci.* **2015**, *282*, 20150614. [CrossRef] [PubMed]
74. Ju, D.; Mathieson, I. The evolution of skin pigmentation associated variation in West Eurasia. *Proc Nat Acad Sci USA* **2021**, *118*, e2009227118. [CrossRef]
75. Gunther, T.; Malmstrom, H.; Svensson, E.M.; Omrak, A.; Sanchez-Quinto, F.; Kilinc, G.M.; Krzewinska, M.; Eriksson, G.; Fraser, M.; Edlund, H.; et al. Population genomics of Mesolithic Scandinavia: Investigating early postglacial migration routes and high-latitude adaptation. *PLoS Biol.* **2018**, *16*, e2003703. [CrossRef]
76. Nakagome, S.; Alkorta-Aranburu, G.; Amato, R.; Howie, B.; Peter, B.M.; Hudson, R.R.; Di Rienzo, A. Estimating the ages of selection signals from different epochs in human history. *Mol. Biol. Evol.* **2016**, *33*, 657–669. [CrossRef]
77. Fu, Q.; Posth, C.; Hajdinjak, M.; Petr, M.; Mallick, S.; Fernandes, D.; Furtwangler, A.; Haak, W.; Meyer, M.; Mittnik, A.; et al. The genetic history of Ice Age Europe. *Nature* **2016**, *534*, 200–205. [CrossRef]
78. Prufer, K.; Racimo, F.; Patterson, N.; Jay, F.; Sankararaman, S.; Sawyer, S.; Heinze, A.; Renaud, G.; Sudmant, P.H.; de Filippo, C.; et al. The complete genome sequence of a Neanderthal from the Altai Mountains. *Nature* **2014**, *505*, 43–49. [CrossRef]
79. Peyregne, S.; Slon, V.; Mafessoni, F.; de Filippo, C.; Hajdinjak, M.; Nagel, S.; Nickel, B.; Essel, E.; Le Cabec, A.; Wehrberger, K.; et al. Nuclear DNA from two early Neandertals reveals 80,000 years of genetic continuity in Europe. *Sci. Adv.* **2019**, *5*, eaaw5873. [CrossRef]
80. Petr, M.; Paabo, S.; Kelso, J.; Vernot, B. Limits of long-term selection against Neandertal introgression. *Proc. Natl. Acad. Sci. USA* **2019**, *116*, 1639–1644. [CrossRef]
81. Villalba-Mouco, V.; van de Loosdrecht, M.S.; Posth, C.; Mora, R.; Martinez-Moreno, J.; Rojo-Guerra, M.; Salazar-Garcia, D.C.; Royo-Guillen, J.I.; Kunst, M.; Rougier, H.; et al. Survival of late Pleistocene hunter-gatherer ancestry in the Iberian peninsula. *Curr. Biol.* **2019**, *29*, 1169–1177.e7. [CrossRef] [PubMed]
82. Mathieson, I.; Lazaridis, I.; Rohland, N.; Mallick, S.; Patterson, N.; Roodenberg, S.A.; Harney, E.; Stewardson, K.; Fernandes, D.; Novak, M.; et al. Genome-wide patterns of selection in 230 ancient Eurasians. *Nature* **2015**, *528*, 499–503. [CrossRef] [PubMed]
83. Mathieson, I.; Alpaslan-Roodenberg, S.; Posth, C.; Szecsenyi-Nagy, A.; Rohland, N.; Mallick, S.; Olalde, I.; Broomandkhoshbacht, N.; Candilio, F.; Cheronet, O.; et al. The genomic history of southeastern Europe. *Nature* **2018**, *555*, 197–203. [CrossRef] [PubMed]
84. Haak, W.; Lazaridis, I.; Patterson, N.; Rohland, N.; Mallick, S.; Llamas, B.; Brandt, G.; Nordenfelt, S.; Harney, E.; Stewardson, K.; et al. Massive migration from the steppe was a source for Indo-European languages in Europe. *Nature* **2015**, *522*, 207–211. [CrossRef] [PubMed]
85. Damgaard, P.B.; Marchi, N.; Rasmussen, S.; Peyrot, M.; Renaud, G.; Korneliussen, T.; Moreno-Mayar, J.V.; Pedersen, M.W.; Goldberg, A.; Usmanova, E.; et al. 137 ancient human genomes from across the Eurasian steppes. *Nature* **2018**, *557*, 369–374. [CrossRef] [PubMed]
86. Mays, S.; Prowse, T.; George, M.; Brickley, M. Latitude, urbanization, age, and sex as risk factors for vitamin D deficiency disease in the Roman Empire. *Am. J. Phys. Anthropol.* **2018**, *167*, 484–496. [CrossRef]
87. Kuan, V.; Martineau, A.R.; Griffiths, C.J.; Hypponen, E.; Walton, R. DHCR7 mutations linked to higher vitamin D status allowed early human migration to northern latitudes. *BMC Evol. Biol.* **2013**, *13*, 144. [CrossRef]
88. Wang, T.J.; Zhang, F.; Richards, J.B.; Kestenbaum, B.; van Meurs, J.B.; Berry, D.; Kiel, D.P.; Streeten, E.A.; Ohlsson, C.; Koller, D.L.; et al. Common genetic determinants of vitamin D insufficiency: A genome-wide association study. *Lancet* **2010**, *376*, 180–188. [CrossRef]
89. Jiang, X.; O'Reilly, P.F.; Aschard, H.; Hsu, Y.H.; Richards, J.B.; Dupuis, J.; Ingelsson, E.; Karasik, D.; Pilz, S.; Berry, D.; et al. Genome-wide association study in 79,366 European-ancestry individuals informs the genetic architecture of 25-hydroxyvitamin D levels. *Nat. Commun.* **2018**, *9*, 260. [CrossRef]
90. Revez, J.A.; Lin, T.; Qiao, Z.; Xue, A.; Holtz, Y.; Zhu, Z.; Zeng, J.; Wang, H.; Sidorenko, J.; Kemper, K.E.; et al. Genome-wide association study identifies 143 loci associated with 25 hydroxyvitamin D concentration. *Nat. Commun.* **2020**, *11*, 1647. [CrossRef]
91. Hardy, A. Commentary: Bread and alum, syphilis and sunlight: Rickets in the nineteenth century. *Int. J. Epidemiol.* **2003**, *32*, 337–340. [CrossRef] [PubMed]
92. Zhang, M.; Shen, F.; Petryk, A.; Tang, J.; Chen, X.; Sergi, C. "English Disease": Historical notes on rickets, the bone–lung link and child neglect issues. *Nutrients* **2016**, *8*, 722. [CrossRef] [PubMed]

93. Huang, S.J.; Wang, X.H.; Liu, Z.D.; Cao, W.L.; Han, Y.; Ma, A.G.; Xu, S.F. Vitamin D deficiency and the risk of tuberculosis: A meta-analysis. *Drug Des. Devel. Ther.* **2017**, *11*, 91–102. [CrossRef] [PubMed]
94. Brace, S.; Diekmann, Y.; Booth, T.J.; van Dorp, L.; Faltyskova, Z.; Rohland, N.; Mallick, S.; Olalde, I.; Ferry, M.; Michel, M.; et al. Ancient genomes indicate population replacement in Early Neolithic Britain. *Nat. Ecol. Evol.* **2019**, *3*, 765–771. [CrossRef]
95. Holick, M.F. Vitamin D deficiency. *N. Engl. J. Med.* **2007**, *357*, 266–281. [CrossRef] [PubMed]

 MDPI

Review

Vitamin D and Systems Biology

Shahid Hussain [1], Clayton Yates [2,3,4] and Moray J. Campbell [1,*,†]

1 Division of Pharmaceutics and Pharmaceutical Chemistry, College of Pharmacy, The Ohio State University, Columbus, OH 43210, USA
2 Department of Biology and Center for Cancer Research, Tuskegee University, Tuskegee, AL 36088, USA
3 Department of Pathology, Johns Hopkins University School of Medicine, Baltimore, MD 21287, USA
4 Department of Oncology Sidney Kimmel Comprehensive Cancer Center, Johns Hopkins University School of Medicine, Baltimore, MD 21287, USA
* Correspondence: campbell.1933@osu.edu
† Current address: Department of Biomedical Sciences, Cedars-Sinai Medical Center, Los Angeles, CA 90048, USA.

Abstract: The biological actions of the vitamin D receptor (VDR) have been investigated intensively for over 100 years and has led to the identification of significant insights into the repertoire of its biological actions. These were initially established to be centered on the regulation of calcium transport in the colon and deposition in bone. Beyond these well-known calcemic roles, other roles have emerged in the regulation of cell differentiation processes and have an impact on metabolism. The purpose of the current review is to consider where applying systems biology (SB) approaches may begin to generate a more precise understanding of where the VDR is, and is not, biologically impactful. Two SB approaches have been developed and begun to reveal insight into VDR biological functions. In a top-down SB approach genome-wide scale data are statistically analyzed, and from which a role for the VDR emerges in terms of being a hub in a biological network. Such approaches have confirmed significant roles, for example, in myeloid differentiation and the control of inflammation and innate immunity. In a bottom-up SB approach, current biological understanding is built into a kinetic model which is then applied to existing biological data to explain the function and identify unknown behavior. To date, this has not been applied to the VDR, but has to the related ERα and identified previously unknown mechanisms of control. One arena where applying top-down and bottom-up SB approaches may be informative is in the setting of prostate cancer health disparities.

Keywords: vitamin D receptor; systems biology; cell differentiation; prostate cancer

Citation: Hussain, S.; Yates, C.; Campbell, M.J. Vitamin D and Systems Biology. *Nutrients* **2022**, *14*, 5197. https://doi.org/10.3390/nu14245197

Academic Editor: Bruce W. Hollis

Received: 26 September 2022
Accepted: 1 December 2022
Published: 7 December 2022

1. Systems Biology and Biomedicine

1.1. The Opportunities of Applying Systems Biology Approaches in Biomedical Research

Much of the discovery in biomedicine has been centered on the classic paradigms of reductionist biology, in which phenotypes are interpreted as the interaction of either single or small groups of molecules. Across biomedical sciences, there are many remarkable examples of how this approach has led to the discovery of drivers of human disease, and equally remarkable examples of new therapies designed to target these drivers. For example, high-profile examples of this reductionist approach include the discovery of the oncogenic role of the BCR-ABL fusion gene in the etiology of chronic myeloid leukemia [1] and the development of a targeting kinase inhibitor such as Imatinib [2].

This is a striking example of the so-called bench-to-bedside research and represents one of the earliest examples of precision medicine. At times, however, this reductionist approach can appear limited both theoretically and clinically as the etiology of Imatinib-resistant phenotypes only too well demonstrates [3]. Across cancers and other disease phenotypes, a stumbling block can be identifying single strong disease driver mechanisms, which in turn accurately predict drug sensitivities and therapeutic effectiveness. Therefore, delivering a fuller prediction of disease drivers and therapeutic vulnerabilities may require developing different methodologies. Ideally, any methodologies would also take advantage of the ever-increasing stream of high-dimensional biological data to inform diagnosis and prognosis. Perhaps these approaches (SB and reductionist) can actually be highly symbiotic.

Systems biology (SB) aims to apply mathematical approaches to build a predicative and quantitative model of biological systems, with the goal to use model predictions to define specific physiological or pathophysiological states and outcomes. The models derived from such SB approaches applied to experimentally derived biological data aim to identify the dynamic behavior of networks that are the center of cellular behaviors [4,5]. These approaches are readily scalable and not restricted by scope. SB approaches can be applied to discrete cell signaling systems, such as gene regulatory networks and signal transduction cascades, to cell–cell interactions, tissue organization, organismal behavior, and to complex multi-organism interactions as seen within, for example, the function of the gut and even complete ecosystems [6].

Furthermore, the models built by SB approaches aim to define the functioning of living organisms not solely by looking at the constituent molecules such as DNA, RNA, proteins, and metabolites but rather by the process-level biological systems they constitute, such as mitosis and metabolic control [7,8]. A key hallmark of these models is that they capture the states of the system in a predicative and quantitative manner and can be exploited to drive novel understanding. As these in silico models can be interrogated rapidly, multiple components of any given model can be dissected to reveal unintuitive findings [9–11].

A consequence of the integrated nature of biological signaling is the emergent complexity, which underpins human health. For example, the dexterous control of transcriptional networks is derived from a high ratio of transcription factors to regulated mRNA or miRNA targets. This ratio, combined with large regulatory regions, results in unparalleled plasticity over the choice, amplitude, and period of transcription [12,13]. More specifically, transcription factor modules recognize, interpret and sustain histone modifications and ultimately establish boundaries between transcriptionally rich euchromatin and transcriptionally restricted heterochromatin. These boundaries are cemented further by the regulation of CpG island methylation; these two processes are dynamically intertwined. Again, from an SB perspective, it is reasoned that modeling these transcriptional processes will help to explain the highly integrated nature and robustness of normal transcriptional control, whereby the processes have redundancies such that they do not radically alter in response to external signals and do not fail when a single component is altered. By contrast, transcriptional networks in cancer cells display a loss of transcriptional plasticity and do not display the full breadth of signaling capacities. The evolution of the malignant transcriptome is seen clearly in the nuclear receptor and MYC superfamilies (reviewed in [14]).

The concept of SB builds on so-called "holistic biology" developed in the 1960s and coupled with informatics theory and modeling approaches developed through the 20th century. However, the application of SB methodologies and the expansion of SB concepts in biomedicine has been boosted since the early 2000s by the technological advances in the development of high dimensional data approaches frequently derived from next-generation sequencing technologies coupled with bioinformatic approaches. This progression from the 1950s to the current state has been profoundly catalyzed by the human genome project, with its draft sequence published in 2001, and a final reference genome published in 2022 [15] alongside other reference genomes across the animal and plant kingdom. In this manner, the combination of high throughput experimental approaches in the wet lab coupled with complex statistical analyses and computational methods in the dry lab have led to a more comprehensive characterization of multiple diverse organisms, and a shift of focus from molecules to their interactions and the networks they form [16]. This sophistication and power of prediction are most likely set to increase when SB models also include a spatio-temporal characterization of cell behavior, including the dynamics of how molecules are exchanged between compartments and exported from the cell. Again, modeling these aspects of cell behavior has been massively impacted by an explosion in single-cell and spatial technologies.

There are several well-justified advantages of applying SB approaches, which have the genuine potential to complement and extend the reductionist paradigm. Expressing the complete interactions within biological systems in mathematical terms reduces the impact of biases introduced by focusing on single proteins, and instead can reveal under-explored control points in the system. Furthermore, expressing biological systems in the context of processes, rather than well-understood individual components, has the potential to assimilate more readily new high dimensional data generated in biological experiments. Consequently, it is reasonably anticipated that significant, novel, and unpredicted strides will be made in understanding the control and responsiveness of biological events, which in turn can generate new insights into disease susceptibility and therapeutic opportunities [17,18].

1.2. Systems Biology Builds upon an Asymptotic Recursion between Wet Lab and Dry Lab

At its core, SB approaches have an asymptotic recursion between research activities in the wet lab and dry lab. Models are built in the dry lab that reflects the biological observations in the wet lab and are used to generate predictions for how the system can behave. Such predictions are then formulated into testable interventions in the wet lab to generate data for model refinement in the dry lab. It is this oscillation between wet and dry labs that is so potentially powerful, but at the same time so daunting. Broadly, two strategies have emerged to develop such models, by either top-down or bottom-up approaches (Figure 1).

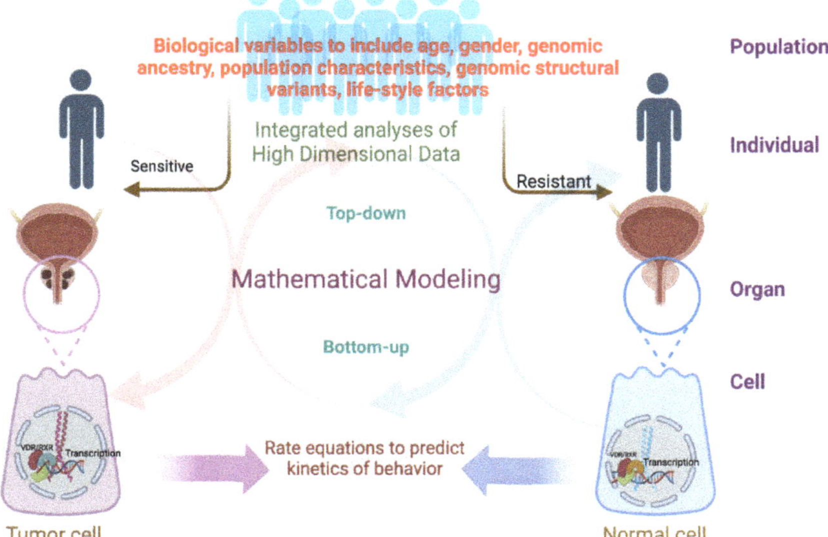

Figure 1. A workflow for top-down and bottoms-up system biology (SB) approaches to address prostate cancer health disparities. For top-down SB there is a plethora of high dimensional data including germline structural variants, population characteristics as serum-borne, and other clinical data that can be integrated through a range of approaches to deliver insights into how the prostate could either be either sensitive or resistant to the growth restraint properties of the VDR. From the bottoms-up perspective, more focused kinetic models can be developed for how the VDR functions and is disrupted between normal and tumor cells. Combining these approaches has the potential to develop a holistic understanding of how the VDR functions in the prostate gland and how this is impacted by genomic ancestry.

In the top-down approach, large datasets are interrogated with statistical methods to find patterns in the data with which to derive predictions on the system organization [19]. The models in top-down SB are phenomenological, meaning they are not directly mechanistically based and do not require knowledge about relationships between different molecular components. Identification of significant associations is used to develop a hypothesis on the nature of the molecule interactions and associations identified. These approaches naturally lend themselves to omics-derived data and develop hypotheses to be tested by wet lab analysis [20]. It is worth noting that the identified associations that are significant may not be causal, and in fact, themselves may not be true. Top-down approaches are often used with subsystems that have not yet been characterized to a high level of mechanistic detail and approaches such as Bayesian modeling are appropriate due to the missingness in the data from biological regulatory networks [21].

By contrast, the bottom-up approach begins with the hypothesis of biological mechanism and formulating equations on how the components of the system interact and then running simulations to generate predictions. This approach relies on experimental studies to determine the kinetic and chemical properties of the components, and starts with formulating system behaviors in rate equations, for example, expressed as differential equations, of the constituting parts of each system. These formulations are then integrated to predict the system behavior, with the goal to combine pathway models into a model for a larger system [21]. The data on the system under study are subjected to perturbations in the cell context and models are refined from the data. An example of a bottom-up approach is the silicon cell program where computational replicas of actual pathways are made to calculate system behavior [22].

The power of these approaches arises from being able to identify and direct experiments to test fundamental questions of a biological system. From the top-down perspective, these questions include identifying in a genome-wide manner the interactions of all components in a system to define the metabolic control that ultimately brings about cell, tissue, or organism behavior. Arising from this approach questions can be asked of a system. For example, what is the interconnectedness of the system and how does that change between health and disease states, or in different development or differentiation states? Which hubs in such networks are central and which are peripheral? Again, how does hub distribution shift in disease and development transitions? How do changes in gene and protein expression combine to control metabolism? How do germline or somatic structural variants change network topology and metabolic flux?

Similarly, from the bottom-up perspective, a model is curated from known biological interactions and expressed in mathematical terms to test behavior and make new predictions. For example, with any signaling system how is activation controlled and silenced? Given the ubiquitous role of enzymes in biology, a common question is how does changing the kinetics of activating/de-activating enzymes lead to altered signal amplitude? To what extent do germline or somatic structural variants in enzymes, or co-factors modulate enzymatic capacity and what effect does that exert on signal strength? More broadly, how are different signaling systems integrated and what is the quantitative effect of convergence? Does crosstalk generate synergistic events in regulatory complexes or are they merely convergent downstream at endpoints? What is the magnitude of control at each step and is there an order to control ranging from find-tuning to signal-independent activation? Finally, questions can be asked of the system in disease or development states and tested for pharmacological relevance including identifying which targets may be impactful but have deleterious side effects.

2. The Opportunities and Challenges of Applying Systems Biology Approaches to Studying the Vitamin D Receptor

A systems-level appreciation for vitamin D signaling has existed for several centuries, given that a description of rickets was first described in the 17th century [23]; rickets arises from impaired bone mineralization due to insufficient signaling via the vitamin D receptor (NR1I1/VDR) (reviewed in [24]).

The VDR is a Type II member of the nuclear receptor superfamily, which binds the active hormone 1α,25$(OH)_2D_3$. Several features of how the VDR functions are important from an SB perspective. A feature of Type II receptors such as the VDR is that independent of ligand the receptor is significantly associated with the genome bound to cis-regulatory elements (CRE) and may exert repressive effects, which is reversed by ligand activation leading to genomic redistribution and transactivation [25–28]. Additionally, of interest from a systems perspective is that VDR interacts in distinct ways with a number of proteins. Firstly, it heterodimerizes with the RXRs (NR2B1/RXRα, NR2B2/RXRβ, NR2B3/RXRγ), which is also a central dimer partner for other Type II receptors including those retinoic acid receptors (NR1B1/RARα, NR1B2/RARβ, NR1B3/RARγ), and the peroxisome proliferator-activated receptors (NR1C1/PPARα, NR1C2/PPARβ, NR1C3/PPARγ) [reviewed in [24]]. Secondly, the VDR interacts with a range of coregulator proteins such as coactivators and corepressors that exert antagonist roles in the control of local chromatin structure, as well as components of the SWI/SNF complexes to remodel nucleosome positioning and other enzymes such as helicases [29,30] and splicing factors [31] to also facilitate transcription. The VDR appears to participate in protein–protein–DNA interactions, for example with other transcription factors, in a trans-regulatory mechanism to regulate transcription. Finally, there is emerging evidence for the VDR, like other nuclear receptors [32,33], to interact functionally with long non-coding RNAs (lncRNA) to control gene expression by directly affecting the DNA environment or other RNA binding proteins [34]. Indeed one such lncRNA, steroid receptor activator (SRA) coimmunoprecipitates with the coactivator NCOA1/SRC1 and may function more broadly as a scaffold for NR complexes (reviewed in [32]).

Finally, of central importance to the complete VDR signaling system is the generation of the ligand, 1α,25$(OH)_2D_3$, the levels of which are dynamically controlled in terms of synthesis of the precursor 25$(OH)D_3$ which is held in a relatively tight range in the serum, and signs of deficiency in the VDR system can be seen when this serum level is diminished. The various metabolic steps that lead to the generation of 25$(OH)D_3$, the active ligand 1α,25$(OH)_2D_3$, and a range of metabolites that ultimately lead to its catabolism are all tightly controlled by enzymatic reactions both in an endocrine and tissue-specific local intracrine manner.

2.1. Top-Down Approaches Applied to VDR Biology

From a top-down SB perspective, it is interesting to ask in what context the VDR itself, or coregulators, or ligand generation are identified as a central hub in gene networks associated with different cell phenotypes and disease states, and from a bottom-up SB perspective, it is attractive to develop models that accurately capture the cycling interactions of the VDR with different co-factor and predict how this relates to diverse transcriptional outputs, again across cell phenotypes and disease states.

High dimensional data approaches have been applied to ask questions about what the transcriptional effects are of adding 1α,25$(OH)_2D_3$, or deleting the VDR. From a top-down SB perspective, the question of VDR function is alternatively phrased to identify biological circumstances where the VDR is identified as a hub in the transcriptional network independent of any knowledge of the system. In this manner, analyses of across myeloid cells [35], granulocytes [36], and megakaryocytes [37] have identified significant control functions for the VDR to regulate specific cell differentiation outcomes. Interestingly, these studies support some of the earliest studies that explored the functions of 1α,25$(OH)_2D_3$ and revealed its capacity to initiate differentiation of leukemia cells [38,39]. These findings might suggest a role for the VDR to be mutated or deleted in leukemia and although there were numerous candidate studies of the VDR expression and genomic integrity, large-scale genomic approaches, for example in the TCGA leukemia cohort [40], suggest that the VDR is neither distorted to act as cancer-driver, nor has it been therapeutically exploited to date.

Together, these findings suggest that whilst the VDR is biologically impactful in the normal differentiation, for example, the process to monocytes, it is not itself so frequently disrupted in leukemia such that when it is mutated it acts in an oncogenic manner to disrupt normal progenitor differentiation, and it cannot be targeted to induce leukemia cell differentiation in patients. At first glance, these findings may appear contradictory, but most likely reflect the nature of redundancy in the system, and the role of VDR to undergo diverse protein interactions. Specifically, the VDR is known to interact with CEBPs [41] to induce leukemia cell differentiation [41], and indeed the CEBPs are a master regulator in the physiological transcriptional module containing the VDR to regulate monocyte differentiation [35]. However, CEPPs interact with multiple different nuclear receptors to regulate cell fates [42–45], and redundancy in the system limits the impact that altered VDR function alone can exert on cell fates. Finally, it is interesting to note that CEBPA mutations are in the top 33 mutational events in the TCGA leukemia cohort [40].

These identified roles for the VDR in monocyte differentiation reflect a broader function for the receptor to signal in differentiation processes [46] and immune phenotypes. Alongside the roles of vitamin D to prevent rickets, in the late 19th century the concept emerged of sunlight therapy, known as heliotherapy, to fight tuberculosis and other infections [47]. Indeed, unbiased transcriptomic analyses of the impact of *Mycobacterium tuberculosis* demonstrated a role for Toll-like receptors to initiate a signal transduction cascade in macrophages that upregulated the VDR and induced an antimicrobial response. Intriguingly, this response appeared to be dampened in people with African genomic ancestry perhaps associated with low serum levels of 25(OH)D_3 [48]. Similarly, *Mycobacterium tuberculosis* infection of lung cells identified VDR as a key regulated transcription factor [49]. Another infection of interest is the coronaviruses and in 2013 the lung cell response towards severe acute respiratory syndrome (SARS) arising from a coronavirus infection again identified the VDR as a key regulator of response [50]. More recently, COVID-19 infection of T cells from bronchoalveolar lavage demonstrated a role for the VDR to trigger super-enhancer activation and the control of innate immunity [51]. An unbiased identification of the role of the VDR in the regulation of innate immunity and immune phenotypes has also been identified by GWAS [52–55].

Viewed from the perspective of unbiased top-down statistical analyses has revealed critical roles for the VDR in innate immunity, control of inflammation, regulation of differentiation, and metabolic control [56–58], but not in some of the other phenotypes that are investigated at the pre-clinical and candidate level, notably including cancer; this was established in a previous structured literature search (reviewed in [59]). That is, no cancer-associated GWAS identifies structural variants in the VDR associated with cancer risk, and none of the TCGA papers (over 10,000 tumor samples across more than 30 tumor types) identify frequent VDR-associated structural variants or significantly altered expression that associates with cancer clinical phenotypes.

GWAS studies have continued to evolve in terms of the complexity of study design [60] and cohort sizes and different genomic ancestry, and methods have also developed to finding interactions between variants and also considering how non-coding variants withing regulatory regions may determine how efficiently a given transcription factor can function; so-called a cis-expression quantitative trait loci (eQTL). Certainly, there is evidence for significant enrichment of germline structural variants in VDR binding sites associated with genes that regulate immune phenotypes [61]. Likewise, there is evidence for such eQTLs are significantly impacted by genomic ancestry, for example in the case of the ex vivo response towards 1α,25$(OH)_2D_3$ in normal colon cells [62].

These methodological developments and the ever-increasing repertoire of high dimensional data are likely to increase the understanding of VDR and vitamin D signaling phenotypes and will most likely become ever more important to develop new hypotheses over how signaling occurs and where it is most biologically relevant. Perhaps this is readily illustrated by the application of Mendelian randomization methodology to test the factors such as levels of serum 25(OH)D_3 (either measured or predicted) and their causal relationship with different clinical phenotypes. In this manner 25(OH)D_3 levels and components of the VDR axis are significantly casual in multiple sclerosis, hand grip strength, bacterial infection, and type 2 diabetes [63–66], but not for the prevention of COVID-19 infection, birth weight, or colon cancer [58,67,68].

2.2. Bottom-Up Approaches Applied to VDR Biology

Applying a bottom-up SB approach to VDR-dependent gene regulation requires constructing a kinetic model that captures the function of different VDR-containing complexes and the quantitative regulation patterns of VDR target genes. Such a model would combine spatial-temporal measurements for the generation of 1α,25(OH)$_2$$D_3$ in a target tissue, VDR genomic interactions, such as initial concentrations in the nucleus of the proteins involved, their half-life, and their associations, to explain how these actions predict the downstream transcriptional impact. Model refinement and validation can then be undertaken by empirical measurements of the cycling of VDR genomic interactions and transcriptional outputs.

The working assumption is that the transcription factor complex contains enzymes that remodel chromatin to allow signaling to the RNA polymerase complex and the initiation of transcription. In the case of a transcription factor such as the VDR, which has nuclear residence independent of ligand exposure, the complex exists in two states of genomic interaction, namely with and without ligand, and in each with qualitatively and quantitatively different cofactors, and also in terms of distribution (distal or proximal). The output of the model will then be to predict the kinetics of mRNA accumulation in a manner that reflects the different ligand-dependent and independent states, and spatial association to target genes. Of course, these components are most likely too simple to predict mRNA accumulation accurately. However, even such a simple model for different genes and the VDR could begin to define how much of the difference between mRNA accumulation could be explained by merely considering spatial and temporal factors for transcription factor residence on a gene. Model components could then be added to consider other well-understood biological components such as disease states, and chromatin states, as well as the impact of DNA helicases required for transcription, or even the role of non-coding RNA to impact mRNA accumulation.

To date, such an approach has been applied to several transcription factors in human cells. Within the nuclear receptor superfamily, kinetic modeling of ERα predicted a role of receptor phosphorylation to act as an underexplored feedback mechanism to predict RNA accumulation and was validated by empirical measurements [69]. Outside of nuclear receptors, other human transcription factor-centered models include describing the separate actions of NF-κB [70,71] and p53 [72], and their collective oscillatory behavior [73]. Separately, signal transduction events have also been modeled quite extensively, including the transduction events for RAS [74,75] and its interaction with other regulatory events such as the circadian clock [76] and EGFR signaling [77].

These approaches are powerful, and it is perhaps surprising why they have not been more widely applied to other nuclear receptors beyond the ERα, for example to the clinically relevant VDR. Several impediments no doubt include the time, resources, and interdisciplinary expertise required for model development and refinement. One impediment is establishing collaborations that bring insight from the wet and dry labs. To be successful these collaborations require sufficient data density of dynamic cellular events, such as precise measurements of the concentration of key proteins and RNA molecules in a cell, the frequency of transcription factor genomic interactions, and levels of mRNA accumulation to justify modelling in mathematical terms. Furthermore, these challenges are sometimes impeded by an uneasy relationship between the biological and mathematical communities [5]. The incentive for these intensive modelling efforts, most likely involving interdisciplinary collaborations, is that once such a model is built it has the potential to be translated to other transcription factor genome interactions, or across cell types, or disease states and allow speculation on the biological processes without having to undertake time-intensive and costly experiments in the wet lab.

3. Prostate Cancer and Health Disparities; An Exemplar of the Opportunities Arising from Systems Biology Approaches in Biomedicine

Men of African genomic ancestry in American (African American, (AA)) experience higher risks of developing more aggressive prostate cancer (PCa) than European American (EA) counterparts [78–81], which reflects underlying genetic [82–85] and epigenetic [86–91] drivers and biopsychosocial processes [89,92,93]. The role of the glucocorticoid receptor (GR) as a primary target for stress response has more recently been examined in the context of cancer (reviewed in [94]) and offers a potential functional explanation for how stress can be an accelerant of PCa in AA men [94–97]. A role for VDR signaling appears to be more high profile in the etiology of AA rather than EA PCa. Given that UVB radiation degrades folic acid as well as catalyzing vitamin D synthesis, a strong inverse correlation between skin pigmentation and latitude has arisen during ancestral adaptation [98,99], and amongst AA PCa patients there are significant associations between low serum vitamin D_3 levels and incidence and progression risks of PCa [100–115]. Furthermore, although vitamin D_3 supplementation in the VITAL cohort [116,117] had no overall impact on cancer incidence, the AA participants experienced a suggestive 23% ($p = 0.07$) reduction in cancer risk, indicating that larger cohorts may be more informative [118–120]. Strikingly, vitamin D_3 supplementation in AA and EA PCa patients only significantly modulated prostate gene expression in the AA patients [121], associated with the control of inflammation. This was supported by our recent study [122] in AA and EA cell models which have identified qualitatively and quantitatively distinct VDR actions in terms of VDR protein–protein interactions and more frequent VDR genomic interactions associated with significantly distinct target gene expression.

These data support the concept that one cell function that appears to be most highly responsive in a manner that reflects genomic ancestry is the role of VDR to modulate innate immunity [48,123–126], and reflecting this there are correlative findings that support a relationship between $25(OH)D_3$ levels and immune-modulatory factors such as interleukin (IL)-6 [127,128]. IL6 release into the serum is associated with activated macrophages and is elevated in Ghanaian and AA men [129–131]. Together these studies suggest that the biology of the prostate is the most sensitive VDR signaling in AA men, and more impacted by inadequate VDR signaling either as a result of molecular mechanisms or environments of low $25(OH)D_3$. One biologically impactful consequence of this is the loss of control of inflammatory signals. These studies also highlight that African ancestry is itself divergent, and is also further modified by admixture, for example in the AA population (refs).

The Potential Application of SB Approaches to Prostate Health and Disease

There are significant knowledge gaps in identifying how genomic ancestry, the environment, and lifestyle choices may combine to impact prostate health and disease. The application of SB modeling could be used to generate biologically plausible hypotheses to test specific relationships in these complex interactions. These knowledge gaps include whether germline structural variants impact the generation of $1\alpha,25(OH)_2D_3$, and the extent of relationships between serum $25(OH)D_3$ levels and PCa in a manner that reflects genomic ancestry, and if, and to what extent, this is impacted by admixture. In experimental models, it is emerging that the VDR genomic interactions differ between EA and AA prostate cell models (ref), and recently AR genomic interactions also appear different in EA and AA PCa samples. However, it is unclear if either the underlying sequence, for example at enhancer binding sites, or other mechanisms are impacting why these different genomic interactions arise. Finally, given the multi-parameter process of cis-regulatory interactions that drive transcription through to processed RNA and translation to protein, it is also unclear how genomic ancestry impacts these processes potentially in an emergent manner.

Together, these knowledge gaps can be combined in a hypothesis; namely that PCa risks in AA men reflect the interplay of genomic ancestry, including admixture, coupled with altered environmental signals that combine to drive qualitatively and quantitatively distinct functioning of the VDR and potentially other transcription factors and underscore the health disparities in PCa. Potentially, this is also an attractive setting to develop a series of SB models with which to test these interactions in silico and generate predictions to validate in vitro and in vivo (Figure 1).

This provides an opportunity for both top-down and bottom-up SB modeling. From the top-down perspective, high dimensional data are available to test develop models that test how genomic ancestry significantly determines the associations between serum 25(OH)D_3 levels [118–120] and PCa structural variants [132,133], epigenetic states [86–91] and gene expression [87,91,134] and how this significantly relates to serum inflammatory markers. More specifically, Mendelian randomization approaches would be appropriate to test how germline structural variants are associated with serum 25(OH)D_3 levels depending on genomic ancestry.

To complement these approaches, other top-down strategies applied to high dimensional data sets could then determine epigenomes such as CpG methylation, histone modification, and chromatin accessibility [135,136] and transcriptome relationships in AA and EA PCa patients. Construction of a spatial matrix from epigenomic data, co-incidence with prostate cancer-specific enhancers [137,138], will be binned by accounting for orientation and distance to gene features. Machine learning approaches can then be applied to identify and test the significance of relationships between this matrix and PCa transcriptomes. For example, bootstrapping approaches can test the associations of the epigenome and patterns of observed gene expression compared to simulated data by random sampling [139]. In parallel, other approaches such as the Pareto optimization algorithm [140] can define the ordered correlation of relationships between the cistrome matrix and genesets, of which the strongest (positive or negative) are of greatest biological significance. Comparing the results of both approaches within and across genomic ancestry will reveal more significant cistrome–transcriptome features that are potentially driving PCa health disparities. Lasso and ridge regression can then be performed on the most significant cistrome–transcriptome relationship genes to identify gene expression patterns that predict clinical outcomes in publicly available data [141,142].

To meet these top-down approaches, bottom-up methods can be applied to build kinetic models that take into consideration the amount of the VDR and its key interacting factors in AA and EA prostate cells to simulate its ligand-dependent and independent genomic states and distribution across epigenetic states. The goal of this modeling would be to provide biological plausibility for why the VDR appears to be significantly transcriptionally divergent between AA and EA PCa. Such predictions would form the justification for validation in the wet lab using AA prostate cell line models and patient-derived xenografts. Finally, such modeling may well justify tailoring recommendations for healthy vitamin D serum levels that reflect genomic ancestry.

4. Summary

The VDR signaling system is physiologically impactful and when its functions are altered it is associated with several disease states. Systems biology approaches to the analyses of VDR functions have been applied in several top-down studies that have identified significant VDR functions. To date, however, no bottom-up approaches to kinetic modeling have been applied. The incentive in biomedicine to continue to apply SB approaches would be to develop insight into what are the most prominent contexts where the VDR system exerts biological impact and in turn how this can be exploited in diagnostic, prognostic, or therapeutic contexts. For example, the application of SB approaches, both from a top-down and bottom-up perspective, has the potential to be impactful in the context of prostate cancer health disparities.

Author Contributions: Conceptualization C.Y. and M.J.C.; Writing and Review S.H., C.Y. and M.J.C.; Funding acquisition C.Y. and M.J.C.; All authors have read and agreed to the published version of the manuscript.

Funding: M.J.C. acknowledges support in part from the Department of Defense Congressionally Directed Medical Research Programs (W81XWH-20-1-0373; W81XWH-21-1-0555), C.Y. acknowledges support in part from the Department of Defense Congressionally Directed Medical Research Programs (W81XWH-21-1-0850; W81XWH-18-1-0589); from U54-MD007585-26 (NIH/NIMHD), U54 CA118623 (NIH/NCI). S.H. acknowledges support from a Pelotonia Post-Doctoral Fellowship. M.J.C. and S.H. also acknowledge the National Institute of Health Cancer Center Support Grant (P30CA016058) to the OSUCCC The James.

Institutional Review Board Statement: Not applicable.

Informed Consent Statement: Not applicable.

Data Availability Statement: Not applicable.

Conflicts of Interest: The authors declare no conflict of interest.

References

1. Rowley, J.D. Letter: A new consistent chromosomal abnormality in chronic myelogenous leukaemia identified by quinacrine fluorescence and Giemsa staining. *Nature* **1973**, *243*, 290–293. [CrossRef] [PubMed]
2. Druker, B.J.; Lydon, N.B. Lessons learned from the development of an abl tyrosine kinase inhibitor for chronic myelogenous leukemia. *J. Clin. Investig.* **2000**, *105*, 3–7. [CrossRef]
3. Bixby, D.; Talpaz, M. Seeking the causes and solutions to imatinib-resistance in chronic myeloid leukemia. *Leukemia* **2011**, *25*, 7–22. [CrossRef] [PubMed]
4. Cassman, M. Barriers to progress in systems biology. *Nature* **2005**, *438*, 1079. [CrossRef] [PubMed]
5. Vera, J.; Lischer, C.; Nenov, M.; Nikolov, S.; Lai, X.; Eberhardt, M. Mathematical Modelling in Biomedicine: A Primer for the Curious and the Skeptic. *Int. J. Mol. Sci.* **2021**, *22*, 547. [CrossRef] [PubMed]
6. Haran, T.K.; Keren, L. From genes to modules, from cells to ecosystems. *Mol. Syst. Biol.* **2022**, *18*, e10726.
7. Angione, C. Human Systems Biology and Metabolic Modelling: A Review-From Disease Metabolism to Precision Medicine. *Biomed. Res. Int.* **2019**, *2019*, 8304260. [CrossRef]
8. Kuenzi, B.M.; Ideker, T. A census of pathway maps in cancer systems biology. *Nat. Rev. Cancer* **2020**, *20*, 233–246. [CrossRef]
9. Wynn, M.L.; Consul, N.; Merajver, S.D.; Schnell, S. Logic-based models in systems biology: A predictive and parameter-free network analysis method. *Integr. Biol.* **2012**, *4*, 1323–1337. [CrossRef]
10. Germain, R.N. Will Systems Biology Deliver Its Promise and Contribute to the Development of New or Improved Vaccines? What Really Constitutes the Study of "Systems Biology" and How Might Such an Approach Facilitate Vaccine Design. *Cold Spring Harb. Perspect. Biol.* **2018**, *10*, a033308. [CrossRef]
11. Petrasek, D. Systems biology: The case for a systems science approach to diabetes. *J. Diabetes Sci. Technol.* **2008**, *2*, 131–134. [CrossRef] [PubMed]
12. Okamura, H.; Yamaguchi, S.; Yagita, K. Molecular machinery of the circadian clock in mammals. *Cell Tissue Res.* **2002**, *309*, 47–56. [CrossRef] [PubMed]
13. Tan, D.; Chen, R.; Mo, Y.; Gu, S.; Ma, J.; Xu, W.; Lu, X.; He, H.; Jiang, F.; Fan, W.; et al. Quantitative control of noise in mammalian gene expression by dynamic histone regulation. *Elife* **2021**, *10*, e65654. [CrossRef] [PubMed]
14. Thorne, J.L.; Campbell, M.J.; Turner, B.M. Transcription factors, chromatin and cancer. *Int. J. Biochem. Cell Biol.* **2009**, *41*, 164–175. [CrossRef] [PubMed]
15. Nurk, S.; Koren, S.; Rhie, A.; Rautiainen, M.; Bzikadze, A.V.; Mikheenko, A.; Vollger, M.R.; Altemose, N.; Uralsky, L.; Gershman, A.; et al. The complete sequence of a human genome. *Science* **2022**, *376*, 44–53. [CrossRef] [PubMed]
16. Barabasi, A.L.; Oltvai, Z.N. Network biology: Understanding the cell's functional organization. *Nat. Rev. Genet.* **2004**, *5*, 101–113. [CrossRef] [PubMed]
17. Carlberg, C.; Dunlop, T.W. An integrated biological approach to nuclear receptor signaling in physiological control and disease. *Crit. Rev. Eukaryot. Gene Expr.* **2006**, *16*, 1–22. [CrossRef]
18. Flores, M.; Glusman, G.; Brogaard, K.; Price, N.D.; Hood, L. P4 medicine: How systems medicine will transform the healthcare sector and society. *Per. Med.* **2013**, *10*, 565–576. [CrossRef]
19. Sobie, E.A.; Lee, Y.S.; Jenkins, S.L.; Iyengar, R. Systems biology—Biomedical modeling. *Sci. Signal* **2011**, *4*, tr2. [CrossRef]
20. Westerhoff, H.V.; Palsson, B.O. The evolution of molecular biology into systems biology. *Nat. Biotechnol.* **2004**, *22*, 1249–1252. [CrossRef]
21. Bruggeman, F.J.; Westerhoff, H.V. The nature of systems biology. *Trends Microbiol.* **2007**, *15*, 45–50. [CrossRef] [PubMed]
22. Snoep, J.L. The Silicon Cell initiative: Working towards a detailed kinetic description at the cellular level. *Curr. Opin. Biotechnol.* **2005**, *16*, 336–343. [CrossRef] [PubMed]
23. Ellis, H. Daniel Whistler: English physician who published the first book on rickets in 1645. *Br. J. Hosp. Med.* **2019**, *80*, 51. [CrossRef] [PubMed]
24. Carlberg, C.; Campbell, M.J. Vitamin D receptor signaling mechanisms: Integrated actions of a well-defined transcription factor. *Steroids* **2013**, *78*, 127–136. [CrossRef]

25. Seuter, S.; Neme, A.; Carlberg, C. Characterization of genomic vitamin D receptor binding sites through chromatin looping and opening. *PLoS ONE* **2014**, *9*, e96184. [CrossRef] [PubMed]
26. Satoh, J.; Tabunoki, H. Molecular network of chromatin immunoprecipitation followed by deep sequencing-based vitamin D receptor target genes. *Mult. Scler.* **2013**, *19*, 1035–1045. [CrossRef] [PubMed]
27. Ramagopalan, S.V.; Heger, A.; Berlanga, A.J.; Maugeri, N.J.; Lincoln, M.R.; Burrell, A.; Handunnetthi, L.; Handel, A.E.; Disanto, G.; Orton, S.M.; et al. A ChIP-seq defined genome-wide map of vitamin D receptor binding: Associations with disease and evolution. *Genome Res.* **2010**, *20*, 1352–1360. [CrossRef]
28. Meyer, M.B.; Goetsch, P.D.; Pike, J.W. VDR/RXR and TCF4/beta-catenin cistromes in colonic cells of colorectal tumor origin: Impact on c-FOS and c-MYC gene expression. *Mol. Endocrinol.* **2012**, *26*, 37–51. [CrossRef]
29. Gonzalez-Duarte, R.J.; Cazares-Ordonez, V.; Diaz, L.; Ortiz, V.; Larrea, F.; Avila, E. The expression of RNA helicase DDX5 is transcriptionally upregulated by calcitriol through a vitamin D response element in the proximal promoter in SiHa cervical cells. *Mol. Cell Biochem.* **2015**, *410*, 65–73. [CrossRef]
30. Wagner, M.; Rid, R.; Maier, C.J.; Maier, R.H.; Laimer, M.; Hintner, H.; Bauer, J.W.; Onder, K. DDX5 is a multifunctional co-activator of steroid hormone receptors. *Mol. Cell Endocrinol.* **2012**, *361*, 80–91. [CrossRef]
31. MacDonald, P.N.; Dowd, D.R.; Zhang, C.; Gu, C. Emerging insights into the coactivator role of NCoA62/SKIP in Vitamin D-mediated transcription. *J. Steroid Biochem. Mol. Biol.* **2004**, *89–90*, 179–186. [CrossRef] [PubMed]
32. Foulds, C.E.; Panigrahi, A.K.; Coarfa, C.; Lanz, R.B.; O'Malley, B.W. Long Noncoding RNAs as Targets and Regulators of Nuclear Receptors. *Curr. Top Microbiol. Immunol.* **2016**, *394*, 143–176. [PubMed]
33. Lanz, R.B.; Razani, B.; Goldberg, A.D.; O'Malley, B.W. Distinct RNA motifs are important for coactivation of steroid hormone receptors by steroid receptor RNA activator (SRA). *Proc. Natl. Acad. Sci. USA* **2002**, *99*, 16081–16086. [CrossRef] [PubMed]
34. Statello, L.; Guo, C.J.; Chen, L.L.; Huarte, M. Gene regulation by long non-coding RNAs and its biological functions. *Nat. Rev. Mol. Cell Biol.* **2021**, *22*, 96–118. [CrossRef]
35. Novershtern, N.; Subramanian, A.; Lawton, L.N.; Mak, R.H.; Haining, W.N.; McConkey, M.E.; Habib, N.; Yosef, N.; Chang, C.Y.; Shay, T.; et al. Densely interconnected transcriptional circuits control cell states in human hematopoiesis. *Cell* **2011**, *144*, 296–309. [CrossRef]
36. Novikova, S.; Tikhonova, O.; Kurbatov, L.; Farafonova, T.; Vakhrushev, I.; Lupatov, A.; Yarygin, K.; Zgoda, V. Omics Technologies to Decipher Regulatory Networks in Granulocytic Cell Differentiation. *Biomolecules* **2021**, *11*, 907. [CrossRef]
37. Fuhrken, P.G.; Chen, C.; Apostolidis, P.A.; Wang, M.; Miller, W.M.; Papoutsakis, E.T. Gene Ontology-driven transcriptional analysis of CD34+ cell-initiated megakaryocytic cultures identifies new transcriptional regulators of megakaryopoiesis. *Physiol. Genom.* **2008**, *33*, 159–169. [CrossRef]
38. Yetgin, S.; Ozsoylu, S. Myeloid metaplasia in vitamin D deficiency rickets. *Scand. J. Haematol.* **1982**, *28*, 180–185. [CrossRef]
39. Koeffler, H.P. Induction of differentiation of human acute myelogenous leukemia cells: Therapeutic implications. *Blood* **1983**, *62*, 709–721. [CrossRef]
40. Tyner, J.W.; Tognon, C.E.; Bottomly, D.; Wilmot, B.; Kurtz, S.E.; Savage, S.L.; Long, N.; Schultz, A.R.; Traer, E.; Abel, M.; et al. Functional genomic landscape of acute myeloid leukaemia. *Nature* **2018**, *562*, 526–531. [CrossRef]
41. Ji, Y.; Studzinski, G.P. Retinoblastoma protein and CCAAT/enhancer-binding protein beta are required for 1,25-dihydroxyvitamin D3-induced monocytic differentiation of HL60 cells. *Cancer Res.* **2004**, *64*, 370–377. [CrossRef] [PubMed]
42. Shiozaki, Y.; Miyazaki-Anzai, S.; Keenan, A.L.; Miyazaki, M. MEF2D-NR4A1-FAM134B2-mediated reticulophagy contributes to amino acid homeostasis. *Autophagy* **2022**, *18*, 1049–1061. [CrossRef]
43. Marchwicka, A.; Marcinkowska, E. Regulation of Expression of CEBP Genes by Variably Expressed Vitamin D Receptor and Retinoic Acid Receptor alpha in Human Acute Myeloid Leukemia Cell Lines. *Int. J. Mol. Sci.* **2018**, *19*, 1918. [CrossRef] [PubMed]
44. Mansure, J.J.; Nassim, R.; Chevalier, S.; Szymanski, K.; Rocha, J.; Aldousari, S.; Kassouf, W. A novel mechanism of PPAR gamma induction via EGFR signalling constitutes rational for combination therapy in bladder cancer. *PLoS ONE* **2013**, *8*, e55997. [CrossRef]
45. Tang, Q.Q.; Otto, T.C.; Lane, M.D. CCAAT/enhancer-binding protein beta is required for mitotic clonal expansion during adipogenesis. *Proc. Natl. Acad. Sci. USA* **2003**, *100*, 850–855. [CrossRef] [PubMed]
46. Zolotarenko, A.; Chekalin, E.; Mesentsev, A.; Kiseleva, L.; Gribanova, E.; Mehta, R.; Baranova, A.; Tatarinova, T.V.; Piruzian, E.S.; Bruskin, S. Integrated computational approach to the analysis of RNA-seq data reveals new transcriptional regulators of psoriasis. *Exp. Mol. Med.* **2016**, *48*, e268. [CrossRef]
47. Greenhalgh, I.; Butler, A.R. Sanatoria revisited: Sunlight and health. *J. R. Coll. Phys. Edinb.* **2017**, *47*, 276–280. [CrossRef]
48. Liu, P.T.; Stenger, S.; Li, H.; Wenzel, L.; Tan, B.H.; Krutzik, S.R.; Ochoa, M.T.; Schauber, J.; Wu, K.; Meinken, C.; et al. Toll-like receptor triggering of a vitamin D-mediated human antimicrobial response. *Science* **2006**, *311*, 1770–1773. [CrossRef]
49. Mvubu, N.E.; Pillay, B.; Gamieldien, J.; Bishai, W.; Pillay, M. Canonical pathways, networks and transcriptional factor regulation by clinical strains of Mycobacterium tuberculosis in pulmonary alveolar epithelial cells. *Tuberculosis* **2016**, *97*, 73–85. [CrossRef]
50. Sims, A.C.; Tilton, S.C.; Menachery, V.D.; Gralinski, L.E.; Schafer, A.; Matzke, M.M.; Webb-Robertson, B.J.; Chang, J.; Luna, M.L.; Long, C.E.; et al. Release of severe acute respiratory syndrome coronavirus nuclear import block enhances host transcription in human lung cells. *J. Virol.* **2013**, *87*, 3885–3902. [CrossRef]
51. Chauss, D.; Freiwald, T.; McGregor, R.; Yan, B.; Wang, L.; Nova-Lamperti, E.; Kumar, D.; Zhang, Z.; Teague, H.; West, E.E.; et al. Autocrine vitamin D signaling switches off pro-inflammatory programs of TH1 cells. *Nat. Immunol.* **2022**, *23*, 62–74. [CrossRef]

52. Adolphe, C.; Xue, A.; Fard, A.T.; Genovesi, L.A.; Yang, J.; Wainwright, B.J. Genetic and functional interaction network analysis reveals global enrichment of regulatory T cell genes influencing basal cell carcinoma susceptibility. *Genome Med.* **2021**, *13*, 19. [CrossRef] [PubMed]
53. Jostins, L.; Ripke, S.; Weersma, R.K.; Duerr, R.H.; McGovern, D.P.; Hui, K.Y.; Lee, J.C.; Schumm, L.P.; Sharma, Y.; Anderson, C.A.; et al. Host-microbe interactions have shaped the genetic architecture of inflammatory bowel disease. *Nature* **2012**, *491*, 119–124. [CrossRef]
54. Wang, J.; Thingholm, L.B.; Skieceviciene, J.; Rausch, P.; Kummen, M.; Hov, J.R.; Degenhardt, F.; Heinsen, F.A.; Ruhlemann, M.C.; Szymczak, S.; et al. Genome-wide association analysis identifies variation in vitamin D receptor and other host factors influencing the gut microbiota. *Nat. Genet.* **2016**, *48*, 1396–1406. [CrossRef]
55. Tavasolian, F.; Hosseini, A.Z.; Soudi, S.; Naderi, M.; Sahebkar, A. A Systems Biology Approach for miRNA-mRNA Expression Patterns Analysis in Rheumatoid Arthritis. *Comb. Chem. High Throughput. Screen* **2021**, *24*, 195–212. [CrossRef] [PubMed]
56. Montasser, M.E.; Aslibekyan, S.; Srinivasasainagendra, V.; Tiwari, H.K.; Patki, A.; Bagheri, M.; Kind, T.; Barupal, D.K.; Fan, S.; Perry, J.; et al. An Amish founder population reveals rare-population genetic determinants of the human lipidome. *Commun. Biol.* **2022**, *5*, 334. [CrossRef] [PubMed]
57. Skinkyte-Juskiene, R.; Kogelman, L.J.A.; Kadarmideen, H.N. Transcription Factor Co-expression Networks of Adipose RNA-Seq Data Reveal Regulatory Mechanisms of Obesity. *Curr. Genom.* **2018**, *19*, 289–299. [CrossRef] [PubMed]
58. Bao, M.H.; Luo, H.Q.; Chen, L.H.; Tang, L.; Ma, K.F.; Xiang, J.; Dong, L.P.; Zeng, J.; Li, G.Y.; Li, J.M. Impact of high fat diet on long non-coding RNAs and messenger RNAs expression in the aortas of ApoE(-/-) mice. *Sci. Rep.* **2016**, *6*, 34161. [CrossRef]
59. Campbell, M.J. Bioinformatic approaches to interrogating vitamin D receptor signaling. *Mol. Cell Endocrinol.* **2017**, *453*, 3–13. [CrossRef]
60. Vuckovic, D.; Bao, E.L.; Akbari, P.; Lareau, C.A.; Mousas, A.; Jiang, T.; Chen, M.H.; Raffield, L.M.; Tardaguila, M.; Huffman, J.E.; et al. The Polygenic and Monogenic Basis of Blood Traits and Diseases. *Cell* **2020**, *182*, 1214–1231.e1211. [CrossRef]
61. Singh, P.K.; van den Berg, P.R.; Long, M.D.; Vreugdenhil, A.; Grieshober, L.; Ochs-Balcom, H.M.; Wang, J.; Delcambre, S.; Heikkinen, S.; Carlberg, C.; et al. Integration of VDR genome wide binding and GWAS genetic variation data reveals co-occurrence of VDR and NF-kappaB binding that is linked to immune phenotypes. *BMC Genom.* **2017**, *18*, 132. [CrossRef] [PubMed]
62. Alleyne, D.; Witonsky, D.B.; Mapes, B.; Nakagome, S.; Sommars, M.; Hong, E.; Muckala, K.A.; Di Rienzo, A.; Kupfer, S.S. Colonic transcriptional response to 1alpha,25(OH)2 vitamin D3 in African- and European-Americans. *J. Steroid Biochem. Mol. Biol.* **2017**, *168*, 49–59. [CrossRef] [PubMed]
63. Colak, Y.; Nordestgaard, B.G.; Afzal, S. Low vitamin D and risk of bacterial pneumonias: Mendelian randomisation studies in two population-based cohorts. *Thorax* **2021**, *76*, 468–478. [CrossRef] [PubMed]
64. Mazidi, M.; Davies, I.G.; Penson, P.; Rikkonen, T.; Isanejad, M. Lifetime serum concentration of 25-hydroxyvitamin D 25(OH) is associated with hand grip strengths: Insight from a Mendelian randomisation. *Age Ageing* **2022**, *51*, afac079. [CrossRef]
65. Mokry, L.E.; Ross, S.; Ahmad, O.S.; Forgetta, V.; Smith, G.D.; Goltzman, D.; Leong, A.; Greenwood, C.M.; Thanassoulis, G.; Richards, J.B. Vitamin D and Risk of Multiple Sclerosis: A Mendelian Randomization Study. *PLoS Med.* **2015**, *12*, e1001866. [CrossRef]
66. Zheng, J.S.; Luan, J.; Sofianopoulou, E.; Sharp, S.J.; Day, F.R.; Imamura, F.; Gundersen, T.E.; Lotta, L.A.; Sluijs, I.; Stewart, I.D.; et al. The association between circulating 25-hydroxyvitamin D metabolites and type 2 diabetes in European populations: A meta-analysis and Mendelian randomisation analysis. *PLoS Med.* **2020**, *17*, e1003394. [CrossRef]
67. He, Y.; Timofeeva, M.; Farrington, S.M.; Vaughan-Shaw, P.; Svinti, V.; Walker, M.; Zgaga, L.; Meng, X.; Li, X.; Spiliopoulou, A.; et al. Exploring causality in the association between circulating 25-hydroxyvitamin D and colorectal cancer risk: A large Mendelian randomisation study. *BMC Med.* **2018**, *16*, 142. [CrossRef]
68. Thompson, W.D.; Tyrrell, J.; Borges, M.C.; Beaumont, R.N.; Knight, B.A.; Wood, A.R.; Ring, S.M.; Hattersley, A.T.; Freathy, R.M.; Lawlor, D.A. Association of maternal circulating 25(OH)D and calcium with birth weight: A mendelian randomisation analysis. *PLoS Med.* **2019**, *16*, e1002828. [CrossRef]
69. Tian, D.; Solodin, N.M.; Rajbhandari, P.; Bjorklund, K.; Alarid, E.T.; Kreeger, P.K. A kinetic model identifies phosphorylated estrogen receptor-alpha (ERalpha) as a critical regulator of ERalpha dynamics in breast cancer. *FASEB J.* **2015**, *29*, 2022–2031. [CrossRef]
70. Cheong, R.; Hoffmann, A.; Levchenko, A. Understanding NF-kappaB signaling via mathematical modeling. *Mol. Syst. Biol.* **2008**, *4*, 192. [CrossRef]
71. Bullock, M.E.; Moreno-Martinez, N.; Miller-Jensen, K. A transcriptional cycling model recapitulates chromatin-dependent features of noisy inducible transcription. *PLoS Comput. Biol.* **2022**, *18*, e1010152. [CrossRef] [PubMed]
72. Jimenez, A.; Lu, D.; Kalocsay, M.; Berberich, M.J.; Balbi, P.; Jambhekar, A.; Lahav, G. Time-series transcriptomics and proteomics reveal alternative modes to decode p53 oscillations. *Mol. Syst. Biol.* **2022**, *18*, e10588. [CrossRef] [PubMed]
73. Ihekwaba, A.E.; Sedwards, S. Communicating oscillatory networks: Frequency domain analysis. *BMC Syst. Biol.* **2011**, *5*, 203. [CrossRef] [PubMed]
74. Markevich, N.I.; Moehren, G.; Demin, O.V.; Kiyatkin, A.; Hoek, J.B.; Kholodenko, B.N. Signal processing at the Ras circuit: What shapes Ras activation patterns? *Syst. Biol.* **2004**, *1*, 104–113. [CrossRef]

75. Stelniec-Klotz, I.; Legewie, S.; Tchernitsa, O.; Witzel, F.; Klinger, B.; Sers, C.; Herzel, H.; Bluthgen, N.; Schafer, R. Reverse engineering a hierarchical regulatory network downstream of oncogenic KRAS. *Mol. Syst. Biol.* **2012**, *8*, 601. [CrossRef]
76. Relogio, A.; Thomas, P.; Medina-Perez, P.; Reischl, S.; Bervoets, S.; Gloc, E.; Riemer, P.; Mang-Fatehi, S.; Maier, B.; Schafer, R.; et al. Ras-mediated deregulation of the circadian clock in cancer. *PLoS Genet.* **2014**, *10*, e1004338. [CrossRef]
77. Davies, A.E.; Pargett, M.; Siebert, S.; Gillies, T.E.; Choi, Y.; Tobin, S.J.; Ram, A.R.; Murthy, V.; Juliano, C.; Quon, G.; et al. Systems-Level Properties of EGFR-RAS-ERK Signaling Amplify Local Signals to Generate Dynamic Gene Expression Heterogeneity. *Cell Syst.* **2020**, *11*, 161–175.e165. [CrossRef]
78. Patel, H.D.; Doshi, C.P.; Koehne, E.L.; Hart, S.; Van Kuiken, M.; Quek, M.L.; Flanigan, R.C.; Gupta, G.N. African American Men have Increased Risk of Prostate Cancer Detection Despite Similar Rates of Anterior Prostatic Lesions and PI-RADS Grade on Multiparametric Magnetic Resonance Imaging. *Urology* **2022**, *163*, 132–137. [CrossRef]
79. Powell, I.J. Epidemiology and pathophysiology of prostate cancer in African-American men. *J. Urol.* **2007**, *177*, 444–449. [CrossRef]
80. Moul, J.W.; Sesterhenn, I.A.; Connelly, R.R.; Douglas, T.; Srivastava, S.; Mostofi, F.K.; McLeod, D.G. Prostate-specific antigen values at the time of prostate cancer diagnosis in African-American men. *JAMA* **1995**, *274*, 1277–1281. [CrossRef]
81. Sanchez-Ortiz, R.F.; Troncoso, P.; Babaian, R.J.; Lloreta, J.; Johnston, D.A.; Pettaway, C.A. African-American men with nonpalpable prostate cancer exhibit greater tumor volume than matched white men. *Cancer* **2006**, *107*, 75–82. [CrossRef] [PubMed]
82. Fiorica, P.N.; Schubert, R.; Morris, J.D.; Abdul Sami, M.; Wheeler, H.E. Multi-ethnic transcriptome-wide association study of prostate cancer. *PLoS ONE* **2020**, *15*, e0236209. [CrossRef] [PubMed]
83. Hoffmann, T.J.; Van Den Eeden, S.K.; Sakoda, L.C.; Jorgenson, E.; Habel, L.A.; Graff, R.E.; Passarelli, M.N.; Cario, C.L.; Emami, N.C.; Chao, C.R.; et al. A large multiethnic genome-wide association study of prostate cancer identifies novel risk variants and substantial ethnic differences. *Cancer Discov.* **2015**, *5*, 878–891. [CrossRef] [PubMed]
84. Helfand, B.T.; Roehl, K.A.; Cooper, P.R.; McGuire, B.B.; Fitzgerald, L.M.; Cancel-Tassin, G.; Cornu, J.N.; Bauer, S.; Van Blarigan, E.L.; Chen, X.; et al. Associations of prostate cancer risk variants with disease aggressiveness: Results of the NCI-SPORE Genetics Working Group analysis of 18,343 cases. *Hum. Genet.* **2015**, *134*, 439–450. [CrossRef] [PubMed]
85. Mori, J.O.; White, J.; Elhussin, I.; Duduyemi, B.M.; Karanam, B.; Yates, C.; Wang, H. Molecular and pathological subtypes related to prostate cancer disparities and disease outcomes in African American and European American patients. *Front. Oncol.* **2022**, *12*, 928357. [CrossRef]
86. Awasthi, S.; Berglund, A.; Abraham-Miranda, J.; Rounbehler, R.J.; Kensler, K.; Serna, A.; Vidal, A.; You, S.; Freeman, M.R.; Davicioni, E.; et al. Comparative Genomics Reveals Distinct Immune-oncologic Pathways in African American Men with Prostate Cancer. *Clin. Cancer Res.* **2021**, *27*, 320–329. [CrossRef]
87. Yuan, J.; Kensler, K.H.; Hu, Z.; Zhang, Y.; Zhang, T.; Jiang, J.; Xu, M.; Pan, Y.; Long, M.; Montone, K.T.; et al. Integrative comparison of the genomic and transcriptomic landscape between prostate cancer patients of predominantly African or European genetic ancestry. *PLoS Genet.* **2020**, *16*, e1008641. [CrossRef]
88. Hardiman, G.; Savage, S.J.; Hazard, E.S.; Wilson, R.C.; Courtney, S.M.; Smith, M.T.; Hollis, B.W.; Halbert, C.H.; Gattoni-Celli, S. Systems analysis of the prostate transcriptome in African-American men compared with European-American men. *Pharmacogenomics* **2016**, *17*, 1129–1143. [CrossRef]
89. Yamoah, K.; Asamoah, F.A.; Abrahams, A.O.D.; Awasthi, S.; Mensah, J.E.; Dhillon, J.; Mahal, B.A.; Gueye, S.M.; Jalloh, M.; Farahani, S.J.; et al. Prostate tumors of native men from West Africa show biologically distinct pathways-A comparative genomic study. *Prostate* **2021**, *81*, 1402–1410. [CrossRef]
90. Koga, Y.; Song, H.; Chalmers, Z.R.; Newberg, J.; Kim, E.; Carrot-Zhang, J.; Piou, D.; Polak, P.; Abdulkadir, S.A.; Ziv, E.; et al. Genomic Profiling of Prostate Cancers from Men with African and European Ancestry. *Clin. Cancer Res.* **2020**, *26*, 4651–4660. [CrossRef]
91. Rayford, W.; Beksac, A.T.; Alger, J.; Alshalalfa, M.; Ahmed, M.; Khan, I.; Falagario, U.G.; Liu, Y.; Davicioni, E.; Spratt, D.E.; et al. Comparative analysis of 1152 African-American and European-American men with prostate cancer identifies distinct genomic and immunological differences. *Commun. Biol.* **2021**, *4*, 670. [CrossRef]
92. Liu, W.; Zheng, S.L.; Na, R.; Wei, L.; Sun, J.; Gallagher, J.; Wei, J.; Resurreccion, W.K.; Ernst, S.; Sfanos, K.S.; et al. Distinct Genomic Alterations in Prostate Tumors Derived from African American Men. *Mol. Cancer Res.* **2020**, *18*, 1815–1824. [CrossRef]
93. Magi-Galluzzi, C.; Tsusuki, T.; Elson, P.; Simmerman, K.; LaFargue, C.; Esgueva, R.; Klein, E.; Rubin, M.A.; Zhou, M. TMPRSS2-ERG gene fusion prevalence and class are significantly different in prostate cancer of Caucasian, African-American and Japanese patients. *Prostate* **2011**, *71*, 489–497. [CrossRef] [PubMed]
94. Kerkvliet, C.P.; Truong, T.H.; Ostrander, J.H.; Lange, C.A. Stress sensing within the breast tumor microenvironment: How glucocorticoid receptors live in the moment. *Essays Biochem.* **2021**, *65*, 971–983. [CrossRef] [PubMed]
95. Lee, M.J.; Rittschof, C.C.; Greenlee, A.J.; Turi, K.N.; Rodriguez-Zas, S.L.; Robinson, G.E.; Cole, S.W.; Mendenhall, R. Transcriptomic analyses of black women in neighborhoods with high levels of violence. *Psychoneuroendocrinology* **2021**, *127*, 105174. [CrossRef]
96. Woods-Burnham, L.; Stiel, L.; Martinez, S.R.; Sanchez-Hernandez, E.S.; Ruckle, H.C.; Almaguel, F.G.; Stern, M.C.; Roberts, L.R.; Williams, D.R.; Montgomery, S.; et al. Psychosocial Stress, Glucocorticoid Signaling, and Prostate Cancer Health Disparities in African American Men. *Cancer Health Disparities* **2020**, *4*, e1–e30.
97. Woods-Burnham, L.; Cajigas-Du Ross, C.K.; Love, A.; Basu, A.; Sanchez-Hernandez, E.S.; Martinez, S.R.; Ortiz-Hernandez, G.L.; Stiel, L.; Duran, A.M.; Wilson, C.; et al. Glucocorticoids Induce Stress Oncoproteins Associated with Therapy-Resistance in African American and European American Prostate Cancer Cells. *Sci. Rep.* **2018**, *8*, 15063. [CrossRef]

98. Hochberg, Z.; Templeton, A.R. Evolutionary perspective in skin color, vitamin D and its receptor. *Hormones* **2010**, *9*, 307–311. [CrossRef]
99. Jablonski, N.G. The evolution of human skin colouration and its relevance to health in the modern world. *J. R. Coll. Phys. Edinb.* **2012**, *42*, 58–63. [CrossRef]
100. Yao, S.; Hong, C.C.; Bandera, E.V.; Zhu, Q.; Liu, S.; Cheng, T.D.; Zirpoli, G.; Haddad, S.A.; Lunetta, K.L.; Ruiz-Narvaez, E.A.; et al. Demographic, lifestyle, and genetic determinants of circulating concentrations of 25-hydroxyvitamin D and vitamin D-binding protein in African American and European American women. *Am. J. Clin. Nutr.* **2017**, *105*, 1362–1371. [CrossRef]
101. Yao, S.; Haddad, S.A.; Hu, Q.; Liu, S.; Lunetta, K.L.; Ruiz-Narvaez, E.A.; Hong, C.C.; Zhu, Q.; Sucheston-Campbell, L.; Cheng, T.Y.; et al. Genetic variations in vitamin D-related pathways and breast cancer risk in African American women in the AMBER consortium. *Int. J. Cancer* **2016**, *138*, 2118–2126. [CrossRef] [PubMed]
102. Mishra, D.K.; Wu, Y.; Sarkissyan, M.; Sarkissyan, S.; Chen, Z.; Shang, X.; Ong, M.; Heber, D.; Koeffler, H.P.; Vadgama, J.V. Vitamin D receptor gene polymorphisms and prognosis of breast cancer among African-American and Hispanic women. *PLoS ONE* **2013**, *8*, e57967. [CrossRef] [PubMed]
103. Neuhouser, M.L.; Bernstein, L.; Hollis, B.W.; Xiao, L.; Ambs, A.; Baumgartner, K.; Baumgartner, R.; McTiernan, A.; Ballard-Barbash, R. Serum vitamin D and breast density in breast cancer survivors. *Cancer Epidemiol. Biomark. Prev.* **2010**, *19*, 412–417. [CrossRef]
104. John, E.M.; Schwartz, G.G.; Koo, J.; Wang, W.; Ingles, S.A. Sun exposure, vitamin D receptor gene polymorphisms, and breast cancer risk in a multiethnic population. *Am. J. Epidemiol.* **2007**, *166*, 1409–1419. [CrossRef] [PubMed]
105. Layne, T.M.; Weinstein, S.J.; Graubard, B.I.; Ma, X.; Mayne, S.T.; Albanes, D. Serum 25-hydroxyvitamin D, vitamin D binding protein, and prostate cancer risk in black men. *Cancer* **2017**, *123*, 2698–2704. [CrossRef] [PubMed]
106. Nelson, S.M.; Batai, K.; Ahaghotu, C.; Agurs-Collins, T.; Kittles, R.A. Association between Serum 25-Hydroxy-Vitamin D and Aggressive Prostate Cancer in African American Men. *Nutrients* **2016**, *9*, 12. [CrossRef]
107. Batai, K.; Murphy, A.B.; Nonn, L.; Kittles, R.A. Vitamin D and Immune Response: Implications for Prostate Cancer in African Americans. *Front. Immunol.* **2016**, *7*, 53. [CrossRef]
108. Murphy, A.B.; Nyame, Y.; Martin, I.K.; Catalona, W.J.; Hollowell, C.M.; Nadler, R.B.; Kozlowski, J.M.; Perry, K.T.; Kajdacsy-Balla, A.; Kittles, R. Vitamin D deficiency predicts prostate biopsy outcomes. *Clin. Cancer Res.* **2014**, *20*, 2289–2299. [CrossRef]
109. Richards, Z.; Batai, K.; Farhat, R.; Shah, E.; Makowski, A.; Gann, P.H.; Kittles, R.; Nonn, L. Prostatic compensation of the vitamin D axis in African American men. *JCI Insight* **2017**, *2*, e91054. [CrossRef]
110. Kidd, L.C.; Paltoo, D.N.; Wang, S.; Chen, W.; Akereyeni, F.; Isaacs, W.; Ahaghotu, C.; Kittles, R. Sequence variation within the 5′ regulatory regions of the vitamin D binding protein and receptor genes and prostate cancer risk. *Prostate* **2005**, *64*, 272–282. [CrossRef]
111. Murphy, A.B.; Kelley, B.; Nyame, Y.A.; Martin, I.K.; Smith, D.J.; Castaneda, L.; Zagaja, G.J.; Hollowell, C.M.; Kittles, R.A. Predictors of serum vitamin D levels in African American and European American men in Chicago. *Am. J. Mens. Health* **2012**, *6*, 420–426. [CrossRef] [PubMed]
112. Paller, C.J.; Kanaan, Y.M.; Beyene, D.A.; Naab, T.J.; Copeland, R.L.; Tsai, H.L.; Kanarek, N.F.; Hudson, T.S. Risk of prostate cancer in African-American men: Evidence of mixed effects of dietary quercetin by serum vitamin D status. *Prostate* **2015**, *75*, 1376–1383. [CrossRef] [PubMed]
113. Batai, K.; Kittles, R.A. Can vitamin D supplementation reduce prostate cancer disparities? *Pharmacogenomics* **2016**, *17*, 1117–1120. [CrossRef] [PubMed]
114. Taksler, G.B.; Cutler, D.M.; Giovannucci, E.; Smith, M.R.; Keating, N.L. Ultraviolet index and racial differences in prostate cancer incidence and mortality. *Cancer* **2013**, *119*, 3195–3203. [CrossRef]
115. Steck, S.E.; Arab, L.; Zhang, H.; Bensen, J.T.; Fontham, E.T.; Johnson, C.S.; Mohler, J.L.; Smith, G.J.; Su, J.L.; Trump, D.L.; et al. Association between Plasma 25-Hydroxyvitamin D, Ancestry and Aggressive Prostate Cancer among African Americans and European Americans in PCaP. *PLoS ONE* **2015**, *10*, e0125151. [CrossRef]
116. King, L.; Dear, K.; Harrison, S.L.; van der Mei, I.; Brodie, A.M.; Kimlin, M.G.; Lucas, R.M. Investigating the patterns and determinants of seasonal variation in vitamin D status in Australian adults: The Seasonal D Cohort Study. *BMC Public Health* **2016**, *16*, 892. [CrossRef] [PubMed]
117. Diffey, B.L. Modelling the seasonal variation of vitamin D due to sun exposure. *Br. J. Dermatol.* **2010**, *162*, 1342–1348. [CrossRef]
118. Manson, J.E.; Cook, N.R.; Lee, I.M.; Christen, W.; Bassuk, S.S.; Mora, S.; Gibson, H.; Albert, C.M.; Gordon, D.; Copeland, T.; et al. Marine n-3 Fatty Acids and Prevention of Cardiovascular Disease and Cancer. *N. Engl. J. Med.* **2019**, *380*, 23–32. [CrossRef] [PubMed]
119. Bassuk, S.S.; Chandler, P.D.; Buring, J.E.; Manson, J.E.; Group, V.R. The VITamin D and OmegA-3 TriaL (VITAL): Do Results Differ by Sex or Race/Ethnicity? *Am. J. Lifestyle Med.* **2021**, *15*, 372–391. [CrossRef]
120. Chandler, P.D.; Chen, W.Y.; Ajala, O.N.; Hazra, A.; Cook, N.; Bubes, V.; Lee, I.M.; Giovannucci, E.L.; Willett, W.; Buring, J.E.; et al. Effect of Vitamin D3 Supplements on Development of Advanced Cancer: A Secondary Analysis of the VITAL Randomized Clinical Trial. *JAMA Netw. Open* **2020**, *3*, e2025850. [CrossRef]
121. Marshall, D.T.; Savage, S.J.; Garrett-Mayer, E.; Keane, T.E.; Hollis, B.W.; Horst, R.L.; Ambrose, L.H.; Kindy, M.S.; Gattoni-Celli, S. Vitamin D3 supplementation at 4000 international units per day for one year results in a decrease of positive cores at repeat biopsy in subjects with low-risk prostate cancer under active surveillance. *J. Clin. Endocrinol. Metab.* **2012**, *97*, 2315–2324. [CrossRef] [PubMed]

122. Siddappa, M.; Hussain, S.; Wani, S.A.; Tang, H.; Gray, J.S.; Jafari, H.; Wu, H.; Long, M.D.; Elhussin, I.; Karanam, B.; et al. Vitamin D receptor cistrome-transcriptome analyses establishes quantitatively distinct receptor genomic interactions in African American prostate cancer regulated by BAZ1A. *bioRxiv* **2022**. [CrossRef]
123. Singh, A.K.; Prakash, S.; Garg, R.K.; Jain, P.; Kumar, R.; Jain, A. Polymorphisms in vitamin D receptor, toll-like receptor 2 and Toll-Like receptor 4 genes links with Dengue susceptibility. *Bioinformation* **2021**, *17*, 506–513. [CrossRef] [PubMed]
124. He, L.; Zhou, M.; Li, Y.C. Vitamin D/Vitamin D Receptor Signaling Is Required for Normal Development and Function of Group 3 Innate Lymphoid Cells in the Gut. *iScience* **2019**, *17*, 119–131. [CrossRef] [PubMed]
125. Konya, V.; Czarnewski, P.; Forkel, M.; Rao, A.; Kokkinou, E.; Villablanca, E.J.; Almer, S.; Lindforss, U.; Friberg, D.; Hoog, C.; et al. Vitamin D downregulates the IL-23 receptor pathway in human mucosal group 3 innate lymphoid cells. *J. Allergy Clin. Immunol.* **2018**, *141*, 279–292. [CrossRef] [PubMed]
126. Gombart, A.F.; Borregaard, N.; Koeffler, H.P. Human cathelicidin antimicrobial peptide (CAMP) gene is a direct target of the vitamin D receptor and is strongly up-regulated in myeloid cells by 1,25-dihydroxyvitamin D3. *FASEB J.* **2005**, *19*, 1067–1077. [CrossRef]
127. Liao, S.L.; Lai, S.H.; Tsai, M.H.; Hua, M.C.; Yeh, K.W.; Su, K.W.; Chiang, C.H.; Huang, S.Y.; Kao, C.C.; Yao, T.C.; et al. Maternal Vitamin D Level Is Associated with Viral Toll-Like Receptor Triggered IL-10 Response but Not the Risk of Infectious Diseases in Infancy. *Mediat. Inflamm.* **2016**, *2016*, 8175898. [CrossRef]
128. Ojaimi, S.; Skinner, N.A.; Strauss, B.J.; Sundararajan, V.; Woolley, I.; Visvanathan, K. Vitamin D deficiency impacts on expression of toll-like receptor-2 and cytokine profile: A pilot study. *J. Transl. Med.* **2013**, *11*, 176. [CrossRef]
129. Darko, S.N.; Yar, D.D.; Owusu-Dabo, E.; Awuah, A.A.; Dapaah, W.; Addofoh, N.; Salifu, S.P.; Awua-Boateng, N.Y.; Adomako-Boateng, F. Variations in levels of IL-6 and TNF-alpha in type 2 diabetes mellitus between rural and urban Ashanti Region of Ghana. *BMC Endocr. Disord.* **2015**, *15*, 50. [CrossRef]
130. Mensah, G.I.; Addo, K.K.; Tetteh, J.A.; Sowah, S.; Loescher, T.; Geldmacher, C.; Jackson-Sillah, D. Cytokine response to selected MTB antigens in Ghanaian TB patients, before and at 2 weeks of anti-TB therapy is characterized by high expression of IFN-gamma and Granzyme B and inter- individual variation. *BMC Infect. Dis.* **2014**, *14*, 495. [CrossRef]
131. Minas, T.Z.; Candia, J.; Dorsey, T.H.; Baker, F.; Tang, W.; Kiely, M.; Smith, C.J.; Zhang, A.L.; Jordan, S.V.; Obadi, O.M.; et al. Serum proteomics links suppression of tumor immunity to ancestry and lethal prostate cancer. *Nat. Commun.* **2022**, *13*, 1759. [CrossRef]
132. Chen, F.; Darst, B.F.; Madduri, R.K.; Rodriguez, A.A.; Sheng, X.; Rentsch, C.T.; Andrews, C.; Tang, W.; Kibel, A.S.; Plym, A.; et al. Validation of a multi-ancestry polygenic risk score and age-specific risks of prostate cancer: A meta-analysis within diverse populations. *Elife* **2022**, *11*, e78304. [CrossRef] [PubMed]
133. Conti, D.V.; Darst, B.F.; Moss, L.C.; Saunders, E.J.; Sheng, X.; Chou, A.; Schumacher, F.R.; Olama, A.A.A.; Benlloch, S.; Dadaev, T.; et al. Trans-ancestry genome-wide association meta-analysis of prostate cancer identifies new susceptibility loci and informs genetic risk prediction. *Nat. Genet.* **2021**, *53*, 65–75. [CrossRef] [PubMed]
134. Berchuck, J.E.; Adib, E.; Abou Alaiwi, S.; Dash, A.K.; Shin, J.N.; Lowder, D.; McColl, C.; Castro, P.; Carelli, R.; Benedetti, E.; et al. The prostate cancer androgen receptor cistrome in African American men associates with upregulation of lipid metabolism and immune response. *Cancer Res.* **2022**, *82*, 2848–2859. [CrossRef]
135. Ernst, J.; Kellis, M. Chromatin-state discovery and genome annotation with ChromHMM. *Nat. Protoc.* **2017**, *12*, 2478–2492. [CrossRef]
136. Ernst, J.; Kellis, M. ChromHMM: Automating chromatin-state discovery and characterization. *Nat. Methods* **2012**, *9*, 215–216. [CrossRef]
137. Long, M.D.; Jacobi, J.J.; Singh, P.K.; Llimos, G.; Wani, S.A.; Rowsam, A.M.; Rosario, S.R.; Hoogstraat, M.; Linder, S.; Kirk, J.; et al. Reduced NCOR2 expression accelerates androgen deprivation therapy failure in prostate cancer. *Cell Rep.* **2021**, *37*, 110109. [CrossRef]
138. Pomerantz, M.M.; Qiu, X.; Zhu, Y.; Takeda, D.Y.; Pan, W.; Baca, S.C.; Gusev, A.; Korthauer, K.D.; Severson, T.M.; Ha, G.; et al. Prostate cancer reactivates developmental epigenomic programs during metastatic progression. *Nat. Genet.* **2020**, *52*, 790–799. [CrossRef] [PubMed]
139. Long, M.D.; van den Berg, P.R.; Russell, J.L.; Singh, P.K.; Battaglia, S.; Campbell, M.J. Integrative genomic analysis in K562 chronic myelogenous leukemia cells reveals that proximal NCOR1 binding positively regulates genes that govern erythroid differentiation and Imatinib sensitivity. *Nucleic Acids Res.* **2015**, *43*, 7330–7348. [CrossRef] [PubMed]
140. Cao, Y.; Kitanovski, S.; Hoffmann, D. intePareto: An R package for integrative analyses of RNA-Seq and ChIP-Seq data. *BMC Genom.* **2020**, *21*, 802. [CrossRef]
141. Liu, J.; Lei, B.; Yu, X.; Li, Y.; Deng, Y.; Yang, G.; Li, Z.; Liu, T.; Ye, L. Combining Immune-Related Genes For Delineating the Extracellular Matrix and Predicting Hormone Therapy and Neoadjuvant Chemotherapy Benefits In Breast Cancer. *Front. Immunol.* **2022**, *13*, 888339. [CrossRef] [PubMed]
142. Frost, H.R.; Amos, C.I. Gene set selection via LASSO penalized regression (SLPR). *Nucleic Acids Res.* **2017**, *45*, e114. [CrossRef] [PubMed]

MDPI

Review

Vitamin D and Its Target Genes

Carsten Carlberg [1,2]

[1] Institute of Animal Reproduction and Food Research, Polish Academy of Sciences, PL-10-748 Olsztyn, Poland; c.carlberg@pan.olsztyn.pl

[2] Institute of Biomedicine, School of Medicine, University of Eastern Finland, FI-70211 Kuopio, Finland

Abstract: The vitamin D metabolite 1α,25-dihydroxyvitamin D_3 is the natural, high-affinity ligand of the transcription factor vitamin D receptor (VDR). In many tissues and cell types, VDR binds in a ligand-dependent fashion to thousands of genomic loci and modulates, via local chromatin changes, the expression of hundreds of primary target genes. Thus, the epigenome and transcriptome of VDR-expressing cells is directly affected by vitamin D. Vitamin D target genes encode for proteins with a large variety of physiological functions, ranging from the control of calcium homeostasis, innate and adaptive immunity, to cellular differentiation. This review will discuss VDR's binding to genomic DNA, as well as its genome-wide locations and interaction with partner proteins, in the context of chromatin. This information will be integrated into a model of vitamin D signaling, explaining the regulation of vitamin D target genes.

Keywords: vitamin D; VDR; target genes; chromatin; epigenome; transcriptome; vitamin D signaling

Citation: Carlberg, C. Vitamin D and Its Target Genes. *Nutrients* **2022**, *14*, 1354. https://doi.org/10.3390/nu14071354

Academic Editor: William B. Grant

Received: 6 March 2022
Accepted: 23 March 2022
Published: 24 March 2022

Publisher's Note: MDPI stays neutral with regard to jurisdictional claims in published maps and institutional affiliations.

1. Introduction

For 100 years, the term "vitamin D" has been used [1] for a molecule, the deficiency of which can lead to bone malformations, such as rickets [2]. More than 50 years ago, evidence accumulated that vitamin D_3 acts via its metabolites 25-hydroxvitamin D_3 (25(OH)D_3) and, in particular, via the nuclear hormone 1α,25-dihydroxyvitamin D_3 (1,25(OH)$_2$D$_3$) [3]. The emerging endocrinology of vitamin D was completed through the identification of vitamin D-binding proteins [4,5] and the cloning of VDR in different species [6,7]. VDR turned out to be an endocrine member of the nuclear receptor superfamily [8,9], suggesting that (similar to the steroid hormones estradiol, testosterone, progesterone, cortisol and mineralocorticoids, the vitamin A derivative all-*trans* retinoic acid and the thyroid hormone triiodothyronine) 1,25(OH)$_2$D$_3$ acts at nanomolar, or even picomolar, concentrations, as a direct regulator of specific target genes [10–12], in VDR-expressing tissues and cell types (www.proteinatlas.org/ENSG00000111424-VDR/tissue, accessed on 5 March 2022). The nearly ubiquitous expression of the VDR gene supports findings obtained during the past 30 years, that vitamin D regulates not only calcium homeostasis [13], but also immunity, cell growth and differentiation, as well as energy metabolism [14–16].

The genomic actions of vitamin D depend on the activation of VDR by 1,25(OH)$_2$D$_3$ and involve changes in the epigenome, leading to changes in the transcriptome and proteome. Therefore, in a typical in vitro vitamin D stimulation experiment, where supra-physiological concentrations of 10–100 nM 1,25(OH)$_2$D$_3$ are applied to a cell culture model, it takes (due to the time needed for RNA and protein synthesis) a few hours before the physiological effects of the nuclear hormone can be observed [17]. This is in contrast to the so-called non-genomic actions of vitamin D that happen within seconds to minutes and do not to involve VDR and changes in gene expression [18,19]. However, the timing of vitamin D signaling may not be of critical importance, since the physiology of vitamin D and its metabolites aims towards homeostasis, i.e., in vivo, there are no larger fluctuations in the concentrations of 1,25(OH)$_2$D$_3$ [20]. Thus, in net effect, the physiology of 1,25(OH)$_2$D$_3$ largely overlaps with the actions of the transcription factor VDR.

This review will outline VDR's DNA-binding modes, genome-wide locations, as well as its interaction with chromatin components and other partner proteins. This will provide the basis of a model of vitamin D signaling that explains the mode of action of vitamin D target genes and allows for their classification.

2. VDR: A Transcription Factor Activated by Vitamin D

Transcription factors are proteins that are able to bind sequence-specifically to genomic DNA and interact with other nuclear proteins, which modulates the activity of RNA polymerase II and mRNA production [21,22]. Some of the approximately 1600 transcription factors encoded by the human genome are constitutively active and regulated primarily by their expression, while most of them are activated by extra- and intracellular signals. A few of these signal-dependent transcription factors are located either in a latent form in the cytosol and activated through translocation into the nucleus, or—most of them—are found in the nucleus and modulated in their activity by post-translational modifications, such as phosphorylation or acetylation. Furthermore, some members of the nuclear receptor superfamily have an additional mechanism of activation, which is a ligand-induced conformational change on the surface of their ligand-binding domain (LBD) [23,24].

The inner surface of VDR's LBD forms a ligand-binding pocket, where 40, mostly non-polar, amino acids snugly enclose the molecule 1,25$(OH)_2D_3$, so that it binds with an affinity of 0.1 nM [25]. This is a very high affinity, even in comparison with other nuclear receptors [26]. Ligand binding changes VDR's interaction profile with many of the more than 50 nuclear proteins that have been reported to cooperate with the receptor [27]. Some of these VDR-interacting proteins function as co-repressors, such as NCOR1 (nuclear receptor corepressor 1) [28] or COPS2 (COP9 signalosome subunit 2, also called ALIEN) [29], co-activators of the NCOA (nuclear receptor coactivator) family [30], or members of the Mediator complex, such as MED1 [31–33] (Figure 1). Other VDR partner proteins are chromatin-modifying enzymes, such as histone acetyltransferases (HATs) [30], histone deacetylases (HDACs) [29], lysine demethylases, such as KDM6B [34] and KDM1A [35] or chromatin remodeling proteins, such as BRD (bromodomain-containing) 7 and 9 [36]. The large variety of its protein interaction partners suggests that VDR is a dynamic member of a large nuclear protein complex [37] (Section 5).

Many in vitro studies indicated that VDR binds efficiently to genomic DNA in a complex with the nuclear receptor retinoid X receptor (RXR) [38–40]. The preferred binding sites for the heterodimeric VDR-RXR complex are RGKTSA (R=A or G, K=G or T, S=C or G) sequence motifs, arranged as a direct repeat with three spacing nucleotides (DR3) [41–43] (Figure 1). However, these genomic binding sites need to be accessible to VDR complexes; i.e., they need to be located within open chromatin, which is referred to as euchromatin (Section 3). Chromatin immunoprecipitation sequencing (ChIP-seq) is a next-generation sequencing method that is able to determine, in an unbiased fashion, the genome-wide binding pattern, the so-called cistrome, of a transcription factor, such as VDR. ChIP-seq for VDR had been performed in many cellular models (Section 4), which confirmed that DR3-type binding sites are the most enriched sequence motifs below the summits ($\pm$100 bp) of VDR peaks [44]. However, depending on the threshold settings of DNA motif-finding algorithms, such as HOMER [45], only 10–20% of all VDR binding sites contain DR3-type motifs [44,46]. Thus, in a genome-wide perspective, not all VDR-containing nuclear complexes contact genomic DNA via DR3-type binding sites, by some distance.

The low percentage of DR3-binding VDR complexes detected by ChIP-seq also suggests that there are a number of scenarios in which VDR acts independently of RXR. VDR may use other nuclear proteins as alternative cooperative binding partners on genomic DNA [47–49] or may bind indirectly to DNA, such as "backpack", to other transcription factors [50]. For example, below VDR ChIP-seq peak binding sites for the pioneer transcription factor PU.1 (purine-rich box-1) are enriched [44]. In fact, in THP-1 monocytic leukemia cells, the presence of PU.1 is observed on two-thirds of VDR's genomic binding sites [51]. This makes sense, since PU.1, VDR and the pioneer factor CEBPα (CCAAT enhancer binding protein alpha) are the key transcription factors directing the differentiation of myeloid progenitor cells into monocytes and granulocytes, during the process of hematopoiesis [52]. Interestingly, THP-1 cells CEBPα [53], GABPα (GA-binding protein transcription factor alpha) [54] and ETS1 (ETS proto-oncogene 1, transcription factor) [55] also co-locate with VDR binding sites and act as pioneer factors for vitamin D signaling (Figure 1). Furthermore, in osteoblasts, vitamin D signaling is enhanced by the pioneer factors CEBPα and RUNX2 (RUNX family transcription factor 2) [56], while in T cells, this is mediated by the transcription factor BACH2 (BTB domain and CNC homolog 2) [57]. Thus, VDR uses help from RXR, but also from many other transcription factors, in order to form functional complexes with genomic DNA.

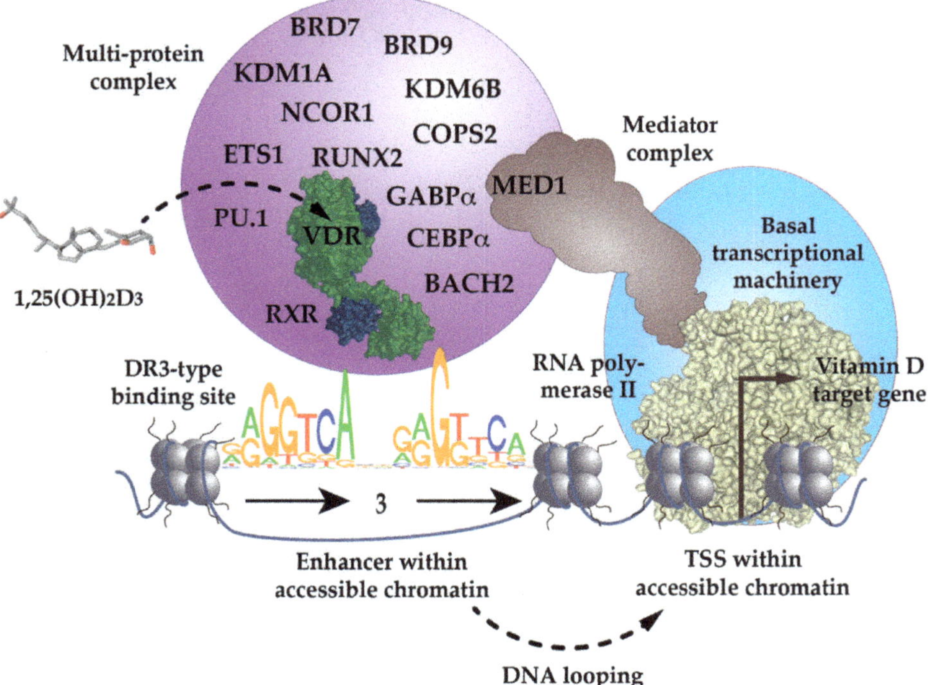

Figure 1. VDR as the key ligand-inducible component of a multi-protein complex. VDR is a part of a multi-protein complex that, e.g., contains co-receptors (RXR), pioneer factors (PU.1, CEBPα, GABPα, ETS1, RUNX2, BACH2), chromatin modifiers (KDM1A, KDM6B), chromatin remodelers (BRD7, BRD9), co-activators (MED1) and co-repressors (NCOR1, COPS2). The complex is activated through the binding of $1,25(OH)_2D_3$ to VDR and attaches preferentially to DR3-type binding sites within enhancer regions. The mediator complex connects the activated VDR complex with the RNA polymerase II waiting on transcription start site (TSS) regions of vitamin D target genes. In most cases the linear distance of enhancer and TSS region are multiple kb, so that the intervening genomic DNA forms a regulatory loop. In this way the expression of the vitamin D target genes is either increased or decreased.

3. Vitamin D Target Gene Regulation in the Context of Chromatin

Each of the approximately 20,000 protein-coding genes of the human genome has one or multiple transcription start sites (TSSs). The latter are core promoter regions, to which RNA polymerase II is directed by general transcription factors, such as TBP (TATA box binding protein) [58]. Depending on a direct interaction of this so-called basal transcriptional machinery, with signal-dependent transcription factors (Section 2) or an indirect interaction via members of the Mediator complex [59], the RNA polymerase modulates the transcription rate of the respective gene, i.e., its expression (mRNA level) increases or decreases. Transcription factors, such as VDR, bind to enhancer regions [60], which are stretches of genomic DNA that contain specific binding sites (Section 2), for one or multiple transcription factors. The interaction of the protein complexes formed on TSS and enhancer regions is facilitated by DNA looping, in a so-called regulatory loop (Figure 1). Therefore, enhancers are equally likely found upstream and downstream of TSS regions [61]. However, both types of genomic regions need to be located within the same TAD (topologically associating domain), in order to efficiently interact. TADs are far larger loops of genomic DNA than regulatory loops, with a size of hundreds of kb to a few Mb [62]. They subdivide the human genome into at least 2000 units, which are functionally insulated from each other [63] (Figure 2). The borders of TADs are defined by the binding of the chromatin-organizing protein CTCF (CCCTC-binding factor) [64,65], forming together with cohesin and other proteins, so-called TAD anchors [66]. The interaction of the VDR-bound enhancers with genomic regions outside of a TAD are prevented by these insulating TAD borders. This is the reason why genes are regulated almost exclusively by enhancers that are located within the same TAD. This also implies that the linear distance between VDR-bound enhancers and TSS region(s) cannot be larger than the size of the respective TAD.

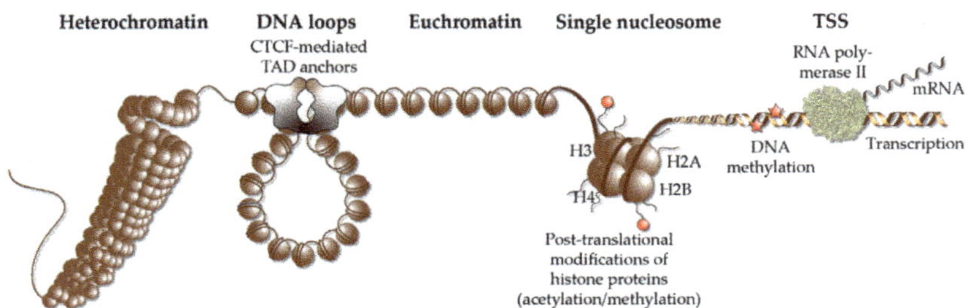

Figure 2. Elements of chromatin. Different elements of chromatin are shown, such as densely packed heterochromatin, DNA loops, such as TADs that are anchored by CTCF proteins, accessible euchromatin, the structure of a single nucleosome, chromatin modification via histone acetylation and methylation as well DNA methylation and a TSS, from which RNA polymerase II starts gene transcription into mRNA.

In a repeating unit of 200 bp, genomic DNA is packaged around nucleosomes, which are complexes of two copies of each of the histone proteins H2A, H2B, H3 and H4 (Figure 2). The complex of nucleosomes and genomic DNA is referred to as chromatin and can be interpreted as the physical expression of the epigenome [67]. Chromatin regions largely differ in their degree of packaging [68], so that transcription factors are limited in their access to genomic binding sites [69]. The accessibility of chromatin can be monitored genome-wide by the methods of DNase-seq (DNase I hypersensitivity sequencing) [70], FAIRE-seq (formaldehyde-assisted isolation of regulatory elements sequencing) [71] and ATAC-seq (assay for transposase-accessible chromatin using sequencing) [72]. The vast majority of the genome is covered in a cell- and tissue-specific fashion by densely packed heterochromatin, which is largely inaccessible, in order to prevent the unintentional activation of genes. In contrast, in an average differentiated cell, genomic DNA is accessible at less than 200,000 loci (representing only some 10% of the whole genome) that primarily comprise TSS and enhancer regions [61]. This has a key impact on vitamin D signaling, since VDR binds exclusively to accessible enhancer regions (Section 4) and activates only those genes, the TSS region of which are located within euchromatin.

The distribution of eu- and heterochromatin (including their specific epigenetic markers) of a given cell is referred to as its epigenetic landscape or epigenome [73]. The epigenome is determined by patterns of DNA methylation, post-translational modification of histone tails and 3-dimensional chromatin organization [74]. It depends on the activity of chromatin-modifying enzymes, such as DNA methyltransferases (DNMTs), which add methyl groups to cytosines within genomic DNA, TET (Tet methylcytosine dioxygenase) proteins that initiate DNA methylation, or HATs, HDACs, lysine methyltransferases (KTMs) and KDMs, which add or remove acetyl and methyl groups to histones [75]. Histone acetylation is generally associated with transcriptional activation, but it is not important which exact amino acid is acetylated. In contrast, for histone methylation, the exact residue and its degree of methylation (mono-, di- or tri-methylation) is critical. Furthermore, the function of chromatin-remodeling enzymes is to shift or evict nucleosomes, in an ATP-dependent fashion. The projects ENCODE (www.encodeproject.org, accessed on 5 March 2022) [61] and Roadmap Epigenomics (www.roadmapepigenomics.org, accessed on 5 March 2022) [76] systematically assessed the epigenomes of more than 100 human cell lines, as well as primary cells, and serve as a reference for the epigenome of non-stimulated human tissues and cell types. However, in contrast to the static genome, the epigenome dynamically responds to intra- and extracellular signals, since chromatin modifying enzymes are often the endpoints of transduction cascades of peptide hormones, cytokines and growth factors [77]. Thus, the response of the epigenome to signals, such as $1,25(OH)_2D_3$, is even more important than its ground state.

Nuclear hormones, such as $1,25(OH)_2D_3$, affect the epigenome via direct interaction of their receptors with chromatin-modifying enzymes (Section 2), as well as through up- or down-regulating the genes encoding for chromatin modifiers. In this way:

1. Vitamin D affects histone markers for active chromatin, such as H3K27ac (acetylated histone H3 at lysine 27), and for TSS regions, such as H3K4me3 (tri-methylated histone H3 at lysine 4) [53,57,78];
2. VDR initiates the demethylation of its binding sites via interaction with TET2 [79];
3. The accessibility of thousands of VDR-binding enhancer and TSS regions is affected by $1,25(OH)_2D_3$ [80,81];
4. The binding of CTCF, to more than 1000 of its genomic sites, is modulated by $1,25(OH)_2D_3$ [82];
5. The organization of some 400 TADs is dependent on $1,25(OH)_2D_3$ [82], i.e., vitamin D affects the 3-dimensional chromatin structure.

Thus, there are multiple ways by which $1,25(OH)_2D_3$ modulates the epigenome of its target tissues. Interestingly, some $1,25(OH)_2D_3$-modulated chromatin loci, such as TSS regions, open only 2 h after ligand stimulation, while most sites take 24 h to reach maximal accessibility [78,80]. This suggests that many effects of vitamin D on the epigenome are secondary, i.e., they are mediated by genes and proteins that are primary vitamin D targets [83,84]. Nevertheless, the vitamin D-triggered effects on the epigenome facilitate the looping of VDR-bound enhancers, towards accessible TSS regions within the same TAD [85]. This assembly of enhancer and TSS regions enables the formation of a large protein complex, containing VDR, nuclear adaptor proteins, chromatin-modifying enzymes and RNA polymerase II, modulating gene transcription (Section 2) (Figure 1).

4. Genome-Wide Location of VDR

In human cellular systems, the VDR cistrome had been determined in B cells (GM10855 and GM10861) [86], T cells [87], macrophages (lipopolysaccharide-polarized THP-1) cells [44], peripheral blood dendritic cells [79], colorectal cancer cells (LS180) [88], prostate epithelial cells (RWPE1) [89], hepatic stellate cells (LX2) [90] and human kidney tissue [91]. However, the VDR cistrome was studied in most detail in monocytes (undifferentiated THP-1 cells) [46,92]. For comparison, the mouse VDR cistrome was obtained in pre-adipocytes (3T3-L1) [93], osteocytic cells (IDG-SW3) [94], pre-osteoblastic and differentiated osteoblastic cells (MC3T3-E1) [95], as well as in mouse intestine [96], mouse kidney [91] and bone-marrow-derived mesenchymal stem cells [56]. In the presence of $1,25(OH)_2D_3$, the VDR cistrome comprises 5000–20,000 genomic loci, which represents a 2- to 10-fold increase compared to respective unstimulated cells. However, this implies that the VDR cistrome contains a lower number of persistent loci that remain constantly occupied, while their occupancy significantly changes after stimulation with a ligand [46]. These persistent VDR binding sites are the primary contact points of the human genome with 1,25(OH)$_2$D$_3$ and are considered as "hotspots" for vitamin D signaling (Section 6). These sites coordinate the functional consequences of ligand stimulation over time, i.e., these sites best represent the spatio-temporal response of the (epi)genome to the extracellular changes in vitamin D levels. In addition, there are transient VDR-binding loci that modulate the response of the epigenome to vitamin D and support persistent VDR sites. Thus, the genome-wide, ligand-induced binding of VDR, to its preferred loci, is the most prominent of all the epigenome-wide effects of vitamin D.

5. Model of Vitamin D Signaling

The here-presented model of vitamin D signaling describes the sequential activation of vitamin D target genes [97]. Typically, protein-coding target genes of vitamin D are considered, but there are also some non-coding RNA genes that are known to be modulated by vitamin D [98–100]. VDR does not act as an isolated protein but functions in the context of a larger, dynamically composed protein complex that contains RXR, other possible co-receptors, pioneer factors, such as PU.1, CEBPα, GABPα, ETS1, RUNX2 and BACH2, co-factors, chromatin modifiers and chromatin remodelers (Figure 1).

1. The pioneer factors within the complex may take the first contact to enhancer regions. With the help of chromatin-remodeling proteins, they optimize the access of VDR to suitable binding motifs within the enhancer region, including the demethylation of genomic DNA.
2. Chromatin modifiers within the complex then leave marks, such as H3K27ac, to the local chromatin region.
3. Although VDR may not be the first protein of the complex making contact with the enhancer region, its specific activation by 1,25(OH)$_2$D$_3$ drives the activity of the other members of the complex. This may explain the epigenetic effects of 1,25(OH)$_2$D$_3$, such as chromatin opening, histone marks and the recruitment of pioneer factors.
4. When the complex is established on the enhancer region, DNA looping events to TSS regions within the same TAD region, which are complexed with a basal transcriptional machinery, become stabilized. Via 1,25(OH)$_2$D$_3$-triggered effects on CTCF-dependent TAD anchor formation, this also affects the structure of the whole TAD.

In terms of net effect, the epigenome-modulating functions of vitamin D will increase or decrease the activity of RNA polymerase II, so that the mRNA expression of the respective gene(s) changes. On persistent VDR binding sites, i.e., enhancer with residual VDR binding, even in the absence of a ligand, the above-described multi-step process happens more efficiently, explaining why these loci are the primary sites of 1,25(OH)$_2$D$_3$-dependent gene expression.

6. Vitamin D Target Genes

Vitamin D target genes are detected via a statistically significant change in their expression, within a given time frame (often 24 h), after ligand stimulation. Long-time known vitamin D target genes, such as *BGLAP* (bone gamma-carboxyglutamate protein, also called osteocalcin) [101] or *CAMP* (cathelicidin antimicrobial peptide) [102], were deduced based on the observation of the physiological effects of vitamin D, e.g., on calcium homeostasis or the defense against microbe infection, respectively. On the level of mRNA changes, vitamin D target genes were analyzed, initially, by single gene approaches, using northern blotting or quantitative PCR. After the completion of the human genome, microarrays became popular, which are able to detect the transcriptome-wide effects of vitamin D stimulation in one assay [103,104]. However, for some 10 years, RNA sequencing (RNA-seq) has been the method of choice for describing the vitamin D-dependent transcriptome [105]. Studies in a large number of cellular models led to a tremendous increase in the number of putative vitamin D target genes. For example, a microarray in THP-1 cells reported, in one experimental setting, 3372 significantly (false discovery rate (FDR) < 0.05) regulated genes [92], while in another microarray, in the same cellular model, 4532 genes passed the statistical threshold [106]. Interestingly, 1227 genes were common and more than half of them (695) were confirmed by RNA-seq [46,80]. Thus, in a given cellular model, it is more likely that a few hundred genes respond to vitamin D than thousands of gene candidates. Interestingly, a meta-analysis of transcriptome-wide investigations of vitamin D target genes in 94 different human and mouse cell models resulted in only two common targets, *CYP24A1* (cytochrome P450 family 24 subfamily A member 1) and *CLMN* (calmin) [107]. Thus, vitamin D target genes are largely tissue specific.

Time course analysis of vitamin D target genes allows one to classify them into rapidly responding (4–8 h) "primary" target genes and delayed-reacting "secondary" targets (Figure 3A). Primary target genes are directly regulated by $1,25(OH)_2D_3$-activated VDR, as described in the model of vitamin D signaling (Section 5); i.e., these genes need to have, within the same TAD, a VDR-binding enhancer. In contrast, secondary target genes may be regulated by transcription factors, co-factors or chromatin modifiers, which are encoded by primary vitamin D target genes (Figure 3A). Suitable proteins encoded by primary target genes are the transcription factors BCL6 (B-cell CLL/lymphoma 6), NFE2 (nuclear factor, erythroid 2), POU4F2 (POU class 4 homeobox 2) and ELF4 (E74-like factor 4) in THP-1 cells [83], as well as IRF5 (interferon regulatory factor 5), MAFF (MAF BZIP transcription factor F), MYCL (MYCL proto-oncogene, BHLH transcription factor), NFXL1 (nuclear transcription factor, X-box binding-like 1) and TFEC (transcription factor EC), as well as the transcriptional co-regulators MAMLD1 (mastermind-like domain-containing 1), PPARGC1B (PPARG coactivator 1 beta), SRA1 (steroid receptor RNA activator 1) and ZBTB46 (zinc finger and BTB domain-containing 46) in human PBMCs (peripheral blood mononuclear cells) [108]. In this way, secondary target genes do not have to carry a VDR-binding enhancer within their TADs. For example, although the time course study in human PBMCs used a strict statistical approach (threshold testing applying a fold change (FC) > 1.5 and 2), 662 vitamin D-responding genes were identified (FDR < 0.05), 179 of which are primary and 483 secondary targets [108]. An alternative classification of the same set of genes suggests that 293 of them are direct and 369 indirect targets of vitamin D (Figure 3B). Irrespective of the timing of their response to $1,25(OH)_2D_3$, the expression change of direct targets is driven by VDR-bound enhancers, while indirect targets are primarily stabilized in their expression by $1,25(OH)_2D_3$ and its receptor, against an up- or down-regulation by other transcription factors and/or their epigenetic effects.

The model of vitamin D signaling (Section 5) illustrates the mechanisms of up-regulation of primary vitamin D target genes. However, the majority of vitamin D target genes are down-regulated, in particular, when cells are stimulated with $1,25(OH)_2D_3$ for 24 h or longer. The down-regulation of a gene by vitamin D is only possible when the gene is first up-regulated by other transcription factors or epigenetic effects, mediated by chromatin modifiers. The most, likely mechanism of down-regulation of a gene is to block one or several of its up-regulating factors. Accordingly, the majority of down-regulated target genes could be classified as indirect targets, i.e., vitamin D counteracts their up-regulation rather than prominently down-regulating their expression [108]. For example, VDR antagonizes pro-inflammatory transcription factors, such as NFAT, AP1 and NFκB, in immune cells [109]. However, since each gene is up-regulated by an individual set of transcription factors and chromatin modifiers, there are also individual mechanisms of its down-regulation. Thus, there is no general model for describing the mechanism of the down-regulation of vitamin D target genes.

The expression of the majority of vitamin D target genes is less than 5-fold, up- or down-regulated (after a stimulation for 24 h with $1,25(OH)_2D_3$); i.e., only a few genes respond with huge expression changes to vitamin D. For example, in THP-1 cells, the top five are the up-regulated genes *CYP24A1*, *CAMP*, *TSPAN18* (tetraspanin 18), *CD14* (CD14 molecule) and *FBP1* (fructose-bisphosphatase 1), with an FC of expression ranging from 47 to 402 [80]. In human PBMCs, *CYP24A1* and *CAMP* also have an FC of 409 and 158, respectively, the top up-regulated vitamin D target genes, but *STEAP4* (STEAP4 metalloreductase), *NRG1* (neuregulin 1) and *CXCL10* (C-X-C motif chemokine ligand 10) show an FC of 471, 450 and 158, respectively, meaning comparable levels of down-regulation of their expression [108]. This prominent down-regulation of expression is more remarkable than the up-regulation, since it is far easier to increase a very low basal expression than a down-regulation of a highly expressed gene. Nevertheless, one should remember that these in vitro $1,25(OH)_2D_3$ experiments are designed for maximal effects and do not reflect the reality of the endocrinology of vitamin D in vivo [110,111].

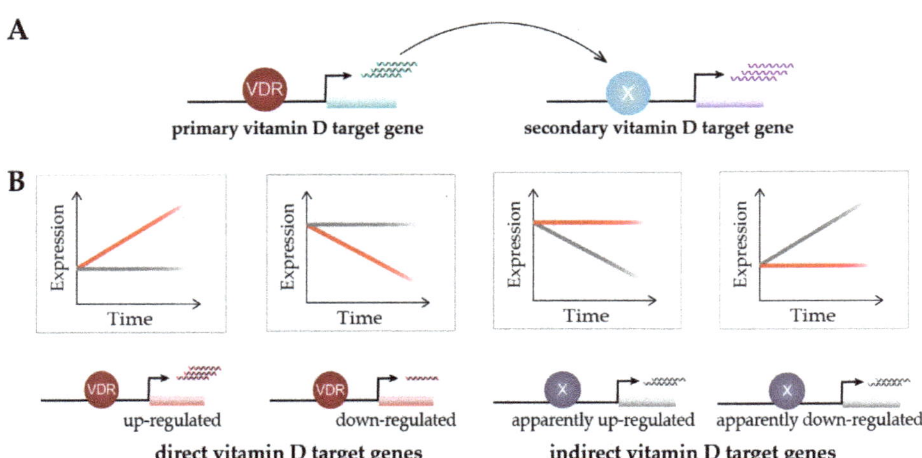

Figure 3. Classification of vitamin D target genes. Primary vitamin D target genes are directly regulated by VDR, while secondary vitamin D targets are controlled by transcriptional regulators that are encoded by primary targets (**A**). Time course analysis allows to differentiate vitamin D target genes in four different types based on cause and direction of expression change [108] (**B**). This suggests an alternative view on vitamin D signaling: $1,25(OH)_2D_3$ either directly induces or reduces the expression of its target genes via VDR or prevents their expression change mediated by other factors. Red and grey lines indicate gene's expression level in the presence or absence of $1,25(OH)_2D_3$, respectively.

7. Functional Profile of Vitamin D Target Genes

The most important thing to consider in the analysis of lists of hundreds of vitamin D target genes is the identification of the underlying biological processes. This is often evaluated by gene ontology analysis, where the list of target genes is assessed for statistically significant enrichment of a predefined list of terms, concerning (i) the molecular function, i.e., the molecular activity of a gene, (ii) the biological process, i.e., the cellular or physiological role carried out by a gene in the context of other genes, and (iii) the cellular component, i.e., the location where the gene's product functions in the cell. For example, in THP-1 cells, the biological processes "neutrophil activation", "inflammatory response", "neutrophil degranulation", "negative regulation of T cell proliferation" and "positive regulation of cytokine secretion", are most significantly associated with the list of 695 vitamin D target genes [84]. This fits with the known key functions of vitamin D in monocytes [112]. Since monocytes are the most vitamin D responsive cell fraction of PBMCs, gene ontology analysis of the 662 vitamin D target genes in this primary tissue suggests similar functions, such as neutrophil degranulation", "inflammatory response", "cytokine-mediated signaling pathway", "extracellular matrix organization" and "positive regulation of angiogenesis" [108]. For example, vitamin D down-regulates 10 of 12 *HLA* (human leukocyte antigen) class II genes and five *S100A* (S100 calcium-binding protein A) genes encoding for alarmins, as well as modulating the expression of six members of the *CXCL* gene family, encoding for chemokines [108]. The resulting vitamin D-triggered immune tolerance leads to the induction of regulatory T cells, which down-regulate the activity of other cells in the immune system [57,79]. This is the central mechanism for how vitamin D dampens chronic inflammation and autoimmunity in diseases, such as inflammatory bowel disease [113] and multiple sclerosis [114].

VDR is expressed in many cell types (Section 1), but most of them differ majorly in their respective epigenetic landscape and, thus, expression of vitamin D target genes. Therefore, there are a large variety of biological processes being regulated by 1,25(OH)$_2$D$_3$, i.e., in summary of all VDR-expressing tissues, vitamin D has rather ubiquitous functions.

8. Conclusions and Future View

Although vitamin D was discovered through its critical role in calcium homeostasis, being essential for proper bone formation, vitamin D signaling is studied most intensively, nowadays, in the immune system [115]. Therefore, the present molecular understanding of vitamin D signaling (Section 5) is mainly based on the integration of data obtained in immune cells [97].

The main challenge for future investigations of vitamin D target genes is their analysis in a human in vivo setting. The first studies involving a transcriptome- and epigenome-wide analysis, performed directly after isolation of PBMCs from vitamin D$_3$-supplemented human donors, have already begun [116–118]. These, and similar types of investigations, may provide us with a further enhanced understanding of the action of VDR and its target genes, in particular, in the context of immunity.

Funding: Financial support was provided by the WELCOME2—ERA Chair European Union's Horizon2020 research and innovation program under grant agreement no. 952601.

Acknowledgments: Many thanks to Andrea Hanel for critical comments for the manuscript.

Conflicts of Interest: The author declare no conflict of interest.

References

1. McMollum, E.V.; Simmonds, N.; Becker, J.E.; Shipley, P.G. Studies on experimental rickets: An experimental demonstration of the existence of a vitamin which promotes calcium deposition. *J. Biol. Chem.* **1922**, *52*, 293–298.
2. Holick, M.F. Resurrection of vitamin D deficiency and rickets. *J. Clin. Investig.* **2006**, *116*, 2062–2072. [CrossRef] [PubMed]
3. Renkema, K.Y.; Alexander, R.T.; Bindels, R.J.; Hoenderop, J.G. Calcium and phosphate homeostasis: Concerted interplay of new regulators. *Ann. Med.* **2008**, *40*, 82–91. [PubMed]
4. Tsai, H.C.; Norman, A.W. Studies on calciferol metabolism. 8. Evidence for a cytoplasmic receptor for 1,25-dihydroxyvitamin D$_3$ in the intestinal mucosa. *J. Biol. Chem.* **1973**, *248*, 5967–5975. [PubMed]
5. Brumbaugh, P.F.; Hughes, M.R.; Haussler, M.R. Cytoplasmic and nuclear binding components for 1α,25-dihydroxyvitamin D$_3$ in chick parathyroid glands. *Proc. Natl. Acad. Sci. USA* **1975**, *72*, 4871–4875.
6. McDonnell, D.P.; Mangelsdorf, D.J.; Pike, J.W.; Haussler, M.R.; O'Malley, B.W. Molecular cloning of complementary DNA encoding the avian receptor for vitamin D. *Science* **1987**, *235*, 1214–1217.
7. Baker, A.R.; McDonnell, D.P.; Hughes, M.; Crisp, T.M.; Mangelsdorf, D.J.; Haussler, M.R.; Pike, J.W.; Shine, J.; O'Malley, B. Cloning and expression of full-length cDNA encoding human vitamin D receptor. *Proc. Natl. Acad. Sci. USA* **1988**, *85*, 3294–3298.
8. Evans, R.; Mangelsdorf, D. Nuclear Receptors, RXR, and the Big Bang. *Cell* **2014**, *157*, 255–266. [CrossRef]
9. Evans, R.M. The nuclear receptor superfamily: A rosetta stone for physiology. *Mol. Endocrinol.* **2005**, *19*, 1429–1438. [CrossRef]
10. Carlberg, C.; Polly, P. Gene regulation by vitamin D$_3$. *Crit. Rev. Eukaryot. Gene Expr.* **1998**, *8*, 19–42.
11. Haussler, M.R.; Haussler, C.A.; Jurutka, P.W.; Thompson, P.D.; Hsieh, J.C.; Remus, L.S.; Selznick, S.H.; Whitfield, G.K. The vitamin D hormone and its nuclear receptor: Molecular actions and disease states. *J. Endocrinol.* **1997**, *154*, S57–S73. [PubMed]
12. Pike, J.W. Vitamin D$_3$ receptors: Structure and function in transcription. *Annu. Rev. Nutr.* **1991**, *11*, 189–216. [PubMed]
13. Veldurthy, V.; Wei, R.; Oz, L.; Dhawan, P.; Jeon, Y.H.; Christakos, S. Vitamin D, calcium homeostasis and aging. *Bone Res.* **2016**, *4*, 16041. [CrossRef]
14. Muller, V.; de Boer, R.J.; Bonhoeffer, S.; Szathmary, E. An evolutionary perspective on the systems of adaptive immunity. *Biol. Rev. Camb. Philos. Soc.* **2018**, *93*, 505–528. [CrossRef] [PubMed]
15. Vanherwegen, A.S.; Gysemans, C.; Mathieu, C. Vitamin D endocrinology on the cross-road between immunity and metabolism. *Mol. Cell. Endocrinol.* **2017**, *453*, 52–67. [CrossRef] [PubMed]
16. Cortes, M.; Chen, M.J.; Stachura, D.L.; Liu, S.Y.; Kwan, W.; Wright, F.; Vo, L.T.; Theodore, L.N.; Esain, V.; Frost, I.M.; et al. Developmental vitamin D availability impacts hematopoietic stem cell production. *Cell Rep.* **2016**, *17*, 458–468. [CrossRef]
17. Carlberg, C. Vitamin D genomics: From in vitro to in vivo. *Front. Endocrinol.* **2018**, *9*, 250. [CrossRef]
18. Khanal, R.C.; Nemere, I. The ERp57/GRp58/1,25D3-MARRS receptor: Multiple functional roles in diverse cell systems. *Curr. Med. Chem.* **2007**, *14*, 1087–1093. [CrossRef]

19. Zmijewski, M.A.; Carlberg, C. Vitamin D receptor(s): In the nucleus but also at membranes? *Exp. Dermatol.* **2020**, *29*, 876–884. [CrossRef]
20. van Etten, E.; Stoffels, K.; Gysemans, C.; Mathieu, C.; Overbergh, L. Regulation of vitamin D homeostasis: Implications for the immune system. *Nutr. Rev.* **2008**, *66*, S125–S134. [CrossRef]
21. Vaquerizas, J.M.; Kummerfeld, S.K.; Teichmann, S.A.; Luscombe, N.M. A census of human transcription factors: Function, expression and evolution. *Nat. Rev. Genet.* **2009**, *10*, 252–263. [PubMed]
22. Carlberg, C.; Molnár, F. Transcription factors. In *Mechanisms of Gene Regulation*, 2nd ed.; Springer: Berlin/Heidelberg, Germany, 2016; pp. 57–73.
23. Huang, P.; Chandra, V.; Rastinejad, F. Structural overview of the nuclear receptor superfamily: Insights into physiology and therapeutics. *Annu. Rev. Physiol.* **2010**, *72*, 247–272. [PubMed]
24. Carlberg, C.; Molnár, F. Switching genes on and off: The example of nuclear receptors. In *Mechanisms of Gene Regulation*, 2nd ed.; Springer: Berlin/Heidelberg, Germany, 2016; pp. 95–108.
25. Molnár, F.; Peräkylä, M.; Carlberg, C. Vitamin D receptor agonists specifically modulate the volume of the ligand-binding pocket. *J. Biol. Chem.* **2006**, *281*, 10516–10526. [CrossRef] [PubMed]
26. Weikum, E.R.; Liu, X.; Ortlund, E.A. The nuclear receptor superfamily: A structural perspective. *Protein Sci.* **2018**, *27*, 1876–1892. [CrossRef]
27. Molnár, F. Structural considerations of vitamin D signaling. *Front. Physiol.* **2014**, *5*, 191. [CrossRef]
28. Tagami, T.; Lutz, W.H.; Kumar, R.; Jameson, J.L. The interaction of the vitamin D receptor with nuclear receptor corepressors and coactivators. *Biochem. Biophys. Res. Commun.* **1998**, *253*, 358–363.
29. Polly, P.; Herdick, M.; Moehren, U.; Baniahmad, A.; Heinzel, T.; Carlberg, C. VDR-Alien: A novel, DNA-selective vitamin D_3 receptor-corepressor partnership. *FASEB J.* **2000**, *14*, 1455–1463.
30. Herdick, M.; Carlberg, C. Agonist-triggered modulation of the activated and silent state of the vitamin D_3 receptor by interaction with co-repressors and co-activators. *J. Mol. Biol.* **2000**, *304*, 793–801.
31. Rachez, C.; Lemon, B.D.; Suldan, Z.; Bromleigh, V.; Gamble, M.; Näär, A.M.; Erdjument-Bromage, H.; Tempst, P.; Freedman, L.P. Ligand-dependent transcription activation by nuclear receptors requires the DRIP complex. *Nature* **1999**, *398*, 824–828.
32. Belorusova, A.Y.; Bourguet, M.; Hessmann, S.; Chalhoub, S.; Kieffer, B.; Cianferani, S.; Rochel, N. Molecular determinants of MED1 interaction with the DNA bound VDR-RXR heterodimer. *Nucleic Acids Res.* **2020**, *48*, 11199–11213. [CrossRef]
33. Yuan, C.-X.; Ito, M.; Fondell, J.D.; Fu, Z.-Y.; Roeder, R.G. The TRAP220 component of a thyroid hormone receptor-associated protein (TRAP) coactivator complex interacts directly with nuclear receptors in a ligand-dependent fashion. *Proc. Natl. Acad. Sci. USA* **1998**, *95*, 7939–7944. [PubMed]
34. Pereira, F.; Barbachano, A.; Silva, J.; Bonilla, F.; Campbell, M.J.; Munoz, A.; Larriba, M.J. KDM6B/JMJD3 histone demethylase is induced by vitamin D and modulates its effects in colon cancer cells. *Hum. Mol. Genet.* **2011**, *20*, 4655–4665. [CrossRef] [PubMed]
35. Battaglia, S.; Karasik, E.; Gillard, B.; Williams, J.; Winchester, T.; Moser, M.T.; Smiraglia, D.J.; Foster, B.A. LSD1 dual function in mediating epigenetic corruption of the vitamin D signaling in prostate cancer. *Clin. Epigenet.* **2017**, *9*, 82. [CrossRef]
36. Wei, Z.; Yoshihara, E.; He, N.; Hah, N.; Fan, W.; Pinto, A.F.M.; Huddy, T.; Wang, Y.; Ross, B.; Estepa, G.; et al. Vitamin D switches BAF complexes to protect beta cells. *Cell* **2018**, *173*, 1135–1149.e15. [CrossRef]
37. Cui, X.; Pertile, R.; Eyles, D.W. The vitamin D receptor (VDR) binds to the nuclear matrix via its hinge domain: A potential mechanism for the reduction in VDR mediated transcription in mitotic cells. *Mol. Cell. Endocrinol.* **2018**, *472*, 18–25. [CrossRef]
38. Carlberg, C.; Bendik, I.; Wyss, A.; Meier, E.; Sturzenbecker, L.J.; Grippo, J.F.; Hunziker, W. Two nuclear signalling pathways for vitamin D. *Nature* **1993**, *361*, 657–660. [CrossRef]
39. Sone, T.; Ozono, K.; Pike, J.W. A 55-kilodalton accessory factor facilitates vitamin D receptor DNA binding. *Mol. Endocrinol.* **1991**, *5*, 1578–1586.
40. Liao, J.; Ozano, K.; Sone, T.; McDonnell, D.P.; Pike, J.W. Vitamin D receptor requires a nuclear protein and 1,25-dihydroxyvitamin D_3. *Proc. Natl. Acad. Sci. USA* **1990**, *87*, 9751–9755.
41. Shaffer, P.L.; Gewirth, D.T. Structural analysis of RXR-VDR interactions on DR3 DNA. *J. Steroid Biochem. Mol. Biol.* **2004**, *89–90*, 215–219.
42. Umesono, K.; Murakami, K.K.; Thompson, C.C.; Evans, R.M. Direct repeats as selective response elements for the thyroid hormone, retinoic acid, and vitamin D_3 receptors. *Cell* **1991**, *65*, 1255–1266.
43. Ozono, K.; Liao, J.; Kerner, S.A.; Scott, R.A.; Pike, J.W. The vitamin D-responsive element in the human osteocalcin gene. Association with a nuclear proto-oncogene enhancer. *J. Biol. Chem.* **1990**, *265*, 21881–21888. [PubMed]
44. Tuoresmäki, P.; Väisänen, S.; Neme, A.; Heikkinen, S.; Carlberg, C. Patterns of genome-wide VDR locations. *PLoS ONE* **2014**, *9*, e96105. [CrossRef]
45. Heinz, S.; Benner, C.; Spann, N.; Bertolino, E.; Lin, Y.C.; Laslo, P.; Cheng, J.X.; Murre, C.; Singh, H.; Glass, C.K. Simple combinations of lineage-determining transcription factors prime cis-regulatory elements required for macrophage and B cell identities. *Mol. Cell* **2010**, *38*, 576–589. [PubMed]
46. Neme, A.; Seuter, S.; Carlberg, C. Selective regulation of biological processes by vitamin D based on the spatio-temporal cistrome of its receptor. *Biochim. Biophys. Acta* **2017**, *1860*, 952–961. [CrossRef]
47. Schräder, M.; Bendik, I.; Becker-Andre, M.; Carlberg, C. Interaction between retinoic acid and vitamin D signaling pathways. *J. Biol. Chem.* **1993**, *268*, 17830–17836. [PubMed]

48. Schräder, M.; Müller, K.M.; Carlberg, C. Specificity and flexibility of vitamin D signaling.: Modulation of the activation of natural vitamin D response elements by thyroid hormone. *J. Biol. Chem.* **1994**, *269*, 5501–5504.
49. Schräder, M.; Müller, K.M.; Nayeri, S.; Kahlen, J.P.; Carlberg, C. VDR-T_3R receptor heterodimer polarity directs ligand sensitivity of transactivation. *Nature* **1994**, *370*, 382–386.
50. Carlberg, C.; Molnár, F. Vitamin D receptor signaling and its therapeutic implications: Genome-wide and structural view. *Can. J. Physiol. Pharmacol.* **2015**, *93*, 311–318. [CrossRef]
51. Seuter, S.; Neme, A.; Carlberg, C. Epigenomic PU.1-VDR crosstalk modulates vitamin D signaling. *Biochim. Biophys. Acta* **2017**, *1860*, 405–415. [CrossRef]
52. Novershtern, N.; Subramanian, A.; Lawton, L.N.; Mak, R.H.; Haining, W.N.; McConkey, M.E.; Habib, N.; Yosef, N.; Chang, C.Y.; Shay, T.; et al. Densely interconnected transcriptional circuits control cell states in human hematopoiesis. *Cell* **2011**, *144*, 296–309. [CrossRef]
53. Nurminen, V.; Neme, A.; Seuter, S.; Carlberg, C. Modulation of vitamin D signaling by the pioneer factor CEBPA. *Biochim. Biophys. Acta* **2019**, *1862*, 96–106. [CrossRef]
54. Seuter, S.; Neme, A.; Carlberg, C. ETS transcription factor family member GABPA contributes to vitamin D receptor target gene regulation. *J. Steroid Biochem. Mol. Biol.* **2018**, *177*, 46–52. [CrossRef] [PubMed]
55. Warwick, T.; Schulz, M.H.; Gunther, S.; Gilsbach, R.; Neme, A.; Carlberg, C.; Brandes, R.P.; Seuter, S. A hierarchical regulatory network analysis of the vitamin D induced transcriptome reveals novel regulators and complete VDR dependency in monocytes. *Sci. Rep.* **2021**, *11*, 6518. [CrossRef] [PubMed]
56. Meyer, M.B.; Benkusky, N.A.; Sen, B.; Rubin, J.; Pike, J.W. Epigenetic plasticity drives adipogenic and osteogenic differentiation of marrow-derived mesenchymal stem cells. *J. Biol. Chem.* **2016**, *291*, 17829–17847. [CrossRef]
57. Chauss, D.; Freiwald, T.; McGregor, R.; Yan, B.; Wang, L.; Nova-Lamperti, E.; Kumar, D.; Zhang, Z.; Teague, H.; West, E.E.; et al. Autocrine vitamin D signaling switches off pro-inflammatory programs of TH1 cells. *Nat. Immunol.* **2022**, *23*, 62–74. [CrossRef]
58. Forrest, A.R.R.; Kawaji, H.; Rehli, M.; Kenneth Baillie, J.; de Hoon, M.J.L.; Haberle, V.; Lassmann, T.; Kulakovskiy, I.V.; Lizio, M.; Itoh, M.; et al. A promoter-level mammalian expression atlas. *Nature* **2014**, *507*, 462–470. [CrossRef]
59. Soutourina, J. Transcription regulation by the Mediator complex. *Nat. Rev. Mol. Cell Biol.* **2018**, *19*, 262–274. [CrossRef]
60. Andersson, R.; Gebhard, C.; Miguel-Escalada, I.; Hoof, I.; Bornholdt, J.; Boyd, M.; Chen, Y.; Zhao, X.; Schmidl, C.; Suzuki, T.; et al. An atlas of active enhancers across human cell types and tissues. *Nature* **2014**, *507*, 455–461. [CrossRef]
61. ENCODE-Project-Consortium; Bernstein, B.E.; Birney, E.; Dunham, I.; Green, E.D.; Gunter, C.; Snyder, M. An integrated encyclopedia of DNA elements in the human genome. *Nature* **2012**, *489*, 57–74. [CrossRef]
62. Ali, T.; Renkawitz, R.; Bartkuhn, M. Insulators and domains of gene expression. *Curr. Opin. Genet. Dev.* **2016**, *37*, 17–26. [CrossRef]
63. Dixon, J.R.; Selvaraj, S.; Yue, F.; Kim, A.; Li, Y.; Shen, Y.; Hu, M.; Liu, J.S.; Ren, B. Topological domains in mammalian genomes identified by analysis of chromatin interactions. *Nature* **2012**, *485*, 376–380. [CrossRef] [PubMed]
64. Phillips, J.E.; Corces, V.G. CTCF: Master weaver of the genome. *Cell* **2009**, *137*, 1194–1211. [CrossRef] [PubMed]
65. Ghirlando, R.; Felsenfeld, G. CTCF: Making the right connections. *Genes Dev.* **2016**, *30*, 881–891. [CrossRef] [PubMed]
66. Felsenfeld, G.; Burgess-Beusse, B.; Farrell, C.; Gaszner, M.; Ghirlando, R.; Huang, S.; Jin, C.; Litt, M.; Magdinier, F.; Mutskov, V.; et al. Chromatin boundaries and chromatin domains. *Cold Spring Harb. Symp. Quant. Biol.* **2004**, *69*, 245–250. [CrossRef] [PubMed]
67. Carlberg, C.; Molnár, F. The impact of chromatin. In *Mechanisms of Gene Regulation*, 2nd ed.; Springer: Berlin/Heidelberg, Germany, 2016.
68. Smith, Z.D.; Meissner, A. DNA methylation: Roles in mammalian development. *Nat. Rev. Genet.* **2013**, *14*, 204–220. [CrossRef]
69. Bell, O.; Tiwari, V.K.; Thoma, N.H.; Schubeler, D. Determinants and dynamics of genome accessibility. *Nat. Rev. Genet.* **2011**, *12*, 554–564. [CrossRef]
70. Song, L.; Crawford, G.E. DNase-seq: A high-resolution technique for mapping active gene regulatory elements across the genome from mammalian cells. *Cold Spring Harb. Protoc.* **2010**, *2010*, pdb.prot5384. [CrossRef]
71. Giresi, P.G.; Kim, J.; McDaniell, R.M.; Iyer, V.R.; Lieb, J.D. FAIRE (Formaldehyde-Assisted Isolation of Regulatory Elements) isolates active regulatory elements from human chromatin. *Genome Res.* **2007**, *17*, 877–885.
72. Buenrostro, J.D.; Giresi, P.G.; Zaba, L.C.; Chang, H.Y.; Greenleaf, W.J. Transposition of native chromatin for fast and sensitive epigenomic profiling of open chromatin, DNA-binding proteins and nucleosome position. *Nat. Methods* **2013**, *10*, 1213–1218. [CrossRef]
73. Bird, A. Perceptions of epigenetics. *Nature* **2007**, *447*, 396–398. [CrossRef]
74. Carlberg, C.; Molnár, F. The epigenome. In *Mechanisms of Gene Regulation*, 2nd ed.; Springer: Berlin/Heidelberg, Germany, 2016.
75. Carlberg, C.; Molnár, F. Chromatin modifiers. In *Mechanisms of Gene Regulation*, 2nd ed.; Springer: Berlin/Heidelberg, Germany, 2016.
76. Roadmap Epigenomics, C.; Kundaje, A.; Meuleman, W.; Ernst, J.; Bilenky, M.; Yen, A.; Heravi-Moussavi, A.; Kheradpour, P.; Zhang, Z.; Wang, J.; et al. Integrative analysis of 111 reference human epigenomes. *Nature* **2015**, *518*, 317–330. [CrossRef]
77. Carlberg, C.; Molnár, F. *Human Epigenetics: How Science Works*; Springer: New York, NY, USA, 2019.
78. Nurminen, V.; Neme, A.; Seuter, S.; Carlberg, C. The impact of the vitamin D-modulated epigenome on VDR target gene regulation. *Biochim. Biophys. Acta* **2018**, *1861*, 697–705. [CrossRef]

79. Catala-Moll, F.; Ferrete-Bonastre, A.G.; Godoy-Tena, G.; Morante-Palacios, O.; Ciudad, L.; Barbera, L.; Fondelli, F.; Martinez-Caceres, E.M.; Rodriguez-Ubreva, J.; Li, T.; et al. Vitamin D receptor, STAT3, and TET2 cooperate to establish tolerogenesis. *Cell Rep.* **2022**, *38*, 110244. [CrossRef] [PubMed]
80. Seuter, S.; Neme, A.; Carlberg, C. Epigenome-wide effects of vitamin D and their impact on the transcriptome of human monocytes involve CTCF. *Nucleic Acids Res.* **2016**, *44*, 4090–4104. [CrossRef]
81. Seuter, S.; Pehkonen, P.; Heikkinen, S.; Carlberg, C. Dynamics of 1α,25-dihydroxyvitamin D-dependent chromatin accessibility of early vitamin D receptor target genes. *Biochim. Biophys. Acta* **2013**, *1829*, 1266–1275. [CrossRef]
82. Neme, A.; Seuter, S.; Carlberg, C. Vitamin D-dependent chromatin association of CTCF in human monocytes. *Biochim. Biophys. Acta* **2016**, *1859*, 1380–1388. [CrossRef]
83. Nurminen, V.; Neme, A.; Ryynanen, J.; Heikkinen, S.; Seuter, S.; Carlberg, C. The transcriptional regulator BCL6 participates in the secondary gene regulatory response to vitamin D. *Biochim. Biophys. Acta* **2015**, *1849*, 300–308. [CrossRef]
84. Nurminen, V.; Seuter, S.; Carlberg, C. Primary vitamin D target genes of human monocytes. *Front. Physiol.* **2019**, *10*, 194. [CrossRef]
85. Carlberg, C.; Campbell, M.J. Vitamin D receptor signaling mechanisms: Integrated actions of a well-defined transcription factor. *Steroids* **2013**, *78*, 127–136. [CrossRef]
86. Ramagopalan, S.V.; Heger, A.; Berlanga, A.J.; Maugeri, N.J.; Lincoln, M.R.; Burrell, A.; Handunnetthi, L.; Handel, A.E.; Disanto, G.; Orton, S.M.; et al. A ChIP-seq defined genome-wide map of vitamin D receptor binding: Associations with disease and evolution. *Genome Res.* **2010**, *20*, 1352–1360.
87. Handel, A.E.; Sandve, G.K.; Disanto, G.; Berlanga-Taylor, A.J.; Gallone, G.; Hanwell, H.; Drablos, F.; Giovannoni, G.; Ebers, G.C.; Ramagopalan, S.V. Vitamin D receptor ChIP-seq in primary $CD4^+$ cells: Relationship to serum 25-hydroxyvitamin D levels and autoimmune disease. *BMC Med.* **2013**, *11*, 163. [CrossRef]
88. Meyer, M.B.; Goetsch, P.D.; Pike, J.W. VDR/RXR and TCF4/beta-catenin cistromes in colonic cells of colorectal tumor origin: Impact on c-FOS and c-MYC gene expression. *Mol. Endocrinol.* **2012**, *26*, 37–51. [CrossRef] [PubMed]
89. Fleet, J.C.; Kovalenko, P.L.; Li, Y.; Smolinski, J.; Spees, C.; Yu, J.G.; Thomas-Ahner, J.M.; Cui, M.; Neme, A.; Carlberg, C.; et al. Vitamin D signaling suppresses early prostate carcinogenesis in TgAPT121 mice. *Cancer Prev. Res.* **2019**, *12*, 343–356. [CrossRef]
90. Ding, N.; Yu, R.T.; Subramaniam, N.; Sherman, M.H.; Wilson, C.; Rao, R.; Leblanc, M.; Coulter, S.; He, M.; Scott, C.; et al. A vitamin D receptor/SMAD genomic circuit gates hepatic fibrotic response. *Cell* **2013**, *153*, 601–613. [CrossRef]
91. Meyer, M.B.; Lee, S.M.; Carlson, A.H.; Benkusky, N.A.; Kaufmann, M.; Jones, G.; Pike, J.W. A chromatin-based mechanism controls differential regulation of the cytochrome P450 gene Cyp24a1 in renal and non-renal tissues. *J. Biol. Chem.* **2019**, *294*, 14467–14481. [CrossRef]
92. Heikkinen, S.; Väisänen, S.; Pehkonen, P.; Seuter, S.; Benes, V.; Carlberg, C. Nuclear hormone 1α,25-dihydroxyvitamin D_3 elicits a genome-wide shift in the locations of VDR chromatin occupancy. *Nucleic Acids Res.* **2011**, *39*, 9181–9193. [CrossRef]
93. Siersbaek, R.; Rabiee, A.; Nielsen, R.; Sidoli, S.; Traynor, S.; Loft, A.; La Cour Poulsen, L.; Rogowska-Wrzesinska, A.; Jensen, O.N.; Mandrup, S. Transcription factor cooperativity in early adipogenic hotspots and super-enhancers. *Cell Rep.* **2014**, *7*, 1443–1455. [CrossRef]
94. St John, H.C.; Bishop, K.A.; Meyer, M.B.; Benkusky, N.A.; Leng, N.; Kendziorski, C.; Bonewald, L.F.; Pike, J.W. The osteoblast to osteocyte transition: Epigenetic changes and response to the vitamin D_3 hormone. *Mol. Endocrinol.* **2014**, *28*, 1150–1165. [CrossRef]
95. Meyer, M.B.; Benkusky, N.A.; Lee, C.H.; Pike, J.W. Genomic determinants of gene regulation by 1,25-dihydroxyvitamin D_3 during osteoblast-lineage cell differentiation. *J. Biol. Chem.* **2014**, *289*, 19539–19554. [CrossRef]
96. Lee, S.M.; Riley, E.M.; Meyer, M.B.; Benkusky, N.A.; Plum, L.A.; DeLuca, H.F.; Pike, J.W. 1,25-Dihydroxyvitamin D_3 controls a cohort of vitamin D receptor target genes in the proximal intestine that Is enriched for calcium-regulating components. *J. Biol. Chem.* **2015**, *290*, 18199–18215. [CrossRef]
97. Carlberg, C. Molecular endocrinology of vitamin D on the epigenome level. *Mol. Cell. Endocrinol.* **2017**, *453*, 14–21. [CrossRef] [PubMed]
98. Zeljic, K.; Supic, G.; Magic, Z. New insights into vitamin D anticancer properties: Focus on miRNA modulation. *Mol. Genet. Genom.* **2017**, *292*, 511–524. [CrossRef] [PubMed]
99. Wang, L.; Zhou, S.; Guo, B. Vitamin D Suppresses Ovarian Cancer Growth and Invasion by Targeting Long Non-Coding RNA CCAT2. *Int. J. Mol. Sci.* **2020**, *21*, 2334. [CrossRef]
100. Zuo, S.; Wu, L.; Wang, Y.; Yuan, X. Long Non-coding RNA MEG3 Activated by Vitamin D Suppresses Glycolysis in Colorectal Cancer via Promoting c-Myc Degradation. *Front. Oncol.* **2020**, *10*, 274. [CrossRef]
101. Lian, J.B.; Glimcher, M.J.; Roufosse, A.H.; Hauschka, P.V.; Gallop, P.M.; Cohen-Solal, L.; Reit, B. Alterations of the gamma-carboxyglutamic acid and osteocalcin concentrations in vitamin D-deficient chick bone. *J. Biol. Chem.* **1982**, *257*, 4999–5003.
102. Gombart, A.F.; Borregaard, N.; Koeffler, H.P. Human cathelicidin antimicrobial peptide (CAMP) gene is a direct target of the vitamin D receptor and is strongly up-regulated in myeloid cells by 1,25-dihydroxyvitamin D_3. *FASEB J.* **2005**, *19*, 1067–1077.
103. Lin, R.; Nagai, Y.; Sladek, R.; Bastien, Y.; Ho, J.; Petrecca, K.; Sotiropoulou, G.; Diamandis, E.P.; Hudson, T.J.; White, J.H. Expression profiling in squamous carcinoma cells reveals pleiotropic effects of vitamin D_3 analog EB1089 signaling on cell proliferation, differentiation, and immune system regulation. *Mol. Endocrinol.* **2002**, *16*, 1243–1256. [CrossRef]
104. Campbell, M.J. Vitamin D and the RNA transcriptome: More than mRNA regulation. *Front. Physiol.* **2014**, *5*, 181. [CrossRef]

105. Craig, T.A.; Zhang, Y.; McNulty, M.S.; Middha, S.; Ketha, H.; Singh, R.J.; Magis, A.T.; Funk, C.; Price, N.D.; Ekker, S.C.; et al. Research resource: Whole transcriptome RNA sequencing detects multiple 1α,25-dihydroxyvitamin D_3-sensitive metabolic pathways in developing zebrafish. *Mol. Endocrinol.* **2012**, *26*, 1630–1642. [CrossRef]
106. Verway, M.; Bouttier, M.; Wang, T.T.; Carrier, M.; Calderon, M.; An, B.S.; Devemy, E.; McIntosh, F.; Divangahi, M.; Behr, M.A.; et al. Vitamin D induces interleukin-1beta expression: Paracrine macrophage epithelial signaling controls M. tuberculosis infection. *PLoS Pathog.* **2013**, *9*, e1003407. [CrossRef]
107. Dimitrov, V.; Barbier, C.; Ismailova, A.; Wang, Y.; Dmowski, K.; Salehi-Tabar, R.; Memari, B.; Groulx-Boivin, E.; White, J.H. Vitamin D-regulated gene expression profiles: Species-specificity and cell-specific effects on metabolism and immunity. *Endocrinology* **2021**, *162*, bqaa218. [CrossRef] [PubMed]
108. Hanel, A.; Carlberg, C. Time-resolved gene expression analysis monitors the regulation of inflammatory mediators and attenuation of adaptive immune response by vitamin D. *Int. J. Mol. Sci.* **2022**, *23*, 911. [CrossRef] [PubMed]
109. Zeitelhofer, M.; Adzemovic, M.Z.; Gomez-Cabrero, D.; Bergman, P.; Hochmeister, S.; N'Diaye, M.; Paulson, A.; Ruhrmann, S.; Almgren, M.; Tegner, J.N.; et al. Functional genomics analysis of vitamin D effects on $CD4^+$ T cells in vivo in experimental autoimmune encephalomyelitis. *Proc. Natl. Acad. Sci. USA* **2017**, *114*, E1678–E1687. [CrossRef] [PubMed]
110. DeLuca, H.F. Overview of general physiologic features and functions of vitamin D. *Am. J. Clin. Nutr.* **2004**, *80*, 1689S–1696S.
111. Norman, A.W. From vitamin D to hormone D: Fundamentals of the vitamin D endocrine system essential for good health. *Am. J. Clin. Nutr.* **2008**, *88*, 491S–499S.
112. Prietl, B.; Treiber, G.; Pieber, T.R.; Amrein, K. Vitamin D and immune function. *Nutrients* **2013**, *5*, 2502–2521. [CrossRef]
113. Limketkai, B.N.; Mullin, G.E.; Limsui, D.; Parian, A.M. Role of vitamin D in inflammatory bowel disease. *Nutr. Clin. Pract.* **2017**, *32*, 337–345. [CrossRef]
114. Munger, K.L.; Levin, L.I.; Hollis, B.W.; Howard, N.S.; Ascherio, A. Serum 25-hydroxyvitamin D levels and risk of multiple sclerosis. *JAMA* **2006**, *296*, 2832–2838. [CrossRef]
115. Carlberg, C. Genome-wide (over)view on the actions of vitamin D. *Front. Physiol.* **2014**, *5*, 167. [CrossRef]
116. Carlberg, C.; Seuter, S.; Nurmi, T.; Tuomainen, T.P.; Virtanen, J.K.; Neme, A. In vivo response of the human epigenome to vitamin D: A proof-of-principle study. *J. Steroid Biochem. Mol. Biol.* **2018**, *180*, 142–148. [CrossRef]
117. Neme, A.; Seuter, S.; Malinen, M.; Nurmi, T.; Tuomainen, T.P.; Virtanen, J.K.; Carlberg, C. In vivo transcriptome changes of human white blood cells in response to vitamin D. *J. Steroid Biochem. Mol. Biol.* **2019**, *188*, 71–76. [CrossRef] [PubMed]
118. Shirvani, A.; Kalajian, T.A.; Song, A.; Holick, M.F. Disassociation of vtamin D's calcemic activity and non-calcemic genomic activity and idividual responsiveness: A randomized controlled double-blind clinical tial. *Sci. Rep.* **2019**, *9*, 17685. [CrossRef] [PubMed]

MDPI

Review

Vitamin D and Its Receptor from a Structural Perspective

Natacha Rochel [1,2,3,4]

1 Integrated Structural Biology Department, Institut de Génétique et de Biologie Moléculaire et Cellulaire, 67404 Illkirch, France; rochel@igbmc.fr
2 Centre National de la Recherche Scientifique, UMR7104, 67404 Illkirch, France
3 Institut National de la Santé et de la Recherche Médicale (INSERM), U1258, 67404 Illkirch, France
4 Université de Strasbourg, 67404 Illkirch, France

Abstract: The activities of 1α,25-dihydroxyvitamin D3, $1{,}25D_3$, are mediated via its binding to the vitamin D receptor (VDR), a ligand-dependent transcription factor that belongs to the nuclear receptor superfamily. Numerous studies have demonstrated the important role of $1{,}25D_3$ and VDR signaling in various biological processes and associated pathologies. A wealth of information about ligand recognition and mechanism of action by structural analysis of the VDR complexes is also available. The methods used in these structural studies were mainly X-ray crystallography complemented by NMR, cryo-electron microscopy and structural mass spectrometry. This review aims to provide an overview of the current knowledge of VDR structures and also to explore the recent progress in understanding the complex mechanism of action of $1{,}25D_3$ from a structural perspective.

Keywords: vitamin D; vitamin D receptor; 3D structure; structural analysis; molecular recognition; protein–ligand interactions; coregulators

Citation: Rochel, N. Vitamin D and Its Receptor from a Structural Perspective. *Nutrients* **2022**, *14*, 2847. https://doi.org/10.3390/nu14142847

Academic Editor: Carsten Carlberg

Received: 15 June 2022
Accepted: 6 July 2022
Published: 12 July 2022

1. Introduction

Upon sun exposure, the secosteroid prohormone vitamin D is two-step hydroxylated in the liver and the kidney into the biologically hormonal form, 1α,25-dihydroxyvitamin D3 ($1{,}25D_3$) [1]. Most effects of $1{,}25D_3$ are mediated via its binding to the vitamin D receptor (VDR, also termed NR1I1), a ligand-dependent transcription factor (TF) that belongs to the nuclear receptor (NR) superfamily. VDR was cloned and sequenced in mammalian and avian species in the late 1980′s [2–5]. VDR acts as a heterodimer with one of the three retinoid X receptor (RXR) isotypes (RXRα, NR2B1; RXRβ, NR2B2; and RXRγ, NR2B3). VDR and its ligand control calcium metabolism, cell growth, differentiation, anti-proliferation, apoptosis and adaptive/innate immune responses [6–8]. VDR is expressed in the different tissues of the body and many of these tissues were not originally considered target tissues for $1{,}25D_3$. Conversion of $25D_3$ to $1{,}25D_3$ also occurs in many non-renal tissues and cells, including the skin, parathyroid glands, bone cells, both cardiovascular and immune cells, and many others [9,10]. The discovery of these autocrine/paracrine activities of $1{,}25D_3$ has markedly increased our appreciation of the wide effects of $1{,}25D_3$. An additional layer of complexity came with the identification of alternative vitamin D pathways and the major enzymes involved that produce other natural vitamin D metabolites, with some of them showing potent VDR activation [11,12]. Deregulation of VDR function may lead to severe diseases, such as cancers, psoriasis, rickets, renal osteodystrophy, and autoimmunity disorders (multiple sclerosis, rheumatoid arthritis, inflammatory bowel diseases, type I diabetes) [13]. Despite the large number of potential applications, clinical use of the native hormone $1{,}25D_3$ is limited by calcification of soft tissues (hypercalcemia). However, synthetic analogs have been developed and some of the highly active and non-calcemic VDR ligands have found clinical applications in the standard topical treatment of psoriasis, secondary hyperparathyroidism or osteoporosis [14,15].

In this review, we discuss the general mechanism of action of VDR and what we have learned from structural studies of VDR complexes from isolated domains to full length proteins focusing on 1,25D$_3$ action. Recent studies that aimed to describe how coregulators interact with VDR will also be discussed. These studies have significantly advanced our understanding on the molecular mechanism by which 1,25D$_3$ mediates transcription regulation. However, important questions remain regarding the basic mechanism of the cell-specific action of ligands and possible cross-talks with other NRs and TFs. Finally, the ongoing effort to characterize the structures, dynamics and relationships with the function of large VDR coregulatory complexes will be discussed.

2. Structure and Mechanism of Activation of the Vitamin D Receptor

VDR shares two main features with other members of the NR superfamily, principal mechanism of action and structural organization. Simplified regulation of target gene expression by VDR can be presented as follows (Figure 1): VDR forms a heterodimer with RXR and binds to specific DNA sequences of controlled genes called VDR response element (VDRE). These VDREs are often localized thousands of base pairs from the coding region of the regulated gene [16,17]. Typically, in the absence of ligands or in the presence of antagonists, corepressors with histone deacetylase activity are recruited to VDR bound to their target genes, while binding of agonist ligands induces a change in the structure of the NR that allows interaction with coactivators [8,16]. Recruitment of coactivators with enzymatic activities, such as histone acetyl-transferases (HAT), prepares target gene promoters through decondensation of the chromatin. HATs can further be replaced by the mediator complex that provides a link with the basal transcriptional machinery.

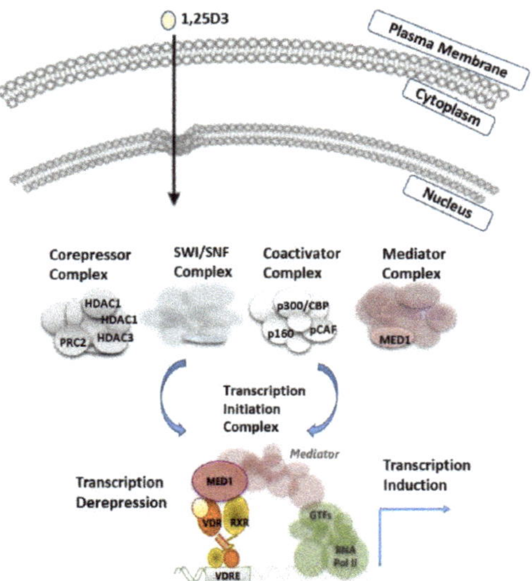

Figure 1. Schematic model for VDR regulation. Upon 1,25D$_3$ binding to VDR, VDR translocates into the nucleus, binds as a heterodimer with RXR to DNA and interacts with various coregulators, leading to the activation of transcription or relief of constitutive repression.

The second feature shared by VDR and other NRs is its structural organization. VDR is a molecule of approximately 50–60 kDa, depending on species. The human VDR has two potential start sites with a common polymorphism (Fok 1) that alters the first ATG start site to ACG, leading to a VDR that is three amino acids shorter (424 AA vs. 427 AA), a polymorphism correlated with reduced bone density [18]. The VDR has a modular organization (Figure 2A) that consists of a short variable and flexible N-terminal domain, a highly conserved DNA-binding domain (DBD), a conserved ligand-binding domain (LBD) and a hinge region connecting the DBD to LBD. In addition, VDR has a unique feature among NRs, with its long insertion region within the LBD that is in a disordered state [19].

Since the first crystal structures of VDR LBD in 2000 [20] and of VDR DBD in 2002 [21] (Figure 2B,C), numerous studies have investigated the isolated domains of VDR, mainly by X-ray crystallography complemented by NMR and hydrogen-deuterium exchange coupled with mass spectrometry (HDX-MS).

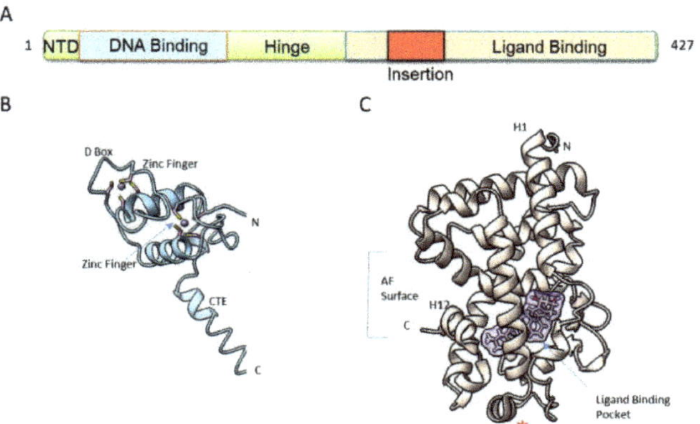

Figure 2. VDR structure. (**A**) Modular organization of VDR. NTD: N-terminal domain. DBD: DNA binding domain. LBD: ligand binding domain. (**B**) Overall structure of hVDR DBD monomer (PDB ID: 1KB4 [21]). Zn atoms are represented by spheres. CTE: C-terminal extension. (**C**) Overall structure of the hVDR LBD bound to 1,25D_3 (PDB ID: 1DB1 [20]). LBP: ligand binding pocket. AF: activation function. The red star corresponds to the position of the truncation of the insertion domain.

3. DNA Binding

The DBD, the most conserved domain in VDR from different species and among the NRs, is comprised of two zinc fingers (Figure 2B). The first zinc finger is important for specific DNA binding to the VDREs, while the second one is involved in heterodimer interaction. Steric constraints of the VDR–RXR complex determine the optimal heterodimer binding site within VDRE as a direct repeat of the sequence RGKTSA (R = A or G, K = G or T, S = C or G), separated by three nucleotides (DR3) [22–24]. RXR binds to the upstream half site, while VDR binds to the downstream site. The crystal structure of the VDR DBD homodimer on DR3 [21] has revealed that the key interactions between the VDR DBD and the DNA, including four conserved residues in the recognition helix, Glu42, Lys45, Arg49 and Arg50, make sequence-specific base contacts in the major groove of the half-site. ChIP-seq studies in various cell types (see references [25–31] among others) have confirmed that the DR3 are the most enriched motifs upon ligand treatment but that represent only 10–20% of all VDR binding sites and most of VDREs spread over the whole genome [17]. A prerequisite for VDR DNA binding is the accessibility of the binding site through the action of pioneer factors and coactivators to open the chromatin and to modify chromatin topology [17,24,25].

4. Ligand Binding

NR is a highly dynamic scaffold protein and VDR LBD (Figure 2C) in a similar way to other NRs, is dynamic and only stabilized into a fixed conformation upon ligand binding. The dynamics of ligand binding process by VDR has been investigated by NMR [32] and HDX-MS [33–35]. These studies revealed that the entire C-terminal of VDR LBD in its apo state is very dynamic, with 80% of amide hydrogen exchange. The region forming the ligand binding pocket (LBP) (Figure 2C) also showed a high exchange rate, while the central layer of the α-helical sandwich appeared to be protected. Binding of 1,25D_3 has been shown to lead to significant protection from hydrogen amide exchange, not only for the LBP but also in regions remote from the LBP.

Detailed information on the binding mode of 1,25D$_3$ has been obtained by the elucidation of the crystal structure by X-ray crystallography of its complex with the human VDR LBD [20]. For the crystallization of the hVDR–1,25D$_3$ complex, a truncated form of the hVDR LBD was used that lacks the insertion domain (Figure 2A). This region is characterized by poor sequence conservation between VDR family members, is predicted to be disordered and does not play a major role in receptor selectivity for 1,25D$_3$ [19,36]. The general fold of VDR LBD (PDB IDs: 1DB1 and 7QPP) consists of a three-layered α-helical sandwich composed of twelve helices (H1 to H12), three two-turn helices (H3n, H4n and Hx) and a three-stranded β-sheet (Figure 3A). The LBP is surrounded by helices H2, H3, H5, H6, H7, H10 an H12. The residues of each of β-sheet strands also form contacts with the ligand.

The ligand occupies 56% of the volume of the LBP (697 Å^3) with some water molecules near the position 2 of the A-ring of 1,25D$_3$ (Figure 3B). The ligand adopts a chair B conformation with the 19-methylene "up" and the 1α-OH and 3β-OH groups in equatorial and axial orientations, respectively, while the aliphatic chain at position 17 of the D-ring adopts an extended conformation. The ligand is anchored in the LBP through three pair of hydrogen bonds formed between three hydroxyl groups of the ligand and polar residues: 1-OH group with Ser237 (H3) and Arg274 (H5), 3-OH group with Ser278 (H5) and Tyr143 (loop H1-H2) and 25-OH group with His305 (loop H6-H7) and His397 (H11) (Figure 3B). In addition, the ligand interacts with the hydrophobic residues lining the LBP (Figure 3C).

The LBD also contains the regions necessary for heterodimerization to RXR, comprising H9 and H10 and the loop 8–9, and for coactivator interaction through the activation function AF-2 formed by H3, H4 and H12. Helix 12 closed the LBP and is stabilized by two interactions with the ligand. Helix 12 is also stabilized by several hydrophobic contacts with residues of H3, H5 and H11 and two polar interactions with residues of H3 and H4. Some of these residues contact the ligand, thus indicating an additional indirect ligand-control of the position of helix H12.

The structures of 1,25D$_3$ in the complex with VDR of other species were also described for *Rattus norvegicus* (rVDR) [37], *Danio rerio* (zebrafish, zVDRα) [38] and *Petromyzon marinus* (sea lamprey (l)) [39], the most basal vertebrate showing the most divergent VDR sequence. In the case of the rVDR LBD complex (PDB ID: 1RK3), the same truncation of the large insertion region connecting helices H1 to H3 as for hVDR was applied. For the zVDR (PDB ID: 2HC4) and lVDR (PDB ID: 7QPI), the wild-type LBD was used. The binding mode of 1,25D$_3$ to VDR LBP is similar in all VDR structures, indicating a conserved ligand selectivity of VDRs across vertebrate species. While the differences are small between the structures of h, r and z VDR LBDs and primarily involve the loops, some significant differences are observed for lVDR around the linker regions between H11-H12 and H9-H10. These differences explain the weaker AF-2 stabilization and weaker RXR dimerization [39], and consequently the lower efficacy to activate lVDR compared to higher vertebrate VDRs [40,41]. Interpretation of the VDR structures and VDR–ligand interaction differences in a phylogenetic context allow for significant progress towards understanding the molecular activities. Indeed, structural and sequence co-evolution analysis allow to identify the conserved residues to maintain their functional integrity, and pairs of amino acids that coevolved to accommodate novel functionalities. Coevolving residues are located in H9 and in the insertion domain [39], accounting for the increased sensitivity to RXR and coregulators during evolution and leading to the increase in transactivation responses to 1,25D$_3$ and VDR to be fully activated by 1,25D$_3$ and to respond to lithocholic acid in higher vertebrates, facilitating novel functions.

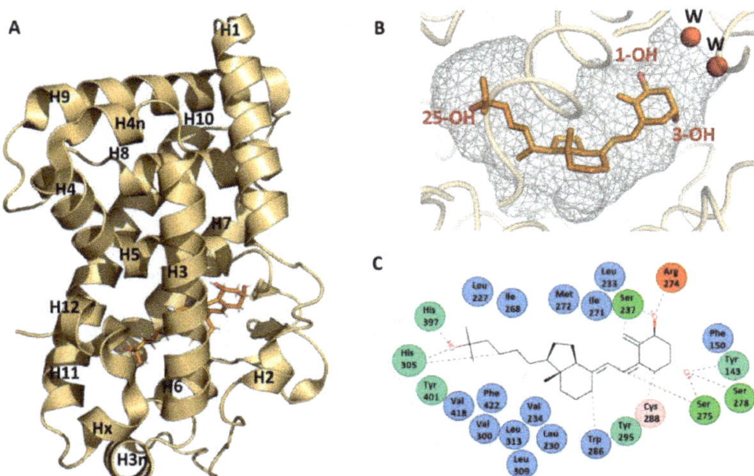

Figure 3. $1,25D_3$ recognition by VDR. (**A**) Overall structure of the hVDR LBD. The hVDR LBD bound to $1,25D_3$ is composed of 15 helices and 1 β-sheet (PDB ID: 1DB1 and 7QPP) [20,39]. (**B**) Conformation of $1,25D_3$ in the VDR LBP shown as a grey surface. (**C**) Interaction map of $1,25D_3$ in the LBP of hVDR.

This structural knowledge has been pivotal in the comprehension of VDR action and to understand the impact of mutations found in hereditary vitamin D-resistant rickets (HVDRR) patients showing a phenotype connected with softening and weakening of the bones [42,43]. The alterations in the VDR gene are caused by mutations that result in suboptimal gene regulatory responses, despite the presence of $1,25D_3$ in the body. To date, 12 missense mutations have been identified in hVDR DBD, all associated with alopecia [43]. In the LBD, 24 amino acids of the hVDR were found mutated in HVDRR patients, some associated with alopecia [43]. The mutations result in alterations of VDR functions as ligand- or DNA-binding, heterodimerization with RXR or coregulator-binding. However, the overall effect of the mutation may result from the combination effects of all four functions at the same time. Among those mutations, the mutations involved residues forming hydrogen bonds with the hydroxyl groups of $1,25D_3$, including Arg274Leu/His [44–46], His305Gln [47] and His397Pro [48], the last one being associated with very severe HVDRR. The discussion of the structural implications of these missense mutations were presented in the studies [49,50].

In addition, the crystal structures of VDR LBD have provided significant information for the characterization and the design of more specific analogs. A large part of the VDR LBP remains unoccupied when bound by $1,25D_3$ (Figure 3B), providing additional space for the fitting of the modified moieties of the hormone and some analogs induce significant conformational changes, enlarging the original LBP (review in [15,51,52]). Not only secosteroidal analogs, but also other VDR ligands with non secosteroidal structures, such as lithocholic acid derivatives and mimics of $1,25D_3$, have been reported. Hundreds of analogs in the complexes with VDR LBD from human, rat or zebrafish species were crystalized. The binding mode of some of these analogs to VDR LBP and their mechanism of action were discussed in [15,51,52]. In addition to the development of molecules targeting the VDR LBP and acting as agonist or antagonist, small molecules that target VDR–coregulator binding and act as protein-protein inhibitors are being developed [53]. Atomic-level understanding allows for the development of compounds with more original chemical structures and more specific action. However, due to the complexity of VDR signaling, these compounds remain largely unsuccessful in reaching therapeutic applications thus far.

5. Structure of Full-Length VDR Complex

Until now, only the VDR LBD monomer has been crystallized. The lower affinity of VDR for RXR compared to other NRs, such as PPAR or RAR [39,54], may explain the difficulty to crystallize the VDR–RXR LBDs heterodimer. DNA binding stabilizes the VDR heterodimer and 1,25D_3 is required for high affinity binding and activation. On other hand, RXR ligand, 9-cis retinoic acid, has been shown to either inhibit [55] or activate [56,57] -1,25D_3 stimulation of gene transcription. In the absence of a full-length heterodimer crystal structure, solution methods using small angle X-ray scattering (SAXS) [58], cryo-electron microscopy (cryoEM) [59] and HDX-MS [60] have provided information on the full length VDR–RXR–DNA complex. These methods render it possible to visualize how the DBD and the LBD/heterodimerization domains are arranged relative to one another and how their binding to ligand, DNA, and coactivators influence one another.

The LBD and DBD domains in the full length structure is structurally conserved, compared to those of previously solved individual domain structures. The solution structures of VDR–RXR show that DBDs and LBDs are separated and positioned asymmetrically (Figure 4). The relative position of the domains and the observed asymmetry of the overall architecture both point to the essential role played by the hinge domains in establishing and maintaining the integrity of the functional structures. While the RXR hinge does not have a well-defined structure, the hinge domain of VDR forms an α-helix that stabilizes the whole complex, thus facilitating the positioning of the LBD and the surface to be accessible by the coregulators [58,59]. The 3D structures of several NRs truncated by their NTDs and in complex with DNA have now been obtained by X-ray crystallography or cryoEM for PPAR–RXR [61], HNF4 [62], LXR–RXR [63], RAR–RXR [64], EcR–USP [65] and AR [66]. These studies suggest that NRs are rather flexible macromolecules, adapting several conformations. In the VDR–RXR complex, even with the relatively separated positioning, there is evidence of long-range allosteric connections between the VDR LBD, DBD and DNA [60]. HDX-MS was used to understand the conformational plasticity and allosteric/dynamic communications in VDR complexes and has revealed cooperative effects between the VDR DBD and LBD to fine tune transcriptional regulation by the ligands and the DNA [67]. Upon 1,25D_3 binding to full VDR–RXR, the differential HDX experiment has revealed a profile very similar to the binding to VDR LBD alone with a stabilization of VDR H12. Interestingly, an increase in solvent exchange in the DBD of VDR was also observed upon ligand binding, indicating that the ligand impact on the DBD conformation. Upon ligand binding, a stabilization of the heterodimer interface was also observed. The HDX profile of 1,25D_3 binding in the presence of RXR ligand was similar to that of 1,25D_3 alone. Allosteric communication is ligand dependent and bidirectional. DNA binding modulates both DBD–DBD and LBD–LBD interactions and influences coactivator recognition and binding. Conformational dynamics indicate that the binding of VDR–RXR to DNA results in significant alterations in the conformation of the LBD within the region important for interactions with coactivators, VDR H12 and RXR H3, and dimer interface H10–H11. Difference in DNA sequences modulate the receptor dynamics in remote regions, such as the coactivator binding surfaces [60,67]. A similar effect was observed for the RAR–RXR–DNA complex [68].

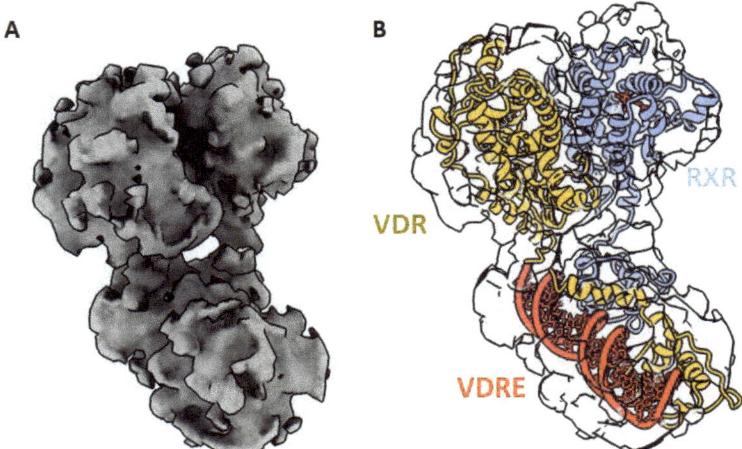

Figure 4. Cryo-EM structure of the VDR–RXR on DNA. (**A**) Electron density map shown as surface. (**B**) Fitted atomic model of the heterodimer in the EM map [59].

6. VDR–Coregulatory Complexes

The successful regulation of transcription by NRs requires the recruitment of coregulators to genomic loci, an event that directly affects the transcriptional rate. Hundreds of NR coregulators have been reported and include coactivators and corepressors [69]. For VDR, the mechanism of coactivation is now well understood, while VDR corepression is less studied. The major VDR coactivators are the NCoAs and mediator complexes [70,71]. The NCoAs (NCoA1, NCoA2 and NCoA3) recruit secondary coactivators CBP/p300 and p/CAF that have histone acetyl transferase activity and interact with VDR in a 1,25D_3-dependent manner [71]. Another important VDR coactivator is MED1, a subunit of the mediator that links the TFs to the transcription machinery [72].

Most of the coactivators that can be recruited to the NRs in a ligand-dependent way contain the conserved LXXLL motif or NR box [73]. An analogous sequence motif (LXXH/IIXXXI/L) was identified in corepressors [74–76]. NR boxes are often located within intrinsically disordered regions of the coactivators and corepressors. NCoAs contain several domains separated by long disordered regions, the N-terminal part contains a highly conserved basic helix-loop-helix (HLH) and a signaling PAS (Per/Arnt/Sim) domain that mediates protein–protein interactions. The NRs interact with NCoAs via the receptor interaction domain (RID) that contains three LXXLL motifs. Each domain in NCoAs specializes in recruiting various TFs or other coregulators of transcription, including protein-modifying enzymes and chromatin remodelers. VDR has been shown to preferentially interact with the second and third LXXLL motifs of the NCoA RID [77]. MED1 contains two NR boxes in its central RID that differentially bind to NRs [78]. The second motif has been shown to preferentially bind to VDR [77]. These coregulators are not specific to VDR, but interact with a large number of other NRs and transcription factors. The mediator complex can be found in large numbers of loci and are known to be involved as super enhancers [79]. In addition, other domains of MED1 outside the RID has been shown to interact with NRs [80].

Multiple biochemical and structural studies have mapped the interaction regions of NRs with the coactivator NR box [73,81–83] and numerous X-ray structures of NR LBDs with bound coregulator LXXLL peptides are nowadays deposited in the PDB. Analogous recognition region in the corepressors, CoR-NR box, binds in the same hydrophobic groove on LBDs. Therefore, the recruitment of coactivators and corepressors is mutually exclusive. The intrinsically disordered properties of the coregulators that are important to adapt and interact with many different transcription factors also limit their studies by X-ray crystallography or cryoEM.

MED1 NR2 as well as NCoA1 NR2, NCoA2 NR2 and NR3 were crystallized with liganded VDR LBD [37,39,84,85], revealing the binding mode of the LXXLL motifs to the VDR LBD. The LXXLL peptide formed a short a-helix that binds to a surface formed by helices H3, H4 and H12 of the LBD. The interaction of MED1 NR2 that buries about 507 Å^2 of the receptor's surface involves the three leucines that are buried within the pocket and surrounded by hydrophobic residues. The peptide is additionally locked through the formation of hydrogen bonds between conserved lysine residue in H3 and a glutamate residue in H12 that define a charge clamp.

Structures of coregulatory complexes with large fragment or full-length coactivators have been less studied due to their dynamic conformations. Several biophysical and solution structural studies in solution have been performed for some NRs bound to coactivator RIDs and recent developments in cryoEM have started to provide structural information on NR–coactivator complexes, including Erα-NcoA3-p300 [86] and AR-NcoA3-p300 [87]. However, NR–coregulator complexes structures have not reached atomic resolution yet, due to the conformational dynamics of the transcriptional complexes. For VDR, mass spectrometry structural methods combined with biophysical and structural methods have provided important details of VDR complex assembly and conformational plasticity and allosteric/dynamic communications within the complexes [60,67,88].

Differential HDX-MS has been used to study the interaction of VDR–RXR with the NCoA1 RID that contains the three LXXLL motifs [60,67]. As expected in the absence of both ligands, no coactivator interaction was observed and the separate addition of the 1,25D$_3$ or RXR ligand, 9-cis retinoic acid, allow the coactivator to interact in a ligand-specific manner and independently. In the presence of 1,25D$_3$, only RXR AF-2 within the full VDR–RXR complex was insensitive to the binding of NCoA1 RID, indicating that it is primarily associated to VDR AF-2. However, RXR and its ligand modulate the interaction when VDR is liganded and NCoA1 RID binding stabilizes not only VDR AF-2 but also RXR H3 and H10-H11 [60]. Helices 3 and 4 of VDR that are part of the coactivator binding cleft cannot be further stabilized, since they achieve maximal stabilization upon 1,25D$_3$ binding. For RXR, the loop between helices 10 and 11 is important in the formation of the hydrophobic groove facilitating coactivator binding. In addition, DNA binding modulates the interaction of NCoA1 with the heterodimer. Mutations of each LXXLL motifs and luciferase transactivation assays suggested that the third motif was associated to VDR and the first one to RXR [60]. Mutations on residues of VDR and RXR involved in the coactivator charge clamp indicate that the coactivator binding surface of each receptor is important for NCoA1 RID (Figure 5A). These data are in contrast with the SAXS data showing that MED1 RID mainly interacts with RXR's heterodimeric partner [58]. Importantly, to fully understand the binding of MED1 to VDR–RXR, it was necessary to use a larger fragment of MED1 that not only contains the RID but also the N-terminal domain (50–660) [89]. Significant differences could be observed in comparison to the NCoA1 RID complex [89] by integrative structural methods combining SAXS, NMR and structural mass spectrometry. Differential HDX-MS and crosslink mass spectrometry confirmed that VDR interaction with MED1 motif NR2 is driving the complex formation but also demonstrated that other VDR–RXR regions outside the VDR AF-2, as well as MED1 regions other than RID, modulate the association and form an extended interaction surface (Figure 5B). Both LXXLL motifs of MED1 were perturbed upon formation of the complex with VDR–RXR, suggesting that while the NR1 box is not accommodated within the classical coactivator binding site, it could be either interacting with an alternative site of the receptors or stabilized allosterically. In addition to RID, the structured N-terminal domain of MED1 is also affected upon binding to VDR–RXR and is likely interacting with both VDR and RXR LBDs; in particular, the MED1 region 243–255 is largely stabilized in the complex with the receptor heterodimer. Crosslink MS also confirmed that this MED1 region is in physical proximity to the RXR. Among other novel MED1-interacting regions within the VDR–RXR heterodimer is the flexible insertion domain in the VDR LBD located between H1 and H3. Conformational changes upon interaction with MED1 was observed by NMR. This effect was not observed in the HDX-MS experiment; however, the observed difference could be attributed to the different temporal resolution of the two methods.

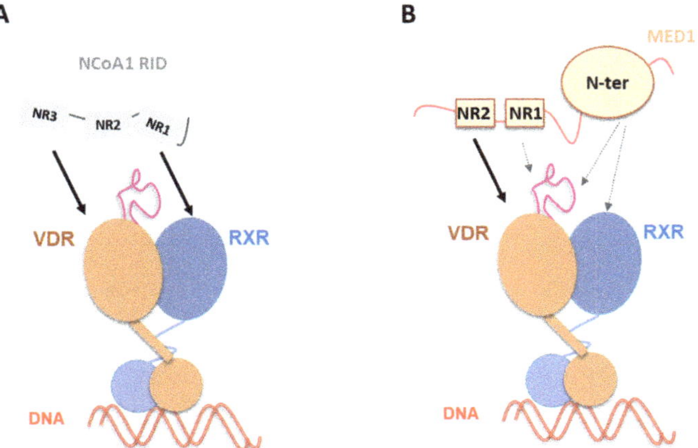

Figure 5. Schematic representation of NCoA1 and MED1 binding to liganded VDR–RXR–DNA complex. (**A**) NCoA1 RID binding to liganded VDR–RXR in complex with and without DNA [60]. The NCoA1 NR3 motif is associated to VDR and the NR1 to RXR. NCoA1 RID binding has been shown to stabilize not only VDR AF-2 but also RXR H3 and H10-H11. (**B**) MED1 (50–660) binding to liganded VDR–RXR–DNA complex [89]. Complex formation is primarily driven by strong ligand-dependent MED1 NR2 binding to the VDR AF-2, but other MED1 regions including NR1 and the structured N-terminal domain are involved in the interaction, as well as alternative sites of the receptors, including VDR insertion domain and RXR.

Differences in the binding modes could serve as molecular determinants of how the NRs discriminate between the coactivators. These studies were performed on large domains of coactivators but future studies should be carried out on the heterodimer complex with full length coactivators, as other domains, as shown for the N-ter domain of MED1, may modulate interactions directly or allosterically.

7. Conclusions and Perspectives

Our understanding of the molecular and structural insights for 1,25D_3 action via its master regulator VDR has continuously advanced in the last twenty years, through the elucidation of the atomic structures of VDR DBD and LBD in complex with 1,25D_3 and analogs. In addition, few structural studies on full length DNA bound VDR complexes have provided information on the allosteric effects driven by other domains and other effectors as DNA and coregulators. Important questions remain regarding the basic mechanisms of cell-specific action of ligands and possible cross-talks with other NRs and TFs. A deeper understanding of the interactions of VDR with specific coregulators will also be essential to better understand VDR action and may likely impact the future of drug development. However, full-length coregulators have large unstructured regions that remain a major hurdle for their structural characterization. CryoEM has started to provide structural information and will be essential to understand the structural dynamics of coregulatory complexes. The new structural biology tools, such as structural mass spectrometry, single molecule cryoEM with new processing tools and 3D classification, and cryo-tomography, will ultimately provide details on the structures of VDR complexes with coregulator complexes in a physiological environment.

Funding: Financial support was provided by the Agence Nationale de la Recherche (ANR-13-BSV8-0024-01 and ANR-21-CE17-0009-01).

Conflicts of Interest: The author declares no conflict of interest.

References

1. Bikle, D.D. Vitamin D: Production, Metabolism and Mechanisms of Action. In *Endotext [Internet]*; Feingold, K.R., Anawalt, B., Boyce, A., Chrousos, G., de Herder, W.W., Dhatariya, K., Dungan, K., Hershman, J.M., Hofland, J., Kalra, S., et al., Eds.; MDText.com, Inc.: South Dartmouth, MA, USA, 2021.
2. McDonnell, D.P.; Mangelsdorf, D.J.; Pike, J.W.; Haussler, M.R.; O'Malley, B.W. Molecular cloning of complementary DNA encoding the avian receptor for vitamin D. *Science* **1987**, *235*, 1214–1217. [CrossRef] [PubMed]
3. Burmester, J.K.; Maeda, N.; DeLuca, H.F. Isolation and expression of rat 1,25-dihydroxyvitamin D3 receptor cDNA. *Proc. Natl. Acad. Sci. USA* **1988**, *85*, 1005–1009. [CrossRef] [PubMed]
4. Peleg, S.; Pike, J.W.; O'Malley, B.W. Characterization of the chromosomal gene of the mouse vitamin D receptor. In *Vitamin D, Molecular Cellular and Clinical Endocrinology*; Norman, A., Schaefer, K., Grigoleit, Herrath, D., Eds.; Walter de Gruyter: New York, NY, USA, 1988; pp. 242–243.
5. Baker, A.R.; McDonnell, D.P.; Hughes, M.; Crisp, T.M.; Mangelsdorf, D.J.; Haussler, M.R.; Pike, J.W.; Shine, J.; O'Malley, B.W. Cloning and expression of full-length cDNA encoding human vitamin D receptor. *Proc. Natl. Acad. Sci. USA* **1988**, *85*, 3294–3298. [CrossRef] [PubMed]
6. Fleet, J.C. The role of vitamin D in the endocrinology controlling calcium homeostasis. *Mol. Cell. Endocrinol.* **2017**, *453*, 36–45. [CrossRef]
7. Goltzman, D. Functions of vitamin D in bone. *Histochem. Cell. Biol.* **2018**, *149*, 305–312. [CrossRef]
8. Bikle, D.; Christakos, S. New aspects of vitamin D metabolism and action—Addressing the skin as source and target. *Nat. Rev. Endocrinol.* **2020**, *16*, 234–252. [CrossRef]
9. Morris, H.A.; Anderson, P.H. Autocrine and paracrine actions of vitamin D. *Clin. Biochem. Rev.* **2010**, *31*, 129–138.
10. Pike, J.W.; Meyer, M.B. The unsettled science of nonrenal calcitriol production and its clinical relevance. *J. Clin. Investig.* **2020**, *130*, 4519–4521. [CrossRef]
11. Tuckey, R.C.; Cheng, C.Y.S.; Slominski, A.T. The serum vitamin D metabolome: What we know and what is still to discover. *J. Steroid Biochem. Mol. Biol.* **2019**, *186*, 4–21. [CrossRef]
12. Bouillon, R.; Bikle, D. Vitamin D Metabolism Revised: Fall of Dogmas. *J. Bone Min. Res.* **2019**, *34*, 1985–1992. [CrossRef]
13. Umar, M.; Sastry, K.S.; Chouchane, A.I. Role of Vitamin D Beyond the Skeletal Function: A Review of the Molecular and Clinical Studies. *Int. J. Mol. Sci.* **2018**, *19*, 1618. [CrossRef] [PubMed]
14. Leyssens, C.; Verlinden, L.; Verstuyf, A. The future of vitamin D analogs. *Front. Physiol.* **2014**, *5*, 122. [CrossRef]
15. Maestro, M.A.; Molnár, F.; Carlberg, C. Vitamin D and Its Synthetic Analogs. *J. Med. Chem.* **2019**, *62*, 6854–6875. [CrossRef]
16. Haussler, M.R.; Whitfield, G.K.; Kaneko, I.; Haussler, C.A.; Hsieh, D.; Hsieh, J.C.; Jurutka, P.W. Molecular mechanisms of vitamin D action. *Calcif. Tissue Int.* **2013**, *92*, 77–98. [CrossRef]
17. Carlberg, C. Vitamin D and Its Target Genes. *Nutrients* **2022**, *4*, 1354. [CrossRef] [PubMed]
18. Morrison, N.A.; Qi, J.C.; Tokita, A.; Kelly, P.J.; Crofts, L.; Nguyen, T.V.; Sambrook, P.N.; Eisman, J.A. Prediction of bone density from vitamin D receptor alleles. *Nature* **1994**, *367*, 284–287. [CrossRef] [PubMed]
19. Rochel, N.; Tocchini-Valentini, G.; Egea, P.F.; Juntunen, K.; Garnier, J.M.; Vihko, P.; Moras, D. Functional and structural characterization of the insertion region in the ligand binding domain of the vitamin D nuclear receptor. *Eur. J. Biochem.* **2001**, *268*, 971–979. [CrossRef]
20. Rochel, N.; Wurtz, J.M.; Mitschler, A.; Klaholz, B.; Moras, D. The crystal structure of the nuclear receptor for vitamin D bound to its natural ligand. *Mol. Cell.* **2000**, *5*, 173–179. [CrossRef]
21. Shaffer, P.L.; Gewirth, D.T. Structural basis of VDR-DNA interactions on direct repeat response elements. *EMBO J.* **2002**, *21*, 2242–2252. [CrossRef]
22. DeLuca, H.F.; Zierold, C. Mechanisms and functions of vitamin D. *Nutr. Rev.* **1998**, *56*, S4–S10. [CrossRef]
23. Carlberg, C. What do we learn from the genome-wide perspective on vitamin D3? *Anticancer. Res.* **2015**, *35*, 1143–1151. [PubMed]
24. Pike, J.W.; Meyer, M.B.; Benkusky, N.A.; Lee, S.M.; St John, H.; Carlson, A.; Onal, M.; Shamsuzzaman, S. Genomic Determinants of Vitamin D-Regulated Gene Expression. *Vitam. Horm.* **2016**, *100*, 21–44. [PubMed]
25. Ramagopalan, S.V.; Heger, A.; Berlanga, A.J.; Maugeri, N.J.; Lincoln, M.R.; Burrell, A.; Handunnetthi, L.; Handel, A.E.; Disanto, G.; Orton, S.M.; et al. A ChIP-seq defined genome-wide map of vitamin D receptor binding: Associations with disease and evolution. *Genome Res.* **2010**, *20*, 1352–1360. [CrossRef]
26. Meyer, M.B.; Goetsch, P.D.; Pike, J.W. VDR/RXR and TCF4/β-catenin cistromes in colonic cells of colorectal tumor origin: Impact on c-FOS and c-MYC gene expression. *Mol. Endocrinol.* **2012**, *26*, 37–51. [CrossRef]
27. Heikkinen, S.; Väisänen, S.; Pehkonen, P.; Seuter, S.; Benes, V.; Carlberg, C. Nuclear hormone 1α,25-dihydroxyvitamin D3 elicits a genome-wide shift in the locations of VDR chromatin occupancy. *Nucleic. Acids Res.* **2011**, *39*, 9181–9193. [CrossRef]
28. Ding, N.; Yu, R.T.; Subramaniam, N.; Sherman, M.H.; Wilson, C.; Rao, R.; Leblanc, M.; Coulter, S.; He, M.; Scott, C.; et al. A vitamin D receptor/SMAD genomic circuit gates hepatic fibrotic response. *Cell* **2013**, *153*, 601–613. [CrossRef]

29. Long, M.D.; Campbell, M.J. Integrative genomic approaches to dissect clinically-significant relationships between the VDR cistrome and gene expression in primary colon cancer. *J. Steroid Biochem. Mol. Biol.* **2017**, *173*, 130–138. [CrossRef] [PubMed]
30. Warwick, T.; Schulz, M.H.; Gilsbach, R.; Brandes, R.P.; Seuter, S. Nuclear receptor activation shapes spatial genome organization essential for gene expression control: Lessons learned from the vitamin D receptor. *Nucleic. Acids Res.* **2022**, *50*, 3745–3763. [CrossRef]
31. Abu El Maaty, M.A.; Grelet, E.; Keime, C.; Rerra, A.I.; Gantzer, J.; Emprou, C.; Terzic, J.; Lutzing, R.; Bornert, J.M.; Laverny, G.; et al. Single-cell analyses unravel cell type-specific responses to a vitamin D analog in prostatic precancerous lesions. *Sci. Adv.* **2021**, *7*, 5982. [CrossRef]
32. Singarapu, K.K.; Zhu, J.; Tonelli, M.; Rao, H.; Assadi-Porter, F.M.; Westler, W.M.; DeLuca, H.F.; Markley, J.L. Ligand-specific structural changes in the vitamin D receptor in solution. *Biochemistry* **2011**, *50*, 11025–11033. [CrossRef]
33. Zhang, J.; Chalmers, M.J.; Stayrook, K.R.; Burris, L.L.; Garcia-Ordonez, R.D.; Pascal, B.D.; Burris, T.P.; Dodge, J.A.; Griffin, P.R. Hydrogen/deuterium exchange reveals distinct agonist/partial agonist receptor dynamics within vitamin D receptor/retinoid X receptor heterodimer. *Structure* **2010**, *18*, 1332–1341. [CrossRef]
34. Kato, A.; Itoh, T.; Anami, Y.; Egawa, D.; Yamamoto, K. Helix12-Stabilization Antagonist of Vitamin D Receptor. *Bioconjug. Chem.* **2016**, *27*, 1750–1761. [CrossRef]
35. Belorusova, A.Y.; Chalhoub, S.; Rovito, D.; Rochel, N. Structural Analysis of VDR Complex with ZK168281 Antagonist. *J. Med. Chem.* **2020**, *63*, 9457–9463. [CrossRef]
36. Krasowski, M.D.; Ai, N.; Hagey, L.R.; Kollitz, E.M.; Kullman, S.W.; Reschly, E.J.; Ekins, S. The evolution of farnesoid X, vitamin D, and pregnane X receptors: Insights from the green-spotted pufferfish (*Tetraodon nigriviridis*) and other non-mammalian species. *BMC Biochem.* **2011**, *12*, 5. [CrossRef]
37. Vanhooke, J.L.; Benning, M.M.; Bauer, C.B.; Pike, J.W.; DeLuca, H.F. Molecular structure of the rat vitamin D receptor ligand binding domain complexed with 2-carbon-substituted vitamin D3 hormone analogues and a LXXLL-containing coactivator peptide. *Biochemistry* **2004**, *43*, 4101–4110. [CrossRef]
38. Ciesielski, F.; Rochel, N.; Moras, D. Adaptability of the Vitamin D nuclear receptor to the synthetic ligand Gemini: Remodelling the LBP with one side chain rotation. *J. Steroid Biochem. Mol. Biol.* **2007**, *103*, 235–242. [CrossRef]
39. Sigüeiro, R.; Bianchetti, L.; Peluso-Iltis, C.; Chalhoub, S.; Dejaegere, A.; Osz, J.; Rochel, N. Advances in Vitamin D Receptor Function and Evolution Based on the 3D Structure of the Lamprey Ligand-Binding Domain. *J. Med. Chem.* **2022**, *65*, 5821–5829. [CrossRef]
40. Reschly, E.J.; Bainy, A.C.D.; Mattos, J.J.; Hagey, L.R.; Bahary, N.; Mada, S.R.; Ou, J.; Venkataramanan, R.; Krasowski, M.D. Functional evolution of the vitamin D and pregnane X receptors. *BMC Evol. Biol.* **2007**, *7*, 222. [CrossRef]
41. Whitfield, G.K.; Dang, H.T.L.; Schluter, S.F.; Bernstein, R.M.; Bunag, T.; Manzon, L.A.; Hsieh, G.; Encinas Dominguez, C.; Youson, J.H.; Haussler, M.R.; et al. Cloning of a functional vitamin D receptor from the lamprey (*Petromyzon marinus*), an ancient vertebrate lacking a calcified skeleton and teeth. *Endocrinology* **2003**, *144*, 2704–2716. [CrossRef]
42. Feldman, D.J.; Malloy, P. Mutations in the vitamin D receptor and hereditary vitamin D-resistant rickets. *Bonekey Rep.* **2014**, *3*, 510. [CrossRef]
43. Acar, S.; Demir, K.; Shi, Y. Genetic causes of rickets. *J. Clin. Res. Pediatr. Endocrinol.* **2017**, *9*, 88–105. [CrossRef] [PubMed]
44. Kristjansson, K.; Rut, A.R.; Hewison, M.; O'Riordan, J.L.; Hughes, M.R. Two mutations in the hormone binding domain of the vitamin D receptor cause tissue resistance to 1,25 dihydroxy vitamin D3. *J. Clin. Investig.* **1993**, *92*, 12–16. [CrossRef] [PubMed]
45. Aljubeh, J.M.; Wang, J.; Al-Remeithi, S.S.; Malloy, P.J.; Feldman, D. Report of two unrelated patients with hereditary vitamin D resistant rickets due to the same novel mutation in the vitamin D receptor. *J. Pediatr. Endocrinol. Metab.* **2011**, *24*, 793–799. [CrossRef] [PubMed]
46. Brar, P.C.; Dingle, E.; Pappas, J.; Raisingani, M. Clinical Phenotype in a Toddler with a Novel Heterozygous Mutation of the Vitamin D Receptor. *Case Rep. Endocrinol.* **2017**, *2017*, 3905905. [CrossRef] [PubMed]
47. Malloy, P.J.; Eccleshall, T.R.; Gross, C.; Van Maldergem, L.; Bouillon, R.; Feldman, D. Hereditary vitamin D resistant rickets caused by a novel mutation in the vitamin D receptor that results in decreased affinity for hormone and cellular hyporesponsiveness. *J. Clin. Investig.* **1997**, *99*, 297–304. [CrossRef]
48. Andary, R.; El-Hage-Sleiman, A.K.; Farhat, T.; Sanjad, S.; Nemer, G. Hereditary vitamin D-resistant rickets in Lebanese patients: The p.R391S and p.H397P variants have different phenotypes. *J. Pediatr. Endocrinol. Metab.* **2017**, *30*, 437–444. [CrossRef]
49. Rochel, N.; Molnár, F. Structural aspects of Vitamin D endocrinology. *Mol. Cell. Endocrinol.* **2017**, *453*, 22–35. [CrossRef]
50. Belorusova, A.Y.; Rochel, N. Structural Basis for Ligand Activity in Vitamin D Receptor. In *Vitamin D*, 5th ed.; Feldman, A., Ed.; Elsevier: Amsterdam, The Netherlands, 2022.
51. Belorusova, A.Y.; Rochel, N. Modulators of vitamin D nuclear receptor: Recent advances from structural studies. *Curr. Top Med. Chem.* **2014**, *14*, 2368–2377. [CrossRef]
52. Yamada, S.; Makishima, M. Structure-activity relationship of nonsecosteroidal vitamin D receptor modulators. *Trends Pharmacol. Sci.* **2014**, *35*, 324–337. [CrossRef]
53. Mutchie, T.R.; Yu, O.B.; Di Milo, E.S.; Arnold, L.A. Alternative binding sites at the vitamin D receptor and their ligands. *Mol. Cell. Endocrinol.* **2019**, *485*, 1–8. [CrossRef]

54. Fadel, L.; Rehó, B.; Volkó, J.; Bojcsuk, D.; Kolostyák, Z.; Nagy, G.; Müller, G.; Simandi, Z.; Hegedüs, É.; Szabó, G.; et al. Agonist binding directs dynamic competition among nuclear receptors for heterodimerization with retinoid X receptor. *J. Biol. Chem.* **2020**, *295*, 10045–10061. [CrossRef] [PubMed]
55. Thompson, P.D.; Jurutka, P.W.; Haussler, C.A.; Whitfield, G.K.; Haussler, M.R. Heterodimeric DNA binding by the vitamin D receptor and retinoid X receptors is enhanced by 1,25-dihydroxyvitamin D3 and inhibited by 9-cis-retinoic acid. Evidence for allosteric receptor interactions. *J. Biol. Chem.* **1998**, *273*, 8483–8491. [CrossRef] [PubMed]
56. Bettoun, D.J.; Burris, T.P.; Houck, K.A.; Buck, D.W.; Stayrook, K.R.; Khalifa, B.; Lu, J.; Chin, W.W.; Nagpal, S. Retinoid X receptor is a nonsilent major contributor to vitamin D receptor-mediated transcriptional activation. *Mol. Endocrinol.* **2003**, *17*, 2320–2328. [CrossRef] [PubMed]
57. Sánchez-Martínez, R.; Castillo, A.I.; Steinmeyer, A.; Aranda, A. The retinoid X receptor ligand restores defective signalling by the vitamin D receptor. *EMBO Rep.* **2006**, *7*, 1030–1034. [CrossRef] [PubMed]
58. Rochel, N.; Ciesielski, F.; Godet, J.; Moman, E.; Roessle, M.; Peluso-Iltis, C.; Moulin, M.; Haertlein, M.; Callow, P.; Mély, Y.; et al. Common architecture of nuclear receptor heterodimers on DNA direct repeat elements with different spacings. *Nat. Struct. Mol. Biol.* **2011**, *18*, 564–570. [CrossRef]
59. Orlov, I.; Rochel, N.; Moras, D.; Klaholz, B.P. Structure of the full human RXR/VDR nuclear receptor heterodimer complex with its DR3 target DNA. *EMBO J.* **2012**, *31*, 291–300. [CrossRef]
60. Zhang, J.; Chalmers, M.J.; Stayrook, K.R.; Burris, L.L.; Wang, Y.; Busby, S.A.; Pascal, B.D.; Garcia-Ordonez, R.D.; Bruning, J.B.; Istrate, M.A.; et al. DNA binding alters coactivator interaction surfaces of the intact VDR-RXR complex. *Nat. Struct. Mol. Biol.* **2011**, *18*, 556–563. [CrossRef]
61. Chandra, V.; Huang, P.; Hamuro, Y.; Raghuram, S.; Wang, Y.; Burris, T.P.; Rastinejad, F. Structure of the intact PPAR-gamma-RXR-nuclear receptor complex on DNA. *Nature* **2008**, *456*, 350–356. [CrossRef]
62. Chandra, V.; Huang, P.; Potluri, N.; Wu, D.; Kim, Y.; Rastinejad, F. Multidomain integration in the structure of the HNF-4α nuclear receptor complex. *Nature* **2013**, *495*, 394–398. [CrossRef]
63. Lou, X.; Toresson, G.; Benod, C.; Suh, J.H.; Philips, K.J.; Webb, P.; Gustafsson, J.A. Structure of the retinoid X receptor α-liver X receptor β (RXRα-LXRβ) heterodimer on DNA. *Nat. Struct. Mol. Biol.* **2014**, *21*, 277–281. [CrossRef]
64. Chandra, V.; Wu, D.; Li, S.; Potluri, N.; Kim, Y.; Rastinejad, F. The quaternary architecture of RARβ-RXRα heterodimer facilitates domain-domain signal transmission. *Nat. Commun.* **2017**, *8*, 868. [CrossRef] [PubMed]
65. Maletta, M.; Orlov, I.; Roblin, P.; Beck, Y.; Moras, D.; Billas, I.M.; Klaholz, B.P. The palindromic DNA-bound USP/EcR nuclear receptor adopts an asymmetric organization with allosteric domain positioning. *Nat. Commun.* **2014**, *5*, 4139. [CrossRef] [PubMed]
66. Wasmuth, E.V.; Broeck, A.V.; LaClair, J.R.; Hoover, E.A.; Lawrence, K.E.; Paknejad, N.; Pappas, K.; Matthies, D.; Wang, B.; Feng, W.; et al. Allosteric interactions prime androgen receptor dimerization and activation. *Mol. Cell.* **2022**, *82*, 2021–2031.e5. [CrossRef] [PubMed]
67. Zheng, J.; Chang, M.R.; Stites, R.E.; Wang, Y.; Bruning, J.B.; Pascal, B.D.; Novick, S.J.; Garcia-Ordonez, R.D.; Stayrook, K.R.; Chalmers, M.J.; et al. HDX reveals the conformational dynamics of DNA sequence specific VDR co-activator interactions. *Nat. Comm.* **2017**, *8*, 923. [CrossRef]
68. Osz, J.; McEwen, A.G.; Bourguet, M.; Przybilla, F.; Peluso-Iltis, C.; Poussin-Courmontagne, P.; Mély, Y.; Cianférani, S.; Jeffries, C.M.; Svergun, D.I.; et al. Structural basis for DNA recognition and allosteric control of the retinoic acid receptors RAR-RXR. *Nucleic. Acids Res.* **2020**, *48*, 9969–9985. [CrossRef]
69. Lonard, D.M.; O'malley, B.W. Nuclear receptor coregulators: Judges, juries, and executioners of cellular regulation. *Mol. Cell.* **2007**, *27*, 691–700. [CrossRef]
70. Pike, J.W.; Christakos, S. Biology and Mechanisms of Action of the Vitamin D Hormone. *Endocrinol. Metab. Clin. N. Am.* **2017**, *46*, 815–843. [CrossRef]
71. Bikle, D.D.; Oda, Y.; Tu, C.L.; Jiang, Y. Novel mechanisms for the vitamin D receptor (VDR) in the skin and in skin cancer. *J. Steroid Biochem. Mol. Biol.* **2015**, *148*, 47–51. [CrossRef]
72. Rachez, C.; Lemon, B.D.; Suldan, Z.; Bromleigh, V.; Gamble, M.; Näär, A.M.; Erdjument-Bromage, H.; Tempst, P.; Freedman, L.P. Ligand-dependent transcription activation by nuclear receptors requires the DRIP complex. *Nature* **1999**, *398*, 824–828. [CrossRef]
73. Heery, D.M.; Kalkhoven, E.; Hoare, S.; Parker, M.G. A signature motif in transcriptional co-activators mediates binding to nuclear receptors. *Nature* **1997**, *387*, 733–736. [CrossRef]
74. Nagy, L.; Kao, H.Y.; Love, J.D.; Li, C.; Banayo, E.; Gooch, J.T.; Krishna, V.; Chatterjee, K.; Evans, R.M.; Schwabe, J.W. Mechanism of corepressor binding and release from nuclear hormone receptors. *Genes Dev.* **1999**, *13*, 3209–3216. [CrossRef] [PubMed]
75. Hu, X.; Lazar, M.A. The CoRNR motif controls the recruitment of corepressors by nuclear hormone receptors. *Nature* **1999**, *402*, 93–96. [CrossRef] [PubMed]
76. Perissi, V.; Staszewski, L.M.; McInerney, E.M.; Kurokawa, R.; Krones, A.; Rose, D.W.; Lambert, M.H.; Milburn, M.V.; Glass, C.K.; Rosenfeld, M.G. Molecular determinants of nuclear receptor-corepressor interaction. *Genes Dev.* **1999**, *13*, 3198–3208. [CrossRef] [PubMed]
77. Teichert, A.; Arnold, L.A.; Otieno, S.; Oda, Y.; Augustinaite, I.; Geistlinger, T.R.; Kriwacki, R.W.; Guy, R.K.; Bikle, D.D. Quantification of the vitamin D receptor-coregulator interaction. *Biochemistry* **2009**, *48*, 1454–1461. [CrossRef]
78. Burakov, D.; Wong, C.W.; Rachez, C.; Cheskis, B.J.; Freedman, L.P. Functional interactions between the estrogen receptor and DRIP205, a subunit of the heteromeric DRIP coactivator complex. *J. Biol. Chem.* **2000**, *275*, 20928–20934. [CrossRef]

79. Whyte, W.A.; Orlando, D.A.; Hnisz, D.; Abraham, B.J.; Lin, C.Y.; Kagey, M.H.; Rahl, P.B.; Lee, T.I.; Young, R.A. Master transcription factors and mediator establish super-enhancers at key cell identity genes. *Cell* **2013**, *153*, 307–319. [CrossRef]
80. Jin, F.; Claessens, F.; Fondell, J.D. Regulation of androgen receptor-dependent transcription by coactivator MED1 is mediated through a newly discovered noncanonical binding motif. *J. Biol. Chem.* **2012**, *287*, 858–870. [CrossRef]
81. Nolte, R.T.; Wisely, G.B.; Westin, S.; Cobb, J.E.; Lambert, M.H.; Kurokawa, R.; Rosenfeld, M.G.; Willson, T.M.; Glass, C.K.; Milburn, M.V. Ligand binding and co-activator assembly of the peroxisome proliferator-activated receptor-gamma. *Nature* **1998**, *395*, 137–143. [CrossRef]
82. Shiau, A.K.; Barstad, D.; Loria, P.M.; Cheng, L.; Kushner, P.J.; Agard, D.A.; Greene, G.L. The structural basis of estrogen receptor/coactivator recognition and the antagonism of this interaction by tamoxifen. *Cell* **1998**, *95*, 927–937. [CrossRef]
83. Darimont, B.D.; Wagner, R.L.; Apriletti, J.W.; Stallcup, M.R.; Kushner, P.J.; Baxter, J.D.; Fletterick, R.J.; Yamamoto, K.R. Structure and specificity of nuclear receptor-coactivator interactions. *Genes Dev.* **1998**, *12*, 3343–3356. [CrossRef]
84. Huet, T.; Laverny, G.; Ciesielski, F.; Molnár, F.; Ramamoorthy, T.G.; Belorusova, A.Y.; Antony, P.; Potier, N.; Metzger, D.; Moras, D.; et al. A vitamin D receptor selectively activated by gemini analogs reveals ligand dependent and independent effects. *Cell. Rep.* **2015**, *10*, 516–526. [CrossRef] [PubMed]
85. Egawa, D.; Itoh, T.; Kato, A.; Kataoka, S.; Anami, Y.; Yamamoto, K. SRC2-3 binds to vitamin D receptor with high sensitivity and strong affinity. *Bioorg. Med. Chem.* **2017**, *25*, 568–574. [CrossRef] [PubMed]
86. Yi, P.; Wang, Z.; Feng, Q.; Chou, C.K.; Pintilie, G.D.; Shen, H.; Foulds, C.E.; Fan, G.; Serysheva, I.; Ludtke, S.J.; et al. Structural and Functional Impacts of ER Coactivator Sequential Recruitment. *Mol. Cell.* **2017**, *67*, 733–743.e4. [CrossRef] [PubMed]
87. Yu, X.; Yi, P.; Hamilton, R.A.; Shen, H.; Chen, M.; Foulds, C.E.; Mancini, M.A.; Ludtke, S.J.; Wang, Z.; O'Malley, B.W. Structural Insights of Transcriptionally Active, Full-Length Androgen Receptor Coactivator Complexes. *Mol. Cell.* **2020**, *79*, 812–823.e4. [CrossRef]
88. Rovito, D.; Belorusova, A.Y.; Chalhoub, S.; Rerra, A.I.; Guiot, E.; Molin, A.; Linglart, A.; Rochel, N.; Laverny, G.; Metzger, D. Cytosolic sequestration of the vitamin D receptor as a therapeutic option for vitamin D-induced hypercalcemia. *Nat. Commun.* **2020**, *11*, 6249. [CrossRef] [PubMed]
89. Belorusova, A.Y.; Bourguet, M.; Hessmann, S.; Chalhoub, S.; Kieffer, B.; Cianférani, S.; Rochel, N. Molecular determinants of MED1 interaction with the DNA bound VDR-RXR heterodimer. *Nucleic. Acids Res.* **2020**, *48*, 11199–11213. [CrossRef]

MDPI

Review

The Centennial Collection of VDR Ligands: Metabolites, Analogs, Hybrids and Non-Secosteroidal Ligands

Miguel A. Maestro [1,*] and Samuel Seoane [2]

[1] Department of Chemistry-CICA, University of A Coruña, Campus da Zapateira, s/n, 15008 A Coruña, Spain
[2] Department of Physiology-CIMUS, University of Santiago, Campus Vida, 15005 Santiago, Spain
* Correspondence: miguel.maestro@udc.es

Abstract: Since the discovery of vitamin D a century ago, a great number of metabolites, analogs, hybrids and nonsteroidal VDR ligands have been developed. An enormous effort has been made to synthesize compounds which present beneficial properties while attaining lower calcium serum levels than calcitriol. This structural review covers VDR ligands published to date.

Keywords: metabolites; analogs; hybrids and VDR nonsecosteroidal ligands

Citation: Maestro, M.A.; Seoane, S. The Centennial Collection of VDR Ligands: Metabolites, Analogs, Hybrids and Non-Secosteroidal Ligands. *Nutrients* **2022**, *14*, 4927. https://doi.org/10.3390/nu14224927

Academic Editor: Federica I. Wolf

Received: 3 October 2022
Accepted: 16 November 2022
Published: 21 November 2022

1. Introduction

Since the chemical structure of vitamin D_3 (**1**, Figure 1 [1–36], cholecalciferol) was established in 1932, successive studies have shown it to be essential in physiological processes. Two hydroxylations of **1** are necessary before attaining its most biologically active form. The first is a 25-hydroxylation, which occurs mainly in the liver and produces the most abundant circulating metabolite, 25-hydroxyvitamin D_3 (**11**, Figure 1, 25-hydroxycholecalciferol, calcidiol, $25OHD_3$) [12]. Subsequently, a second hydroxylation at the 1α position generates the vitamin D hormone, 1α,25-dihydroxyvitamin D_3 (**13**, Figure 1, 1α,25-dihydroxycholecalciferol, calcitriol, $1{,}25(OH)_2D_3$) [14]. This is a pleiotropic hormone that exerts genomic actions by binding to its specific receptor (the vitamin D receptor, VDR), which is present on target cells and found in more than 200 different tissues.

The biological role of $1{,}25(OH)_2D_3$ has been related to calcium and phosphorus homeostasis. However, the effects of vitamin D are not limited to mineral homeostasis, skeletal health maintenance, or immune modulation. In addition, this hormone also has fundamental effects on cellular proliferation and differentiation, regulating genes involved in the cell cycle and apoptosis both in normal and tumor cells. These properties and its wide distribution have led to the study of its effects on various pathologies, such as osteoporosis and cancer, thus arousing interest in the field of health and the pharmaceutical industry. Unfortunately, the therapeutic use of $1{,}25(OH)_2D_3$ also leads to an increase in the concentration of calcium in blood (hypercalcemia), which can cause significant side effects. Therefore, numerous attempts have been made to synthesize noncalcemic analogs of $1{,}25(OH)_2D_3$ for use in health treatment.

In recent decades, structure–function relationships (SARs) have been determined to support the chemical modifications of the secosteroid structure of $1{,}25(OH)_2D_3$. The novel structures' goal is to reduce their calcemic activity in comparison with calcitriol while exerting their interesting biological properties. A huge synthesis effort has been carried out, yielding interesting chemical reviews in this regard [2]. The current review updates the scientific information on the structural library of VDR ligands and incorporates nonsteroidal VDR ligands.

2. Materials and Methods

All compounds contained in this review were collected from published papers and patents. Most of the materials were freely accessible via the Internet, and paper copies were available in other cases. After careful reading, relevant structures were drawn using CHEMDRAW software [3]. No database was generated. A structural analysis of this collection may require future elaboration of a database.

3. Results

We found 1778 VDR compounds, which are displayed chronologically in 31 figures. All of these compounds are ligands that specifically bind to their VDR receptor. This binding allows the interaction of the $1{,}25(OH)_2D_3$-VDR complex with target genes in the cell nucleus, modulating their expression and mediating a biological response. The following color scheme was used in the figures: dark blue corresponds to marketed compounds (Figures 1, 3, 4, 8 and 9), light violet to outstanding compounds with interesting properties (Figures 2, 4–10, 12–15, 17–21 and 23–31), and dark green to non-secosteroidal VDR ligands (Figures 9, 11, 12, 15 and 20–27).

Vitamin D is closely associated with calcium and phosphorus homeostasis. No scientific rational has yet been found for the calcemic properties of a compound in comparison with calcitriol. Therefore, structure–function relationships (SARs) were carried out in order to validate the key modifications in the structure of $1{,}25(OH)_2D_3$ that may alter biological and calcemic properties. After more than 50 years of study, some hints have been obtained. For example, it is known that C-19 methylene deletion yields low calcemic analogs; it is also known that deletion/substitution of the steroidal cycles de-A ring, de-C ring, and/or de-D ring may yield low calcemic analogs. Lowering the calcemic side effects of the vitamin D analogs is important; however, we must not lose sight of other modifications that may increase the antiproliferative and prodifferentiation activity (side-chain modification with extra double and or triple bonds) as well as increase the metabolic stability (fluorine atom incorporation). In summary, the following main structural topics are covered in the current review:

- C-21 Methyl epimerization;
- C-19 Methylene deletion;
- Incorporation of fluorine atoms;
- Deletion/substitution of steroidal cycles: de-A ring, de-C ring, and/or de-D ring;
- C-2 Functionalization;
- C-3 Epimerization;
- Side-chain modification with extra double and/or triple bonds, heteroatoms, and/or branched hydrocarbons.

What is novel in this collection is the incorporation of non-secosteroidal VDR ligands (dark green). In 1999, Boehm [4] hypothesized that "non-secosteroidal VDR ligands might display different profiles of activity and metabolism than do secosteroidal $1{,}25(OH)_2D_3$, analogs, including less calcemic properties, which might render them attractive as both topical and oral pharmaceuticals for treating a variety of diseases. This hypothesis was based in part on the success that nonsteroidal androgen receptor (AR) and estrogen receptor (ER) modulators have had as drugs. Nonsteroidal compounds have been synthesized that modulate the activity of these receptors and show enhanced tissue selectivity in comparison to the steroids".

Figure 1 (1931–1978) [1–36]. Vitamin D_3 (**1**, cholecalciferol) [1] was discovered in 1922, but it was not chemically characterized until 1931. Dihydrotachysterol$_2$ (**5**) [10] was introduced in 1934, and it is still on the market as an antitetanic agent AT-10. In 1968, the most abundant metabolite of vitamin D_3 was discovered as 25-hydroxyvitamin D_3 (**11**, 25-hydroxycholecalciferol) [18], and in 1971, 1α,25-dihydroxyvitamin D_3 **13**, 1α,25-dihydroxycholecalciferol, calcitriol, $1\alpha{,}25(OH)_2D_3$ [21], the vitamin D_3 hormone, was identified. Later, 1α-hydroxyvitamin D_3 (**21**, Alfacalcidiol) [25], a synthetic analog, was marketed for the treatment of secondary hyperparathyroidism (2HPT), renal failure, and osteoporosis.

Figure 2 (1978–1982) [37–52]. 25-Hydroxyvitamin D_3 26(23)-lactones (**58–61**) were discovered in 1980 [50–52], and they behave as antagonists of gene transcription induced by VDR. They were the first compounds discovered to have antagonist properties.

Figure 3 (1982–1987) [53–78]. 26,26,26,27,27,27-Hexafluoro-1α,25-dihydroxyvitamin D_3 (**70**, Falecalcitriol) [60] is used in the treatment of 2HPT and osteoporosis. 1α,25-Dihydroxy-22-oxavitamin D_3 (**100**, Maxacalcitol) [76] is used in 2HPT and psoriasis.

Figure 4 (1987–1991) [78–94]. **111** (Calcipotriol, MC903) [79] is marketed as a treatment with exceptional clinical response in psoriasis. 1α,25-Dihydroxy-22(23)-didehydrovitamin D_3 (**116**) [83] has shown potent antiproliferative activity. 2β-(Hydroxypropoxy)-1α,25-dihydroxyvitamin D_3 (**131,** ED-71) [85] is used in osteoporosis treatment.

Figure 5 (1991–1992) [95–117]. Compound **186** [107] is an important analog functionalized at C-11 that may allow the synthesis of haptens, without disturbing the VDR ligand anchoring groups (1α-OH, 3β-OH and 25-OH).

Figure 6 (1993–1994) [118–136]. Compounds **225** [93] and **208** [108] were independently developed by different research groups and are important analogs functionalized at C-18 and C-11, respectively. They may allow the synthesis of haptens without disturbing the VDR ligand anchoring groups.

Figure 7 (1994–1997) [136–148]. Compounds **308** and **309** [147] present an interesting property by exhibiting only nongenomic rapid effects at physiological concentrations. Moreover, 1α-hydroxyl group addition (**309**) does not alter the sensitivity of nongenomic effects of **308**.

Figure 8 (1997–1999) [149–158]. 1α-Hydroxyvitamin D_2 (**325**, Doxercalciferol) [151] is marketed as a 2HPT treatment. (22*E*,24*E*)-Diene-24,26,27-trishomo-19-nor-1α,25-dihydroxyvitamin D_3 (**348**, Ro 25-8584) [152] represents an outstanding compound inhibiting the proliferation in myeloid leukemia cell lines. When 2-methylene-19-nor-1α,25-dihydroxyvitamin D_3 (**349**, 2MD) [156] is given as oral therapy, it is at least 100 times more potent than 1α,25$(OH)_2D_3$ in stimulating bone mass increase. A randomized clinical trial showed that **349** increased bone turnover but not BMD (bone mass density) in postmenopausal woman with osteopenia.

Figure 9 (1999) [158–168]. 24*R*,25-Dihydroxyvitamin D_3 (**388**, Tacalcitol) [160] is prescribed for psoriasis. 24,26,27-Trishomo-1α,25-dihydroxyvitamin D_3 (**406**, Seocalcitol, EB 1089) [163] acts as a powerful antiproliferative used in breast, colon, or pancreas tumor models.

Figure 10 (2000–2001) [169–182]. 1α-Hydroxy-26(27)-dehydro-25-(butylcarboxylate)-vitamin D_3 (**433**, ZK159222) and 1α-hydroxy-26(27)-dehydro-25-(ethylpropenoate)-vitamin D_3 (**434**, ZK168281) [170] have been identified as VDR antagonists, though **434** is more potent than **433**. Both compounds selectively stabilize an antagonist conformation of the VDR-LBD (ligand-binding domain). 1α,25-Dihydroxy-21-(3-hydroxy-3-methylbutyl)-vitamin D_3 (**435**, Gemini) [171] has emerged as the lead compound with superior gene transcription activity and tumor-cell-line inhibition.

Figure 11 (2001–2002) [183–196]. 1α,25-$(OH)_2$-16-ene-20-epi-23-yne-3-epi-D_3 (**493**), 1α,25$(OH)_2$-16-ene-23-yne-hexafluoro-3-epi-D_3 (**494**), and 1α,25$(OH)_2$-16-ene-3-epi-D_3 (**495**) are potent inducers of apoptosis of HL-60 cells. Their 3-natural (3β-OH) analogs have been shown to be potent modulators of HL-60 cell growth and differentiation [184]. This is the first report to demonstrate that the epimerization of the hydroxyl group at C-3 of the A-ring of 1α,25$(OH)_2D_3$ plays an important modulation role for HL-60 cell differentiation and apoptosis. 2,2-Difluoro-1α,25-dihydroxyvitamin D_3 (**507**) [185] is similar to 1,25$(OH)_2D_3$ in terms of in vitro antiproliferative activity, but it is different in terms of transcriptional activity. In addition, **507** is 2–3 times more transcriptionally active than calcitriol, with similar in vivo calcemic activity. 2,2-Dimethyl-1α,25-dihydroxy-19-norvitamin D_3 (**509**) [186] is 7.5 times less transcriptionally active than calcitriol and considerably less calcemic. Moreover, **509** strongly suppresses parathyroid hormone (PTH) secretion.

Figure 12 (2002) [197–204]. Seco-C-9,11-bisnor-17-methyl-26,26,26,27,27,27-hexafluoro-20-epi-1α,25-dihydroxyvitamin D_3 (**533**, WY1112) [197] and seco-C-9,11,21-trisnor-17-methyl-23(24)-didehydro-26,26,26,27,27,27-hexafluoro-1α,25-dihydroxyvitamin D_3 (**559**, CD578) [198] display high differentiation ratios between antiproliferative and calcemic affects.

26,27-Bishomo-1α-fluoro,25-hydroxy-23-en-vitamin D_3 (**582**, Ro-26-9228) [203] is used for treatment of osteoporosis.

Figure 13 (2003–2004) [205–218]. Dienyne **646** [215] represents the first locked side-chain analog of calcitriol with remarkable VDR transcriptional activity. Lactone **657** [217] showed one order of magnitude higher antagonist activity than lactone **66** (Figure 2).

Figure 14 (2004–2006) [218–223]. Further development in double side-chain vitamin D analogs, the Gemini series, made it possible to assess the steric VDR requirements of drug candidates. Compounds **684–695** **[220]** present two different side chains at C-20 that improve their toxicity profiles and pharmacokinetic drug performance.

Figure 15 (2006–2007) [224–240]. C-20 cyclopropyl vitamin D_3 analog **755** [233] showed high MLR (mixed lymphocyte reaction) activity for the suppression of interferon-γ release with no calcemic activity. Immunomodulatory activity was measured by suppression of interferon-γ release in mixed lymphocyte reaction cells. The inhibition of clonal proliferation was evaluated in the leukemia HL-60, breast cancer MCF-7, prostate PC-3, and LNCaP cell lines. Significant separation of the immunomodulatory activity from hypercalcemic effects (MTD, maximum tolerated dose) was observed. Compound **747** was 2900 times more active and 100 times less hypercalcemic than 1α,25$(OH)_2D_3$, while **755** was 29 times more active and 100 less hypercalcemic. In the breast cancer MCF-7 cell line, compounds **753**, **754**, **755**, and **757** were ten thousand times more active but equally or less hypercalcemic than 1α,25$(OH)_2D_3$. Metabolism of 16-ene-20-cyclopropyl compounds is arrested at the 24-keto stage, which explains the increased biological activity of the 16-ene variants.

Figure 16 (2006–2008) [241–253]. Intensive research activity was carried out on the leading structures with outstanding biological properties, i.e., Gemini compounds **799–803** [246,247]. These studies focused on the structural modifications of Gemini that influenced the differentiation-inducing, antiproliferative, and transcriptional activity of the compounds in human leukemia cells. The cyclopropyl modification at the pro-*R* side chain decreased the activity of the compound compared to 1α,25$(OH)_2D_3$, and further A-ring modifications did not restore this activity. Cyclopropyl modification at the pro-*S* side chain of Gemini increased the VDR-induced transcriptional activity. In addition, privileged VDR antagonists lactones **804–832** and **833–864** [243,244] were studied. The antagonistic activity was markedly affected by the structure of the lactone ring, including length of the alkyl chain and the stereochemistries on the C23 and C24 positions. The VDR binding affinity of the (23*S*,24*S*)-24-alkylated vitamin D_3 lactones increased 2.3–3.7-fold as compared to the unsubstituted lactones **64–67** (Figure 2). The antagonistic activity of (23*S*,24*S*)-isomers were enhanced to be 2.2-,3.5-, 1.8-, and 1.7-fold higher compared to the unsubstituted lactones **64–67** (Figure 2).

Figure 17 (2008–2009) [254–264]. 2-Methylene-19-nor-(20*S*)-1α-hydroxy-bishomopregnacalciferol **942** [20(*S*)-2MbisP] [263] were able to suppress PTH at levels that did not stimulate bone resorption, intestinal calcium, or phosphate absorption and may have potential for use in the treatment of 2HPT in chronic kidney disease.

Figure 18 (2009–2010) [265–270]. Hybrid compounds **1020** (26,27-bis-nor-25-bishomo-19-nor-25′-oxo-25″-methylcarboxamide-1α-hydroxyvitamin D_3) and **1022** (26,27-bis-nor-25-homo-19-nor-25′-(2aminophenyl)-carboxamide-1α-hydroxyvitamin D_3) [270] showed antiproliferative activity against AT84 carcinoma cells; neither of them induced hypercalcemia even at concentrations 100-fold higher than those tolerated for 1,25D. This demonstrates that it is possible to create a wide range of bifunctional molecules that possess VDR agonism and HDACi (histone deacetylases inhibitor) activity. Structural latitude is significant with a wide variety of ZBGs (zinc-binding group) amenable to incorporation into the side chain of vitamin D-like secosteroids. Importantly, several of these molecules function as antiproliferative agents against AT84 cells in vitro, while possessing minimal hypercalcemic activity in vivo.

Figure 19 (2009–2010) [271–283]. Intensive research activity was carried out on Gemini compounds **1053–1069** [273]. Calcitriol was implicated in many cellular functions including cell growth and differentiation. It was shown that Gemini compounds were active in gene transcription induction with enhanced antitumor activity. Fine tuning of their structurally derived biological properties would be required for therapeutic use.

Figure 20 (2010–2012) [284–298]. 25-Diethylphosphite-1α-hydroxy-23(24)-didehydro-vitamin D_3 **1131** [290] was tested for antiproliferative effects on several human and murine tumor cell lines: human squamous cell carcinoma HN12, human glioma T98G, and Kaposi sarcoma SVEC vGPCR cell lines. Furthermore, in human glioma T98G and human squamous cell carcinoma HN12 cell lines, the antiproliferative effects exerted by compound **1131** were greater than those elicited by 1α,25$(OH)_2D_3$. Visual observation of internal animal organs such as liver, duodenum, lungs, and kidneys showed no macroscopic morphological alterations after treatment with this compound. This compound appears to be well tolerated even at high doses. Altogether, these results suggest that compound **1131** exerts considerable antiproliferative activity at nonhypercalcemic dosages and may have therapeutic potential for the treatment of various hyperproliferative disorders. Non-secosteroidal VDR ligand (4-{1-ethyl-1-[4-(2-hydroxy-3,3-dimethyl-butoxy)-3-methyl-phenyl]-propyl}-2-methyl-phenoxy)-hydroxyacetamide **1173** [295] was confirmed to significantly prevent bone loss after daily treatment without inducing hypercalcemia. These types of compound are potent inhibitors of the Hh (Hedgehog) signaling pathway. Studies show that, contrary to secosteroidal hybrids, the optimal location for incorporating the highly hydrophilic hydroxamic acid corresponds to the portion of the molecules that serve as secosteroidal A-ring mimetics. The best hybrid, **1173**, is a full VDR agonist, as assessed by several criteria, and an efficacious antiproliferative agent against both 1,25D-sensitive (SCC25, AT84) and 1α,25$(OH)_2D_3$-resistant (SCC4) squamous carcinoma cell lines. Importantly, the activity in 1α,25$(OH)_2D_3$-resistant SCC4 cells required both the VDR agonism and HDACi activity of **1173**. This study revealed the remarkable flexibility in the conversion of calcitriol analogs into fully integrated bifunctional molecules, suggesting that it may be possible to extend fully integrated bifunctionalization to other pharmacophores.

Figure 21 (2012–2013) [298–313]. 24*S*-Methyl-21-epi-2-methylene-22-oxa-1α,25-dihydroxyvitamin D_3 (**1191**, VS-105) [306] bound to VDR is highly inductive of functional responses in vitro and effectively suppresses PTH in a dose range that does not affect serum calcium in 5/6 NX uremic rats. [6-(4-{1-Ethyl-1-[4-((*E*)-3-ethyl-3-hydroxy-1-pentenyl)-3-methylphenyl]propyl}-2-methylphenyl)pyridin-3-yl]acetic acid (**1218**) [308] showed excellent ability to prevent BMD loss in mature rats in an osteoporosis model, without severe hypercalcemia and with good PK profiling.

Figure 22 (2013–2014) [313–316]. Compounds **1247**–**1301** (non-secosteroidal VDR ligands) [315] were analyzed and presented better therapeutic efficacy when compared to 1α,25$(OH)_2D_3$ in experimental models of cancer and osteoporosis with less induction of hypercalcemia, a major potential adverse effect in the clinical application of VDR ligands. Compounds **1302–1313** [316] were analyzed for their binding affinity and inhibitory activity against CYP24A1 (24-hydroxylase; this mitochondrial protein initiates the degradation of 1α,25$(OH)_2D_3$ by hydroxylation of the side chain), and the imidazole styrylbenzamides **1305–1309** were identified as potent inhibitors of CYP24A1, with similar or greater CYP27B1 (1α-hydroxylase; the protein encoded by this gene it hydroxylates $25OHD_3$ at the 1α-position, producing 1α,25$(OH)_2D_3$) selectivity than standard ketoconazole. Further evaluation of the 3,5-dimethoxy (**1308**) and 3,4,5-trimethoxy derivatives (**1309**) in chronic lymphocytic leukemia cells revealed that cotreatment of 1α,25-dihydroxyvitamin D_3 and inhibitor upregulated GADD45α (growth arrest and DNA damage 45 gen) and CDKN1A (cyclin-dependent kinase inhibitor 1A gen).

Figure 23 (2014) [317–321]. Intensive research activity was carried out on Gemini compounds **1338–1364** [320].

Figure 24 (2014–2015) [322–336]. 1α,20*S*,24*R*-Trihydroxyvitamin D_3 (**1410**) [332] showed a higher degree of activation, anti-inflammatory activity, and antiproliferative activity than vitamin D_3 receptor.

Figure 25 (2015–2017) [337–351]. 1α,25-Dihydroxy-21-(3-hydroxy-3-methyl-1-methylene-butyl)vitamin D_3 (**1428**, UV1) [337] presented potent antitumoral effects over a wide panel of tumor cell lines without inducing hypercalcemia or toxicity in vivo. The first vitamin D analog carrying an *o*-carborane in the side chain **1436** [340] showed that the substitution of hydroxyl group at C-25 by this apolar bulky group was possible. VDR binding was half of calcitriol's, the transcriptional activity was similar, and the calcemic induction was significantly lower. **1436** is an outstanding B-carrier containing 10 boron atoms, which notably bind to VDR, a nuclear receptor. This suggests that **1436** may be interesting as a BNCT (boron neutron capture therapy) drug.

Figure 26 (2017–2018) [351–358]. 1,1′-([4-(3-[4-(3-Hydroxypropoxy)-3-methylphenyl] pentan-3-yl)-1,2-phenylene]bis(oxy))bis(3,3-dimethylbutan-2-ol) (**1503**) [358] displayed efficient inhibitory activity against collagen deposition and fibrotic gene expression in chronic pancreatitis. It also showed physicochemical and pharmacokinetic properties with antitumor activity, highlighting its potential therapeutic applications in cancer treatment.

Figure 27 (2018) [359–364]. (1*R*,3*S*,*Z*)-5-{(*E*)-3-[3-(6-Hydroxy-6-methylheptyl)phenyl] pent-2-en-1-ylidene}-4-methylenecyclohexane-1,3-diol (**1573**) [359] exhibited significant tumor growth inhibition and increased survival in SCID mouse models implanted with MDA-MB-231 breast tumor cells. Des-C-ring aromatic D-ring analog **1587** [363] showed remarkable lack of calcemic activity together with its significant antiproliferative and transcriptional activities in breast cancer cell lines, suggesting the therapeutical potential of **1587** for the treatment of breast tumors.

Figure 28 (2018–2019) [365–378]. 21-nor-17(*S*)-Methyl-20(22),23(24)-didehydro-26,26, 26,27,27,27-hexafluoro-1α,25-dihydroxyvitamin D_3 (**1600**) [368] bound strongly to VDR ligand binding domain and induced VDR transcriptional activity. Hybrid **1619** [371] was found to be a potent inhibitor of Hh (Hedgehog) signaling pathway.

Figure 29 (2019–2020) [379–383]. It is known that 25(OH)D3, down-regulates SREBP (sterol regulatory element-binding protein) independently of VDR. A screening of over 250 vitamin D congeners was carried out for their ability to inhibit the activity of an SREBP-responsive luciferase reporter. This is a VDR-responsive reporter assay. A comparison of the relative activity of the six compounds revealed **1639 [379]** as the VDR-selective activator.

Figure 30 (2020–2022) [384–389]. Des-C-ring aromatic D-ring analogs **1712** and **1713** [373] showed a remarkable lack of calcemic activity together with significant antiproliferative and transcriptional properties in breast cancer cell lines, suggesting a therapeutical potential for **1712** and **1713** in breast tumor treatment.

Figure 31 (2021–2022) [390–392]. KK-052 (**1746**) [391], was found to be the first vitamin D-based SREBP (sterol regulatory element-binding proteins) inhibitor that mitigates hepatic lipid accumulation without calcemic action in mice. KK-052 maintained the ability of 25-hydroxyvitamin D_3 to induce the degradation of SREBP but lacked VDR-mediated activity. KK-052 serves as a valuable compound for interrogating SREBP/SCAP in vivo and may represent an unprecedented translational opportunity for synthetic vitamin D analogs.

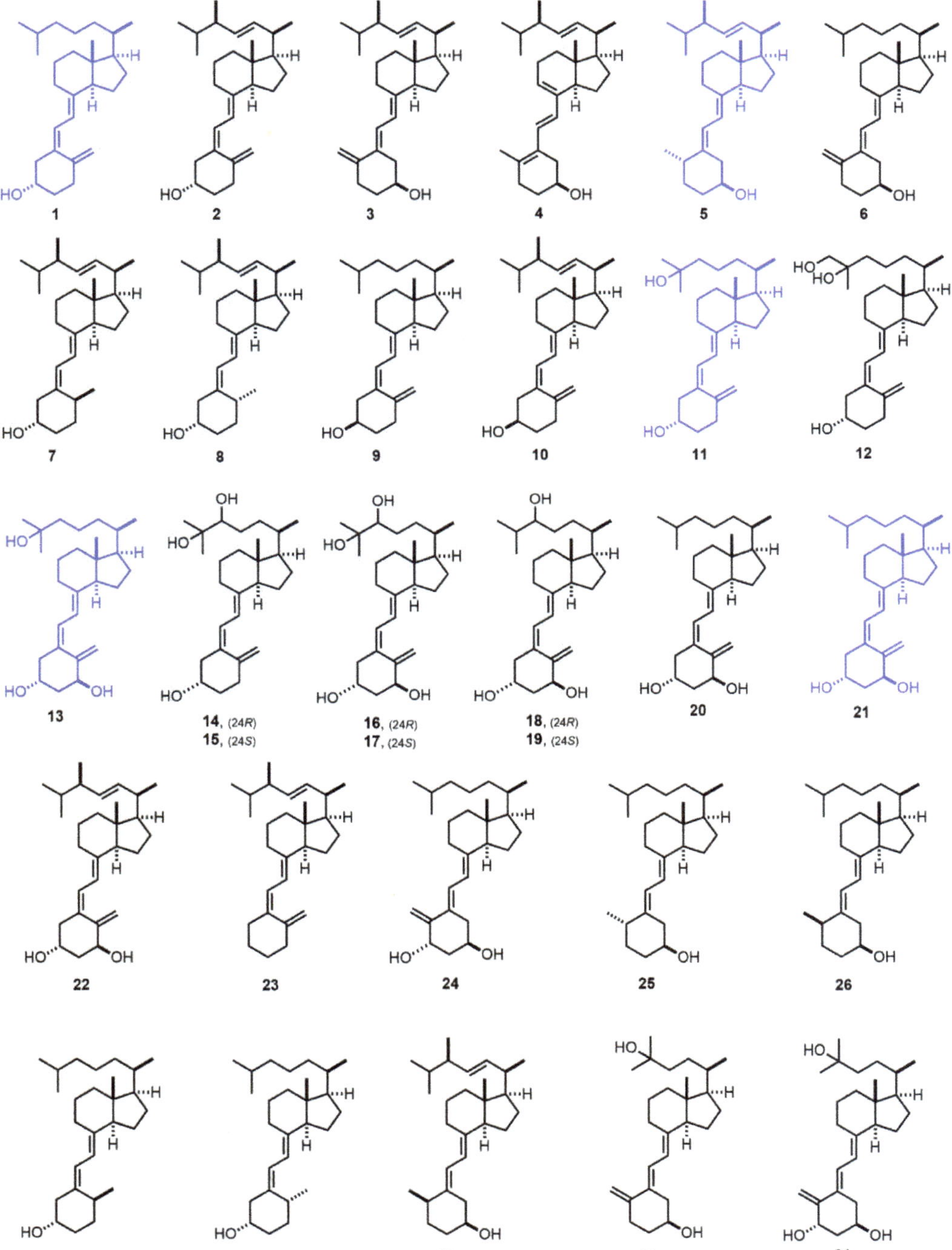

Figure 1. (1931–1978) [1–36].

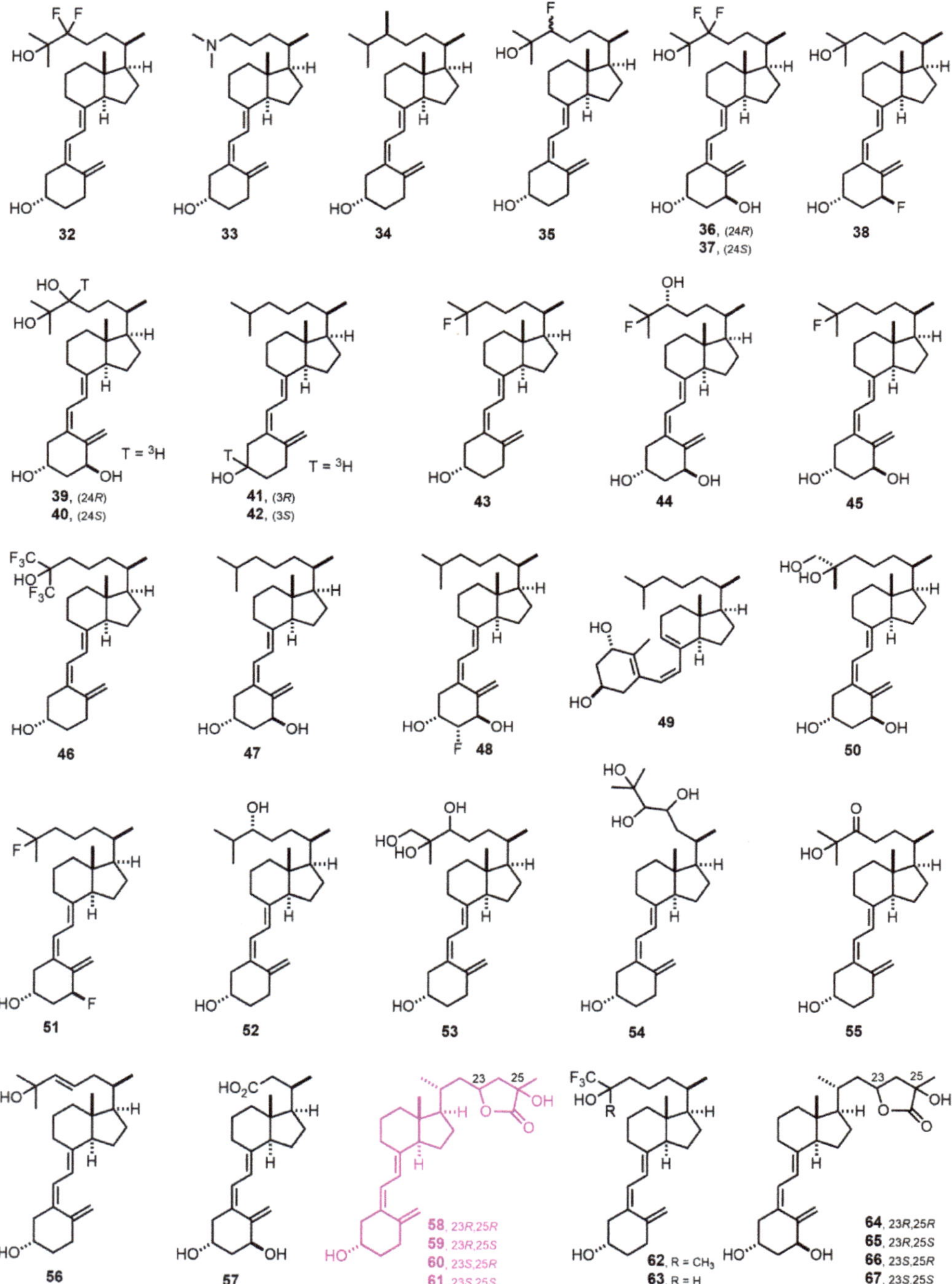

Figure 2. (1978–1982) [37–52].

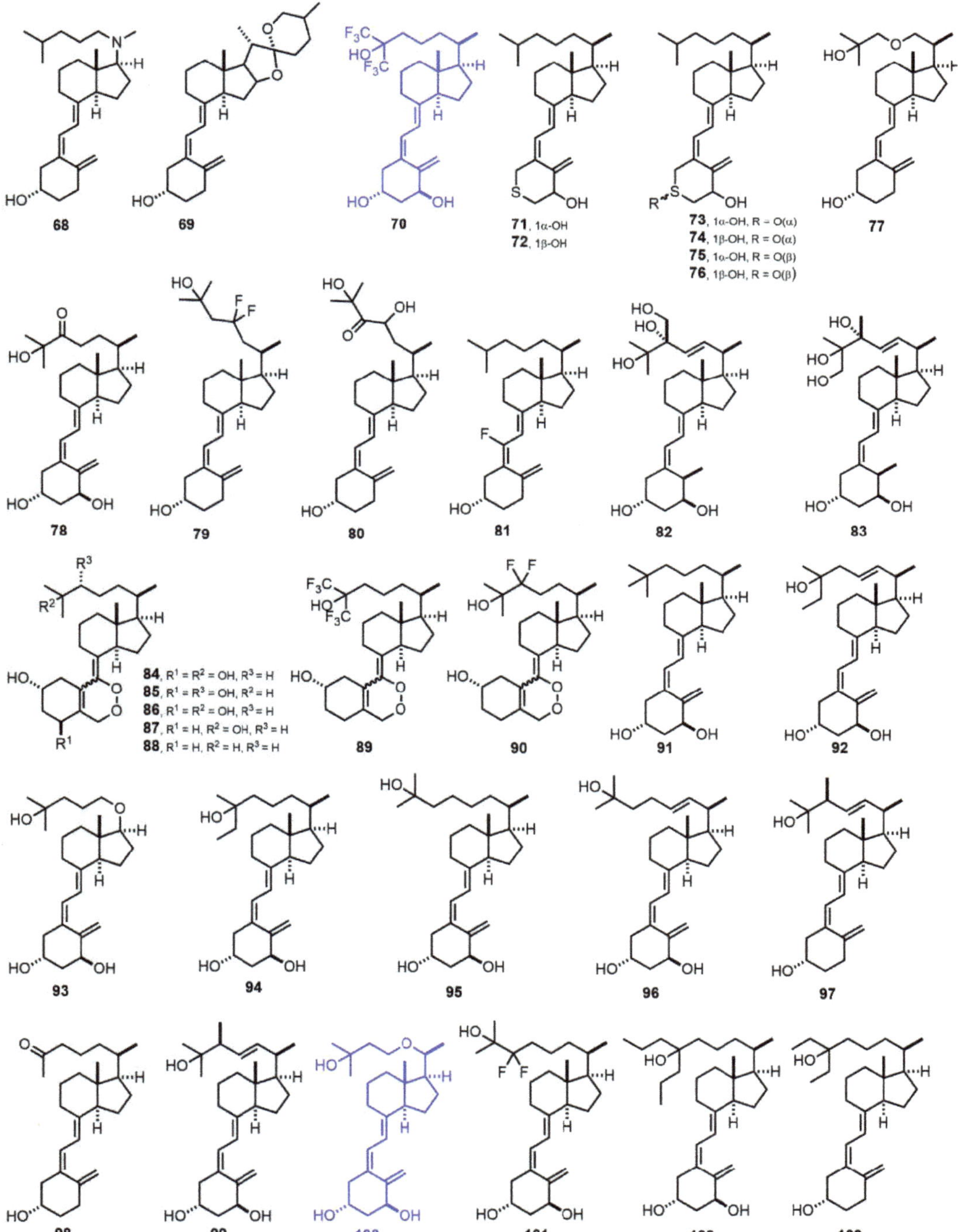

Figure 3. (1982–1987) [53–78].

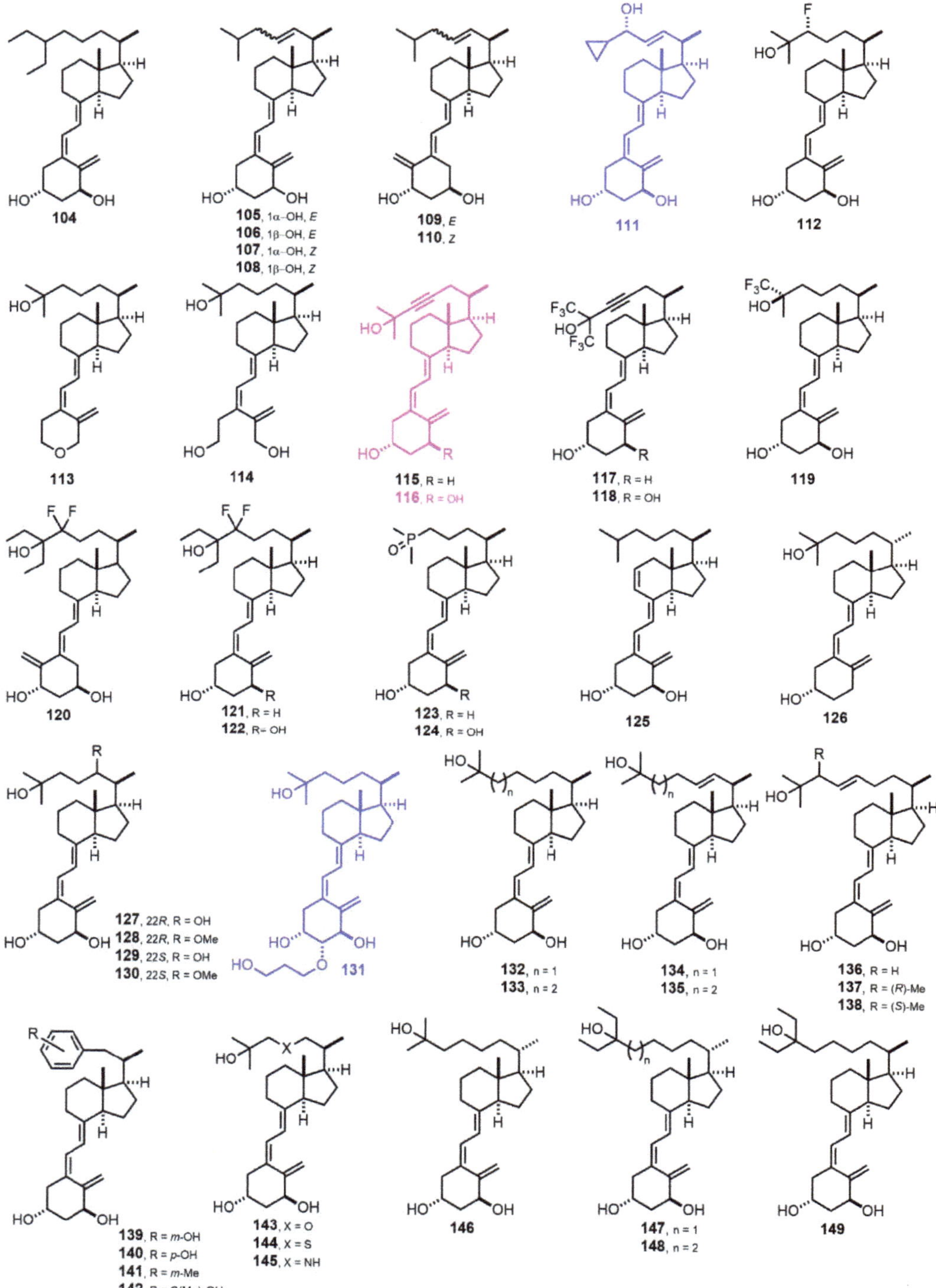

Figure 4. (1987–1991) [78–94].

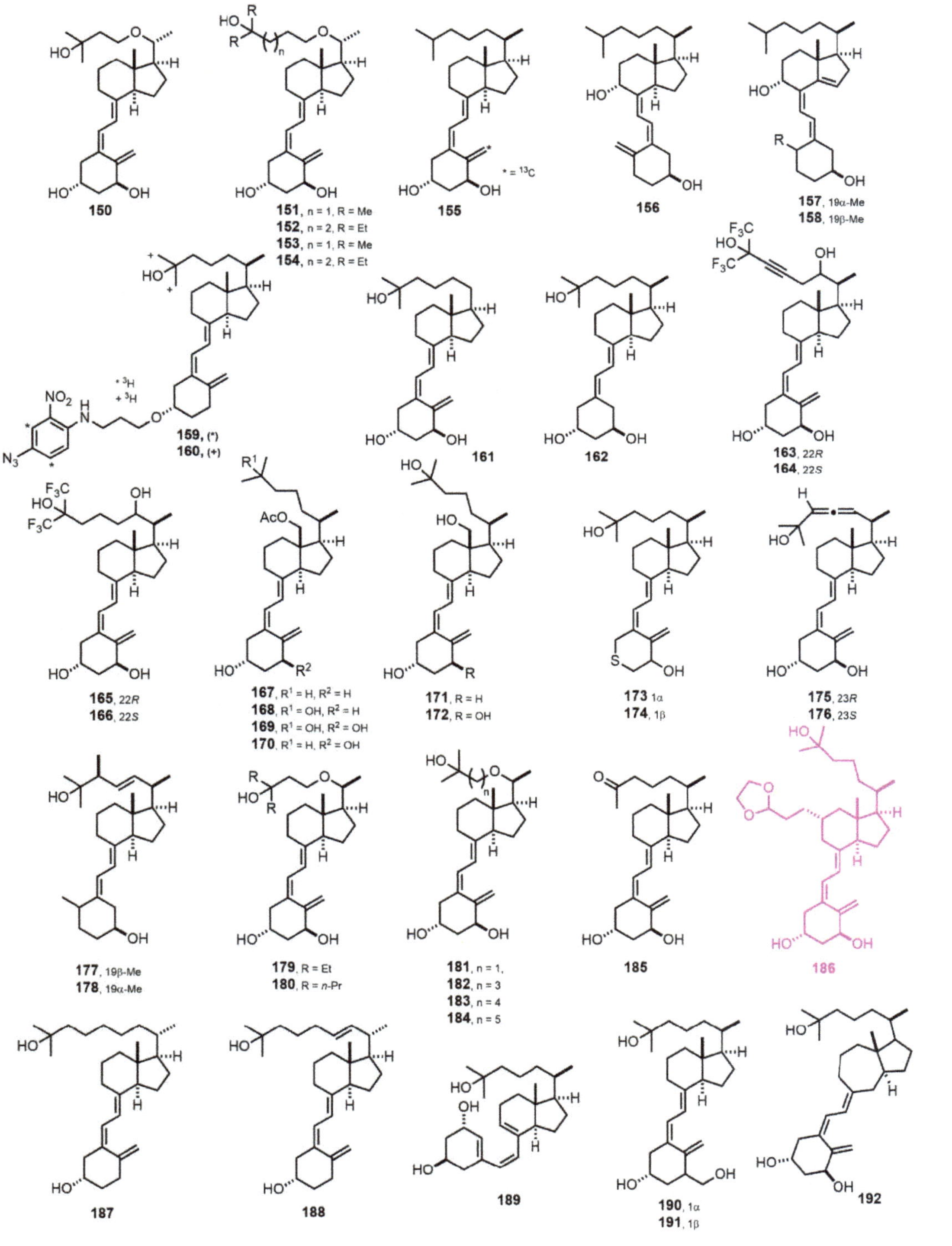

Figure 5. (1991–1992) [95–117].

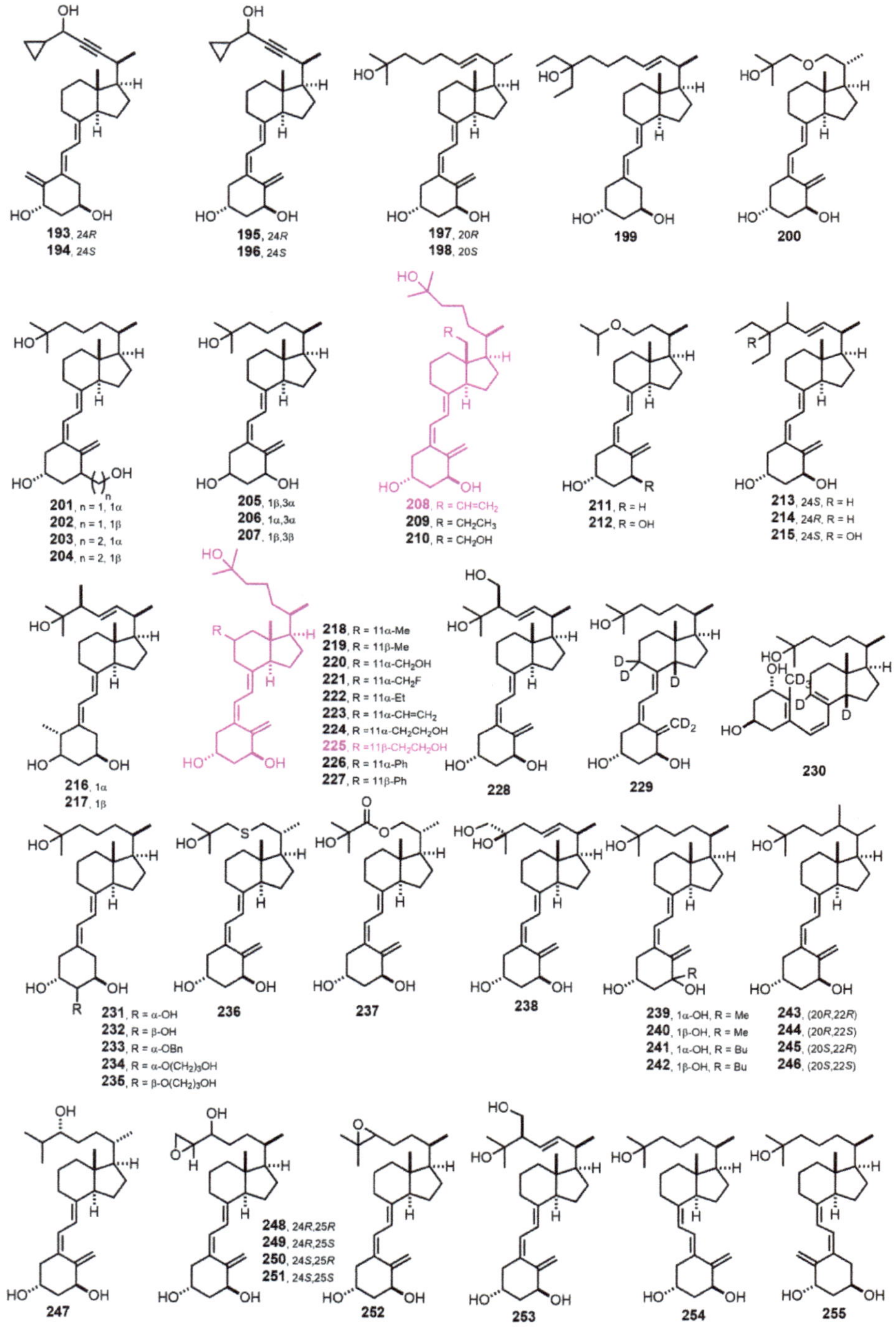

193, 24*R*
194, 24*S*
195, 24*R*
196, 24*S*
197, 20*R*
198, 20*S*
199
200

201, n = 1, 1α
202, n = 1, 1β
203, n = 2, 1α
204, n = 2, 1β
205, 1β,3α
206, 1α,3α
207, 1β,3β
208, R = CH=CH$_2$
209, R = CH$_2$CH$_3$
210, R = CH$_2$OH
211, R = H
212, R = OH
213, 24*S*, R = H
214, 24*R*, R = H
215, 24*S*, R = OH

216, 1α
217, 1β
218, R = 11α-Me
219, R = 11β-Me
220, R = 11α-CH$_2$OH
221, R = 11α-CH$_2$F
222, R = 11α-Et
223, R = 11α-CH=CH$_2$
224, R = 11α-CH$_2$CH$_2$OH
225, R = 11β-CH$_2$CH$_2$OH
226, R = 11α-Ph
227, R = 11β-Ph
228
229
230

231, R = α-OH
232, R = β-OH
233, R = α-OBn
234, R = α-O(CH$_2$)$_3$OH
235, R = β-O(CH$_2$)$_3$OH
236
237
238
239, 1α-OH, R = Me
240, 1β-OH, R = Me
241, 1α-OH, R = Bu
242, 1β-OH, R = Bu
243, (20*R*,22*R*)
244, (20*R*,22*S*)
245, (20*S*,22*R*)
246, (20*S*,22*S*)

247
248, 24*R*,25*R*
249, 24*R*,25*S*
250, 24*S*,25*R*
251, 24*S*,25*S*
252
253
254
255

Figure 6. (1993–1994) [118–136].

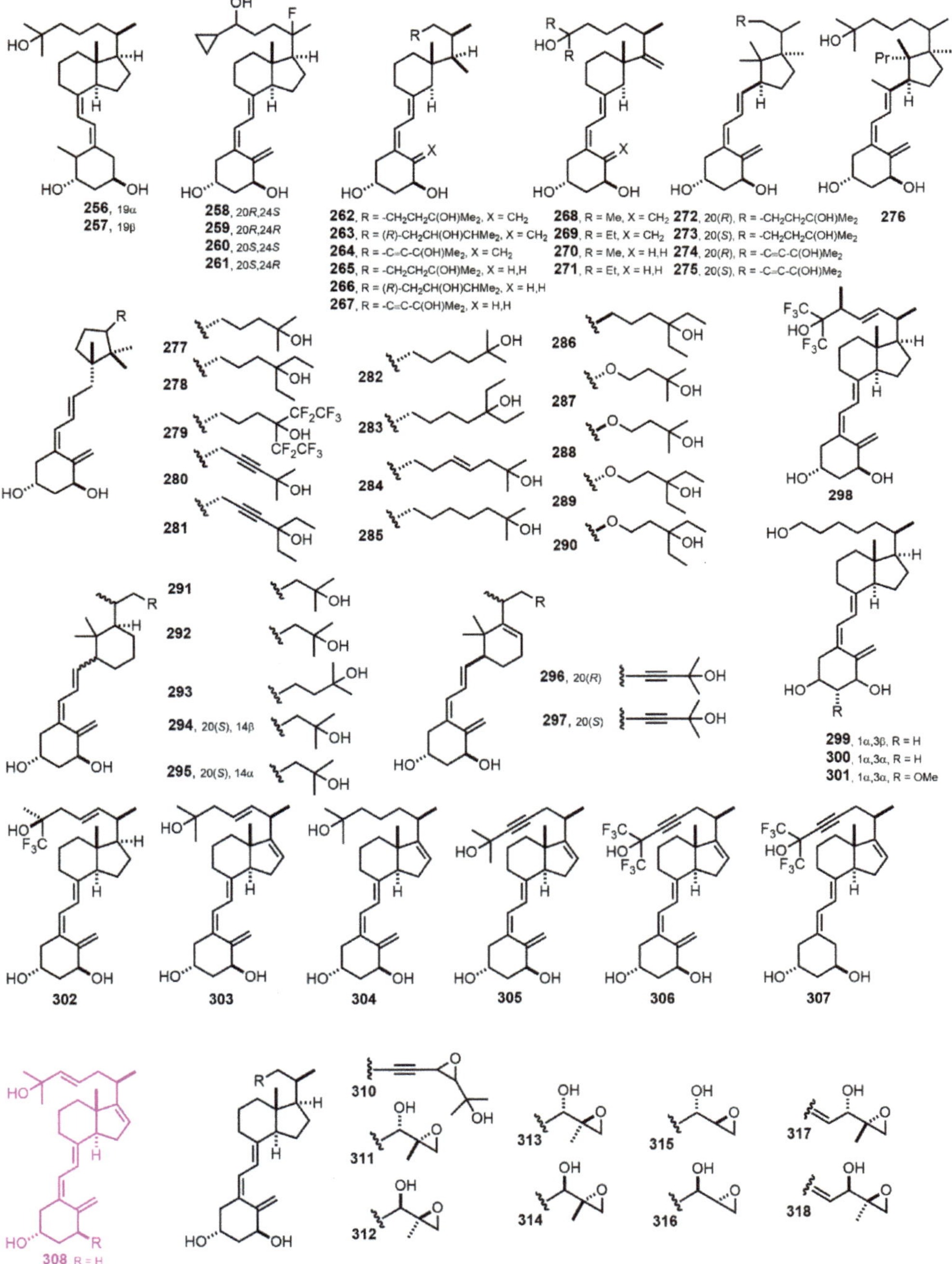

Figure 7. (1994–1997) [136–148].

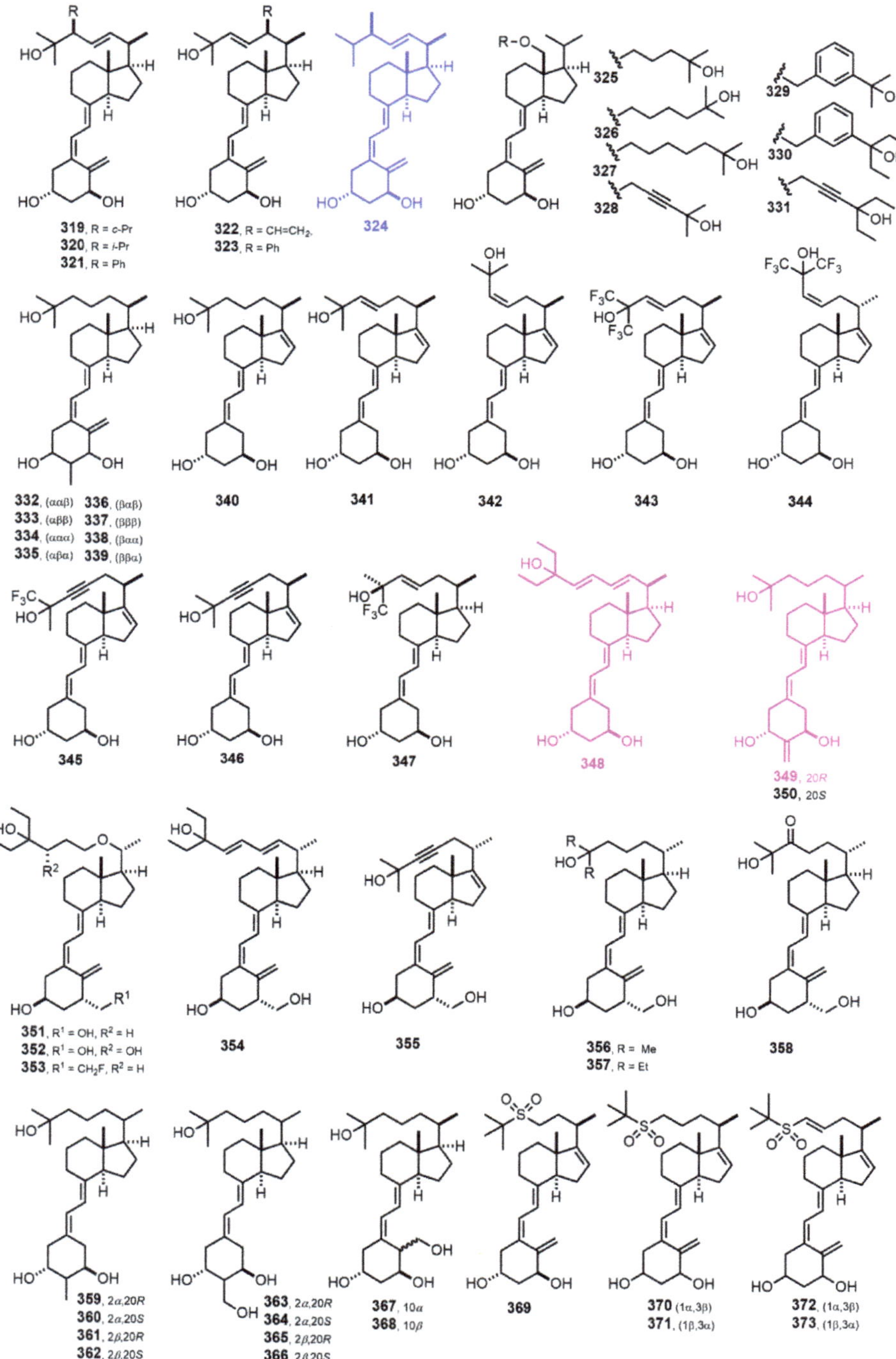

Figure 8. (1997–1999) [149–158].

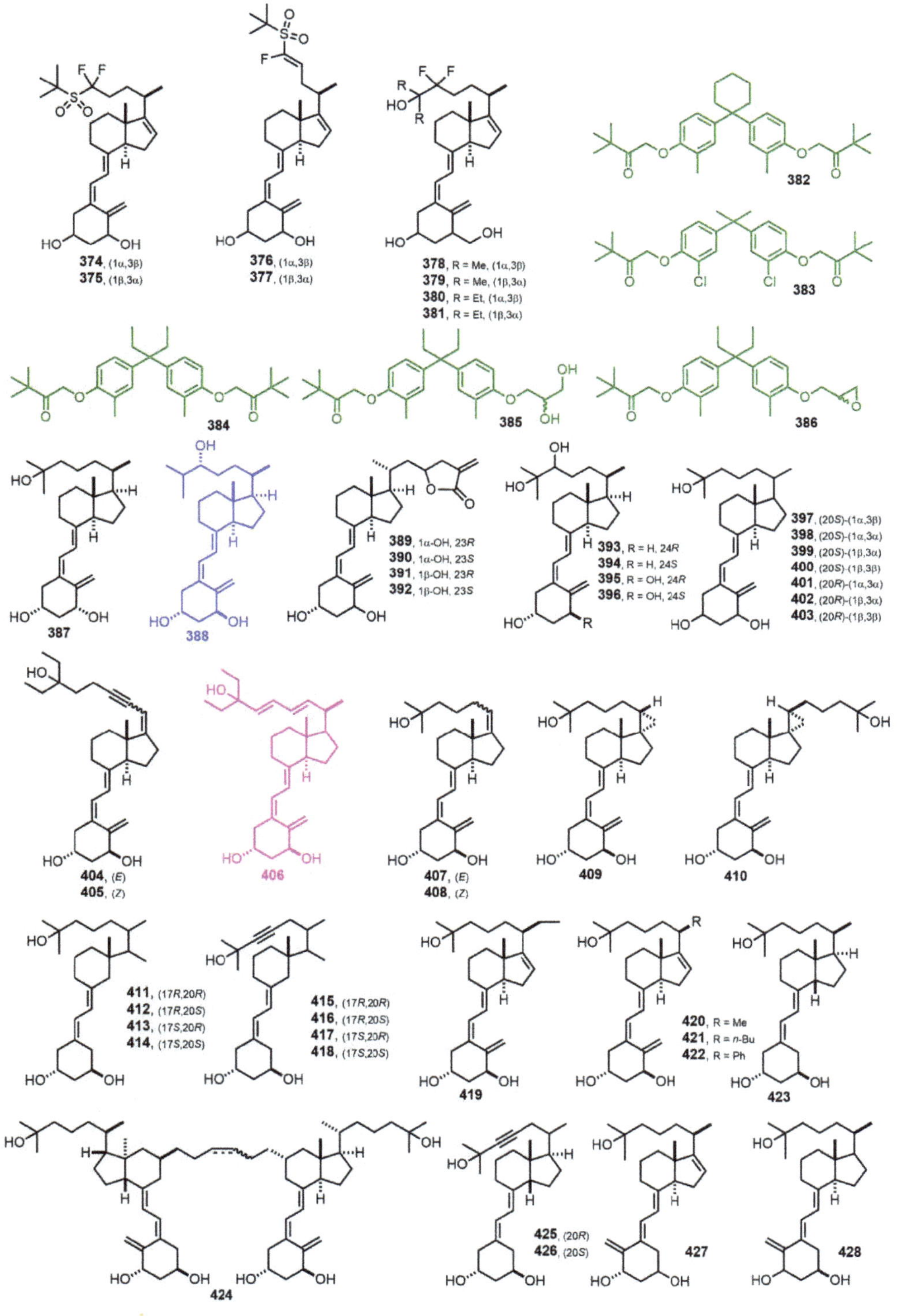

Figure 9. (1999) [158–168].

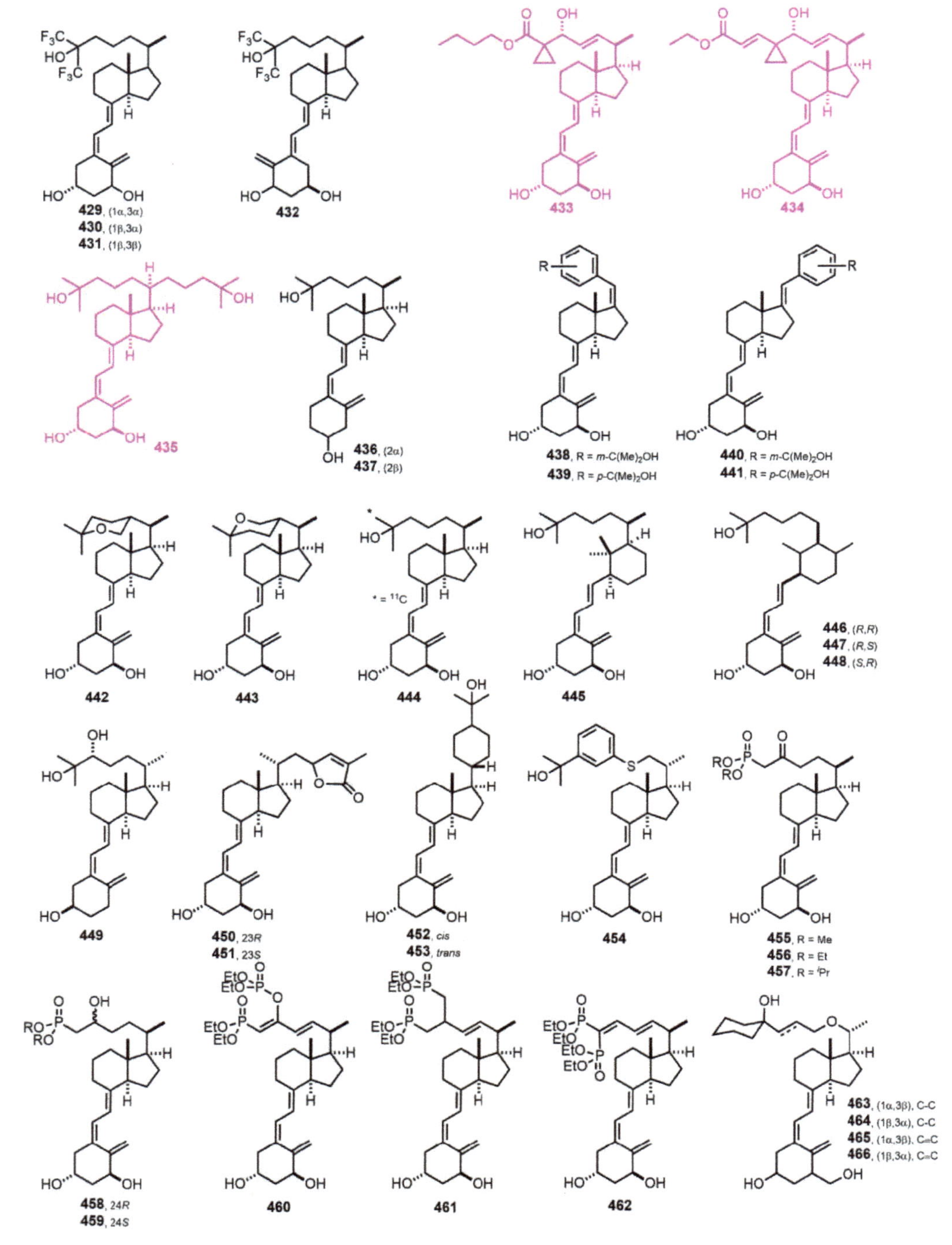

Figure 10. (2000–2001) [169–182].

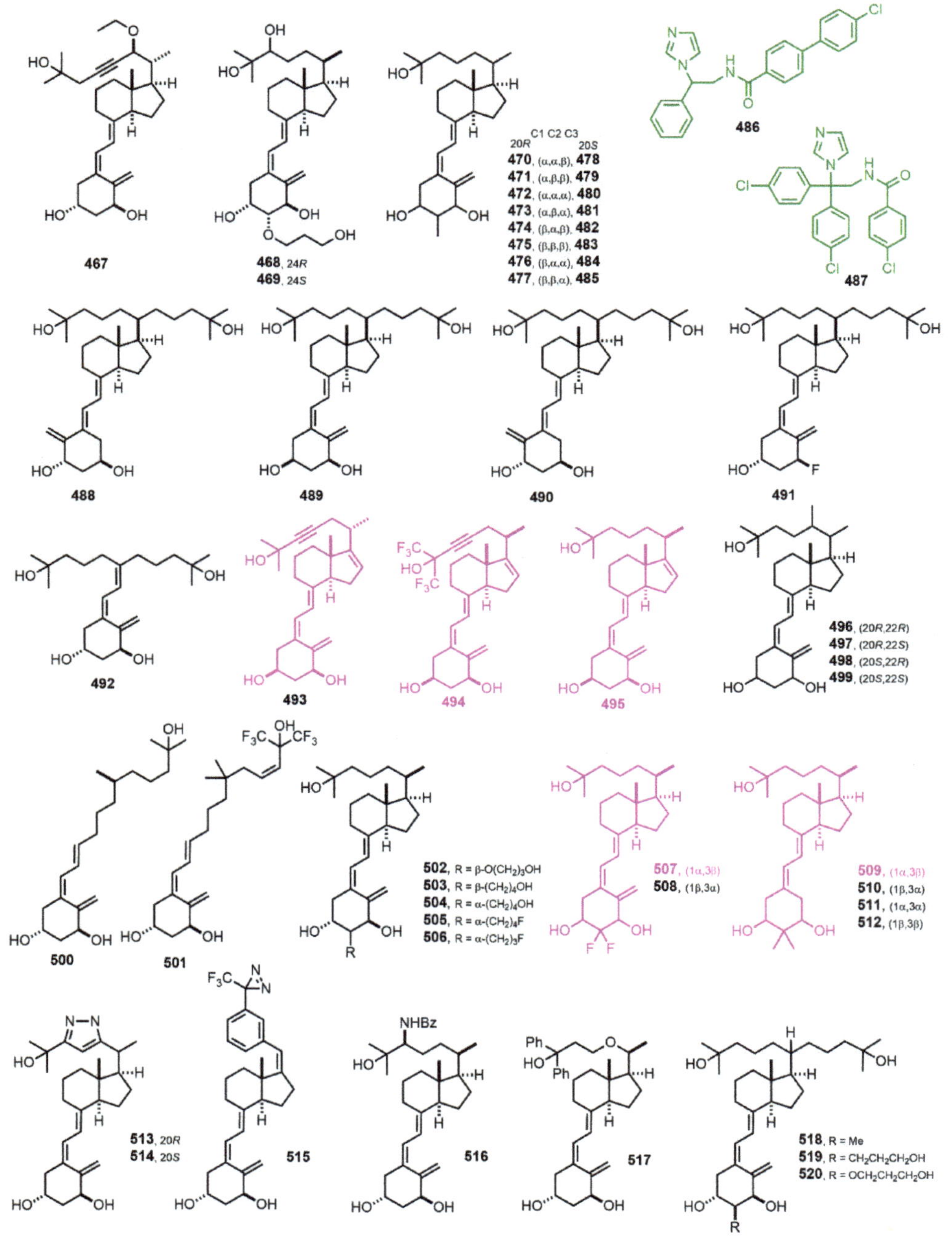

Figure 11. (2001–2002) [183–196].

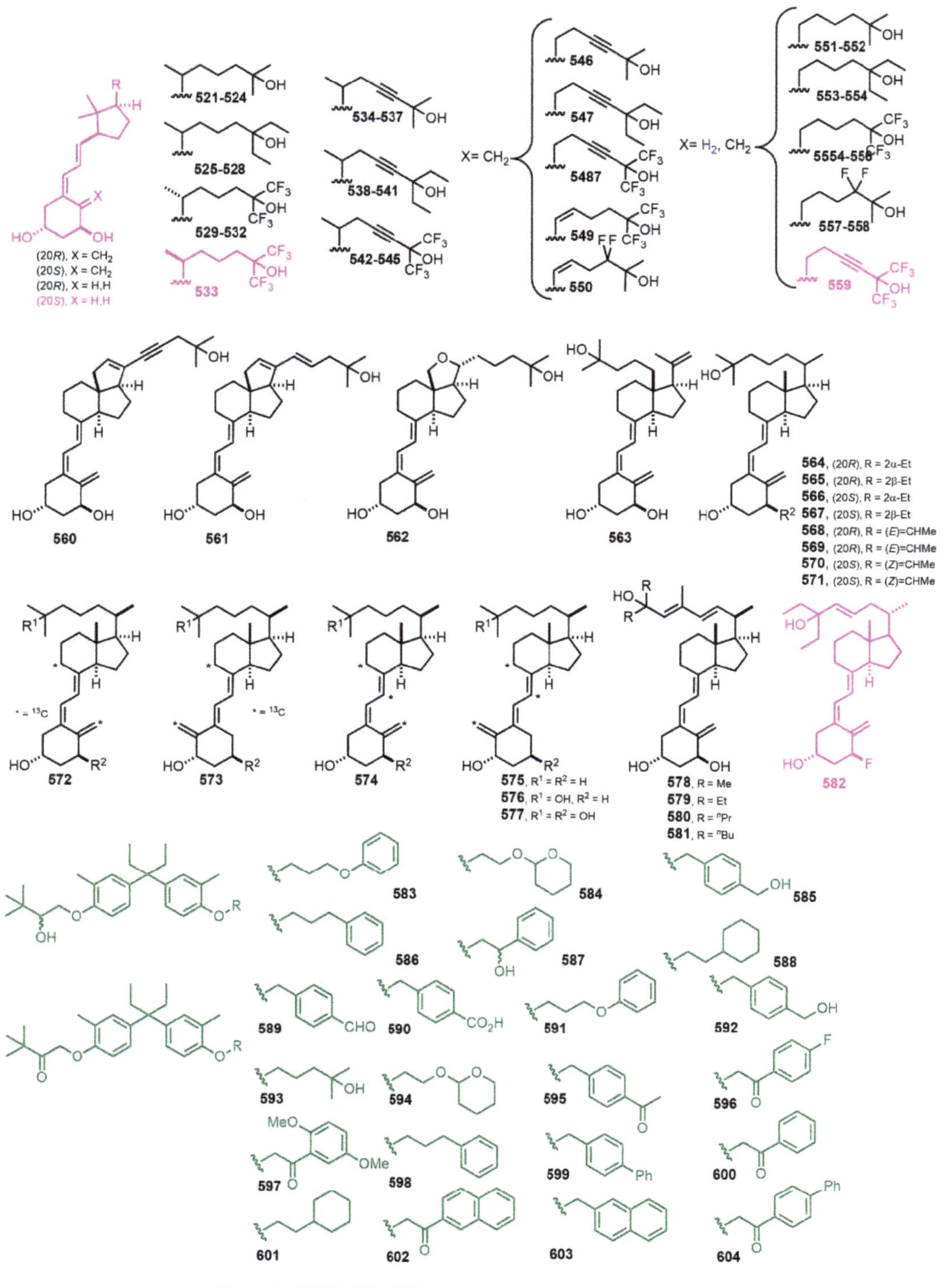

Figure 12. (2002) [197–204].

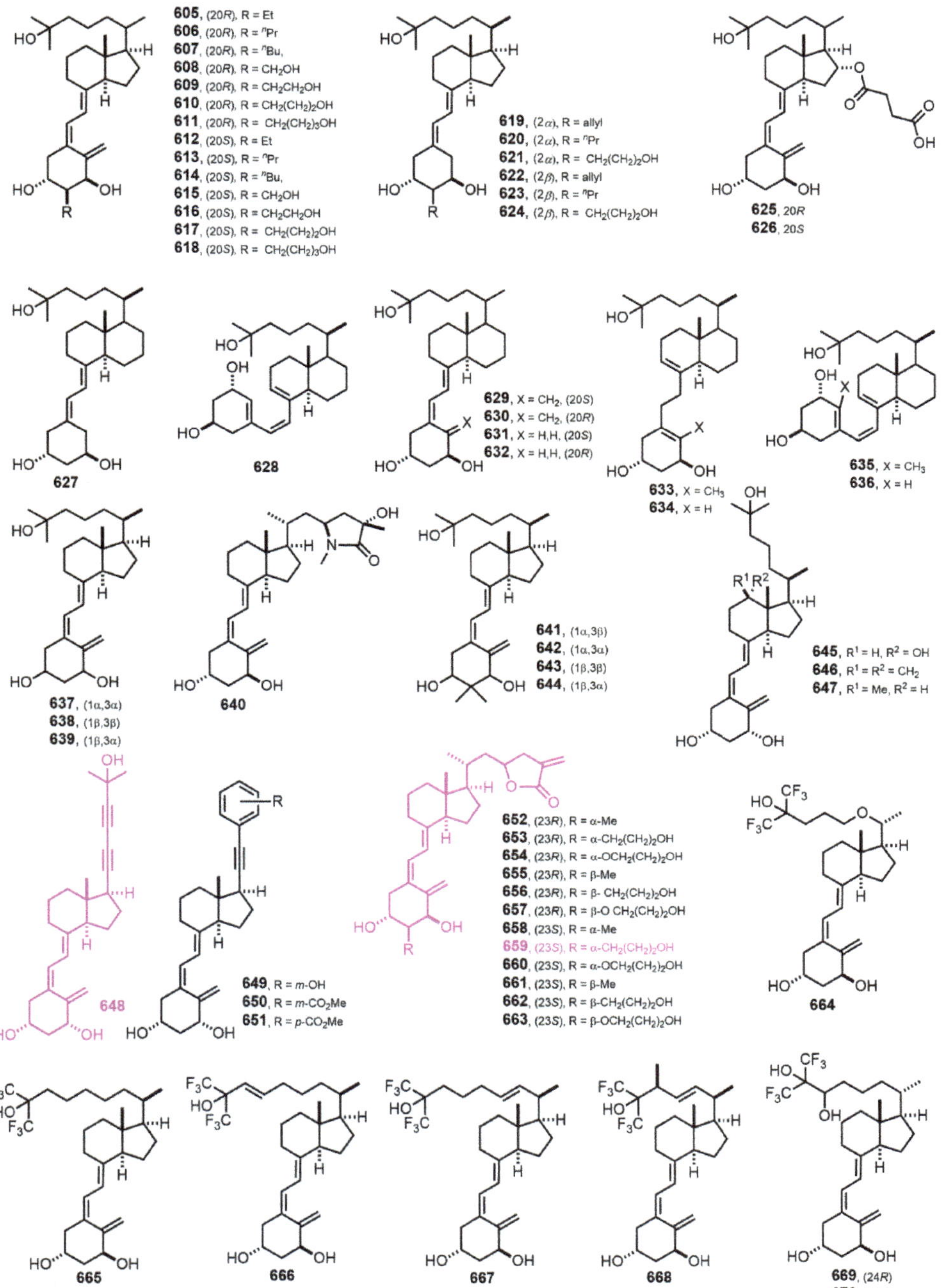

Figure 13. (2003–2004) [205–218].

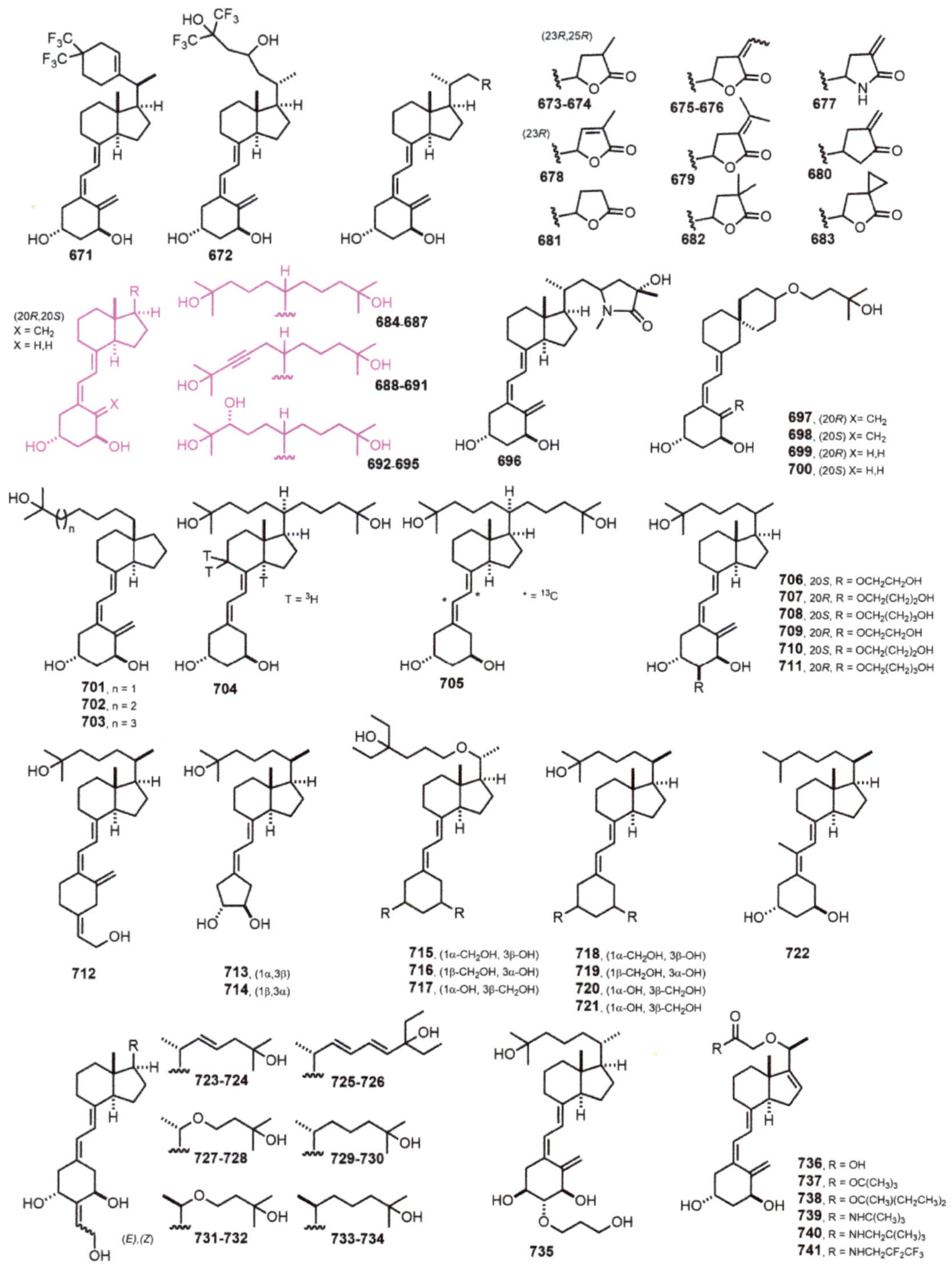

Figure 14. (2004–2006) [218–223].

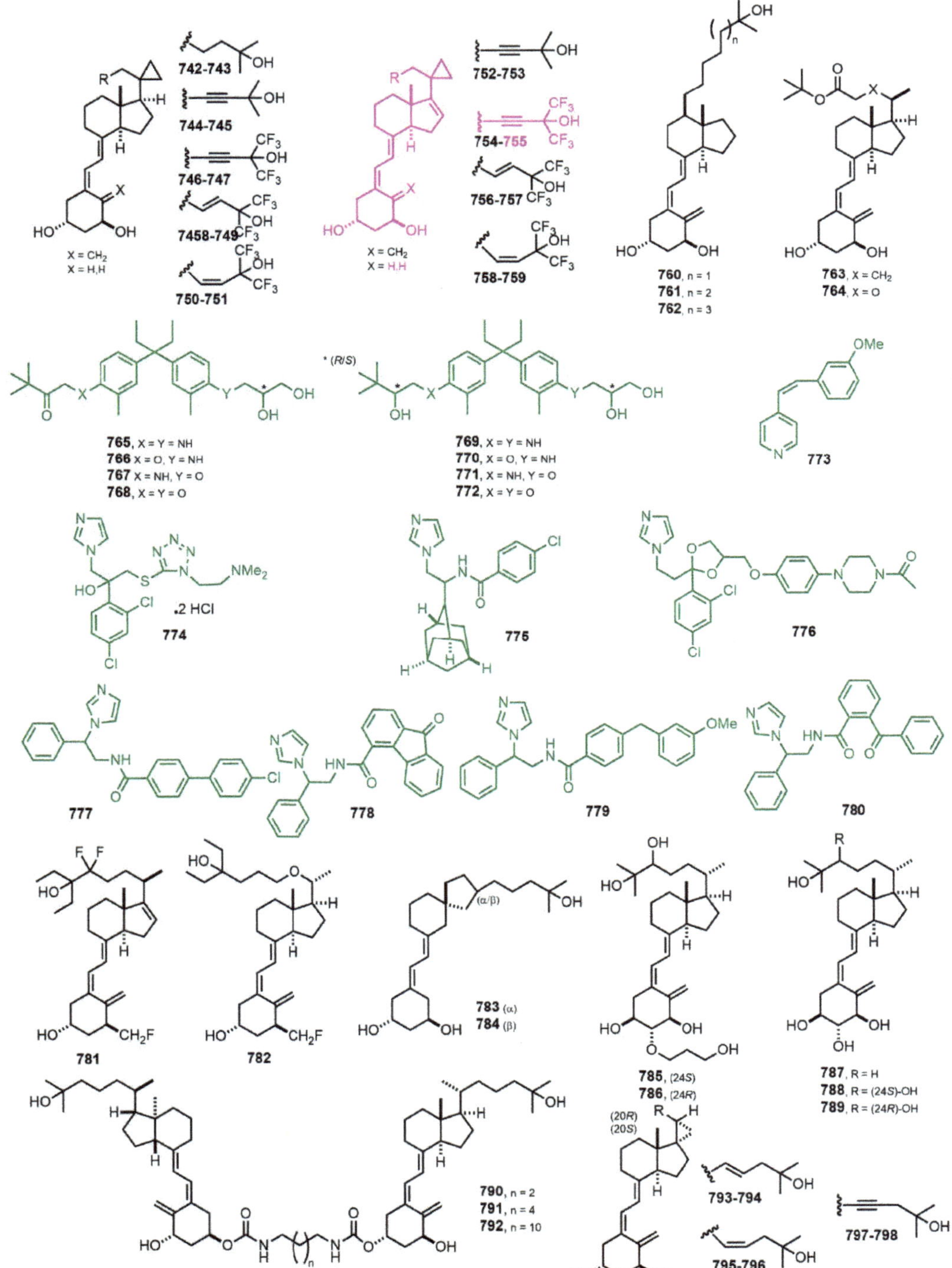

Figure 15. (2006–2007) [224–240].

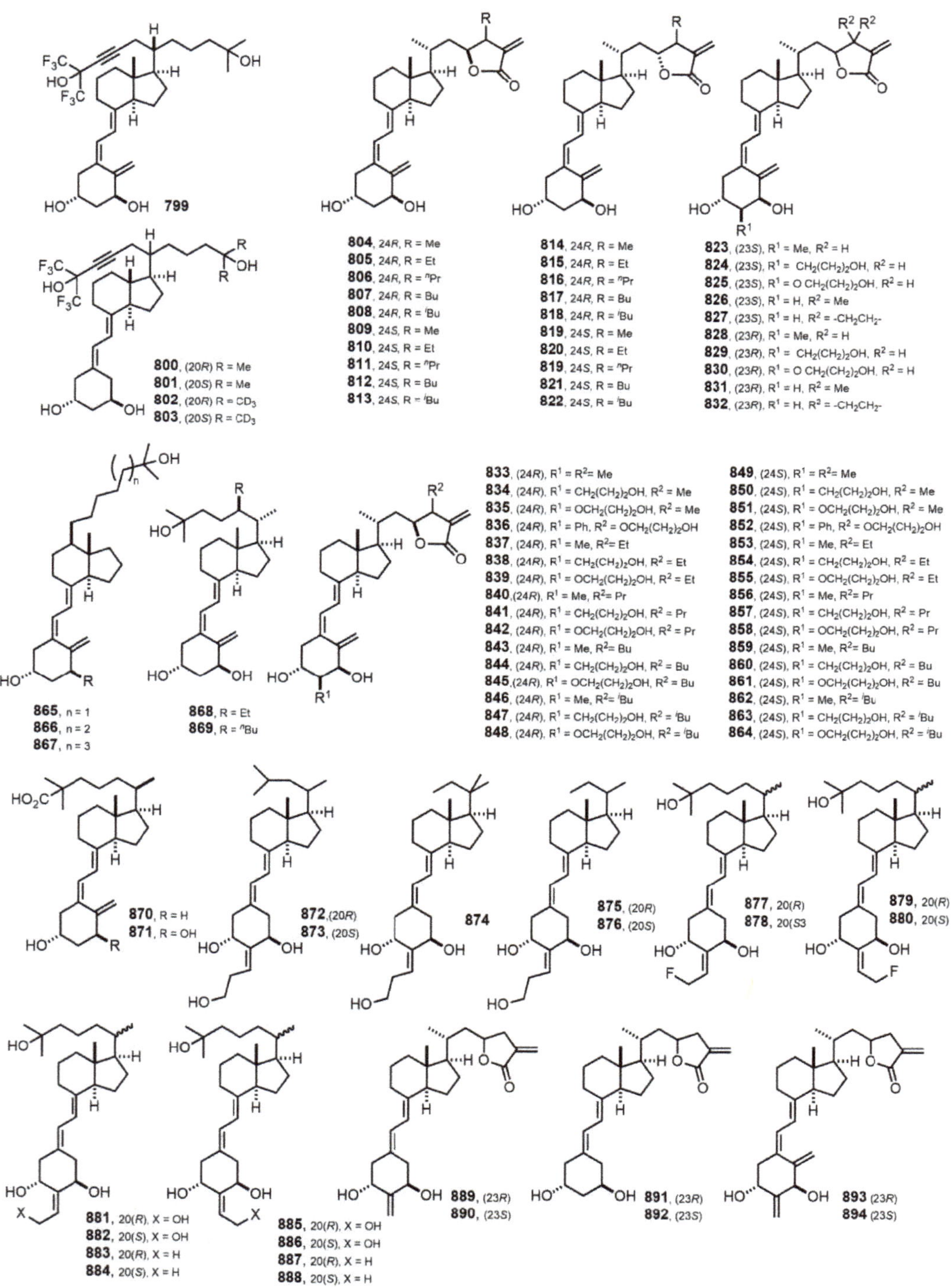

Figure 16. (2006–2008) [241–253].

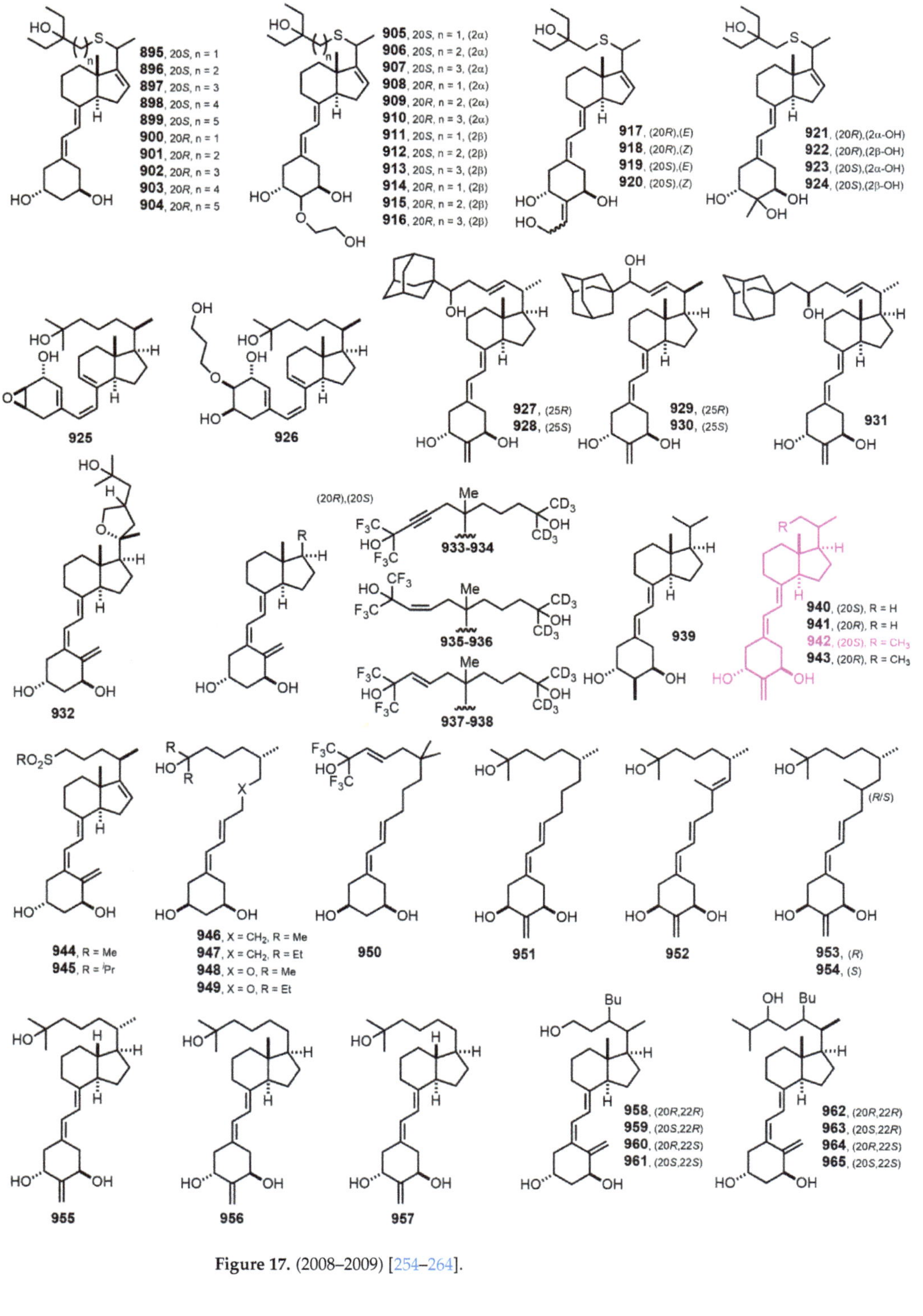

895, 20S, n = 1
896, 20S, n = 2
897, 20S, n = 3
898, 20S, n = 4
899, 20S, n = 5
900, 20R, n = 1
901, 20R, n = 2
902, 20R, n = 3
903, 20R, n = 4
904, 20R, n = 5

905, 20S, n = 1, (2α)
906, 20S, n = 2, (2α)
907, 20S, n = 3, (2α)
908, 20R, n = 1, (2α)
909, 20R, n = 2, (2α)
910, 20R, n = 3, (2α)
911, 20S, n = 1, (2β)
912, 20S, n = 2, (2β)
913, 20S, n = 3, (2β)
914, 20R, n = 1, (2β)
915, 20R, n = 2, (2β)
916, 20R, n = 3, (2β)

917, (20R),(E)
918, (20R),(Z)
919, (20S),(E)
920, (20S),(Z)

921, (20R),(2α-OH)
922, (20R),(2β-OH)
923, (20S),(2α-OH)
924, (20S),(2β-OH)

925 **926**

927, (25R)
928, (25S)

929, (25R)
930, (25S)

931

932

(20R),(20S)

933-934

935-936

937-938

939

940, (20S), R = H
941, (20R), R = H
942, (20S), R = CH_3
943, (20R), R = CH_3

944, R = Me
945, R = ^{i}Pr

946, X = CH_2, R = Me
947, X = CH_2, R = Et
948, X = O, R = Me
949, X = O, R = Et

950 **951** **952**

953, (R)
954, (S)

955 **956** **957**

958, (20R,22R)
959, (20S,22R)
960, (20R,22S)
961, (20S,22S)

962, (20R,22R)
963, (20S,22R)
964, (20R,22S)
965, (20S,22S)

Figure 17. (2008–2009) [254–264].

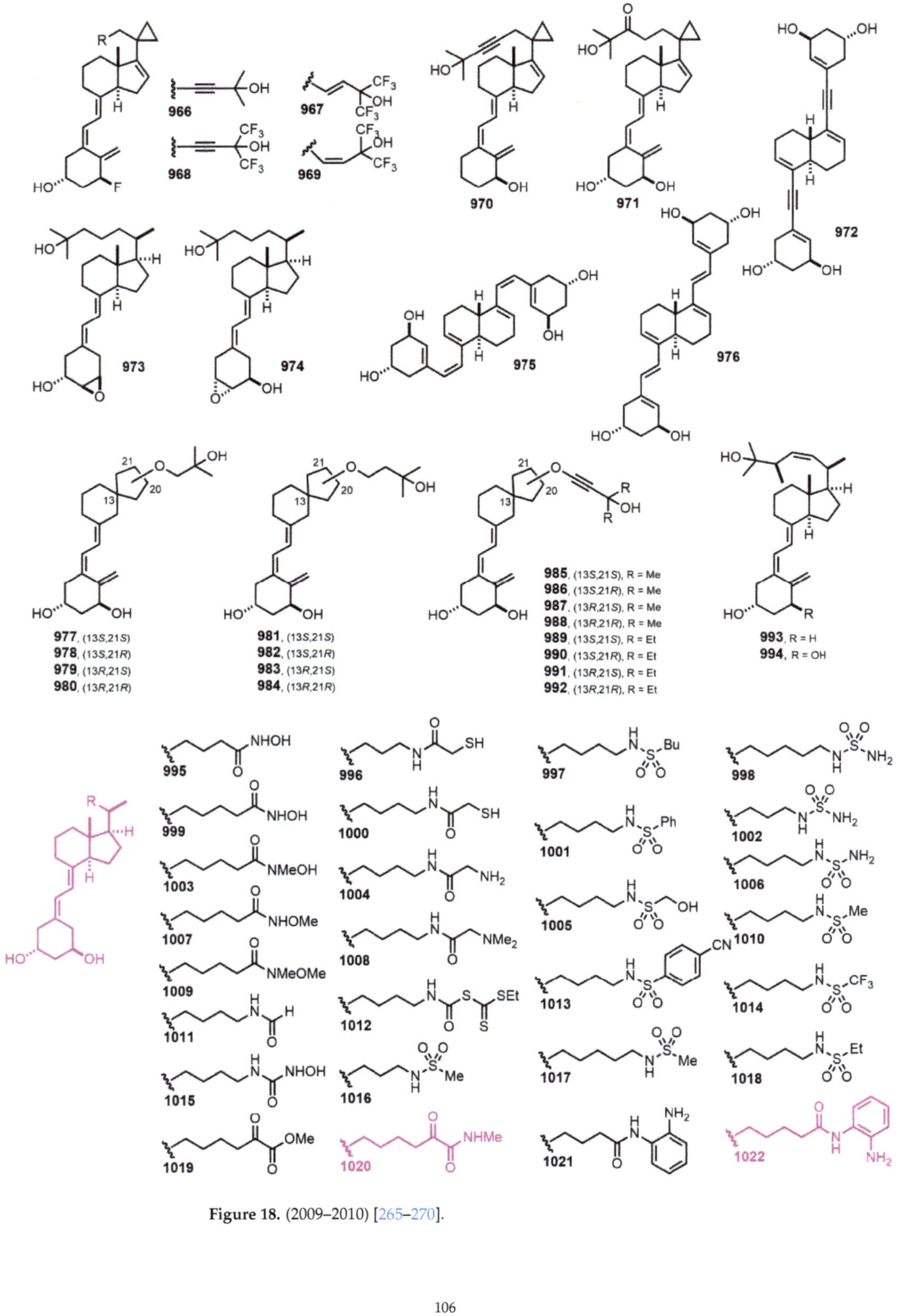

Figure 18. (2009–2010) [265–270].

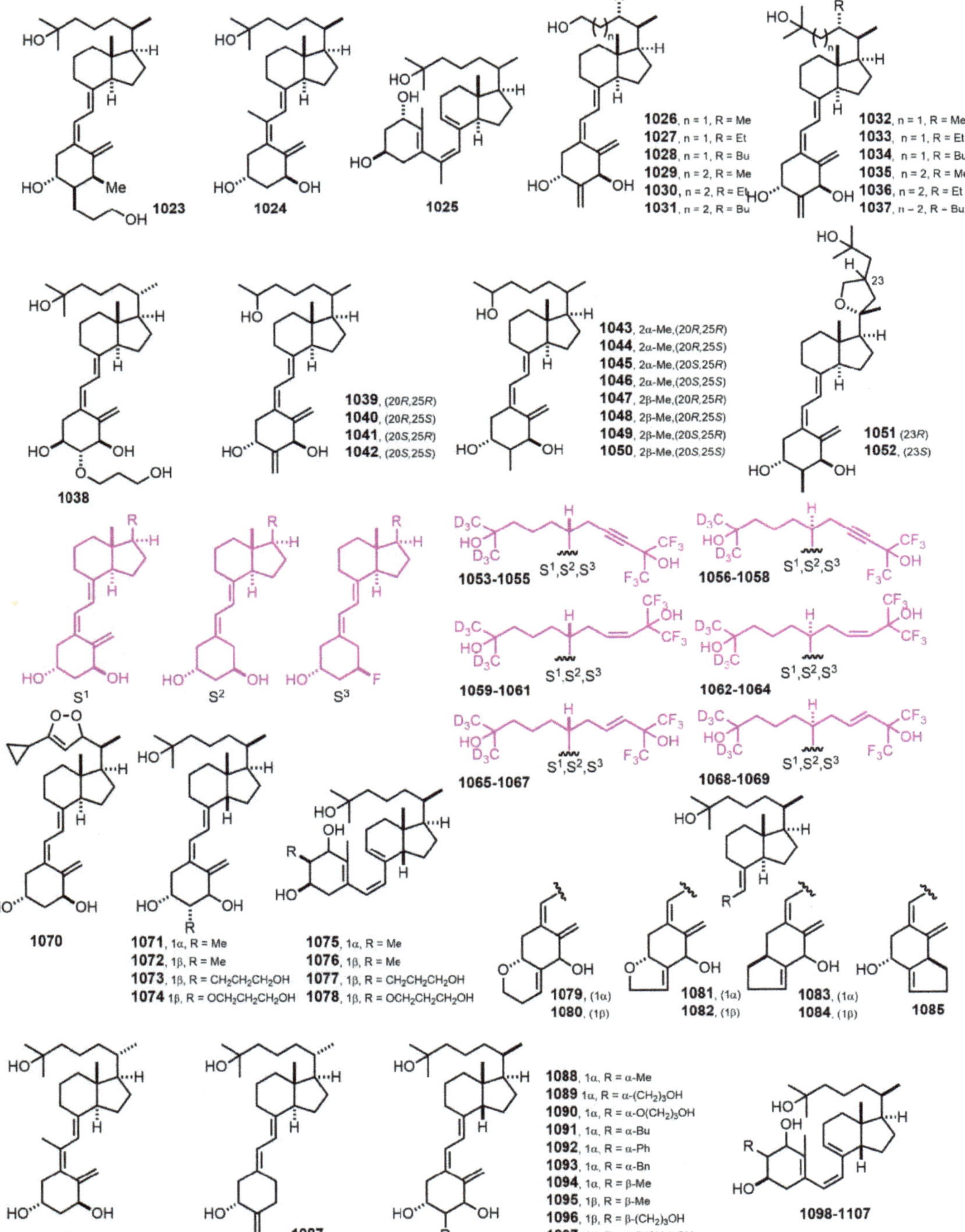

Figure 19. (2009–2010) [271–283].

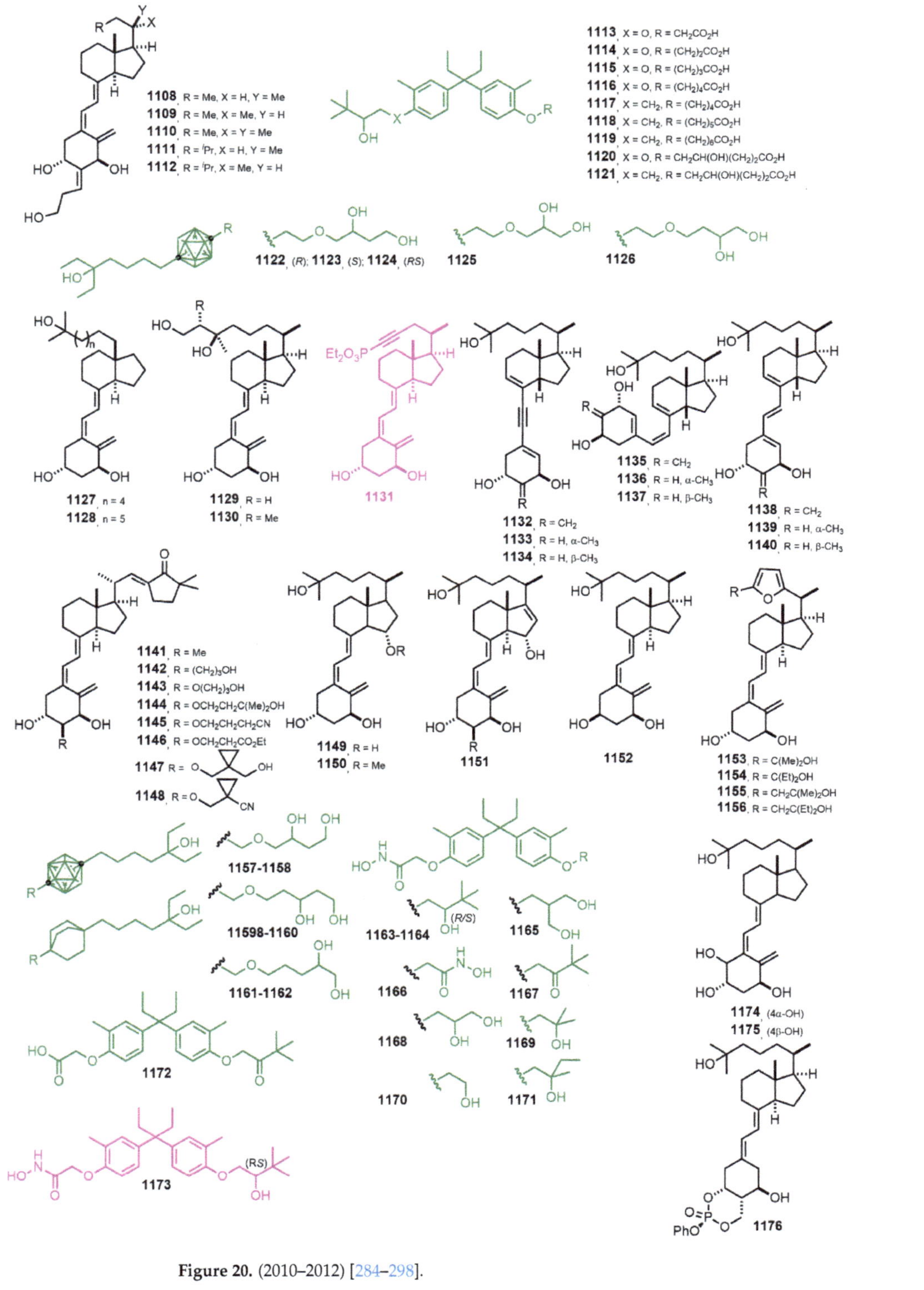

1108, R = Me, X = H, Y = Me
1109, R = Me, X = Me, Y = H
1110, R = Me, X = Y = Me
1111, R = ^{i}Pr, X = H, Y = Me
1112, R = ^{i}Pr, X = Me, Y = H

1113, X = O, R = CH_2CO_2H
1114, X = O, R = $(CH_2)_2CO_2H$
1115, X = O, R = $(CH_2)_3CO_2H$
1116, X = O, R = $(CH_2)_4CO_2H$
1117, X = CH_2, R = $(CH_2)_4CO_2H$
1118, X = CH_2, R = $(CH_2)_5CO_2H$
1119, X = CH_2, R = $(CH_2)_6CO_2H$
1120, X = O, R = $CH_2CH(OH)(CH_2)_2CO_2H$
1121, X = CH_2, R = $CH_2CH(OH)(CH_2)_2CO_2H$

1122, (*R*); **1123**, (*S*); **1124**, (*RS*) **1125** **1126**

1127, n = 4
1128, n = 5

1129, R = H
1130, R = Me

1131

1132, R = CH_2
1133, R = H, α-CH_3
1134, R = H, β-CH_3

1135, R = CH_2
1136, R = H, α-CH_3
1137, R = H, β-CH_3

1138, R = CH_2
1139, R = H, α-CH_3
1140, R = H, β-CH_3

1141, R = Me
1142, R = $(CH_2)_3OH$
1143, R = $O(CH_2)_3OH$
1144, R = $OCH_2CH_2C(Me)_2OH$
1145, R = $OCH_2CH_2CH_2CN$
1146, R = $OCH_2CH_2CO_2Et$
1147 R =
1148, R =

1149, R = H
1150, R = Me

1151

1152

1153, R = $C(Me)_2OH$
1154, R = $C(Et)_2OH$
1155, R = $CH_2C(Me)_2OH$
1156, R = $CH_2C(Et)_2OH$

1157-1158

11598-1160

1161-1162

1163-1164 (*R/S*)

1165

1166

1167

1168

1169

1170

1171

1172

1173

1174, (4α-OH)
1175, (4β-OH)

1176

Figure 20. (2010–2012) [284–298].

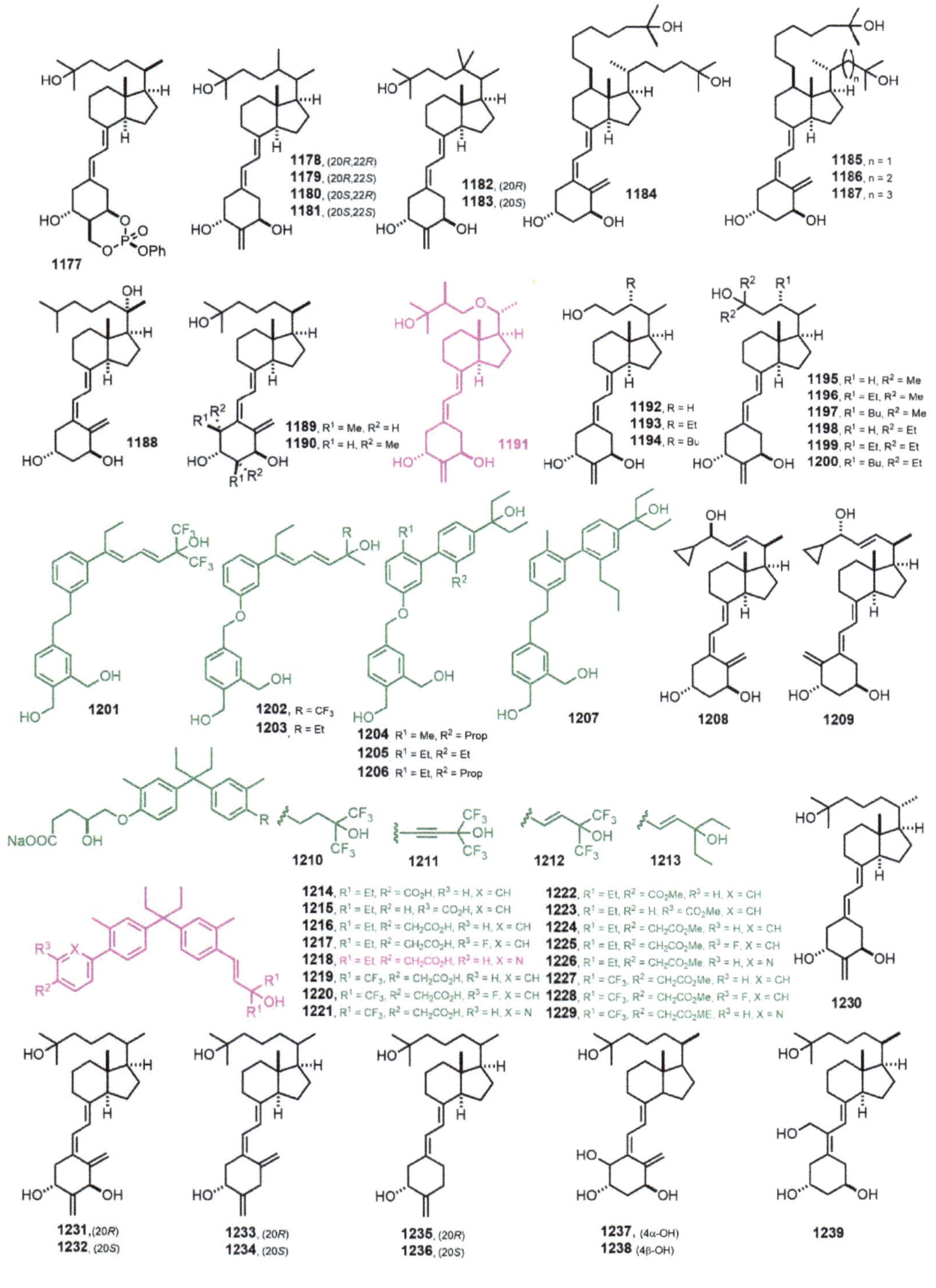

Figure 21. (2012–2013) [298–313].

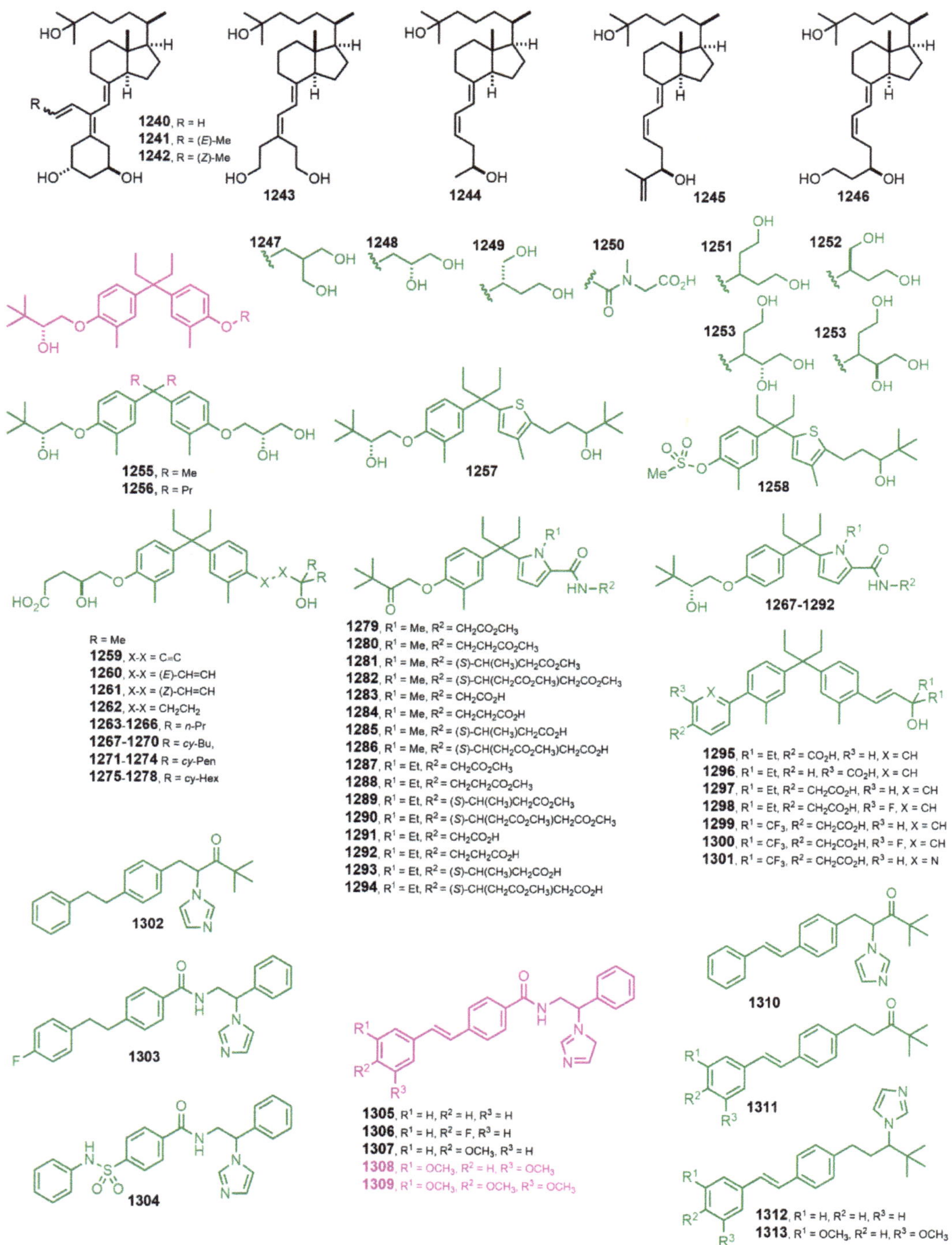

Figure 22. (2013–2014) [313–316].

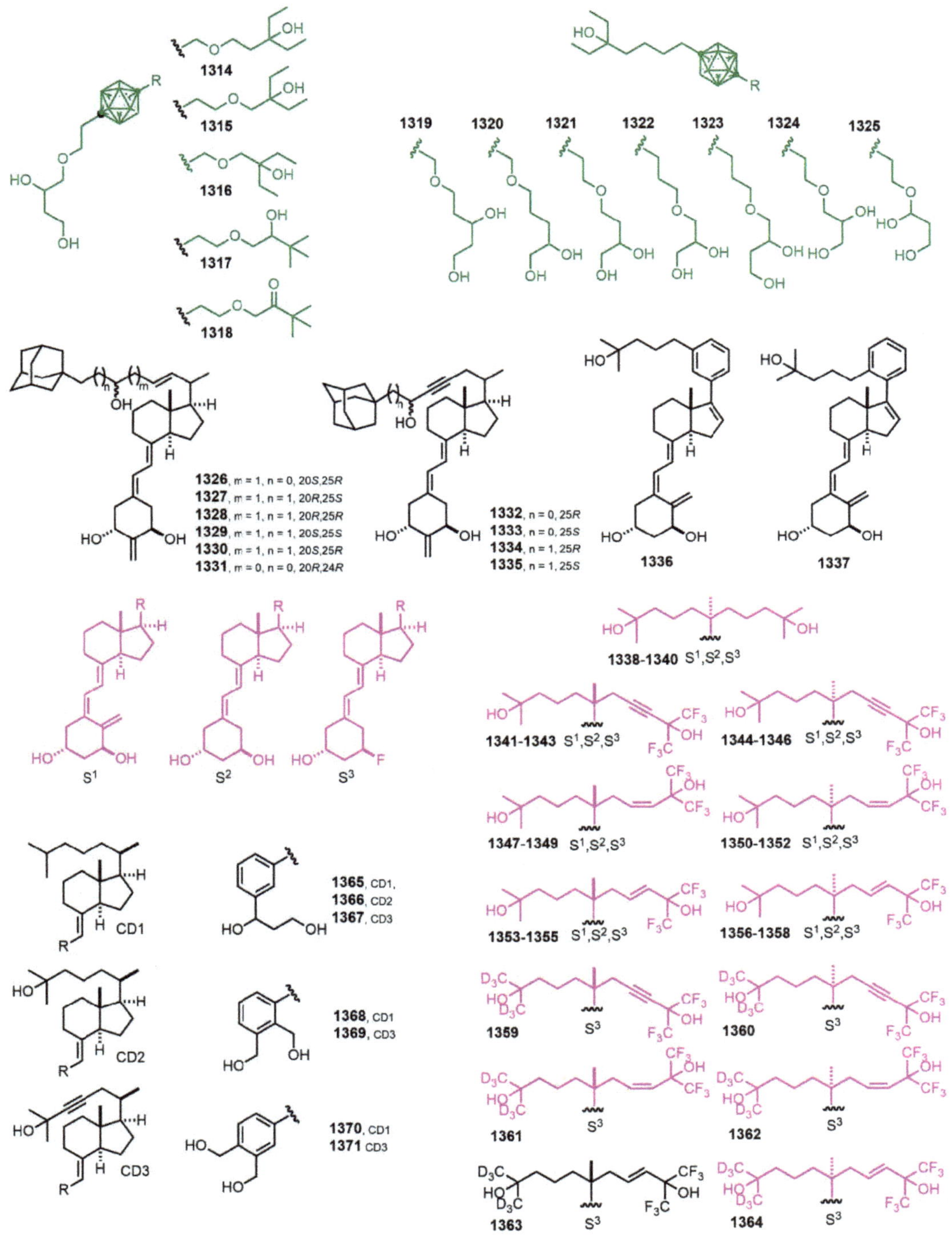

Figure 23. (2014) [317–321].

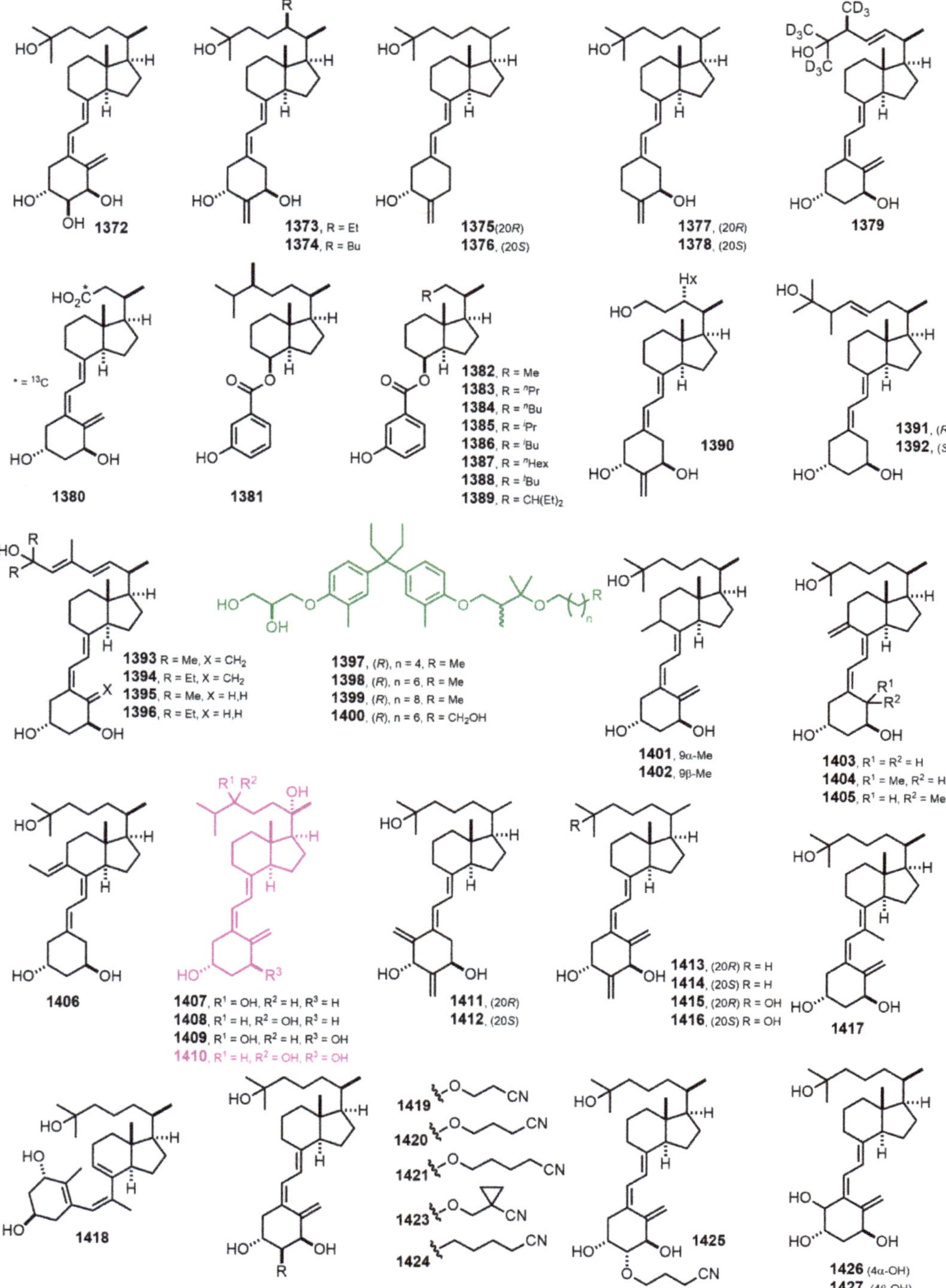

Figure 24. (2014–2015) [322–336].

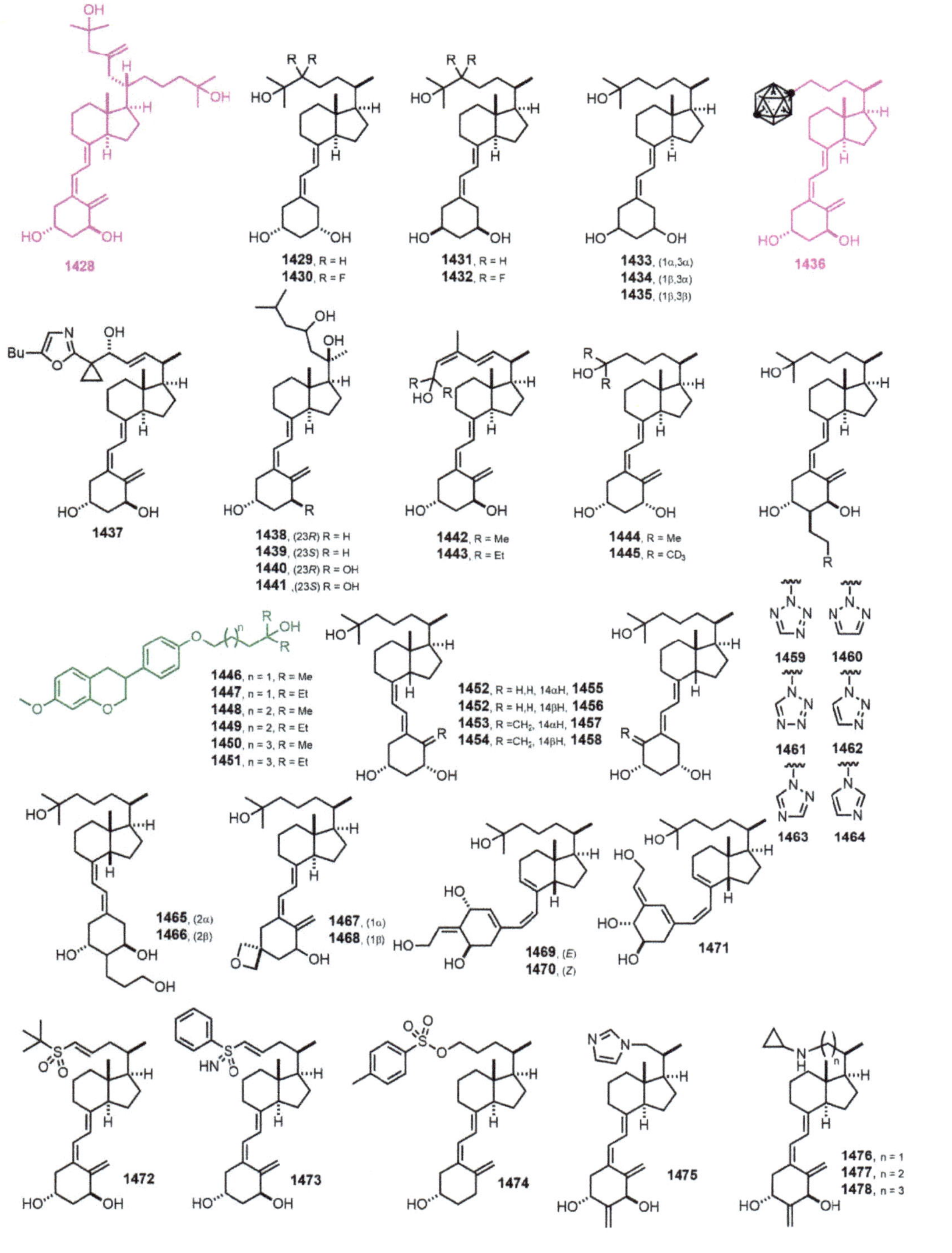

Figure 25. (2015–2017) [337–351].

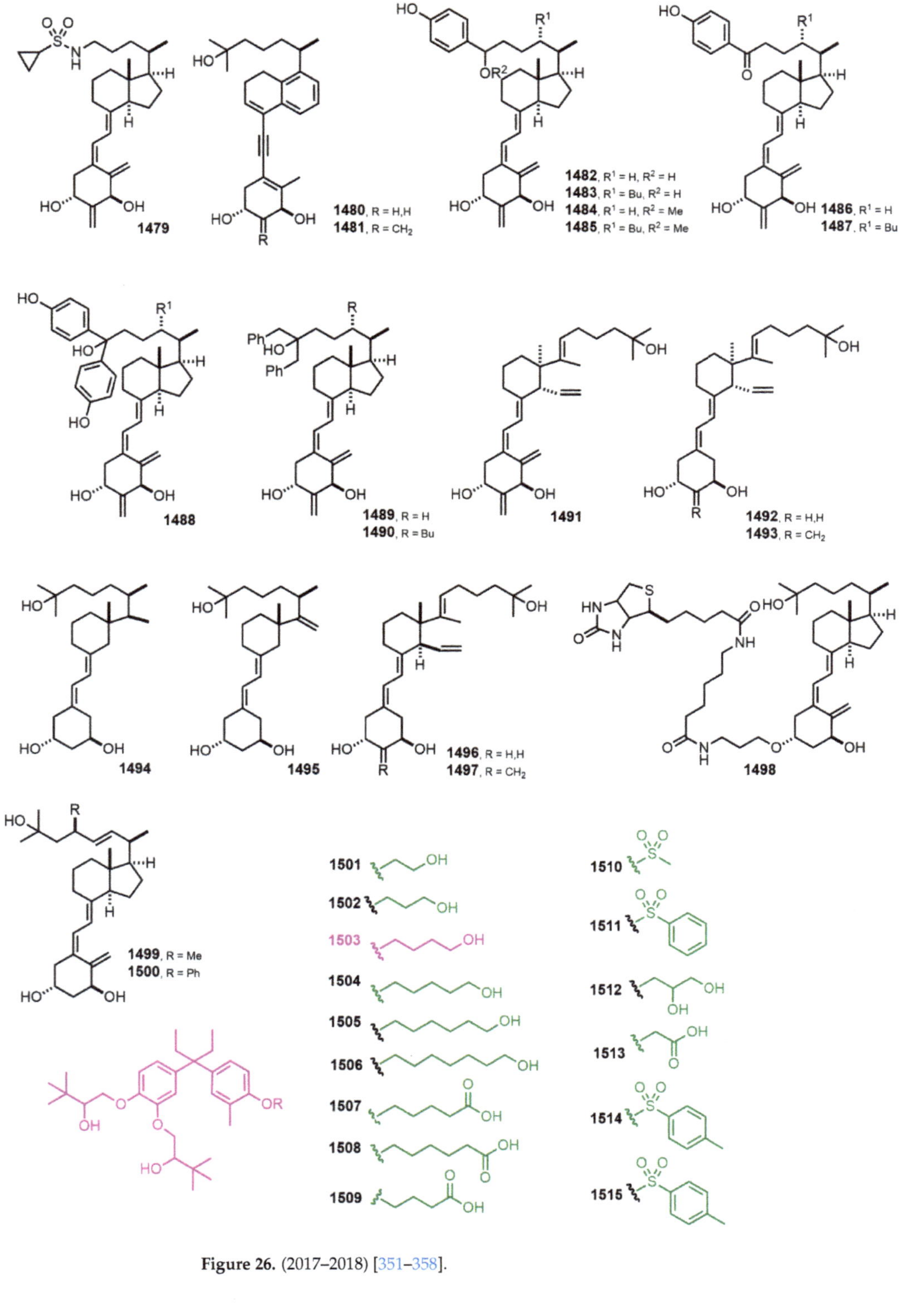

Figure 26. (2017–2018) [351–358].

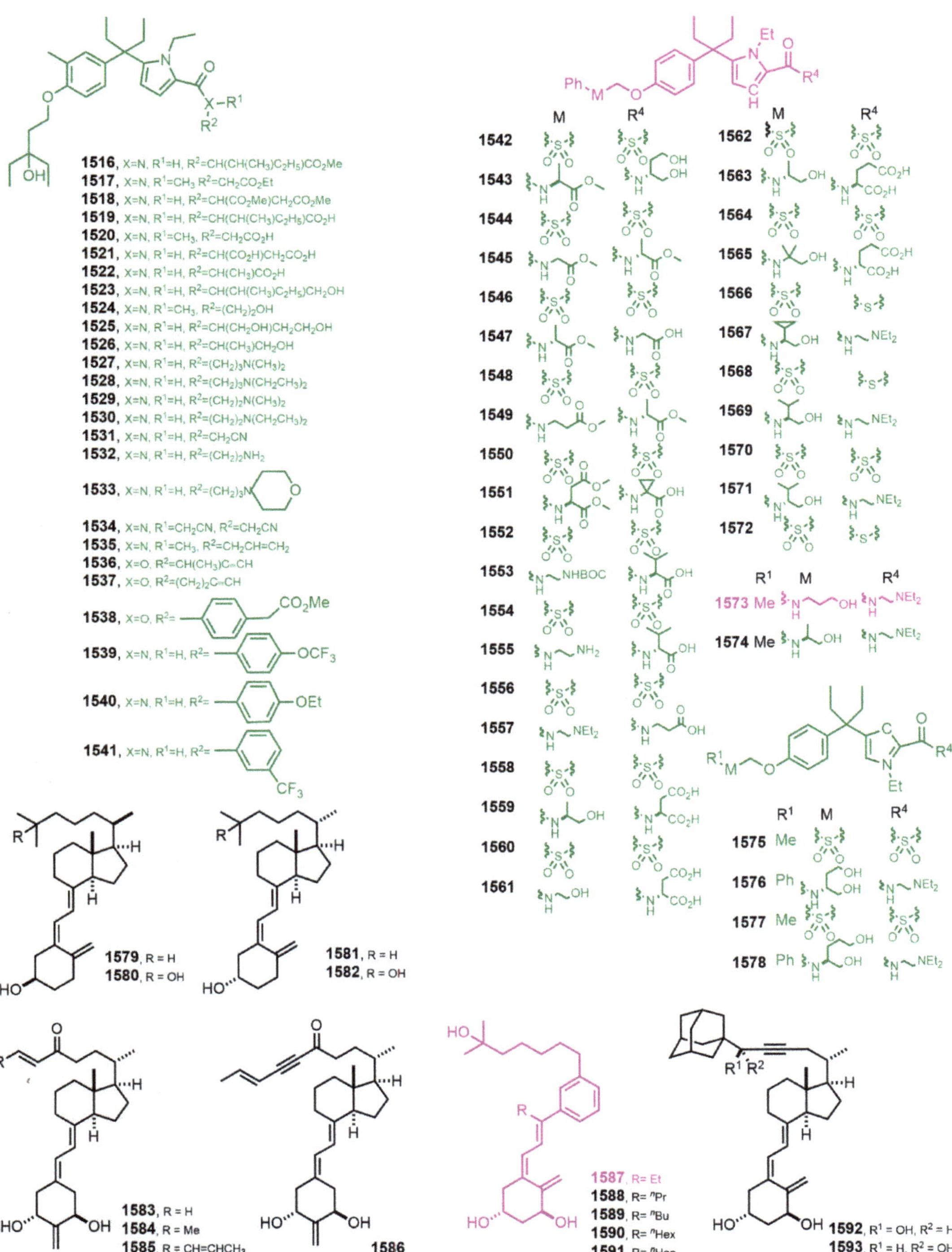

Figure 27. (2018) [359–364].

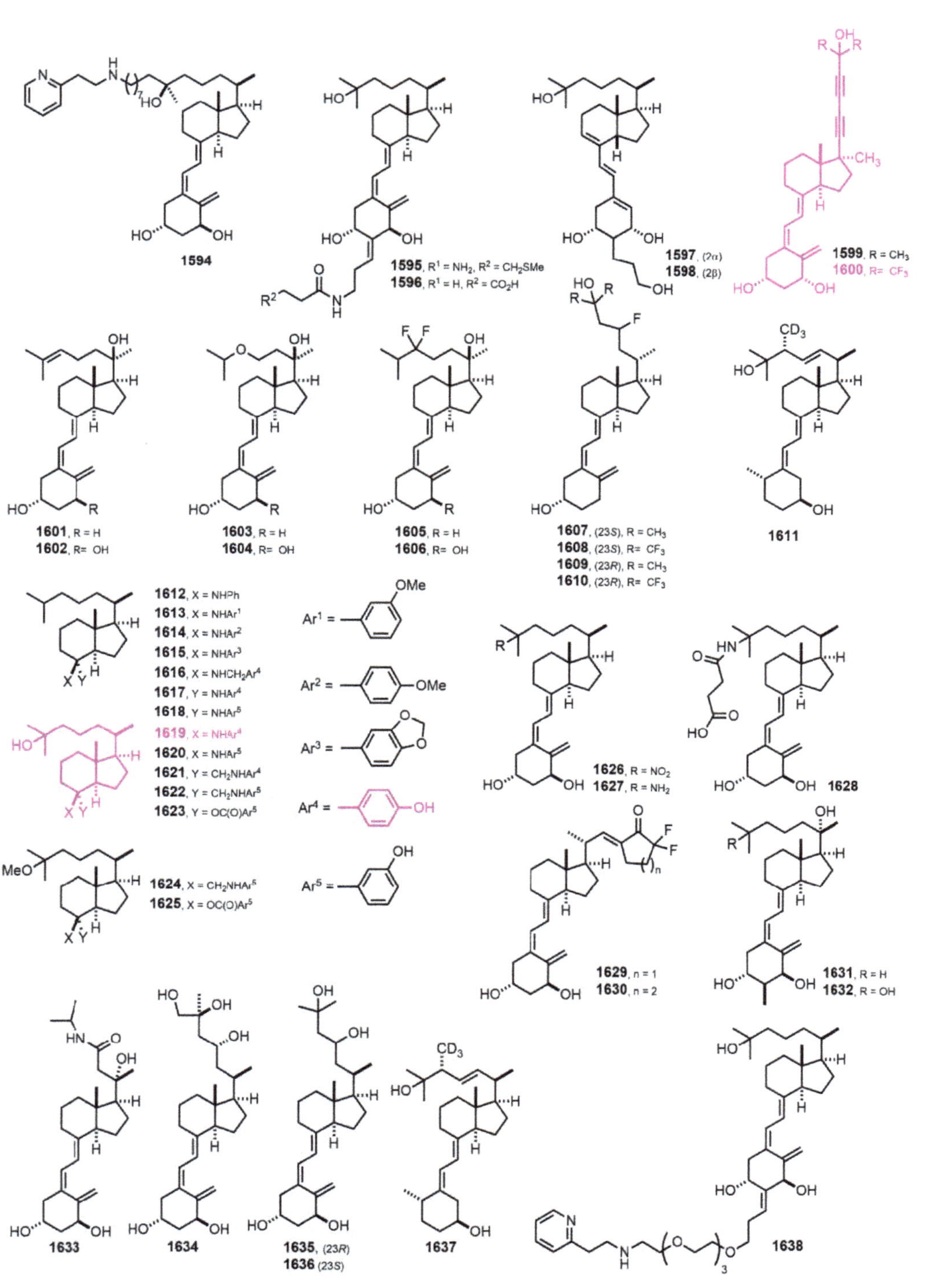

Figure 28. (2018–2019) [365–378].

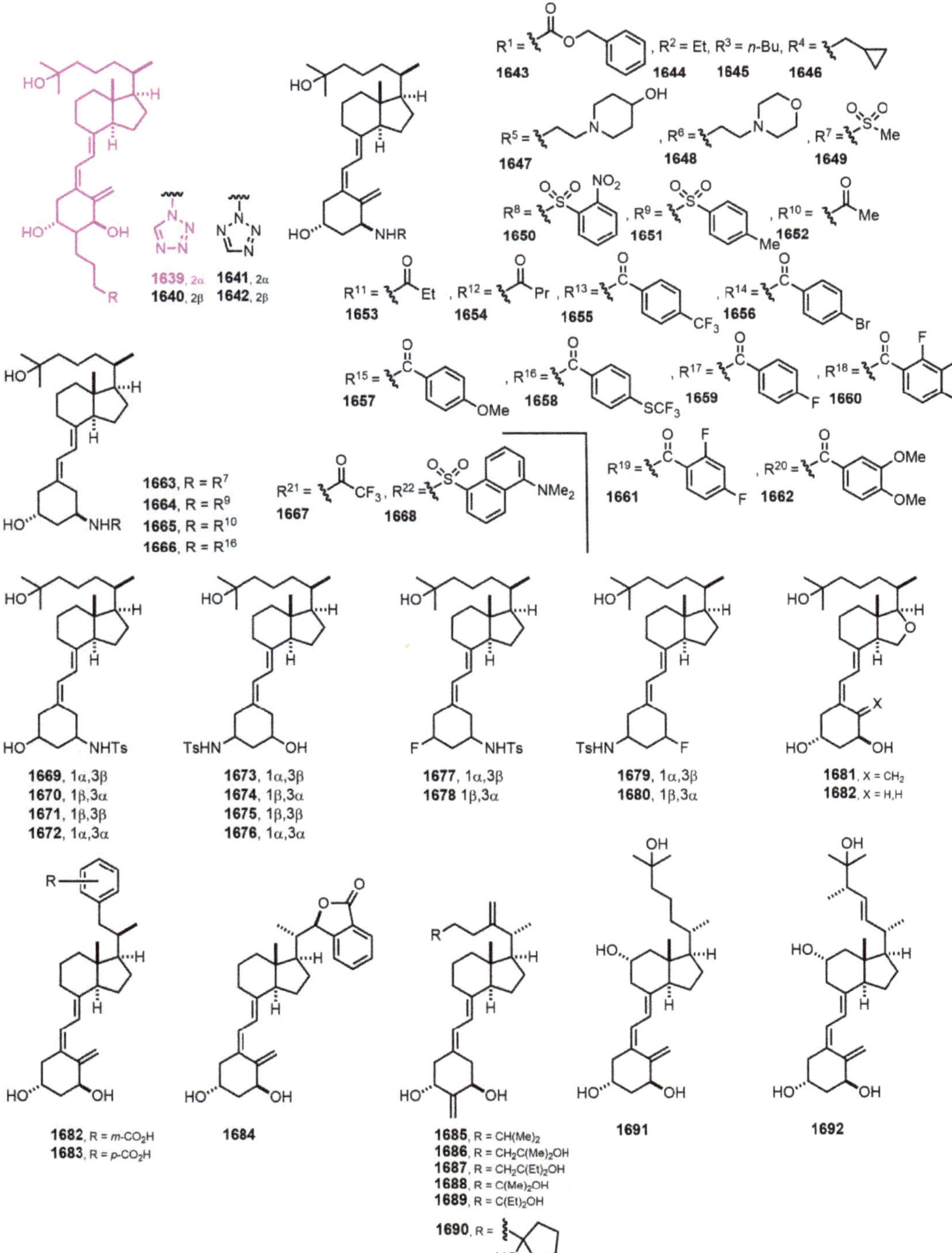

Figure 29. (2019–2020) [379–383].

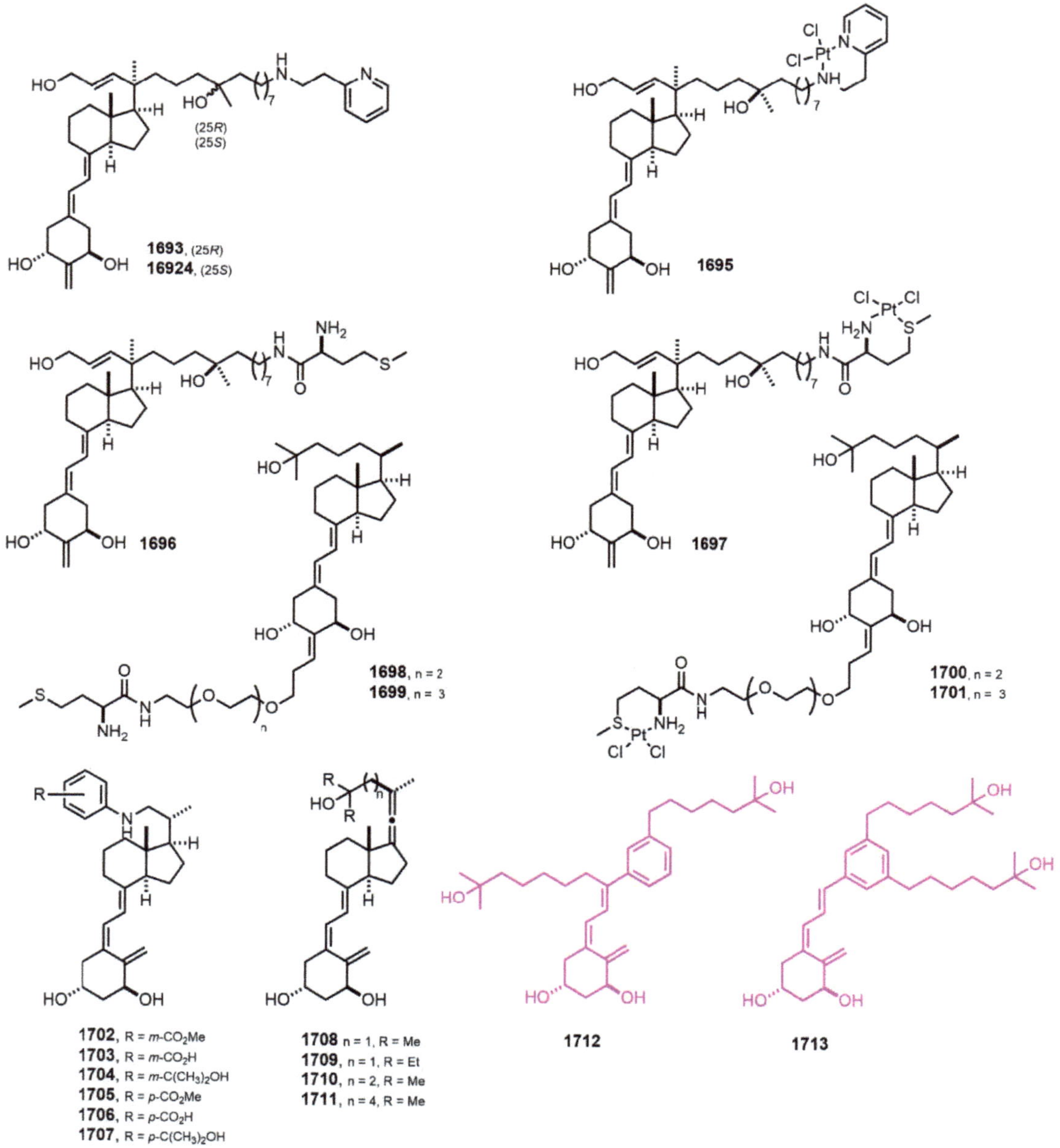

Figure 30. (2020–2022) [384–389].

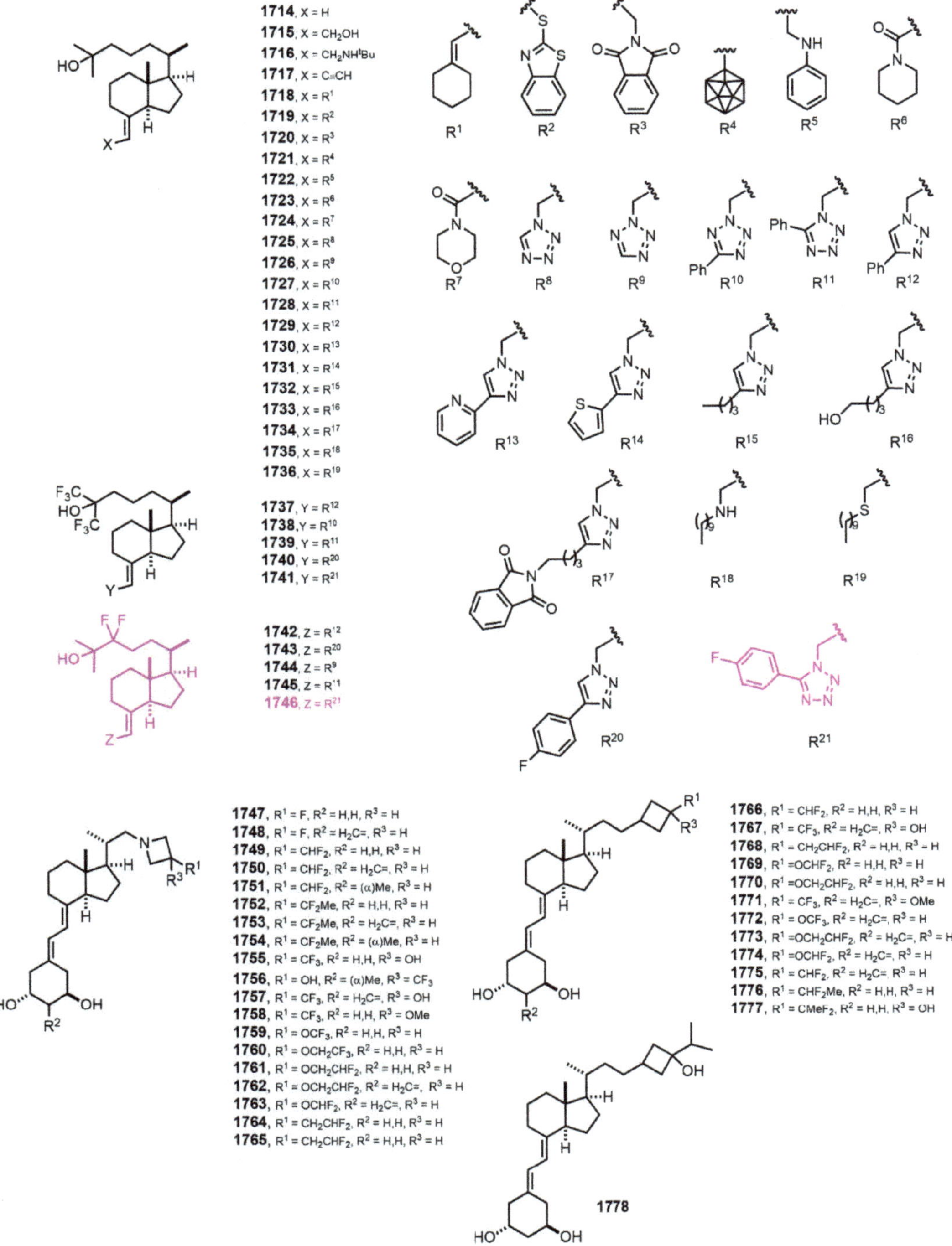

Figure 31. (2021–2022) [390–392].

4. Conclusions

A century has passed since vitamin D was discovered. The structural diversity achieved among vitamin D receptor ligands (1785 ligands involving metabolites, analogs, hybrids, and nonsteroidal ligands). Seeing as vitamin D plays a ubiquitous role in human physiology, VDR ligands have been found to cure or ameliorate the symptoms of various diseases. It is disheartening to note that for more than twenty years no drug based on a VDR ligand (i.e., analogues, hybrids, or nonsteroidal ligands) has been placed on the market because the structural diversity achieved in the VDR ligands might encode new therapies for other illness different than the calcium–phosphorous homeostasis.

Author Contributions: Conceptualization, M.A.M.; methodology, M.A.M.; investigation, M.A.M. and S.S.; writing—original draft preparation, M.A.M.; writing—review and editing, M.A.M. and S.S. All authors have read and agreed to the published version of the manuscript.

Funding: This research received no external funding.

Conflicts of Interest: The authors declare no conflict of interest.

References

1. Askew, F.A.; Bruce, H.M.; Callow, R.K.; Philpot, L., Jr.; Webster, T.A. Crystalline vitamin D. *Nature* **1931**, *128*, 758. [CrossRef]
2. Havinga, E.; Bots, J.P.L. Studies on vitamin D I. The synthesis of vitamin D_3 3 ^{14}C. *Rec. Trav. Chim.* **1954**, *73*, 393–400. [CrossRef]
3. Bouillon, R.; Okamura, W.H.; Norman, A.W. Structure-function relationships in vitamin D endocrine system. *Endocrine Rev.* **1995**, *16*, 200–257. [CrossRef]
4. Zhu, G.-D.; Okamura, W.H. Synthesis of vitamin D (calciferol). *Chem. Rev.* **1995**, *95*, 1877–1952. [CrossRef]
5. Saito, N.; Kittaka, A. Highly potent vitamin D receptor antagonists; design, synthesis and biological evaluation. *ChemBioChem* **2006**, *7*, 1478–1490. [CrossRef]
6. Posner, G.H.; Kahraman, M. Organic chemistry on vitamin D analogues (Deltanoids). *Eur. J. Org. Chem.* **2003**, *2003*, 3889–3895. [CrossRef]
7. PerkinElmer. *CHEM DRAW v.21.0.028 Chemical Drawing Software*; PerkinElmer Informatics, Inc.: Waltham, MA, USA, 2011.
8. Boehm, M.F.; Fitzgerald, P.; Zou, A.; Elgort, M.G.; Bischoff, E.D.; Mere, L.; Mais, D.E.; Bissonnette, R.P.; Heyman, R.A.; Nadzan, A.M.; et al. Novel nonsecosteroidal vitamin D mimics exert VDR-modulating activities with less calcium mobilization than 1,25-dihydroxyvitamin D_3. *Chem. Biol.* **1999**, *6*, 265–275. [CrossRef]
9. Windaus, A.; Linsert, O.; Lütringhaus, A.; Weidlich, G. Über das krystallisierte vitamin D_2. *Ann. Chem.* **1932**, *492*, 226–241. [CrossRef]
10. Jordans, G.H.W. A.T. 10, a new drug against tetany. *Ned. Tijdschr.* **1934**, *78*, 2750–2756.
11. Albright, F.; Sulkowitch, H.W.; Bloomberg, E. A comparison of the effects of vitamin D, dihydrotachysterol (A.T. 10) and parathyroid extract on the disordered metabolism of rickets. *J. Clin. Inv.* **1939**, *18*, 165. [CrossRef]
12. Verloop, A.; Koevoet, A.L.; Havinga, E. Studies on vitamin D compounds and related compounds III. Short communication on the *cis-trans* isomerization of calciferol and properties of *"trans"*-vitamin D_2. *Rec. Trav. Chim.* **1955**, *74*, 1125–1130. [CrossRef]
13. Koevoet, A.L.; Verloop, A.; Havinga, E. Studies on vitamin D compounds and related compounds II. Preliminary communication on the interconversion and the possible *cis-trans* isomerism of previtamin D and tachysterol. *Rec. Trav. Chim.* **1955**, *74*, 788–792. [CrossRef]
14. Westerhof, P.; Keverling Buisman, J.A. Investigations on sterols. VI. The preparation of dihydrotachysterol$_2$. *Rec. Trav. Chim.* **1956**, *75*, 453–462. [CrossRef]
15. Inhoffen, H.H.; Quinkert, G.; Hess, H.-J.; Hirschfeld, H. Studien in der vitamin D-reihe, XXIV. Photo-isomerisierung der trans-vitamin D_2 un D_3 zu den vitaminen D_2 und D_3. *Chem Ber.* **1957**, *90*, 2544–2553. [CrossRef]
16. Westerhof, P.; Keverling Buisman, J.A. Investigations on sterols. IX. Dihydroderivatives of ergocalciferol. *Rec. Trav. Chim.* **1957**, *76*, 679–688. [CrossRef]
17. Inhoffen, H.H.; Irmscher, K.; Hirschfeld, H.; Stache, U.; Kreutzer, A. Partial synthesis of vitamin D_2 and D_3. *J. Chem. Soc.* **1959**, 385–386. [CrossRef]
18. Blunt, J.W.; DeLuca, H.F.; Schnoes, H.K. 25-Hydroxycholecalciferol. A biologically active metabolite of vitamin D_3. *Biochemistry* **1968**, *7*, 3317–3322. [CrossRef]
19. Suda, T.; DeLuca, H.F.; Schnoes, H.K.; Tanaka, Y.; Holick, M.F. 25,26-Dihydroxycholecalciferol, a metabolite of vitamin D_3 with intestinal calcium transport activity. *Biochemistry* **1970**, *9*, 4776–4780. [CrossRef]
20. Redel, J.; Bell, P.; Delbarre, F.; Kodicek, E. Synthese du dihydydroxy-25,26 cholecalciferol, metabolite de la vitamine D_3. *C. R. Acad. Sc. Paris Serie D* **1973**, *276*, 2907–2909.
21. Holick, M.F.; Schnoes, H.K.; DeLuca, H.F. Identification of 1,25-dihydroxycholecalciferol, a form of vitamin D_3 metabolically active in the intestine. *Proc. Nat. Acad. Sci. USA* **1971**, *68*, 803–804. [CrossRef]

22. Norman, A.W.; Myrtle, J.F.; Midgett, R.J.; Noviki, H.G.; Williams, V.; Popják, G. 1,25-Dihydroxycholecalciferol: Identification of the proposed active form of vitamin D_3 in the intestine. *Science* **1971**, *173*, 51–54. [CrossRef] [PubMed]
23. Lawson, D.E.M.; Fraser, D.R.; Kodicek, E.; Morris, H.R.; Dudley, H.W. Calcitriol, identificaction of 1,25-dihydroxycholecalciferol, a new kidney controlling calcium metabolism. *Nature* **1971**, *230*, 228–230. [CrossRef] [PubMed]
24. Lam, H.-Y.; Schnoes, H.K.; DeLuca, H.F.; Chen, T.C. 24,25-Dihydroxyvitamin D_3. Synthesis and biological activity. *Biochemistry* **1973**, *12*, 4851–4855. [CrossRef] [PubMed]
25. Chalmers, T.M.; Hunter, J.O.; Davie, M.W.; Szaz, K.F.; Pelc, B.; Kodicek, E. 1-α-Hydroxycholecalciferol as a substitute for the kidney hormone 1,25-dihydroxycholecalciferol in chronic renal failure. *Lancet* **1973**, *2*, 696–699. [CrossRef]
26. Fürst, A.; Labler, L.; Meier, W.; Pfoertner, K.-H. Synthese von 1α-hydroxycholecalferol. *Helv. Chim. Acta* **1973**, *56*, 1708–1710. [CrossRef]
27. Barton, D.H.R.; Hesse, R.H.; Pechet, M.M.; Rizzardo, E. A convenient synthesis of 1α-hydroxyvitamin D_3. *J. Am. Chem. Soc.* **1973**, *95*, 2748–2749. [CrossRef]
28. Harrison, R.G.; Lythgoe, B.; Wright, P.W. Total synthesis of 1α-hydroxyvitamin D_3. *Tet. Lett.* **1973**, *14*, 3649–3652. [CrossRef]
29. Harrison, R.G.; Lythgoe, B.; Wright, P.W. Calciferol and its relatives. Part XVIII. Total synthesis of 1α-hydroxy-vitamin D3. *J. Chem. Soc. Perkin I* **1974**, 2654–2657. [CrossRef]
30. Lam, H.Y.; Schnoes, H.K.; DeLuca, H.F. 1α-Hydroxyvitamin D_2. Potent synthetic analog of vitamin D_2. *Science* **1974**, *186*, 1038–1040.
31. Ikekewa, N.; Morisaki, M.; Koizumi, N.; Kato, Y.; Takeshita, T. Synthesis of active forms of vitamin D. VIII. Synthesis of [24*R*]- and [24*S*]-1α,24,25-trihydroxyvitamin D_3. *Chem. Pharm. Bull.* **1975**, *23*, 695–697. [CrossRef]
32. Morisaki, M.; Koizumi, N.; Ikekewa, N. Synthesis of active forms of vitamin D. IX. Synthesis of 1α,24-dihydroxycholecalciferol. *J. Chem. Soc. Perkin I* **1975**, 1421–1424. [CrossRef] [PubMed]
33. Lythgoe, B.; Moran, T.A.; Nambudiry, M.E.N.; Ruston, S.; Tideswell, J.; Wright, P.W. Allylic phosphine oxides as precusors of dienes of defined geometry: Synthesis of 3-deoxyvitamin D_2. *Tet. Lett.* **1975**, *44*, 3863–3866. [CrossRef]
34. Okamura, W.H.; Hammond, M.L.; Rego, A.; Norman, A.W.; Wing, R.M. Studies on vitamin D (calciferol) and its analogues. 12. Structural and synthetic studies of 5,6-*trans*-vitamin D_3 and stereoisomers of 10,19-dihydrovitamin D_3 including dihydrotachysterol$_3$. *J. Org. Chem.* **1977**, *42*, 2284–2291. [CrossRef] [PubMed]
35. Mouriño, A.; Blair, P.; Wecksler, W.; Johnson, R.L.; Norman, A.W.; Okamura, W.H. Studies on vitamin D (calciferol) and its analogues. 15. 24-Nor-1α,25-dihydroxyvitamin D_3 and 14-nor-25-hydroxy-5,6-*trans*-vitamin D_3. *J. Med. Chem.* **1978**, *21*, 1025–1029. [CrossRef] [PubMed]
36. Mouriño, A.; Okamura, W.H. Studies on vitamin D (calciferol) and its analogues. 14. On the 10,19-dihydrovitamins related to vitamin D_2 including dihydrotachysterol$_2$. *J. Org. Chem.* **1978**, *43*, 1653–1656. [CrossRef]
37. Yamada, S.; Ohmori, M.; Takayama, H. Synthesis of 24,24-difluoro-1α,25-dihydroxyvitamin D_3. *Chem. Pharm. Bull.* **1979**, *27*, 3196–3198. [CrossRef]
38. Onisko, B.L.; Schnoes, H.K.; DeLuca, H.F. 25-Aza-vitamin D_3, an inhibitor of vitamin D metabolism and action. *J. Biol. Chem.* **1979**, *254*, 3493–3496. [CrossRef]
39. Kocienski, P.J.; Lythgoe, B.; Rutson, S. Calciferol and its relatives. Part 24. A synthesis of vitamin D_4. *J. Chem. Soc. Perkin I* **1979**, 1290–1293. [CrossRef]
40. Kobayahi, Y.; Taguchi, T.; Terada, T. Synthesis of 24,24-difluoro- and 24-fluoro-25-hydroxyvitamin D_3. *Tet. Lett.* **1979**, *20*, 2023–2026. [CrossRef]
41. Yamada, S.; Ohmori, M.; Takayama, H. Synthesis of 24,24-difluoro-25-hydroxyvitamin D_3. *Tet. Lett.* **1979**, *20*, 1859–1862. [CrossRef]
42. Napoli, J.L.; Fivizzani, M.A.; Schnoes, H.K.; DeLuca, H.F. 1-Fluorovitamin D_3, a vitamin D_3 analogue more active on bone-calcium mobilization than on intestinal-calcium transport. *Biochemistry* **1979**, *18*, 1641–1646. [CrossRef] [PubMed]
43. Ishizuka, S.; Bannai, K.; Naruchi, T.; Hashimoto, Y.; Noguchi, T.; Hosoya, N. Studies on the mechanism of action of 1α,24-dihydroxyvitamin D_3. I. Synthesis of 1α,24(*R*)- and 1α,24(*S*)-dihydroxy-[24-^{3}H]- vitamin D_3 and their metabolism in the rat. *J. Biochem.* **1980**, *88*, 87–95. [PubMed]
44. Holick, S.A.; Holick, M.F.; Frommer, J.E.; Henley, J.W.; Lenz, J.A. Synthesis of [3α-^{3}H]-3-epivitamin D_3 and its metabolism in the rat. *Biochem* **1980**, *19*, 3993–3997. [CrossRef]
45. Onisko, B.L.; Schnoes, H.K.; DeLuca, H.F. Inhibitors of the 25-hydroxylation of vitamin D_3 in the rat. *Bioorg. Chem.* **1980**, *9*, 187–198. [CrossRef]
46. Kobayashi, Y.; Taguchi, T.; Kanuma, N. Synthesis of 26,26,26,27,27,27-hexafluoro-25-hydroxyvitamin D_3. *J. Chem Soc. Chem. Comm.* **1980**, 459–460. [CrossRef]
47. Jacobus, D.P.; Jones, H.; Yang, S.S. Cholecalciferol and Dihydrotachysterol3 Derivatives in Metabolic Blocking Drugs. Federal Republic Germany. DE2646240 A1, 28 April 1977.
48. Kocienski, P.J.; Lythgoe, B. Calciferol and its relatives. Part 27. A synthesis of 1α-hydroxyvitamin D_3 by way of 1α-hydroxytachysterol$_3$. *J. Chem. Soc. Perkin I* **1980**, 1400–1404. [CrossRef]
49. Oshida, J.-I.; Morisaki, M.; Ikekawa, N. Synthesis of 2β-fluoro-1α-hydroxyvitamin D_3. *Tet. Lett.* **1980**, *21*, 1755–1756. [CrossRef]
50. Ohmura, N.; Bannai, K.; Yamaguchi, H.; Hashimoto, Y.; Norman, A.W. Isolation of a new metabolite of vitamin D produced in vivo, 1α,25-dihydroxyvitamin D_3-26,23-lactone. *Arch. Biochem. Biophys.* **1980**, *204*, 387–391. [CrossRef]

51. Wichmann, J.K.; Paaren, H.E.; Fivizzani, M.A.; Schnoes, H.K.; DeLuca, H.F. Synthesis of 25-hydroxyvitamin D_3 26,23-lactone. *Tet. Lett.* **1980**, *21*, 4667–4670. [CrossRef]
52. Eguchi, T.; Takatsuto, S.; Hirano, Y.; Ishiguro, M.; Ikekawa, N. Synthesis of four isomers of 25-hydroxyvitamin D_3 26,23-lactone. *Heterocycles* **1982**, *17*, 359–375.
53. Paaren, H.E.; Fivizzani, M.A.; Schnoes, H.K.; DeLuca, H.F. 1α,25-Difluorovitamin D_3: An inert vitamin D analog. *Arch. Biochem. Biophys.* **1981**, *209*, 579–583. [CrossRef]
54. Wichmann, J.; Schnoes, H.K.; DeLuca, H.F. Isolation and identification of 24(*R*)-hydroxyvitamin D_3 from chicks given large doses of vitamin D_3. *Biochemistry* **1981**, *20*, 2350–2353. [CrossRef] [PubMed]
55. Wichmann, J.; Schnoes, H.K.; DeLuca, H.F. 23,24,25-Trihydroxyvitamin D_3, 24,25,26-trihydroxyvitamin D_3, 24-keto-25-hydroxyvitamin D_3 and 23-dehydro-25-hydroxyvitamin D_3. *Biochemistry* **1981**, *20*, 7385–7391. [CrossRef] [PubMed]
56. Esvelt, R.P.; Fivizzani, M.A.; Paaren, H.E.; Schnoes, H.K.; DeLuca, H.F. Synthesis of calcitroic acid, a metabolite of 1α,25-dihydroxycholecalciferol. *J. Org. Chem.* **1981**, *46*, 456–458. [CrossRef]
57. Eguchi, T.; Takatsuto, S.; Ishiguro, M.; Ikekawa, N.; Tanaka, Y.; DeLuca, H.F. Synthesis and determination of configuration of natural 25-hydroxyvitamin D_3 26,23-lactone. *Proc. Nat. Acad. Sci. USA* **1981**, *78*, 6579–6583. [CrossRef]
58. Kobayashi, Y.; Taguchi, T.; Kanuma, N. Synthesis of 26,26,26-trifluoro-25-hydroxy and 27-nor-26,26,26-trifluoro-25-hydroxyvitamin D_3. *Tet. Lett.* **1981**, *22*, 4309–4312. [CrossRef]
59. Matoba, K.; Kondo, K.; Yamazaki, T. Syntheses of vitamin D analogs I. *Chem. Pharm. Bull.* **1982**, *30*, 4593–4596. [CrossRef]
60. Kobayashi, Y.; Taguchi, T.; Mitsihashi, S.; Eguchi, T.; Ohshima, E.; Ikekawa, N. Studies on organic fluorine compounds. XXXIX. Studies on steroids. LXXXIX. Synthesis of 1α,25-dihydroxy-26,26,26,27,27,27-hexafluorovitamin D_3. *Chem. Pharm. Bull.* **1982**, *30*, 4297–4303. [CrossRef]
61. Haces, A.; Okamura, W.H. Heterocalciferols: Novel 3-thia and 3-sulfinyl analogues of 1α-hydroxyvitamin D_3. *J. Am. Chem. Soc.* **1982**, *104*, 6105–6109. [CrossRef]
62. Yamada, S.; Ohmori, M.; Takayama, H.; Takasaki, Y.; Suda, T. Isolation and identification of 1α- and 23-hydroxylated metabolites of 25-hydroxy-24-oxo-vitamin D_3 from in vitro incubates of chick kidney homogenates. *J. Biol. Chem.* **1983**, *258*, 457–463. [CrossRef]
63. Wovkulich, P.M.; Barcelos, F.; Batcho, A.D.; Sereno, J.F.; Baggiolini, E.G.; Hennessy, B.M.; Uskokovic, M.R. Stereoselective total synthesis of 1α,25*S*,26-trihydroxycholecalciferol. *Tetrahedron* **1984**, *40*, 2283–2296. [CrossRef]
64. Barner, R.; Hübscher, J.; Daly, J.J.; Schönholzer, P. Zur Konfiguration des Vitamin D_3-Metaboliten 25,26-Dihydroxycholecalciferol: Synthese von (25*S*,26)- und (25*R*,26)-Dihydroxy-cholecalciferol. *Helv. Chim. Acta* **1981**, *64*, 915–938. [CrossRef]
65. Toh, H.T.; Okamura, W.H. Studies on a convergent route to side-chain analogues of vitamin D: 25-hydroxy-23-oxavitamin D_3. *J. Org. Chem.* **1983**, *48*, 1414–1417. [CrossRef]
66. Midgley, J.M.; Upton, R.M.; Watt, R.A.; Whalley, W.B.; Zhang, X.M. Unsaturated steroids. Part 11. Synthesis of 1α-hydroxy-25-methyl vitamin D_3. *J. Chem. Res., Synopses* **1983**, *11*, 273.
67. Taguchi, T.; Mitsuhashi, S.; Yamanouchi, A.; Kabayashi, Y. Synthesis of 23,23-difluoro-25-hydroxyvitamin D_3. *Tet. Lett.* **1984**, *25*, 4933–4936. [CrossRef]
68. Yamada, S.; Yamamoto, K.; Naito, H.; Suzuki, T.; Ohmori, M.; Takayama, H.; Shiina, Y.; Miyaura, C.; Tanaka, H.; Abe, E.; et al. Synthesis and differentiating action of vitamin D endoperoxides. Singlet oxygen adducts of vitamin D derivatives in human myeloid leukemia cells (HL-60). *J. Med. Chem.* **1985**, *28*, 1148–1153. [CrossRef]
69. Dauben, W.G.; Kohler, B.; Roesle, A. Synthesis of 6-fluorovitamina D_3. *J. Org. Chem.* **1985**, *50*, 2007–2010. [CrossRef]
70. Tanaka, Y.; Sicinski, R.R.; DeLuca, H.F.; Sai, H.; Ikekawa, N. Unique rearrangement of ergosterol side chain in vivo; production of a biologically highly active homologue of 1α,25-dihydroxyvitamin D_3. *Biochemistry* **1986**, *25*, 5512–5518. [CrossRef]
71. Reddy, G.S.; Tserng, K.-Y. Isolation and identification of 1,24,25-trihydroxyvitamin D_2, 1,24,25,28-tetrahydroxyvitamin D_2 and 1,24,25,26-tetrahydroxyvitamin D_2: New metabolites of 1,25-dihydroxyvitamin D_2 produced in rat kidney. *Biochemistry* **1986**, *25*, 5328–5336. [CrossRef]
72. Kubodera, N.; Miyamoto, K.; Ochi, K.; Matsunaga, M. Synthetic studies of vitamin D analogues. VII. Synthesis of 20-oxa-21-norvitamin D_3 analogues. *Chem. Pharm. Bull.* **1986**, *34*, 2286–2289. [CrossRef]
73. Sai, H.; Takatsuto, S.; Ikekawa, N.; Tanaka, I.; DeLuca, H.F. Synthesis of side-chain homologues of 1,25-dihydroxyvitamin D_3 and investigation of their biological activities. *Chem. Pharm. Bull.* **1986**, *34*, 4508–4515. [CrossRef] [PubMed]
74. Sardina, F.J.; Mouriño, A.; Castedo, L. Studies on the synthesis of side-chain hydroxylated metabolites of vitamin D. 2. Stereocontrolled synthesis of 25-hydroxyvitamin D_2. *J. Org. Chem.* **1986**, *51*, 1264–1269. [CrossRef]
75. Mascareñas, J.L.; Mouriño, A.; Castedo, L. Studies on the synthesis of side-chain hydroxylated metabolites of vitamin D. 3. Synthesis of 25-ketovitamin D_3 and 25-hydroxyvitamin D_3. *J. Org. Chem.* **1986**, *51*, 1269–1272. [CrossRef]
76. Murayama, E.; Miyamoto, K.; Kubodera, N.; Mori, T.; Matsunaga, M. Synthetic studies of vitamin D analogues. VIII. Synthesis of 22-oxavitamin D_3 analogues. *Chem. Pharm. Bull.* **1986**, *34*, 4410–4413. [CrossRef] [PubMed]
77. Sicinski, R.R.; DeLuca, H.F.; Schnoes, H.K.; Tanaka, Y.; Smith, C.M. Δ^{22}-Unsaturated analogs of vitamin D_3 and their C(1)-hydroxylated derivatives. *Bioorg. Chem.* **1987**, *15*, 152–166. [CrossRef]
78. Ikekawa, N.; Eguchi, T.; Hara, N.; Takatsuto, S.; Honda, A.; Mori, Y.; Otomo, S. 26,27-Diethyl-1α,25-dihydroxyvitamin D_3 and 24,24-difluoro-24-homo-1α,25-dihydroxyvitamin D_3: Highly potent inducer for differentiation of human leukemia cells HL-60. *Chem. Pharm. Bull.* **1987**, *35*, 4362–4365. [CrossRef]

79. Calverley, M.J. Synthesis of MC 903, a biologically active vitamin D metabolite analogue. *Tetrahedron* **1987**, *43*, 4609–4619. [CrossRef]
80. Eguchi, T.; Sai, H.; Takatsuto, S.; Hara, N.; Ikekawa, N. Synthesis of 26,27-dialkyl analogues of 1α,25-dihydroxyvitamin D_3. *Chem. Pharm. Bull.* **1988**, *36*, 2303–2311. [CrossRef]
81. Shiuey, S.-J.; Partridge, J.J.; Uskokovic, M.R. Triply convergent synthesis of 1α,25-dihydroxy-24(*R*)-fluorocholecalciferol. *J. Org. Chem.* **1988**, *53*, 1040–1046. [CrossRef]
82. Barrack, S.A.; Gibbs, R.A.; Okamura, W.H. Potential inhibitors of vitamin D metabolism: An oxa analogue of vitamin D. *J. Org. Chem.* **1988**, *53*, 1790–1796. [CrossRef]
83. Baggiolini, E.G.; Hennessy, B.M.; Truitt, G.A.; Uskokovic, M.R. Dehydrocholecalciferol Derivatives. U.S. US14532882A, 20 January 1988.
84. Kutner, A.; Perlman, K.L.; Lago, A.; Sicinski, R.R.; Schnoes, H.K.; DeLuca, H.F. Novel convergent synthesis of side-chain modified analogues of 1α,25-dihydroxycholecalciferol and 1α,25-dihydroxyergocalciferol. *J. Org. Chem.* **1988**, *53*, 3450–3457. [CrossRef]
85. Okano, T.; Tsugawa, N.; Masuda, S.; Takeuchi, A.; Kobayashi, T.; Takita, Y.; Nishii, Y. Regulatory activities of 2β-(3-hydroxypropoxy)-1α,25-dihydroxy-vitamin D_3, a novel synthetic vitamin D_3 derivative, on calcium metabolism. *Biochem. Biophys. Res. Commm.* **1989**, *163*, 1444–1449. [CrossRef]
86. Eguchi, T.; Yoshida, M.; Ikekawa, N. Synthesis and biological activities of 22-hydroxy and 22-methoxy derivatives of 1α,25-dihydroxyvitamin D_3: Importance of side chain conformation for biological activities. *Bioorg. Chem.* **1989**, *17*, 294–307. [CrossRef]
87. Dauben, W.G.; Ollmann, R.R., Jr.; Funhoff, A.S.; Neidlein, R. The synthesis of 25-oxo-25-phosphavitamin D_3. *Tett. Lett.* **1989**, *30*, 677–680. [CrossRef]
88. Perlman, K.; Kutner, A.; Prahl, J.; Smith, C.; Inaba, M.; Schnoes, H.K.; DeLuca, H.F. 24-Homologated 1,25-dihydroxyvitamin D_3 compounds: Separation of calcium and cell differentiation activities. *Biochemistry* **1990**, *29*, 190–196. [CrossRef] [PubMed]
89. Gill, H.S.; Londonwski, J.M.; Corradino, R.A.; Zinsmeister, A.R.; Kumar, R. Synthesis and biological activity of novel vitamin D analogues: 24,24-difluoro-25-hydroxy-26,27-dimethylvitamin D_3 and 24,24-difluoro-1α,25-dihydroxy-26,27-dimethyl vitamin D_3. *J. Med. Chem.* **1990**, *33*, 480–490. [CrossRef]
90. Hara, N.; Eguchi, T.; Ikekawa, N.; Ishizuka, S.; Sato, J.-i. Synthesis and biological activity of (22*E*,25*R*)- and (22*E*,25*S*)-22-dehydro-1α,25-dihydroxy-26-methylvitamin D_3. *J. St. Biochem.* **1990**, *35*, 655–664. [CrossRef]
91. Binderup, L.; Latini, S.; Binderup, E.; Bretting, C.; Calverley, M.; Hansen, K. 20-epi-vitamin D_3 analogues: A novel class of potent regulators of cell growth and immune responses. *Biochem. Pharm.* **1991**, *42*, 1569–1575. [CrossRef]
92. Kubodera, N.; Miyamoto, K.; Akiyama, M.; Matsumoto, M.; Mori, T. Synthetic studies of vitamin D analogues. IX. Synthesis and diffrentiation-inducing activity of 1α,25-dihydroxy-23-oxa-, thia-, and azavitamin D_3. *Chem. Pharm. Bull.* **1991**, *39*, 3221–3224. [CrossRef]
93. Figadere, B.; Norman, A.W.; Henry, H.L.; Koeffler, H.P.; Zhou, J.Y.; Okamura, W.H. Arocalciferols: Synthesis and biological evaluation of aromatic side-chain analogues of 1α,25-dihydroxyvitamin D_3. *J. Med. Chem.* **1991**, *34*, 2452–2463. [CrossRef]
94. Okamura, W.H.; Aurrecoechea, J.M.; Gibbs, R.A.; Norman, A.W. Synthesis and biological activity of 9,11-dehydrovitamin D_3 analogues: Stereoselective preparation of 6β-vitamin D vinylallenes and concise enynol synthesis for preparing the A-ring. *J. Org. Chem.* **1989**, *54*, 4072–4083. [CrossRef]
95. Eguchi, T.; Kakinuma, K.; Ikekawa, N. Synthesis of 1α-[19-^{13}C]hydroxyvitamin D_3 and ^{13}C NMR analysis of the conformational equilibrium of the A-ring. *Bioorg. Chem.* **1991**, *19*, 327–332. [CrossRef]
96. Chodynski, M.; Kutner, A. Synthesis of side-chain homologated analogs of 1,25-dihydroxycholecalciferol and 1,25-dihydroxyergocalciferol. *Steroids* **1991**, *56*, 311–315. [CrossRef]
97. Baggiolini, E.G.; Hennessy, B.M.; Shiuey, S.J.; Truitt, G.A.; Uskokovic, M.R. Preparation of Dehydrocholecalciferol Derivatives for Treatment of Hyperproliferative Skin Diseases and Neoplasms and Pharmaceutical Compositions Containing Them. European Patent Organization. EP325279A1, 26 July 1989.
98. Maestro, M.A.; Sardina, F.J.; Castedo, L.; Mouriño, A. Stereoselective synthesis and thermal rearrangement of the first analogue of (7Z)-vitamin D. *J. Org. Chem.* **1991**, *56*, 3582–3587. [CrossRef]
99. Ray, R.; Bouillon, R.; Van Baelen, H.; Holick, M.F. Synthesis of 25-hydroxyvitamin D_3 3β-3′-[*N*-(4-azido-2-nitrophenyl)amino]propyl ether, a second-generation photoaffinity analogue of 25-hydroxyvitamin D_3. Photoaffinity labeling of rat serum vitamin D binding protein. *Biochemistry* **1991**, *30*, 4809–4813. [CrossRef] [PubMed]
100. Perlman, K.L.; Swenson, R.E.; Paaren, H.E.; Schnoes, H.K.; DeLuca, H.F. Novel synthesis of 19-nor-vitamin compounds. *Tet. Lett.* **1991**, *32*, 7663–7666. [CrossRef]
101. Perlman, K.L.; DeLuca, H.F. 1α-Hydroxy-19-nor-vitamina D C-22 aldehyde. A versatile intermediate in the synthesis of side chain modified 1α,25-dihydroxy-19-nor-vitamina D_3. *Tet. Lett.* **1992**, *33*, 2937–2940. [CrossRef]
102. Mascareñas, J.L.; Sarandeses, L.A.; Castedo, L.; Mouriño, A. Palladium-catalysed coupling of vinyl triflates with enynes and its application to the synthesis of 1α,25-dihydroxyvitamin D_3. *Tetrahedron* **1991**, *47*, 3485–3498. [CrossRef]
103. Kubodera, N.; Miyamoto, K.; Matsumoto, M.; Kawanishi, T.; Ohkawa, H.; Mori, T. Synthetic studies of vitamin D analogues. X. Synthesis and biological activity of 1α,25-dihydroxy-21-norvitamin D_3. *Chem. Pharm. Bull.* **1992**, *40*, 648–651. [CrossRef]
104. Iseki, K.; Nagal, T.; Kobayashi, Y. Synthesis of 24-homo-26,26,26,27,27,27-hexafluoro-1α,22.25-trihydroxyvitamin D_3. *Chem. Pharm. Bull.* **1992**, *40*, 1346–1348. [CrossRef]

105. Kubodera, N.; Watanabe, H.; Kawanishi, T.; Matsumoto, M. Synthetic studies of vitamin D analogues. XI. Synthesis and differentiation-inducing activity of 1α,25-dihydroxy-22-oxavitamin D_3 analogues. *Chem. Pharm. Bull.* **1992**, *40*, 1494–1499. [CrossRef] [PubMed]
106. Ohira, Y.; Taguchi, T.; Iseki, K.; Kobayashi, Y. Preparation of (22*S*)- and (22*R*)-24-homo-26,26,26,27,27,27-hexafluoro-1,22,25-trihydroxy-24-yne-vitamin D_3. *Chem. Pharm. Bull.* **1992**, *40*, 1647–1649. [CrossRef] [PubMed]
107. Bouillon, R.; Allewaert, K.; van Leeuwen, P.T.M.; Tan, B.-K.; Xiang, D.Z.; De Clercq, P.; Vandewalle, M.; Pols, H.A.P.; Bos, M.P.; Van Baelen, H.; et al. Structure function analysis of vitamin D analogs with C-ring modifications. *J. Biol. Chem.* **1992**, *267*, 3044–3051. [CrossRef]
108. Okamura, W.H.; Palenzuela, J.A.; Plumet, J.; Midland, M.M. Vitamin D: Structure-function analyses and the design of analogs. *J. Cell Biochem.* **1992**, *49*, 10–18. [CrossRef] [PubMed]
109. Maynard, D.F.; Norman, A.W.; Okamura, W.H. 18-Substituted derivatives of vitamin D: 18-acetoxy-1α,25-dihydroxyvitamin D_3 and related analogues. *J. Org. Chem.* **1992**, *57*, 3214–3217. [CrossRef]
110. Lee, A.S.; Norman, A.W.; Okamura, W.H. 3-Deoxy-3-thia-1α,25-dihydroxyvitamin D_3 and its 1β-epimer: Synthesis and biological evaluation. *J. Org. Chem.* **1992**, *57*, 3846–3854. [CrossRef]
111. Craig, A.S.; Norman, A.W.; Okamura, W.H. Two novel allenic side chain analogues of 1α,25-dihydroxyvitamin D_3. *J. Org. Chem.* **1992**, *57*, 4374–4380. [CrossRef]
112. Maestro, M.A.; Castedo, L.; Mouriño, A. A convergent appoach to the dihydrotachysterol diene system. Application to the synthesis of dihydrotachysterol$_2$ (DHT_2), 25-hydroxydihydrotachysterol$_2$ (25-OH-DHT_2), 10(*R*),19-dihydro-(5*E*)-epivitamina D_2 and 25-hydroxy-10(*R*),19-dihydro-(5*E*)-epivitamina D_2. *J. Org. Chem.* **1992**, *57*, 528–5213. [CrossRef]
113. Torneiro, M.; Fall, Y.; Castedo, L.; Mouriño, A. An efficient route to 1α,25-dihydroxyvitamin D_3 functionalized at C-11. *Tet. Lett.* **1992**, *33*, 105–108. [CrossRef]
114. Vallés, M.J.; Castedo, L.; Mouriño, A. Functionalization of vitamin D metabolites at C-18 and application to the synthesis of 1α,18,25-trihydroxyvitamin D_3 and 18,25-dihydroxyvitamin D_3. *Tet. Lett.* **1992**, *33*, 1503–1506. [CrossRef]
115. Sarandeses, L.A.; Mascareñas, J.L.; Castedo, L.; Mouriño, A. Synthesis of 1α,25-dihydroxy-19-norprevitamin D_3. *Tet. Lett.* **1992**, *33*, 5445–5448. [CrossRef]
116. Posner, G.H.; Nelson, T.D.; Guyton, K.Z.; Kensler, T.W. New vitamin D_3 derivatives with unexpected antiproliferative activity: 1-(hydroxymethyl)-25-hydroxyvitamin D_3 homologs. *J. Med. Chem.* **1992**, *35*, 3280–3287. [CrossRef] [PubMed]
117. Steinmeyer, A.; Neef, G.; Kirsch, G.; Schwarz, K.; Rach, P.; Habery, M.; Thieroff-Ekerdt, R. Synthesis and biological activities of 8(14)a-homocalcitriol. *Steroids* **1992**, *57*, 447–452. [CrossRef]
118. Kutner, A.; Chodynski, M.; Halkes, S.J.; Brugman, J. Novel concurrent synthesis of side-chain analogues of vitamina D_2 and D_3: 24,24-dihomo-25-hydroxycholecalciferol and (22*E*)-22-dehydro-24,24-dihomo-25-hydroxycholecalciferol. *Bioorg. Chem.* **1993**, *21*, 13–23. [CrossRef]
119. Calverley, M.J.; Bretting, C.A.S. 1α,24*S*-Dihydroxy-26,27-cyclo-22-yne-vitamin D_3: Tha side chain triple bond analogue of MC 903 (calcipotriol). *Bioorg. Med. Chem. Lett.* **1993**, *3*, 1841–1844. [CrossRef]
120. Calverley, M.J.; Binderup, L. Synthesis and biological evaluation of MC 1357, a new 20-epi-23-oxa-1α,25-dihydroxy-vitamin D_3 analogue with potent non-classical effects. *Bioorg. Med. Chem. Lett.* **1993**, *3*, 1845–1848. [CrossRef]
121. Batcho, A.D.; Sereno, J.F.; Hennessy, B.M.; Baggiolini, E.G.; Uskokovic, M.R.; Horst, R.L. Total synthesis of 1α,25,28-trihydroxyergocalciferol. *Bioorg. Med. Chem. Lett.* **1993**, *3*, 1821–1824. [CrossRef]
122. Nilsson, K.; Vallés, M.J.; Castedo, L.; Mouriño, A.; Halkes, S.J.; van de Velde, J.P. Synthesis and biological evaluation of 18-substituted analogs of 1α,25-dihydroxyvitamin D_3. *Bioorg. Med. Chem. Lett.* **1993**, *3*, 1855–1858. [CrossRef]
123. Posner, G.H.; Dai, H. 1-(Hydroxyalkyl)-25-hydroxyvitamin D_3 analogs of calcitriol. 1. Synthesis. *Bioorg. Med. Chem. Lett.* **1993**, *3*, 1829–1834. [CrossRef]
124. Allewaert, K.; Van Baelen, H.; Bouillon, R.; Zhao, X.-y.; De Clercq, P.; Vandewalle, M. Synthesis and biological evaluation of 23-oxa-, 23-thia- and 24-oxa-24-oxo-1α,25-dihydroxyvitamin D_3. *Bioorg. Med. Chem. Lett.* **1993**, *3*, 1859–1862. [CrossRef]
125. Sarandeses, L.A.; Vallés, M.J.; Castedo, L.; Mouriño, A. Synthesis of 24-oxavitamin D_3 and 1α-hydroxy-24-oxavitamin D_3. *Tetrahedron* **1993**, *49*, 731–738. [CrossRef]
126. Okabe, M.; Sun, R.-C. An efficient synthesis of (22*E*,25*R*)-1α,25,26-trihydroxy-Δ^{22}-vitamin D_3. *Tet. Lett.* **1993**, *34*, 6533–6536. [CrossRef]
127. Curtin, M.L.; Okamura, W.H. 1α,25-Dihydroxyprevitamin D_3: Synthesis of the 9,14,19,19,19-pentadeuterio derivative and a kinetic study of its [1,7]-sigmatropic shift to 1α,25-dihydroxyvitamin D_3. *J. Am. Chem. Soc.* **1991**, *113*, 6958–6966. [CrossRef]
128. Zhao, X.-y.; De Clercq, P.; Vandewalle, M.; Alkwaert, K.; Van Baelen, I.; Bouillon, R. Synthesis and biological evaluation of some 25,26-epoxy-1α,24-dihydroxyvitamin D_3. *Bioorg. Med. Chem. Lett.* **1993**, *3*, 1863–1867. [CrossRef]
129. Allewaert, K.; Zhao, X.-Y.; Zhao, J.; Glibert, F.; Branisteanu, D.; De Clercq, P.; Vandewalle, M.; Bouillon, R. Biological evaluation of epoxy analogs of 1α,25-dihydroxyvitamin D_3. *Steroids* **1995**, *60*, 324–336. [CrossRef]
130. Choudhry, S.C.; Belica, P.S.; Coffen, D.L.; Focella, A.; Maehr, H.; Manchand, P.S.; Serico, L.; Yang, R.T. Synthesis of a biologically active vitamin D_2 metabolite. *J. Org. Chem.* **1993**, *58*, 1496–1500. [CrossRef]
131. Muralidharan, K.R.; de Lera, A.R.; Isaeff, S.D.; Norman, A.; Okamura, W.H. Studies on the A-ring diastereomers of 1α,25-dihydroxyvitamin D_3. *J. Org. Chem.* **1993**, *58*, 1895–1899. [CrossRef]

132. Sicinski, R.R.; Perlman, K.L.; DeLuca, H.F. Synthesis and biological activity of 2-hydroxy and 2-alkoxy analogs of 1α,25-dihydroxy-19-norvitamin D_3. *J. Med. Chem.* **1994**, *37*, 3730–3738. [CrossRef]
133. Schoerder, N.J.; Trafford, D.J.H.; Cunningham, J.; Jones, G.; Makin, H.L.J. In vivo dihydrotachysterol$_2$ metabolism in normal man: 1α- and 1β-hydroxylation of 25-hydroxytachysterol$_2$ and effects on plasma parathyroid hormone and 1α,25-dihydroxyvitamin D_3 concentrations. *J. Clin. Endocr. Met.* **1994**, *78*, 1841–1847. [CrossRef]
134. Perlman, K.L.; Prahl, J.M.; Smith, C.; Sicinski, R.R.; DeLuca, H.F. 26,27-Dihomo-1α-hydroxy- and 26,27-dihomo-1α,25-hydroxyvitamina D_2 analogs that differ markedly in biological activity in vivo. *J. Biol. Chem.* **1994**, *269*, 24014–24019. [CrossRef]
135. Ishida, H.; Shimizu, M.; Yamamoto, K.; Iwasaki, Y.; Yamada, S. Syntheses of 1-alkyl-1,25-dihydroxyvitamin D_3. *J. Org. Chem.* **1995**, *60*, 1828–1833. [CrossRef]
136. VanAlstyne, E.M.; Norman, A.W.; Okamura, W.H. 7,8-Cis Geometric isomers of the steroid hormone 1α,25-dihydroxyvitamin D_3. *J. Am. Chem. Soc.* **1994**, *116*, 6207–6216. [CrossRef]
137. Okamoto, M.; Fujii, T.; Tanaka, T. The first convergent synthesis of 1α,24(*R*)-dihydroxyvitamin D_3 via diastereoselective isopropylation and alkylative enyne cyclization. *Tetrahedron* **1995**, *51*, 5543–5556. [CrossRef]
138. Schwarz, K.; Neef, G.; Kirsh, G.; Müller-Fahrnow, A.; Steinmeyer, A. Synthesis of 20-fluorovitamin D analogues. *Tetrahedron* **1995**, *51*, 9543–9550. [CrossRef]
139. Iseki, K.; Oishi, S.; Namba, H.; Taguchi, T.; Kobayashi, Y. Preparation and biological activity of 24-epi-26,26,26,27,27,27-hexafluoro-1α,25-dihydroxyvitamin D_2. *Chem. Pharm. Bull.* **1995**, *43*, 1897–1901. [CrossRef]
140. Sabbe, K.; D'Hallewyn, C.; De Clercq, P.; Vandewalle, M.; Bouillon, R.; Verstuyf, A. Synthesis of CD-ring modified 1α,25-dihydroxyvitamin D analogues: E-ring analogues. *Bioorg. Med. Chem. Lett.* **1996**, *6*, 1697–1702. [CrossRef]
141. Zhu, G.-D.; Chen, Y.; Zhou, X.; De Clercq, P.; Vandewalle, M.; Bouillon, R.; Verstuyf, A. Synthesis of CD-ring modified 1α,25-dihydroxyvitamin D analogues: C-ring analogues. *Bioorg. Med. Chem. Lett.* **1996**, *6*, 1703–1708. [CrossRef]
142. Yamamoto, K.; Sun, W.Y.; Ohta, M.; Hamada, K.; DeLuca, H.F.; Yamada, S. Conformationally restricted analogs of 1α,25-dihydroxyvitamin D_3 and its 20-epimer: Compounds for study of the three-dimension structure of vitamin D responsible for binding to the receptor. *J. Med. Chem.* **1996**, *39*, 2727–2737. [CrossRef]
143. Yong, W.; Ling, S.; D'Hallewyn, C.; Van Haver, D.; De Clercq, P.; Vandewalle, M.; Bouillon, R.; Verstuyf, A. Synthesis of CD-ring modified 1α,25-dihydroxyvitamin D analogues: Five membered D-ring analogues. *Bioorg. Med. Chem. Lett.* **1997**, *7*, 923–928. [CrossRef]
144. Linclau, B.; De Clercq, P.; Vandewalle, M.; Bouillon, R.; Verstuyf, A. The synthesis of CD-ring modified 1α,25-dihydroxyvitamin D: Six-membered D-ring analogues I. *Bioorg. Med. Chem. Lett.* **1997**, *7*, 1461–1464. [CrossRef]
145. Linclau, B.; De Clercq, P.; Vandewalle, M.; Bouillon, R.; Verstuyf, A. The synthesis of CD-ring modified 1α,25-dihydroxyvitamin D: Six-membered D-ring analogues II. *Bioorg. Med. Chem. Lett.* **1997**, *7*, 1465–1468. [CrossRef]
146. Scheddin, D.; Mayer, H.; Wittmann, S.; Schönecker, B.; Gliesing, S.; Reichenbächer, M. Synthesis and biological activities of 26-hydroxy-27-nor-derivatives of 1α,25-dihydroxyvitamin D_3. *Steroids* **1996**, *61*, 598–608. [CrossRef]
147. Hedlund, T.E.; Moffatt, K.A.; Uskokovic, M.R.; Miller, G.J. Three synthetic vitamin D analogues induce protate-specific acid phosphate and protate-specific antigen while inhibiting the growth of human prostate cancer cells in a vitamin D receptor-depenent fashion. *Clin. Cancer Res.* **1997**, *3*, 1331–1338. [PubMed]
148. Kabat, M.M.; Burger, W.; Guggino, S.; Hennessy, B.; Iacobelli, J.A.; Takeuchi, K.; Uskokovic, M.R. Total synthesis of 25-hydroxy-16,23*E*-diene vitamin D_3 and 1α,25-dihydroxy-16,23*E*-diene vitamin D_3: Separation of genomic and non-genomic vitamin D activities. *Bioorg. Med. Chem.* **1998**, *6*, 2051–2059. [CrossRef]
149. Asou, H.; Koike, M.; Elstner, E.; Cambel, M.; Le, J.; Uskokovic, M.R.; Kamada, N.; Koeffler, H.P. 19-nor Vitamin D analogs: A new class of potent inhibitors of proliferation and differentiation of human myeloid leukemia cell lines. *Blood* **1998**, *92*, 2441–2449. [CrossRef]
150. Torneiro, M.; Fall, Y.; Castedo, L.; Mouriño, A. A short, efficient copper-mediated synthesis of 1α,25-dihydroxyvitamin D_2 (1α,25-dihydroxyergocalciferol) and C-24 analogs. *J. Org. Chem.* **1997**, *62*, 6344–6352. [CrossRef]
151. Andrews, D.R.; Barton, D.H.R.; Cheng, K.P.; Finet, J.P.; Hesse, R.H.; Johnson, G.; Pechet, M.M. A direct, regio- and stereoselective 1α-hydroxylation of (5*E*)-calciferol derivative. *J. Org. Chem.* **1986**, *51*, 1635–1637. [CrossRef]
152. Konno, K.; Maki, S.; Fujishima, T.; Liu, Z.; Miura, D.; Chokki, M.; Takayama, H. A novel and practical route to A-ring enyne synthon for 1α,25-dihydroxyvitamin D_3 analogues: Synthesis of A-ring diastereomers of 1α,25-dihydroxyvitamin D_3 and 2-methyl-1α,25-dihydroxyvitamin D_3. *Bioorg. Med. Chem. Lett.* **1998**, *8*, 151–156. [CrossRef]
153. Fujishima, T.; Liu, Z.; Miura, D.; Chokki, M.; Ishizuka, S.; Konno, K.; Takayama, H. Synthesis and biological activity of 2-methyl-20-epi-1α,25-dihydroxyvitamin D_3. *Bioorg. Med. Chem. Lett.* **1998**, *8*, 2145–2148. [CrossRef]
154. Grue-Sørensen, G.; Hansen, C.M. New 1α,25-dihydroxyvitamin D_3 analogues with side chains attached to C-18: Synthesis and biological activity. *Bioorg. Med. Chem.* **1998**, *6*, 2029–2039. [CrossRef]
155. Posner, G.H.; Lee, J.K.; White, M.C.; Hutchings, R.H.; Dai, H.; Kachinski, J.L.; Dolan, P.; Kensler, T.W. Antiproliferative hibrid analogs of the hormone 1α,25-dihydroxyvitamin D_3: Design, synthesis, and preliminary biological evaluation. *J. Org. Chem.* **1997**, *62*, 3299–3314. [CrossRef]

156. Sicinski, R.R.; Prahl, J.M.; Smith, C.M.; DeLuca, H.F. New 1α,25-dihydroxy-19-norvitamin D_3 compounds of high biological activity: Synthesis and biological evaluation of 2-hydroxymethyl, 2-methyl, and 2-methylene analogues. *J. Med. Chem.* **1998**, *41*, 4662–4674. [CrossRef] [PubMed]
157. Posner, G.H.; Lee, J.K.; Wang, Q.; Peleg, S.; Burke, M.; Brem, H.; Dolan, P.; Kensler, T.W. Noncalcemic, antiproliferative, transcriptionally active, 24-fluorinated hybrid analogues of the hormone 1α,25-dihydroxyvitamin D_3. Synthesis and preliminary biological evaluation. *J. Med. Chem.* **1998**, *41*, 3008–3014. [CrossRef] [PubMed]
158. Posner, G.H.; Wang, Q.; Han, G.; Lee, J.K.; Crawford, K.; Zand, S.; Brem, H.; Peleg, S.; Dolan, P.; Kensler, T.W. Conceptually new sulfone analogues of the hormone 1α,25-dihydroxyvitamin D_3: Synthesis and preliminary biological evaluation. *J. Med. Chem.* **1999**, *42*, 3425–3435. [CrossRef] [PubMed]
159. Odrzywolska, M.; Chodynski, M.; Zorgdrager, J.; Van de Velde, J.-P.; Kutner, A. Diasteroselective synthesis, binding affinity for vitamin D receptor, and chiral stationary phase chromatography of hydroxy analogs of 1α,25-dihydroxycholecalciferol and 25-hydroxycholecalciferol. *Chirality* **1999**, *11*, 701–706. [CrossRef]
160. Kawashima, H.; Hoshina, K.; Hashimoto, Y.; Takeshita, T.; Ishimoto, S.; Noguchi, T.; Ikekawa, N.; Morisaki, M.; Orimo, H. Biological activity of 1α,24-dihydroxycholecalciferol: A new synthetic analog of the hormonal form of vitamin D. *FEBS Lett.* **1977**, *76*, 177–181. [CrossRef]
161. Miura, D.; Manabe, K.; Gao, Q.; Norman, A.W.; Ishizuka, S. 1α,25-Dihydroxyvitamin D_3-26,23-lactone analogs antagonize differentiation of human leukemia cells (HL-60 cells) but not of human acute promyelocytic leukemia cells (NB4 cells). *FEBS Lett.* **1999**, *460*, 297–302. [CrossRef]
162. Fujishima, T.; Konno, K.; Nakagawa, K.; Kurobe, M.; Okano, T.; Takayama, H. Efficient synthesis and biological evaluation of all A-ring diastereomers of 1α,25-dihydroxyvitamin D_3 and its 20-epimer. *Bioorg. Med. Chem.* **2000**, *8*, 123–134. [CrossRef]
163. El Abdaimi, K.; Dion, N.; Papavasiliou, V.; Cardinal, P.-E.; Binderup, L.; Goltzman, D.; Ste-Marie, L.-G.; Kremer, R. The vitamin D analogue EB1039 prevents skeletal metastasis and prolongs survival time in nude mice transplanted with human breastcancer cells. *Cancer Res.* **2000**, *60*, 4412–4418.
164. Colston, K.W.; Mackay, A.G.; James, S.Y.; Binderup, L.; Chander, S.; Coombes, R.C. EB1089: A new vitamin D analog that inhibits the growth of breast cancer cells in vivo and in vitro. *Biochem. Pharmacol.* **1992**, *44*, 2273–2280. [CrossRef]
165. Zhou, X.; Zhu, G.-D.; Van Haver, D.; Vandewalle, M.; De Clercq, P.; Verstuyf, A.; Bouillon, R. Synthesis, biological activity, and conformational analysis of four *seco*-D-15,19-*bisnor*-1α,25-dihydroxyvitamin D_3 analogues, diastereomeric at C17 and C20. *J. Med. Chem.* **1999**, *42*, 3439–3456. [CrossRef] [PubMed]
166. Rey, M.A.; Martínez-Pérez, J.A.; Fernández-Gacio, A.; Halkes, K.; Fall, Y.; Mouriño, A. New synthetic strategies to vitamin D analogues modified at the side chain and D ring. Synthesis of 1α,25-dihydroxy-16-ene-vitamin D_3 and C-20 analogues. *J. Org. Chem.* **1999**, *64*, 3196–3206. [CrossRef]
167. Pérez Sestelo, J.; Mouriño, A.; Sarandeses, L.A. Design and synthesis of a 1α,25-dihydroxyvitamin D_3 dimer as a potential chemical inducer of vitamin D receptor dimerization. *Org. Lett.* **1999**, *1*, 1005–1007. [CrossRef] [PubMed]
168. Hisatake, J.-I.; Kubota, T.; Hisatake, Y.; Uskokovic, M.; Tomoyasu, S.; Koeffler, H.P. 5,6-*trans*-16-ene-Vitamin D_3: A New Class of Potent Inhibitors of Proliferation of Prostate, Breast, and Myeloid Leukemic Cells. *Cancer Res.* **1999**, *59*, 4023–4029. [PubMed]
169. Ikeda, M.; Takahashi, K.; Dan, A.; Koyama, K.; Kubota, K.; Tanaka, T.; Hayashi, M. Synthesis and biological evaluations of A-ring isomers of 26,26,26,27,27,27-hexafluoro-1α,25-dihydroxyvitamin D_3. *Bioorg. Med. Chem.* **2000**, *8*, 2157–2166. [CrossRef]
170. Herdick, M.; Steinmeyer, A.; Calberg, C. Carboxylic ester antagonist of 1α,25-dihydroxyvitamin D_3 show cell-specific actions. *Chem. Biol.* **2000**, *7*, 885–894. [CrossRef]
171. Norman, A.W.; Manchand, P.S.; Uskokovic, M.R.; Okamura, W.H.; Takeuchi, J.A.; Bishop, J.E.; Hisatake, J.-i.; Koeffler, H.P.; Peleg, S. Characterization of a novel analogue of 1α,25$(OH)_2$-vitamin D_3 with two side chains: Interaction with its nuclear receptor and cellular actions. *J. Med. Chem.* **2000**, *43*, 2719–2730. [CrossRef]
172. Verlinden, L.; Verstuyf, A.; Van Camp, M.; Marcelis, S.; Sabbe, K.; Zhao, X.-Y.; De Clercq, P.; Vandewalle, M.; Bouillon, R. Two novel 14-epi-analogs of 1α,25-dihydroxyvitamin D_3 inhibitthe growth of human breast cancer cells in vitro and in vivo. *Cancer Res.* **2000**, *60*, 2673–2679.
173. Fernández-Gacio, A.; Vitale, C.; Mouriño, A. Synthesis of new aromatic (C17-C20)-locked side-chain analogues of calcitriol (1α,25-dihydroxyvitamin D_3). *J. Org. Chem.* **2000**, *65*, 6978–6983. [CrossRef]
174. Fall, Y.; Fernández, C.; Vitale, C.; Mouriño, A. Stereoselective synthesis of vitamin D analogues with cyclic side chains. *Tet. Lett.* **2000**, *41*, 7323–7326. [CrossRef]
175. Codesido, E.M.; Cid, M.M.; Castedo, L.; Mouriño, A.; Granja, J.R. Synthesis of vitamin D analogues with a 2-hydroxy-3-deoxy ring A. *Tet. Lett.* **2000**, *41*, 5861–5864. [CrossRef]
176. Bonasera, T.A.; Grue-Sørensen, G.; Ortu, G.; Binderup, E.; Bergström, M.; Björkling, F.; Långström, B. The synthesis of [26,26-^{11}C]dihydroxyvitamin D_3, a tracer for positron emission tomography (PET). *Bioorg. Med. Chem.* **2001**, *9*, 3123–3128. [CrossRef]
177. Gabriëls, S.; Van Haver, D.; Vandewalle, M.; De Clercq, P.; Verstuyf, A.; Bouillon, R. Development of analogues of 1α,25-dihydroxyvitamin D_3 with biased side chain orientation: Methylated des-CD-homo analogues. *Chem. Eur. J.* **2001**, *7*, 520–532. [CrossRef]
178. Calverley, M. Novel side chain analogs of 1α,25-dihydroxyvitamin D_3: Design and synthesis of the 21,24-methano derivatives. *Steroids* **2001**, *66*, 249–255. [CrossRef]

179. Fujishima, T.; Zhaopeng, L.; Konno, K.; Nakagawa, K.; Okano, T.; Yamaguchi, K.; Takayama, H. Highly potent cell diffrentiation-inducing analogues of 1α,25-dihydroxyvitamin D_3: Synthesis and biological activity of 2-methyl-1α,25-dihydroxyvitamin D_3 with side-chain modifications. *Bioorg. Med. Chem.* **2001**, *9*, 525–535. [CrossRef]
180. Kamao, M.; Tatematsu, S.; Reddy, G.S.; Hatakeyama, S.; Sugiura, M.; Ohashi, N.; Kubodera, N.; Okano, T. Isolation, identification and biological activity of 24*R*,25-dihydroxy-3-epi-vitamin D_3; a novel metabolite of 24*R*,25-dihydroxyvitamin D_3 produced in rat osteosarcoma (UMR 106). *J. Nutr. Sci. Vitaminol.* **2001**, *47*, 108–115. [CrossRef] [PubMed]
181. Ishizuka, I.; Miura, D.; Ozono, K.; Saito, M.; Eguchi, H.; Chokki, M.; Norman, A.W. (23*S*)- and (23*R*)-25-Dehydro-1α-hydroxyvitamin D_3-26,23-lactone function as antagonist of vitamin D receptor-mediated genomic actions of 1α,25-dihydroxyvitamin D_3. *Steroids* **2001**, *66*, 227–237. [CrossRef]
182. Steinmeyer, A.; Schwarz, K.; Haberey, M.; Langer, G.; Wiesinger, H. Synthesis and biological activities of a new series of secosteroids: Vitamin D phosphonate hybrids. *Steroids* **2001**, *66*, 257–266. [CrossRef]
183. White, M.C.; Burke, M.D.; Peleg, S.; Brem, H.; Posner, G.H. Conformationally restricted hybrid analogues of the hormone of 1α,25-dihydroxyvitamin D_3: Design, synthesis and biological evaluation. *Bioorg. Med. Chem.* **2001**, *9*, 1691–1699. [CrossRef]
184. Mäenpää, P.H.; Väisänen, S.; Jääskeläinen, T.; Ryhänen, S.; Rouvinen, J.; Duchier, C.; Mahonen, A. Vitamin D_3 analogs (MD 1288, KH 1060, EB 1089, GS 1558, and CD 1093): Studies on their mechanism of action. *Steroids* **2001**, *66*, 223–225. [CrossRef]
185. Hatakeyama, S.; Kawase, A.; Uchiyama, Y.; Maeyama, J.; Iwabuchi, Y.; Kubodera, N. Synthesis and biological characterization of 1α,24,25-trihydroxy-2β-(3-hydroxypropoxy)vitamin D_3 (24-hydroxylated ED-71). *Steroids* **2001**, *66*, 267–276. [CrossRef]
186. Takayama, H.; Konno, K.; Fujishima, T.; Maki, S.; Liu, Z.; Miura, D.; Chokki, M.; Ishizuka, S.; Smith, C.; DeLuca, H.F.; et al. Systematic studies on synthesis, structural elucidation and biological evaluation of A-ring diastereomers of 2-methyl-1α,25-dihydroxyvitamin D_3 and 20-epi-2-methyl-1α,25-dihydroxyvitamin D_3. *Steroids* **2001**, *66*, 277–285. [CrossRef]
187. Schuster, I.; Egger, H.; Astecker, N.; Herzig, G.; Schüssler, M.; Vorisek, G. Selective inhibitors of CYP24: Mechanistic tools to explore vitamin D metabolism in human keratinocytes. *Steroids* **2001**, *66*, 451–462. [CrossRef]
188. Bury, Y.; Herdick, M.; Uskokovic, M.R.; Calberg, C. Gene regulatory potential of 1α,25-dihydroxyvitamin D_3 analogues with two side chains. *J. Cell Biochem.* **2001**, *Supp. 36*, 179–190. [CrossRef]
189. Nakagawa, K.; Sowa, Y.; Kurobe, M.; Ozono, K.; Siu-Caldera, M.-L.; Reddy, G.S.; Uskokovic, M.R.; Okano, T. Differential activities of 1α,25-dihydroxy-16-ene-vitamin D_3 analogs and their 3-epimers on human promyelocytic leukemia (HL-60) cell differentaition and apoptosis. *Steroids* **2001**, *66*, 327–337. [CrossRef]
190. Fall, Y.; Fernández, C.; González, V.; Mouriño, A. Stereoselective synthesis of (22*R*)- and (22*S*)-22-methyl-1α,25-dihydroxyvitamin D_3. *Synlett* **2001**, 1567–1568. [CrossRef]
191. Hilpert, H.; Wirz, B. Novel versatile approach to an enantiopure 19-*nor*,*des*-CD vitamin D_3 derivative. *Tetrahedron* **2001**, *57*, 681–694. [CrossRef]
192. Posner, G.H.; Woodard, B.T.; Crawford, K.R.; Peleg, S.; Brown, A.J.; Dolan, P.; Kensler, T.W. 2,2-Disubstituted analogues of the natural hormone 1α,25-dihydroxyvitamin D_3: Chemistry and biology. *Bioorg. Med. Chem.* **2002**, *10*, 2353–2365. [CrossRef]
193. Fall, Y.; Barreiro, C.; Fernández, C.; Mouriño, A. Vitamin D heterocyclic analogues. Part 1: A stereoselective route to CD systems with pyrazole rings on their side chains. *Tet. Lett.* **2002**, *43*, 1433–1436. [CrossRef]
194. Fernández-Gacio, A.; Mouriño, A. Studies on the introduction of a photoreactive aryldiazirine group into the vitamin D skeleton. *Eur. J. Org. Chem.* **2002**, 2529–2534. [CrossRef]
195. Pérez-Sestelo, J.; de Uña, O.; Mouriño, A.; Sarandeses, L.A. Synthesis of the first 24-aminovitamin D_3 derivatives by diastereoselective conjugate addition to a chiral methyleneoxazolidinone in aqueous media. *Synlett* **2002**, 719–722. [CrossRef]
196. Suhara, Y.; Kittaka, A.; Kishimoto, S.; Calverley, M.J.; Fujishima, T.; Saito, N.; Sugiura, T.; Waku, K.; Takayama, H. Synthesis and testing of 2α-modified 1α,25-dihydroxyvitamin D_3 analogues with a double side chain: Marked cell diffrentiation activity. *Bioorg. Med. Chem. Lett.* **2002**, *12*, 3255–3258. [CrossRef]
197. Wu, Y.; Sabbe, K.; De Clercq, P.; Vandewalle, M.; Bouillon, R.; Verstuyf, A. Vitamin D_3: Synthesis of *seco* C-9,11,21-*trisnor*-17-methyl-1α,25-dihydroxy vitamin D_3 analogues. *Bioorg. Med. Chem. Lett.* **2002**, *12*, 1629–1632. [CrossRef]
198. Wu, Y.; De Clercq, P.; Vandewalle, M.; Bouillon, R.; Verstuyf, A. Vitamin D_3: Synthesis of *seco* C-9,11-*bisnor*-17-methyl-1α,25-dihydroxy vitamin D_3 analogues. *Bioorg. Med. Chem. Lett.* **2002**, *12*, 1633–1636. [CrossRef]
199. Chodynski, M.; Wietrzyk, J.; Marcinkowska, E.; Opolski, A.; Szelejewski, W.; Kutner, A. Synthesis and antiproliferative activity of side-chain unsaturated and homologated analogs of 1α,25-dihydroxyvitamin D_2 (24*E*)-(1*S*)-24-dehydro-24a-homo-1,25-dihydroxyergalciferol and congeners. *Steroids* **2002**, *67*, 789–798. [CrossRef]
200. Varela, C.; Nilsson, K.; Torneiro, M.; Mouriño, A. Synthesis of tetracyclic analogues of calcitriol (1α,25-dihydroxyvitamin D_3) with side-chain-locked spatial orientations at C(20). *Helv. Chim. Acta* **2002**, *85*, 3251–3261. [CrossRef]
201. Cornella, I.; Pérz-Sestelo, J.; Mouriño, A.; Sarandeses, L.A. Synthesis of 18-substituted analogues of calcitriol using photochemical remote functionalization. *J. Org. Chem.* **2002**, *67*, 4707–4714. [CrossRef]
202. Okamura, W.H.; Zhu, G.-D.; Hill, D.K.; Thomas, R.J.; Ringe, K.; Borchardt, D.B.; Norman, A.W.; Mueller, L.J. Synthesis and NMR studies of ^{13}C-labeled vitamin D metabolites. *J. Org. Chem.* **2002**, *67*, 1637–1650. [CrossRef]
203. Brandl, M.; Wu, X.; Liu, Y.; Pease, J.; Holper, M.; Hooijmaaijer, E.; Lu, Y.; Wu, P. Chemical reactivity of Ro-26-9228, 1α-fluoro-25-hydroxy-10,23*E*-diene-26,27-*bishomo*-20-*epi*-cholecalciferol in aqueous solution. *J. Pharm. Sci.* **2003**, *92*, 1981–1989. [CrossRef]
204. Swann, S.L.; Bergh, J.; Farach-Carson, M.C.; Ocasio, C.A.; Koh, J.T. Structure-Based design of selective agonists for a rickets-associated mutant of the vitamin D receptor. *J. Am. Chem. Soc.* **2002**, *124*, 13795–13805. [CrossRef]

205. Fujishima, T.; Kojima, Y.; Azumaya, I.; Kittaka, A.; Takayama, H. Design and synthesis of potent vitamin D receptor antagonists with A-ring modificactions: Remarkable effects of 2α-methyl introduction an antagonist activity. *Bioorg. Med. Chem.* **2003**, *11*, 3621–3631. [CrossRef]
206. Honzawa, S.; Suhara, Y.; Nihei, K.-I.; Saito, N.; Kishimoto, S.; Fujishima, T.; Kurihara, M.; Sugiura, T.; Waku, K.; Takayama, H.; et al. Concise synthesis and biological activities of 2α-alkyl and 2α-(ω-hydroxyalkyl)-20-epi-1α,25-dihydroxyvitamin D_3. *Bioorg. Med. Chem. Lett.* **2003**, *13*, 3503–3506. [CrossRef]
207. Verlinden, L.; Verstuyf, A.; Verboven, C.; Eelen, G.; De Ranter, C.; Gao, L.-J.; Chen, Y.-J.; Murad, I.; Choi, M.; Yamamoto, K.; et al. Previtamin D_3 with a *trans*-fused decalin CD-ring has pronounced genomic activity. *J. Biol. Chem.* **2003**, *278*, 35476–35482. [CrossRef] [PubMed]
208. Hanazawa, T.; Koyama, A.; Nakata, K.; Okamoto, S.; Sato, F. New convergent synthesis of 1α,25-dihydroxyvitamin D_3 and its analogues by Suzuki-Miyaura coupling between A-ring and C,D-ring parts. *J. Org. Chem.* **2003**, *68*, 9767–9772. [CrossRef] [PubMed]
209. Blæhr, L.K.A.; Björkling, F.; Calverley, M.J.; Binderup, E.; Begtrup, M. Synthesis of novel hapten derivatives of 1α,25-dihydroxyvitamin D_3 and its 20-epi analogue. *J. Org. Chem.* **2003**, *68*, 1367–1375. [CrossRef]
210. Ono, K.; Yoshida, A.; Saito, N.; Fujishima, T.; Honzawa, S.; Suhara, Y.; Kishimoto, S.; Sugiura, T.; Waku, K.; Takayama, H.; et al. Efficent synthesis of 2-modified 1α,25-dihydroxy-19-norvitamin D_3 with Julia olefination: High potency in induction of the differentiation on HL-60 cells. *J. Org. Chem.* **2003**, *68*, 7407–7415. [CrossRef]
211. Kato, H.; Hashimoto, Y.; Nagasawa, K. Novel heteroatom-containing vitamin D_3 analogs: Efficient synthesis of 1α,25-dihydroxyvitamin D_3-26,23-lactam. *Molecules* **2003**, *8*, 488–499. [CrossRef]
212. Chen, Y.-J.; Gao, L.-J.; Murad, I.; Verstuyf, A.; Verlinden, L.; Verboven, C.; Bouillon, R.; Viterbo, D.; Van Haver, D.; Vandewalle, M.; et al. Synthesis, biological activity, and conformational analysis of CD-ring modified *trans*-decalin 1α,25-dihydroxyvitamin D_3 analogs. *Org. Biomol. Chem.* **2003**, *1*, 257–267. [CrossRef]
213. Fujishima, T.; Kittaka, A.; Yamaoka, K.; Takeyama, K.-i.; Kato, S.; Takayama, H. Synthesis of 2,2-dimethyl-1α,25-dihydroxyvitamin D_3: A-ring structural motif that modulates interactions of vitamin D receptor with transcriptional activators. *Org. Biomol. Chem.* **2003**, *1*, 1863–1869. [CrossRef]
214. González-Avión, X.C.; Mouriño, A. Functionalization at C-12 of 1α,25-dihydroxyvitamin D_3 strongly modulates the affinity for the vitamin D receptor (VDR). *Org. Lett.* **2003**, *5*, 2291–2293. [CrossRef]
215. Pérez-García, X.; Rumbo, A.; Larriba, M.J.; Ordóñez, P.; Muñoz, A.; Mouriño, A. The first locked side-chain analogues of calcitriol (1α,25-dihydroxyvitamin D_3) induce vitamin D receptor transcriptional activity. *Org. Lett.* **2003**, *5*, 4033–4036. [CrossRef] [PubMed]
216. Saito, N.; Matsunaga, T.; Fujishima, T.; Anzai, M.; Saito, H.; Takenouchi, K.; Miura, D.; Takayama, H.; Kittaka, A. Remarkable effect of 2α-modification on the VDR antagonistic activity of 1α-hydroxyvitamin D_3-26,23-lactones. *Org. Biomol. Chem.* **2003**, *1*, 4396–4402. [CrossRef] [PubMed]
217. Saito, N.; Saito, H.; Anzai, M.; Yoshido, A.; Fujishima, T.; Takenouchi, K.; Miura, D.; Ishizuka, I.; Takayama, H.; Kittaka, A. Dramatic enhancement of antagonistic activity on vitamin D receptor: A double functionalization of 1α-hydroxyvitamin D_3-26,23-lactones. *Org. Lett.* **2003**, *5*, 4859–4862. [CrossRef] [PubMed]
218. Unten, S.; Ishihara, M.; Sakagami, H. Relationship between differentiation-inducing activity and hypercalcemic activity of hexafluorotrihydroxyvitamin D derivatives. *Anticancer Res.* **2004**, *24*, 683–690. [PubMed]
219. Kato, H.; Nakano, Y.; Sano, H.; Tanatani, A.; Kobayashi, H.; Shimazawa, R.; Koshino, H.; Hashimoto, Y.; Nagasawa, K. Synthesis of 1α,25-dihydroxyvitamin D_3-26,23-lactams, a novel series of 1α,25-dihydroxyvitamin D_3 antagonist. *Bioorg. Med. Chem. Lett.* **2004**, *14*, 2579–2583. [CrossRef] [PubMed]
220. Maehr, H.; Uskokovic, M.R. Formal desymmetrization of the diastereotopic chains in Gemini calcitriol derivatives with two different side chains at C-20. *Eur. J. Org. Chem.* **2004**, 1703–1713. [CrossRef]
221. Takenouchi, K.; Sogawa, R.; Manabe, K.; Saitoh, H.; Gao, Q.; Miura, D.; Ishizuka, S. Synthesis and structure-activity relationships of TEI-9647 derivatives as vitamin D_3 antagonists. *J. St. Biochem. Mol. Biol.* **2004**, *89–90*, 31–34. [CrossRef]
222. Schepens, W.; Van Haver, D.; Vandewalle, M.; De Clercq, P.; Bouillon, R.; Verstuyf, A. Synthesis and biological activity of 22-oxa CD-modified analogues of 1α,25-dihydroxyvitamin D_3: Spiro[5,5]-undecane CF-ring analogues. *Bioorg, Med. Chem. Lett.* **2004**, *14*, 3889–3892. [CrossRef]
223. Maehr, H.; Uskokovic, M.R.; Adorini, L.; Reddy, G.S. Calcitriol derivatives with two different side chains at C20. II. Diastereo selective synthesis of the metabolically produced 24(*R*)-Gemini. *J. Med. Chem.* **2004**, *47*, 6476–6484. [CrossRef]
224. Momán, E.; Nicoletti, D.; Mouriño, A. Synthesis of novel analogues of 1α,25-dihydroxyvitamin D_3 with side chains at C-18. *J. Org. Chem.* **2004**, *69*, 4615–4625. [CrossRef]
225. Saito, N.; Suhara, Y.; Kurihara, M.; Fujishima, T.; Honzawa, S.; Takayanagi, H.; Kozono, T.; Matsumoto, M.; Ohmori, M.; Miyata, N.; et al. Design and efficient synthesis of 2α-(ω-hydroxyalkoxy)-1α,25-dihydroxyvitamin D_3 analogues, including 2-*epi*-ED-71 and their 20-epimers with HL-60 cell differentiation activity. *J. Org. Chem.* **2004**, *69*, 7463–7471. [CrossRef] [PubMed]
226. Wu, S.-Y.; de Keczer, S.A.; Masjedizadeh, M.R. [^{3}H] and [^{14}C]-Ro0275646, a vitamin D analog. Synth. *Appl Isot. Labl. Comp.* **2004**, *8*, 203–206.

227. Posner, G.H.; Tony Lee, S.H.; Kim, H.J.; Peleg, S.; Dolan, P.; Kensler, T.W. Novel A-ring analogs of the hormone 1α,25-dihydroxyvitamin D_3: Synthesis and preliminary biological evaluation. *Bioorg. Med. Chem.* **2005**, *13*, 2959–2966. [CrossRef] [PubMed]
228. Hatcher, M.A.; Peleg, S.; Dolan, P.; Kensler, T.W.; Sarjeant, A.; Posner, G.H. A-ring hydroxymethyl 19-nor analogs of the natural hormone 1α,25-dihydroxyvitamin D_3: Synthesis and preliminary biological evaluation. *Bioorg. Med. Chem.* **2005**, *13*, 3964–3976. [CrossRef]
229. Gómez-Reino, C.; Vitale, C.; Maestro, M.; Mouriño, A. Pd-Catalyzed carbocyclization-Negishi cross-coupling cascade: A novel approach to 1α,25-dihydroxyvitamin D_3 and analogues. *Org. Lett.* **2005**, *7*, 5885–5887. [CrossRef]
230. Shimizu, M.; Miyamoto, Y.; Kobayashi, E.; Shimazaki, M.; Yamamoto, K.; Reischl, W.; Yamada, S. Synthesis and biological activities of new 1α,25-dihydroxy-19-norvitamin D_3 analogs with modifications in both A-ring and the side chain. *Bioorg. Med. Chem.* **2006**, *14*, 4277–4294. [CrossRef]
231. Shimizu, K.; Kawase, A.; Haneishi, T.; Kato, Y.; Kinoshita, K.; Ohmori, M.; Furuta, Y.; Emura, T.; Kato, N.; Mitsui, T.; et al. Design and evaluation of new antipsoriatic antedrug candidates having 16-en-22-oxa-vitamin D_3 structures. *Bioorg. Med. Chem. Lett.* **2006**, *16*, 3323–3329. [CrossRef]
232. Hatakeyama, S.; Nahashima, S.; Imai, N.; Takahashi, K.; Ishihara, J.; Sugita, A.; Nihei, T.; Saito, H.; Takahashi, F.; Kubodera, N. Synthesis and biological evaluation of a 3-position diastereomer of 1α,25-dihydroxy-2β-(3-hydroxypropoxy)vitamin D_3 (ED-71). *Bioorg. Med. Chem.* **2006**, *14*, 8050–8056. [CrossRef]
233. Uskokovic, M.R.; Manchand, P.; Marezak, S.; Maehr, H.; Jankowski, P.; Adorini, L.; Reddy, G.S. C-20 Cyclopropyl vitamin D analogs. *Curr. Top. Med. Chem.* **2006**, *6*, 1289–1296. [CrossRef]
234. Oves, D.; Fernández, S.; Verlinden, L.; Bouillon, R.; Verstuyf, A.; Ferrero, M.; Gotor, V. Novel A-ring homodimeric C-3-carbamate analogues of 1α,25-dihydroxyvitamin D_3: Synthesis and preliminary biological evaluation. *Bioorg. Med. Chem.* **2006**, *14*, 7512–7519. [CrossRef]
235. Hosoda, S.; Tanatani, A.; Wakabayashi, K.-i.; Makishima, M.; Imai, K.; Miyachi, H.; Nagasawa, K.; Hashimoto, Y. Ligands with a 3,3-diphenylpentane skeleton for nuclear vitamin D and androgen receptors: Dual activities and metabolic activation. *Bioorg Med. Chem.* **2006**, *14*, 5489–5502. [CrossRef] [PubMed]
236. Schuster, I.; Egger, H.; Nussbaumer, P.; Kroemer, T. Inhibitors of vitamin D hydroxylases: Structure-activity relationships. *J. Cell. Biochem.* **2003**, *88*, 372–380. [CrossRef] [PubMed]
237. Peleg, S.; Petersen, K.S.; Suh, B.C.; Dolan, P.; Agoston, E.S.; Kensler, T.W.; Posner, G.H. Low-calcemic, antiproliferatives, 1-difluoromethyl hybrid analogs of the hormone 1α,25-dihydroxyvitamin D_3: Design, synthesis and preliminary biological evaluation. *J. Med. Chem.* **2006**, *49*, 7513–7517. [CrossRef] [PubMed]
238. Schepens, W.; Van Haver, D.; Vandewalle, M.; Bouillon, R.; Verstuyf, A.; De Clercq, P. Synthesis of spiro[4,5]-decane CF-ring analogues of 1α,25-dihydroxyvitamin D_3. *Org. Lett.* **2006**, *8*, 4247–4250. [CrossRef]
239. Ono, Y.; Watanabe, H.; Taira, I.; Takahashi, K.; Ishihara, J.; Hatakeyama, S.; Kubodera, N. Synthesis of putative metabolites of 1α,25-dihydroxy-2β-(3-hydroxypropoxy)vitamin D_3 (ED-71). *Steroids* **2006**, *71*, 529–540. [CrossRef]
240. Riveiros, R.; Rumbo, A.; Sarandeses, L.A.; Mouriño, A. Synthesis and conformational analysis of 17α,21-*cyclo*-22-unsaturated analogues of calcitriol. *J. Org. Chem.* **2007**, *72*, 5477–5485. [CrossRef]
241. Yamamoto, K.; Abe, D.; Yoshimoto, N.; Choi, M.; Yamagishi, K.; Tokiwa, H.; Shimizu, M.; Makishima, M.; Yamada, S. Vitamin D receptor: Ligand recognition and allosteric network. *J. Med. Chem.* **2006**, *49*, 1313–1324. [CrossRef]
242. Lee, H.J.; Wislocki, A.; Goodman, C.; Ji, Y.; Ge, R.; Maehr, H.; Uskokovic, M.R.; Reiss, M.; Suh, N. A vitamin D derivative activates bone morphogenetic protein signaling in MCF10 breast epithelial cells. *Mol. Pharm.* **2006**, *69*, 1840–1848. [CrossRef]
243. Yoshimoto, N.; Inaba, Y.; Yamada, S.; Makishima, M.; Shimizu, M.; Yamamoto, K. 2-Methylene-19-nor-25-dehydro-1α-hydroxyvitamin D_3-26,23-lactones; synthesis, biological activities and molecular basis of passive antagonism. *Bioorg. Med. Chem.* **2008**, *16*, 457–473. [CrossRef]
244. Saito, N.; Matsunaga, T.; Saito, H.; Anzai, M.; Takenouchi, K.; Miura, D.; Namekawa, J.-i.; Ishizuka, S.; Kittaka, A. Further synthetic and biological studies on vitamin D hormone antagonist based on C24-alkylation and C2α-functionalization of 25-dehydro-1α-hydroxyvitamin D_3-26,23-lactones. *J. Med. Chem.* **2006**, *49*, 7063–7075. [CrossRef]
245. González-Avión, X.C.; Mouriño, A.; Rochel, N.; Moras, D. Novel 1α,25-dihydroxyvitamin D_3 analogs with side chain at C12. *J. Med. Chem.* **2006**, *49*, 1509–1516. [CrossRef] [PubMed]
246. Maehr, H.; Uskokovic, M.R.; Adorini, L.; Penna, G.; Mariani, R.; Panina, P.; Passini, N.; Bono, E.; Perego, S.; Biffi, M.; et al. Calcitriol derivatives with two different side chains at C20. III. An epimeric pair of the Gemini family with unprecedented antiproliferative effects on tumor cells and renin mRNA expression inhibition. *J. St. Biochem. Mol. Biol.* **2007**, *103*, 277–281. [CrossRef] [PubMed]
247. Garay, E.; Jankowski, P.; Lizano, P.; Marczak, S.; Maehr, H.; Adorini, L.; Uskokovic, M.R.; Studzinski, G.P. Calcitriol derivatives with two different side chains at C20. Part 4. Further side chain modifications that alter VDR-dependent monocytic differentiation potency inhuman leukemia cells. *Bioorg. Med. Chem.* **2007**, *15*, 4444–4455. [CrossRef] [PubMed]
248. Yamamoto, K.; Inaba, Y.; Yoshimoto, N.; Choi, M.; DeLuca, H.F.; Yamada, S. 22-Alkyl-20-epi-1α,25-dihydroxyvitamin D_3 compounds of superagonistic activity: Syntheses, biological activities and interaction with the receptor. *J. Med. Chem.* **2007**, *50*, 932–939. [CrossRef] [PubMed]

249. Gregorio, C.; Eduardo, S.; Rodrigues, L.C.; Regueira, M.; Fraga, R.; Riveiros, R.; Maestro, M.; Mouriño, A. Synthesis of two carboxylic haptens for raising antibodies to 25-hydroxyvitamin D_3 and 1α,25-dihydroxyvitamin D_3. *J. St. Biochem. Mol. Biol.* **2007**, *103*, 227–230. [CrossRef] [PubMed]
250. Glebocka, A.; Sicinski, R.R.; Plum, L.A.; DeLuca, H.F. 2-(3′-Hydroxypropylidene)-1α-hydroxy-19-norvitamin D compounds with truncated side chain. *J. St. Biochem. Mol. Biol.* **2007**, *103*, 310–315. [CrossRef]
251. Kobayashi, E.; Shimazaki, M.; Miyamoto, Y.; Masuno, H.; Yamamoto, K.; DeLuca, H.F.; Yamada, S.; Shimizu, M. Structure-activity relationships of 19-norvitamin D analogs having a fluoroethylene group at the C-2 position. *Bioorg. Med. Chem.* **2007**, *15*, 1475–1482. [CrossRef]
252. Inaba, Y.; Yamamoto, K.; Yoshimoto, N.; Matsunawa, M.; Uno, S.; Yamada, S.; Makishima, M. Vitamin D_3 derivatives with adamantane or lactone ring side chains are cell type-selective vitamin D receptor modulators. *Mol. Pharm.* **2007**, *71*, 1298–1311. [CrossRef]
253. Chiellini, G.; Grzywack, P.; Plum, L.A.; Barycki, R.; Clagett-Dame, M.; DeLuca, H.F. Synthesis and biological properties of 2-methylene-19-nor-25-dehydro-1α-hydroxyvitamin D_3-26,26-lactones-weak agonists. *Bioorg. Med. Chem.* **2008**, *16*, 8563–8573. [CrossRef]
254. Takaku, H.; Miyamoto, Y.; Asami, S.; Shimazaki, M.; Yamada, S.; Yamamoto, K.; Udagawa, N.; DeLuca, H.F.; Shimizu, M. Synthesis and structure-activity relationships of 16-ene-22-thia-1α,25-dihydroxy-26,27-dimethyl-19-norvitamin D_3 analogs having side chains of different sizes. *Bioorg. Med. Chem.* **2008**, *16*, 1796–1815. [CrossRef]
255. Shimizu, M.; Miyamoto, Y.; Takaku, H.; Matsuo, M.; Nakabayashi, M.; Masuno, H.; Udagawa, N.; DeLuca, H.F.; Ikura, T.; Ito, N. 2-Substituted-16-ene-22-thia-1α,25-dihydroxy-26,27-dimethyl-19-norvitamin D_3 analogs: Synthesis, biological evaluation and crystal structure. *Bioorg. Med. Chem.* **2008**, *16*, 6949–6964. [CrossRef] [PubMed]
256. Sánchez-Abella, L.; Fernández, S.; Verstuyf, A.; Verlinden, L.; Gotor, V. Synthesis and biological activity of previtamin D_3 analogues with A-ring modifications. *Bioorg. Med. Chem.* **2008**, *16*, 10244–10250. [CrossRef] [PubMed]
257. Nakabayashi, M.; Yamada, S.; Yoshimoto, N.; Tanaka, T.; Igarashi, M.; Ikura, T.; Ito, N.; Makishima, M.; Tokiwa, H.; DeLuca, H.F.; et al. Crystal structures of rat vitamin D receptor bound to adamantyl vitamin D analogs: Structural basis for vitamin D receptor antagonism and partial agonism. *J. Med. Chem.* **2008**, *51*, 5320–5329. [CrossRef] [PubMed]
258. Hourai, S.; Rodrigues, L.C.; Antony, P.; Reina-San-Martin, B.; Ciesielski, F.; Magnier, B.C.; Schoonjans, K.; Mouriño, A.; Rochel, N.; Moras, D. Structure-based design of a superagonist ligand for the vitamin D nuclear receptor. *Chem. Biol.* **2008**, *15*, 383–392. [CrossRef] [PubMed]
259. Saito, T.; Okamoto, R.; Haritunians, T.; O'Kelly, J.; Uskokovic, M.; Maehr, H.; Marczak, S.; Jankowski, P.; Badr, R.; Koeffler, H.P. Novel Gemini vitamin D_3 analogs have potent antitumor activity. *J. St. Biochem. Mol. Biol.* **2008**, *112*, 151–156. [CrossRef] [PubMed]
260. Williams, K.B.; DeLuca, H.F. 2-Methylene-19-nor-20(*S*)-1α-hydroxy-bishomopregnacalciferol [(20*S*)-2MbisP], an analog of vitamin D_3 [1,25$(OH)_2D_3$], does not stimulate intestinal phosphate absorption at levels previously shown to suppress parathyroid hormone. *Steroids* **2008**, *73*, 1277–1284. [CrossRef]
261. Usera, A.R.; Dolan, P.; Kensler, T.W.; Posner, G.H. Novel alkyl chain sulfone 1α,25-dihydroxyvitamin D_3 analogs: A comparison of in vitro antiproliferative and in vivo calcemic activities. *Bioorg. Med. Chem.* **2009**, *17*, 5627–5631. [CrossRef]
262. Plonska-Ocypa, K.; Sicinski, R.R.; Plum, L.A.; Grzywacz, P.; Frelek, J.; Clagett-Dame, M.; DeLuca, H.F. 13-Methyl-substituted *des*-CD analogs of (20*S*)-1α,25-dihydroxy-2-methylene-19-norvitamin D_3 (2MD): Synthesis and biological evaluation. *Bioorg. Med. Chem.* **2009**, *17*, 1747–1763. [CrossRef]
263. Barycki, R.; Sicinski, R.R.; Plum, L.A.; Grzywacz, P.; Clagett-Dame, M.; DeLuca, H.F. Removal of the 20-methyl group from 2-methylene-19-nor-20(*S*)-1α-hydroxyvitamin D_3 (2MD) selectively eliminates bone calcium mobilization activity. *Bioorg. Med. Chem.* **2009**, *17*, 7658–7669. [CrossRef]
264. Inaba, Y.; Yoshimoto, N.; Sakamaki, N.; Nakabayashi, T.; Ikura, T.; Tamamura, H.; Ito, N.; Shimizu, M.; Yamamoto, K. A new class of vitamin D analogues that induce structural rearrangement of the ligand-binding pocket of the receptor. *J. Med. Chem.* **2009**, *52*, 1438–1449. [CrossRef]
265. Laverny, G.; Penna, G.; Uskokovic, M.; Marczak, S.; Maehr, H.; Jankowski, P.; Ceailles, C.; Vouros, P.; Smith, B.; Robinson, M.; et al. Synthesis and anti-inflammatory properties of 1α,25-dihydroxy-16-ene-20-cyclopropyl-24-oxovitamin D_3, a hypocalcemic, stable metabolite of 1α,25-dihydroxy-16-ene-20-cyclopropylvitamin D_3. *J. Med. Chem.* **2009**, *52*, 2204–2213. [CrossRef] [PubMed]
266. Sánchez-Abella, L.; Fernández, S.; Verstuyf, A.; Verlinden, L.; Gotor, V.; Ferrero, M. Synthesis, conformational analysis and biological evaluation of 19-*nor*-vitamin D_3 analogues with A-ring modifications. *J. Med. Chem.* **2009**, *52*, 6158–6162. [CrossRef] [PubMed]
267. Minne, G.; Verlinden, L.; Verstuyf, A.; De Clercq, P.J. Synthesis of 1α,25-dihydroxyvitamin D analogues featuring a S_2-symmetric CD-ring core. *Molecules* **2009**, *14*, 894–903. [CrossRef] [PubMed]
268. De Brysser, F.; Verlinden, L.; Verstuyf, A.; De Clercq, P.J. Synthesis of 22-oxaspiro[4.5]decane CD-ring modified analogs of 1α,25-dihydroxyvitamin D_3. *Tet. Lett.* **2009**, *50*, 4174–4177. [CrossRef]
269. Gándara, Z.; Pérez, M.; Pérez-García, X.; Gómez, G.; Fall, Y. Stereoselective synthesis of (22*Z*)-25-hydroxyvitamin D_2 and (22*Z*)-1α,25-dihydroxyvitamin D_2. *Tet. Lett.* **2009**, *50*, 4874–4877. [CrossRef]

270. Lambin, M.; Dabbas, B.; Spingarn, R.; Mendoza-Sanchez, R.; Wang, T.-T.; An, B.-S.; Huang, D.C.; Kremer, R.; White, J.H.; Gleason, J.L. Vitamin D receptor agonist/histone deacetylase inhibitor molecular hybrids. *Bioorg. Med. Chem.* **2010**, *18*, 4119–4137. [CrossRef]
271. Honzawa, S.; Takahashi, N.; Yamashita, A.; Sugiura, T.; Kurihara, M.; Arai, M.A.; Kato, S.; Kittaka, A. Synthesis of a 1α-C-methyl analogue of 25-hydroxyvitamin D_3: Interaction with the mutant vitamin D receptor Arg274Leu. *Tetahedron* **2009**, *65*, 7135–7147. [CrossRef]
272. Ben Shabat, S.; Sintov, A. Substituted Cyclohexylidene-Ethylidene-Octahydro-Indene Compounds. World Patent Organization. WO2009153782 A1, 23 December 2009.
273. Maehr, H.; Lee, H.J.; Perry, B.; Suh, N.; Uskokovic, M.R. Calcitriol derivatives with two different side chain at C-20. V. Potent inhibitors of mammary carcinogenesis and inducers of leukemia differentiation. *J. Med. Chem.* **2009**, *52*, 5505–5519. [CrossRef]
274. Gogoi, P.; Sigüeiro, R.; Eduardo, S.; Mouriño, A. An expeditious route to 1α,25-dihydroxyvitamin D_3 and its analogues by an aqueous tandem palladium-catalyzed A-ring closure and Suzuki coupling to the C/D unit. *Chem. Eur. J.* **2010**, *16*, 1432–1435. [CrossRef]
275. Sakamaki, Y.; Inaba, Y.; Yoshimoto, N.; Yamamoto, K. Potent antagonist for the vitamin D receptor: Vitamin D analogues with simple side chain structure. *J. Med. Chem.* **2010**, *53*, 5813–5826. [CrossRef]
276. Yoshino, M.; Eto, K.; Takahashi, K.; Ishihara, J.; Hatakeyama, S.; Ono, Y.; Saito, H.; Kubodera, N. Synthesis of 20-*epi*-eldecalcitol [20-epi-1α,25-dihydroxy-2β-(3-hydroxypropoxy)vitamin D_3: 20-*epi*-ED-71]. *Heterocycles* **2010**, *81*, 381–394. [CrossRef]
277. Grzywacz, P.; Chiellini, G.; Plum, L.A.; Clagett-Dame, M.; DeLuca, H.F. Removal of the 26-methyl group from 2-methylene-19-nor-1α,25-dihydroxyvitamin D_3 markely reduces in vivo calcemic activity without altering in vitro VDR binding, HL-60 cell differentiation, and transcription. *J. Med. Chem.* **2010**, *53*, 8642–8649. [CrossRef] [PubMed]
278. Antony, P.; Sigüeiro, R.; Huet, T.; Sato, Y.; Ramalanjaona, N.; Rodrigues, L.C.; Mouriño, A.; Rochel, N.; Moras, D. Structure-function relationships and crystal structures of the vitamin D receptor bound 2α-methyl-(20*S*,23*R*)- and 2α-methyl-(20*S*,23*S*)-epoxymethano-1α,25-dihydroxyvitamin D_3. *J. Med. Chem.* **2010**, *53*, 1159–1171. [CrossRef] [PubMed]
279. Sawada, D.; Tsukuda, Y.; Saito, H.; Takagi, K.-i.; Ochiai, E.; Ishizuka, S.; Takenouchi, K.; Kittaka, A. Synthesis of 2β-substituted-14-epi-previtamin D_3 and testing its genomic activity. *J. St. Biochem. Mol. Biol.* **2010**, *121*, 20–24. [CrossRef]
280. Sokolowska, K.; Mouriño, A.; Sicinski, R.R.; Sigüeiro, R.; Plum, L.A.; DeLuca, H.F. Synthesis and biological evaluation of 6-methyl analog of 1α,25-dihydroxyvitamin D_3. *J. St. Biochem. Mol. Biol.* **2010**, *121*, 29–33. [CrossRef]
281. Glebocka, A.; Sicinski, R.R.; Plum, L.A.; DeLuca, H.F. New 1α,25-dihydroxy-19-nor-vitamin D_3 analogs with frozen A-ring conformation. *J. St. Biochem. Mol. Biol.* **2010**, *121*, 46–50. [CrossRef]
282. Sibilska, I.; Barycka, K.M.; Sicinski, R.R.; Plum, L.A.; DeLuca, H.F. 1-Desoxy analog of 2MD: (20*S*)-25-hydroxy-2-methylene-19-norvitamin D_3. *J. St. Biochem. Mol. Biol.* **2010**, *121*, 51–55. [CrossRef]
283. Sawada, D.; Katayama, T.; Tsukuda, Y.; Yamashita, A.; Saito, N.; Saito, H.; Takagi, K.-i.; Ochiai, E.; Ishizuka, S.; Takenouchi, K.; et al. Synthesis of 2α-and 2β-substituted-14-epi-previtamin D_3 and their genomic activity. *Tetrahedron* **2010**, *66*, 5407–5423. [CrossRef]
284. Glebocka, A.; Sicinski, R.R.; Plum, L.A.; DeLuca, H.F. Synthesis and biological activity of 2-(3′-hydroxypropylidene)-1α-hydroxy-19-norvitamin D analogues with shortened alkyl side chains. *J. Med. Chem.* **2011**, *54*, 6832–6842. [CrossRef]
285. Kashiwagi, H.; Ono, Y.; Shimizu, K.; Haneishi, T.; Ito, T.; Iijima, S.; Kobayashi, T.; Ichikawa, T.; Harada, S.; Sato, H.; et al. Novel nonsecosteroidal vitamin D_3 carboxylic acid analogs for osteoporosis and SAR analysis. *Bioorg. Med. Chem.* **2011**, *19*, 4721–4729. [CrossRef]
286. Fujii, S.; Masuno, H.; Taoda, Y.; Kano, A.; Wongmayura, A.; Nakabayashi, M.; Ito, N.; Shimizu, M.; Kawachi, E.; Hirano, T.; et al. Boron cluster-based development of potent nonsecosteroidal vitamin D receptor ligands: Direct observation of hydrophobic interaction between protein surface and carborane. *J. Am. Chem. Soc.* **2011**, *133*, 20933–20941. [CrossRef] [PubMed]
287. Verlinden, L.; Verstuyf, A.; Eelen, G.; Bouillon, R.; Ordóñez-Morán, P.; Larriba, M.J.; Muñoz, A.; Rochel, N.; Sato, Y.; Moras, D.; et al. Synthesis, structure, and biological activity of *des*-side chain analogues of 1α,25-dihydroxyvitamin D_3 with substituents at C18. *ChemMedChem* **2011**, *6*, 788–793. [CrossRef] [PubMed]
288. Sawada, D.; Tsukuda, Y.; Saito, H.; Takimoto-Kamimura, M.; Ochiai, E.; Ishizuka, S.; Takenouchi, K.; Kittaka, A. Development of 14-*epi*-19-nortachysterol and its unprecedented binding configuration for the human vitamin D receptor. *J. Am. Chem. Soc.* **2011**, *133*, 7215–7221. [CrossRef] [PubMed]
289. Regueira, M.A.; Samanta, S.; Malloy, P.J.; Ordóñez-Morán, P.; Resende, D.; Sussman, F.; Muñoz, A.; Mouriño, A.; Feldman, D.; Torneiro, M. Synthesis and biological evaluation of 1α,25-dihydroxyvitamin D_3 analogues hydroxymethylated at C26. *J. Med. Chem.* **2011**, *54*, 3950–3962. [CrossRef] [PubMed]
290. Salomón, D.G.; Grioli, S.M.; Buschiazzo, M.; Mascaró, E.; Vitale, C.; Radivoy, G.; Perez, M.; Fall, Y.; Mesri, E.A.; Curino, A.C.; et al. Novel alkynylphospate analogue of calcitriol with potent antiproliferative effects in cancer cells and lack of calcemic activity. *ACS Med. Chem. Lett.* **2011**, *2*, 503–508. [CrossRef] [PubMed]
291. Saito, H.; Chida, T.; Takagi, K.; Horie, K.; Sawai, Y.; Nakamura, Y.; Harada, Y.; Takenouchi, K.; Kittaka, A. Synthesis of C-2 substituted vitamin D derivatives having ringed side chains and their biological evaluation, especially biological effect on bone by modification at the C-2 poistion. *Org. Biomol. Chem.* **2011**, *9*, 3954–3964. [CrossRef]
292. Shindo, K.; Kumagai, G.; Takano, M.; Sawada, D.; Saito, N.; Saito, H.; Kakuda, S.; Takagi, K.-i.; Ochiai, E.; Horie, K.; et al. New C15-substituted active vitamin D_3. *Org. Lett.* **2011**, *13*, 2852–2855. [CrossRef]

293. Molnár, F.; Sigüeiro, R.; Sato, Y.; Araujo, C.; Schuster, I.; Antony, P.; Peluso, J.; Muller, C.; Mouriño, A.; Moras, D.; et al. 1α,25-$(OH)_2$-3-*epi*-vitamin D_3, a natural metabolite of vitamin D_3: Its synthesis, biological activity and crystal structure with its receptor. *PLoS ONE* **2011**, *6*, e19124. [CrossRef]
294. Wongmayura, A.; Fujii, S.; Ito, S.; Kano, A.; Taoda, Y.; Kawachi, E.; Kagechika, H.; Tanatani, A. Novel vitamin D receptor ligands bearing a spherical hydrophobic core structure-comparison of bicyclic hydrocarbon derivatives with boron cluster derivatives. *Bioorg. Med. Chem. Lett.* **2012**, *22*, 1756–1760. [CrossRef]
295. Fischer, J.; Wang, T.-T.; Kaldre, D.; Rochel, N.; Moras, D.; White, J.H.; Gleason, J.L. Synthetically accessible non-secosteroidal hybrid molecules combining vitamin D receptor agonism and histone deacetylase inhibition. *Chem. Biol.* **2012**, *19*, 963–971. [CrossRef]
296. Fraga, R.; Zacconi, F.; Sussman, F.; Ordóñez-Morán, P.; Muñoz, A.; Huet, T.; Molnar, F.; Moras, D.; Rochel, N.; Maestro, M.; et al. Design, synthesis, evaluation, and structure of vitamin D analogues with furan side chains. *Chem. Eur. J.* **2012**, *18*, 603–612. [CrossRef] [PubMed]
297. Sawada, D.; Tsukuda, Y.; Yasuda, K.; Sakaki, T.; Saito, H.; Kakuda, S.; Takagi, K.-i.; Takenouchi, K.; Chen, T.C.; Reddy, G.S.; et al. Synthesis and biological activities of 1α,4α,25- and 1α,4β,25-trihydroxyvitamin D_3 and their metabolism by human CYP24A1 and UDP-glucuronosyltransferase. *Chem. Pharm. Bull.* **2012**, *60*, 1343–1346. [CrossRef] [PubMed]
298. Sikervar, V.; Fleet, J.C.; Fuchs, P.L. A general approach to the synthesis of enantiopure 19-*nor*-vitamin D_3 and its C-2 phosphate analogs prepared from cyclohexadienyl sulfone. *Chem. Comm.* **2012**, *48*, 9077–9079. [CrossRef] [PubMed]
299. Carballa, D.M.; Rumbo, A.; Torneiro, M.; Maestro, M.; Mouriño, A. Synthesis of (1α)-1,25-dihydroxyvitamin D_3 with a β-postioned seven-carbon side chain at C(12). *Helv. Chim. Acta* **2012**, *95*, 1842–1850. [CrossRef]
300. Flores, A.; Sicinski, R.R.; Grzywacz, P.; Thoden, J.B.; Plum, L.A.; Clagett-Dame, M.; DeLuca, H.F. A 20*S* Combined with 22*R* configuration markedly increases both in vivo and in vitro biological activity of 1α,25-dihydroxy-22-methyl-2-methylene-19-norvitamin D_3. *J. Med. Chem.* **2012**, *55*, 4352–4366. [CrossRef]
301. Carballa, D.M.; Seoane, S.; Zacconi, F.; Pérez, X.; Rumbo, A.; Álvarez-Díaz, S.; Larriba, M.J.; Pérez-Fernández, R.; Muñoz, A.; Maestro, M.; et al. Synthesis and biological evaluation of 1α,25-dihydroxyvitamin D_3 analogues with a long side chain at C12 and short C17 side chains. *J. Med. Chem.* **2012**, *55*, 8642–8656. [CrossRef]
302. Lu, Y.; Chen, J.; Janjetovic, Z.; Michaels, P.; Tang, E.K.Y.; Wang, J.; Tuckey, R.; Slominski, A.T.; Li, W.; Miller, D.D. Design, synthesis, and biological action of (20*R*)-hydroxyvitamin D_3. *J. Med. Chem.* **2012**, *55*, 3573–3577. [CrossRef]
303. Yoshimoto, N.; Sakamaki, Y.; Haeta, M.; Kato, A.; Inaba, Y.; Itoh, T.; Nakabayashi, M.; Itoh, N.; Yamamoto, K. Butyl pocket formation in the vitamin D receptor strongly affects the agonistic or antagonistic behavior of ligands. *J. Med. Chem.* **2012**, *55*, 4373–4381. [CrossRef]
304. Sikervar, V.; Fleet, J.C.; Fuchs, P.L. Fluoride-mediated elimination of allyl sulfones: Application to the synthesis of a 2,4-dimethyl-A-ring vitamin D_3 analogue. *J. Org. Chem.* **2012**, *77*, 5132–5138. [CrossRef]
305. Ciesielski, F.; Sato, Y.; Chebaro, Y.; Moras, D.; Dejaegere, A.; Rochel, N. Structural basis for the accommodation of bis- and tris-aromatic derivatives in vitamin D nuclear receptor. *J. Med. Chem.* **2012**, *55*, 8440–8449. [CrossRef]
306. Chen, B.; Kawai, M.; Wu-Wong, R. Synthesis of VS-105: A novel and potent vitamin receptor agonist with reduced hypercalcemic effects. *Bioorg. Med. Chem. Lett.* **2013**, *23*, 5949–5952. [CrossRef] [PubMed]
307. Milczarek, M.; Chodynski, M.; Filip-Psurska, B.; Martowicz, A.; Krup, M.; Krajewski, K.; Kutner, A.; Wietrzyk, J. Synthesis and biological activity of diastereomeric and geometric analogs of calcipotriol, PRI-2202 and PRI-2205, against human HL-60 leukemia and MCF-7 breast cancer cells. *Cancers* **2013**, *5*, 1355–1378. [CrossRef] [PubMed]
308. Kashiwagi, H.; Ono, Y.; Ohta, M.; Itoh, S.; Ichikawa, F.; Harada, S.; Takeda, S.; Sekiguchi, N.; Ishigai, M.; Takahashi, T. A series of nonsecosteroidal vitamin D receptor agonists for osteoporosis therapy. *Bioorg. Med. Chem.* **2013**, *21*, 1823–1833. [CrossRef] [PubMed]
309. Kashiwagi, H.; Ohta, M.; Ono, Y.; Morikami, K.; Itoh, S.; Sato, H.; Takahashi, T. Effects of fluorines on nonsecosteroidal vitamin D receptor agonists. *Bioorg. Med. Chem.* **2013**, *21*, 712–721. [CrossRef]
310. Sibilska, I.S.; Szybinski, M.; Sicinski, R.R.; Plum, L.A.; DeLuca, H.F. Highly potent 2-methylene analogs of 1α,25-dihydroxyvitamin D_3: Synthesis and biological evaluation. *J. St. Biochem. Mol. Biol.* **2013**, *136*, 9–13. [CrossRef]
311. Sibilska, I.S.; Sicinski, R.R.; Plum, L.A.; DeLuca, H.F. Synthesis and biological activity of 25-hydroxy-2-methylene-vitamin D_3 compounds. *J. St. Biochem. Mol. Biol.* **2013**, *136*, 17–22. [CrossRef]
312. Kulezka, U.; Mouriño, A.; Plum, L.A.; DeLuca, H.F.; Sicinski, R.R. Synthesis of 19-norvitamin D_3 analogs with unnatural triene system. *J. St. Biochem. Mol. Biol.* **2013**, *136*, 23–26. [CrossRef]
313. Sokolowska, K.; Sicinski, R.R.; Plum, L.A.; DeLuca, H.F.; Mouriño, A. Synthesis and biological evaluation of novel 6-substituted analogs of 1α,25-dihydroxy-19-norvitamin D_3. *J. St. Biochem. Mol. Biol.* **2013**, *136*, 30–33. [CrossRef]
314. Glebocka, A.; Sicinski, R.R.; Plum, L.A.; DeLuca, H.F. Ring-A-*seco* analogs of 1α,25-dihydroxy-19-norvitamin D_3. *J. St. Biochem. Mol. Biol.* **2013**, *136*, 39–43. [CrossRef]
315. Yamada, S.; Makishima, M. Structure-activity relationship of nonsecoesteroidal vitamin D receptor modulators. *Trends Phar. Sci.* **2014**, *35*, 324–337. [CrossRef]
316. Ferla, S.; Aboraia, A.S.; Brancale, A.; Pepper, C.J.; Zhu, J.; Ochalek, J.T.; DeLuca, H.F.; Simons, C. Small-molecule inhibitors of 25-hydroxyvitamin D-24-hydroxylase (CYP24A1): Synthesis and biological evaluation. *J. Med. Chem.* **2014**, *57*, 7702–7715. [CrossRef] [PubMed]

317. Fujii, S.; Kano, A.; Songkram, C.; Masuno, H.; Taoda, Y.; Kawachi, E.; Hirano, T.; Tanatani, A.; Kagechika, H. Synthesis and structure-activity relationship of *p*-carborane-based non-secosteroidal vitamin D analogs. *Bioorg. Med. Chem.* **2014**, *22*, 1227–1235. [CrossRef] [PubMed]
318. Kudo, T.; Ishizawa, M.; Maekawa, K.; Nakabayashi, M.; Watari, Y.; Uchida, H.; Tokiwa, H.; Ikura, T.; Ito, N.; Makishima, M.; et al. Combination of triple bond and adamantane ring on the vitamin D side chain produced partial agonists for vitamin D receptor. *J. Med. Chem.* **2014**, *57*, 4073–4087. [CrossRef] [PubMed]
319. Liu, C.; Zhao, G.-D.; Mao, X.; Suenaga, T.; Fujishima, T.; Zhang, C.-M.; Liu, Z.-P. Synthesis and biological evaluation of 1α,25-dihydroxyvitamin D_3 analogues with aromatic side chains attached at C-17. *Eur. J. Med. Chem.* **2014**, *85*, 569–575. [CrossRef]
320. Okamoto, R.; Gery, S.; Kuwayama, Y.; Borregaard, N.; Ho, Q.; Alvarez, R.; Akagi, T.; Liu, G.Y.; Uskokovic, M.R.; Koeffler, H.P. Novel Gemini vitamin D_3 analogs: Large structure/function analysis and ability to induce antimicrobial peptide. *Int. J. Cancer* **2014**, *134*, 207–217. [CrossRef]
321. Thomas, E.; Brion, J.-D.; Peyrat, J.-F. Synthesis and preliminary biological evaluation of new antiproliferative aromatic analogues of 1α,25-dihydroxyvitamin D_3. *Eur. J. Med. Chem.* **2014**, *86*, 381–393. [CrossRef]
322. Takano, M.; Ohya, S.; Yasuda, K.; Nishikawa, M.; Takeguchi, A.; Sawada, D.; Sakaki, T.; Kittaka, A. Synthesis and biological activity of 1α,2α,25-trihydroxyvitamin D_3; active metabolite of 2α-(3-hydroxypropoxy)-1α,25-dihydroxyvitamin D_3 by human CYP3A4. *Chem. Pharm. Bull.* **2014**, *62*, 182–184. [CrossRef]
323. Anami, Y.; Itoh, T.; Egawa, D.; Oshimoto, N.; Yamamoto, K. A mixed population of antagonist and agonist binding conformers in a single crystal explains partial agonism against vitamin D receptor: Active vitamin D analogues with 22*R*-alkyl group. *J. Med. Chem.* **2014**, *57*, 4351–4367. [CrossRef] [PubMed]
324. Sibilska, I.S.; Sicinski, R.R.; Ochalek, J.T.; Plum, L.A.; DeLuca, H.F. Synthesis and biological activity of 25-hydroxy-2-methylene-vitamin D_3 analogues monohydroxylated in the A-ring. *J. Med. Chem.* **2014**, *57*, 8319–8331. [CrossRef]
325. Sigüeiro, R.; Álvarez, A.; Otero, R.; López-Pérez, B.; Carballa, D.; Regueira, T.; González-Berdullas, P.; Seoane, S.; Pérez-Fernández, R.; Mouriño, A.; et al. Synthesis of nonadeuterated 1α,25-dihydroxyvitamin D_2. *J. St. Biochem. Mol. Biol.* **2014**, *144*, 204–206. [CrossRef]
326. Meyer, D.; Rentsch, L.; Marti, R. Effcient and scalable total synthesis of calcitroic acid and its ^{13}C-labeled derivative. *RSC Adv.* **2014**, *4*, 32327–32334. [CrossRef]
327. Banerjee, U.; DeBerardinis, A.M.; Hadden, M.K. Design, synthesis and evaluation of hybrid vitamin D_3 side chain analogues as hedgehog pathway inhibitors. *Bioorg. Med. Chem.* **2015**, *23*, 548–555. [CrossRef] [PubMed]
328. Anami, Y.; Sakamaki, Y.; Itoh, T.; Inaba, Y.; Nakabayashi, M.; Ikura, T.; Ito, N.; Yamamoto, K. Fine tuning of agonistic/antagonistic activity for vitamin D receptor by 22-alkyl chain length of ligands: 22*S*-hexyl compound unexpectedly restored agonistic activity. *Bioorg. Med. Chem.* **2015**, *23*, 7274–7281. [CrossRef] [PubMed]
329. Trynda, J.; Turlej, E.; Milczarek, M.; Pietraszek, A.; Chodynski, M.; Kutner, A.; Wietrzyk, J. Antiproliferative activity and in vivo toxicity of double-point modified analogs of 1,25-dihydroxyergocalciferol. *Int. J. Mol. Sci.* **2015**, *16*, 24873–24894. [CrossRef] [PubMed]
330. Misawa, T.; Yorioka, M.; Demizu, Y.; Noguchi-Yachide, T.; Ohoka, N.; Kurashima-Kinoshita, M.; Motoyoshi, H.; Nojiri, H.; Kittaka, A.; Makishima, M.; et al. Effects of the alkyl side chains and terminal hydrophilicity on vitamin D receptor (VDR) agonist activity based on the diphenylpentane skeleton. *Bioorg. Med. Chem. Lett.* **2015**, *25*, 5362–5366. [CrossRef]
331. Kulesza, U.; Plum, L.A.; DeLuca, H.F.; Mouriño, A.; Sicinski, R.R. Novel 9-alkyl and 9-alkylidene-substituted 1α,25-dihydroxyvitamin D_3 analogues: Synthesis and biological examinations. *J. Med. Chem.* **2015**, *58*, 6237–6247. [CrossRef]
332. Lin, Z.; Marepally, S.R.; Ma, D.; Myers, L.; Postlethwaite, A.E.; Tuckey, R.C.; Chen, C.Y.S.; Kim, T.-K.; Yue, J.; Slominski, A.T.; et al. Chemical synthesis and biological activities of 20,24*S*/*R*-dihydroxyvitamin D_3 epimers and their 1α-hydroxyl derivatives. *J. Med. Chem.* **2015**, *58*, 7881–7887. [CrossRef]
333. Sibilska, I.K.; Szybinski, M.; Sicinski, R.R.; Plum, L.A.; DeLuca, H.F. Synthesis and biological activity of active 2-methylene analogues of calcitriol and related compounds. *J. Med. Chem.* **2015**, *58*, 9653–9662. [CrossRef]
334. Sokolowska, K.; Carballa, D.; Seoane, S.; Pérez-Fernández, R.; Mouriño, A.; Sicinski, R.R. Synthesis and biological activity of two C-7 methyl analogs of vitamin D. *J. Org. Chem.* **2015**, *80*, 165–173. [CrossRef]
335. Saitoh, H.; Watanabe, H.; Kakuda, S.; Takimoto-Kamimura, M.; Takagi, K.; Takeuchi, A.; Takenouchi, K. Synthesis and biological activity of vitamin D_3 derivatives with cyanoalkyl side chain at C-2 position. *J. St. Biochem. Mol. Biol.* **2015**, *148*, 27–30. [CrossRef]
336. Takano, M.; Sawada, D.; Yasuda, K.; Nishikawa, M.; Takeuchi, A.; Takagi, K.-i.; Horie, K.; Reddy, G.S.; Chen, T.C.; Sakaki, T.; et al. Synthesis and metabolic studies of 1α,2α,25-, 1α,4α,25- and 1α,4β,25-trihydroxyvitamin D_3. *J. St. Biochem. Mol. Biol.* **2015**, *148*, 34–37. [CrossRef] [PubMed]
337. Ferronato, M.J.; Salomón, D.G.; Fermento, M.E.; Gandini, N.A.; López Romero, A.; Rivadulla, M.L.; Pérez-García, X.; Gómez, G.; Pérez, M.; Fall, Y.; et al. Vitamin D analogue: Potent antiproliferative effects on cancer cell lines and lack of hypercalcemic activity. *Arch. Pharm. Chem. Life Sci.* **2015**, *348*, 315–329. [CrossRef] [PubMed]
338. Biswas, T.; Akagi, Y.; Usuda, K.; Yasui, K.; Shimizu, I.; Okamoto, M.; Uesugi, M.; Hosokawa, S.; Nagasawa, K. Synthesis of 24,24-difluoro-1,3-*cis*-25-hydroxy-19-norvitamin D_3 derivatives and evaluation of their vitamin D receptor-binding affinity. *Biol. Pharm. Bull.* **2016**, *39*, 1387–1391. [CrossRef] [PubMed]
339. Usuda, K.; Biswas, T.; Akagi, Y.; Yasui, K.; Uesugi, M.; Shimizu, I.; Hosokawa, S.; Nagasawa, K. Synthesis of diastereomers of 1,3-*cis*-25-hydroxy-19-norvitamin D_3. *Chem. Pharm. Bull.* **2016**, *64*, 1190–1195. [CrossRef] [PubMed]

340. Otero, R.; Seoane, S.; Sigüeiro, R.; Belorusova, A.Y.; Maestro, M.A.; Pérez-Fernández, R.; Rochel, N.; Mouriño, A. Carborane-based design of a potent vitamin D receptor agonist. *Chem. Sci.* **2016**, *7*, 1033–1037. [CrossRef] [PubMed]
341. Nijenhuis, T.; van der Eerden, B.; Zügel, U.; Steinmeyer, A.; Weinans, H.; Hoenderop, J.G.J.; van Leeuwen, J.P.T.M.; Bindels, R.J.M. The novel vitamin D analog ZK191784 as an intestine-specific vitamin D antagonist. *FASEB J.* **2016**, *20*, 2171–2173. [CrossRef]
342. Lin, Z.; Marepally, S.R.; Ma, D.; Kim, T.-K.; Oak, A.S.W.; Myers, L.; Tuckey, R.C.; Slominski, A.T.; Miller, D.D.; Li, W. Synthesis and biological evaluation of vitamin D_3 metabolite 20,23*S*-dihydroxyvitamin D_3 and its 23*R* epimer. *J. Med. Chem.* **2016**, *59*, 5102–5108. [CrossRef] [PubMed]
343. Bolla, N.R.; Corcoran, A.; Yasuda, K.; Chodynski, M.; Krajewski, K.; Cmoch, P.; Marcinkowska, E.; Brown, G.; Sakaki, T.; Kutner, A. Synthesis and evaluation of geometric analogs of 1α,25-dihydroxyvitamin D_2 as potential therapeutics. *J. St. Biochem. Mol. Biol.* **2016**, *164*, 50–55. [CrossRef]
344. Loureiro, J.; Maestro, M.A.; Mouriño, A.; Sigüeiro, R. Stereoselective synthesis of 1β,25-dihydroxyvitamin D_3 and its 26,27-hexadeuterated derivative. *J. St. Biochem. Mol. Biol.* **2016**, *164*, 56–58. [CrossRef]
345. Ahmad, M.I.; Raghuvanshi, D.S.; Singh, S.; John, A.A.; Prakash, R.; Nainwat, K.S.; Singh, D.; Tripathi, S.; Sharma, A.; Gupta, A. Design and synthesis of 3-arylbenzopyran based non-steroidal vitamin D_3 mimics as osteogenic agents. *Med. Chem. Commun.* **2016**, *7*, 2381–2394. [CrossRef]
346. Sawada, D.; Kakuda, S.; Kamimura-Takimoto, M.; Takeuchi, A.; Matsumoto, Y.; Kittaka, A. Revisiting the 7,8-*cis*-vitamin D_3 derivatives: Synthesis, evaluating the biological activity, and study of the binding configuration. *Tetrahedron* **2016**, *72*, 2838–2848. [CrossRef]
347. Kittaka, A.; Takano, M.; Saitoh, H. Vitamin D analogs with nitrogen atom at C2 substitution and effect on bone formation. *Vitam. Horm.* **2016**, *100*, 379–394. [CrossRef] [PubMed]
348. Sawada, D.; Ochiai, E.; Takeuchi, A.; Kakuda, S.; Kamimura-Takimoto, M.; Kawagoe, F.; Kittaka, A. Synthesis of 2α- and 2β-(3-hydroxypropyl)-7,8-*cis*-14-*epi*-1α,25-dihydroxy-19-norvitamin D_3 and their biological activity. *J. St. Biochem. Mol. Biol.* **2017**, *173*, 79–82. [CrossRef]
349. Suenaga, T.; Nozaki, T.; Fujishima, T. Synthesis of novel vitamin D_3 analogues having a spiro-oxetane structure. *Vitamins* **2016**, *90*, 109–114.
350. Hernández-Martín, A.; Fernández, S.; Verstuyf, A.; Verlinden, L.; Ferrero, M. A-Ring-modified 2-hydroxyethylidene previtamin D_3 analogues: Synthesis and biological evaluation. *Eur. J. Org. Chem.* **2017**, *2017*, 504–513. [CrossRef]
351. Chiellini, G.; Rapposelli, S.; Nesi, G.; Sestito, S.; Sabatini, M.; Zhu, J.; Massarelli, I.; Plum, L.A.; Clagett-Dame, M.; DeLuca, H.F. Synthesis and biological evaluation of cyclopropylamine vitamin D-like CYP24A1 inhibitors. *ChemistrySelect* **2017**, *2*, 8346–8353. [CrossRef]
352. Szybinski, M.; Brzeminski, P.; Fabisiak, A.; Berkowska, K.; Marcinkowska, E.; Sicinski, R.R. Seco-B-ring steroidal dienynes with aromatic D ring: Design, synthesis and biological evaluation. *Int. J. Mol. Sci.* **2017**, *18*, 2162–2176. [CrossRef]
353. Kato, A.; Yamao, M.; Hashihara, Y.; Ishida, H.; Toh, T.; Yamamoto, K. Vitamin D analogues with a *p*-hydroxyphenyl group at C25 position: Crystal structure of vitamin D receptor ligand-binding domain complexed with the ligand explains the mechanism underlying full antagonistic action. *J. Med. Chem.* **2017**, *60*, 8394–8406. [CrossRef]
354. Szybinski, M.; Sokolowska, K.; Sicinski, R.R.; Plum, L.A.; DeLuca, H.F. D-*seco*-Vitamin D analogs having reversed configurations at C-13 and C-14: Synthesis, dockings studies and biological evaluation. *J. St. Biochem. Mol. Biol.* **2017**, *173*, 57–63. [CrossRef]
355. Szybinski, M.; Sektas, K.; Sicinski, R.R.; Plum, L.A.; Frelek, J.; DeLuca, H.F. Design, synthesis and biological properties of seco-D-ring modified 1α,25-dihydroxyvitamin D_3 analogues. *J. St. Biochem. Mol. Biol.* **2017**, *171*, 144–154. [CrossRef]
356. Kattner, L.; Bernardi, D. An efficient synthesis of 1α,25-dihydroxyvitamin D_3 LC-biotin. *J. St. Biochem. Mol. Biol.* **2017**, *173*, 89–92. [CrossRef] [PubMed]
357. Vinhas, S.; Vázquez, S.; Rodríguez-Borges, J.E.; Sigüeiro, R. A convergent approach to the side-chain homologated of 1α,25-dihydroxyergocalciferol. *J. St. Biochem. Mol. Biol.* **2017**, *173*, 83–85. [CrossRef] [PubMed]
358. Kang, Z.-S.; Wang, C.; Han, X.-L.; Du, J.-J.; Li, Y.-Y.; Zhang, C. Design, synthesis and biological evaluation of non-secosteroidal vitamin D receptor ligand bearing double side chain for the treatment of chronic pancreatitis. *Eur. J. Med. Chem.* **2018**, *146*, 541–553. [CrossRef]
359. Kang, Z.-S.; Wang, C.; Han, X.-L.; Wang, B.; Yuan, H.-L.; Hao, M.; Hou, S.-Y.; Hao, M.-X.; Du, J.-J.; Li, Y.-Y.; et al. Sulfonyl-containing phenyl-pyrrolyl pentane analogues: Novel non-secosteroidal vitamin D receptor modulators with favorable physicochemical properties, pharmacokinetic properties and anti-tumor activity. *Eur. J. Med. Chem.* **2018**, *157*, 1174–1191. [CrossRef] [PubMed]
360. Hao, M.; Hou, S.; Xue, L.; Yuan, H.; Zhu, L.; Wang, C.; Wang, B.; Tang, C.; Zhang, C. Further developments of the phenyl-pyrrolyl pentane series of nonsteroidal vitamin D receptor modulators as anticancer agents. *J. Med. Chem.* **2018**, *61*, 3059–3075. [CrossRef] [PubMed]
361. Carballa, D.; Sigüeiro, R.; Rodríguez-Docampo, Z.; Zacconi, F.; Maestro, M.A.; Mouriño, A. Stereoselective palladium-catalyzed approach to vitamin D_3 derivatives in protic medium. *Chem. Eur. J.* **2018**, *24*, 3314–3320. [CrossRef] [PubMed]
362. Yoshizawa, M.; Itoh, T.; Hori, T.; Kato, A.; Anami, Y.; Yoshimoto, N.; Yamamoto, K. Identification of the histidine residue in vitamin D receptor that covalently binds to electrphilic ligands. *J. Med. Chem.* **2018**, *61*, 6339–6349. [CrossRef] [PubMed]
363. Gogoi, P.; Seoane, S.; Sigüeiro, R.; Guiberteau, T.; Maestro, M.A.; Pérez-Fernández, R.; Rochel, N.; Mouriño, A. Aromatic-based design of highly active and non calcemic vitamin D receptor agonists. *J. Med. Chem.* **2018**, *61*, 4928–4937. [CrossRef]

364. Otero, R.; Ishizawa, M.; Numoto, N.; Ikura, T.; Ito, N.; Tokiwa, H.; Mouriño, A.; Makishima, M.; Yamada, S. 25*S*-Adamantyl-23-yne-26,27-dinor-1α,25-dihydroxyvitamin D_3: Synthesis, tissue selective biological activities, and X-ray crystal structural analysis of its vitamin D complex. *J. Med. Chem.* **2018**, *61*, 6658–6673. [CrossRef]
365. Brzeminski, P.; Fabisiak, A.; Sektas, K.; Berkowska, K.; Marcinkowska, E.; Sicinski, R.R. Synthesis of 19-norcalcitriol analogs with elongated side chain. *J. St. Biochem. Mol. Biol.* **2018**, *177*, 231–234. [CrossRef]
366. Fabisiak, A.; Brzeminski, P.; Berkowska, K.; Marcinkowska, E.; Sicinski, R.R. Synthesis of 19-norcalcitriol analogs with alkylidene moieties at C-2 based on succinic acid and L-methionine. *J. St. Biochem. Mol. Biol.* **2018**, *177*, 235–239. [CrossRef] [PubMed]
367. Sawada, D.; Kakuda, S.; Takeuchi, A.; Kawagoe, F.; Takimoto-Kamimura, M.; Kittaka, A. Effects of 2-substitution on 14-*epi*-19-nortachysterol-mediated biological events; based on synthesis and X-ray co-crystallographic analysis with the human vitamin D receptor. *Org. Biomol. Chem.* **2018**, *16*, 2448–2455. [CrossRef] [PubMed]
368. Sigüeiro, R.; Maestro, M.A.; Mouriño, A. Synthesis of side-chain locked analogs of 1α,25-dihydroxyvitamin D_3 bearing a C17 methyl group. *Org. Lett.* **2018**, *20*, 2641–2644. [CrossRef] [PubMed]
369. Lin, Z.; Marepally, S.R.; Goh, E.S.Y.; Chen, C.Y.S.; Janjetovic, Z.; Kim, T.-k.; Miller, D.D.; Postlehwaite, A.E.; Slominski, A.T.; Tuckey, R.C.; et al. Investigation of 20*S*-hydroxyvitamin D_3 analogs and their 1α-OH derivatives as potent vitamin D receptor agonists with anti-inflammatory activities. *Sci. Rep.* **2018**, *8*, 1478–1489. [CrossRef] [PubMed]
370. Kawagoe, F.; Yasuda, K.; Mototani, S.; Sugiyama, T.; Uesugi, M.; Sakaki, T.; Kittaka, A. Synthesis and CYP24A1-dependent metabolism of 23-fluorinated vitamin D_3 analogues. *ACS Omega* **2019**, *4*, 11332–11337. [CrossRef] [PubMed]
371. Maschinot, C.A.; Chau, L.Q.; Wechsler-Reya, R.J.; Hadden, M.K. Synthesis and evaluating of third generation vitamin D_3 analogues as inhibitors of Hedgehog signaling. *Eur. J. Med. Chem.* **2019**, *162*, 495–506. [CrossRef] [PubMed]
372. Sigüeiro, R.; Loureiro, J.; González-Berdullas, P.; Mouriño, A.; Maestro, M.A. Synthesis of 28,28,28-trideutero-25-hydroxydihydrotachysterol$_2$. *J. St. Biochem. Mol. Biol.* **2019**, *185*, 248–250. [CrossRef]
373. Ferronato, M.J.; Obiol, D.J.; Alonso, E.N.; Guevara, J.A.; Grioli, S.M.; Mascaró, M.; Rivadulla, M.L.; Martínez, A.; Gómez, G.; Fall, Y.; et al. Synthesis of a novel analog of calcitriol and its biological evaluation as antitumor agent. *J. St. Biochem. Mol. Biol.* **2019**, *185*, 118–136. [CrossRef]
374. Wang, W.; Zhao, G.-D.; Cui, J.-i.; Li, M.-Q.; Liu, Z.-P. Synthesis of 1α,25-dihydroxyvitamin D_3 Analogues with α,α-difluorocycloketone at the CD-ring side chains and their biological properties in ovariectomized rats. *J. St. Biochem. Mol. Biol.* **2019**, *186*, 66–73. [CrossRef]
375. Kawagoe, F.; Sugiyama, T.; Yasuda, K.; Uesugi, M.; Sakaki, T.; Kittaka, A. Concise synthesis of 23-hydroxylated vitamin D_3 metabolites. *J. St. Biochem. Mol. Biol.* **2019**, *186*, 161–168. [CrossRef]
376. Fujishima, T.; Suenaga, T.; Kawahata, M.; Yamaguchi, K. Synthesis and characterization of 20-hydroxyvitamin D_3 with the A-ring modification. *J. St. Biochem. Mol. Biol.* **2019**, *187*, 27–33. [CrossRef] [PubMed]
377. Kawagoe, F.; Mototani, S.; Yasuda, K.; Nagasawa, K.; Uesugi, M.; Sakaki, T.; Kittaka, A. Introduction of fluorine atoms to vitamin D_3 side-chain and synthesis of 24,24-difluoro-25-hydroxyvitamin D_3. *J. St. Biochem. Mol. Biol.* **2019**, *195*, 105477. [CrossRef] [PubMed]
378. Fabisiak, A.; Brzeminski, P.; Berkowska, K.; Marcinkowska, E.; Sicinski, R.R. Synthesis of 19-norcalcitriol with pegylated alkylidene chains at C-2. *J. St. Biochem. Mol. Biol.* **2019**, *185*, 251–255. [CrossRef] [PubMed]
379. Nagata, A.; Akagi, Y.; Asano, L.; Kotake, K.; Kawagoe, F.; Mendoza, A.; Masoud, S.S.; Usuda, K.; Takemoto, Y.; Kittaka, A.; et al. Synthetic chemical probes that dissect vitamin D activities. *ACS Chem. Biol.* **2019**, *14*, 2851–2858. [CrossRef]
380. Ibe, K.; Yamada, T.; Okamoto, S. Synthesis and vitamin D receptor affinity of 16-oxavitamin D_3 analogues. *Org. Biomol. Chem.* **2019**, *17*, 10188–10200. [CrossRef]
381. Fujishima, T.; Komatsu, T.; Takao, Y.; Yonamine, W.; Suenaga, T.; Isono, H.; Morikawa, M.; Takaguchi, K. Design and concise synthesis of novel vitamin D analogues bearing a functionalized aromatic ring on the side chain. *Tetrahedron* **2019**, *75*, 1098–1106. [CrossRef]
382. Sibilska-Kaminski, I.K.; Sicinski, R.R.; Plum, L.A.; DeLuca, H.F. Synthesis and biological activity of 2,22-dimethylene analogues of 19-norcalcitriol and related compounds. *J. Med. Chem.* **2020**, *63*, 7355–7368. [CrossRef]
383. Katner, L.; Rauch, E. Efficient synthesis of 3-TBDMS-11α,25-dihydroxyvitamin D_3 and D_2 ethers. *J. St. Biochem. Mol. Biol.* **2020**, *200*, 105638. [CrossRef]
384. Brzeminski, P.; Fabisiak, A.; Berkowska, K.; Rárova, L.; Marcinkowska, E.; Sicinski, R.R. Synthesis of Gemini analogs of 19-norcalcitriol and their platinum(II) complexes. *Bioorg. Chem.* **2020**, *100*, 103883. [CrossRef]
385. Fabisiak, A.; Brzeminski, P.; Berkowska, K.; Rárova, L.; Marcinkowska, E.; Sicinski, R.R. Design, synthesis and biological evaluation of novel 2-alkylidene 19-norcalcitriol analogs. *Bioorg. Chem.* **2020**, *100*, 104013. [CrossRef]
386. Sibilska-Kaminski, I.K.; Fabisiak, A.; Brzeminski, P.; Plum, L.A.; Sicinski, R.R.; DeLuca, H.F. Novel superagonist analogs of 2-methylene calcitriol: Design, molecular docking, synthesis and biological evaluation. *Bioorg. Chem.* **2022**, *118*, 105416. [CrossRef] [PubMed]
387. Obelleiro, A.; Gómez-Bouzó, U.; Gómez, G.; Fall, Y.; Santalla, H. Design and efficient synthesis of novel vitamin D analogues bearing an aniline moiety in their side chains. *Tet. Lett.* **2020**, *61*, 152493. [CrossRef]
388. Fraga, R.; Len, K.; Lutzing, R.; Laverny, G.; Loureiro, J.; Maestro, M.A.; Rochel, N.; Rodriguez-Borges, E.; Mouriño, A. Design, synthesis, evaluation and structure of allenic 1α,25-dihydroxyvitamin D_3 analogs with locked mobility at C17. *Chem. Eur. J.* **2021**, *27*, 13384–13389. [CrossRef] [PubMed]

389. Seoane, S.; Gogoi, P.; Zárate-Ruiz, A.; Peluso-Iltis, C.; Peters, S.; Guiberteau, T.; Maestro, M.A.; Pérez-Fernández, R.; Rochel, N.; Mouriño, A. Design, synthesis, biological activity and structural analysis of novel des-C—Ring aromatic-D-ring analogues of 1α,25-dihydroxyvitamin D_3. *J. Med. Chem.* **2022**, *65*, 13112–13124. [CrossRef] [PubMed]
390. Ibe, K.; Nakada, H.; Ohgami, M.; Yamada, T.; Okamoto, S. Design, synthesis, and properties of *des*-D-ring interphenylene derivatives of 1α,25-dihydroxyvitamin D_3. *Eur. J. Med. Chem.* **2022**, *243*, 114795. [CrossRef]
391. Kawagoe, F.; Mendoza, A.; Hayata, Y.; Asano, L.; Kotake, K.; Mototani, S.; Kawamura, S.; Kurosaki, S.; Akagi, Y.; Takemoto, Y.; et al. Discovery of a vitamin D receptor-silent vitamin D derivative that impairs regulatory element-binding protein in vivo. *J. Med. Chem.* **2021**, *64*, 5689–5709. [CrossRef]
392. Saito, H.; Horie, K.; Suga, A.; Kaibara, Y.; Mashiko, T. Vitamin D Derivative Having Cyclic Amine in Side Chain. World Patent Organization. WO2022059684A1, 24 March 2022.

MDPI

Review

Nongenomic Activities of Vitamin D

Michał A. Żmijewski

Department of Histology, Faculty of Medicine, Medical University of Gdańsk, PL-80211 Gdańsk, Poland; mzmijewski@gumed.edu.pl; Tel.: +48-58-3491455

Abstract: Vitamin D shows a variety of pleiotropic activities which cannot be fully explained by the stimulation of classic pathway- and vitamin D receptor (VDR)-dependent transcriptional modulation. Thus, existence of rapid and nongenomic responses to vitamin D was suggested. An active form of vitamin D (calcitriol, $1,25(OH)_2D_3$) is an essential regulator of calcium–phosphate homeostasis, and this process is tightly regulated by VDR genomic activity. However, it seems that early in evolution, the production of secosteroids (vitamin-D-like steroids) and their subsequent photodegradation served as a protective mechanism against ultraviolet radiation and oxidative stress. Consequently, direct cell-protective activities of vitamin D were proven. Furthermore, calcitriol triggers rapid calcium influx through epithelia and its uptake by a variety of cells. Subsequently, protein disulfide-isomerase A3 (PDIA3) was described as a membrane vitamin D receptor responsible for rapid nongenomic responses. Vitamin D was also found to stimulate a release of secondary massagers and modulate several intracellular processes—including cell cycle, proliferation, or immune responses—through wingless (WNT), sonic hedgehog (SSH), STAT1-3, or NF-kappaB pathways. Megalin and its coreceptor, cubilin, facilitate the import of vitamin D complex with vitamin-D-binding protein (DBP), and its involvement in rapid membrane responses was suggested. Vitamin D also directly and indirectly influences mitochondrial function, including fusion–fission, energy production, mitochondrial membrane potential, activity of ion channels, and apoptosis. Although mechanisms of the nongenomic responses to vitamin D are still not fully understood, in this review, their impact on physiology, pathology, and potential clinical applications will be discussed.

Keywords: vitamin D; VDR; nongenomic response; PDIA3; ultraviolet radiation; megalin; mitochondria; vitamin D analogs

Citation: Żmijewski, M.A. Nongenomic Activities of Vitamin D. *Nutrients* **2022**, *14*, 5104. https://doi.org/10.3390/nu14235104

Academic Editor: Stefan Pilz

Received: 30 October 2022
Accepted: 24 November 2022
Published: 1 December 2022

1. Vitamin D—Overview

Vitamin D is present in almost forms of life, from phytoplankton to humans, and is considered one of the oldest hormones on Earth [1–3]. In humans, the classic endocrine function of this powerful secosteroid is regulation of calcium–phosphorus homeostasis in order to maintain proper function of the body [4]. Initially, vitamin D has been described as a vitamin because it can be acquired from the diet, mainly from fatty fish, fish oil, milk products, and some mushrooms. However, for humans, the major and a natural source of vitamin D is its photochemical production by 7-dehydrocholesterol (7-DHC) [5]. This process is stimulated by ultraviolet type B (UVB) radiation (280–320 nm) and takes place in the basal layer of epidermis of the skin [6,7]. To be exact, photodegradation of the B-ring of 7-DHC resulted in production of three isomers: previtamin D, tachysterol, and lumisterol. Those byproducts are subjected to time-dependent thermal conversion to vitamin D, which—after release from the cells—enters the circulatory system and is transported to all organs by vitamin-D-binding protein (DBP) [8–10]. Regardless of the source, vitamin D requires two subsequent hydroxylations to achievement its full hormonal activity [3,5,11]. First, 25-hydroxylase (CYP2R1) facilitates the production of 25-hydroxyvitamin D_3 ($25(OH)D_3$) in the liver's hepatocytes. This is the major metabolite of vitamin D, and its serum level is widely used as an indicator of vitamin D status [3,12–15]. The second hydroxylation takes place in proximal tubules of the kidney and requires activity of 1α-hydroxylase (CYP27B1). The final product is a fully active hormone—1,25(OH)$_2$D$_3$, calcitriol [16,17]. Serum levels of vitamin D are tightly regulated on two levels. Firstly, an excess of sun (ultraviolet radiation type B) leads to photodegradation of vitamin D and the formation of suprasterols [18]. Secondly, elevated levels of 1,25(OH)$_2$D$_3$ stimulate the expression of CYP24A1, which is 24-hydroxylase and is responsible for the catabolism (deactivation) of both 25(OH)D$_3$ and 1,25(OH)$_2$D$_3$ [6]. Several other metabolites of vitamin D have been described in recent years, including 3-epi analogs [19–21], lactones [22], and CYP1A1 (CYP450scc) metabolites [23–26]. However, their biological function still is far from being understood.

1,25(OH)$_2$D$_3$ regulates a variety of cellular processes thought activation of a nuclear receptor—VDR. It binds to VDR, which forms a complex with its coreceptor RXR and is subsequently translocated into the nucleus. Upon activation, the VDR-RXR complex regulates expression from hundreds to more than 3000 genes in the human genome depending on cell type and physiological conditions [27–29]. Interestingly, recent studies have shown that not only 1,25(OH)$_2$D$_3$ can affect the expression of genes. 25(OH)D$_3$ also binds to VDR but with an affinity approximately 1000 times lower compared to 1,25(OH)$_2$D$_3$. Transcriptomic analyses of human peripheral blood mononuclear cells (PBMCs) revealed similar patterns of genes affected by 25(OH)D$_3$ treatment at 1000 or 10,000 nM concentrations compared to 10 nM 1,25(OH)$_2$D$_3$ [30,31]. Similar overlapping patterns of expression were also observed in the prostate cancer cell line LNCaP [32]. Intestinally, the metallothionein 2A gene was identified as a unique target for 1,25(OH)$_2$D$_3$ but not for 25(OH)D$_3$ [33]. Other nonclassic metabolites of vitamin D, such as 20(OH)D$_3$ and 20,23(OH)$_2$D$_3$, not only interact with VDR but also target other transcription factors, including RORα and RORγ [34,35] or AhR [28]. These experiments led to the detection of unique genes regulated by alternative nuclear receptors. Furthermore, it seems that the genes coding transcription factors (TFs) are activated by 1,25(OH)$_2$D$_3$, which in turn regulates the expression of additional sets of genes [36], which could be described as secondary genomic response. Recently, regulatory effects of 1,25(OH)$_2$D$_3$ on the expression of microRNAs and long noncoding RNAs have been uncovered as a potential anticancer mechanism [37,38].

The main target of vitamin D is the regulation of the calcium–phosphate homeostasis, which contributes to maintaining proper bone health [1,11,13,39–41]. Furthermore, the presence of VDR is already well documented in the cells of various organs, and it was established that proper function of musculoskeletal, immune, nervous, and cardiovascular systems as well as regeneration of epithelial barriers strongly depends on vitamin D. Vitamin D deficiency (25(OH)D$_3$ level below 20 ng/mL) was not only described as a risk factor for rickets or osteoporosis [39,42] but also impairs the function of the immune system [43–47] and its response to pathogens, including influenza viruses [48] and coronaviruses (SARS-CoV-2) [49]. Vitamin D supplementation resulting in adequate level of vitamin D (preferably 30 ng/mL of 25(OH)D$_3$ in the serum) improves muscle performance and reduces falls in vitamin-D-deficient older adults [50]. It also has a neuroprotective role [51,52] and prevents and reduces outcomes of psoriasis [53]. Furthermore, it was suggested that even higher levels of vitamin D (40–60 ng/mL of 25(OH)D$_3$ in the serum) are beneficial, reducing the occurrence and severity of multiple types of cancer [23,54–60] and displaying neuroprotective [61] and cardioprotective properties [62,63]. Consequently, high doses of vitamin D were found to be beneficial in the prevention or treatment of several diseases, including cancer [23,54,56,58,59,64], neurodegenerative disorders, autoimmune diseases [43,45,65,66], psoriasis [67,68], and preeclampsia [69]. This must be of special interest because vitamin D deficiency is still a global problem [21,39,55,70–72]. Furthermore, seasonal changes in vitamin D levels that were observed pointed to the necessity of its proper supplementation, especially in the winter season [12,13,72].

In spite of the well-established role of VDR in the biological activity of vitamin D, not all effects of this powerful hormone could be liked to VDR-driven regulation of the expression of the genes. Firstly, some proven physiological effects can be observed within seconds or minutes after stimulation with vitamin D. These do not include activation of gene expression and subsequent protein synthesis, which take at least a few. Thus, the so-called rapid nongenomic responses to vitamin D have been described. The presence of membrane receptor for vitamin D has been suggested as an analogy to other steroid hormones [73]. Interestingly, in spite of the fact that VDR is a nuclear receptor, its involvement was not fully excluded. However, rapid responses do not rely on its transcriptional activity [74]. In this review, I will discuss different aspects of alternatives to the genomic activities of vitamin D. Figure 1 visualizes our current understanding of the variety of cellular processes activated by vitamin D and its metabolites.

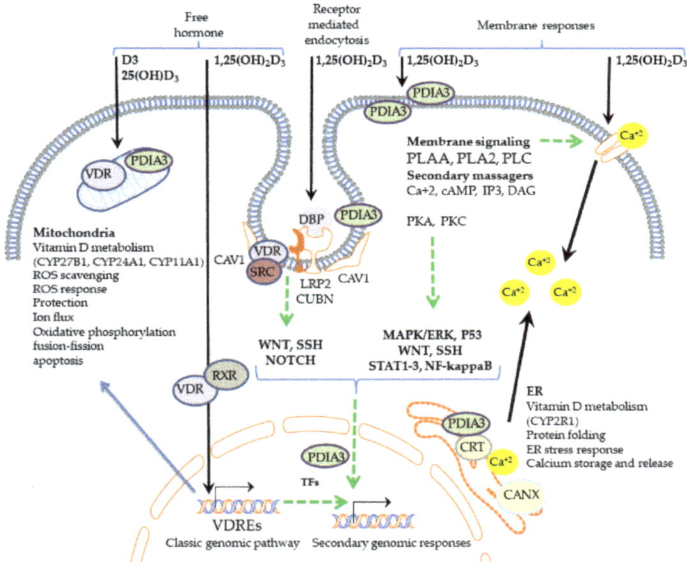

Figure 1. Vitamin-D-activated pathways. All secosteroids, including vitamin D_3, 25(OH)D_3, 1,25(OH)$_2D_3$, and other derivatives, enter the cell directly though the plasma membrane according to the free hormone hypothesis. Alternatively, 1,25(OH)$_2D_3$ bound to DBP may be imported by the megalin/cubulin complex (LRP2-CUBN complex) via receptor-mediated endocytosis. Inside the cell, 1,25(OH)$_2D_3$ binds the VDR-RXR complex, which is translocated into the nucleus and modulates the expression of targeted genes. Vitamin D metabolism takes place in the mitochondria, including activation of 25(OH)D_3 by CYP27B1, resulting in formation of 1,25(OH)$_2D_3$ and inactivation of both metabolites by CYP24A1. In mitochondria, vitamin D may act as ROS scavengers, thus protecting the mitochondria from damage. Vitamin D metabolites may also directly influence ion transport, oxidative phosphorylation, and ROS response or may do so indirectly through the genomic pathway (blue arrow). PDIA3 mediates 1,25(OH)$_2D_3$-dependent membrane-signaling cascades. PDIA3 was detected on both sides of the cell membrane, on the endoplasmic reticulum, and in the cytoplasm, and its translocation to the nucleus was also postulated. PDIA3 mediates 1,25(OH)2D3-dependent activation of PLAA, PLA2, and PLC and the opening of Ca^{+2} and Pi (NaPi II a,c) channels. These result in the rapid accumulation of secondary messengers, including DAG, IP3, cAMP, and Ca^{+2}, followed by PKA, PKC, or CAMK2G activation and the alteration of several downstream targets, including MAPK pathways. VDR was found also to be colocalized with CAV1 and SRC in caveolae. Its membrane activity was linked with the downregulation of WNT, SSH, and NOTCH signaling pathways. Alone or together with VDR after stimulation with 1,25(OH)$_2D_3$, PDIA3 modulates the transcription factors NF-kappaB, STAT3, and P53. At the endoplasmic reticulum, PDIA3—together with calreticulin (CRT) and calnexin (CANX)—facilitates protein folding. 1,25(OH)$_2D_3$, through VDR transcriptional activity, induces several transcription factors (TFs) which are involved in secondary genomic responses to this multipotent hormone.

2. Direct Effects of Vitamin D on Calcium Transport

The idea of alternative pathways activated by vitamin D arose from an observation of rapid (1–10 min) influx of calcium induced by calcitriol in osteogenic sarcoma cell line ROS 17/2.8 [75] or myocytes isolated from chicken embryonic heart [76]. Consequently, transcaltachia was described as a rapid—measured in seconds to minutes—resorption of calcium by enterocytes in response to calcitriol [77]. Interestingly, the rapid influx of calcium requires as low as picomolar concentrations of calcitriol (0.13 nM [78] or 0.3 nM [79]). Later, activation of rapid nongenomic responses to calcitriol was confirmed in vivo in chicken [80], in cultured mice chondrocytes with VDR knockout, and in osteoblastic cell line ROS 24/1 deprived of VDR expression [81]. The time frame (seconds to minutes) of observed calcium transport excludes involvement of the genomic pathway of response to vitamin D. Thus, the idea of membrane response was suggested.

3. Transmembrane Transport of Vitamin D

It is still presumed, according to free hormone hypothesis, that vitamin D can freely diffuse through the membranes and does not require transporter or channel to enter the cells [82], as can other lipophilic steroids. However, only ~0.04% of 1,25(OH)$_2$D$_3$ is considered to be a free hormone in circulation. Furthermore, the major circulating metabolite of vitamin D–25(OH)D$_3$ is mainly carried by vitamin-D-binding protein (DBP; ~85%) and albumin (~15%). Thus, only a fraction of 25(OH)D$_3$ is free (~0.03%) and can be directly absorbed by the target cells. In an addition, the affinity of 1,25(OH)$_2$D$_3$ to DBP is 10 to 100 times lower in comparison to 25(OH)D$_3$ but is sufficient to be its main transporter [8]. This also raises an important question: How do vitamin D metabolites bound with high affinity to DBP become freed and enter into cells? It is well established that other hormones, including steroids and thyroid hormones are bound with corresponding globulins in circulation and require carrier proteins for efficient cellular uptake of their complexes [83]. Surprisingly, early studies revealed that deletion of DBP did not result in vitamin D deficiency under normal vitamin D supplementation, which strongly supported the free hormone hypothesis [84]. On the other hand, a deletion of megalin (LRP2) 600-kDa transmembrane glycoprotein, which functions as an endocytic receptor, results in vitamin D deficiency. Later studies reviled that megalin and two additional proteins—cubilin (CUBN) and disabled-2 (DAB2)—are required for effective endocytosis of DBP/vitamin D complexes. It was shown that deletion or deactivation of any of those proteins results in vitamin D deficiency, hypocalcaemia, and subsequent deterioration of bone structure and function [85–87].

Although expression of megalin, as well as megalin-mediated endocytosis of DBP protein, was shown in several tissues, including muscle [88], mammary gland [89], placenta, prostate, colon epithelium [90], and skin tissues, the best-studied model is epithelium of the proximal convoluted tubules of the kidney [91]. It was shown that the megalin complex is responsible for rapid and efficient resorption of small proteins, including DBP binding vitamin D, and this mechanism is necessary to maintain an optimal level of vitamin D in circulation. Interestingly, epithelial cells of proximal convoluted tubules express CYP27B1 and CYP24A1. Thus, the kidney is the primary location of both activation and deactivation of vitamin D [85,86,92]. Recent studies also postulated the role of megalin in membrane to mitochondria trafficking of various proteins (angiotensin II, stanniocalcin-1, and TGF-β) [93,94], which raises the question of whether the same mechanism could be utilized by vitamin D in complex with DBP. In summary, the role of the megalin-mediated transport of DBP-bound vitamin D metabolite in a rapid membrane response to this secosteroid and its mitochondrial targeting is currently under discussion and intensive investigation.

4. Membrane-Bound Receptor(s) and Targets for Vitamin D

As opposed to its genomic counterparts, nongenomic vitamin D responses appear rapidly (within a range of seconds or minutes) and are not susceptible to actinomycin D or cycloheximide. Therefore, they do not require expression of genes or protein synthesis [95].

Thus, a direct interaction of this secosteroid with membrane-bound or intracellular macromolecules was suggested as an acceptable explanation [96,97]. In fact, several steroids and other small regulatory molecules bind to membrane receptors in addition to their nuclear receptors, activating rapid responses [73,97]. Thus, in an analogy, an involvement of a G-coupled membrane receptor (GPCR) in vitamin D signaling was suggested, and its presence initially confirmed in membranous cell extracts [76,98]. Furthermore, the existence of membrane-associated activities of vitamin D was supported by observation of a rapid and efficient mobilization of the secondary messengers, including cAMP and calcium, triggered by $1,25(OH)_2D_3$ [75,79]. Other studies documented fast (15–300 s) activation of phospholipase C (PLC) and calcium influx in response to the stimulus in the ROS 24/1 osteoblastic cell line [98]. It must be underscored that VDR involvement should be excluded because ROS 24/1 cells lack VDR expression. Similar studies on cultured chondrocytes derived from VDR knockout mice confirmed this observation [81]. Unfortunately, no receptor of the GPCR or receptor tyrosine kinase families dedicated to vitamin D or its metabolites has been identified so far. Nevertheless, the early studies of Nemere et al. on transcaltachia [99] resulted not only in the discovery of rapid response to vitamin D and its metabolites but also in the detection of a membrane-associated protein that binds to radiolabeled $1,25(OH)_2D_3$ (KD-value 0.72 nM) [100]. The protein was initially described as 1,25D3-MARRS (membrane-associated rapid response to steroid), but it is also known as PDIA3 (protein disulfide isomerase family A member 3), GRP58, and ERp57 [101–108]. PDIA3 interacts with calreticulin (CALR) and calnexin (CANX) and plays a crucial role in the folding and export of proteins from the endoplasmic reticulum [80,109]. PDIA3 activity is essential for the proper function of the immune [74,110] and musculoskeletal systems [102,111] as well as for mammary gland growth and development [112]. Several groups have reported colocalization of PDIA3 with the cell membrane, cytoplasm, mitochondria, and even the nucleus [113,114]. Most importantly, PDIA3 was found to be involved in a rapid intestinal uptake of calcium and phosphate. Thus, its involvement in the classic function of $1,25(OH)_2D_3$ was postulated [79,80,100,105,107]. It was also confirmed that $1,25(OH)_2D_3$ binds to the a' domain of PDIA3, with an estimated KD of 1 nM [115]. Site-directed mutagenesis studies revealed that the catalytic side of PDIA3 (C406) and its calreticulin (CALR) interaction site (K214 and R282) are required for the rapid response to $1,25(OH)_2D_3$ [116]. Unfortunately, the nature of interaction of PDIA3 with vitamin D metabolites is not fully understood. The binding side for $1,25(OH)_2D_3$ was not fully characterized because only a partial crystal structure of PDIA3 is available [117]. However, several studies have strongly supported involvement of PDIA3 in vitamin D signaling. For example, although a deletion of PDIA3 (PDIA3+/-) is lethal in mice [108,110], its partial silencing ($PDIA3^{+/-}$) resulted in aberration in skeletal development [118], thus at least indirectly linking PDIA3 activity to calcium homoeostasis and vitamin D. Furthermore, targeted disruption of the PDIA3 gene results in an inhibition of rapid calcium transport and attenuates PKA or PKC signaling, thus impairing rapid nongenomic responses to $1,25(OH)_2D_3$ [107,108]. In an addition, PDIA3 was also shown to interact with PLA2-activating protein (PLAA), which led to activation of phospholipaseA2 (PLA2), while targeted disruption of the PDIA3 gene attenuates cellular Ca^{+2} influx through L-type Ca^{+2} channels [119]. Other studies pointed out that uptake of calcium by the ER through the Ca^{+2} pump SERCA2b is regulated by PDIA3 and vitamin D [120]. It should be recalled, however, that the deletion of VDR also impairs bone formation [118], but it seems that PDIA3 but not VDR is essential for fast response to vitamin D, including activation of protein kinase C (PKC) signaling pathway [121] or a protection against UV-induced thymine dimer formation and DNA damage [122]. Furthermore, chondrocytes with a VDR knockdown but expressing PDIA3 showed a rapid increase in PKC levels and accelerated proteoglycan production in response to $1,25(OH)_2D_3$. Thus, a distinct role for VDR and PDIA3 in $1,25(OH)_2D_3$-mediated proliferation control of rat growth plate chondrocytes was postulated [81].

Association of PDIA3 with cell membrane signaling is strongly supported by observation of its interaction within caveolin-1 [123], which is the main scaffolding protein of caveolae (small invaginations of the plasma membrane which play important role in membrane singling and transmembrane trafficking). This observation at least partially explains the nature of PDIA3-driven rapid response to vitamin D through the association of PDIA3 with cell membrane caveolae and the activation of downstream signaling, including calcium mobilization, IP3, DAG, and cAMP production. PDIA3 was also found to regulate other intracellular pathways. For example, it was shown that 1,25(OH)$_2$D$_3$ triggers PDIA3-mediated rapid signaling cascade via CAMK2G (calcium/calmodulin dependent protein kinase II gamma), PLA2, PLC, and PKC, followed by an activation of mitogen-activated protein kinases (MAPK1 and MAPK2) [111,124]. Similarly, the PDIA3 membrane complex takes part in the rapid activation of WNT5A (Wnt family member 5A) by 1,25(OH)$_2$D$_3$-stimulated calcium influx and generation of secondary messengers, such as cAMP and/or IP3 [101]. Finally, some reports suggested nuclear localization of PDIA3 [114,125]. It is well established that PDIA3 has a noncanonic ER retention signal, Q/KEDL, but the presence of a Lys-rich nuclear localization signal was also suggested [111]. However, other studies suggested that TNFα, but not 1,25(OH)$_2$D3, triggers fast translocation of PDIA3 to the nucleus [126]. In an addition, an active form of vitamin D was shown to trigger translocation of the complexes of PDIA3, with other transcription factors including STAT3 [127–129] and NF-κB [130]. Furthermore, binding of PDIA3 to DNA in close proximity to the STAT3 binding site was shown, and its regulatory role in the expression of at least a subset of STAT3-dependent genes was postulated [129]. Interestingly, other groups also showed that PDIA3 binds to specific fragments of DNA [104,131] and is involved in the regulation and expression of genes, including MSH6, TMEM126A, LRBA, and ETS1 [104]. Finally, PDIA3 was also found to support trafficking of retinoic acid receptor alpha (RARA) into the nucleus and subsequent degradation in Sertoli cells [132]. However, the effect of vitamin D on this process still requires verification.

In summary, the nature of direct or indirect interactions of 1,25(OH)$_2$D$_3$ with PDIA3 associated with membranes is still not fully understood since the biding side was not fully characterized. However, there is growing evidence that PDIA3 modulates the response to vitamin D. Furthermore, it is still under debate whether 1,25(OH)$_2$D$_3$ triggers translocation of PDIA3 into the nucleus and whether PDIA3 can act as a transcription factor or if PDIA3 only facilitates the trafficking of other transcription factors [133].

5. Direct or Indirect Regulation of Ion Transport by Vitamin D Metabolites

Keeping in mind that one of the best-described fast nongenomic targets of vitamin D is calcium influx, it could be speculated that vitamin D and its derivatives bind directly to ion channels or proteins affecting ion transport across the membranes. For example, the binding of sulfated and glucoronated derivatives of 25(OH)D$_3$ to the multidrug resistance proteins SLCO1B1 (OATP1B1) and SLCO1B3 (OATP1B3) was described [134]. On the other hand, it is well established that 1,25(OH)$_2$D$_3$, through the genomic pathway, regulates expression of cell membrane transient receptor potential cation channels (TRPV1, 5, and 6), which at least in part explains the impact of vitamin D on calcium influx and cell proliferation [95,135–137]. Furthermore, 1,25(OH)$_2$D$_3$ increases expression of K^+ two-pore domain channel subfamily K member 3 (KCNK3 expression) [138] but decreases the level of mRNA for potassium channels KCNH1 (Kv10.1) [139,140] and TASK-1 [138]. The electrophysiological studies on conductivity of ion channels showed that 1,25(OH)$_2$D$_3$, at 100 nM concentration, acts as a mild agonist of the TRPV1 channel. In an addition, 25OHD and 1,25(OH)$_2$D can also act as inhibitors of capsaicin-induced TRPV1 activity [136,141]. Molecular docking studies suggested that 1,25(OH)$_2$D$_3$ shared the capsaicin binding site at the vanilloid binding pocket of the TRPV1 [136]. Other studies suggested that the fast effect of 1,25(OH)$_2$D$_3$ on an ion transport may be mediated through activation of L-type calcium channels. Calcitriol triggered an increase in the L-type calcium current and the fast transient outward K^+ current in myocytes. Furthermore, fast response to vitamin D depended on protein kinase A (PKA) and was absent in myocytes derived from mice with VDR knockout. Subsequent, intracellular Ca^{2+} mobilization resulted in ventricular myocytes' contractility [142,143]. Finally, activation of chloride channels by vitamin D was found to be involved in the protection of human skin (ex vivo) from UV irradiation [144]. Interestingly, the modulation of mitochondrial potassium channels by $1,25(OH)_2D_3$ has been shown recently [145] (the effects of vitamin D on mitochondria are discussed in Chapter 8). These intriguing electrophysiological observations require further investigations in order to provide more mechanistic details and the physiological significance of a direct impact of vitamin D on ion transport.

6. Nongenomic Activity of VDR

It must be underscored that the time frame (seconds to minutes) of rapid membrane responses to vitamin D virtually excludes the involvement of the transcriptional activity of VDR. Furthermore, multiple studies on the cell lines or model organisms with deletion of VDR have strongly suggested that rapid responses to vitamin D do not require the presence of vitamin D receptor (VDR). Thus, it seems that PDIA3, but not VDR, regulates the nongenomic activities of vitamin D [119,121,122]. However, the requirement of VDR in nongenomic activities of vitamin D may be strongly affected by the experimental model used in the study (e.g., cell type specific, type of assay). Furthermore, direct or indirect interaction between VDR and PDIA3 should also be considered. Indeed, several studies reported a membrane localization of VDR as well as colocalization with PDAI3 and caveolin 1 (CAV1) [146]. Interestingly, some reports have suggested that PDIA3 may serve as molecular chaperone for VDR [115,116]. However, as it was postulated above, at least some effects of $1,25(OH)_2D_3$, such as rapid activation of PKC, are also observed in cells lacking VDR and thus are VDR-independent. On the other hand, a mutation within the isomerase catalytic side and removal of the KDEL motive (ER retention signal) of PDIA3 abolish this activity [116].

VDR was also detected in mitochondrial membranes [147–149] or even lipid droplets [150]. Interestingly, the existence of a second ligand-binding domain of VDR responsible for membrane signaling was postulated [27,151]. Although rapid activation of PLA2 was attributed to PDIA3, some other effects—such as $1,25(OH)_2D_3$ activation of the SRC (SRC proto-oncogene, nonreceptor tyrosine kinase) WNT pathway [152–155], sonic hedgehog signaling molecule (SHH) [156–160], and NOTCH [161,162] signaling pathways [155]—were found to be VDR-dependent. Interestingly, VDR and NF-kB share DNA binding sites that may explain the influence of vitamin D on immune response [163]. Taken together, VDR may be involved in the activation of at least some of the nongenomic activities triggered by vitamin D, most probably in cooperation with PDIA3 [101,102,111,116,124,132,146,164,165].

7. Is Vitamin D as a Scavenger of Free Radicals or Their Source?

It is established that UVB is a major factor contributing to skin malignancy, but at the same time, it is essential for epidermal vitamin D production [60]. It could be hypothesized that during evolution, vitamin D may be adopted in direct or indirect response to radiation. Alternatively, UVB-driven vitamin D production may represent its primary function, which protected primitive aquatic organisms from UV or oxidative stress long before development of its endocrine function as a calcium–phosphate regulator.

The concept of a direct involvement of vitamin D in response to UV originated from a simple question: Can one overdose on vitamin D by sun tanning? This issue was addressed more than 40 years ago by Micheal Holick et al. in the classic paper [18]. It was shown that prolonged exposure to UVB did not lead to excessive accumulation of vitamin D. In fact, surplus of vitamin D was fast and efficiently removed by further structural rearrangements involving production of 5,6-transvitamin D_3, suprasterols I and II in the skin [18]. Further studies revealed that irradiation of 5,7-dienes lead also to the formation of 5,7,9(11)-trienes with probable generation of a singlet oxygen, which may act as a photosensitizer [166,167]. Finally, it was shown that extensive irradiation of vitamin D and its analogs (such as isotachysterol) resulted in the formation of large groups of hydroxyl-, peroxy derivatives [18,168,169]. Detection of nonclassical products of UVB triggered photolysis of 5,7-dienes (7DHC and its natural or synthetic derivatives) might represent a natural mechanism of regulation of vitamin D levels as well as a cell protective mechanism. However, it is still not fully understood whether those generally unstable byproducts may be acceptors or also donors of reactive oxygen species (ROS). For example, accumulation of 5,7-dienes and 5,7,9(11)-trienes is important in the etiology of a 7Δ-reductase deficiency syndrome (Smith-Lemli-Opitz syndrome—SLOS) [166,167,170]. In this syndrome, the lack of the key enzyme converting 7DHC to cholesterol results in multiple metabolic defects with accumulation of 7DHC and its steroidal derivatives, including cholesta-5,7,9(11)-trien-3β-ol (9DDHC) [167] and pregna-5,7-diene-3b,17a,20-triol [170]. It was postulated that oxidation of 7DHC and 9DDHC leads to severe photosensitivity in SLOS patients with potential production of highly reactive singlet oxygen [166,171]. Nevertheless, it is still not known whether the potential phototoxicity of 7DHC derivatives is unique to SLOS patients or if it could be a general property.

Solar UV radiation reaching human skin damages cells and tissues directly or indirectly through the production of ROS, which are exemplified by superoxide and hydrogen peroxide [60,172,173]. UVB also induces DNA lesions, including formation of cyclobutane pyrimidine dimers (CPDs) [174], while UVB-generated ROS target guanine, leading to the subsequent generation of 8-oxo-7,8-dihydro-20-deoxyguanosine (8-OHdG) [55,175]. Unrepaired UV-induced DNA damage results in the accumulation of mutations, which contribute strongly to the formation of skin neoplasms [1,55,60]. On the other hand, it is well established that vitamin D protects skin cells from UVB-induced damage (including DNA damage and induction of inflammation), as was shown on various models, including in vitro, mice models, and human subjects [175–178]. Thus, it was postulated that cutaneous production of vitamin D protects skin cells from UV-driven carcinogenesis. The mechanism requires vitamin-D-induced upregulation of the genes involved in ROS response through VDR-mediated genomic pathways with additional activation of NRF2 [179]. However, PDIA3 was also shown to participate in cells' protection against the formation of thymine dimers after irradiation with UV [122]. The effect was associated with activation of PKC signaling and calcium influx [180]. In addition, $1,25(OH)_2D_3$ protected cells from UVR-induced thymine dimers via stimulation of chloride currents [144]. Photoprotective effects of $1,25(OH)_2D_3$ were also associated with an increase in p53 tumor-suppressor protein translocation to the nucleus and a decrease in the level of nitric oxide species (NOS). Vitamin D can also affect endothelial function through the nongenomic pathway through the induction of PDIA3-mediated calcium, cAMP, Akt, and PKC downstream signaling, which in turn affect endothelial NO synthesis (eNOS) activity [181]. Finally, the potential UV-protective effect of $1,25(OH)_2D_3$ and other vitamin D derivatives and analogues could be attributed to direct impact on mitochondria bioenergetics. It was postulated that the process of DNA repair induced by UV requires substantial energy expenditure, and $1,25(OH)_2D_3$ and its analogs may directly affect mitochondrial activities (discussed below).

In summary, although vitamin D photoproduction may represent the primary and still-existing mechanism of cell protection against UVB and ROS/RNS, it must be underscored that vitamin D hormonal activity, namely VDR-mediated induction of the genes involved in UV and ROS/RNS, responses is also crucial for cell survival. On the other hand, potential involvement of 7DHC derivatives with triene moiety, such as 9DDHC in ROS generation (pro-oxidative), may be unique to SLOS.

8. Mitochondria as a Target for Vitamin D

Mitochondria are not only the powerhouse of the cell but also a key organelle to vitamin D metabolism, including activation, modification, and inactivation by the cytochromes of the P450 family (CYP27A1, CYP27B1, CYP24A1, and CYP11A1) [6,182,183]. All of those enzymes are located in the inner mitochondrial membrane. Thus, at least theoretically, all steps of the activation of vitamin D including hydroxylation in position 25 (CYP27A1) and position 1 (CYP27B1), as well as deactivation by hydroxylation at C-24, could be conducted within mitochondria. Although it must be acknowledged that CYP2R1 located in the endoplasmic reticulum is the major enzyme responsible for 25-hydoxylation, CYP27A1 also shows such activity [182,184]. Interestingly, CYP11A1 (CYP450scc—the major enzyme of steroidogenesis responsible for conversion of cholesterol to pregnenolone) was shown to also metabolize vitamin D, tachysterol, and lumisterol, which opens a new route of secosteroidogenesis [185–187]. There is no doubt that mitochondria are the direct target for vitamin D and its metabolites, but the mechanism of an entry, a regulation of vitamin D metabolism, or its direct impact on mitochondrial biogenesis or oxidative stress still requires an in-depth investigation. As was discussed previously, megalin may be responsible for vitamin D import to the cell. Interestingly, recent studies have suggested that megalin may also be involved in the translocation of some molecules directly to the mitochondria. Furthermore, mitochondrial megalin was found to be involved in the defense against ROS, and its knockout impairs mitochondrial respiration and glycolysis [93]. Megalin itself was colocalised with mitochondria in association with stanniocalcin-1 and SIRT3, which are involved in a defense against ROS [93]. Interestingly, it was shown that megalin is involved in the intracellular traffic and mitochondrial import of angiotensin II, stanniocalcin-1, and TGF-β [93], which raises the question of whether vitamin D and its metabolites, in concert with DBP, could be acquired and target mitochondria accordingly.

An active form of vitamin D also has indirect impact on mitochondria function through the VDR-dependent regulation of expression of genes involved in the oxidative stress response [145,183,188]. By using transcriptomic analyses focused on mitochondrial genes, it was discovered that 1,25$(OH)_2D_3$ inhibits expression of the several genes involved in oxidative phosphorylation and fusion/fission and upregulates genes involved in mitophagy and ROS defense [188]. In seems that the activation of selected genes' coding proteins in response to ROS by 1,25$(OH)_2D_3$ is mediated by nuclear-factor-erythroid 2-related factor 2 (NRF2) [179,188]. NRF2 is a key transcription factor that can bind to antioxidant response elements (AREs) on DNA and initiate transcription of the genes responsible for the protection of cells against oxidative stress associated with diabetic neuropathy [188].

An involvement of VDR in mitochondrial physiology was confirmed by observation; deletion of VDR affects mitochondrial membrane potential, enhances ROS production [147], and results in subsequent mitochondrial damage [189]. Removal of VDR also resulted in an increased expression of elements of the respiratory chain, such as cyclooxygenase 2 and 4 (COX2 and 4), the ATP synthase subunits (6MT-ATP6 and ATP5B) [147], and subunits II and IV of cytochrome c oxidase [134]. Interestingly, VDR was also found in mitochondria, but in contrast to its nuclear translocation, its import to mitochondria is most probably ligand-independent [148]. VDR was found to interact with the mitochondrial permeability transition pore (PTP), the voltage-dependent anion-selective channel (VDAC). Recently, it was shown that vitamin D regulates the activity of mitochondrial calcium channels. Furthermore, VDR may also directly impact cholesterol and vitamin D transport through the interaction with the steroidogenic acute regulatory (StAR) protein [148]. The mitochondrial localization of VDR was linked to the redirection tricarboxylic acid cycle (TCA) towards biosynthesis, which is the common feature of neoplastic rather than fully differentiated cells [189]. Thus, it seems that VDR may affect mitochondrial function both through the activation of classic genomic pathways and also directly inside of mitochondria, including interaction with the mitochondrial genome. Interestingly, the protective anti-ROS properties of 1,25$(OH)_2D_3$ were shown on isolated mitochondria. Thus, the genomic effects of VDR activation may not be fully required [190]. It must also be underscored that 1,25$(OH)_2D_3$ may protect mitochondria from potential toxic insults, as it was shown that it can attenuate the cytotoxicity induced by aluminum phosphide via inhibiting mitochondrial dysfunction and oxidative stress in isolated rat cardiomyocytes [191].

9. Clinical Implications of Nongenomic Pathways Activated by Vitamin D

It is well established that 20 ng/mL of 25(OH)D_3 in the serum is an adequate concentration for bone health. However, several studies and recommendations have suggested a concentration of vitamin D >30 ng/mL as optimal in order to provide extraskeletal benefits. Recently, it was postulated that even higher serum concentrations of vitamin D should be considered. For example, it was calculated that 37 ng/mL during pregnancy decreases the chance of complications in patients with risk of preeclampsia to values characteristic for the entire population [192]. Even higher concentrations (40–50 ng/mL) were proposed to be beneficial for extraskeletal outcomes of vitamin D, including its anticancer properties [193]. This leads to the significant change in recommendation concerning vitamin supplementation (800 IU vs. 4000 IU) and very high doses (50,000–100,000 IU) are recommended for patients with a severe vitamin D deficiency. However, potential hypercalcemia and hypercaluria must be considered as a potential side-effect of such as supplementation. However, several studies revealed that vitamin-D-induced hypercalcemia is observed usually in patients with defects in vitamin D metabolism (e.g., patients with *CYP24A1* mutation) [194]. This raises the very interesting question of whether an activation of the alternative nongenomic pathways is responsible for the additional phenotypical effects of vitamin D. It was shown that 1,25(OH)$_2D_3$ binds to VDR with an affinity of 0.1 nmol/L and to PDIA3 with an estimated Kd of 1 nmol/L [195]. Thus, it could be speculated that higher doses of vitamin D supplementation and higher (30 ng/mL or higher) serum concentrations of 25(OH)D_3 are essential for an effective activation of nongenomic pathways. Nevertheless, this speculation requires in-depth laboratory and clinical investigations.

The description of transcaltachia was historically the first indication of existence of nongenomic pathways activated by 1,25(OH)$_2D_3$ [100]. This rapid vitamin-D-stimulated transport of calcium through intestinal endothelium is an important factor regulating the level of calcium in the body. Another clinical implication of nongenomic pathways is direct scavenging of UV-generated ROS and RNS observed as early as 15–30 min after treatment of irradiated keratinocytes with 1,25(OH)$_2D_3$ (please see [73] for discussion). In an addition to the clear UV-protective effects of vitamin D (see also chapter 7), it was postulated that vitamin D protects against skin photocarcinogenesis [57]. 1,25(OH)$_2D_3$ also affects ERK, PARP-1, and p53 activities and takes part in DNA protection and repair [73]. It is well established that increased production of ROS and deregulation of cellular energetics is a characteristic feature of cancer progression and drug resistance [196]. Thus, vitamin D may directly inhibit carcinogenesis through ROS scavenging. In fact, is also know that activation of VDR-dependent genomic pathway protects cells by activation of transcription of the genes involved in the response to ROS and DNA-repair [197]. Involvement of genomic pathways in the anticancer activities of vitamin D and its derives is also supported by observation that expression of VDR expression decreases with cancer progression [24]. However, it was also postulated that the extraskeletal benefits of vitamin D, including effects on cardiovascular and autoimmune diseases and cancer, are observed at relatively high concentrations [198]. Indeed, it was recently estimated that an increase in the serum concentration of 25(OH)D from 10 to 80 ng/mL would decrease cancer incidence rates by $70 \pm 10\%$ [60]. Thus, it is tempting to suggest that higher doses of vitamin D are required for the activation of alternative pathways, but this still remains to be confirmed.

Another, striking nongenomic activity of 1,25(OH)$_2D_3$ was shown on models of collagen or ADP-induced platelet aggregation. Interestingly, the inhibitory effect of 1,25(OH)$_2D_3$ on platelet aggregation varied depending on the state of diabetes [199]. Although VDR was detected in platelets and associated with mitochondria, because platelets naturally lack cell nuclei [149], only activation of nongenomic pathways should be considered. In an addition, it was also shown that glycemic control was inversely associated with high platelet aggregation and low levels of 25(OH)$_2$D in diabetic patients [199]. These findings provide important clinical implication of alternative pathways activated by vitamin D.

The impact of PDIA3 and alternative pathways activated by vitamin D was also shown on an experimental model of cholangiopathy induced by Abcb4 knockout in mice. Firstly, additional silencing of VDR (VDR and Abcb4 double knockout) promotes a proinflammatory phenotype in cholangiocytes. Interestingly, vitamin D—or its analog, calcipotriol—treatment resulted in PDIA3-dependent reduction of inflammatory response [200]. Interestingly, PDIA3 was also found to be essential for the autolysosomal degradation of *Helicobacter pylori* [201]. Thus, activation of membrane response with potential involvement of PDIA3 may play a role in the regulation of immune response, including response against pathogens.

In melanoma, it was found that decreased expression of VDR correlates with poor prognosis. On the other hand, overexpression of PDIA3 was shown to be a poor survival factor in patients with clear cell renal cell carcinoma—ccRCC [127]. Thus, it seems that proper functioning of classic VDR-dependent pathways as well as activation of alternative pathways is essential in order to maintain body homeostasis as well as to protect from potential harmful insults, such as UVB or pathogen infection.

In summary, although genomic pathways activated by VDR still remain the major targets for vitamin D, with clear clinical implications (e.g., rickets), the existence of alternative pathways explains fast responses to vitamin D, and its direct impact on mitochondria and other intracellular processes. Unfortunately, the pleiotropic effects of vitamin D genomic activities, and the potential involvement of VDR in nongenomic regulation, hinder the studies of vitamin D membrane signaling, its effects on mitochondria, and the precise determination of the role of PDIA3 (Table 1).

Table 1. Genomic and membrane pathways activated by vitamin D.

	Genomic	Membrane
Time course	Delayed response (hours–days)	Fast response (minutes–hours)
Primary Mechanisms	$1,25(OH)_2D_3$ binds to VDR, which is translocated to the nucleus together with RXR, where the complex binds vitamin-D-responding elements (VDRE) of DNA.	$1,25(OH)_2D_3$ affect activity of membrane proteins including PDIA3 and direct or indirect activity of ion channels. ROS scavenging.
Primary effects	Transcriptional regulation of up to 3000 genes resulting in inhibition of cell proliferation, induction of cell differentiation, immunomodulation, UVB response, ROS response, and alteration of mitochondrial function.	Membrane responses with activation of secondary messengers (calcium, cAMP, IP3, DAG). Transcaltachia, modulation of mitochondrial bioenergetics. Direct cell protection against UV, ROS, and pathogens.
Secondary mechanisms	Alteration of the expression of several TFs results in activation of secondary genomic responses.	Activation of secondary messages may activate MAPK/ERK, P53, WNT, SSH, STAT1-3, and NF-kappaB genomic activities. PDIA3 was found in the nucleus. Thus, its function as transcriptional regulator is considered.
Secondary effects	A dissection of primary from secondary genomic effects still requires careful investigation. Potential secondary impact of membrane signaling on genes expression adds an additional complexity to vitamin D response.	Modulation of activity of signaling pathways MAPK/ERK, P53, WNT, SSH, STAT1-3, and NF-kappaB results in modulation of cell physiology as well as activation of secondary genomic responses.
Role of VDR	VDR act as a transcription factor.	VDR may be engaged in membrane signaling.
Role of PDIA3	Potential secondary genomic effect is considered as PDIA3 was found in the nucleus.	PDIA3 is involved in initiation of membrane signaling.
Role in calcium regulation	Change in the expression of the genes responsible for calcium homeostasis	Rapid direct, or indirect effect on calcium transport (transcaltachia)
Effect on mitochondria	Indirect though alteration of the expression of mitochondria related genes	Potential direct effect on mitochondrial proteins (including cytochromes P450) and potassium ion channels from the inner mitochondrial membrane
UVB protection	Indirect through activation of stress response genes and DNA repair mechanisms	Potential direct ROS scavenging
Immunomodulation	Inhibition of B cell differentiation and immunoglobulin secretion, shift from a Th1 to a Th2 response, induction of T regulatory cells, decrease of expression of proinflammatory cytokines, and stimulation of expression of antimicrobial peptides.	Regulation of calcium influx may affect immune cell physiology.

10. Beyond 1,25(OH)$_2$D$_3$

Although it is not the major scope of this review, potential biological activities of vitamin D metabolites or analogs must be acknowledged. For years, it has been believed that calcitriol is the only active metabolite of vitamin D. In fact, the binding of 25(OH)D$_3$ or 24,25(OH)$_2$D$_3$ to VDR is at least 100–1000 times weaker in comparison to 1,25(OH)$_2$D$_3$. Relatively recently, some alternative metabolic routes for vitamin D have been described, and some unique biological futures of those secosteroids have been proposed. In silico docking experiments showed the variable binding affinities to VDR of various alternative vitamin D metabolites or analogs. However, 1,25(OH)2D3 remained its main unquestioned ligand responsible for activation of the classic genomic response. It seems that alternative metabolites may alter or tune up or down transcription response to vitamin D. For example, 20(OH)D$_3$ and 20,23(OH)$_2$D$_3$ generated by CYP450scc (CYP11A1) inhibit the activity of key factor in immune response NF-kB [202] or stimulate ROS response through NRF2 [179,187] similarly to 1,25(OH)2D3. However, it is believed that in addition to VDR, they may also target other transcription factors, such as RORα and RORγ [34,35] or AhR [28]. For many years, lumisterol and tachysterol we considered nonfunctional isomers of vitamin D, but recent studies have proposed their biological activity and further metabolism, which opens a new chapter in the field of secostroids [187,203]. Thus, it seems that 1,25(OH)$_2$D$_3$ is not the only functional metabolite of vitamin D and that other secosteroids may help in the fine-tuning of vitamin D response or even replace 1,25(OH)$_2$D$_3$, but it seems that higher concentrations are required.

11. Conclusions

Keeping the pluripotent properties of vitamin D in mind, the presence of alternative intracellular pathways may help to explain a variety of responses [204]. VDR genomic activity of course plays the central role in 1,25(OH)$_2$D$_3$ signaling. However, it must be noted that from the evolutionary perspective, some primal functions, such as protection from ultraviolet radiation, should be considered. On the other hand, membrane response, including calcium mobilization, may provide a fast route of action for this powerful secosteroid. The detailed mechanisms underlying membrane nongenomic responses still remain to be elucidated. However, the role of PDIA3, as well as that of membrane-associated VDR, should be considered. Furthermore, direct or indirect interaction of vitamin D metabolites with other proteins including megalin, ion channels, and as their impact on mitochondria may help us to understand the versatility of its phenotypic effects [64,186,205]. However, further in vivo studies and randomized controlled trials [206] are required to confirm physiological and clinical importance of alternative pathways of vitamin D signaling.

Funding: M.A.Z. was funded by the NCN grant 2017/25/B/NZ3/00431.

Institutional Review Board Statement: Not applicable.

Informed Consent Statement: Not applicable.

Data Availability Statement: Not applicable.

Conflicts of Interest: The author declares no conflict of interest.

References

1. Carlberg, C. Vitamin D in the Context of Evolution. *Nutrients* **2022**, *14*, 3018. [CrossRef]
2. Hanel, A.; Carlberg, C. Vitamin D and evolution: Pharmacologic implications. *Biochem. Pharmacol.* **2020**, *173*, 113595. [CrossRef]
3. Holick, M.F. Vitamin D: Evolutionary, physiological and health perspectives. *Curr. Drug Targets* **2011**, *12*, 4–18. [CrossRef] [PubMed]
4. Wacker, M.; Holick, M.F. Sunlight and Vitamin D: A global perspective for health. *Derm.-Endocrinol.* **2013**, *5*, 51–108. [CrossRef] [PubMed]
5. Holick, M.F.; Frommer, J.E.; McNeill, S.C.; Richtand, N.M.; Henley, J.W.; Potts, J.T., Jr. Photometabolism of 7-dehydrocholesterol to previtamin D3 in skin. *Biochem. Biophys. Res. Commun.* **1977**, *76*, 107–114. [CrossRef] [PubMed]

6. Bikle, D.; Christakos, S. New aspects of vitamin D metabolism and action—Addressing the skin as source and target. *Nat. Rev. Endocrinol.* **2020**, *16*, 234–252. [CrossRef]
7. Bikle, D.D. Vitamin D and the skin: Physiology and pathophysiology. *Rev. Endocr. Metab. Disord.* **2012**, *13*, 3–19. [CrossRef]
8. Bikle, D.D.; Schwartz, J. Vitamin D Binding Protein, Total and Free Vitamin D Levels in Different Physiological and Pathophysiological Conditions. *Front. Endocrinol.* **2019**, *10*, 317. [CrossRef]
9. Bouillon, R.; Schuit, F.; Antonio, L.; Rastinejad, F. Vitamin D Binding Protein: A Historic Overview. *Front. Endocrinol.* **2019**, *10*, 910. [CrossRef]
10. Holick, M.F.; Clark, M.B. The photobiogenesis and metabolism of vitamin D. *Fed. Proc.* **1978**, *37*, 2567–2574.
11. Fleet, J.C. The role of vitamin D in the endocrinology controlling calcium homeostasis. *Mol. Cell. Endocrinol.* **2017**, *453*, 36–45. [CrossRef] [PubMed]
12. Płudowski, P.; Karczmarewicz, E.; Bayer, M.; Carter, G.; Chlebna-Sokół, D.; Czech-Kowalska, J.; Dębski, R.; Decsi, T.; Dobrzańska, A.; Franek, E.; et al. Practical guidelines for the supplementation of vitamin D and the treatment of deficits in Central Europe—Recommended vitamin D intakes in the general population and groups at risk of vitamin D deficiency. *Endokrynol. Pol.* **2013**, *64*, 319–327. [CrossRef] [PubMed]
13. Pludowski, P.; Holick, M.F.; Grant, W.B.; Konstantynowicz, J.; Mascarenhas, M.R.; Haq, A.; Povoroznyuk, V.; Balatska, N.; Barbosa, A.P.; Karonova, T.; et al. Vitamin D supplementation guidelines. *J. Steroid Biochem. Mol. Biol.* **2018**, *175*, 125–135. [CrossRef] [PubMed]
14. Grant, W.B.; Boucher, B.J.; Bhattoa, H.P.; Lahore, H. Why vitamin D clinical trials should be based on 25-hydroxyvitamin D concentrations. *J. Steroid Biochem. Mol. Biol.* **2018**, *177*, 266–269. [CrossRef]
15. Amon, U.; Baier, L.; Yaguboglu, R.; Ennis, M.; Holick, M.F.; Amon, J. Serum 25-hydroxyvitamin D levels in patients with skin diseases including psoriasis, infections, and atopic dermatitis. *Derm.-Endocrinol.* **2018**, *10*, e1442159. [CrossRef]
16. Takeyama, K.; Kitanaka, S.; Sato, T.; Kobori, M.; Yanagisawa, J.; Kato, S. 25-Hydroxyvitamin D3 1alpha-hydroxylase and vitamin D synthesis. *Science* **1997**, *277*, 1827–1830. [CrossRef]
17. Holick, M.F.; Uskokovic, M.; Henley, J.W.; MacLaughlin, J.; Holick, S.A.; Potts, J.T. The photoproduction of 1 alpha,25-dihydroxyvitamin D3 in skin: An approach to the therapy of vitamin-D-resistant syndromes. *N. Engl. J. Med.* **1980**, *303*, 349–354. [CrossRef]
18. Webb, A.R.; DeCosta, B.R.; Holick, M.F. Sunlight regulates the cutaneous production of vitamin D3 by causing its photodegradation. *J. Clin. Endocrinol. Metab.* **1989**, *68*, 882–887. [CrossRef]
19. Tuckey, R.C.; Tang, E.K.Y.; Maresse, S.R.; Delaney, D.S. Catalytic properties of 25-hydroxyvitamin D3 3-epimerase in rat and human liver microsomes. *Arch. Biochem. Biophys.* **2019**, *666*, 16–21. [CrossRef]
20. Berger, S.E.; Van Rompay, M.I.; Gordon, C.M.; Goodman, E.; Eliasziw, M.; Holick, M.F.; Sacheck, J.M. Investigation of the C-3-epi-25(OH)D. *Appl. Physiol. Nutr. Metab.* **2018**, *43*, 259–265. [CrossRef]
21. Kmieć, P.; Minkiewicz, I.; Sworczak, K.; Żmijewski, M.A.; Kowalski, K. Vitamin D status including 3-epi-25(OH)D3 among adult patients with thyroid disorders during summer months. *Endokrynol. Pol.* **2018**, *69*, 653–660. [CrossRef] [PubMed]
22. Jones, G.; Kaufmann, M. Diagnostic Aspects of Vitamin D: Clinical Utility of Vitamin D Metabolite Profiling. *JBMR Plus* **2021**, *5*, e10581. [CrossRef] [PubMed]
23. Chen, J.; Tang, Z.; Slominski, A.T.; Li, W.; Żmijewski, M.A.; Liu, Y. Vitamin D and its analogs as anticancer and anti-inflammatory agents. *Eur. J. Med. Chem.* **2020**, *207*, 112738. [CrossRef] [PubMed]
24. Slominski, A.T.; Brożyna, A.A.; Skobowiat, C.; Zmijewski, M.A.; Kim, T.K.; Janjetovic, Z.; Oak, A.S.; Jozwicki, W.; Jetten, A.M.; Mason, R.S.; et al. On the role of classical and novel forms of vitamin D in melanoma progression and management. *J. Steroid Biochem. Mol. Biol.* **2018**, *177*, 159–170. [CrossRef] [PubMed]
25. Tuckey, R.C.; Li, W.; Zjawiony, J.K.; Zmijewski, M.A.; Nguyen, M.N.; Sweatman, T.; Miller, D.; Slominski, A. Pathways and products for the metabolism of vitamin D3 by cytochrome P450scc. *FEBS J.* **2008**, *275*, 2585–2596. [CrossRef]
26. Slominski, A.T.; Semak, I.; Zmijewski, M.A.; Sweatman, T.; Janjetovic, Z.; Wortsman, J.; Tuckey, R.C. Sequential metabolism of provitamin D3 (7-dehydrocholesterol) to 5,7-diene products in the adrenal gland. *J. Investig. Dermatol.* **2007**, *127*, 443.
27. Haussler, M.R.; Jurutka, P.W.; Mizwicki, M.; Norman, A.W. Vitamin D receptor (VDR)-mediated actions of $1\alpha,25(OH)_2$ vitamin D_3: Genomic and non-genomic mechanisms. *Best Pract. Res. Clin. Endocrinol. Metab.* **2011**, *25*, 543–559. [CrossRef]
28. Slominski, A.T.; Kim, T.K.; Janjetovic, Z.; Brożyna, A.A.; Żmijewski, M.A.; Xu, H.; Sutter, T.R.; Tuckey, R.C.; Jetten, A.M.; Crossman, D.K. Differential and Overlapping Effects of $20,23(OH)_2D3$ and $1,25(OH)_2D3$ on Gene Expression in Human Epidermal Keratinocytes: Identification of AhR as an Alternative Receptor for $20,23(OH)_2D3$. *Int. J. Mol. Sci.* **2018**, *19*, 3072. [CrossRef]
29. Carlberg, C. Vitamin D and Its Target Genes. *Nutrients* **2022**, *14*, 1354. [CrossRef]
30. Hanel, A.; Veldhuizen, C.; Carlberg, C. Gene-Regulatory Potential of 25-Hydroxyvitamin D. *Front. Nutr.* **2022**, *9*, 910601. [CrossRef]
31. Hanel, A.; Bendik, I.; Carlberg, C. Transcriptome-Wide Profile of 25-Hydroxyvitamin D. *Nutrients* **2021**, *13*, 4100. [CrossRef]
32. Ellfolk, M.; Norlin, M.; Gyllensten, K.; Wikvall, K. Regulation of human vitamin D(3) 25-hydroxylases in dermal fibroblasts and prostate cancer LNCaP cells. *Mol. Pharmacol.* **2009**, *75*, 1392–1399. [CrossRef]
33. Susa, T.; Iizuka, M.; Okinaga, H.; Tamamori-Adachi, M.; Okazaki, T. Without 1α-hydroxylation, the gene expression profile of 25(OH)D. *Sci. Rep.* **2018**, *8*, 9024. [CrossRef] [PubMed]

34. Slominski, A.T.; Kim, T.K.; Takeda, Y.; Janjetovic, Z.; Brozyna, A.A.; Skobowiat, C.; Wang, J.; Postlethwaite, A.; Li, W.; Tuckey, R.C.; et al. RORalpha and RORgamma are expressed in human skin and serve as receptors for endogenously produced noncalcemic 20-hydroxy- and 20,23-dihydroxyvitamin D. *FASEB J.* **2014**, *28*, 2775–2789. [CrossRef] [PubMed]
35. Slominski, A.T.; Kim, T.K.; Hobrath, J.V.; Oak, A.S.W.; Tang, E.K.Y.; Tieu, E.W.; Li, W.; Tuckey, R.C.; Jetten, A.M. Endogenously produced nonclassical vitamin D hydroxy-metabolites act as "biased" agonists on VDR and inverse agonists on RORα and RORγ. *J. Steroid Biochem. Mol. Biol.* **2017**, *173*, 42–56. [CrossRef] [PubMed]
36. Warwick, T.; Schulz, M.H.; Günther, S.; Gilsbach, R.; Neme, A.; Carlberg, C.; Brandes, R.P.; Seuter, S. A hierarchical regulatory network analysis of the vitamin D induced transcriptome reveals novel regulators and complete VDR dependency in monocytes. *Sci. Rep.* **2021**, *11*, 6518. [CrossRef] [PubMed]
37. Gallardo Martin, E.; Cousillas Castiñeiras, A. Vitamin D modulation and microRNAs in gastric cancer: Prognostic and therapeutic role. *Transl. Cancer Res.* **2021**, *10*, 3111–3127. [CrossRef]
38. Shahrzad, M.K.; Gharehgozlou, R.; Fadaei, S.; Hajian, P.; Mirzaei, H.R. Vitamin D and Non-coding RNAs: New Insights into the Regulation of Breast Cancer. *Curr. Mol. Med.* **2021**, *21*, 194–210. [CrossRef]
39. Charoenngam, N.; Ayoub, D.; Holick, M.F. Nutritional rickets and vitamin D deficiency: Consequences and strategies for treatment and prevention. *Expert Rev. Endocrinol. Metab.* **2022**, *17*, 351–364. [CrossRef]
40. Veldurthy, V.; Wei, R.; Oz, L.; Dhawan, P.; Jeon, Y.H.; Christakos, S. Vitamin D, calcium homeostasis and aging. *Bone Res.* **2016**, *4*, 16041. [CrossRef]
41. Wierzbicka, J.; Piotrowska, A.; Żmijewski, M.A. The renaissance of vitamin D. *Acta Biochim. Pol.* **2014**, *61*, 679–686. [CrossRef] [PubMed]
42. Holick, M.F. The vitamin D deficiency pandemic: Approaches for diagnosis, treatment and prevention. *Rev. Endocr. Metab. Disord.* **2017**, *18*, 153–165. [CrossRef] [PubMed]
43. Yamamoto, E.A.; Jørgensen, T.N. Relationships Between Vitamin D, Gut Microbiome, and Systemic Autoimmunity. *Front. Immunol.* **2019**, *10*, 3141. [CrossRef] [PubMed]
44. Holmes, E.A.; Rodney Harris, R.M.; Lucas, R.M. Low Sun Exposure and Vitamin D Deficiency as Risk Factors for Inflammatory Bowel Disease, With a Focus on Childhood Onset. *Photochem. Photobiol.* **2019**, *95*, 105–118. [CrossRef]
45. Mak, A. The Impact of Vitamin D on the Immunopathophysiology, Disease Activity, and Extra-Musculoskeletal Manifestations of Systemic Lupus Erythematosus. *Int. J. Mol. Sci.* **2018**, *19*, 2355. [CrossRef]
46. Medrano, M.; Carrillo-Cruz, E.; Montero, I.; Perez-Simon, J.A. Vitamin D: Effect on Haematopoiesis and Immune System and Clinical Applications. *Int. J. Mol. Sci.* **2018**, *19*, 2663. [CrossRef]
47. Trochoutsou, A.I.; Kloukina, V.; Samitas, K.; Xanthou, G. Vitamin-D in the Immune System: Genomic and Non-Genomic Actions. *Mini Rev. Med. Chem.* **2015**, *15*, 953–963. [CrossRef]
48. Gruber-Bzura, B.M. Vitamin D and Influenza-Prevention or Therapy? *Int. J. Mol. Sci.* **2018**, *19*, 2419. [CrossRef]
49. Charoenngam, N.; Shirvani, A.; Holick, M.F. Vitamin D and Its Potential Benefit for the COVID-19 Pandemic. *Endocr. Pract.* **2021**, *27*, 484–493. [CrossRef]
50. Giustina, A.; Bouillon, R.; Dawson-Hughes, B.; Ebeling, P.R.; Lazaretti-Castro, M.; Lips, P.; Marcocci, C.; Bilezikian, J.P. Vitamin D in the older population: A consensus statement. *Endocrine* 2022. [CrossRef]
51. Cui, X.; Gooch, H.; Petty, A.; McGrath, J.J.; Eyles, D. Vitamin D and the brain: Genomic and non-genomic actions. *Mol. Cell. Endocrinol.* **2017**, *453*, 131–143. [CrossRef] [PubMed]
52. Eyles, D.W. Vitamin D: Brain and Behavior. *JBMR Plus* **2021**, *5*, e10419. [CrossRef] [PubMed]
53. Owczarczyk-Saczonek, A.; Purzycka-Bohdan, D.; Nedoszytko, B.; Reich, A.; Szczerkowska-Dobosz, A.; Bartosińska, J.; Batycka-Baran, A.; Czajkowski, R.; Dobrucki, I.T.; Dobrucki, L.W.; et al. Pathogenesis of psoriasis in the "omic" era. Part III. Metabolic disorders, metabolomics, nutrigenomics in psoriasis. *Postepy Dermatol. Alergol.* **2020**, *37*, 452–467. [CrossRef] [PubMed]
54. Carlberg, C.; Velleuer, E. Vitamin D and the risk for cancer: A molecular analysis. *Biochem. Pharmacol.* **2022**, *196*, 114735. [CrossRef]
55. Holick, M.F. Sunlight, UV Radiation, Vitamin D, and Skin Cancer: How Much Sunlight Do We Need? *Adv. Exp. Med. Biol.* **2020**, *1268*, 19–36. [CrossRef]
56. Ferrer-Mayorga, G.; Larriba, M.J.; Crespo, P.; Muñoz, A. Mechanisms of action of vitamin D in colon cancer. *J. Steroid Biochem. Mol. Biol.* **2019**, *185*, 1–6. [CrossRef]
57. Reichrath, J.; Saternus, R.; Vogt, T. Endocrine actions of vitamin D in skin: Relevance for photocarcinogenesis of non-melanoma skin cancer, and beyond. *Mol. Cell. Endocrinol.* **2017**, *453*, 96–102. [CrossRef]
58. De Smedt, J.; Van Kelst, S.; Boecxstaens, V.; Stas, M.; Bogaerts, K.; Vanderschueren, D.; Aura, C.; Vandenberghe, K.; Lambrechts, D.; Wolter, P.; et al. Vitamin D supplementation in cutaneous malignant melanoma outcome (ViDMe): A randomized controlled trial. *BMC Cancer* **2017**, *17*, 562. [CrossRef]
59. Piotrowska, A.; Wierzbicka, J.; Żmijewski, M.A. Vitamin D in the skin physiology and pathology. *Acta Biochim. Pol.* **2016**, *63*, 17–29. [CrossRef]
60. Muñoz, A.; Grant, W.B. Vitamin D and Cancer: An Historical Overview of the Epidemiology and Mechanisms. *Nutrients.* **2022**, *14*, 1448. [CrossRef]
61. Moretti, R.; Morelli, M.E.; Caruso, P. Vitamin D in Neurological Diseases: A Rationale for a Pathogenic Impact. *Int. J. Mol. Sci.* **2018**, *19*, 2245. [CrossRef] [PubMed]

62. Dziedzic, E.A.; Grant, W.B.; Sowińska, I.; Dąbrowski, M.; Jankowski, P. Small Differences in Vitamin D Levels between Male Cardiac Patients in Different Stages of Coronary Artery Disease. *J. Clin. Med.* **2022**, *11*, 779. [CrossRef] [PubMed]
63. Boucher, B.J.; Grant, W.B. Difficulties in designing randomised controlled trials of vitamin D supplementation for reducing acute cardiovascular events and in the analysis of their outcomes. *Int. J. Cardiol. Heart Vasc.* **2020**, *29*, 100564. [CrossRef]
64. Duffy, M.J.; Murray, A.; Synnott, N.C.; O'Donovan, N.; Crown, J. Vitamin D analogues: Potential use in cancer treatment. *Crit. Rev. Oncol. Hematol.* **2017**, *112*, 190–197. [CrossRef]
65. Tabatabaeizadeh, S.A.; Avan, A.; Bahrami, A.; Khodashenas, E.; Esmaeili, H.; Ferns, G.A.; Abdizadeh, M.F.; Ghayour-Mobarhan, M. High Dose Supplementation of Vitamin D Affects Measures of Systemic Inflammation: Reductions in High Sensitivity C-Reactive Protein Level and Neutrophil to Lymphocyte Ratio (NLR) Distribution. *J. Cell. Biochem.* **2017**, *118*, 4317–4322. [CrossRef]
66. Rosen, Y.; Daich, J.; Soliman, I.; Brathwaite, E.; Shoenfeld, Y. Vitamin D and autoimmunity. *Scand. J. Rheumatol.* **2016**, *45*, 439–447. [CrossRef] [PubMed]
67. Armstrong, A.W.; Read, C. Pathophysiology, Clinical Presentation, and Treatment of Psoriasis: A Review. *JAMA* **2020**, *323*, 1945–1960. [CrossRef]
68. Finamor, D.C.; Sinigaglia-Coimbra, R.; Neves, L.C.; Gutierrez, M.; Silva, J.J.; Torres, L.D.; Surano, F.; Neto, D.J.; Novo, N.F.; Juliano, Y.; et al. A pilot study assessing the effect of prolonged administration of high daily doses of vitamin D on the clinical course of vitiligo and psoriasis. *Derm.-Endocrinol.* **2013**, *5*, 222–234. [CrossRef]
69. Suárez-Varela, M.M.; Uçar, N.; Peraita-Costa, I.; Huertas, M.F.; Soriano, J.M.; Llopis-Morales, A.; Grant, W.B. Vitamin D-Related Risk Factors for Maternal Morbidity during Pregnancy: A Systematic Review. *Nutrients* **2022**, *14*, 3166. [CrossRef]
70. Płudowski, P.; Ducki, C.; Konstantynowicz, J.; Jaworski, M. Vitamin D status in Poland. *Pol. Arch. Med. Wewn.* **2016**, *126*, 530–539. [CrossRef]
71. Kmieć, P.; Żmijewski, M.; Waszak, P.; Sworczak, K.; Lizakowska-Kmieć, M. Vitamin D deficiency during winter months among an adult, predominantly urban, population in Northern Poland. *Endokrynol. Pol.* **2014**, *65*, 105–113. [CrossRef] [PubMed]
72. Kmieć, P.; Żmijewski, M.; Lizakowska-Kmieć, M.; Sworczak, K. Widespread vitamin D deficiency among adults from northern Poland (54° N) after months of low and high natural UVB radiation. *Endokrynol. Pol.* **2015**, *66*, 30–38. [CrossRef] [PubMed]
73. Thiebaut, C.; Vlaeminck-Guillem, V.; Trédan, O.; Poulard, C.; Le Romancer, M. Non-genomic signaling of steroid receptors in cancer. *Mol. Cell. Endocrinol.* **2021**, *538*, 111453. [CrossRef]
74. Khanal, R.C.; Nemere, I. The ERp57/GRp58/1,25D3-MARRS receptor: Multiple functional roles in diverse cell systems. *Curr. Med. Chem.* **2007**, *14*, 1087–1093. [CrossRef]
75. Civitelli, R.; Kim, Y.S.; Gunsten, S.L.; Fujimori, A.; Huskey, M.; Avioli, L.V.; Hruska, K.A. Nongenomic activation of the calcium message system by vitamin D metabolites in osteoblast-like cells. *Endocrinology* **1990**, *127*, 2253–2262. [CrossRef] [PubMed]
76. Selles, J.; Boland, R. Evidence on the participation of the 3′,5′-cyclic AMP pathway in the non-genomic action of 1,25-dihydroxy-vitamin D3 in cardiac muscle. *Mol. Cell. Endocrinol.* **1991**, *82*, 229–235. [CrossRef] [PubMed]
77. Fleet, J.C.; Schoch, R.D. Molecular mechanisms for regulation of intestinal calcium absorption by vitamin D and other factors. *Crit. Rev. Clin. Lab. Sci.* **2010**, *47*, 181–195. [CrossRef]
78. Sterling, T.M.; Nemere, I. 1,25-dihydroxyvitamin D3 stimulates vesicular transport within 5 s in polarized intestinal epithelial cells. *J. Endocrinol.* **2005**, *185*, 81–91. [CrossRef]
79. Tunsophon, S.; Nemere, I. Protein kinase C isotypes in signal transduction for the 1,25D3-MARRS receptor (ERp57/PDIA3) in steroid hormone-stimulated phosphate uptake. *Steroids* **2010**, *75*, 307–313. [CrossRef]
80. Nemere, I.; Farach-Carson, M.C.; Rohe, B.; Sterling, T.M.; Norman, A.W.; Boyan, B.D.; Safford, S.E. Ribozyme knockdown functionally links a $1,25(OH)_2D_3$ membrane binding protein ($1,25D_3$-MARRS) and phosphate uptake in intestinal cells. *Proc. Natl. Acad. Sci. USA* **2004**, *101*, 7392–7397. [CrossRef]
81. Boyan, B.D.; Sylvia, V.L.; McKinney, N.; Schwartz, Z. Membrane actions of vitamin D metabolites 1alpha,$25(OH)_2D_3$ and 24R,$25(OH)_2D_3$ are retained in growth plate cartilage cells from vitamin D receptor knockout mice. *J. Cell. Biochem.* **2003**, *90*, 1207–1223. [CrossRef]
82. Bikle, D.D. The Free Hormone Hypothesis: When, Why, and How to Measure the Free Hormone Levels to Assess Vitamin D, Thyroid, Sex Hormone, and Cortisol Status. *JBMR Plus* **2021**, *5*, e10418. [CrossRef]
83. Willnow, T.E.; Nykjaer, A. Cellular uptake of steroid carrier proteins—Mechanisms and implications. *Mol. Cell. Endocrinol.* **2010**, *316*, 93–102. [CrossRef] [PubMed]
84. Safadi, F.F.; Thornton, P.; Magiera, H.; Hollis, B.W.; Gentile, M.; Haddad, J.G.; Liebhaber, S.A.; Cooke, N.E. Osteopathy and resistance to vitamin D toxicity in mice null for vitamin D binding protein. *J. Clin. Investig.* **1999**, *103*, 239–251. [CrossRef]
85. Nykjaer, A.; Fyfe, J.C.; Kozyraki, R.; Leheste, J.R.; Jacobsen, C.; Nielsen, M.S.; Verroust, P.J.; Aminoff, M.; de la Chapelle, A.; Moestrup, S.K.; et al. Cubilin dysfunction causes abnormal metabolism of the steroid hormone 25(OH) vitamin D(3). *Proc. Natl. Acad. Sci. USA* **2001**, *98*, 13895–13900. [CrossRef] [PubMed]
86. Leheste, J.R.; Rolinski, B.; Vorum, H.; Hilpert, J.; Nykjaer, A.; Jacobsen, C.; Aucouturier, P.; Moskaug, J.O.; Otto, A.; Christensen, E.I.; et al. Megalin knockout mice as an animal model of low molecular weight proteinuria. *Am. J. Pathol.* **1999**, *155*, 1361–1370. [CrossRef] [PubMed]

87. Leheste, J.R.; Melsen, F.; Wellner, M.; Jansen, P.; Schlichting, U.; Renner-Müller, I.; Andreassen, T.T.; Wolf, E.; Bachmann, S.; Nykjaer, A.; et al. Hypocalcemia and osteopathy in mice with kidney-specific megalin gene defect. *FASEB J.* **2003**, *17*, 247–249. [CrossRef] [PubMed]
88. Abboud, M.; Puglisi, D.A.; Davies, B.N.; Rybchyn, M.; Whitehead, N.P.; Brock, K.E.; Cole, L.; Gordon-Thomson, C.; Fraser, D.R.; Mason, R.S. Evidence for a specific uptake and retention mechanism for 25-hydroxyvitamin D (25OHD) in skeletal muscle cells. *Endocrinology* **2013**, *154*, 3022–3030. [CrossRef]
89. Rowling, M.J.; Kemmis, C.M.; Taffany, D.A.; Welsh, J. Megalin-mediated endocytosis of vitamin D binding protein correlates with 25-hydroxycholecalciferol actions in human mammary cells. *J. Nutr.* **2006**, *136*, 2754–2759. [CrossRef]
90. Ternes, S.B.; Rowling, M.J. Vitamin D transport proteins megalin and disabled-2 are expressed in prostate and colon epithelial cells and are induced and activated by all-trans-retinoic acid. *Nutr. Cancer* **2013**, *65*, 900–907. [CrossRef]
91. Negri, A.L. Proximal tubule endocytic apparatus as the specific renal uptake mechanism for vitamin D-binding protein/25-(OH)D3 complex. *Nephrology* **2006**, *11*, 510–515. [CrossRef] [PubMed]
92. Gao, Y.; Zhou, S.; Luu, S.; Glowacki, J. Megalin mediates 25-hydroxyvitamin D. *FASEB J.* **2019**, *33*, 7684–7693. [CrossRef] [PubMed]
93. Sheikh-Hamad, D.; Holliday, M.; Li, Q. Megalin-Mediated Trafficking of Mitochondrial Intracrines: Relevance to Signaling and Metabolism. *J. Cell. Immunol.* **2021**, *3*, 364–369. [PubMed]
94. Li, Q.; Lei, F.; Tang, Y.; Pan, J.S.; Tong, Q.; Sun, Y.; Sheikh-Hamad, D. Megalin mediates plasma membrane to mitochondria cross-talk and regulates mitochondrial metabolism. *Cell. Mol. Life Sci.* **2018**, *75*, 4021–4040. [CrossRef] [PubMed]
95. Taparia, S.; Fleet, J.C.; Peng, J.B.; Wang, X.D.; Wood, R.J. 1,25-Dihydroxyvitamin D and 25-hydroxyvitamin D—Mediated regulation of TRPV6 (a putative epithelial calcium channel) mRNA expression in Caco-2 cells. *Eur. J. Nutr.* **2006**, *45*, 196–204. [CrossRef] [PubMed]
96. Schwartz, N.; Verma, A.; Bivens, C.B.; Schwartz, Z.; Boyan, B.D. Rapid steroid hormone actions via membrane receptors. *Biochim. Biophys. Acta* **2016**, *1863*, 2289–2298. [CrossRef]
97. Gerdes, D.; Christ, M.; Haseroth, K.; Notzon, A.; Falkenstein, E.; Wehling, M. Nongenomic Actions of Steroids—From the Laboratory to Clinical Implications. *J. Pediatr. Endocrinol. Metab.* **2000**, *13*, 853–878. [CrossRef]
98. Baran, D.T.; Ray, R.; Sorensen, A.M.; Honeyman, T.; Holick, M.F. Binding characteristics of a membrane receptor that recognizes 1α25-dihydroxyvitamin D3 and its epimer, 1β,25-dihydroxyvitamin D3. *J. Cell. Biochem.* **1994**, *56*, 510–517. [CrossRef]
99. Nemere, I.; Yoshimoto, Y.; Norman, A.W. Calcium Transport in Perfused Duodena from Normal Chicks: Enhancement within Fourteen Minutes of Exposure to 1,25-Dihydroxyvitamin D_3*. *Endocrinology* **1984**, *115*, 1476–1483. [CrossRef]
100. Nemere, I.; Dormanen, M.C.; Hammond, M.W.; Okamura, W.H.; Norman, A.W. Identification of a specific binding protein for 1 alpha,25-dihydroxyvitamin D3 in basal-lateral membranes of chick intestinal epithelium and relationship to transcaltachia. *J. Biol. Chem.* **1994**, *269*, 23750–23756. [CrossRef]
101. Doroudi, M.; Olivares-Navarrete, R.; Boyan, B.D.; Schwartz, Z. A review of 1α,25(OH)$_2$D$_3$ dependent Pdia3 receptor complex components in Wnt5a non-canonical pathway signaling. *J. Steroid Biochem. Mol. Biol.* **2015**, *152*, 84–88. [CrossRef]
102. Doroudi, M.; Plaisance, M.C.; Boyan, B.D.; Schwartz, Z. Membrane actions of 1α,25(OH)$_2$D$_3$ are mediated by Ca^{2+}/calmodulin-dependent protein kinase II in bone and cartilage cells. *J. Steroid Biochem. Mol. Biol.* **2015**, *145*, 65–74. [CrossRef] [PubMed]
103. Nemere, I.; Garbi, N.; Winger, Q. The 1,25D$_3$-MARRS receptor/PDIA3/ERp57 and lifespan. *J. Cell. Biochem.* **2014**, *116*, 380–385. [CrossRef] [PubMed]
104. Aureli, C.; Gaucci, E.; Arcangeli, V.; Grillo, C.; Eufemi, M.; Chichiarelli, S. ERp57/PDIA3 binds specific DNA fragments in a melanoma cell line. *Gene* **2013**, *524*, 390–395. [CrossRef] [PubMed]
105. Nemere, I.; Garbi, N.; Hammerling, G.; Hintze, K.J. Role of the 1,25D3-MARRS receptor in the 1,25(OH)2D3-stimulated uptake of calcium and phosphate in intestinal cells. *Steroids* **2012**, *77*, 897–902. [CrossRef]
106. Chen, J.; Olivares-Navarrete, R.; Wang, Y.; Herman, T.R.; Boyan, B.D.; Schwartz, Z. Protein-disulfide Isomerase-associated 3 (Pdia3) Mediates the Membrane Response to 1,25-Dihydroxyvitamin D3 in Osteoblasts. *J. Biol. Chem.* **2010**, *285*, 37041–37050. [CrossRef] [PubMed]
107. Nemere, I.; Garbi, N.; Hämmerling, G.J.; Khanal, R.C. Intestinal Cell Calcium Uptake and the Targeted Knockout of the 1,25D3-MARRS (Membrane-associated, Rapid Response Steroid-binding) Receptor/PDIA3/Erp57. *J. Biol. Chem.* **2010**, *285*, 31859–31866. [CrossRef] [PubMed]
108. Wang, Y.; Chen, J.; Lee, C.S.; Nizkorodov, A.; Riemenschneider, K.; Martin, D.; Hyzy, S.; Schwartz, Z.; Boyan, B.D. Disruption of Pdia3 gene results in bone abnormality and affects 1α,25-dihydroxy-vitamin D3-induced rapid activation of PKC. *J. Steroid Biochem. Mol. Biol.* **2010**, *121*, 257–260. [CrossRef]
109. Nemere, I.; Safford, S.E.; Rohe, B.; DeSouza, M.M.; Farach-Carson, M.C. Identification and characterization of 1,25D3-membrane-associated rapid response, steroid (1,25D3-MARRS) binding protein. *J. Steroid Biochem. Mol. Biol.* **2004**, *89–90*, 281–285. [CrossRef]
110. Garbi, N.; Tanaka, S.; Momburg, F.; Hämmerling, G.J. Impaired assembly of the major histocompatibility complex class I peptide-loading complex in mice deficient in the oxidoreductase ERp57. *Nat. Immunol.* **2005**, *7*, 93–102. [CrossRef]
111. Doroudi, M.; Chen, J.; Boyan, B.D.; Schwartz, Z. New insights on membrane mediated effects of 1α,25-dihydroxy vitamin D3 signaling in the musculoskeletal system. *Steroids* **2014**, *81*, 81–87. [CrossRef]
112. Wilkin, A.M.; Harnett, A.; Underschultz, M.; Cragg, C.; Meckling, K.A. Role of the ERp57 protein (1,25D3-MARRS receptor) in murine mammary gland growth and development. *Steroids* **2018**, *135*, 63–68. [CrossRef] [PubMed]

113. Hettinghouse, A.; Liu, R.; Liu, C.-J. Multifunctional molecule ERp57: From cancer to neurodegenerative diseases. *Pharmacol. Ther.* **2018**, *181*, 34–48. [CrossRef] [PubMed]
114. Wasiewicz, T.; Szyszka, P.; Cichorek, M.; Janjetovic, Z.; Tuckey, R.C.; Slominski, A.T.; Zmijewski, M.A. Antitumor Effects of Vitamin D Analogs on Hamster and Mouse Melanoma Cell Lines in Relation to Melanin Pigmentation. *Int. J. Mol. Sci.* **2015**, *16*, 6645–6667. [CrossRef]
115. Gaucci, E.; Raimondo, D.; Grillo, C.; Cervoni, L.; Altieri, F.; Nittari, G.; Eufemi, M.; Chichiarelli, S. Analysis of the interaction of calcitriol with the disulfide isomerase ERp57. *Sci. Rep.* **2016**, *6*, 37957. [CrossRef]
116. Chen, J.; Lobachev, K.S.; Grindel, B.J.; Farach-Carson, M.C.; Hyzy, S.L.; El-Baradie, K.B.; Olivares-Navarrete, R.; Doroudi, M.; Boyan, B.D.; Schwartz, Z. Chaperone properties of pdia3 participate in rapid membrane actions of 1α,25-dihydroxyvitamin d3. *Mol. Endocrinol.* **2013**, *27*, 1065–1077. [CrossRef] [PubMed]
117. Kozlov, G.; Maattanen, P.; Schrag, J.D.; Pollock, S.; Cygler, M.; Nagar, B.; Thomas, D.; Gehring, K. Crystal Structure of the bb$'$ Domains of the Protein Disulfide Isomerase ERp57. *Structure* **2006**, *14*, 1331–1339. [CrossRef]
118. Wang, Y.; Nizkorodov, A.; Riemenschneider, K.; Lee, C.S.D.; Olivares-Navarrete, R.; Schwartz, Z.; Boyan, B.D. Impaired Bone Formation in Pdia3 Deficient Mice. *PLoS ONE* **2014**, *9*, e112708. [CrossRef]
119. Yang, W.S.; Yu, H.; Kim, J.J.; Lee, M.J.; Park, S.K. Vitamin D-induced ectodomain shedding of TNF receptor 1 as a nongenomic action: D3 vs. D2 derivatives. *J. Steroid Biochem. Mol. Biol.* **2016**, *155*, 18–25. [CrossRef]
120. Li, Y.; Camacho, P. Ca^{2+}-dependent redox modulation of SERCA 2b by ERp57. *J. Cell Biol.* **2003**, *164*, 35–46. [CrossRef]
121. Khanal, R.; Nemere, I. Membrane receptors for vitamin D metabolites. *Crit. Rev. Eukaryot. Gene Expr.* **2007**, *17*, 31–47. [CrossRef] [PubMed]
122. Sequeira, V.B.; Rybchyn, M.S.; Tongkao-On, W.; Gordon-Thomson, C.; Malloy, P.J.; Nemere, I.; Norman, A.W.; Reeve, V.E.; Halliday, G.M.; Feldman, D.; et al. The role of the vitamin D receptor and ERp57 in photoprotection by 1α,25-dihydroxyvitamin D3. *Mol. Endocrinol.* **2012**, *26*, 574–582. [CrossRef] [PubMed]
123. Egger, A.N.; Rajabi-Estarabadi, A.; Williams, N.M.; Resnik, S.R.; Fox, J.D.; Wong, L.L.; Jozic, I. The importance of caveolins and caveolae to dermatology: Lessons from the caves and beyond. *Exp. Dermatol.* **2019**, *29*, 136–148. [CrossRef] [PubMed]
124. Doroudi, M.; Schwartz, Z.; Boyan, B.D. Membrane-mediated actions of 1,25-dihydroxy vitamin D3: A review of the roles of phospholipase A2 activating protein and Ca^{2+}/calmodulin-dependent protein kinase II. *J. Steroid Biochem. Mol. Biol.* **2015**, *147*, 81–84. [CrossRef] [PubMed]
125. Chichiarelli, S.; Altieri, F.; Paglia, G.; Rubini, E.; Minacori, M.; Eufemi, M. ERp57/PDIA3: New insight. *Cell. Mol. Biol. Lett.* **2022**, *27*, 1–18. [CrossRef]
126. Grindel, B.J.; Rohe, B.; Safford, S.E.; Bennett, J.J.; Farach-Carson, M.C. Tumor necrosis factor-α treatment of HepG2 cells mobilizes a cytoplasmic pool of ERp57/1,25D_3-MARRS to the nucleus. *J. Cell. Biochem.* **2011**, *112*, 2606–2615. [CrossRef] [PubMed]
127. Hu, W.; Zhang, L.; Li, M.X.; Shen, J.; Liu, X.D.; Xiao, Z.G.; Wu, D.L.; Ho, I.H.T.; Wu, J.C.Y.; Cheung, C.K.Y.; et al. Vitamin D3 activates the autolysosomal degradation function against Helicobacter pylori through the PDIA3 receptor in gastric epithelial cells. *Autophagy* **2019**, *15*, 707–725. [CrossRef]
128. Guo, G.G.; Patel, K.; Kumar, V.; Shah, M.; Fried, V.A.; Etlinger, J.D.; Sehgal, P.B. Association of the Chaperone Glucose-Regulated Protein 58 (GRP58/ER-60/ERp57) with Stat3 in Cytosol and Plasma Membrane Complexes. *J. Interf. Cytokine Res.* **2002**, *22*, 555–563. [CrossRef]
129. Chichiarelli, S.; Gaucci, E.; Ferraro, A.; Grillo, C.; Altieri, F.; Cocchiola, R.; Arcangeli, V.; Turano, C.; Eufemi, M. Role of ERp57 in the signaling and transcriptional activity of STAT3 in a melanoma cell line. *Arch. Biochem. Biophys.* **2009**, *494*, 178–183. [CrossRef]
130. Wu, W.; Beilhartz, G.; Roy, Y.; Richard, C.L.; Curtin, M.; Brown, L.; Cadieux, D.; Coppolino, M.; Farach-Carson, M.C.; Nemere, I.; et al. Nuclear translocation of the 1,25D3-MARRS (membrane associated rapid response to steroids) receptor protein and NFκB in differentiating NB4 leukemia cells. *Exp. Cell Res.* **2010**, *316*, 1101–1108. [CrossRef]
131. Grillo, C.; D'Ambrosio, C.; Consalvi, V.; Chiaraluce, R.; Scaloni, A.; Maceroni, M.; Eufemi, M.; Altieri, F. DNA-binding Activity of the ERp57 C-terminal Domain Is Related to a Redox-dependent Conformational Change. *J. Biol. Chem.* **2007**, *282*, 10299–10310. [CrossRef]
132. Doroudi, M.; Olivares-Navarrete, R.; Hyzy, S.L.; Boyan, B.D.; Schwartz, Z. Signaling components of the 1α,25$(OH)_2D_3$-dependent Pdia3 receptor complex are required for Wnt5a calcium-dependent signaling. *Biochim. Biophys. Acta* **2014**, *1843*, 2365–2375. [CrossRef] [PubMed]
133. Zhu, L.; Santos, N.C.; Kim, K.H. Disulfide isomerase glucose-regulated protein 58 is required for the nuclear localization and degradation of retinoic acid receptor α. *Reproduction* **2010**, *139*, 717–731. [CrossRef]
134. Gao, C.; Liao, M.Z.; Han, L.W.; Thummel, K.E.; Mao, Q. Hepatic Transport of 25-Hydroxyvitamin D. *Drug Metab. Dispos.* **2018**, *46*, 581–591. [CrossRef]
135. Chen, Y.; Liu, X.; Zhang, F.; Liao, S.; He, X.; Zhuo, D.; Huang, H.; Wu, Y. Vitamin D receptor suppresses proliferation and metastasis in renal cell carcinoma cell lines via regulating the expression of the epithelial Ca^{2+} channel TRPV5. *PLoS ONE* **2018**, *13*, e0195844. [CrossRef]
136. Long, W.; Fatehi, M.; Soni, S.; Panigrahi, R.; Philippaert, K.; Yu, Y.; Kelly, R.; Boonen, B.; Barr, A.; Golec, D.; et al. Vitamin D is an endogenous partial agonist of the transient receptor potential vanilloid 1 channel. *J. Physiol.* **2020**, *598*, 4321–4338. [CrossRef] [PubMed]

137. Long, W.; Johnson, J.; Kalyaanamoorthy, S.; Light, P. TRPV1 channels as a newly identified target for vitamin D. *Channels* **2021**, *15*, 360–374. [CrossRef] [PubMed]
138. Callejo, M.; Mondejar-Parreño, G.; Morales-Cano, D.; Barreira, B.; Esquivel-Ruiz, S.; Olivencia, M.A.; Manaud, G.; Perros, F.; Duarte, J.; Moreno, L.; et al. Vitamin D deficiency downregulates TASK-1 channels and induces pulmonary vascular dysfunction. *Am. J. Physiol. Lung Cell. Mol. Physiol.* **2020**, *319*, L627–L640. [CrossRef] [PubMed]
139. García-Becerra, R.; Díaz, L.; Camacho, J.; Barrera, D.; Ordaz-Rosado, D.; Morales, A.; Ortiz, C.S.; Avila, E.; Bargallo, E.; Arrecillas, M.; et al. Calcitriol inhibits Ether-à go-go potassium channel expression and cell proliferation in human breast cancer cells. *Exp. Cell Res.* **2010**, *316*, 433–442. [CrossRef]
140. Avila, E.; García-Becerra, R.; Rodríguez-Rasgado, J.A.; Díaz, L.; Ordaz-Rosado, D.; Zügel, U.; Steinmeyer, A.; Barrera, D.; Halhali, A.; Larrea, F.; et al. Calcitriol down-regulates human ether a go-go 1 potassium channel expression in cervical cancer cells. *Anticancer Res.* **2010**, *30*, 2667–2672.
141. Tripathy, B.; Majhi, R.K. TRPV1 channel as the membrane vitamin D receptor: Solving part of the puzzle. *J. Physiol.* **2020**, *598*, 5601–5603. [CrossRef] [PubMed]
142. Tamayo, M.; Manzanares, E.; Bas, M.; Martín-Nunes, L.; Val-Blasco, A.; Jesús Larriba, M.; Fernández-Velasco, M.; Delgado, C. Calcitriol (1,25-dihydroxyvitamin D. *Heart Rhythm* **2017**, *14*, 432–439. [CrossRef] [PubMed]
143. Tamayo, M.; Martin-Nunes, L.; Val-Blasco, A.; Piedras, M.J.; Larriba, M.J.; Gómez-Hurtado, N.; Fernández-Velasco, M.; Delgado, C. Calcitriol, the Bioactive Metabolite of Vitamin D, Increases Ventricular K^+ Currents in Isolated Mouse Cardiomyocytes. *Front. Physiol.* **2018**, *9*, 1186. [CrossRef]
144. Sequeira, V.B.; Rybchyn, M.S.; Gordon-Thomson, C.; Tongkao-On, W.; Mizwicki, M.T.; Norman, A.W.; Reeve, V.E.; Halliday, G.M.; Mason, R.S. Opening of chloride channels by 1α,25-dihydroxyvitamin D3 contributes to photoprotection against UVR-induced thymine dimers in keratinocytes. *J. Investig. Dermatol.* **2013**, *133*, 776–782. [CrossRef]
145. Olszewska, A.M.; Sieradzan, A.K.; Bednarczyk, P.; Szewczyk, A.; Żmijewski, M.A. Mitochondrial potassium channels: A novel calcitriol target. *Cell. Mol. Biol. Lett.* **2022**, *27*, 3. [CrossRef] [PubMed]
146. Chen, J.; Doroudi, M.; Cheung, J.; Grozier, A.L.; Schwartz, Z.; Boyan, B.D. Plasma membrane Pdia3 and VDR interact to elicit rapid responses to 1α,25(OH)(2)D(3). *Cell. Signal.* **2013**, *25*, 2362–2373. [CrossRef] [PubMed]
147. Ricca, C.; Aillon, A.; Bergandi, L.; Alotto, D.; Castagnoli, C.; Silvagno, F. Vitamin D Receptor Is Necessary for Mitochondrial Function and Cell Health. *Int. J. Mol. Sci.* **2018**, *19*, 1672. [CrossRef] [PubMed]
148. Silvagno, F.; Consiglio, M.; Foglizzo, V.; Destefanis, M.; Pescarmona, G. Mitochondrial translocation of vitamin D receptor is mediated by the permeability transition pore in human keratinocyte cell line. *PLoS ONE* **2013**, *8*, e54716. [CrossRef]
149. Silvagno, F.; De Vivo, E.; Attanasio, A.; Gallo, V.; Mazzucco, G.; Pescarmona, G. Mitochondrial localization of vitamin D receptor in human platelets and differentiated megakaryocytes. *PLoS ONE* **2010**, *5*, e8670. [CrossRef]
150. Filipović, N.; Bočina, I.; Restović, I.; Grobe, M.; Kretzschmar, G.; Kević, N.; Mašek, T.; Vitlov Uljević, M.; Jurić, M.; Vukojević, K.; et al. Ultrastructural characterization of vitamin D receptors and metabolizing enzymes in the lipid droplets of the fatty liver in rat. *Acta Histochem.* **2020**, *122*, 151502. [CrossRef]
151. Mizwicki, M.T.; Menegaz, D.; Yaghmaei, S.; Henry, H.L.; Norman, A.W. A molecular description of ligand binding to the two overlapping binding pockets of the nuclear vitamin D receptor (VDR): Structure-function implications. *J. Steroid Biochem. Mol. Biol.* **2010**, *121*, 98–105. [CrossRef]
152. Tapia, C.; Suares, A.; De Genaro, P.; González-Pardo, V. In vitro studies revealed a downregulation of Wnt/β-catenin cascade by active vitamin D and TX 527 analog in a Kaposi's sarcoma cellular model. *Toxicol. In Vitro* **2020**, *63*, 104748. [CrossRef] [PubMed]
153. Muralidhar, S.; Filia, A.; Nsengimana, J.; Poźniak, J.; O'Shea, S.J.; Diaz, J.M.; Harland, M.; Randerson-Moor, J.A.; Reichrath, J.; Laye, J.P.; et al. Vitamin D-VDR Signaling Inhibits Wnt/β-Catenin-Mediated Melanoma Progression and Promotes Antitumor Immunity. *Cancer Res.* **2019**, *79*, 5986–5998. [CrossRef] [PubMed]
154. Tang, L.; Fang, W.; Lin, J.; Li, J.; Wu, W.; Xu, J. Vitamin D protects human melanocytes against oxidative damage by activation of Wnt/β-catenin signaling. *Lab. Investig.* **2018**, *98*, 1527–1537. [CrossRef] [PubMed]
155. Larriba, M.J.; Gonzalez-Sancho, J.M.; Bonilla, F.; Munoz, A. Interaction of vitamin D with membrane-based signaling pathways. *Front. Physiol.* **2014**, *5*, 60. [CrossRef]
156. Bikle, D.D.; Jiang, Y.; Nguyen, T.; Oda, Y.; Tu, C.L. Disruption of Vitamin D and Calcium Signaling in Keratinocytes Predisposes to Skin Cancer. *Front. Physiol.* **2016**, *7*, 296. [CrossRef]
157. Bandera Merchan, B.; Morcillo, S.; Martin-Nuñez, G.; Tinahones, F.J.; Macías-González, M. The role of vitamin D and VDR in carcinogenesis: Through epidemiology and basic sciences. *J. Steroid Biochem. Mol. Biol.* **2017**, *167*, 203–218. [CrossRef]
158. Hadden, M.K. Hedgehog and Vitamin D Signaling Pathways in Development and Disease. *Vitam. Horm.* **2016**, *100*, 231–253. [CrossRef]
159. Lisse, T.S.; Saini, V.; Zhao, H.; Luderer, H.F.; Gori, F.; Demay, M.B. The vitamin D receptor is required for activation of cWnt and hedgehog signaling in keratinocytes. *Mol. Endocrinol.* **2014**, *28*, 1698–1706. [CrossRef]
160. Teichert, A.E.; Elalieh, H.; Elias, P.M.; Welsh, J.; Bikle, D.D. Overexpression of hedgehog signaling is associated with epidermal tumor formation in vitamin D receptor-null mice. *J. Investig. Dermatol.* **2011**, *131*, 2289–2297. [CrossRef]
161. Wang, H.; Wang, X.; Xu, L.; Zhang, J.; Cao, H. A molecular sub-cluster of colon cancer cells with low VDR expression is sensitive to chemotherapy, BRAF inhibitors and PI3K-mTOR inhibitors treatment. *Aging* **2019**, *11*, 8587–8603. [CrossRef] [PubMed]

162. Olsson, K.; Saini, A.; Strömberg, A.; Alam, S.; Lilja, M.; Rullman, E.; Gustafsson, T. Evidence for Vitamin D Receptor Expression and Direct Effects of 1α,25(OH)$_2$D$_3$ in Human Skeletal Muscle Precursor Cells. *Endocrinology* **2016**, *157*, 98–111. [CrossRef] [PubMed]
163. Singh, P.K.; van den Berg, P.R.; Long, M.D.; Vreugdenhil, A.; Grieshober, L.; Ochs-Balcom, H.M.; Wang, J.; Delcambre, S.; Heikkinen, S.; Carlberg, C.; et al. Integration of VDR genome wide binding and GWAS genetic variation data reveals co-occurrence of VDR and NF-κB binding that is linked to immune phenotypes. *BMC Genom.* **2017**, *18*, 132. [CrossRef]
164. Doroudi, M.; Boyan, B.D.; Schwartz, Z. Rapid 1α,25(OH)$_2$D$_3$ membrane-mediated activation of Ca^{2+}/calmodulin-dependent protein kinase II in growth plate chondrocytes requires Pdia3, PLAA and caveolae. *Connect. Tissue Res.* **2014**, *55* (Suppl. S1), 125–128. [CrossRef]
165. Doroudi, M.; Schwartz, Z.; Boyan, B.D. Phospholipase A2 activating protein is required for 1α,25-dihydroxyvitamin D3 dependent rapid activation of protein kinase C via Pdia3. *J. Steroid Biochem. Mol. Biol.* **2012**, *132*, 48–56. [CrossRef]
166. Chignell, C.F.; Kukielczak, B.M.; Sik, R.H.; Bilski, P.J.; He, Y.Y. Ultraviolet A sensitivity in Smith-Lemli-Opitz syndrome: Possible involvement of cholesta-5,7,9(11)-trien-3 beta-ol. *Free Radic. Biol. Med.* **2006**, *41*, 339–346. [CrossRef] [PubMed]
167. De Fabiani, E.; Caruso, D.; Cavaleri, M.; Galli Kienle, M.; Galli, G. Cholesta-5,7,9(11)-trien-3 beta-ol found in plasma of patients with Smith-Lemli-Opitz syndrome indicates formation of sterol hydroperoxide. *J. Lipid Res.* **1996**, *37*, 2280–2287. [CrossRef]
168. Jin, X.; Yang, X.; Yang, L.; Liu, Z.-L.; Zhang, F. Autoxidation of isotachysterol. *Tetrahedron* **2004**, *60*, 2881–2888. [CrossRef]
169. Zmijewski, M.A.; Li, W.; Chen, J.; Kim, T.K.; Zjawiony, J.K.; Sweatman, T.W.; Miller, D.D.; Slominski, A.T. Synthesis and photochemical transformation of 3beta,21-dihydroxypregna-5,7-dien-20-one to novel secosteroids that show anti-melanoma activity. *Steroids* **2011**, *76*, 193–203. [CrossRef]
170. Guo, L.W.; Wilson, W.K.; Pang, J.; Shackleton, C.H. Chemical synthesis of 7- and 8-dehydro derivatives of pregnane-3,17alpha,20-triols, potential steroid metabolites in Smith-Lemli-Opitz syndrome. *Steroids* **2003**, *68*, 31–42. [CrossRef]
171. Valencia, A.; Kochevar, I.E. Ultraviolet A induces apoptosis via reactive oxygen species in a model for Smith-Lemli-Opitz syndrome. *Free Radic. Biol. Med.* **2006**, *40*, 641–650. [CrossRef] [PubMed]
172. Jablonski, N.G.; Chaplin, G. The roles of vitamin D and cutaneous vitamin D production in human evolution and health. *Int. J. Paleopathol.* **2018**, *23*, 54–59. [CrossRef] [PubMed]
173. de Jager, T.L.; Cockrell, A.E.; Du Plessis, S.S. Ultraviolet Light Induced Generation of Reactive Oxygen Species. *Adv. Exp. Med. Biol.* **2017**, *996*, 15–23. [CrossRef] [PubMed]
174. Ravanat, J.L.; Douki, T.; Cadet, J. Direct and indirect effects of UV radiation on DNA and its components. *J. Photochem. Photobiol. B* **2001**, *63*, 88–102. [CrossRef] [PubMed]
175. De Silva, W.G.M.; Han, J.Z.R.; Yang, C.; Tongkao-On, W.; McCarthy, B.Y.; Ince, F.A.; Holland, A.J.A.; Tuckey, R.C.; Slominski, A.T.; Abboud, M.; et al. Evidence for Involvement of Nonclassical Pathways in the Protection from UV-Induced DNA Damage by Vitamin D-Related Compounds. *JBMR Plus* **2021**, *5*, e10555. [CrossRef] [PubMed]
176. Lin, Y.; Cao, Z.; Lyu, T.; Kong, T.; Zhang, Q.; Wu, K.; Wang, Y.; Zheng, J. Single-cell RNA-seq of UVB-radiated skin reveals landscape of photoaging-related inflammation and protection by vitamin D. *Gene* **2022**, *831*, 146563. [CrossRef]
177. Reichrath, J.; Rass, K. Ultraviolet damage, DNA repair and vitamin D in nonmelanoma skin cancer and in malignant melanoma: An update. *Adv. Exp. Med. Biol.* **2014**, *810*, 208–233.
178. Scott, J.F.; Das, L.M.; Ahsanuddin, S.; Qiu, Y.; Binko, A.M.; Traylor, Z.P.; Debanne, S.M.; Cooper, K.D.; Boxer, R.; Lu, K.Q. Oral Vitamin D Rapidly Attenuates Inflammation from Sunburn: An Interventional Study. *J. Investig. Dermatol.* **2017**, *137*, 2078–2086. [CrossRef]
179. Chaiprasongsuk, A.; Janjetovic, Z.; Kim, T.K.; Jarrett, S.G.; D'Orazio, J.A.; Holick, M.F.; Tang, E.K.Y.; Tuckey, R.C.; Panich, U.; Li, W.; et al. Protective effects of novel derivatives of vitamin D. *Redox Biol.* **2019**, *24*, 101206. [CrossRef]
180. Donati, S.; Palmini, G.; Aurilia, C.; Falsetti, I.; Miglietta, F.; Iantomasi, T.; Brandi, M.L. Rapid Nontranscriptional Effects of Calcifediol and Calcitriol. *Nutrients* **2022**, *14*, 1291. [CrossRef]
181. Janjusevic, M.; Gagno, G.; Fluca, A.L.; Padoan, L.; Beltrami, A.P.; Sinagra, G.; Moretti, R.; Aleksova, A. The peculiar role of vitamin D in the pathophysiology of cardiovascular and neurodegenerative diseases. *Life Sci.* **2022**, *289*, 120193. [CrossRef] [PubMed]
182. Zhu, J.G.; Ochalek, J.T.; Kaufmann, M.; Jones, G.; Deluca, H.F. CYP2R1 is a major, but not exclusive, contributor to 25-hydroxyvitamin D production in vivo. *Proc. Natl. Acad. Sci. USA* **2013**, *110*, 15650–15655. [CrossRef] [PubMed]
183. Slominski, A.T.; Li, W.; Kim, T.K.; Semak, I.; Wang, J.; Zjawiony, J.K.; Tuckey, R.C. Novel activities of CYP11A1 and their potential physiological significance. *J. Steroid Biochem. Mol. Biol.* **2015**, *151*, 25–37. [CrossRef] [PubMed]
184. Tuckey, R.C.; Li, W.; Ma, D.; Cheng, C.Y.S.; Wang, K.M.; Kim, T.K.; Jeayeng, S.; Slominski, A.T. CYP27A1 acts on the pre-vitamin D3 photoproduct, lumisterol, producing biologically active hydroxy-metabolites. *J. Steroid Biochem. Mol. Biol.* **2018**, *181*, 1–10. [CrossRef] [PubMed]
185. Slominski, A.T.; Kim, T.K.; Li, W.; Postlethwaite, A.; Tieu, E.W.; Tang, E.K.Y.; Tuckey, R.C. Detection of novel CYP11A1-derived secosteroids in the human epidermis and serum and pig adrenal gland. *Sci. Rep.* **2015**, *5*, 14875. [CrossRef]
186. Slominski, A.T.; Janjetovic, Z.; Fuller, B.E.; Zmijewski, M.A.; Tuckey, R.C.; Nguyen, M.N.; Sweatman, T.; Li, W.; Zjawiony, J.; Miller, D.; et al. Products of vitamin D3 or 7-dehydrocholesterol metabolism by cytochrome P450scc show anti-leukemia effects, having low or absent calcemic activity. *PLoS ONE* **2010**, *5*, e9907. [CrossRef]

187. Slominski, A.T.; Chaiprasongsuk, A.; Janjetovic, Z.; Kim, T.K.; Stefan, J.; Slominski, R.M.; Hanumanthu, V.S.; Raman, C.; Qayyum, S.; Song, Y.; et al. Photoprotective Properties of Vitamin D and Lumisterol Hydroxyderivatives. *Cell Biochem. Biophys.* **2020**, *78*, 165–180. [CrossRef]
188. Quigley, M.; Rieger, S.; Capobianco, E.; Wang, Z.; Zhao, H.; Hewison, M.; Lisse, T.S. Vitamin D Modulation of Mitochondrial Oxidative Metabolism and mTOR Enforces Stress Adaptations and Anticancer Responses. *JBMR Plus* **2022**, *6*, e10572. [CrossRef]
189. Consiglio, M.; Destefanis, M.; Morena, D.; Foglizzo, V.; Forneris, M.; Pescarmona, G.; Silvagno, F. The vitamin D receptor inhibits the respiratory chain, contributing to the metabolic switch that is essential for cancer cell proliferation. *PLoS ONE* **2014**, *9*, e115816. [CrossRef]
190. Demonacos, C.V.; Karayanni, N.; Hatzoglou, E.; Tsiriyiotis, C.; Spandidos, D.A.; Sekeris, C.E. Mitochondrial genes as sites of primary action of steroid hormones. *Steroids* **1996**, *61*, 226–232. [CrossRef]
191. Mersa, A.; Atashbar, S.; Ahvar, N.; Salimi, A. 1,25-dihydroxyvitamin D3 prevents deleterious effects of erythromycin on mitochondrial function in rat heart isolated mitochondria. *Clin. Exp. Pharmacol. Physiol.* **2020**, *47*, 1554–1563. [CrossRef] [PubMed]
192. Mirzakhani, H.; Litonjua, A.A.; McElrath, T.F.; O'Connor, G.; Lee-Parritz, A.; Iverson, R.; Macones, G.; Strunk, R.C.; Bacharier, L.B.; Zeiger, R.; et al. Early pregnancy vitamin D status and risk of preeclampsia. *J. Clin. Investig.* **2016**, *126*, 4702–4715. [CrossRef] [PubMed]
193. Grant, W.B.; Al Anouti, F.; Boucher, B.J.; Dursun, E.; Gezen-Ak, D.; Jude, E.B.; Karonova, T.; Pludowski, P. A Narrative Review of the Evidence for Variations in Serum 25-Hydroxyvitamin D Concentration Thresholds for Optimal Health. *Nutrients* **2022**, *14*, 639. [CrossRef] [PubMed]
194. Pilz, S.; Theiler-Schwetz, V.; Pludowski, P.; Zelzer, S.; Meinitzer, A.; Karras, S.N.; Misiorowski, W.; Zittermann, A.; März, W.; Trummer, C. Hypercalcemia in Pregnancy Due to CYP24A1 Mutations: Case Report and Review of the Literature. *Nutrients* **2022**, *14*, 2518. [CrossRef] [PubMed]
195. Zmijewski, M.A.; Carlberg, C. Vitamin D receptor(s): In the nucleus but also at membranes? *Exp. Dermatol.* **2020**, *29*, 876–884. [CrossRef]
196. Kumari, S.; Badana, A.K.; Murali Mohan, G.; Shailender, G.; Malla, R. Reactive Oxygen Species: A Key Constituent in Cancer Survival. *Biomark. Insights* **2018**, *13*, 1177271918755391. [CrossRef] [PubMed]
197. Berridge, M.J. Vitamin D: A custodian of cell signalling stability in health and disease. *Biochem. Soc. Trans.* **2015**, *43*, 349–358. [CrossRef]
198. Wimalawansa, S.J. Non-musculoskeletal benefits of vitamin D. *J. Steroid Biochem. Mol. Biol.* **2018**, *175*, 60–81. [CrossRef]
199. Sultan, M.; Twito, O.; Tohami, T.; Ramati, E.; Neumark, E.; Rashid, G. Vitamin D diminishes the high platelet aggregation of type 2 diabetes mellitus patients. *Platelets* **2019**, *30*, 120–125. [CrossRef]
200. Gonzalez-Sanchez, E.; El Mourabit, H.; Jager, M.; Clavel, M.; Moog, S.; Vaquero, J.; Ledent, T.; Cadoret, A.; Gautheron, J.; Fouassier, L.; et al. Cholangiopathy aggravation is caused by VDR ablation and alleviated by VDR-independent vitamin D signaling in ABCB4 knockout mice. *Biochim. Biophys. Acta Mol. Basis Dis.* **2021**, *1867*, 166067. [CrossRef]
201. Liu, Y.; Wang, J.X.; Nie, Z.Y.; Wen, Y.; Jia, X.J.; Zhang, L.N.; Duan, H.J.; Shi, Y.H. Upregulation of ERp57 promotes clear cell renal cell carcinoma progression by initiating a STAT3/ILF3 feedback loop. *J. Exp. Clin. Cancer Res.* **2019**, *38*, 439. [CrossRef] [PubMed]
202. Janjetovic, Z.; Zmijewski, M.A.; Tuckey, R.C.; DeLeon, D.A.; Nguyen, M.N.; Pfeffer, L.M.; Slominski, A.T. 20-Hydroxycholecalciferol, product of vitamin D3 hydroxylation by P450scc, decreases NF-kappaB activity by increasing IkappaB alpha levels in human keratinocytes. *PLoS ONE* **2009**, *4*, e5988. [CrossRef]
203. Slominski, A.T.; Kim, T.K.; Slominski, R.M.; Song, Y.; Janjetovic, Z.; Podgorska, E.; Reddy, S.B.; Raman, C.; Tang, E.K.Y.; Fabisiak, A.; et al. Metabolic activation of tachysterol. *FASEB J.* **2022**, *36*, e22451. [CrossRef]
204. Carlberg, C. The physiology of vitamin D-far more than calcium and bone. *Front. Physiol.* **2014**, *5*, 335. [CrossRef]
205. Shirvani, A.; Kalajian, T.A.; Song, A.; Holick, M.F. Disassociation of Vitamin D's Calcemic Activity and Non-calcemic Genomic Activity and Individual Responsiveness: A Randomized Controlled Double-Blind Clinical Trial. *Sci. Rep.* **2019**, *9*, 17685. [CrossRef]
206. Grant, W.B.; Boucher, B.J.; Al Anouti, F.; Pilz, S. Comparing the Evidence from Observational Studies and Randomized Controlled Trials for Nonskeletal Health Effects of Vitamin D. *Nutrients* **2022**, *14*, 3811. [CrossRef]

MDPI

Review

Vitamin D-Mediated Regulation of Intestinal Calcium Absorption

James C. Fleet

Department of Nutritional Sciences, University of Texas, Austin, TX 78723, USA; james.fleet@austin.utexas.edu

Abstract: Vitamin D is a critical regulator of calcium and bone homeostasis. While vitamin D has multiple effects on bone and calcium metabolism, the regulation of intestinal calcium (Ca) absorption efficiency is a critical function for vitamin D. This is necessary for optimal bone mineralization during growth, the protection of bone in adults, and the prevention of osteoporosis. Intestinal Ca absorption is regulated by 1,25 dihydroxyvitamin D ($1,25(OH)_2$ D), a hormone that activates gene transcription following binding to the intestinal vitamin D receptor (VDR). When dietary Ca intake is low, Ca absorption follows a vitamin-D-regulated, saturable pathway, but when dietary Ca intake is high, Ca absorption is predominately through a paracellular diffusion pathway. Deletion of genes that mediate vitamin D action (i.e., VDR) or production (CYP27B1) eliminates basal Ca absorption and prevents the adaptation of mice to low-Ca diets. Various physiologic or disease states modify vitamin-D-regulated intestinal absorption of Ca (enhanced during late pregnancy, reduced due to menopause and aging).

Keywords: diet; transcellular; absorption; diffusion; intestine; homeostasis; parathyroid hormone

Citation: Fleet, J.C. Vitamin D-Mediated Regulation of Intestinal Calcium Absorption. *Nutrients* **2022**, *14*, 3351. https://doi.org/10.3390/nu14163351

Academic Editor: Carsten Carlberg

Received: 18 July 2022
Accepted: 10 August 2022
Published: 16 August 2022

1. Introduction

It has now been 100 years since E.V. McCollum first identified a fat-soluble compound in food that supported bone growth and prevented rickets; he called this compound vitamin D. In the intervening century, many scientists have contributed to our understanding for how vitamin D regulates the physiology of calcium (Ca) metabolism. For example, in 1937 Nicolaysen showed that vitamin D is critical for intestinal Ca absorption [1] while later studies by Pansu et al. [2] and Sheikh et al. [3] showed that vitamin D deficiency significantly reduces intestinal Ca absorption. The critical breakthrough defining the mechanism used by vitamin D to regulate Ca metabolism came in the early 1970′s when research by Holick et al. [4] and Norman et al. [5] isolated the active metabolite of vitamin D, 1,25 dihydroxyvitamin D_3 ($1,25(OH)_2D$), from the intestine. Shortly thereafter, Brumbaugh and Haussler [6] discovered the nuclear receptor for $1,25(OH)_2D$, the vitamin D receptor (VDR), in intestinal mucosa as well. Since then, we've learned the detailed mechanism used by $1,25(OH)_2D$ to regulate gene expression [7] while global and conditional VDR knockout mice have allow us to study the function of vitamin D in Ca/bone metabolism and in other physiologic systems. As part of this effort, my research group has shown that the single most important role for vitamin D during growth is the regulation of intestinal Ca absorption [8] but that $1,25(OH)_2D$ signaling through the VDR has a broad array of target genes in the intestine [9]. Because of the central role that vitamin D plays in the regulation of intestinal Ca absorption, this review provides a critical starting point for anyone who wants to understand the physiologic importance of vitamin D.

2. Vitamin D Has a Critical Physiologic Role for Protecting Bone through the Regulation of Intestinal Ca Absorption

Bone mass is lost when dietary Ca intake is inadequate and so one usually thinks of bone when the term "Ca homeostasis" is used. However, Ca homeostasis is not regulated to maintain bone integrity. Instead, bone, the parathyroid gland, intestine, and kidney make up a multi-tissue axis that work together to maintain serum Ca within a narrow range (8.9–10.2 mg/dL). As a result, after a meal intestinal Ca absorption is a signal that disturbs and elevates serum Ca while bone formation and resorption, along with renal Ca excretion, respond to fluxes in serum Ca in an attempt to limit perturbations in serum Ca. This coordination was shown clearly by Bronner and Aubert who used Ca kinetics in rat models to show how the body adapts to habitual low Ca intake by increasing Ca absorption efficiency, reducing renal Ca loss, and mobilizing Ca from the bone (i.e., resorption) [10]. Pansu et al. [11] showed the impact of dietary Ca intake on intestinal Ca absorption directly when they found that feeding rats a 0.17%, low Ca diet for 5 weeks increased duodenal Ca absorption efficiency by increasing the saturable component of transport (Vmax increased 55%). Similarly, Norman et al. [12] found that in adult humans, feeding a diet with 300 mg of Ca/day for 8 weeks increased Ca absorption efficiency by 43% compared to subjects consuming 1600 mg Ca/day. These studies, and others like them, led to the identification of 1,25 dihydroxyvitamin D (1,25(OH)$_2$ D) and parathyroid hormone (PTH) as the major hormonal regulators of Ca homeostasis.

When dietary Ca is habitually low, there is a transient reduction in serum Ca that is sensed at the parathyroid gland through the Ca sensing receptor (CaSR). This cell surface receptor mediates signals into the parathyroid gland to increase the production and release of PTH into the circulation—a condition called nutritional secondary hyperparathyroidism. PTH has several important functions in Ca metabolism. First, it regulates renal production of 1,25(OH)$_2$ D by inducing the CYP27B1 gene that encodes the enzyme 25 hydroxyvitamin D-1α hydroxylase [13] and it suppresses expression of the CYP24A1 gene that encodes the 25 hydroxyvitamin D-24 hydroxylase [14,15]. 1,25(OH)$_2$ D released from the kidney is the most important regulator of increased intestinal Ca absorption. Of course, this physiologic adaptation has limits so that if the degree of dietary Ca deprivation is too great, the adaptation of intestinal Ca absorption will not be sufficient to compensate. This case, 1,25(OH)$_2$ D and PTH will both promote bone resorption by stimulating osteoclastic activity while also enhancing renal Ca reabsorption in the proximal renal tubule. Collectively, this physiological adaptation can protect serum Ca but at the expense of bone mass.

The central role for vitamin D as a regulator of intestinal Ca absorption has been known for more than 80 years [1,16]. In vitamin D deficient animals intestinal Ca absorption efficiency falls by >75% [2]. Similarly, dialysis patients with low circulating 1,25(OH)$_2$ D levels also have low intestinal Ca absorption [3]. Finally, in elderly adults, secondary hyperparathyroidism can maintain serum 1,25(OH)$_2$ D (and Ca absorption) until serum 25-hydroxyvitamin D (25(OH)D) levels fall to $\leq$10 nmol/L, at which point there is not enough substrate to convert to 1,25(OH)$_2$ D [17].

A challenging concept for many people examining Ca absorption is that it is not a single process but the sum of events that occur through two routes, a transcellular, saturable pathway and non-saturable, paracellular diffusion pathway [11,18–20] (see Figure 1). The relative importance of these two routes depends upon a person's habitual Ca intake. When Ca intake is low, like most adult American women [21], the saturable pathways predominates while when Ca intake is high the bulk of absorption occurs through the diffusional route. Absorption through these two routes can be modeled mathematically using a Michaelis–Menten-like equation that contains a linear component that models diffusion (Figure 1). The saturable transport pathway comprises three parts, apical membrane Ca entry, transcellular diffusion, and basolateral membrane extrusion. The apical membrane transport occurs down a concentration gradient while basolateral membrane extrusion is against a concentration gradient and requires energy [22]. The saturable pathway is present in the proximal small intestine (duodenum and jejunum), cecum, and colon [23–27] but is absent in the ileum [2]. Studies in rat duodenum [2] and in differentiated monolayers of the human intestinal cell line Caco-2 [28], show that 1,25(OH)$_2$ D acts on the saturable transport component where it increases the Vmax (maximal absorptive capacity) but not Km (the affinity of the process for Ca). This suggests that 1,25(OH)$_2$ D increases the production of intestinal Ca transporters (which we'll discuss below) but that this increase has limits.

In contrast, the passive Ca movement across the intestinal barrier occurs at ~13% of the luminal Ca level per hour in humans [20]) and is seen in all segments of the intestine. There is some evidence that the non-saturable portion of Ca absorption in the human ileum is also vitamin D sensitive; the slope of the non-saturable transport pathway is reduced in chronic renal disease patients, and it returns to normal after $1{,}25(OH)_2$ D injection [20].

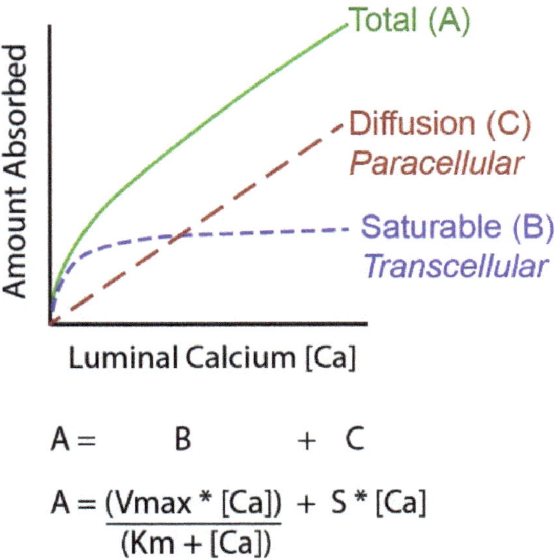

Figure 1. A Mathematical model of intestinal calcium (Ca) absorption. By studying Ca absorption over a range of luminal Ca levels it has been shown that the total amount of Ca absorbed across the intestinal barrier can be described as a curvilinear function. Total transport (A) is the sum of a saturable component (likely transcellular, B) that can be defined by the Michaelis–Menten equation and a diffusional process (C) that is defined by a straight line. [Ca] = luminal Ca concentration; S = the slope of the diffusional component; Vmax = the maximum transport rate seen for the saturable transport component; Km = the luminal concentration of the mineral at $\frac{1}{2}$ the Vmax.

In normal healthy adults, the Km for the saturable component of Ca absorption from the small intestine of adults is 3.3 mM, a concentration met by 265 mg Ca in a meal (calculated from data in [19,20]). As a result, when a person eats a meal with ~400 mg (1/3 the RDA for Ca), saturable Ca transport is about 60% of total Ca absorption. However, the apparent efficiency of total Ca absorption falls as the meal Ca intake level is increased and the diffusional component of absorption takes a larger role. Normally, the amount of Ca absorbed in each intestinal segment is determined by: (a) the presence of the saturable and non-saturable pathways, (b) the residence time in the segment, and (c) the solubility of Ca within the segment (e.g., lower Ca solubility at higher pH [29–31]) (Figure 2). Although $1{,}25(OH)_2$ D clearly increases intestinal Ca absorption efficiency [2,32,33], some have argued that elevated serum 25(OH)D levels might also regulate intestinal Ca absorption. This is not supported by several large, well-designed studies that show no benefit of increasing serum 25(OH)D levels beyond 50 nmol/L [34–37].

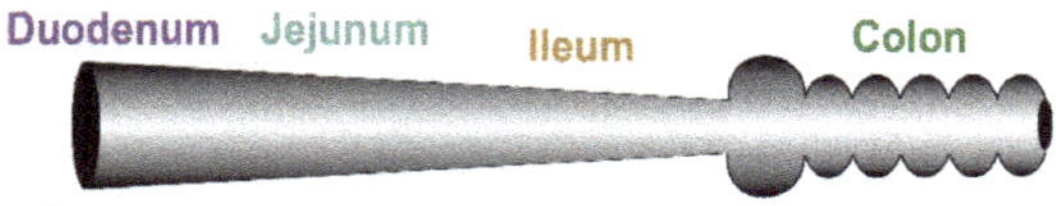

	Duodenum	Jejunum	Ileum	Colon
Saturable Transport	+++	++	-	++
Diffusion	+	+	+	+
VDR	+++	+++	+++	+++
Calbindin D_{9k}/TRPV6	+++	++	-	++
PMCA1b	+	+	+	+
Claudin 2/12	+/-	++	+++	+/-
Ca solubility	+++	++	+	+
Residence Time	+	++	++++	++
Total Ca Absorption (%)	8	17	65	10

Figure 2. Critical factors that influence intestinal calcium (Ca) absorption. Many factors influence net Ca absorption and distinct Ca absorption mechanisms used in various intestinal segments. See Bronner and Pansu [38] for a discussion of solubility and transit time as factors affecting Ca absorption. The number of "+" signs reflects the magnitude of the parameter across tissues while a "-" sign indicates that the parameter is absent in a segment.

There are many studies that show adequate dietary Ca and vitamin D, and therefore total intestinal Ca absorption, is necessary for adequate bone growth. Deficiency of either Ca or vitamin D in growing children or animals causes nutritional rickets characterized by under-mineralized bone and low bone mass [39,40]. This is consistent with the concept that bone matrix cannot mineralize in the absence of mineral. However, net Ca absorption (which reflects both transport routes) is positively correlated to Ca balance in children, reflecting the critical role for Ca absorption in optimizing peak bone mass [41]. Consistent with this idea, we have shown that efficiency of Ca absorption through the saturable, vitamin-D-regulated pathway is significantly positively correlated with femoral trabecular bone volume/total volume in a genetically diverse population of 11 inbred mouse lines [42]. In addition, Patel et al. [43] reported that femur neck BMD was significantly positively correlated with Ca absorption efficiency in adult men. Several studies also indicate the high intestinal Ca absorption efficiency can protect against femoral bone loss in mice fed low Ca diets [44] or reduce the risk of osteoporotic hip fracture in women with low dietary Ca intake [45]. Collectively, these data show that both adequate Ca intake and genetically programed high intestinal Ca absorption efficiency are necessary to build and protect strong bones.

3. Vitamin D Effects on Intestinal Ca Absorption Are Mediated through the VDR

$1,25(OH)_2$ D regulates Ca metabolism and intestinal Ca absorption by regulating gene transcription, a process that requires binding of the hormone to the Vitamin D Receptor (VDR), a nuclear receptor that is a ligand-activated transcription factor [7,46]. A number of studies have shown the critical importance of VDR for the regulation of Ca and bone metabolism. For example, children with inactivating mutations in the VDR gene (i.e., type II genetic rickets) have defects in Ca metabolism that include lower intestinal Ca absorption efficiency [47]. Similarly, VDR knockout mice have severe defects in bone growth and mineralization as well as a >70% reduction in Ca absorption efficiency [48,49]. While the gross phenotype of VDR knockout mice is abnormal bone, several lines of evidence indicate that the most important role for VDR in Ca/bone metabolism during growth is the control of intestinal Ca absorption. First, the phenotype of the intestine-specific VDR knockout mouse is identical to dietary Ca deficiency (osteomalacia, reduced serum Ca, elevated serum 1,25(OH)$_2$ D levels) [50]. Conversely, feeding high Ca/high phosphate/high lactose "rescue" diets that promote passive/diffusional intestinal Ca absorption can prevent the abnormal bone and Ca metabolism phenotype of VDR knockout mice [51]. Finally, experiments from my lab showed that the VDR knockout mouse phenotype (e.g., hypocalcemia, elevated serum PTH, low bone mineral density) can be completely prevented by intestine epithelial cell-specific, transgenic expression of VDR that normalizes intestinal Ca absorption [8].

Several studies show that lower intestinal VDR levels disrupt the physiologic response to 1,25(OH)$_2$D. In VDR KO mice, low level, intestine-specific transgenic VDR expression (10% of wild-type values) was insufficient to maintain normal intestinal Ca absorption [52]. Meanwhile, my lab has shown that a 50% reduction in intestinal VDR levels blunts 1,25(OH)$_2$ D-regulated intestinal Ca absorption efficiency [53]. Consistent with the idea that reduced VDR function impairs intestinal Ca absorption, several studies have shown that Ca absorption efficiency is reduced in people with the longer, less transcriptionally active "f" allele of the Fok I restriction fragment length polymorphism [54–56]. Collectively these data support the hypothesis that variations in VDR level or function can influence vitamin-D-regulated intestinal Ca absorption as well as optimal intestinal responses to the increased serum 1,25(OH)$_2$ D levels that accompany dietary Ca restriction.

4. Molecular Models of Ca Absorption

Ion microscopy reveals that Ca can enter at the apical membrane and flow through the absorptive epithelial cell in 20 min [57]. However, during vitamin D deficiency Ca becomes trapped in the region just below the microvilli. Treating vitamin D deficient chicks with 1,25(OH)$_2$ D reverses this effect starting 2–4 h after treatment [58], consistent with the induction of gene expression mediated through the VDR. In 1986 Bronner et al. [59] critically reviewed transport data from a wide variety of well-controlled mechanistic studies, and from this analysis built the facilitated diffusion model (Figure 3, with protein distributions across the intestinal tract in Figure 2). In the first step of this model, brush border membrane uptake of Ca is mediated by an apical membrane Ca channel, which was later identified as the transient receptor potential cation channel vanilloid family member 6 (TRPV6, originally called CaT1 or ECAC2) [60]. TRPV6 gene expression is strongly regulated by 1,25(OH)$_2$ D in the duodenum of mice [33,61] and in Caco-2 cells [62] and this induction is mediated by VDR binding enhancers upstream from the transcription start site [63,64]. Induction of TRPV6 mRNA precedes the increase in duodenal Ca absorption that occurs following 1,25(OH)$_2$ D injection [33]. While initial studies suggested that 1,25(OH)$_2$ D-mediated Ca absorption was not reduced in TRPV6 knockout mice [65,66], later studies showed that TRPV6 knockout mice [66], and mice with a D541A variant TRPV6 that inactivates Ca movement through the channel [67], had a blunted ability to increase Ca absorption in response to feeding a low Ca diet. In addition, my lab has shown that intestine-specific transgenic expression of TRPV6 increases Ca absorption efficiency and that this prevents the abnormalities in bone/Ca metabolism of VDR knockout mice [68].

An alternative model for apical membrane Ca uptake during Ca absorption is that Ca flows through the L-type Ca channel Ca$_v$1.3 (Figure 3), a transporter activated by glucose-induced membrane depolarization following a meal (reviewed in [69]). However, several studies do not support a physiologic role of Ca$_v$1.3 for vitamin-D-regulated Ca absorption in growing mice [70,71]. In contrast, other studies suggest that Ca$_v$1.3 may have a prolactin-regulated role in transcellular Ca transport during lactation [72] and contributes to Ca absorption prior to the development of vitamin-D-regulated Ca absorption in the neonatal mouse [73].

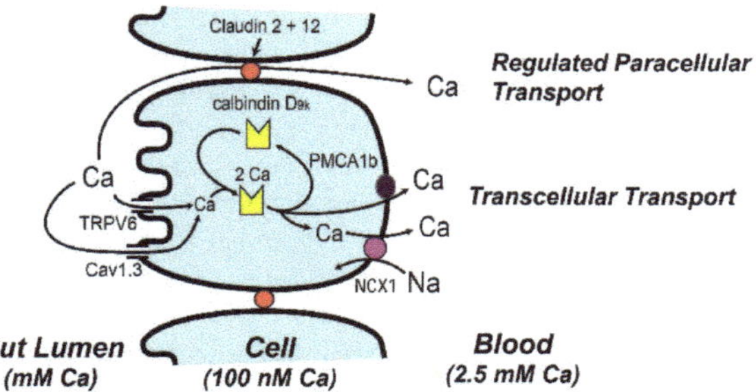

Figure 3. A Model describing intestinal calcium (Ca) absorption. The transcellular absorption pathway is described by the facilitated diffusion model while a regulated paracellular transport mechanisms mediated by claudin 2 and claudin 12 provides selectivity for Ca movement through the tight junction complex. For details of how vitamin D regulates various aspects of these models refer to the text.

The central player in the facilitated diffusion model is calbindin-D, a cytoplasmic Ca binding protein [59] found in intestine (the 9 kd form, calbindin D_{9k}) and the kidney (the 28 kd form, calbindin D_{28k}) [74] (Figures 2 and 3). This was based on studies that show: (a) intestinal calbindin D_{9k} protein levels are positively correlated to Ca absorption [59], (b) intestinal calbindin D_{9k} levels are significantly reduced in vitamin D deficient animals and in VDR knockout mice [48,75], (c) 1,25$(OH)_2$ D injections increase intestinal calbindin D_{9k} levels [76], and (d) theophylline-mediated inhibition of Ca binding to calbindin D_{9k} disrupts intestinal Ca absorption [77]. These observations led to the hypothesis that calbindin D acts as a ferry for intracellular Ca movement during Ca absorption [57,78]. In contrast, other studies indicate that calbindin D_{9k} is not essential for intestinal Ca absorption but may instead act as an intracellular Ca buffer that protects cells from increases in intracellular Ca during Ca absorption. For example, neither basal nor 1,25$(OH)_2$ D-induced Ca absorption are reduced in calbindin-D_{9k} null mice [66,79]. In contrast, 1,25$(OH)_2$ D-induced Ca absorption is reduced by 60% in calbindin-D_{9k}/TRPV6 double knockout mice [66], suggesting the interaction of TRPV6 and calbindin-D_{9k} has a special role in Ca absorption. Another observation that suggests elevated calbindin levels alone are not sufficient to drive intestinal Ca absorption is that calbindin-D protein remains elevated in the intestine even after 1,25$(OH)_2$ D-induced Ca absorption returns to normal in chicks [80] and mice [33]. Finally, we have observed that intestinal calbindin-D_{9k} levels increase in intestine-specific TRPV6 transgenic mice with elevated intestinal Ca absorption efficiency even in VDR knockout mice [68]. This suggests that calbindin-D_{9k} is an intracellular Ca buffer that increases in response to elevated transcellular Ca absorption and that it is not induced to act as a facilitator of transcellular Ca movement.

The final step in the facilitated diffusion model is the extrusion of Ca from the cell. This is an energy dependent process [22] mediated by the plasma membrane Ca ATPase 1b (PMCA1b) [81,82] (Figures 2 and 3). Deletion of the PMCA1b gene (Atp2b1), or the 4.1R protein that stabilizes PMCA1b in the basolateral membrane, reduces basal and 1,25$(OH)_2$ D-induced intestinal Ca absorption [83,84]. While some suggest that the basolateral extrusion of Ca is also be mediated by a sodium-Ca exchanger [85], disrupting the sodium gradient necessary for sodium-Ca exchange did not block duodenal Ca transport [22].

Most intestinal Ca absorption research has focused on vitamin-D-regulated saturable Ca transport but several studies have shown that vitamin D signaling can increase Ca diffusion across the jejunum and ileum [20,86]. Tudpor et al. [87] found that 1,25$(OH)_2$ D induced Ca ion movement across the intestinal barrier by a solvent drag mechanism that may involve charge selectivity of the tight junction. This is similar to the role of paracellin 1 (aka claudin 16), a tight junction protein that regulates ion-specific movement of magnesium and Ca in the kidney [88]. Consistent with this idea, 1,25$(OH)_2$ D treatment significantly increased claudin-2 and 12 mRNA levels in Caco-2 cells and siRNA against these claudins reduced Ca permeability across Caco-2 cell monolayers [89]. In vivo, claudin 2/claudin 12 double knockout mice (but not single KO mice) have reduced Ca absorption across the colon but not small intestine [90]. Claudin-2 and -12 expression is highest in the distal small intestine [91] but essentially absent from the duodenum. A role for claudins in paracellular Ca transport may explain why the non-saturable component of ileal Ca absorption was reduced in chronic kidney disease patients with low serum 1,25$(OH)_2$ D levels [20].

Two other models have been presented to explain vitamin D-mediated, transcellular intestinal Ca absorption: vesicular transport and transcaltachia. The vesicular transport model is an alternative to the role proposed for calbindin-D as a Ca ferry during transcellular intestinal Ca absorption. This is based on the observation that 1,25$(OH)_2$ D treatment increased the activity and cycling of lysosomes [92,93], that Ca accumulates within brush border membrane endosomes [94] and in lysosomes [95] during Ca absorption, and that disrupting lysosomal pH prevents lysosomal Ca accumulation and blocks Ca absorption [95,96]. While these data suggest that vesicular movement may be a legitimate pathway for uptake and movement of Ca through the intestinal epithelial cell, it is not clear what makes the vesicular transport pathway specific for Ca. Transcaltachia is a mode of Ca transport that occurs within minutes of exposing the basolateral side of enterocytes to physiologic levels of 1,25$(OH)_2$ D. Transcaltachia has been directly demonstrated in the perfused chick duodenum [97]. Some data suggests transcaltachia results from 1,25$(OH)_2$ D binding to a unique, alternative ligand binding pocket [98,99] in VDR within caveolae [100], i.e., a novel non-nuclear role for the receptor. Other data suggest the basolateral membrane protein mediating transcaltachia is a multi-functional Membrane Associated Rapid Response Steroid receptor (MARRS). However, while intestine-specific deletion of MARRS in mice reduced cellular 1,25$(OH)_2$ D binding, disrupted 1,25$(OH)_2$ D regulated Ca and phosphate uptake into enterocytes [101,102] and reduced basal Ca absorption in by 30% [103], these reports have not reported the physiological impact of MARRS deletion on bone density. Additionally, the rapid fluxes in serum 1,25$(OH)_2$ D needed for transcaltachia have not been reported during the consumption of Ca-rich meals when transcaltachia would have to occur for the physiologic benefit of the process to be realized. As such, transcaltachia is not a generally accepted mechanism for vitamin-D-regulated intestinal Ca absorption.

5. Physiologic Regulation of Vitamin D-Mediated Intestinal Ca Absorption

As it has been described above, the major physiologic condition where vitamin D signaling is engaged to regulate intestinal Ca absorption is the habitual consumption of a low Ca diet. However, there are a number of other physiologic states that affect vitamin D metabolism or action to influence intestinal Ca absorption, i.e., growth and development, pregnancy/lactation, and aging.

The bulk of mechanistic studies on intestinal Ca absorption have been conducted in growing 2–3-month-old rodents, but recent studies in mice indicate that vitamin D-mediated Ca absorption is not important prior to weaning [73]. This is not completely surprising as VDR is not expressed prior to 14d postnatally in rodents [104,105]. Ca absorption studies in premature infants also suggests that Ca absorption during late development is vitamin D independent [106]. Research suggests that during childhood and adolescence, growth hormone (GH) and its physiologic mediator insulin-like growth factor I (IGF-1) promote intestinal Ca absorption in two ways. The first effect is through activation of renal CYP27B1 and the elevation of serum 1,25$(OH)_2$ D levels [107]. However, in adult animals, GH treatment increases intestinal Ca absorption without significantly increasing serum 1,25$(OH)_2$ D levels [108]. It does so by modulating intestinal VDR levels and increasing cell sensitivity to 1,25$(OH)_2$ D [109]. The effect of GH on Ca absorption is likely mediated through IGF-1 but the intestinal actions of IGF-1 may also be independent of vitamin D signaling [49,110].

Dietary Ca requirements increase significantly during the third trimester of pregnancy and during lactation to meet the needs of the fetus and term infant. While pregnancy causes a vitamin D-independent increase in Ca absorption whose mechanism is not clearly understood [111–114], during late pregnancy serum 1,25(OH)$_2$ D levels and intestinal Ca absorption are both elevated [115]. This is because of PTH-independent 1,25(OH)$_2$ D production by the placenta [116]. Ca absorption is also regulated during lactation in rodents (but not humans) but this is due to a prolactin-dependent mechanism [117,118]. However, prolactin cooperates with 1,25(OH)$_2$D$_3$ to regulate intestinal Ca transport and the expression of TRPV6 and calbindin-D$_{9k}$ in rats [119], suggesting prolactin acts together with 1,25(OH)$_2$D$_3$ to increase active intestinal Ca absorption.

Aging reduces Ca absorption efficiency [120–125]. Yet, despite the fact that age-associated Ca malabsorption was discovered 50 ago, we still do not know the molecular mechanism underlying this phenomenon. Lower serum 1,25(OH)$_2$D levels in the elderly has been reported in some studies [126,127] but not others [128]. In fact, some research indicates that serum 1,25(OH)$_2$ D is higher in older subjects even though fractional Ca absorption is not changed [128–130]. Similar age-associated intestinal resistance to 1,25(OH)$_2$ D signaling has been formally demonstrated in rats [124] and humans [125]. Some evidence suggests that this phenomenon may be caused by lower intestinal VDR levels [130–132] but after adulthood is reached, age-related declines in intestinal VDR content are modest (−20%) [130,131] or non-existent [124]. Consistent with the lack of an impact of age on VDR expression, we recently reported that the open chromatin regions that control the expression of the intestinal VDR gene are not different between 3- and 21-mo-old mice, [104]. Thus, while my research group [53] has shown that a 50% reduction in intestinal VDR level blunts the intestinal response to elevated serum 1,25(OH)$_2$D levels, the inconsistency in the reports on the impact of age on intestinal VDR levels suggests that other mechanisms may contribute to age-associated intestinal resistance to vitamin D.

Another aspect of aging that could negatively impact vitamin D metabolism or intestinal regulation of Ca absorption is the decline in sex hormone levels. Consistent with this, estrogen loss severely disrupts Ca metabolism in post-menopausal women, including reducing Ca absorption [133,134]. While estrogen signaling directly regulates intestinal Ca absorption [135–137], it also enhances the intestinal responsiveness to 1,25(OH)$_2$ D [138]. Some [109,139,140], but not all [141], studies report that low estrogen levels reduce intestinal VDR levels and that this is responsible for intestinal vitamin D-resistance following estrogen loss. In prepubertal boys, testosterone therapy increased intestinal Ca absorption by 61% [142] and this was accompanied higher serum IGF-1 levels that might influence vitamin D metabolism. As men age, both Ca absorption efficiency and serum levels of the sulfated form of the testosterone prohormone DHEA, dehydroepiandrosterone sulphate (DHEAS), fall significantly [143]. However, the change in Ca absorption was independent of changes in serum 1,25(OH)$_2$ D, suggesting changes in androgen signaling do not alter vitamin D metabolism. It is not known if testosterone regulates intestinal VDR levels.

6. Conclusions

A large amount of data supports the conclusion that adequate vitamin D status and adequate production of the metabolite 1,25(OH)$_2$ D are needed to support transcellular, saturable intestinal Ca absorption. 1,25(OH)$_2$ D regulates intestinal biology by activating the VDR to stimulate gene expression. This increases the maximum capacity of the Ca transport system by increasing levels of a transporter that mediates saturable, transcellular Ca transport. The saturable, vitamin-D-regulated component of intestinal Ca absorption plays a significant role in maintaining Ca absorption efficiency because in most individuals, serum 1,25(OH)$_2$ D levels are elevated by habitually low dietary Ca intake that is common in the general population. The exact mechanism that describes Ca movement through the enterocyte is still in question. The facilitated diffusion model is the best characterized mechanism but there are some inconsistencies in the model that must be resolved through additional research. In contrast, research supports a model for regulated paracellular Ca movement through tight junctions that predominates across the ileum and in the early postnatal period. Overall, the data suggest that vitamin D signaling regulates intestinal Ca absorption by different mechanisms that are segment specific.

Funding: This research received no external funding.

Conflicts of Interest: The authors have no conflict to declare.

References

1. Nicolaysen, R. Studies upon the mode of action of vitamin D. III. the influence of vitamin D on the absorption of calcium and phosphorus in the rat. *Biochem. J.* **1937**, *37*, 122–129. [CrossRef] [PubMed]
2. Pansu, D.; Bellaton, C.; Roche, C.; Bronner, F. Duodenal and ileal calcium absorption in the rat and effects of vitamin D. *Am. J. Physiol.* **1983**, *244*, G695–G700. [CrossRef] [PubMed]
3. Sheikh, M.S.; Ramirez, A.; Emmett, M.; Santa, A.C.; Schiller, L.R.; Fordtran, J.S. Role of vitamin D-dependent and vitamin D-independent mechanisms in absorption of food calcium. *J. Clin. Investig.* **1988**, *81*, 126–132. [CrossRef] [PubMed]
4. Holick, M.F.; Schnoes, H.K.; DeLuca, H.F.; Suda, T.; Cousins, R.J. Isolation and identification of 1,25-dihydroxycholecalciferol. A metabolite of vitamin D active in intestine. *Biochemistry* **1971**, *10*, 2799–2804.
5. Norman, A.W.; Myrtle, J.F.; Midgett, R.J.; Nowicki, H.G.; Williams, V.; Popjak, G. 1,25-dihydroxycholecalciferol: Identification of the proposed active form of vitamin D3 in the intestine. *Science* **1971**, *173*, 51–54. [CrossRef]
6. Brumbaugh, P.F.; Haussler, M.R. Nuclear and cytoplasmic receptors for 1,25-dihydroxycholecalciferol in intestinal mucosa. *Biochem. Biophys. Res. Commun.* **1973**, *51*, 74–80. [CrossRef]
7. Haussler, M.R.; Whitfield, G.K.; Haussler, C.A.; Hsieh, J.C.; Thompson, P.D.; Selznick, S.H.; Dominguez, C.E.; Jurutka, P.W. The nuclear vitamin D receptor: Biological and molecular regulatory properties revealed. *J. Bone Miner. Res.* **1998**, *13*, 325–349. [CrossRef]
8. Xue, Y.B.; Fleet, J.C. Intestinal Vitamin D Receptor Is Required for Normal Calcium and Bone Metabolism in Mice. *Gastroenterology* **2009**, *136*, 1317–1327. [CrossRef]
9. Aita, R.A.D.; Hassan, S.; Hur, J.; Pellon-Cardenas, O.; Cohen, E.; Chen, L.; Shroyer, N.; Christakos, S.; Verzi, M.; Fleet, J.C. Genomic Analysis of 1,25-dihydroxyvitamin D3 Action in Mouse Intestine Reveals Compartment and Segment-Specific Gene Regulatory Effects. *J. Biol. Chem.* **2022**, *298*, 102213. [CrossRef]
10. Bronner, F.; Aubert, J.P. Bone metabolism and regulation of the blood calcium level in rats. *Am. J. Physiol.* **1965**, *209*, 887–890. [CrossRef]
11. Pansu, D.; Bellaton, C.; Bronner, F. Effect of Ca intake on saturable and nonsaturable components of duodenal Ca transport. *Am. J. Physiol.* **1981**, *240*, 32–37. [CrossRef] [PubMed]
12. Norman, D.A.; Fordtran, J.S.; Brinkley, L.J.; Zerwekh, J.E.; Nicar, M.J.; Strowig, S.M.; Pak, C.Y. Jejunal and ileal adaptation to alterations in dietary calcium: Changes in calcium and magnesium absorption and pathogenetic role of parathyroid hormone and 1,25-dihydroxyvitamin D. *J. Clin. Investig.* **1981**, *67*, 1599–1603. [CrossRef] [PubMed]
13. Favus, M.J.; Walling, M.W.; Kimberg, D.V. Effects of dietary calcium restriction and chronic thyroparathyroidectomy on the metabolism of (3H)25-hydroxyvitamin D3 and the active transport of calcium by rat intestine. *J. Clin. Investig.* **1974**, *53*, 1139–1148. [CrossRef]
14. Zierold, C.; Mings, J.A.; DeLuca, H.F. Regulation of 25-hydroxyvitamin D3-24-hydroxylase mRNA by 1,25-dihydroxyvitamin D3 and parathyroid hormone. *J. Cell. Biochem.* **2003**, *88*, 234–237. [CrossRef]
15. Armbrecht, H.J.; Boltz, M.A.; Hodam, T.L. PTH increases renal 25(OH)D3-1alpha -hydroxylase (CYP1alpha) mRNA but not renal 1,25(OH)2D3 production in adult rats. *Am. J. Physiol. Ren. Physiol.* **2003**, *284*, F1032–F1036. [CrossRef] [PubMed]
16. Gershoff, S.N.; Hegsted, D.M. Effect of vitamin D and Ca:P ratios on chick gastrointestinal tract. *Am. J. Physiol.* **1956**, *187*, 203–206. [CrossRef] [PubMed]
17. Need, A.G.; O'Loughlin, P.D.; Morris, H.A.; Coates, P.S.; Horowitz, M.; Nordin, B.E. Vitamin D Metabolites and Calcium Absorption in Severe Vitamin D Deficiency. *J. Bone Miner. Res.* **2008**, *23*, 1859–1863. [CrossRef] [PubMed]
18. Wasserman, R.H.; Taylor, A.N. Some aspects of the intestinal absorption of calcium, with special reference to vitamin D. In *Mineral Metabolism, An Advanced Treatise*; Comar, C.L., Bronner, F., Eds.; Academic Press: New York, NY, USA, 1969; Volume 3, pp. 321–403.
19. Heaney, R.P.; Saville, P.D.; Recker, R.R. Calcium absorption as a function of calcium intake. *J. Lab. Clin. Med.* **1975**, *85*, 881–890.
20. Sheikh, M.S.; Schiller, L.R.; Fordtran, J.S. In vivo intestinal absorption of calcium in humans. *Miner. Electrolyte Metab.* **1990**, *16*, 130–146.
21. Wallace, T.C.; Reider, C.; Fulgoni, V.L., 3rd. Calcium and vitamin D disparities are related to gender, age, race, household income level, and weight classification but not vegetarian status in the United States: Analysis of the NHANES 2001–2008 data set. *J. Am. Coll. Nutr.* **2013**, *32*, 321–330. [CrossRef]
22. Favus, M.J.; Angeid-Backman, E.; Breyer, M.D.; Coe, F.L. Effects of trifluoperazine, ouabain, and ethacrynic acid on intestinal calcium. *Am. J. Physiol.* **1983**, *244*, G111–G115. [CrossRef]
23. Favus, M.J.; Kathpalia, S.C.; Coe, F.L. Kinetic characteristics of calcium absorption and secretion by rat colon. *Am. J. Physiol.* **1981**, *240*, G350–G354. [CrossRef] [PubMed]

24. Grinstead, W.C.; Pak, C.Y.C.; Krejs, G.J. Effect of 1,25-Dihydroxyvitamin-D3 on Calcium-Absorption in the Colon of Healthy Humans. *Am. J. Physiol.* **1984**, *247*, G189–G192. [CrossRef] [PubMed]
25. Karbach, U.; Feldmeier, H. The cecum is the site with the highest calcium absorption in rat intestine. *Dig. Dis. Sci.* **1993**, *38*, 1815–1824. [CrossRef] [PubMed]
26. Dhawan, P.; Veldurthy, V.; Yehia, G.; Hsaio, C.; Porta, A.; Kim, K.I.; Patel, N.; Lieben, L.; Verlinden, L.; Carmeliet, G.; et al. Transgenic Expression of the Vitamin D Receptor Restricted to the Ileum, Cecum, and Colon of Vitamin D Receptor Knockout Mice Rescues Vitamin D Receptor-Dependent Rickets. *Endocrinology* **2017**, *158*, 3792–3804. [CrossRef]
27. Jiang, H.; Horst, R.L.; Koszewski, N.J.; Goff, J.P.; Christakos, S.; Fleet, J.C. Targeting 1,25(OH)2D-mediated calcium absorption machinery in proximal colon with calcitriol glycosides and glucuronides. *J. Steroid Biochem. Mol. Biol.* **2020**, *198*, 105574. [CrossRef]
28. Giuliano, A.R.; Wood, R.J. Vitamin D-regulated calcium transport in Caco-2 cells: Unique in vitro model. *Am. J. Physiol.* **1991**, *260*, G207–G212. [CrossRef]
29. Fordtran, J.S.; Locklear, T.W. Ionic constituents and osmolality of gastric and small-intestinal fluids after eating. *Am. J. Dig. Dis.* **1966**, *11*, 503–521. [CrossRef]
30. Marcus, C.S.; Lengemann, F.W. Absorption of Ca 45 and Sr 85 from solid and liquid food at various levels of the alimentary tract of the rat. *J. Nut.* **1962**, *77*, 155–160. [CrossRef]
31. Duflos, C.; Bellaton, C.; Pansu, D.; Bronner, F. Calcium solubility, intestinal sojourn time and paracellular permeability codetermine passive calcium absorption in rats. *J. Nutr.* **1995**, *125*, 2348–2355. [CrossRef]
32. Boyle, I.T.; Gray, R.W.; Omdahl, J.L.; DeLuca, H.F.; Schilling, R.F. The mechanism of adaptation of intestinal calcium absorption to low dietary calcium. *J. Lab. Clin. Med.* **1971**, *78*, 813.
33. Song, Y.; Peng, X.; Porta, A.; Takanaga, H.; Peng, J.B.; Hediger, M.A.; Fleet, J.C.; Christakos, S. Calcium transporter 1 and epithelial calcium channel messenger ribonucleic acid are differentially regulated by 1,25 dihydroxyvitamin D3 in the intestine and kidney of mice. *Endocrinology* **2003**, *144*, 3885–3894. [CrossRef]
34. Abrams, S.A.; Hicks, P.D.; Hawthorne, K.M. Higher serum 25-hydroxyvitamin D levels in school-age children are inconsistently associated with increased calcium absorption. *J. Clin. Endocrinol. Metab.* **2009**, *94*, 2421–2427. [CrossRef] [PubMed]
35. Abrams, S.A.; Hawthorne, K.M.; Chen, Z. Supplementation with 1000 IU vitamin D/d leads to parathyroid hormone suppression, but not increased fractional calcium absorption, in 4-8-y-old children: A double-blind randomized controlled trial. *Am. J. Clin. Nutr.* **2013**, *97*, 217–223. [CrossRef] [PubMed]
36. Lewis, R.D.; Laing, E.M.; Hill Gallant, K.M.; Hall, D.B.; McCabe, G.P.; Hausman, D.B.; Martin, B.R.; Warden, S.J.; Peacock, M.; Weaver, C.M. A randomized trial of vitamin D(3) supplementation in children: Dose-response effects on vitamin D metabolites and calcium absorption. *J. Clin. Endocrinol. Metab.* **2013**, *98*, 4816–4825. [CrossRef] [PubMed]
37. Hansen, K.E.; Johnson, R.E.; Chambers, K.R.; Johnson, M.G.; Lemon, C.C.; Vo, T.N.; Marvdashti, S. Treatment of Vitamin D Insufficiency in Postmenopausal Women: A Randomized Clinical Trial. *JAMA Intern. Med.* **2015**, *175*, 1612–1621. [CrossRef] [PubMed]
38. Bronner, F.; Pansu, D. Nutritional aspects of calcium absorption. *J. Nutr.* **1999**, *129*, 9–12. [CrossRef]
39. Sempos, C.T.; Durazo-Arvizu, R.A.; Fischer, P.R.; Munns, C.F.; Pettifor, J.M.; Thacher, T.D. Serum 25-hydroxyvitamin D requirements to prevent nutritional rickets in Nigerian children on a low-calcium diet-a multivariable reanalysis. *Am. J. Clin. Nutr.* **2021**, *114*, 231–237. [CrossRef]
40. Fischer, P.R.; Almasri, N.I. Nutritional rickets-Vitamin D and beyond. *J. Steroid Biochem. Mol. Biol.* **2022**, *219*, 106070. [CrossRef]
41. Matkovic, V.; Fontana, D.; Tominac, C.; Goel, P.; Chesnut, C.H. Factors that influence peak bone mass formation: A study of calcium balance and the inheritance of bone mass in adolescent females. *Am. J. Clin. Nutr.* **1990**, *52*, 878–888. [CrossRef]
42. Replogle, R.A.; Li, Q.; Wang, L.; Zhang, M.; Fleet, J.C. Gene-by-Diet Interactions Influence Calcium Absorption and Bone Density in Mice. *J. Bone Miner. Res.* **2014**, *29*, 657–665. [CrossRef] [PubMed]
43. Patel, M.B.; Makepeace, A.E.; Jameson, K.A.; Masterson, L.M.; Holt, R.I.; Swaminathan, R.; Javaid, M.K.; Cooper, C.; Arden, N.K. Weight in infancy and adult calcium absorption as determinants of bone mineral density in adult men: The hertfordshire cohort study. *Calcif. Tissue Int.* **2012**, *91*, 416–422. [CrossRef] [PubMed]
44. Reyes Fernandez, P.C.; Replogle, R.A.; Wang, L.; Zhang, M.; Fleet, J.C. Novel Genetic Loci Control Calcium Absorption and Femur Bone Mass as Well as Their Response to Low Calcium Intake in Male BXD Recombinant Inbred Mice. *J. Bone Miner. Res.* **2016**, *31*, 994–1002. [CrossRef] [PubMed]
45. Ensrud, K.E.; Duong, T.; Cauley, J.A.; Heaney, R.P.; Wolf, R.L.; Harris, E.; Cummings, S.R. Low fractional calcium absorption increases the risk for hip fracture in women with low calcium intake. Study of Osteoporotic Fractures Research Group. *Ann. Intern. Med.* **2000**, *132*, 345–353. [CrossRef]
46. Haussler, M.R.; Norman, A.W. Chromosomal receptor for a vitamin D metabolite. *Proc. Natl. Acad. Sci. USA* **1969**, *62*, 155–162. [CrossRef] [PubMed]
47. Tiosano, D.; Hadad, S.; Chen, Z.; Nemirovsky, A.; Gepstein, V.; Militianu, D.; Weisman, Y.; Abrams, S.A. Calcium absorption, kinetics, bone density, and bone structure in patients with hereditary vitamin D-resistant rickets. *J. Clin. Endocrinol. Metab.* **2011**, *96*, 3701–3709. [CrossRef]
48. Van Cromphaut, S.J.; Dewerchin, M.; Hoenderop, J.G.; Stockmans, I.; Van Herck, E.; Kato, S.; Bindels, R.J.; Collen, D.; Carmeliet, P.; Bouillon, R.; et al. Duodenal calcium absorption in vitamin D receptor-knockout mice: Functional and molecular aspects. *Proc. Natl. Acad. Sci. USA* **2001**, *98*, 13324–13329. [CrossRef] [PubMed]

49. Song, Y.; Kato, S.; Fleet, J.C. Vitamin D Receptor (VDR) Knockout Mice Reveal VDR-Independent Regulation of Intestinal Calcium Absorption and ECaC2 and Calbindin D9k mRNA. *J. Nutr.* **2003**, *133*, 374–380. [CrossRef]
50. Lieben, L.; Masuyama, R.; Torrekens, S.; Van Looveren, R.; Schrooten, J.; Baatsen, P.; Lafage-Proust, M.H.; Dresselaers, T.; Feng, J.Q.; Bonewald, L.F.; et al. Normocalcemia is maintained in mice under conditions of calcium malabsorption by vitamin D-induced inhibition of bone mineralization. *J. Clin. Investig.* **2012**, *122*, 1803–1815. [CrossRef]
51. Amling, M.; Priemel, M.; Holzmann, T.; Chapin, K.; Rueger, J.M.; Baron, R.; Demay, M.B. Rescue of the skeletal phenotype of vitamin D receptor ablated mice in the setting of normal mineral ion homeostasis: Formal histomorphometric and biomechanical analysis. *Endocrinology* **1999**, *140*, 4982–4987. [CrossRef]
52. Lieben, L.; Verlinden, L.; Masuyama, R.; Torrekens, S.; Moermans, K.; Schoonjans, L.; Carmeliet, P.; Carmeliet, G. Extra-intestinal calcium handling contributes to normal serum calcium levels when intestinal calcium absorption is suboptimal. *Bone* **2015**, *81*, 502–512. [CrossRef]
53. Song, Y.; Fleet, J.C. Intestinal Resistance to 1,25 Dihydroxyvitamin D in Mice Heterozygous for the Vitamin D Receptor Knockout Allele. *Endocrinology* **2007**, *148*, 1396–1402. [CrossRef] [PubMed]
54. Ames, S.K.; Ellis, K.J.; Gunn, S.K.; Copeland, K.C.; Abrams, S.A. Vitamin D receptor gene Fok1 polymorphisms predicts calcium absorption and bone mineral density in children. *J. Bone Miner. Res.* **1999**, *14*, 740–746. [CrossRef]
55. Jurutka, P.W.; Remus, L.S.; Whitfield, K.; Thompson, P.D.; Hsieh, J.C.; Zitzer, H.; Tavakkoli, P.; Galligan, M.A.; Dang, H.T.L.; Haussler, C.A.; et al. The polymorphic N terminus in human vitamin D receptor isoforms influences trascriptional activity by modulating interaction with transcription factor IIB. *Mol. Endocrinol.* **2000**, *14*, 401–420. [CrossRef]
56. Huang, Z.W.; Dong, J.; Piao, J.H.; Li, W.D.; Tian, Y.; Xu, J.; Yang, X.G. Relationship between the absorption of dietary calcium and the Fok I polymorphism of VDR gene in young women. *Zhonghua Yu Fang Yi Xue Za Zhi* **2006**, *40*, 75–78. [PubMed]
57. Chandra, S.; Fullmer, C.S.; Smith, C.A.; Wasserman, R.H.; Morrison, G.H. Ion microscopic imaging of calcium transport in the intestinal tissue of vitamin D-deficient and vitamin D-replete chickens: A 44Ca stable isotope study. *Proc. Natl. Acad. Sci. USA* **1990**, *87*, 5715–5719. [CrossRef] [PubMed]
58. Fullmer, C.S.; Chandra, S.; Smith, C.A.; Morrison, G.H.; Wasserman, R.H. Ion microscopic imaging of calcium during 1,25-dihydroxyvitamin D-mediated intestinal absorption. *Histochem. Cell. Biol.* **1996**, *106*, 215–222. [CrossRef]
59. Bronner, F.; Pansu, D.; Stein, W.D. An analysis of intestinal calcium transport across the rat intestine. *Am. J. Physiol.* **1986**, *250*, G561–G569. [CrossRef]
60. Peng, J.B.; Chen, X.Z.; Berger, U.V.; Vassilev, P.M.; Tsukaguchi, H.; Brown, E.M.; Hediger, M.A. Molecular cloning and characterization of a channel-like transporter mediated intestinal calcium absorption. *J. Biol. Chem.* **1999**, *274*, 22739–22746. [CrossRef]
61. Meyer, M.B.; Zella, L.A.; Nerenz, R.D.; Pike, J.W. Characterizing early events associated with the activation of target genes by 1,25-dihydroxyvitamin D3 in mouse kidney and intestine in vivo. *J. Biol. Chem.* **2007**, *282*, 22344–22352. [CrossRef]
62. Fleet, J.C.; Eksir, F.; Hance, K.W.; Wood, R.J. Vitamin D-inducible calcium transport and gene expression in three Caco-2 cell lines. *Am. J. Physiol.* **2002**, *283*, G618–G625. [CrossRef] [PubMed]
63. Lee, S.M.; Riley, E.M.; Meyer, M.B.; Benkusky, N.A.; Plum, L.A.; DeLuca, H.F.; Pike, J.W. 1,25-Dihydroxyvitamin D3 Controls a Cohort of Vitamin D Receptor Target Genes in the Proximal Intestine That Is Enriched for Calcium-regulating Components. *J. Biol. Chem.* **2015**, *290*, 18199–18215. [CrossRef] [PubMed]
64. Meyer, M.B.; Watanuki, M.; Kim, S.; Shevde, N.K.; Pike, J.W. The human transient receptor potential vanilloid type 6 distal promoter contains multiple vitamin D receptor binding sites that mediate activation by 1,25-dihydroxyvitamin D3 in intestinal cells. *Mol. Endocrinol.* **2006**, *20*, 1447–1461. [CrossRef] [PubMed]
65. Kutuzova, G.D.; Sundersingh, F.; Vaughan, J.; Tadi, B.P.; Ansay, S.E.; Christakos, S.; DeLuca, H.F. TRPV6 is not required for 1alpha,25-dihydroxyvitamin D3-induced intestinal calcium absorption in vivo. *Proc. Natl. Acad. Sci. USA* **2008**, *105*, 19655–19659. [CrossRef] [PubMed]
66. Benn, B.S.; Ajibade, D.; Porta, A.; Dhawan, P.; Hediger, M.; Peng, J.B.; Jiang, Y.; Oh, G.T.; Jeung, E.B.; Lieben, L.; et al. Active intestinal calcium transport in the absence of transient receptor potential vanilloid type 6 and calbindin-D9k. *Endocrinology* **2008**, *149*, 3196–3205. [CrossRef]
67. Woudenberg-Vrenken, T.E.; Lameris, A.L.; Weissgerber, P.; Olausson, J.; Flockerzi, V.; Bindels, R.J.; Freichel, M.; Hoenderop, J.G. Functional TRPV6 channels are crucial for transepithelial Ca2+ absorption. *Am. J. Physiol. Gastrointest. Liver Physiol.* **2012**, *303*, G879–G885. [CrossRef]
68. Cui, M.; Li, Q.; Johnson, R.; Fleet, J.C. Villin promoter-mediated transgenic expression of transient receptor potential cation channel, subfamily V, member 6 (TRPV6) increases intestinal calcium absorption in wild-type and vitamin D receptor knockout mice. *J. Bone Miner. Res.* **2012**, *27*, 2097–2107. [CrossRef]
69. Kellett, G.L. Alternative perspective on intestinal calcium absorption: Proposed complementary actions of Ca(v)1.3 and TRPV6. *Nutr. Rev.* **2011**, *69*, 347–370. [CrossRef]
70. Reyes-Fernandez, P.C.; Fleet, J.C. Luminal glucose does not enhance active intestinal calcium absorption in mice: Evidence against a role for Ca(v)1.3 as a mediator of calcium uptake during absorption. *Nutr. Res.* **2015**, *35*, 1009–1015. [CrossRef]
71. Li, J.; Zhao, L.; Ferries, I.K.; Jiang, L.; Desta, M.Z.; Yu, X.; Yang, Z.; Duncan, R.L.; Turner, C.H. Skeletal phenotype of mice with a null mutation in Cav 1.3 L-type calcium channel. *J. Musculoskelet. Neuronal Interact.* **2010**, *10*, 180–187.

72. Nakkrasae, L.I.; Thongon, N.; Thongbunchoo, J.; Krishnamra, N.; Charoenphandhu, N. Transepithelial calcium transport in prolactin-exposed intestine-like Caco-2 monolayer after combinatorial knockdown of TRPV5, TRPV6 and Ca(v)1.3. *J. Physiol. Sci.* **2010**, *60*, 9–17. [CrossRef]
73. Beggs, M.R.; Lee, J.J.; Busch, K.; Raza, A.; Dimke, H.; Weissgerber, P.; Engel, J.; Flockerzi, V.; Alexander, R.T. TRPV6 and Cav1.3 Mediate Distal Small Intestine Calcium Absorption Before Weaning. *Cell. Mol. Gastroenterol. Hepatol.* **2019**, *8*, 625–642. [CrossRef] [PubMed]
74. Christakos, S.; Gill, R.; Lee, S.; Li, H. Molecular aspects of the calbindins. *J. Nutr.* **1992**, *122*, 678–682. [CrossRef]
75. Wasserman, R.H.; Taylor, A.N. Vitamin D3-induced calcium-binding proteins in chick intestinal mucosa. *Science* **1966**, *252*, 791–793. [CrossRef]
76. Bronner, F.; Buckley, M. The molecular nature of 1,25-(OH)2-D3-induced calcium-binding protein biosynthesis in the rat. *Adv. Exp. Med. Biol.* **1982**, *151*, 355–360. [PubMed]
77. Pansu, D.; Bellaton, C.; Roche, C.; Bronner, F. Theophylline inhibits active Ca transport in rat intestine by inhibiting Ca binding by CaBP. *Prog. Clin. Biol. Res.* **1988**, *252*, 115–120.
78. Feher, J.J.; Fullmer, C.S.; Wasserman, R.H. Role of facilitated diffusion of calcium by calbindin in intestinal calcium absorption. *Am. J. Physiol.* **1992**, *262*, C517–C526. [CrossRef] [PubMed]
79. Akhter, S.; Kutuzova, G.D.; Christakos, S.; DeLuca, H.F. Calbindin D9k is not required for 1,25-dihydroxyvitamin D3-mediated Ca2+ absorption in small intestine. *Arch. Biochem. Biophys.* **2007**, *460*, 227–232. [CrossRef]
80. Spencer, R.; Charman, M.; Wilson, P.W.; Lawson, D.E.M. The relationship between vitamin D-stimulated calcium transport and intestinal calcium-binding protein in the chicken. *Biochem. J.* **1978**, *170*, 93–101. [CrossRef]
81. Wasserman, R.H.; Smith, C.A.; Brindak, M.E.; Detalamoni, N.; Fullmer, C.S.; Penniston, J.T.; Kumar, R. Vitamin-D and mineral deficiencies increase the plasma membrane calcium pump of chicken intestine. *Gastroenterology* **1992**, *102*, 886–894. [CrossRef]
82. Cai, Q.; Chandler, J.S.; Wasserman, R.H.; Kumar, R.; Penniston, J.T. Vitamin D and adaptation to dietary calcium and phosphate deficiencies increase intestinal plasma membrane calcium pump gene expression. *Proc. Natl. Acad. Sci. USA* **1993**, *90*, 1345–1349. [CrossRef]
83. Liu, C.; Weng, H.; Chen, L.; Yang, S.; Wang, H.; Debnath, G.; Guo, X.; Wu, L.; Mohandas, N.; An, X. Impaired intestinal calcium absorption in protein 4.1R-deficient mice due to altered expression of plasma membrane calcium ATPase 1b (PMCA1b). *J. Biol. Chem.* **2013**, *288*, 11407–11415. [CrossRef] [PubMed]
84. Ryan, Z.C.; Craig, T.A.; Filoteo, A.G.; Westendorf, J.J.; Cartwright, E.J.; Neyses, L.; Strehler, E.E.; Kumar, R. Deletion of the intestinal plasma membrane calcium pump, isoform 1, Atp2b1, in mice is associated with decreased bone mineral density and impaired responsiveness to 1, 25-dihydroxyvitamin D3. *Biochem. Biophys. Res. Commun.* **2015**, *467*, 152–156. [CrossRef] [PubMed]
85. van Corven, E.J.; Roche, C.; Van Os, C.H. Distribution of Ca2+-ATPase, ATP-dependent Ca2+-transport, calmodulin and vitamin D-dependent Ca2+-binding protein along the villus-crypt axis in rat duodenum. *Biochim. Biophys. Acta* **1985**, *820*, 274–282. [CrossRef]
86. Karbach, U. Paracellular Calcium Transport Across the Small Intestine. *J. Nutr.* **1992**, *122*, 672–677. [CrossRef] [PubMed]
87. Tudpor, K.; Teerapornpuntakit, J.; Jantarajit, W.; Krishnamra, N.; Charoenphandhu, N. 1,25-dihydroxyvitamin d(3) rapidly stimulates the solvent drag-induced paracellular calcium transport in the duodenum of female rats. *J. Physiol. Sci.* **2008**, *58*, 297–307. [CrossRef]
88. Simon, D.B.; Lu, Y.; Choate, K.A.; Velazquez, H.; Al-Sabban, E.; Praga, M.; Casari, G.; Bettinelli, A.; Colussi, G.; Rodriquez-Soriano, J.; et al. Paracellin-1, a renal tight junction protein required for paracellular Mg 2+ resorption. *Science* **1999**, *285*, 103–106. [CrossRef]
89. Fujita, H.; Sugimoto, K.; Inatomi, S.; Maeda, T.; Osanai, M.; Uchiyama, Y.; Yamamoto, Y.; Wada, T.; Kojima, T.; Yokozaki, H.; et al. Tight junction proteins claudin-2 and-12 are critical for vitamin D-dependent Ca2+ absorption between enterocytes. *Mol. Biol. Cell.* **2008**, *19*, 1912–1921. [CrossRef]
90. Beggs, M.R.; Young, K.; Pan, W.; O'Neill, D.D.; Saurette, M.; Plain, A.; Rievaj, J.; Doschak, M.R.; Cordat, E.; Dimke, H.; et al. Claudin-2 and claudin-12 form independent, complementary pores required to maintain calcium homeostasis. *Proc. Natl. Acad. Sci. USA* **2021**, *118*, e2111247118. [CrossRef]
91. Fujita, H.; Chiba, H.; Yokozaki, H.; Sakai, N.; Sugimoto, K.; Wada, T.; Kojima, T.; Yamashita, T.; Sawada, N. Differential expression and subcellular localization of claudin-7, -8, -12, -13, and -15 along the mouse intestine. *J. Histochem. Cytochem.* **2006**, *54*, 933–944. [CrossRef]
92. Davis, W.L.; Jones, R.G. Lysosomal proliferation in rachitic avian intestinal absorptive cells following 1,25-dihydroxycholecalciferol. *Tissue Cell.* **1982**, *14*, 585–595. [CrossRef]
93. Nemere, I.; Szego, C.M. Early actions of parathyroid hormone and 1,25-dihydroxycholecalciferol on isolated epithelial cells from rat intestine: 1. Limited lysosomal enzyme release and calcium uptake. *Endocrinology* **1981**, *108*, 1450–1462. [CrossRef] [PubMed]
94. Nemere, I.; Norman, A.W. 1,25-Dihydroxyvitamin D3-mediated vesicular transport of calcium in intestine: Time-course studies. *Endocrinology* **1988**, *122*, 2962–2969. [CrossRef] [PubMed]
95. Nemere, I.; Leathers, V.; Norman, A.W. 1, 25 dihydroxyvitamin D3-mediated intestinal calcium transport. Biochemical identification of lysozomes containing calcium and calcium-binding protein (calbindin-D 28k). *J. Biol. Chem.* **1986**, *261*, 16106–16114. [CrossRef]
96. Favus, M.J.; Tembe, V.; Tanklefsky, M.D.; Ambrosic, K.A.; Nellans, H.N. Effects of quinacrine on calcium active transport by rat intestinal epithelium. *Am. J. Physiol.* **1989**, *257*, G818–G822. [CrossRef]

97. Nemere, I.; Yoshimoto, Y.; Norman, A.W. Calcium transport in perfused duodena from normal chicks: Enhancement within fourteen minutes of exposure to 1,25 dihydroxyvitamin D3. *Endocrinology* **1984**, *115*, 1476–1483. [CrossRef]
98. Mizwicki, M.T.; Keidel, D.; Bula, C.M.; Bishop, J.E.; Zanello, L.P.; Wurtz, J.M.; Moras, D.; Norman, A.W. Identification of an alternative ligand-binding pocket in the nuclear vitamin D receptor and its functional importance in 1alpha,25(OH)2-vitamin D3 signaling. *Proc. Natl. Acad. Sci. USA* **2004**, *101*, 12876–12881. [CrossRef]
99. Norman, A.W.; Bishop, J.E.; Bula, C.M.; Olivera, C.J.; Mizwicki, M.T.; Zanello, L.P.; Ishida, H.; Okamura, W.H. Molecular tools for study of genomic and rapid signal transduction responses initiated by 1 alpha,25(OH)(2)-vitamin D(3). *Steroids* **2002**, *67*, 457–466. [CrossRef]
100. Huhtakangas, J.A.; Olivera, C.J.; Bishop, J.E.; Zanello, L.P.; Norman, A.W. The vitamin D receptor is present in caveolae-enriched plasma membranes and binds 1 alpha,25(OH)(2)-vitamin D-3 in vivo and in vitro. *Mol. Endocrinol.* **2004**, *18*, 2660–2671. [CrossRef]
101. Nemere, I.; Garbi, N.; Hammerling, G.J.; Khanal, R.C. Intestinal cell calcium uptake and the targeted knockout of the 1,25D3-MARRS (membrane-associated, rapid response steroid-binding) receptor/PDIA3/Erp57. *J. Biol. Chem.* **2010**, *285*, 31859–31866. [CrossRef]
102. Nemere, I.; Garcia-Garbi, N.; Hammerling, G.J.; Winger, Q. Intestinal cell phosphate uptake and the targeted knockout of the 1,25D(3)-MARRS receptor/PDIA3/ERp57. *Endocrinology* **2012**, *153*, 1609–1615. [CrossRef] [PubMed]
103. Nemere, I.; Garbi, N.; Hammerling, G.; Hintze, K.J. Role of the 1,25D(3)-MARRS receptor in the 1,25(OH)(2)D(3)-stimulated uptake of calcium and phosphate in intestinal cells. *Steroids* **2012**, *77*, 897–902. [CrossRef] [PubMed]
104. Fleet, J.C.; Aldea, D.; Chen, L.; Christakos, S.; Verzi, M. Regulatory domains controlling high intestinal vitamin D receptor gene expression are conserved in mouse and human. *J. Biol. Chem.* **2022**, *298*, 101616. [CrossRef] [PubMed]
105. Pierce, E.A.; DeLuca, H.F. Regulation of the intestinal 1,25-dihydroxyvitamin D3 receptor during neonatal development in the rat. *Arch. Biochem. Biophys.* **1988**, *261*, 241–249. [CrossRef]
106. Bronner, F.; Salle, B.L.; Putet, G.; Rigo, J.; Senterre, J. Net calcium absorption in premature infants: Results of 103 metabolic balance studies. *Am. J. Clin. Nutr.* **1992**, *56*, 1037–1044. [CrossRef]
107. Zoidis, E.; Gosteli-Peter, M.; Ghirlanda-Keller, C.; Meinel, L.; Zapf, J.; Schmid, C. IGF-I and GH stimulate Phex mRNA expression in lungs and bones and 1,25-dihydroxyvitamin D(3) production in hypophysectomized rats. *Eur. J. Endocrinol.* **2002**, *146*, 97–105. [CrossRef]
108. Fleet, J.C.; Bruns, M.E.; Hock, J.M.; Wood, R.J. Growth hormone and parathyroid hormone stimulate intestinal calcium absorption in aged female rats. *Endocrinology* **1994**, *134*, 1755–1760. [CrossRef]
109. Chen, C.; Noland, K.A.; Kalu, D.N. Modulation of intestinal vitamin D receptor by ovariectomy, estrogen and growth hormone. *Mech. Ageing Dev.* **1997**, *99*, 109–122. [CrossRef]
110. Fatayerji, D.; Mawer, E.B.; Eastell, R. The role of insulin-like growth factor I in age-related changes in calcium homeostasis in men. *J. Clin. Endocrinol. Metab.* **2000**, *85*, 4657–4662. [CrossRef]
111. Quan-Sheng, D.; Miller, S.C. Calciotrophic hormone levels and calcium absorption during pregnancy in rats. *Am. J. Physiol.* **1989**, *257*, E118–E123. [CrossRef]
112. Halloran, B.P.; DeLuca, H.F. Calcium-Transport in Small-Intestine During Pregnancy and Lactation. *Am. J. Physiol.* **1980**, *239*, E64–E68. [CrossRef] [PubMed]
113. Brommage, R.; Baxter, D.C.; Gierke, L.W. Vitamin D-independent intestinal calcium and phosphorus absorption during reproduction. *Am. J. Physiol.* **1990**, *259*, 631–638. [CrossRef] [PubMed]
114. Fudge, N.J.; Kovacs, C.S. Pregnancy up-regulates intestinal calcium absorption and skeletal mineralization independently of the vitamin D receptor. *Endocrinology* **2010**, *151*, 886–895. [CrossRef] [PubMed]
115. Ritchie, L.D.; Fung, E.B.; Halloran, B.P.; Turnlund, J.R.; Van Loan, M.D.; Cann, C.E.; King, J.C. A longitudinal study of calcium homeostasis during human pregnancy and lactation and after resumption of menses. *Am. J. Clin. Nutr.* **1998**, *67*, 693–701. [CrossRef] [PubMed]
116. Breslau, N.A.; Zerwekh, J.E. Relationship of estrogen and pregnancy to calcium homeostasis in pseudohypoparathyroidism. *J. Clin. Endocrinol. Metab.* **1986**, *62*, 45–51. [CrossRef]
117. Robinson, C.J.; Spanos, E.; James, M.F.; Pike, J.W.; Haussler, M.R.; Makeen, A.M.; Hillyard, C.J.; MacIntyre, I. Role of Prolactin in Vitamin-D Metabolism and Calcium-Absorption During Lactation in the Rat. *J. Endocrinol.* **1982**, *94*, 443–453. [CrossRef]
118. Pahuja, D.N.; DeLuca, H.F. Stimulation of Intestinal Calcium-Transport and Bone Calcium Mobilization by Prolactin in Vitamin-D-Deficient Rats. *Science* **1981**, *214*, 1038–1039. [CrossRef] [PubMed]
119. Ajibade, D.V.; Dhawan, P.; Fechner, A.J.; Meyer, M.B.; Pike, J.W.; Christakos, S. Evidence for a role of prolactin in calcium homeostasis: Regulation of intestinal transient receptor potential vanilloid type 6, intestinal calcium absorption, and the 25-hydroxyvitamin D(3) 1alpha hydroxylase gene by prolactin. *Endocrinology* **2010**, *151*, 2974–2984. [CrossRef]
120. Avioli, L.V.; McDonald, J.E.; Lee, S.W. The influence of age on the intestinal absorption of 47Ca in women and its relation to 47Ca absorption in postmenopausal osteoporosis. *J. Clin. Investig.* **1965**, *44*, 1960–1967. [CrossRef]
121. Bullamore, J.R.; Gallagher, J.C.; Wilkinson, R.; Nordin, B.E.C.; Marshall, D.H. Effect of age on calcium absorption. *Lancet* **1970**, *2*, 535–537. [CrossRef]
122. Armbrecht, H.J.; Zenser, T.V.; Bruns, M.E.; Davis, B.B. Effect of age on intestinal calcium absorption and adaptation to dietary calcium. *Am. J. Physiol.* **1979**, *236*, E769–E774. [CrossRef] [PubMed]
123. Ireland, P.; Fordtran, J.S. Effect of dietary calcium and age on jejunal calcium absorption in humans studied by intestinal perfusion. *J. Clin. Investig.* **1973**, *52*, 2672–2681. [CrossRef] [PubMed]

124. Wood, R.J.; Fleet, J.C.; Cashman, K.; Bruns, M.E.; DeLuca, H.F. Intestinal calcium absorption in the aged rat: Evidence of intestinal resistance to 1,25(OH)2 vitamin D. *Endocrinology* **1998**, *139*, 3843–3848. [CrossRef] [PubMed]
125. Pattanaungkul, S.; Riggs, B.L.; Yergey, A.L.; Vieira, N.E.; O'Fallon, W.M.; Khosla, S. Relationship of intestinal calcium absorption to 1,25-dihydroxyvitamin D [1,25(OH)2D] levels in young versus elderly women: Evidence for age- related intestinal resistance to 1,25(OH)2D action. *J. Clin. Endocrinol. Metab.* **2000**, *85*, 4023–4027. [CrossRef] [PubMed]
126. Gallagher, J.C.; Riggs, B.L.; Eisman, J.; Hamstra, A.; Arnaud, S.B.; DeLuca, H.F. Intestinal calcium absorption and serum vitamin D metabolites in normal subjects and osteoporotic patients: Effect of age and dietary calcium. *J. Clin. Investig.* **1979**, *64*, 729–736. [CrossRef]
127. Fujisawa, Y.; Kida, K.; Matsuda, H. Role of change in vitamin D metabolism with age in calcium and phosphorus metabolism in normal human subjects. *J. Clin. Endocrinol. Metab.* **1984**, *59*, 719–726. [CrossRef]
128. Scopacasa, F.; Wishart, J.M.; Horowitz, M.; Morris, H.A.; Need, A.G. Relation between calcium absorption and serum calcitriol in normal men: Evidence for age-related intestinal resistance to calcitriol. *Eur. J. Clin. Nutr.* **2004**, *58*, 264–269. [CrossRef]
129. Eastell, R.; Yergey, A.L.; Vieira, N.E.; Cedel, S.L.; Kumar, R.; Riggs, B.L. Interrelationship among vitamin-D metabolism, true calcium absorption, parathyroid function, and age in women—Evidence of an age-related intestinal resistance to 1,25-dihydroxyvitamin-D action. *J. Bone Min. Res.* **1991**, *6*, 125–132. [CrossRef]
130. Ebeling, P.R.; Sandgren, M.E.; Dimagno, E.P.; Lane, A.W.; DeLuca, H.F.; Riggs, B.L. Evidence of an age-related decrease in intestinal responsiveness to vitamin-D—Relationship between serum 1,25-dihydroxyvitamin-D3 and intestinal vitamin-D receptor concentrations in normal women. *J. Clin. Endocrinol. Metab.* **1992**, *75*, 176–182.
131. Takamoto, S.; Seino, Y.; Sacktor, B.; Liang, C.T. Effect of age on duodenal 1,25-dihydroxyvitamin D-3 receptors in Wistar rats. *Biochim. Biophys. Acta* **1990**, *1034*, 22–28. [CrossRef]
132. Horst, R.L.; Goff, J.P.; Reinhardt, T.A. Advancing age results in reduction of intestinal and bone 1,25 dihydroxyvitamin D receptor. *Endocrinology* **1990**, *126*, 1053–1057. [CrossRef] [PubMed]
133. Heaney, R.P.; Recker, R.R.; Saville, P.D. Menopausal changes in calcium balance performance. *J. Lab. Clin. Med.* **1978**, *92*, 953–963. [CrossRef] [PubMed]
134. Riggs, B.L.; Khosla, S.; Melton, L.J., III. Sex steroids and the construction and conservation of the adult skeleton. *Endocr. Rev.* **2002**, *23*, 279–302. [CrossRef] [PubMed]
135. Ten Bolscher, M.; Netelenbos, J.C.; Barto, R.; Van Buuren, L.M.; Van der vijgh, W.J. Estrogen regulation of intestinal calcium absorption in the intact and ovariectomized adult rat. *J. Bone Miner. Res.* **1999**, *14*, 1197–1202. [CrossRef]
136. Van Abel, M.; Hoenderop, J.G.; van der Kemp, A.W.; van Leeuwen, J.P.; Bindels, R.J. Regulation of the Epithelial Ca^{2+} Channels in Small Intestine as Studied by Quantitative mRNA Detection. *Am. J. Physiol. Gastrointest. Liver Physiol.* **2003**, *285*, G78–G85. [CrossRef]
137. Van Cromphaut, S.J.; Rummens, K.; Stockmans, I.; Van Herck, E.; Dijcks, F.A.; Ederveen, A.; Carmeliet, P.; Verhaeghe, J.; Bouillon, R.; Carmeliet, G. Intestinal calcium transporter genes are upregulated by estrogens and the reproductive cycle through vitamin D receptor-independent mechanisms. *J. Bone Miner. Res.* **2003**, *18*, 1725–1736. [CrossRef]
138. Gennari, C.; Agnusdei, D.; Nardi, P.; Civitelli, R. Estrogen preserves a normal intestinal responsiveness to 1,25-dihydroxyvitamin D3 in oophorectomized women. *J. Clin. Endocrinol. Metab.* **1990**, *71*, 1288–1293. [CrossRef]
139. Arjmandi, B.H.; Holis, B.W.; Kalu, D.N. In vivo effect of 17 b-estradiol on intestinal calcium absorption in rats. *Bone Miner.* **1994**, *26*, 181–189. [CrossRef]
140. Liel, Y.; Shany, S.; Smirnoff, P.; Schwartz, B. Estrogen increases 1,25-dihydroxyvitamin D receptors expression and bioresponse in the rat duodenal mucosa. *Endocrinology* **1999**, *140*, 280–285. [CrossRef]
141. Colin, E.M.; Van Den Bemd, G.J.; Van Aken, M.; Christakos, S.; De Jonge, H.R.; Deluca, H.F.; Prahl, J.M.; Birkenhager, J.C.; Buurman, C.J.; Pols, H.A.; et al. Evidence for involvement of 17beta-estradiol in intestinal calcium absorption independent of 1,25-dihydroxyvitamin D3 level in the Rat. *J. Bone Miner. Res.* **1999**, *14*, 57–64. [CrossRef]
142. Mauras, N.; Haymond, M.W.; Darmaun, D.; Vieira, N.E.; Abrams, S.A.; Yergey, A.L. Calcium and protein kinetics in prepubertal boys. Positive effects of testosterone. *J. Clin. Investig.* **1994**, *93*, 1014–1019. [CrossRef] [PubMed]
143. Chen, R.Y.; Nordin, B.E.; Need, A.G.; Scopacasa, F.; Wishart, J.; Morris, H.A.; Horowitz, M. Relationship between calcium absorption and plasma dehydroepiandrosterone sulphate (DHEAS) in healthy males. *Clin. Endocrinol.* **2008**, *69*, 864–869. [CrossRef] [PubMed]

MDPI

Article

Vitamin D and Diseases of Mineral Homeostasis: A *Cyp24a1* R396W Humanized Preclinical Model of Infantile Hypercalcemia Type 1

René St-Arnaud [1,2,3,4,*], Alice Arabian [1], Dila Kavame [1,2], Martin Kaufmann [5,6] and Glenville Jones [5]

1 Research Centre, Shriners Hospital for Children-Canada, Montreal, QC H4A 0A9, Canada
2 Department of Human Genetics, Faculty of Medicine and Health Sciences, McGill University, Montreal, QC H3A 0C7, Canada
3 Department of Surgery, Faculty of Medicine and Health Sciences, McGill University, Montreal, QC H3G 1A4, Canada
4 Department of Medicine, Faculty of Medicine and Health Sciences, McGill University, Montreal, QC H3A 1A1, Canada
5 Department of Biomedical and Molecular Sciences, Queen's University, Kingston, ON K7L 3N6, Canada
6 Department of Surgery, Queen's University, Kingston, ON K7L 3N6, Canada
* Correspondence: rst-arnaud@shriners.mcgill.ca; Tel.: +1-514-282-7155; Fax: +1-514-842-5581

Abstract: Infantile hypercalcemia type 1 (HCINF1), previously known as idiopathic infantile hypercalcemia, is caused by mutations in the 25-hydroxyvitamin D 24-hydroxylase gene, *CYP24A1*. The R396W loss-of-function mutation in *CYP24A1* is the second most frequent mutated allele observed in affected HCINF1 patients. We have introduced the site-specific R396W mutation within the murine *Cyp24a1* gene in knock-in mice to generate a humanized model of HCINF1. On the C57Bl6 inbred background, homozygous mutant mice exhibited high perinatal lethality with 17% survival past weaning. This was corrected by crossbreeding to the CD1 outbred background. Mutant animals had hypercalcemia in the first week of life, developed nephrolithiasis, and had a very high 25(OH)D_3 to 24,25$(OH)_2D_3$ ratio which is a diagnostic hallmark of the HCINF1 condition. Expression of the mutant *Cyp24a1* allele was highly elevated while *Cyp27b1* expression was abrogated. Impaired bone fracture healing was detected in CD1-R396$^{w/w}$ mutant animals. The augmented lethality of the C57Bl6-R396W strain suggests an influence of distinct genetic backgrounds. Our data point to the utility of unique knock-in mice to probe the physiological ramifications of CYP24A1 variants in isolation from other biological and environmental factors.

Keywords: infantile hypercalcemia type 1; idiopathic infantile hypercalcemia; CYP24A1; 25-hydroxyvitamin D 24-hydroxylase; hypercalcemia; 24,25-dihydroxyvitamin D_3; fracture repair

Citation: St-Arnaud, R.; Arabian, A.; Kavame, D.; Kaufmann, M.; Jones, G. Vitamin D and Diseases of Mineral Homeostasis: A *Cyp24a1* R396W Humanized Preclinical Model of Infantile Hypercalcemia Type 1. *Nutrients* **2022**, *14*, 3221. https://doi.org/10.3390/nu14153221

Academic Editor: Carsten Carlberg

Received: 22 July 2022
Accepted: 4 August 2022
Published: 6 August 2022

1. Introduction

Idiopathic infantile hypercalcemia (IIH) was first reported by Lightwood in the 1950s in the United Kingdom. An elevation in the number of IIH cases was observed following the introduction of an increased prophylactic dose of vitamin D in infant formula and milk products [1,2]. The infants were found to have high levels of serum and urinary excretion of calcium, physiological levels of serum phosphate and magnesium, and occasionally low serum alkaline phosphatase (ALP). Reducing the vitamin D dose by half lowered the observed incidence. Because some infants were not affected by treatment, Lightwood later hypothesized that the disease was caused by fluctuation in infants' sensitivity to vitamin D.

The general manifestations of this rare disease are high serum calcium and high urinary calcium loss, failure to thrive, vomiting, constipation, polyuria with dehydration, and Ca deposits in the kidney, with susceptibility to renal complications such as nephrolithiasis or nephrocalcinosis. Affected infants may also manifest muscular hypotonia and lethargy, which leads to growth impairment, with delays in mental and mobility development [3].

Affected adults can also manifest some extra-renal pathologies, such as Ca deposits in the joints or the cornea, and low bone mineral density with osteoporosis, mainly due to enhanced osteoclastic activity [4,5]. The genetic cause of IIH remained unknown until Schlingmann and coworkers identified it in 2011. They found inactivating mutations in *CYP24A1* that drive the disease evolution [6].

Since mutations in the *SLC34A1* (solute carrier family 34 member 1, a type II sodium-phosphate cotransporter) gene cause a similar clinical presentation [7], the nomenclature has been changed to infantile hypercalcemia type 1 for *CYP24A1* mutations (HCINF1, OMIM 143880) and infantile hypercalcemia type 2 for *SLC34A1* affected patients (HCINF2, OMIM 616963).

To date, more than 50 disease-causing mutations of *CYP24A1* have been described in the literature [8–13]. Alterations of the *CYP24A1* sequence can impact CYP24A1 structure and activity in different ways. Mutations are predicted to alter the substrate or heme binding site, the binding of adrenodoxin, access of the substrate or exit of the product, or generally to affect protein folding [6,14–17].

The currently available *Cyp24a1* knockout strain [18] developed more than 20 years ago exhibits hypercalcemia, hypercalciuria, and nephrocalcinosis and could represent a model of HCINF1. Supporting this view, it was shown that surviving *Cyp24a1*-null animals possess a much-reduced ability to clear a bolus dose of $[1\beta\text{-}^3H]1\alpha,25\text{-}(OH)_2D_3$ compared with wild-type littermates [19]. However, it should be noted that the *Cyp24a1*-deficient strain was generated with a selection cassette inserted in the opposite orientation to replace the heme-binding exon [18]. The *Cyp24a1*-deficient mice from this strain does not express any *Cyp24a1* message [18]. This contrasts with the situation in patients who, in many cases, express the mutated *CYP24A1* mRNA and protein [16]. Moreover, some of the *CYP24A1* mutations identified in HCINF1 patients are hypomorphic mutations, demonstrating residual or altered activity [6]. There is thus a need to generate improved preclinical animal models of infantile hypercalcemia type 1 based on human mutations to better understand the effects of dysfunctional vitamin D metabolism in HCINF1 patients.

We have introduced the site-specific R396W mutation within the murine *Cyp24a1* gene in knock-in mice. The R396W loss-of-function mutation is the second most frequent mutated allele observed in affected HCINF1 patients [8]. Profiling vitamin D metabolites using a sensitive LC-MS/MS-based assay demonstrated that the R396W strain is a valid preclinical model of HCINF1.

2. Materials and Methods

2.1. Generation of the R396W Knock-in Strain

All animal procedures were reviewed and approved under Animal Use Protocol number 7470 by the Shriners Hospitals for Children—Canada Institutional Animal Care and Use Committee and followed the guidelines of the Canadian Council on Animal Care. Mice were maintained in an environmentally controlled barrier animal facility with a 12 h light, 12 h dark cycle, and had access to mouse chow and water ad libitum.

The R396W knock-in embryonic stem cells were generated by Cyagen Biosciences (Santa Clara, CA, USA) under contract. A confirmed targeted clone was injected into C57Bl6/N blastocysts. Chimeras produced by blastocyst injection of targeted embryonic stem cells transmitted the knock-in mutation to their progeny and heterozygous mice were interbred to obtain all three genotypes (wild-type: C57Bl6-R396$^{+/+}$; heterozygous: C57Bl6-R396$^{+/W}$; homozygous mutant: C57Bl6-R396$^{W/W}$). For phenotyping, mice were sacrificed at postnatal Day 7 (P7) and at 4 months of age.

The C57Bl6-R396W strain was also outbred to CD1 mice to generate the CD1-R396W strain. The phenotype of these mice was analyzed at 3 months of age.

2.2. Genotyping

All mice were genotyped using genomic DNA isolated from carcasses found in cages prior to weaning or from ear punches at 3 weeks of age. For the R396W strains, the different genotypes were determined using a custom TaqMan single-nucleotide polymorphism (SNP) genotyping assay from Applied Biosystems (Life Technologies, Foster City, CA, USA) with *Cyp24a1* primers (forward: 5′-GCTTACCCCAAGTGTGCCATT-3′ and reverse: 5′-CCAGAACGGTTGGCTTGTC-3′) flanking the single-point mutation site. The 396R SNP primer (VIC-conjugated) was 5′-AAGGGTCCGAGTTGTG-3′ and the 396W SNP primer (FAM-conjugated) was 5′-AAGGGTCCAAGTTGTG-3′ (mutated nucleotide underlined).

2.3. Survival Analysis

Cages were monitored daily until weaning (day 21) and when found, cadavers/remains were collected for genotyping. The percent survival curves per genotype were graphed using GraphPad Prism (San Diego, CA, USA) version 7.04.

2.4. Serum Biochemistry

Total serum calcium was measured using an automated analyzer. Serum levels of vitamin D metabolites were assayed by LC-MS/MS following 4-(2-(6,7-dimethoxy-4-methyl-3,4-dihydroquinoxalinyl)ethyl)-1,2,4-triazoline-3,5-dione (DMEQ-TAD) derivatization as described previously [20]. Briefly, serum aliquots and calibrators were diluted 1:3 with water and spiked with internal standards. Proteins were precipitated by sequentially adding 0.1 M HCl, 0.2 M zinc sulfate, and methanol, with vortexing after the addition of each component. Tubes were centrifuged 10 min at 12,000× *g* and supernatants were transferred to borosilicate glass tubes. Organic extraction was carried out by adding equal volumes of hexane and methyl tertiary butyl ether with vortexing after the addition of each component. The upper organic phase was transferred into LC-MS/MS sample vials and evaporated under nitrogen flow. Dried residues were derivatized by addition of 0.1 mg/mL DMEQ-TAD dissolved in ethyl acetate for 30 min at room temperature in the dark, then a second time for 60 min. The reaction was stopped by addition of ethanol, samples were dried and redissolved in 60:40 methanol:water running buffer. LC-MS/MS analysis was performed using an Acquity UPLC connected in line with a Xevo TQ-S mass spectrometer in electrospray positive mode (Waters). Chromatographic separations were achieved using a BEH-Phenyl UPLC column (1.7 μm, 2.1 × 50 mm) (Waters) and methanol/water-based gradient solvent system. Simultaneous assay of $1,25(OH)_2D_3$ and $1,24,25(OH)_3D_3$ was possible based on cross-reactivity of an anti-$1,25(OH)_2D_3$ antibody slurry (Immundiagnostik, Manchester, NH) with $1,24,25(OH)_3D_3$. The serum was incubated with 100 μL of anti-$1,25(OH)_2D_3$ antibody slurry [21] for 2h at room temperature with orbital shaking at 1200 rpm. The slurry was isolated by vacuum filtration and vitamin D metabolites were eluted, derivatized with DMEQ-TAD, and separated using a longer LC step as previously described [22,23].

2.5. Microcomputed Tomography of Whole Kidneys

Kidneys were harvested from mice at P7 and fixed in 4% paraformaldehyde diluted in PBS for 24 h, washed two to three times, and then stored in 70% ethanol. Kidneys were scanned with high-resolution microcomputed tomography using a SkyScan model 1272 scanner (Bruker, Kontich, Belgium).

2.6. Gene Expression Monitoring

Kidneys were harvested at P7 and 4 months, dissected free of surrounding tissue, immersed in RNA later (Ambion, Austin, TX, USA), and stored at −80 °C until ready for testing. Quantitative gene expression was assessed by real-time reverse transcriptase polymerase chain reaction (RT-qPCR). Briefly, kidneys were homogenized in 1 mL TRIzol reagent (Invitrogen, Carlsbad, CA, USA) and total RNA was isolated according to the manufacturer's protocol. One (1) μg of RNA was reverse transcribed to cDNA using High-Capacity cDNA Reverse Transcription Kit (Life Technologies Applied Biosystems, Waltham, MA, USA). Real-time qPCR using TaqMan universal PCR master mix and gene-specific Taqman assays for *Cyp24a1* (Mm00487244_m1), *Cyp27b1* (Mm01165918_m1), vitamin D receptor (VDR, Mm01309608_m1), *Casr* (Mm00443375_m1),

Calbindin D9k (*S100g*, Mm00486654_m1), Calbindin D28k (*Calb1*, Mm00486645_m1), *Trpv5* (Mm01166030_m1), *Trpv6* (Mm00499069_m1), *Npt2a* (*Slc34a1*, Mm00441450_m1), *Npt2b* (*Slc34a2*, Mm01215846_m1), *Npt2c* (*Slc34a3*, Mm00551746_m1), *Npt3* (*Slc17a2*, Mm00522866_m1), *Nkcc2* (*Slc12a1*, Mm01275821_m1), and *Nhe3* (*Slc9a3*, Mm01352473_m1) was performed using a QuantStudio 7 real-time PCR system (Life Technologies Applied Biosystems). The assay was performed in triplicate. Relative quantification of mRNA was performed according to the comparative C_t method and normalized to housekeeping genes.

2.7. Intramedullary Rodded Tibial Osteotomy

The surgical procedure was performed on mice under general isoflurane anesthesia as described in Martineau et al. [24]. Briefly, an incision was performed above the right knee to free the patellar ligament from lateral tissue. A 25G spinal needle wire guide was inserted down the medullary canal through a 26G needle pushed through the tibial plateau. The wire guide was bent at a right angle, cut at the tibial plateau, and secured by a mattress suture on either side of the patellar ligament after the needle was pulled out. The tibial shaft was cut using micro scissors about 2–3 mm above the tibiofibular junction. Topical analgesics were applied at the wound site, then the skin was sutured. Carprofen was provided at the time of surgery and for the following 48 h post-osteotomy. The mice were sacrificed using isoflurane and CO_2 at day 10 (D10) post-surgery.

2.8. Three-Point Bending Assays

For three-point bending assays, callus tibiae were thawed overnight at room temperature and tested for mechanical properties using an Instron model 5943 single-column table frame machine (Instron, Norwood, MA, USA). The fractured bones rested on two fulcra set 6 mm apart. Load-sensing cell was applied to the widest part of the callus. Raw output used for comparison was strength (load at break, in N).

2.9. Statistical Analysis

Statistical analyses were performed using GraphPad Prism version 7.04. Statistical tests involved 2-tailed *t*-tests for gene expression monitoring, 1-way ANOVA followed by Dunnett's or Tukey's post hoc test for calcemia and three-point bending assays, respectively, or 2-way ANOVA with Sidak's post hoc test for vitamin D metabolites measurements. The statistical significance threshold was set at a *p*-value of less than 0.05.

3. Results

3.1. Severe Postnatal Lethality in C57Bl6-R396$^{w/w}$ Mutant Animals

The R396W knock-in targeting construct was electroporated into C57Bl6/N embryonic stem cells. The targeting event was confirmed using both PCR screening and Southern blotting. One (1) correctly targeted clone was injected into C57Bl6/N blastocysts, which allowed to derive germline-transmitting chimeras on the homogeneous C57Bl6/N genetic background. Heterozygous progeny was mated inter se to obtain all three genotypes (wild-type: C57Bl6-R396$^{+/+}$; heterozygous: C57Bl6-R396$^{+/w}$; homozygous mutant: C57Bl6-R396$^{w/w}$).

The mutant allele was transmitted at the expected Mendelian frequency, but we detected severe postnatal lethality in the homozygous mutant animals. We observed a significant number of dead pups of the C57Bl6-R396$^{w/w}$ genotype from birth, which culminated in a 17% survival rate of homozygous mutant pups from postnatal day 14 onwards (Figure 1). This is a striking difference from the 50% postnatal lethality measured in the global *Cyp24a1*-deficient strain maintained on a mixed genetic background for more than 20 years [18].

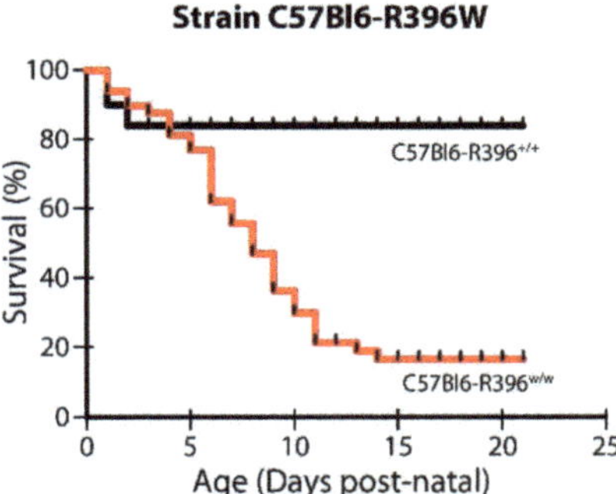

Figure 1. Postnatal lethality in C57Bl6-R396$^{W/W}$ mutant pups. Cages were monitored daily until weaning (day 21) and cadavers were genotyped. Percent survival for wild-type and homozygous mutants is shown.

3.2. Early Postnatal Hypercalcemia in C57Bl6-R396$^{w/w}$ Mutant Animals

We measured serum calcium levels in wild-type, heterozygous and homozygous mutant pups during the second postnatal week (P7 to P10). C57Bl6-R396$^{W/W}$ mutant animals had significantly elevated circulating calcium levels at P7 compared to wild-type and heterozygous control littermates (Figure 3.3). This was accompanied by the marked formation of kidney stones (nephrolithiasis) (Figure 3). We did not investigate for signs of extra renal calcification. Circulating serum levels in surviving mutant pups past P7 were within the normal range (Supplementary Figure S1).

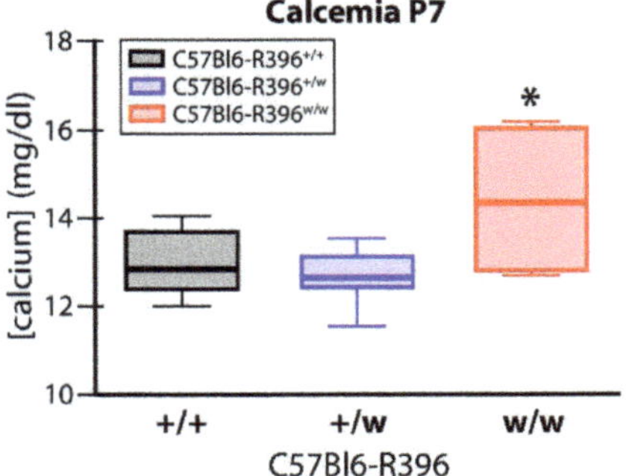

Figure 2. Serum calcium levels in C57Bl6-R396 littermates at postnatal day 7. Total serum calcium was measured using an automated analyzer. +/+, C57Bl6-R396$^{+/+}$; +/w, C57Bl6-R396$^{+/W}$; w/w, C57Bl6-R396$^{W/W}$. *, $p < 0.05$ by one-way ANOVA and Dunnett's post hoc test.

3.3. Gene Expression Monitoring

We compared kidney tissue gene expression in C57Bl6-R396$^{W/W}$ mutants and wild-type littermates at P7 and 4 months. The R396W mutant *Cyp24a1* allele was highly overexpressed (30-fold increase) in pups (Figure 4A); this was reduced in surviving adult mutant animals but still remained almost 10-fold higher as compared to wild-type littermates (Figure 4B). As observed in the global *Cyp24a1*-deficient animals [18,19], *Cyp27b1* expression was markedly inhibited at all ages (Figure 4A,B). Vitamin D receptor expression was moderately but significantly overexpressed (Figure 4A,B).

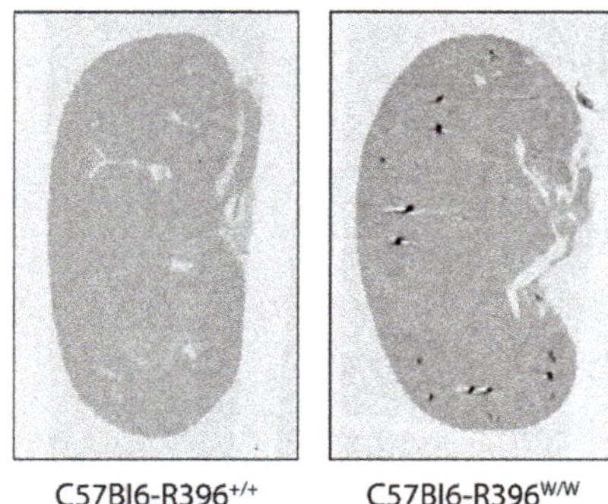

Figure 3. Nephrolithiasis in C57Bl6-R396$^{w/w}$ mutant pups. Kidneys were harvested at P7 and imaged by microCT.

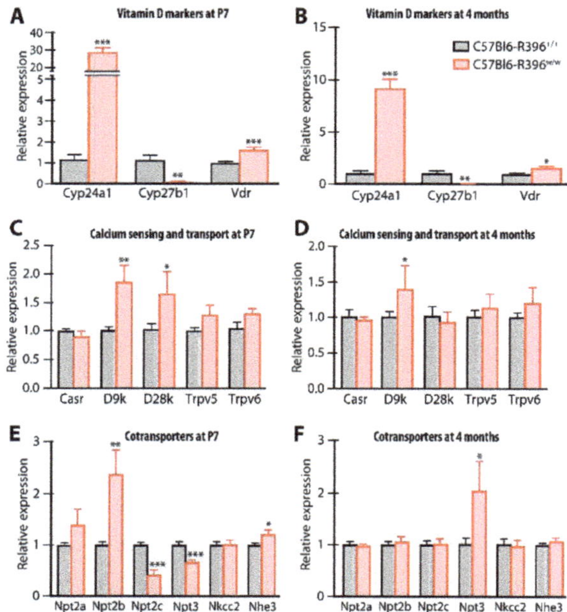

Figure 4. Gene expression monitoring in newborn and adult C57Bl6-R396 littermates. Quantitative gene expression was assessed by RT-qPCR with TaqMan probes on mRNA extracted from kidneys. *, $p < 0.05$; **, $p < 0.01$; ***, $p < 0.001$, double-sided *t*-tests for each gene. (**A**) Vitamin D markers at P7. (**B**) Vitamin D markers at 4 months. (**C**) Calcium sensing and transport at P7. (**D**) Calcium sensing and transport at 4 months. (**E**) Cotransporters at P7. (**F**) Cotransportersat 4 months.

The calcium transporters calbindin D9k and calbindin D28k were significantly overexpressed in mutant pups and this normalized for D28k in adults (Figure 4C,D). There were no differences in the expression levels of the calcium-sensing receptor (*Casr*) or the calcium channels *Trpv5* or *6* between genotypes (Figure 4C,D).

In newborn C57Bl6-R396$^{w/w}$ mutants, we measured the increased expression of the sodium-dependent phosphate transport protein 2b (Npt2b, *Slc34a2*) while related transporters Npt2c (*Slc34a3*) and Npt3 (*Slc17a2*) were expressed at lower levels in mutant animals compared to wild-type littermates (Figure 4E). With the exception of Npt3, which was overexpressed in adult homozygous mutant mice, the expression of co-transporters normalized as animals aged to adulthood (Figure 4F).

3.4. CD1-R396W Strain

We crossed the C57Bl6-R396W strain to CD1 outbred mice in an attempt to decrease the extreme perinatal lethality associated with carrying the R396W mutant allele on a homogeneous genetic background. In this mixed genetic background, survival of the CD1-R396$^{W/W}$ mutants was improved to 86% (Figure 5). This allowed us to compare vitamin D metabolites between the strains and examine additional phenotypic manifestations.

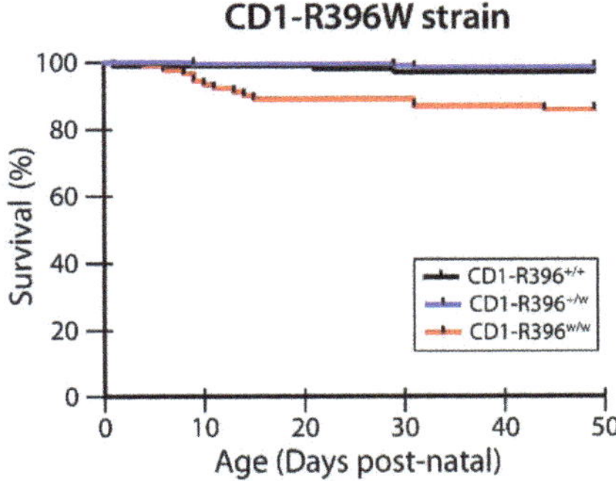

Figure 5. Survival of CD1-R396$^{W/W}$ mutant pups. Cages were monitored daily until postnatal day 49 and cadavers found in cages were genotyped. Percent survival for all 3 genotypes is shown.

3.5. Vitamin D Metabolites

We used a sensitive LC-MS/MS-based assay to profile multiple serum vitamin D metabolites [20] in the global *Cyp24a1*-null mice, the C57Bl6-R396W strain, and the CD1-R396W line. The mutant genotypes of all three strains exhibited a five- to six-fold increase in circulating 25(OH)D_3 compared to control genotypes (Table 1). Surprisingly, levels of 24,25(OH)$_2D_3$ remained detectable but were reduced by 3- to 5.4-fold in mutant mice. These changes led to a very high 25(OH)D_3 to 24,25(OH)$_2D_3$ ratio in homozygote mutants which is a diagnostic hallmark of the infantile hypercalcemia type 1 condition [20]. Concentrations of 25(OH)D_3-26,23-lactone were at the lower limit of detection of the assay, in accord with the demonstrated 23-hydroxylation activity of CYP24A1 [25]. Levels of 1,24,25(OH)$_3D_3$ were also significantly reduced in homozygous mutant animals sporting the R396W mutation (Table 1). Despite the much reduced CYP24A1 catabolic enzyme activity in mutant mice, the serum concentration of the hormone 1,25(OH)$_2D_3$ remained normal (Table 1) presumably due to much reduced *Cyp27b1* expression noted in Figure 4A,B. This adaptation is presumably necessary to permit the viability of the surviving animals.

3.6. Impaired Bone Fracture Healing in CD1-R396$^{w/w}$ Mice

The improved survival of CD1-R396$^{W/W}$ mutants permitted their skeletal analysis. There were no detectable steady-state skeletal phenotypic manifestations in the mutant animals as shown by normal trabecular bone volume, trabecular separation, thickness, and separation (Supplementary Figure S2A–D). Since we have observed impaired bone fracture repair in global *Cyp24a1*-deficient mice [24], we challenged the CD1-R396$^{W/W}$ mutant animals to heal surgical osteotomies as a model of bone fracture recovery. The intramedullary rodded immobilized fracture surgery was performed on the left tibia at 3 months of age. The analyses of the fractured bones were performed on post-surgery day 10 (D10). The tibiae with calluses were dissected and the intramedullary nail was carefully removed. The biomechanical properties of the repaired calluses were evaluated using the three-point bending test.

At D10 post-surgery, the analysis of the callus showed a significant decrease in load at break (strength) in heterozygous and homozygous mutant mice compared to wild-type littermates (Figure 6).

Table 1. Vitamin D metabolites in preclinical models of infantile hypercalcemia type 1.

Strain/Genotype	Sex	$25(OH)D_3$ (ng/mL)	$24,25(OH)_2D_3$ (ng/mL)	Ratio 25:24,25	$1,24,25(OH)_3D_3$ (pg/mL)	$25(OH)D_3$-26,23-lactone (ng/mL)	$1,25(OH)_2D_3$ (pg/mL)
Cyp24a1$^{+/-}$	m	-	-	-	-	-	-
	f	17.21 ± 2.36	6.68 ± 1.05	2.60 ± 0.10	72.80 ± 4.20	3.84 ± 0.72	27.80 ± 4.20
Cyp24a1$^{-/-}$	m	-	-	-	-	-	-
	f	101.29 ± 20.61 ***	1.45 ± 0.11 ***	70.60 ± 15.7 #	n.d.	0.06 ± 0.02 #	34.00 ± 6.20
C57bl6-R396$^{+/+}$	m	19.59 ± 2.90	6.71 ± 0.76	2.91 ± 0.33	57.77 ± 10.88	2.76 ± 0.32	27.39 ± 3.35
	f	18.59 ± 2.50	7.38 ± 0.96	2.53 ± 0.25	66.53 ± 13.18	3.16 ± 0.89	21.28 ± 5.96
C57bl6-R396$^{w/w}$	m	110.73 ± 15.40 #	2.59 ± 0.44 #	43.10 ± 4.08 #	17.81 ± 3.53 #	0.07 ± 0.03 #	37.06 ± 8.85
	f	106.16 ± 26.73 #	2.12 ± 0.44 #	49.71 ± 4.37 #	11.09 ± 5.35 #	0.05 ± 0.02 #	37.17 ± 12.12 **
CD1-R396$^{+/+}$	m	17.23 ± 3.77	6.31 ± 1.89	2.94 ± 1.07	74.66 ± 34.78	2.24 ± 0.70	45.83 ± 22.74
	f	18.24 ± 4.14	8.80 ± 2.14	2.08 ± 0.19	55.19 ± 17.50	2.56 ± 0.67	27.43 ± 9.77
CD1-R396$^{w/w}$	m	81.90 ± 19.39 #	2.26 ± 0.53 #	36.58 ± 5.35 #	15.18 ± 5.84 #	0.06 ± 0.05 #	45.12 ± 16.57
	f	99.16 ± 13.59 #	2.23 ± 0.39 #	44.90 ± 5.53 #	7.49 ± 1.95 #	0.03 ± 0.01 #	29.24 ± 7.98

Results are means ± S.D. Ratio 25:24,25, ratio of $25(OH)_2D_3$ concentrations over $24,25(OH)_2D_3$ concentrations. n.d., non-detectable. **, $p < 0.01$; ***, $p < 0.001$; #, $p < 0.0001$ vs. corresponding wild-type sex within strain by 2-way ANOVA and Sidak's post hoc test.

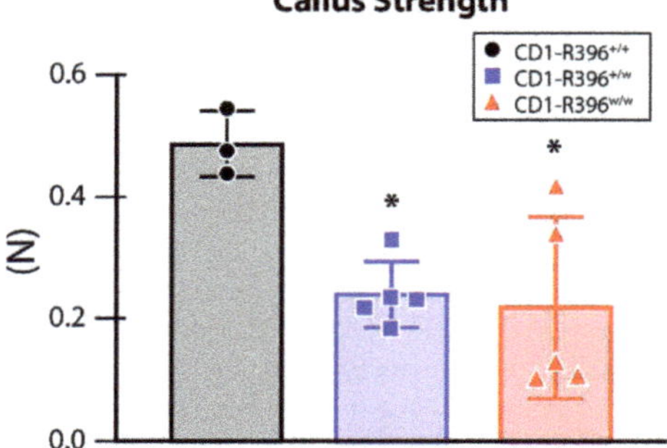

Figure 6. Impaired bone fracture healing in CD1-R396$^{w/w}$ mutant animals. Callus strength (load at break) was calculated from the three-point bending test. *, $p < 0.05$ by one-way ANOVA with Tukey's post hoc test.

4. Discussion

The global inactivation of *Cyp24a1* in mice confirmed the physiological importance of CYP24A1 in the catabolism of $1,25(OH)_2D_3$ [18,19] and elucidated the mechanism of action of $24,25(OH)_2D_3$ in bone fracture healing [24]. As a preclinical model of infantile hypercalcemia type 1, however, it left to be desired, since the mutation involved replacing the heme-binding exon of *Cyp24a1* by the PGK-neo selection cassette [26] in the opposite transcriptional orientation [18], a genetic change that is significantly different from the mutated alleles identified in patients suffering from the disease (reviewed in [8]). We set out to generate a mouse strain expressing a mutated allele of *Cyp24a1* that corresponds to what has been described in patients and decided on one of the first identified mutations [6] which turns out to be the second most frequent mutated allele observed in affected HCINF1 patients, R396W [8].

We first established the mutation on a homogeneous genetic background, and this resulted in a dramatic rate of perinatal lethality. The elevated calcium concentrations in the first week of life may have contributed to the mortality rate. Despite significant changes in the expression of sodium-phosphate cotransporters, circulating phosphate concentrations were not affected by the mutation (Supplementary Figure S3), similar to what is observed in patients [27,28]. The differences in the expression of calcium and phosphate transporters observed soon after birth normalized in the surviving population that reached adulthood.

As reported for the global *Cyp24a1*-deficient mice [19], the expression of the vitamin D 1α-hydroxylase gene, *Cyp27b1*, was completely inhibited in young and older C57Bl6-R396$^{w/w}$ mutants, most likely as a protection mechanism to avoid hypervitaminosis. Such an explanation is consistent with the fact that young and older C57Bl6-R396$^{w/w}$ mutants have serum 1,25(OH)$_2$D$_3$ levels within the normal range. Of note was the very high expression of the mutated *Cyp24a1* allele in homozygous mutants. This contrasts with the measurable but diminished expression of *CYP24A1* in affected patients [16].

The R396W mutation is a loss-of-function mutation [6], yet we measured circulating levels of 24,25(OH)$_2$D$_3$ in both the inbred and outbred R396W mutated strains. These measurable levels of dihydroxylated metabolite could represent the activity of other cytochrome P450s or be due to the migration of interfering dihydroxyvitamin D metabolites with similar retention time and mass spectral characteristics in LC-MS/MS. We estimate this latter probability as very low. On the other hand, it has been reported that cytochrome P450s distinct from CYP24A1 can synthesize 24,25(OH)$_2$D$_3$, but it remains unclear if this activity, measured in vitro [29], has relevance in vivo. At any rate, the ability to detect circulating levels of 24,25(OH)$_2$D$_3$ in CYP24A1-deficient mice or patients allows for calculating the ratio of the metabolite to its precursor 25(OH)D$_3$, namely serum 25(OH)D$_3$:24,25(OH)$_2$D$_3$ [20,30–32]. Clinically, this ratio is a more reliable indicator of HCINF1 due to *CYP24A1* mutations than serum 24,25(OH)$_2$D$_3$ alone because it eliminates the possibility that the patient might have a low serum 24,25(OH)$_2$D$_3$ level due to vitamin D deficiency. The high 25(OH)D$_3$:24,25(OH)$_2$D$_3$ ratio measured in R396W homozygote mutants on both inbred and outbred backgrounds aligns with the reports from over 100 HCINF1 patients [20,32]. We thus posit that the R396W knock-in strains are valid preclinical models of HCINF1.

Another phenotypic manifestation of the R396W mouse mutation that mimics what is observed in patients is the presence of kidney stones. Indeed, hypercalciuria and the risk of renal stones continue in most subjects with *CYP24A1* mutations throughout adult life [28,33,34]. In patients, the disorder may in fact first manifest as hypercalciuria and painful kidney stones in adulthood, without childhood symptoms [16,35–38]. Based on gene frequency studies, Nesterova and colleagues estimate that the frequency of kidney stones in the general population due to *CYP24A1* mutations might be as high as 4–20% [16].

The perinatal lethality phenotype of *Cyp24a1* homozygous mutants exhibits significant variable penetrance depending on the genetic background upon which the mutation is established. Crossbreeding the R396W mutation on an outbred CD1 background most adequately represents the human condition with no obvious complications unless challenged. Exploiting the differential penetrance of the perinatal lethality between the 3 strains of *Cyp24a1* deficiency available could allow for the identification of modifier gene(s) [39] impacting vitamin D and mineral ion homeostasis.

The *Cyp24a1*-deficient mouse strain [18] was an invaluable tool to study the physiological role of 24,25(OH)$_2$D$_3$ in mammalian fracture repair. *Cyp24a1*$^{-/-}$ mice show suboptimal endochondral ossification during fracture repair with smaller calluses that exhibit inferior biomechanical properties [24]. The strength of the repairing callus was reduced in heterozygous and homozygous CD1-R396$^{w/w}$ mutant mice compared to wild-type littermates. This change was the only phenotypic manifestation that we observed in heterozygous carriers of the mutation. The impaired bone fracture healing observed in CD1-R396$^{w/w}$ mutant animals establishes that two distinct preclinical models of *Cyp24a1* deficiency exhibit bone fracture repair defects. These results support the need to study bone fracture healing in patients with HCINF1.

5. Conclusions

Introducing the site-specific R396W mutation within the murine *Cyp24a1* gene in knock-in mice generated a valid preclinical model of infantile hypercalcemia type 1 with the characteristic high 25(OH)D_3:24,25$(OH)_2D_3$ ratio that is a diagnostic hallmark of the condition.

Supplementary Materials: The following supporting information can be downloaded at: https://www.mdpi.com/article/10.3390/nu14153221/s1, Supplementary Figure S1. Serum calcium levels in C57Bl6-R396W littermates, Supplementary Figure S2. Steady-state trabecular parameters in CD1-R396W knock-in strain, Supplementary Figure S3. Serum phosphate levels in surviving wild-type and mutant C57Bl6-R396W mice at 3 months of age.

Author Contributions: A.A., D.K. and M.K. generated data. M.K., G.J. and R.S.-A. participated in data analysis and interpretation. G.J. and R.S.-A. obtained the funding and participated in the conception and design of the study. R.S.-A. wrote the manuscript. All authors have read and agreed to the published version of the manuscript.

Funding: This work was supported by grant No. ERA-132931 from ERA-Net for Research Programs on Rare Diseases to G.J. and R.St-A. and by NIH grant R01 AR070544 to R.St-A.

Institutional Review Board Statement: The animal study protocol was approved by the Institutional Animal Care and Use Committee of Shriners Hospitals for Children—Canada (animal use protocol number 7470 approved 2022-03-01) as well as by Queen's University Animal Care Committee (animal use protocol number 1837 approved 2022-04-18).

Informed Consent Statement: Not applicable.

Data Availability Statement: Not applicable.

Acknowledgments: Through a Queen's University and Waters Corporation agreement, Waters provided the LC-MS/MS instrument used in this study. We thank Mia Esser, Louise Marineau, and Alexandria Norquay for expert animal care. We also thank Paige Conrad, Melanie Vig, and Linore Berezin for assistance with the vitamin D metabolite analysis. Mark Lepik prepared the publication-ready final version of the figures.

Conflicts of Interest: The authors declare no conflict of interest.

References

1. Lightwood, R.; Stapleton, T. Idiopathic hypercalcaemia in infants. *Lancet* **1953**, *265*, 255–256. [CrossRef]
2. Stapleton, T.; Macdonald, W.B.; Lightwood, R. The pathogenesis of idiopathic hypercalcemia in infancy. *Am. J. Clin. Nutr.* **1957**, *5*, 533–542. [CrossRef] [PubMed]
3. Gut, J.; Kutilek, S. Idiopathic infantile hypercalcaemia in 5-month old girl. *Prague Med. Rep.* **2011**, *112*, 124–131. [PubMed]
4. Ferraro, P.M.; Minucci, A.; Primiano, A.; De Paolis, E.; Gervasoni, J.; Persichilli, S.; Naticchia, A.; Capoluongo, E.; Gambaro, G. A novel CYP24A1 genotype associated to a clinical picture of hypercalcemia, nephrolithiasis and low bone mass. *Urolithiasis* **2017**, *45*, 291–294. [CrossRef] [PubMed]
5. Figueres, M.L.; Linglart, A.; Bienaime, F.; Allain-Launay, E.; Roussey-Kessler, G.; Ryckewaert, A.; Kottler, M.L.; Hourmant, M. Kidney function and influence of sunlight exposure in patients with impaired 24-hydroxylation of vitamin D due to CYP24A1 mutations. *Am. J. Kidney Dis.* **2015**, *65*, 122–126. [CrossRef]
6. Schlingmann, K.P.; Kaufmann, M.; Weber, S.; Irwin, A.; Goos, C.; John, U.; Misselwitz, J.; Klaus, G.; Kuwertz-Broking, E.; Fehrenbach, H.; et al. Mutations in CYP24A1 and idiopathic infantile hypercalcemia. *N. Engl. J. Med.* **2011**, *365*, 410–421. [CrossRef]
7. Schlingmann, K.P.; Ruminska, J.; Kaufmann, M.; Dursun, I.; Patti, M.; Kranz, B.; Pronicka, E.; Ciara, E.; Akcay, T.; Bulus, D.; et al. Autosomal-Recessive Mutations in SLC34A1 Encoding Sodium-Phosphate Cotransporter 2A Cause Idiopathic Infantile Hypercalcemia. *J. Am. Soc. Nephrol.* **2016**, *27*, 604–614. [CrossRef]
8. Cappellani, D.; Brancatella, A.; Morganti, R.; Borsari, S.; Baldinotti, F.; Caligo, M.A.; Kaufmann, M.; Jones, G.; Marcocci, C.; Cetani, F. Hypercalcemia due to CYP24A1 mutations: A systematic descriptive review. *Eur. J. Endocrinol.* **2022**, *186*, 137–149. [CrossRef]
9. Hanna, C.; Potretzke, T.A.; Cogal, A.G.; Mkhaimer, Y.G.; Tebben, P.J.; Torres, V.E.; Lieske, J.C.; Harris, P.C.; Sas, D.J.; Milliner, D.S.; et al. High Prevalence of Kidney Cysts in Patients With CYP24A1 Deficiency. *Kidney Int. Rep.* **2021**, *6*, 1895–1903. [CrossRef]

10. Macdonald, C.; Upton, T.; Hunt, P.; Phillips, I.; Kaufmann, M.; Florkowski, C.; Soule, S.; Jones, G. Vitamin D supplementation in pregnancy: A word of caution. Familial hypercalcaemia due to disordered vitamin D metabolism. *Ann. Clin. Biochem.* **2020**, *57*, 186–191. [CrossRef]
11. Molin, A.; Lemoine, S.; Kaufmann, M.; Breton, P.; Nowoczyn, M.; Ballandonne, C.; Coudray, N.; Mittre, H.; Richard, N.; Ryckwaert, A.; et al. Overlapping Phenotypes Associated With CYP24A1, SLC34A1, and SLC34A3 Mutations: A Cohort Study of Patients With Hypersensitivity to Vitamin D. *Front. Endocrinol. (Lausanne)* **2021**, *12*, 736240. [CrossRef] [PubMed]
12. Rousseau-Nepton, I.; Jones, G.; Schlingmann, K.; Kaufmann, M.; Zuijdwijk, C.S.; Khatchadourian, K.; Gupta, I.R.; Pacaud, D.; Pinsk, M.N.; Mokashi, A.; et al. CYP24A1 and SLC34A1 Pathogenic Variants Are Uncommon in a Canadian Cohort of Children with Hypercalcemia or Hypercalciuria. *Horm. Res. Paediatr.* **2021**, *94*, 124–132. [CrossRef] [PubMed]
13. Sun, Y.; Shen, J.; Hu, X.; Qiao, Y.; Yang, J.; Shen, Y.; Li, G. CYP24A1 Variants in Two Chinese Patients with Idiopathic Infantile Hypercalcemia. *Fetal Pediatr. Pathol.* **2019**, *38*, 44–56. [CrossRef]
14. De Paolis, E.; Scaglione, G.L.; De Bonis, M.; Minucci, A.; Capoluongo, E. CYP24A1 and SLC34A1 genetic defects associated with idiopathic infantile hypercalcemia: From genotype to phenotype. *Clin. Chem. Lab. Med.* **2019**, *57*, 1650–1667. [CrossRef] [PubMed]
15. Ji, H.F.; Shen, L. CYP24A1 mutations in idiopathic infantile hypercalcemia. *N. Engl. J. Med.* **2011**, *365*, 1741. [PubMed]
16. Nesterova, G.; Malicdan, M.C.; Yasuda, K.; Sakaki, T.; Vilboux, T.; Ciccone, C.; Horst, R.; Huang, Y.; Golas, G.; Introne, W.; et al. 1,25-(OH)2D-24 Hydroxylase (CYP24A1) Deficiency as a Cause of Nephrolithiasis. *Clin. J. Am. Soc. Nephrol.* **2013**, *8*, 649–657. [CrossRef]
17. Gigante, M.; Santangelo, L.; Diella, S.; Caridi, G.; Argentiero, L.; D'Alessandro, M.M.; Martino, M.; Stea, E.D.; Ardissino, G.; Carbone, V.; et al. Mutational Spectrum of CYP24A1 Gene in a Cohort of Italian Patients with Idiopathic Infantile Hypercalcemia. *Nephron* **2016**, *133*, 193–204. [CrossRef] [PubMed]
18. St-Arnaud, R.; Arabian, A.; Travers, R.; Barletta, F.; Raval-Pandya, M.; Chapin, K.; Depovere, J.; Mathieu, C.; Christakos, S.; Demay, M.B.; et al. Deficient mineralization of intramembranous bone in vitamin D-24-hydroxylase-ablated mice is due to elevated 1,25-dihydroxyvitamin D and not to the absence of 24,25-dihydroxyvitamin D. *Endocrinology* **2000**, *141*, 2658–2666. [CrossRef]
19. Masuda, S.; Byford, V.; Arabian, A.; Sakai, Y.; Demay, M.B.; St-Arnaud, R.; Jones, G. Altered pharmacokinetics of 1alpha,25-dihydroxyvitamin D3 and 25-hydroxyvitamin D3 in the blood and tissues of the 25-hydroxyvitamin D-24-hydroxylase (Cyp24a1) null mouse. *Endocrinology* **2005**, *146*, 825–834. [CrossRef]
20. Kaufmann, M.; Gallagher, J.C.; Peacock, M.; Schlingmann, K.P.; Konrad, M.; DeLuca, H.F.; Sigueiro, R.; Lopez, B.; Mourino, A.; Maestro, M.; et al. Clinical utility of simultaneous quantitation of 25-hydroxyvitamin D and 24,25-dihydroxyvitamin D by LC-MS/MS involving derivatization with DMEQ-TAD. *J. Clin. Endocrinol. Metab.* **2014**, *99*, 2567–2574. [CrossRef]
21. Laha, T.J.; Strathmann, F.G.; Wang, Z.; de Boer, I.H.; Thummel, K.E.; Hoofnagle, A.N. Characterizing antibody cross-reactivity for immunoaffinity purification of analytes prior to multiplexed liquid chromatography-tandem mass spectrometry. *Clin. Chem.* **2012**, *58*, 1711–1716. [CrossRef]
22. Kaufmann, M.; Morse, N.; Molloy, B.J.; Cooper, D.P.; Schlingmann, K.P.; Molin, A.; Kottler, M.L.; Gallagher, J.C.; Armas, L.; Jones, G. Improved Screening Test for Idiopathic Infantile Hypercalcemia Confirms Residual Levels of Serum 24,25-(OH)2 D3 in Affected Patients. *J. Bone Miner. Res.* **2017**, *32*, 1589–1596. [CrossRef] [PubMed]
23. Meyer, M.B.; Benkusky, N.A.; Kaufmann, M.; Lee, S.M.; Onal, M.; Jones, G.; Pike, J.W. A kidney-specific genetic control module in mice governs endocrine regulation of the cytochrome P450 gene Cyp27b1 essential for vitamin D3 activation. *J. Biol. Chem.* **2017**, *292*, 17541–17558. [CrossRef] [PubMed]
24. Martineau, C.; Naja, R.P.; Husseini, A.; Hamade, B.; Kaufmann, M.; Akhouayri, O.; Arabian, A.; Jones, G.; St-Arnaud, R. Optimal bone fracture repair requires 24R,25-dihydroxyvitamin D3 and its effector molecule FAM57B2. *J. Clin. Investig.* **2018**, *128*, 3546–3557. [CrossRef] [PubMed]
25. Beckman, M.J.; Tadikonda, P.; Werner, E.; Prahl, J.; Yamada, S.; DeLuca, H.F. Human 25-hydroxyvitamin D3-24-hydroxylase, a multicatalytic enzyme. *Biochemistry* **1996**, *35*, 8465–8472. [CrossRef] [PubMed]
26. Rudnicki, M.A.; Braun, T.; Hinuma, S.; Jaenisch, R. Inactivation of MyoD in mice leads to up-regulation of the myogenic HLH gene Myf-5 and results in apparently normal muscle development. *Cell* **1992**, *71*, 383–390. [CrossRef]
27. Pronicka, E.; Ciara, E.; Halat, P.; Janiec, A.; Wojcik, M.; Rowinska, E.; Rokicki, D.; Pludowski, P.; Wojciechowska, E.; Wierzbicka, A.; et al. Biallelic mutations in CYP24A1 or SLC34A1 as a cause of infantile idiopathic hypercalcemia (IIH) with vitamin D hypersensitivity: Molecular study of 11 historical IIH cases. *J. Appl. Genet.* **2017**, *58*, 349–353. [CrossRef]
28. Pronicka, E.; Rowinska, E.; Kulczycka, H.; Lukaszkiewicz, J.; Lorenc, R.; Janas, R. Persistent hypercalciuria and elevated 25-hydroxyvitamin D3 in children with infantile hypercalcaemia. *Pediatr. Nephrol.* **1997**, *11*, 2–6. [CrossRef]
29. Sawada, N.; Sakaki, T.; Ohta, M.; Inouye, K. Metabolism of vitamin D(3) by human CYP27A1. *Biochem. Biophys. Res. Commun.* **2000**, *273*, 977–984. [CrossRef]
30. Cools, M.; Goemaere, S.; Baetens, D.; Raes, A.; Desloovere, A.; Kaufman, J.M.; De Schepper, J.; Jans, I.; Vanderschueren, D.; Billen, J.; et al. Calcium and bone homeostasis in heterozygous carriers of CYP24A1 mutations: A cross-sectional study. *Bone* **2015**, *81*, 89–96. [CrossRef]
31. Ketha, H.; Kumar, R.; Singh, R.J. LC-MS/MS for Identifying Patients with CYP24A1 Mutations. *Clin. Chem.* **2016**, *62*, 236–242. [CrossRef] [PubMed]

32. Molin, A.; Baudoin, R.; Kaufmann, M.; Souberbielle, J.C.; Ryckewaert, A.; Vantyghem, M.C.; Eckart, P.; Bacchetta, J.; Deschenes, G.; Kesler-Roussey, G.; et al. CYP24A1 Mutations in a Cohort of Hypercalcemic Patients: Evidence for a Recessive Trait. *J. Clin. Endocrinol. Metab.* **2015**, *100*, E1343–E1352. [CrossRef] [PubMed]
33. Huang, J.; Coman, D.; McTaggart, S.J.; Burke, J.R. Long-term follow-up of patients with idiopathic infantile hypercalcaemia. *Pediatr. Nephrol.* **2006**, *21*, 1676–1680. [CrossRef]
34. Nguyen, M.; Boutignon, H.; Mallet, E.; Linglart, A.; Guillozo, H.; Jehan, F.; Garabedian, M. Infantile hypercalcemia and hypercalciuria: New insights into a vitamin D-dependent mechanism and response to ketoconazole treatment. *J. Pediatr.* **2010**, *157*, 296–302. [CrossRef] [PubMed]
35. Dauber, A.; Nguyen, T.T.; Sochett, E.; Cole, D.E.; Horst, R.; Abrams, S.A.; Carpenter, T.O.; Hirschhorn, J.N. Genetic defect in CYP24A1, the vitamin D 24-hydroxylase gene, in a patient with severe infantile hypercalcemia. *J. Clin. Endocrinol. Metab.* **2012**, *97*, E268–E274. [CrossRef]
36. Streeten, E.A.; Zarbalian, K.; Damcott, C.M. CYP24A1 mutations in idiopathic infantile hypercalcemia. *N. Engl. J. Med.* **2011**, *365*, 1741–1742.
37. Dinour, D.; Beckerman, P.; Ganon, L.; Tordjman, K.; Eisenstein, Z.; Holtzman, E.J. Loss-of-function mutations of CYP24A1, the vitamin D 24-hydroxylase gene, cause long-standing hypercalciuric nephrolithiasis and nephrocalcinosis. *J. Urol.* **2013**, *190*, 552–557. [CrossRef]
38. Jones, G.; Prosser, D.E.; Kaufmann, M. 25-Hydroxyvitamin D-24-hydroxylase (CYP24A1): Its important role in the degradation of vitamin D. *Arch. Biochem. Biophys.* **2012**, *523*, 9–18. [CrossRef]
39. Nadeau, J.H. Modifier genes in mice and humans. *Nat. Rev. Genet.* **2001**, *2*, 165–174. [CrossRef]

MDPI

Review

Vitamin D and Bone: A Story of Endocrine and Auto/Paracrine Action in Osteoblasts

Marjolein van Driel * and Johannes P. T. M. van Leeuwen

Department of Internal Medicine, Room Ee585c, Erasmus MC, Dr. Molewaterplein 40, 3015 GD Rotterdam, The Netherlands
* Correspondence: m.vandriel@erasmusmc.nl; Tel.: +31-107033046

Abstract: Despite its rigid structure, the bone is a dynamic organ, and is highly regulated by endocrine factors. One of the major bone regulatory hormones is vitamin D. Its renal metabolite 1α,25-OH2D3 has both direct and indirect effects on the maintenance of bone structure in health and disease. In this review, we describe the underlying processes that are directed by bone-forming cells, the osteoblasts. During the bone formation process, osteoblasts undergo different stages which play a central role in the signaling pathways that are activated via the vitamin D receptor. Vitamin D is involved in directing the osteoblasts towards proliferation or apoptosis, regulates their differentiation to bone matrix producing cells, and controls the subsequent mineralization of the bone matrix. The stage of differentiation/mineralization in osteoblasts is important for the vitamin D effect on gene transcription and the cellular response, and many genes are uniquely regulated either before or during mineralization. Moreover, osteoblasts contain the complete machinery to metabolize active 1α,25-OH2D3 to ensure a direct local effect. The enzyme 1α-hydroxylase (*CYP27B1*) that synthesizes the active 1α,25-OH2D3 metabolite is functional in osteoblasts, as well as the enzyme 24-hydroxylase (*CYP24A1*) that degrades 1α,25-OH2D3. This shows that in the past 100 years of vitamin D research, 1α,25-OH2D3 has evolved from an endocrine regulator into an autocrine/paracrine regulator of osteoblasts and bone formation.

Keywords: vitamin D metabolism; vitamin D receptor; bone; osteoblasts; differentiation and mineralization

Citation: van Driel, M.; van Leeuwen, J.P.T.M. Vitamin D and Bone: A Story of Endocrine and Auto/Paracrine Action in Osteoblasts. *Nutrients* **2023**, *15*, 480. https://doi.org/10.3390/nu15030480

Academic Editor: Carsten Carlberg

Received: 2 December 2022
Revised: 11 January 2023
Accepted: 12 January 2023
Published: 17 January 2023

1. Introduction

The skeleton plays a fundamental role in the human body by providing structural support and allowing movement. Moreover, it has a protective role for vital internal organs and stem cells, is a source for mineral and growth factors, and is the center of regulatory pathways. Bone is highly dynamic and undergoes continuous remodeling throughout life; it can repair itself. To illustrate this, damaged or (micro)fractured areas are removed by osteoclastic bone resorption, which is followed by new bone formation by osteoblasts (bone remodeling). Bone formation is characterized by secretion of an extracellular proteinaceous matrix, which is subsequently mineralized. Bone remodeling is tightly controlled by an interplay of local, bone and bone marrow-derived factors (e.g., cytokines, growth factors, chemokines) and endocrine factors. One of these endocrine factors is the seco-steroid 1α,25-dihydroxyvitamin D_3 (1α,25-OH_2D_3). 1α,25-OH_2D_3 can affect bone in a direct as well as an indirect manner [1–3]. The indirect effect occurs via control of calcium reabsorption in the kidney and absorption in the intestine, as well as via control of parathyroid hormone production. Although rickets and osteomalacia were prevented in vitamin D receptor (*VDR*) knockout mice fed with a rescue diet that contained high levels of calcium and phosphorus, not all bone changes were rescued, indicating the importance of a direct role for 1α,25-OH_2D_3 in bone metabolism [4–6]. The presence of VDRs in cells of the osteoblast lineage [7,8] enables direct effects of 1α,25-OH_2D_3 on bone metabolism. *VDR* expression in osteoblasts can be regulated by 1α,25-OH_2D_3 itself, as well as by other factors including parathyroid hormone, glucocorticoids, transforming growth factor-β, and epidermal growth factor [9–13]. Transgenic mice specifically overexpressing the *VDR* in osteoblasts have increased trabecular bone volume and increased bone strength, supporting an anabolic effect of 1α,25-OH_2D_3 [14]. This observation was confirmed in a study using mice with a different genetic background [15]. Interestingly, a study with global *VDR* knockout mice [5] knockout mice reported a similar phenotype, with increased trabecular thickness and increased osteoid volume and osteoblast numbers, suggesting an inhibitory effect of 1α,25-OH_2D_3 on bone formation. This was supported by data from an osteoblast-specific *VDR* knockout mouse study [16]. In this latter study, the bone effect appeared to be via reduced bone resorption. The effects on bone may be related to overall levels of calcium intake [17], but whether this explains the apparent opposite effects in murine studies remains to be established. Nevertheless, these observations support a direct effect of 1α,25-OH_2D_3 on bone metabolism via osteoblasts. There is less consensus on VDR expression in osteoclasts. Genomic deletion of the *VDR* in osteoclasts did not impact the positive effect of a 1α,25-OH_2D_3 analog (eldecalcitol) on bone mass [7]. This is supported by Verlinden et al., who reported that *VDR*s in osteoclast precursors are not essential to maintain bone homeostasis [18]. It was concluded that 1α,25-OH_2D_3 regulates osteoclasts indirectly via cells of the osteoblast lineage. In the current review, we will focus on 1α,25-OH_2D_3 in osteoblast function and bone metabolism.

2. Literature Search Strategy

We built on our pre-existing literature database and expanded this with a new search from 2016 until October 2022. With the support of the Erasmus MC Medical Library Literature Search Service, the search strategy was developed and executed. Supplemental Figure S1 shows in detail the search strings used. In this way, we obtained a list of 2713 publications on vitamin D. From this dataset, we excluded 2583 clinical and (genetic) epidemiological association studies and focused on 128 bone-related molecular and cellular studies. Two publications appeared to be retracted after the search was performed.

3. Osteoblasts

Osteoblasts originate from mesenchymal stromal cells via a tightly controlled differentiation process. The eventual fate of osteoblasts is three-fold, either to become lining cells that cover the bone surface, or to become embedded in the extracellular matrix as osteocytes, or to die via apoptosis.

3.1. Proliferation and Apoptosis

The data on 1α,25-OH_2D_3 effects on osteoblast proliferation are variable. Inhibition [19–27], as well as stimulation [20,28] or no effect [29,30] on the proliferation of osteoblasts of mouse, rat, and human origins have been reported. Effects on cell viability [31] and apoptosis [32,33] have also been documented. Although different directions in effect have been observed, these data demonstrate direct effects of 1α,25-OH_2D_3 on osteoblast proliferation and survival. The direction of effect may depend on the timing of treatment, dosage, origin, and environment of the osteoblasts [27,34–36].

3.2. Differentiation

Immature mesenchymal stromal cells differentiate into osteoblasts that produce extracellular matrix proteins, enzymes, and matrix vesicles involved in the mineralization of the extracellular matrix produced (Figure 1). It has been demonstrated that 1α,25-OH_2D_3 impacts all of these processes [3,37,38]. 1α,25-OH_2D_3 stimulation of differentiation has been shown in all in vitro studies using human osteoblasts, human mesenchymal stem cells, and osteogenic-induced pluripotent stem cells [30,39–46]. Most studies with rat osteoblasts resemble these studies using human osteoblasts and show increased differentiation [29,47,48]. Studies with mouse osteoblasts are more diverse. These studies show inhibition [49,50], as well as stimulation of osteoblast differentiation by 1α,25-OH_2D_3 [51]. The definitive explanation for the discrepancies in 1α,25-OH_2D_3 effects between, on the one hand, mouse osteoblast cultures, and on the other hand, between mouse and human/rat osteoblast cultures, is absent; however, several explanations can be put forward. The source of osteoblasts may play a role. Different sites of the skeleton differ in origin and bone formation, such as enchondral (long bones) and intramembranous (calvaria) sites. 1α,25-OH_2D_3 did not affect osteoblasts from cortical bone, and inhibited differentiation of calvaria-derived cells [52,53]. Furthermore, within one skeletal element, differences in osteoblast regulation are observed. A recent study reported differences between periosteal- and bone-marrow-derived osteoblasts in cortical bone [54]. Whether this fully explains the diverse effects observed is not clear, but it shows the importance of origin for the eventual activity and regulation. This may also relate to stage of osteoblast differentiation, donor age, culture conditions, etc., which have been shown to relate to 1α,25-OH_2D_3 action [17,47,55,56]. Furthermore, differences may be species-intrinsic, and may have a genomic explanation. 1α,25-OH_2D_3 increases *RUNX2* and *BGLAP* (osteocalcin) gene expressions in human osteoblasts, while in murine osteoblasts, 1α,25-OH_2D_3 treatment inhibits the gene expressions of *RUNX2* and *BGLAP* [43,57–61].

A picture that emerges from all in vitro osteoblast data is that the osteoblast (micro)environment is a determinant of the eventual outcome of 1α,25-OH_2D_3 action. The extracellular milieu (growth factors, cytokines, matrix proteins, ions (calcium/phosphate), and other signaling molecules) and the intracellular milieu (e.g., the insulin-like growth factor binding protein-6) are important for the eventual effect of 1α,25-OH_2D_3 [62,63]. For example, interactions with transforming growth factor-β, insulin-like growth factor, bone morphogenetic proteins, and interferon have been demonstrated [64–69]. Consequently, the absence or presence of these, but potentially other factors as well, can modulate 1α,25-OH_2D_3 action and determine the eventual response. An example of interaction with other intracellular regulatory pathways is Wnt signaling. Wnt signaling plays an important role in osteoblast differentiation and bone formation. An interplay between 1α,25-OH_2D_3 and Wnt signaling has been described [70–74].

Osteoblast differentiation, bone matrix production, and mineralization, as part of bone formation, are high energy-demanding processes [75–77]. Regulation of energy metabolism impacts osteoblast differentiation and bone formation [78–80]. Vitamin D and energy metabolism have been discussed in relation to obesity and metabolic syndrome [81] and to cancer [82–84], but data on vitamin D and energy metabolism in the context of osteoblast differentiation remain limited. Forkhead Box O (*FoxO*) transcription factors are regulated by 1α,25-OH_2D_3 in murine MC3T3 osteoblasts. *FoxO3a* is upregulated, *FoxO1* is downregulated, and *FoxO4* is unchanged after 1α,25-OH_2D_3 treatment. si-RNA knockdown of the *FoxOs* did not change 1α,25-OH_2D_3 inhibition of proliferation [85]. Unfortunately, the effect on differentiation was not reported. Changes in *FoxO* expression were coupled to increase in reactive oxygen species accumulation, which may be linked to cellular metabolism and bone formation [75,80,86]. Glucose, insulin, and 1α,25-OH_2D_3 regulation of osteoblast proliferation, alkaline phosphatase activity, and production of (uncarboxylated) osteocalcin have been studied in isolated rat osteoblasts, but unfortunately, no coupling to mineralization was made [87]. Nevertheless, these data, together with those on interactions between vitamin D and *PPAR*γ signaling in osteoblast differentiation [88], support that control of energy metabolism can be a vitamin D target in bone formation and mineralization.

3.3. Mineralization

Mineralization can be divided into two phases. In the first phase, formation of hydroxyapatite (HA) crystals takes place in nano-sized extracellular matrix vesicles produced by osteoblasts. In the second phase, HA is propagated outside these vesicles, with a resulting buildup of mineral in the extracellular matrix [89,90]. Calcium and phosphate concentrations increase inside matrix vesicles via involvement of specific proteins, and when the solubility product of calcium and phosphate is exceeded, mineral deposits are formed inside the extracellular vesicles and the second phase of mineralization starts with the release of the preformed HA crystals [90,91]. Proteomic analyses of extracellular matrix vesicles revealed many proteins with a potential role in mineralization [92,93]. Gene profiling studies also identified novel regulators of osteoblast matrix mineralization [94].

Mineralization is controlled by a balanced action of promoters and inhibitors. Alkaline phosphatase and bone sialoprotein are important promoters [95,96]. Alkaline phosphatase increases the phosphate concentration in matrix vesicles by hydrolyzing inorganic pyrophosphate. Pyrophosphate is an inhibitor of mineralization; consequently, alkaline phosphatase also decreases the level of this inhibitor. Pyrophosphatase phosphodiesterase 1 (NPP1, encoded by the gene *ENPP1*) and ankylosis protein (ANK) are involved in inhibiting mineralization. NPP1 generates pyrophosphate, and the transmembrane protein ANK allows pyrophosphate to pass through the plasma membrane to the extracellular matrix; thus, HA formation is inhibited in the extracellular vesicles [97,98]. $1\alpha,25\text{-}OH_2D_3$ stimulates mineralization via direct action on osteoblasts [68,88,99]. $1\alpha,25\text{-}OH_2D_3$ can influence the mineralization process via gene expression and matrix vesicle production. Gene expression profiling studies demonstrated that the $1\alpha,25\text{-}OH_2D_3$ effect is not likely primarily due to changes in the expression of extracellular matrix genes, and thereby to changes in composition of the extracellular matrix [99]. Studies on the expression and production of procollagen type I by human osteoblasts showed stimulation [100,101] as well as no effect [101–104], or inhibition [105].

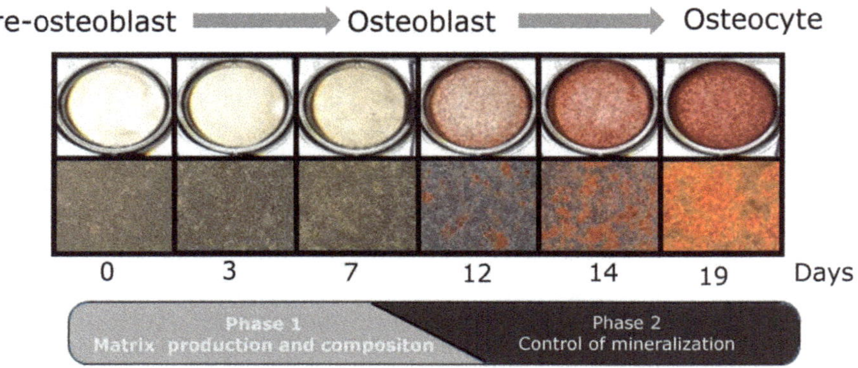

Figure 1. Alizarin red staining of osteoblast culture exemplifying the pre-mineralization and mineralization phases. Red staining shows mineralization. Details on cell culture and Alizarin red staining procedures can be found in Woeckel et al. [99]. Adapted with permission from Eijken, M., Koedam, M., van Driel, M., Buurman, C.J., Pols, H.A.P., van Leeuwen J.P.T.M. The essential role of glucocorticoids for proper human osteoblast differentiation and matrix mineralization. Mol Cell Endocrinol 2006, 248(1–2):87–93. https://doi.org/10.1016/j.mce.2005.11.034. 2006, J.P.T.M. van Leeuwen.

It is postulated that vitamin D may enhance mineralization by stimulating both NPP1, generating pyrophosphate, and alkaline phosphatase, generating phosphate from pyrophosphate [106]. This involves acceleration of the production of alkaline phosphatase-positive matrix vesicles, leading to enhanced formation and deposition of HA crystals, and eventually mineralization. This direct effect of vitamin D occurred in the period prior to the onset of mineralization, and also involved accelerated extracellular matrix maturation [99]. Interestingly, treatment with vitamin D after initiation of mineralization did not affect mineralization. This supports the above-described osteoblast differentiation stage dependency of the $1\alpha,25\text{-}OH_2D_3$ effect. A study by Yajima et al. described the significance of $1\alpha,25\text{-}OH_2D_3$ for osteocytic perilacunar mineralization [107].

$1\alpha,25\text{-}OH_2D_3$ also directly stimulates the production of inhibitors of mineralization. VDR-dependent $1\alpha,25\text{-}OH_2D_3$ expression of *ENPP1* and *ANK* in murine osteoblasts led to an increase in the mineralization inhibitor pyrophosphate [108]. $1\alpha,25\text{-}OH_2D_3$ also stimulates activin A expression in human osteoblasts. Treatment with the activin A blocker follistatin enhanced vitamin-D-induced mineralization of human osteoblasts [109]. 1α,25-OH_2D_3 also increases the expression of osteopontin, which is shown to inhibit mineralization. These observations may provide a fine-tuning mechanism to prevent excessive mineralization of bone. 1α,25-OH_2D_3 induction of carboxylated osteocalcin may be in line with this. 1α,25-OH_2D_3-stimulated mineralization is enhanced by blocking osteocalcin carboxylation by warfarin [109]. The interaction of 1α,25-OH_2D_3 with other factors, as described above, also holds for mineralization, for example, the interaction with DKK1, the inhibitor of Wnt signaling [74].

The counterbalance of bone formation and mineralization by osteoblasts is bone resorption by osteoclasts. In the healthy skeleton, these processes are in balance, securing healthy and strong bones. The osteoblasts/osteocytes are the major regulators of osteoclast formation and action via production of the stimulating factor RANKL, and the RANKL inhibitor, osteoprotegerin (OPG). 1α,25-OH_2D_3 influences the RANKL/OPG ratio, and thereby also impacts bone resorption [110–113]. 1α,25-OH_2D_3 is involved at both the bone formation and the bone resorption sides of the balance, and is an important player in maintaining healthy bones via direct effects on bone, in addition to indirect effects via calcium and phosphate homeostasis [114].

3.4. Gene Expression

The basis of all cellular effects of 1α,25-OH_2D_3 involves VDR-mediated transcriptional regulation. The VDR is a member of the nuclear receptor family. Upon binding to 1α,25-OH_2D_3, the VDR heterodimerizes with the retinoic X receptor (RXR), and binds as a dimer to the vitamin D response element (VDRE) in the DNA to regulate gene expression [115]. Over the years, many studies have unraveled the molecular fundamentals of 1α,25-OH_2D_3 transcriptional regulation. Examples and information can be found in these publications and references therein [116–118]. In a previous publication, we discussed 1α,25-OH_2D_3 and gene transcription in osteoblasts [38]. This will not be repeated or discussed in detail in this review.

A factor that may determine the transcriptional effect of 1α,25-OH_2D_3 effect is not only the basal level of gene expression [51,119], but also the stage of osteoblast differentiation [99]. Studies with rat osteoblasts in the early 1990s showed already that effects of 1α,25-OH_2D_3 on osteoblasts may depend on the osteoblast differentiation phase [119]. An example is the 1α,25-OH_2D_3 stimulation of phosphaturic hormone fibroblast growth factor 23 (FGF23) only in late-stage differentiation osteoblasts and osteocytes [120,121]. FGF23 is a hormone that acts in the kidney to enhance phosphate excretion, and suppresses 1α,25-OH_2D_3 synthesis by inhibiting 1α-hydroxylase (CYP27B1), forming an important loop in the regulation of mineralization [122,123]. Vitamin D signaling in osteocytes [124] is further supported by the 1α,25-OH_2D_3 regulation of *PHEX* (phosphate-regulating neutral endopeptidase, X linked), which suppresses FGF23 transcription [125].

The various osteoblast differentiation stages actually reflect different functional stages of the osteoblast, e.g., proliferation, extracellular matrix production, mineralizing and mechanosensing. It is important to keep in mind the osteoblast differentiation stage when studying 1α,25-OH_2D_3 effects, as this may be an important determinant of the eventual effect (e.g., stimulation or inhibition) on gene transcription and subsequent cellular responses and bone formation. The relationship between the osteoblast differentiation stage and 1α,25-OH_2D_3 gene expression control was shown by Woeckel et al. [99]. 1α,25-OH_2D_3 changed the expression of different sets of genes in the phase before the onset of mineralization, and during the mineralization. For this review, we performed a reanalysis of this gene profiling study [99] with the 2022 updated annotation. Comparison of transcripts regulated (i.e., two-fold up or down) in the phase before and after the start of mineralization (Figure 1) demonstrated that only 2.5% (18 out of the 721 regulated transcripts) were regulated in both phases (Table 1). The gene symbols of the transcripts regulated in both phases are shown in Table 2. To focus in more detail on phase-specific gene expression, we next selected the transcripts that were uniquely regulated in either the pre-mineralization or in the mineralization phase [99]. In this regard, the transcripts should be at least two-fold up- or downregulated in one phase (either pre-mineralization or mineralization phase), and not regulated (fold change on average between 0.8 and 1.2) in the other phase (either the mineralization or pre-mineralization phase). Table 3 shows the number of transcripts uniquely regulated in either of these phases, and Table 4 reports the gene symbols belonging to these transcripts. This binary comparison of pre-mineralization and mineralization phases is not absolute and does not mean that further zooming in on specific phases of osteoblast differentiation will not reveal other sets of vitamin-D-regulated genes. However, it does support the notion that vitamin D gene regulation during osteoblast differentiation and mineralization displays temporal dynamics, and it does show that for proper interpretation of vitamin D effects, knowledge on the differentiation and functional stage of cells and tissues is important. This knowledge can explain the apparent differences in 1α,25-OH_2D_3 effects that have been reported.

Table 1. Number of transcripts on average that are 2-fold up- or downregulated in the pre-mineralization or mineralization phase of human osteoblasts *.

Condition	# of Genes UP	# of Genes DOWN
Pre-mineralization phase	155	164
Mineralization phase	166	236
In both phases	10	8

* Experimental procedures and culture conditions of human osteoblasts (SV-HFO) are described in Woeckel et al. [99]. Two-fold change is based on the average expression at the timepoints in the pre-mineralization or mineralization period.

Table 2. Gene symbols of transcripts that are 2-fold upregulated or downregulated in both the pre-mineralization and mineralization phases of human osteoblasts (i.e., 10 and 8 in both phases in Table 1) *.

Upregulated	Downregulated
ABCC3	*AGAP10*
CYP24A1	*CCL20*
MAGEE1	*DDIT3*
RARRES2	*GRK4*
RICH2	*LOC727869*
SLC25A45	*NFE2L2*
SULT1C2	*ODF1*
THBD	*TSC22D2*
TMEM180	
TOX3	

* Experimental procedures and culture conditions of human osteoblasts (SV-HFO) are described in Woeckel et al. [99]. Two-fold change is based on the average expression at the timepoints in the pre-mineralization or mineralization period.

Table 3. Number of transcripts uniquely 2-fold up- or downregulated in either the pre-mineralization or in the mineralization phase of human osteoblasts *.

Condition	# of Genes UP	# of Genes DOWN
Pre-mineralization phase	65	66
Mineralization phase	77	100

* Experimental procedures and culture conditions of human osteoblasts (SV-HFO) are described in Woeckel et al. [99]. The 2-fold and 0.8–1.2-fold change is based on the average expression at the timepoints in the pre-mineralization or mineralization period.

Table 4. Overview of transcript gene names that are uniquely 2-fold up- or downregulated in either the pre-mineralization or mineralization phase of human osteoblasts *.

Pre-Mineralization Phase				Mineralization Phase			
Upregulated		Downregulated		Upregulated		Downregulated	
AQR	RAB9BP1	ADAM22	RARA	ABCD4	MYH11	AASDH	MOSPD1
ARHGEF7	RLTPR	ADORA1	RBM	AKAP13	NFIX	ABCD3	MRPS23
ATAT1	SARDH	ATF7IP2	RIMKLB	ANKRD11	ORC5L	ABT1	MS4A1
ATG16L1	SHISA8	BAGE	SLC19A1	APIP	PCDHB3	ACTR3C	MTUS2
ATP1A4	SLC38A11	BRS3	SLC26A7	ARHGDIB	PDLIM5	ANUBL1	NCRNA00188
BCL11A	SZT2	BRWD1	SLC3A1	ASH1L	PDZRN4	AP5S1	NDRG2
BMF	TEX9	BST2	SNRPN	ATM	PGAP1	B4GALNT2	NDUFB7
BMP15	TMEM120B	C1orf68	TBK1	BNC2	PLEKHG2	C11orf65	NRAP
C15orf48	TMEM33	CACNA1A	TFAP4	BPTF	PPP4R4	C14orf156	NUDT14
C2orf27A	UBE2G2	CCDC144C	THPO	BRD4	PRPF18	C14orf2	OGFR
C3orf20	UBXN10	CSF2RA	TMPRSS15	CAP1	PTGES	C17orf104	PANK2
C8orf34	UNC13C	CTNS	TRIB3	CCDC67	PTGS1	C4orf36	PAPPA
CCDC124	ZC3H12A-DT	DEFB132	TRMT2A	CCDC76	RAB3IP	CCL5	PAX8
COL24A1	ZNF668	EDA	TTBK2	CD14	RASAL2	CCT2	PIP5K1A
CTU2	ZNF703	ERCC6L2	ZNF396	CLCN4	RG9MTD2	CNOT2	PLCH1
DCTN2		FAM219A	ZNF93	CROCCL1	SERTAD4	COX7C	PMCH
DOCK6		FCGR2C		DCLK3	SMARCA4	CSRP2BP	PML
DST		FLJ10213		DPP4	SRGAP1	DAZL	POLE4
DUSP28		FSD1L		EGFR	SRRM2	DBI	POLR2K
EPG5		GAS2		EP300	SULF1	DCUN1D1	PTPRA
EYA2		GLIPR1		FAM102A	TBC1D13	DNAH1	RHEB
GABRB3		GPR155		FAM186A	TBL1X	DUSP16	RPAIN
GNRHR		HM13		FAM20C	UGGT2	EEF1D	RPL13
HCRTR2		ICA1		FGF7	VCAN	EGFL8	RPL14
HIST1H4C		KLHL36		FLJ11292	ZNF397	EHD1	RPL34
HSPB7		KLK7		FLJ13773	ZNF462	ELP6	RPS11
IL1RN		LEKR1		FOXP2	ZSWIM1	ESPNL	SEMA6D
KCNJ15		LELP1		GABRA5		EXOG	SHLD1
LOC100131283		LIN28B		H2AFY		EXOSC2	SLC10A7
LOC148987		LOC100286895		HMCN1		FABP4	SLC9A5
LOC149351		LOC100287114		HOXA6		FAM126A	SNAP23
LOC285205		LOC283854		HSPA12A		FAM27A	SNCAIP
LOC645591		LOC285692		IL17C		FAXC	SNTG1
LOC728903		LOC390595		INTS4		FUT7	STEEP1
LOC780529		LOC440944		KCNAB1		GOSR1	STK32A
LRRC46		MAN1A2		KCNG3		GPR39	STMN3
LYZL6		MAPRE3		KRTAP3-3		GSN	SUPT16H
MGC42157		MGC12916		LOC100127980		HCG4P6	TAL1
MRS2		MRPL19		LOC100128640		IRGQ	TBC1D8
NCOR2		MSR1		LOC100131993		KCNIP3	TEN1
NOX4		MYO10		LOC283682		KY	TLK1
NTRK2		NR2E3		LOC285500		LOC100133109	TWF1

Table 4. *Cont.*

Pre-Mineralization Phase		Mineralization Phase		
Upregulated	Downregulated	Upregulated	Downregulated	
OR1J4	NUP210L	LOC388210	LOC100287911	TXNIP
PDE1A	OTX2	LOC441461	LOC100289246	UHRF1BP1L
PENK	PCLO	MAGEB18	LOC338862	UQCRB
PGM2L1	PKP2	MARK2	LOC643749	UQCRQ
PHC3	PLXNA2	MEGF10	LPAR5	VMA21
POU2F1	POU2F2	MGAT5B	MATR3	WFDC21P
PRRG2	PRLR	MLXIP	MMP16	XAF1
PTCD3	RAD54L2	MS4A6A	MMP17	ZNF880

* Experimental procedures and culture conditions of human osteoblasts (SV-HFO) are described in Woeckel et al. [99].

4. Vitamin D Metabolism

Metabolism, synthesis of the active form of 1α,25-OH_2D_3 as well as its inactivation, has been an important research topic since the identification of vitamin D. This has contributed to the understanding of the initiation and termination actions of vitamin D and its endocrine function. Figure 2 shows the classical vitamin D metabolism pathway. Serum levels of 1α,25-OH_2D_3 are determined by the activity of the renal enzyme 1α-hydroxylase (CYP27B1). 24-Hydroxylase (CYP24A1) is the first step of a 1α,25-OH_2D_3 inactivation cascade present in all target tissues. In the next sections, we discuss CYP27B1 and CYP24A1 in osteoblasts.

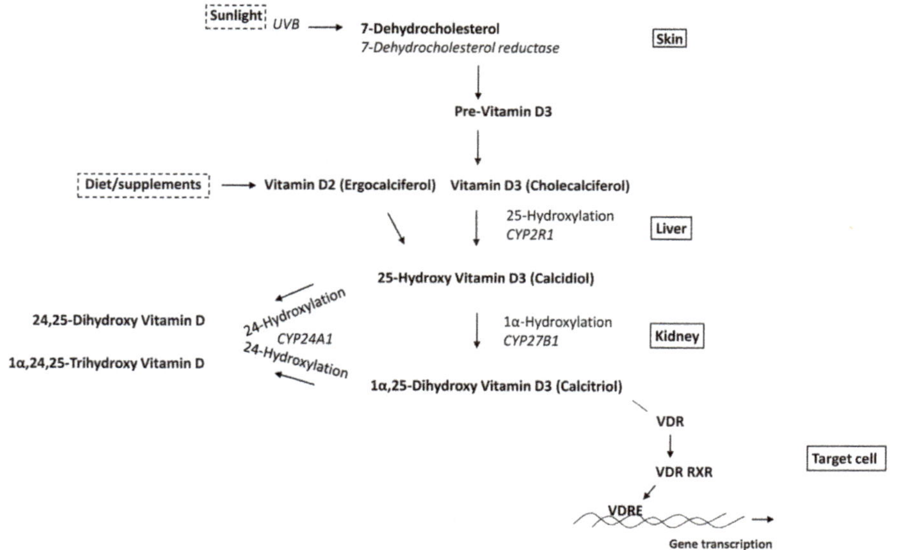

Figure 2. Schematic representation of classic vitamin D metabolism and signaling pathway. Either from sunlight or food, vitamin D is converted via enzymatic reactions in the liver and kidney into its active metabolite, 1α,25-OH_2D_3, which binds to the VDR. Gene activation follows after binding of the vitamin D/receptor complex to vitamin D response elements (VDREs) in target genes.

4.1. *CYP27B1*

In the late 1970s and early 1980s, reports were already coming out that in tissues other than the kidney, 1α,25-OH_2D_3 can be synthesized. Cells isolated from chicken calvaria [126] and human osteosarcoma cells, as well as bone cells isolated from an ileac crest biopsy [127], can produce 1α,25-OH_2D_3.. Its functional significance in human osteoblasts was shown by the fact that inhibition of 1α-hydroxylase activity by ketoconazole blocked the 25(OH)D_3 induction of *CYP24A1* and osteocalcin expression [30]. This was supported by studies on siRNA silencing in human osteoblasts [19,46]. Additional evidence came from a study showing the importance of CYP27B1 for proliferation and osteogenic differentiation of human mesenchymal stromal cells (MSCs) [128,129]. MSCs of older donors had reduced *CYP27B1* expression and resistance to 25(OH)D_3 regulation of osteoblast differentiation [130]. Broader tissue distribution of extra renal CYP27B1 expression beyond bone was recently summarized by Bikle et al. [131].

However, renal synthesis is still considered the major contributor to circulating 1α,25-OH_2D_3 levels. Only in diseases such as sarcoidosis extra is renal synthesis sufficient to contribute to circulating levels. The presence of 1α,25-OH_2D_3 synthesis within bone provides a means to explain the associations of bone phenotypes and other parameters with circulating 25(OH)D_3 and not with 1α,25-OH_2D_3, as discussed by Anderson and colleagues [132,133]. Pharmacokinetic differences between locally produced 1α,25-OH_2D_3 from 25(OH)D_3 and added 1α,25-OH_2D_3 have been suggested from a cell culture study [134]. Further studies, in particular, in vivo studies, are needed for full appreciation of the impact of an autocrine/paracrine role of 1α,25-OH_2D_3.

Observations that the vitamin-D-binding protein receptors cubulin and megalin, as well as the vitamin D_3 25-hydroxylase genes *CYP2R1* and *CYP3A4*, are also expressed in human osteoblasts, supports an autocrine/paracrine role [19,30,131].

Renal CYP27B1 is tightly controlled by factors such as parathyroid hormone (PTH) and fibroblast growth factor-23 (FGF23), which are involved in calcium and phosphate homeostasis. Extrarenal CYP27B1 expression is differently regulated, and probably involves other factors and tissue specificity [135]. For example, PTH and ambient calcium do not regulate *CYP27B1* in human osteoblasts [30], while 1α,25-OH_2D_3 reduces *CYP27B1* expression in human MSCs similar as in the kidney [136]. Several growth factors and cytokines can regulate CYP27B1 expression. IGF-I increases *CYP27B1* expression in human MSCs [136]. Interleukin-1 stimulates while interferon-β reduces *CYP27B1* expression in human osteoblasts [30,69]. The earlier described impact of the osteoblast differentiation stage on 1α,25-OH_2D_3 action can also be translated to expression of CYP27B1. *CYP27B1* expression is increased by 25(OH)D_3 in human MSCs [136], but not in mature osteoblasts [30].

4.2. *CYP24A1*

The first step in the degradation cascade of 1α,25-OH_2D_3 is hydroxylation at the C-24 position by 24-hydroxylase (CYP24A1) [137]. CYP24A1 is expressed in all vitamin D target cells, and its expression is very rapidly and strongly increased after 1α,25-OH_2D_3 binding to VDRs [138–141]. The VDR level is tightly linked to the induction of CYP24A1 expression and 24-hydroxylase activity and, consequently, degradation of 1α,25-OH_2D_3. Thus, the homologous upregulation of VDRs concomitantly induces the inactivation of 1α,25-OH_2D_3, and thereby limits its effect [142,143]. Hydroxylation at the C-24 position of 1α,25-OH_2D_3 or 25(OH)D_3 alone does not immediately lead to an inactive vitamin D molecule. Henry and Norman demonstrated in the 1970s the functional significance of 24,25$(OH)_2$D3 for normal chicken egg hatchability and calcium and phosphorus homeostasis [144,145]. The effects of 24,25$(OH)_2D_3$ on bone metabolism were shown in human, chicken, rat, and mouse studies. 24,25$(OH)_2D_3$, synergistically with PTH, directly stimulates mineralization, and 24,25$(OH)_2D_3$ decreases the number and size of resorption sites on the bone surface [146,147]. 24,25$(OH)_2D_3$ restores and accelerates the bone mineral apposition rate in vitamin-D-deficient and in parathryoidectomized rats [147]. 24,25$(OH)_2D_3$ did not change bone histomorphometric parameters in ovariectomized rats [148], but 24,25$(OH)_2D_3$, and not 1α,25-OH_2D_3, increased bone strength [149].

Several studies focused on 24-hydroxylated vitamin D molecules and fracture healing. 24,25$(OH)_2D_3$ binds to fracture calluses [150], and improves fracture healing [151–153]. Serum 24,25$(OH)_2D_3$ levels were found to correlate with fracture healing in chicken [151], but not in a small human study in 1978 [154]. However, a study on pre-dialysis renal insufficiency patients supported a direct, i.e., PTH-independent, functional role of $24,25(OH)_2D_3$ in human bone. $24,25(OH)_2D_3$, together with $1\alpha,25\text{-}(OH)_2D_3$, preserved the osteoblast perimeter and improved mineralization, while $1\alpha,25\text{-}(OH)_2D_3$ alone was ineffective [155]. A direct effect on bone, in particular osteoblasts, is supported by in vitro studies showing that, similarly to $1\alpha,25\text{-}OH_2D_3$, $24,25(OH)_2D_3$ has direct effects on human osteoblast differentiation [45]. Knowing that 24-hydroxylation per se does not lead to inactivation of vitamin D molecules, it is important to understand target tissue/target cell dynamics of the next steps in the degradation cascade. Control of the velocity of the subsequent steps in the degradation pathway can be a means to regulate vitamin D action in target tissues/cells. Together, these data on CYP24A1 and the biological activities of $24,25(OH)_2D_3$ add to the notion of an auto/paracrine vitamin D regulatory system in bone. This system is most likely not restricted to bone and may also be present in other tissues.

5. Conclusions

This review revealed that the central role for vitamin D in bone physiology is directed via osteoblasts and depends on their stage of development. VDRs and the vitamin-D-metabolizing enzymes CYP27B1 and CYP24A1, known from the vitamin D endocrine system, are present and functional in osteoblasts. This uncovers a direct local role for $1\alpha,25\text{-}OH_2D_3$ vitamin D in osteoblast function, and expands the vitamin D action profile from endocrine regulation of calcium and phosphate homeostasis to an auto/paracrine regulatory network in bone. Several target-tissue-derived factors (growth factors, cytokines), intracellular signaling cascades (Wnt), and functional states of the osteoblast interact with this auto/paracrine network and determine the eventual response. In this way, vitamin D controls the proliferation, apoptosis, differentiation, and mineralization of osteoblasts, as well as their gene profile and interaction with other factors that maintain healthy bone. Moreover, even local degradation products of vitamin D metabolism ($24,25(OH)_2D_3$) have a beneficial contribution to osteoblast function. Together, these observations underscore the importance of contextual knowledge (molecular and cellular) in order to fully understand and appreciate the effects of vitamin D on bone cells.

This warrants research for the next 100 years: future studies may focus on assessing tissue levels of vitamin D metabolites in addition to circulating levels, and study functionality of the complete metabolic profile of vitamin D.

Supplementary Materials: The following supporting information can be downloaded at https://www.mdpi.com/article/10.3390/nu15030480/s1, Figure S1: Literature search strategy performed October 2022.

Author Contributions: M.v.D.: writing—review and editing; J.P.T.M.v.L.: writing—review and editing. All authors have read and agreed to the published version of the manuscript.

Funding: This review received no external funding.

Institutional Review Board Statement: Not applicable.

Informed Consent Statement: Not applicable.

Data Availability Statement: Not applicable.

Acknowledgments: We thank Jeroen van de Peppel from the Erasmus MC Department of Internal Medicine for support with reanalysis of gene expression data, and we are very grateful to Wichor Bramer from the Erasmus MC Medical Library for his support in developing and updating the literature search strategies.

Conflicts of Interest: The authors declare no conflict of interest.

References

1. Datta, H.K.; Ng, W.-F.; Walker, J.A.; Tuck, S.P.; Varanasi, S.S. The cell biology of bone metabolism. *J. Clin. Pathol.* **2008**, *61*, 577–587. [CrossRef] [PubMed]
2. Bikle, D.D. Vitamin D and Bone. *Curr. Osteoporos. Rep.* **2012**, *10*, 151–159. [CrossRef] [PubMed]
3. Christakos, S.; Li, S.; DeLa Cruz, J.; Verlinden, L.; Carmeliet, G. Vitamin D and Bone. *Handb. Exp. Pharmacol.* **2020**, *262*, 47–63. [CrossRef] [PubMed]
4. Li, Y.C.; Amling, M.; Pirro, A.E.; Priemel, M.; Meuse, J.; Baron, R.; Delling, G.; Demay, M.B. Normalization of Mineral Ion Homeostasis by Dietary Means Prevents Hyperparathyroidism, Rickets, and Osteomalacia, But Not Alopecia in Vitamin D Receptor-Ablated Mice. *Endocrinology* **1998**, *139*, 4391–4396. [CrossRef]
5. Amling, M.; Priemel, M.; Holzmann, T.; Chapin, K.; Rueger, J.M.; Baron, R.; DeMay, M.B. Rescue of the Skeletal Phenotype of Vitamin D Receptor-Ablated Mice in the Setting of Normal Mineral Ion Homeostasis: Formal Histomorphometric and Biomechanical Analyses. *Endocrinology* **1999**, *140*, 4982–4987. [CrossRef]
6. Panda, D.K.; Miao, D.; Bolivar, I.; Li, J.; Huo, R.; Hendy, G.N.; Goltzman, D. Inactivation of the 25-Hydroxyvitamin D 1α-Hydroxylase and Vitamin D Receptor Demonstrates Independent and Interdependent Effects of Calcium and Vitamin D on Skeletal and Mineral Homeostasis. *J. Biol. Chem.* **2004**, *279*, 16754–16766. [CrossRef]
7. Nakamichi, Y.; Udagawa, N.; Horibe, K.; Mizoguchi, T.; Yamamoto, Y.; Nakamura, T.; Hosoya, A.; Kato, S.; Suda, T.; Takahashi, N. VDR in Osteoblast-Lineage Cells Primarily Mediates Vitamin D Treatment-Induced Increase in Bone Mass by Suppressing Bone Resorption. *J. Bone Miner. Res.* **2017**, *32*, 1297–1308. [CrossRef]
8. Zarei, A.; Morovat, A.; Javaid, K.; Brown, C.P. Vitamin D receptor expression in human bone tissue and dose-dependent activation in resorbing osteoclasts. *Bone Res.* **2016**, *4*, 16030. [CrossRef]
9. Pols, H.A.P.A.; van Leeuwen, J.P.T.M.P.; Schilte, J.P.P.; Visser, T.J.J.; Birkenhäger, J.C.C. Heterologous up-regulation of the 1,25-dihydroxyvitamin D3 receptor by parathyroid hormone (PTH) and PTH-like peptide in osteoblast-like cells. *Biochem. Biophys. Res. Commun.* **1988**, *156*, 588–594. [CrossRef]
10. Van Leeuwen, J.P.; Birkenhäger, J.C.; Buurman, C.J.; Bemd, G.J.V.D.; Bos, M.P.; A Pols, H. Bidirectional regulation of the 1,25-dihydroxyvitamin D3 receptor by phorbol ester-activated protein kinase-C in osteoblast-like cells: Interaction with adenosine 3′,5′-monophosphate-induced up-regulation of the 1,25-dihydroxyvitamin D3 receptor. *Endocrinology* **1992**, *130*, 2259–2266. [CrossRef]
11. van Leeuwen, J.P.; Pols, H.A.; Schilte, J.P.; Visser, T.J.; Birkenhäger, J.C. Modulation by epidermal growth factor of the basal $1,25(OH)_2D_3$ receptor level and the heterologous up-regulation of the $1,25(OH)_2D_3$ receptor in clonal osteoblast-like cells. *Calcif. Tissue Int.* **1991**, *49*, 35–42. [CrossRef]
12. Godschalk, M.; Levy, J.R.; Downs, R.W., Jr. Glucocorticoids decrease vitamin D receptor number and gene expression in human osteosarcoma cells. *J. Bone Miner. Res.* **1992**, *7*, 21–27. [CrossRef]
13. Reinhardt, T.A.; Horst, R.L. Parathyroid Hormone Down-Regulates 1,25-Dihydroxyvitamin D Receptors (VDR) and VDR Messenger Ribonucleic Acid in Vitro and Blocks Homologous Up-Regulation of VDR in Vivo. *Endocrinology* **1990**, *127*, 942–948. [CrossRef]
14. Gardiner, E.M.; Baldock, P.A.; Thomas, G.P.; Sims, N.A.; Henderson, N.K.; Hollis, B.; White, C.P.; Sunn, K.L.; Morrison, N.A.; Walsh, W.R.; et al. Increased formation and decreased resorption of bone in mice with elevated vitamin D receptor in mature cells of the osteoblastic lineage. *FASEB J.* **2000**, *14*, 1908–1916. [CrossRef]
15. Triliana, R.; Lam, N.N.; Sawyer, R.K.; Atkins, G.J.; Morris, H.A.; Anderson, P.H. Skeletal characterization of an osteoblast-specific vitamin D receptor transgenic (ObVDR-B6) mouse model. *J. Steroid Biochem. Mol. Biol.* **2016**, *164*, 331–336. [CrossRef]
16. Yamamoto, Y.; Yoshizawa, T.; Fukuda, T.; Shirode-Fukuda, Y.; Yu, T.; Sekine, K.; Sato, T.; Kawano, H.; Aihara, K.-I.; Nakamichi, Y.; et al. Vitamin D Receptor in Osteoblasts Is a Negative Regulator of Bone Mass Control. *Endocrinology* **2013**, *154*, 1008–1020. [CrossRef]
17. Lieben, L.; Carmeliet, G. The delicate balance between vitamin D, calcium and bone homeostasis: Lessons learned from intestinal- and osteocyte-specific VDR null mice. *J. Steroid Biochem. Mol. Biol.* **2013**, *136*, 102–106. [CrossRef]
18. Verlinden, L.; Janssens, I.; Doms, S.; Vanhevel, J.; Carmeliet, G.; Verstuyf, A. Vdr expression in osteoclast precursors is not critical in bone homeostasis. *J. Steroid Biochem. Mol. Biol.* **2019**, *195*, 105478. [CrossRef]
19. Atkins, G.J.; Anderson, P.; Findlay, D.M.; Welldon, K.J.; Vincent, C.; Zannettino, A.; O'Loughlin, P.D.; Morris, H.A. Metabolism of vitamin D3 in human osteoblasts: Evidence for autocrine and paracrine activities of 1α,25-dihydroxyvitamin D3. *Bone* **2007**, *40*, 1517–1528. [CrossRef]
20. van den Bemd, G.J.; Pols, H.A.; Birkenhäger, J.C.; Kleinekoort, W.M.; van Leeuwen, J.P. Differential effects of 1,25-dihydroxyvitamin D3-analogs on osteoblast-like cells and on in vitro bone resorption. *J. Steroid Biochem. Mol. Biol.* **1995**, *55*, 337–346. [CrossRef]
21. Döhla, J.; Kuuluvainen, E.; Gebert, N.; Amaral, A.; Englund, J.I.; Gopalakrishnan, S.; Konovalova, S.; Nieminen, A.I.; Salminen, E.S.; Torregrosa Muñumer, R.; et al. Metabolic determination of cell fate through selective inheritance of mitochondria. *Nat. Cell Biol.* **2022**, *24*, 148–154. [CrossRef] [PubMed]
22. Eelen, G.; Verlinden, L.; Van Camp, M.; Mathieu, C.; Carmeliet, G.; Bouillon, R.; Verstuyf, A. Microarray analysis of 1α,25-dihydroxyvitamin D3-treated MC3T3-E1 cells. *J. Steroid Biochem. Mol. Biol.* **2004**, *89–90*, 405–407. [CrossRef] [PubMed]

23. Chen, T.L.; Cone, C.M.; Feldman, D. Effects of 1α,25-dihydroxyvitamin D3 and glucocorticoids on the growth of rat and mouse osteoblast-like bone cells. *Calcif. Tissue Int.* **1983**, *35*, 806–811. [CrossRef] [PubMed]
24. Murray, S.S.; A Glackin, C.; Murray, E.J.B. Variation in 1,25-dihydroxyvitamin D3 regulation of proliferation and alkaline phosphatase activity in late-passage rat osteoblastic cell lines. *J. Steroid Biochem. Mol. Biol.* **1993**, *46*, 227–233. [CrossRef] [PubMed]
25. Kanatani, M.; Sugimoto, T.; Fukase, M.; Chihara, K. Effect of 1.25-Dihydroxyvitamin D3 on the Proliferation of Osteoblastic MC3T3-E1 Cells by Modulating the Release of Local Regulators from Monocytes. *Biochem. Biophys. Res. Commun.* **1993**, *190*, 529–535. [CrossRef]
26. Rubin, J.; Fan, X.; Thornton, D.; Bryant, R.; Biskobing, D. Regulation of Murine Osteoblast Macrophage Colony-Stimulating Factor Production by 1,25(OH)$_2$D$_3$. *Calcif. Tissue Int.* **1996**, *59*, 291–296. [CrossRef]
27. Maehata, Y.; Takamizawa, S.; Ozawa, S.; Kato, Y.; Sato, S.; Kubota, E.; Hata, R.-I. Both direct and collagen-mediated signals are required for active vitamin D3-elicited differentiation of human osteoblastic cells: Roles of osterix, an osteoblast-related transcription factor. *Matrix Biol.* **2006**, *25*, 47–58. [CrossRef]
28. Ishida, H.; Bellows, C.G.; E Aubin, J.; Heersche, J.N. Characterization of the 1,25-(OH)$_2$D$_3$-induced inhibition of bone nodule formation in long-term cultures of fetal rat calvaria cells. *Endocrinology* **1993**, *132*, 61–66. [CrossRef]
29. Wang, D.; Song, J.; Ma, H. An in vitro Experimental Insight into the Osteoblast Responses to Vitamin D3 and Its Metabolites. *Pharmacology* **2018**, *101*, 225–235. [CrossRef]
30. van Driel, M.; Koedam, M.; Buurman, C.J.J.; Hewison, M.; Chiba, H.; Uitterlinden, A.G.; Pols, H.A.P.; Van Leeuwen, J.P.T.M. Evidence for auto/paracrine actions of vitamin D in bone: 1a-hydroxylase expression and activity in human bone cells. *FASEB J.* **2006**, *20*, 2417–2419. [CrossRef]
31. Shi, Y.-C.; Worton, L.; Esteban, L.; Baldock, P.; Fong, C.; Eisman, J.A.; Gardiner, E.M. Effects of continuous activation of vitamin D and Wnt response pathways on osteoblastic proliferation and differentiation. *Bone* **2007**, *41*, 87–96. [CrossRef]
32. Hansen, C.M.; Hansen, D.; Holm, P.K.; Binderup, L. Vitamin D compounds exert anti-apoptotic effects in human osteosarcoma cells in vitro. *J. Steroid Biochem. Mol. Biol.* **2001**, *77*, 1–11. [CrossRef]
33. Thompson, L.; Wang, S.; Tawfik, O.; Templeton, K.; Tancabelic, J.; Pinson, D.; Anderson, H.C.; Keighley, J.; Garimella, R. Effect of 25-hydroxyvitamin D$_3$ and 1 α,25 dihydroxyvitamin D3 on differentiation and apoptosis of human osteosarcoma cell lines. *J. Orthop. Res.* **2011**, *30*, 831–844. [CrossRef]
34. Van Leeuwen, J.P.T.M.; van Driel, M.; van den Bemd, G.J.C.M.; Pols, H.A.P. Vitamin D Control of Osteoblast Function and Bone Extracellular Matrix Mineralization. *Crit. Rev. Eukaryot. Gene Expr.* **2001**, *11*, 199–226. [CrossRef]
35. Skjødt, H.; Gallagher, J.A.; Beresford, J.N.; Couch, M.; Poser, J.W.; Russell, R.G.G. Vitamin D metabolites regulate osteocalcin synthesis and proliferation of human bone cells in vitro. *J. Endocrinol.* **1985**, *105*, 391–396. [CrossRef]
36. Uranoa, T.; Hosoib, T.; Shirakic, M.; Toyoshimad, H.; Ouchia, Y.; Inoue, S. Possible Involvement of the p57Kip2 Gene in Bone Metabolism. *Biochem. Biophys. Res. Commun.* **2000**, *269*, 422–426. [CrossRef]
37. van Driel, M.; van Leeuwen, J.P. Vitamin D endocrinology of bone mineralization. *Mol. Cell. Endocrinol.* **2017**, *453*, 46–51. [CrossRef]
38. van de Peppel, J.; van Leeuwen, J.P.T.M. Vitamin D and gene networks in human osteoblasts. *Front. Physiol.* **2014**, *5*, 137. [CrossRef]
39. Zhou, S.; Glowacki, J.; Kim, S.W.; Hahne, J.; Geng, S.; Mueller, S.M.; Shen, L.; Bleiberg, I.; LeBoff, M.S. Clinical characteristics influence in vitro action of 1,25-dihydroxyvitamin D$_3$in human marrow stromal cells. *J. Bone Miner. Res.* **2012**, *27*, 1992–2000. [CrossRef]
40. Geng, S.; Zhou, S.; Bi, Z.; Glowacki, J. Vitamin D metabolism in human bone marrow stromal (mesenchymal stem) cells. *Metabolism* **2013**, *62*, 768–777. [CrossRef]
41. Piek, E.; Sleumer, L.S.; van Someren, E.P.; Heuver, L.; de Haan, J.R.; de Grijs, I.; Gilissen, C.; Hendriks, J.M.; van Ravestein-van Os, R.I.; Bauerschmidt, S.; et al. Osteo-transcriptomics of human mesenchymal stem cells: Accelerated gene expression and osteoblast differentiation induced by vitamin D reveals c-MYC as an enhancer of BMP2-induced osteogenesis. *Bone* **2010**, *46*, 613–627. [CrossRef] [PubMed]
42. Kato, H.; Ochiai-Shino, H.; Onodera, S.; Saito, A.; Shibahara, T.; Azuma, T. Promoting effect of 1,25(OH)$_2$ vitamin D3 in osteogenic differentiation from induced pluripotent stem cells to osteocyte-like cells. *Open Biol.* **2015**, *5*, 140201. [CrossRef] [PubMed]
43. Prince, M.; Banerjee, C.; Javed, A.; Green, J.; Lian, J.B.; Stein, G.S.; Bodine, P.V.; Komm, B.S. Expression and regulation of Runx2/Cbfa1 and osteoblast phenotypic markers during the growth and differentiation of human osteoblasts. *J. Cell. Biochem.* **2001**, *80*, 424–440. [CrossRef] [PubMed]
44. Jorgensen, N.R.; Henriksen, Z.; Sorensen, O.H.; Civitelli, R. Dexamethasone, BMP-2, and 1,25-dihydroxyvitamin D enhance a more differentiated osteoblast phenotype: Validation of an in vitro model for human bone marrow-derived primary osteoblasts. *Steroids* **2004**, *69*, 219–226. [CrossRef] [PubMed]
45. van Driel, M.; Koedam, M.; Buurman, C.J.J.; Roelse, M.; Weyts, F.; Chiba, H.; Uitterlinden, A.G.G.; Pols, H.A.P.; Van Leeuwen, J.P.T.M. Evidence that both 1α,25-dihydroxyvitamin D3 and 24-hydroxylated D3 enhance human osteoblast differentiation and mineralization. *J. Cell. Biochem.* **2006**, *99*, 922–935. [CrossRef]
46. van der Meijden, K.; Lips, P.; van Driel, M.; Heijboer, A.C.; Schulten, E.A.J.M.; Heijer, M.D.; Bravenboer, N. Primary Human Osteoblasts in Response to 25-Hydroxyvitamin D3, 1,25-Dihydroxyvitamin D3 and 24R,25-Dihydroxyvitamin D3. *PLoS ONE* **2014**, *9*, e110283. [CrossRef]

47. van Driel, M.; van Leeuwen, J.P.T.M. Vitamin D endocrine system and osteoblasts. *BoneKEy Rep.* **2014**, *3*, 493. [CrossRef]
48. Siggelkow, H.; Rebenstorff, K.; Kurre, W.; Niedhart, C.; Engel, I.; Schulz, H.; Atkinson, M.J.; Hüfner, M. Development of the osteoblast phenotype in primary human osteoblasts in culture: Comparison with rat calvarial cells in osteoblast differentiation. *J. Cell. Biochem.* **1999**, *75*, 22–35. [CrossRef]
49. Weitzmann, M.N.; Yamaguchi, M. High dose 1,25$(OH)_2D_3$ inhibits osteoblast mineralization in vitro. *Int. J. Mol. Med.* **2012**, *29*, 934–938. [CrossRef]
50. Chen, Y.-C.; Ninomiya, T.; Hosoya, A.; Hiraga, T.; Miyazawa, H.; Nakamura, H. 1α,25-Dihydroxyvitamin D3 inhibits osteoblastic differentiation of mouse periodontal fibroblasts. *Arch. Oral Biol.* **2012**, *57*, 453–459. [CrossRef]
51. Matsumoto, T.; Igarashi, C.; Takeuchi, Y.; Harada, S.; Kikuchi, T.; Yamato, H.; Ogata, E. Stimulation by 1,25-Dihydroxyvitamin D3 of in vitro mineralization induced by osteoblast-like MC3T3-E1 cells. *Bone* **1991**, *12*, 27–32. [CrossRef]
52. Yang, D.; Atkins, G.J.; Turner, A.G.; Anderson, P.H.; Morris, H.A. Differential effects of 1,25-dihydroxyvitamin D on mineralisation and differentiation in two different types of osteoblast-like cultures. *J. Steroid Biochem. Mol. Biol.* **2012**, *136*, 166–170. [CrossRef]
53. Shevde, N.K.; Plum, L.A.; Clagett-Dame, M.; Yamamoto, H.; Pike, J.W.; DeLuca, H.F. A potent analog of 1α,25-dihydroxyvitamin D3 selectively induces bone formation. *Proc. Natl. Acad. Sci. USA* **2002**, *99*, 13487–13491. [CrossRef]
54. Jeffery, E.C.; Mann, T.L.; Pool, J.A.; Zhao, Z.; Morrison, S.J. Bone marrow and periosteal skeletal stem/progenitor cells make distinct contributions to bone maintenance and repair. *Cell Stem Cell* **2022**, *29*, 1547–1561.e6. [CrossRef]
55. Eisman, J.A.; Bouillon, R. Vitamin D: Direct effects of vitamin D metabolites on bone: Lessons from genetically modified mice. *BoneKEy Rep.* **2014**, *3*, 499. [CrossRef]
56. Yan, X.-Z.; Yang, W.; Yang, F.; Kersten-Niessen, M.; Jansen, J.A.; Both, S.K. Effects of Continuous Passaging on Mineralization of MC3T3-E1 Cells with Improved Osteogenic Culture Protocol. *Tissue Eng. Part C Methods* **2014**, *20*, 198–204. [CrossRef]
57. Viereck, V.; Siggelkow, H.; Tauber, S.; Raddatz, D.; Schutze, N.; Hüfner, M. Differential regulation of Cbfa1/Runx2 and osteocalcin gene expression by vitamin-D3, dexamethasone, and local growth factors in primary human osteoblasts. *J. Cell. Biochem.* **2002**, *86*, 348–356. [CrossRef]
58. Zhang, R.; Ducy, P.; Karsenty, G. 1,25-Dihydroxyvitamin D3 Inhibits Osteocalcin Expression in Mouse through an Indirect Mechanism. *J. Biol. Chem.* **1997**, *272*, 110–116. [CrossRef]
59. Lian, J.B.; Shalhoub, V.; Aslam, F.; Frenkel, B.; Green, J.; Hamrah, M.; Stein, G.S.; Stein, J.L. Species-Specific Glucocorticoid and 1,25-Dihydroxyvitamin D Responsiveness in Mouse MC3T3-E1 Osteoblasts: Dexamethasone Inhibits Osteoblast Differentiation and Vitamin D Down-Regulates Osteocalcin Gene Expression. *Endocrinology* **1997**, *138*, 2117–2127. [CrossRef]
60. Drissi, H.; Pouliot, A.; Koolloos, C.; Stein, J.L.; Lian, J.B.; Stein, G.S.; van Wijnen, A.J. 1,25-$(OH)_2$-Vitamin D3 Suppresses the Bone-Related Runx2/Cbfa1 Gene Promoter. *Exp. Cell Res.* **2002**, *274*, 323–333. [CrossRef]
61. Thomas, G.P.; Bourne, A.; Eisman, J.A.; Gardiner, E.M. Species-Divergent Regulation of Human and Mouse Osteocalcin Genes by Calciotropic Hormones. *Exp. Cell Res.* **2000**, *258*, 395–402. [CrossRef] [PubMed]
62. Cui, J.; Ma, C.; Qiu, J.; Ma, X.; Wang, X.; Chen, H.; Huang, B. A novel interaction between insulin-like growth factor binding protein-6 and the vitamin D receptor inhibits the role of vitamin D3 in osteoblast differentiation. *Mol. Cell. Endocrinol.* **2011**, *338*, 84–92. [CrossRef] [PubMed]
63. Ito, N.; Findlay, D.M.; Anderson, P.H.; Bonewald, L.F.; Atkins, G.J. Extracellular phosphate modulates the effect of 1α,25-dihydroxy vitamin D3 (1,25D) on osteocyte like cells. *J. Steroid Biochem. Mol. Biol.* **2013**, *136*, 183–186. [CrossRef] [PubMed]
64. Staal, A.; Birkenhäger, J.C.; Pols, H.A.; Buurman, C.J.; Vink-van Wijngaarden, T.; Kleinekoort, W.M.; van den Bemd, G.J.; van Leeuwen, J.P. Transforming growth factor beta-induced dissociation between vitamin D receptor level and 1,25-dihydroxyvitamin D3 action in osteoblast-like cells. *Bone Miner.* **1994**, *26*, 27–42. [CrossRef] [PubMed]
65. Staal, A.; Geertsma-Kleinekoort, W.M.C.; Van Den Bemd, G.J.C.M.; Buurman, C.J.; Birkenhäger, J.C.; Pols, H.A.P.; Van Leeuwen, J.P.T.M. Regulation of Osteocalcin Production and Bone Resorption by 1,25-Dihydroxyvitamin D3 in Mouse Long Bones: Interaction with the Bone-Derived Growth Factors TGF-β and IGF-I. *J. Bone Miner. Res.* **1998**, *13*, 36–43. [CrossRef] [PubMed]
66. Staal, A.; VanWijnen, A.; Desai, R.; Pols, H.; Birkenhager, J.; Deluca, H.; Denhardt, D.; Stein, J.; Van Leeuwen, J.; Stein, G.; et al. Antagonistic effects of transforming growth factor-beta on vitamin D3 enhancement of osteocalcin and osteopontin transcription: Reduced interactions of vitamin D receptor/retinoid X receptor complexes with vitamin E response elements. *Endocrinology* **1996**, *137*, 2001–2011. [CrossRef]
67. Wergedal, J.E.; Matsuyama, T.; Strong, D.D. Differentiation of normal human bone cells by transforming growth factor-β and 1,25$(OH)_2$ vitamin D3. *Metabolism* **1992**, *41*, 42–48. [CrossRef]
68. Chen, J.; Dosier, C.R.; Park, J.H.; De, S.; Guldberg, R.E.; Boyan, B.D.; Schwartz, Z. Mineralization of three-dimensional osteoblast cultures is enhanced by the interaction of 1α,25-dihydroxyvitamin D3 and BMP2 via two specific vitamin D receptors. *J. Tissue Eng. Regen. Med.* **2013**, *10*, 40–51. [CrossRef]
69. Woeckel, V.J.J.; Koedam, M.; van de Peppel, J.; Chiba, H.; van der Eerden, B.C.J.; van Leeuwen, J.P.T.M. Evidence of vitamin D and interferon-β cross-talk in human osteoblasts with 1α,25-dihydroxyvitamin D3 being dominant over interferon-β in stimulating mineralization. *J. Cell. Physiol.* **2012**, *227*, 3258–3266. [CrossRef]
70. Fretz, J.A.; Zella, L.A.; Kim, S.; Shevde, N.K.; Pike, J.W. 1,25-Dihydroxyvitamin D3 induces expression of the Wnt signaling co-regulator LRP5 via regulatory elements located significantly downstream of the gene's transcriptional start site. *J. Steroid Biochem. Mol. Biol.* **2007**, *103*, 440–445. [CrossRef]

71. Haussler, M.R.; Haussler, C.A.; Whitfield, G.K.; Hsieh, J.-C.; Thompson, P.D.; Barthel, T.K.; Bartik, L.; Egan, J.B.; Wu, Y.; Kubicek, J.L.; et al. The nuclear vitamin D receptor controls the expression of genes encoding factors which feed the "Fountain of Youth" to mediate healthful aging. *J. Steroid Biochem. Mol. Biol.* **2010**, *121*, 88–97. [CrossRef]
72. Guler, E.; Baripoglu, Y.E.; Alenezi, H.; Arikan, A.; Babazade, R.; Unal, S.; Duruksu, G.; Alfares, F.S.; Yazir, Y.; Oktar, F.N.; et al. Vitamin D3/vitamin K2/magnesium-loaded polylactic acid/tricalcium phosphate/polycaprolactone composite nanofibers demonstrated osteoinductive effect by increasing Runx2 via Wnt/β-catenin pathway. *Int. J. Biol. Macromol.* **2021**, *190*, 244–258. [CrossRef]
73. Doroudi, M.; Olivares-Navarrete, R.; Hyzy, S.L.; Boyan, B.D.; Schwartz, Z. Signaling components of the 1α,25$(OH)_2D_3$-dependent Pdia3 receptor complex are required for Wnt5a calcium-dependent signaling. *Biochim. Biophys. Acta BBA Mol. Cell Res.* **2014**, *1843*, 2365–2375. [CrossRef]
74. Jo, S.; Yoon, S.; Lee, S.Y.; Kim, S.Y.; Park, H.; Han, J.; Choi, S.H.; Han, J.-S.; Yang, J.-H.; Kim, T.-H. DKK1 Induced by 1,25D3 Is Required for the Mineralization of Osteoblasts. *Cells* **2020**, *9*, 236. [CrossRef]
75. Chen, C.-T.; Shih, Y.-R.V.; Kuo, T.K.; Lee, O.K.; Wei, Y.-H. Coordinated Changes of Mitochondrial Biogenesis and Antioxidant Enzymes during Osteogenic Differentiation of Human Mesenchymal Stem Cells. *Stem Cells* **2008**, *26*, 960–968. [CrossRef]
76. Weivoda, M.M.; Chew, C.K.; Monroe, D.G.; Farr, J.N.; Atkinson, E.J.; Geske, J.R.; Eckhardt, B.; Thicke, B.; Ruan, M.; Tweed, A.J.; et al. Identification of osteoclast-osteoblast coupling factors in humans reveals links between bone and energy metabolism. *Nat. Commun.* **2020**, *11*, 87. [CrossRef]
77. Shen, L.; Hu, G.; Karner, C.M. Bioenergetic Metabolism In Osteoblast Differentiation. *Curr. Osteoporos. Rep.* **2022**, *20*, 53–64. [CrossRef]
78. Pal, S.; Singh, M.; Porwal, K.; Rajak, S.; Das, N.; Rajput, S.; Trivedi, A.K.; Maurya, R.; Sinha, R.A.; Siddiqi, M.I.; et al. Adiponectin receptors by increasing mitochondrial biogenesis and respiration promote osteoblast differentiation: Discovery of isovitexin as a new class of small molecule adiponectin receptor modulator with potential osteoanabolic function. *Eur. J. Pharmacol.* **2021**, *913*, 174634. [CrossRef]
79. Bruedigam, C.; Eijken, M.; Koedam, M.; Pols, H.A.P.; van Leeuwen, J.P.T. New insights into peroxisome proliferatoractivated receptor gamma action: Stimulation of human osteoblast differentiation. *Calcif. Tissue Int.* **2007**, *80*, S73.
80. Bruedigam, C.; Eijken, M.; Koedam, M.; van de Peppel, J.; Drabek, K.; Chiba, H.; van Leeuwen, J.P.T.M. A New Concept Underlying Stem Cell Lineage Skewing That Explains the Detrimental Effects of Thiazolidinediones on Bone. *Stem Cells* **2010**, *28*, 916–927. [CrossRef]
81. Bouillon, R.; Carmeliet, G.; Lieben, L.; Watanabe, M.; Perino, A.; Auwerx, J.; Schoonjans, K.; Verstuyf, A. Vitamin D and energy homeostasis—Of mice and men. *Nat. Rev. Endocrinol.* **2013**, *10*, 79–87. [CrossRef] [PubMed]
82. Abu el Maaty, M.A.; Wölfl, S. Vitamin D as a Novel Regulator of Tumor Metabolism: Insights on Potential Mechanisms and Implications for Anti-Cancer Therapy. *Int. J. Mol. Sci.* **2017**, *18*, 2184. [CrossRef] [PubMed]
83. Sheeley, M.P.; Andolino, C.; Kiesel, V.A.; Teegarden, D. Vitamin D regulation of energy metabolism in cancer. *Br. J. Pharmacol.* **2021**, *179*, 2890–2905. [CrossRef] [PubMed]
84. Zhang, P.; Schatz, A.; Adeyemi, B.; Kozminski, D.; Welsh, J.; Tenniswood, M.; Wang, W.-L.W. Vitamin D and testosterone co-ordinately modulate intracellular zinc levels and energy metabolism in prostate cancer cells. *J. Steroid Biochem. Mol. Biol.* **2019**, *189*, 248–258. [CrossRef]
85. Eelen, G.; Verlinden, L.; Meyer, M.B.; Gijsbers, R.; Pike, J.W.; Bouillon, R.; Verstuyf, A. 1,25-Dihydroxyvitamin D3 and the aging-related Forkhead Box O and Sestrin proteins in osteoblasts. *J. Steroid Biochem. Mol. Biol.* **2013**, *136*, 112–119. [CrossRef]
86. Komarova, S.V.; Ataullakhanov, F.I.; Globus, R.K. Bioenergetics and mitochondrial transmembrane potential during differentiation of cultured osteoblasts. *Am. J. Physiol. Cell. Physiol.* **2000**, *279*, C1220–C1229. [CrossRef]
87. Wu, Y.-Y.; Yu, T.; Zhang, X.-H.; Liu, Y.-S.; Li, F.; Wang, Y.-Y.; Wang, Y.-Y.; Gong, P. 1,25$(OH)_2D_3$ inhibits the deleterious effects induced by high glucose on osteoblasts through undercarboxylated osteocalcin and insulin signaling. *J. Steroid Biochem. Mol. Biol.* **2012**, *132*, 112–119. [CrossRef]
88. Woeckel, V.J.; Bruedigam, C.; Koedam, M.; Chiba, H.; van der Eerden, B.C.; van Leeuwen, J.P. 1α,25-Dihydroxyvitamin D3 and rosiglitazone synergistically enhance osteoblast-mediated mineralization. *Gene* **2012**, *512*, 438–443. [CrossRef]
89. Ali, S.Y.; Sajdera, S.W.; Anderson, H.C. Isolation and Characterization of Calcifying Matrix Vesicles from Epiphyseal Cartilage. *Proc. Natl. Acad. Sci. USA* **1970**, *67*, 1513–1520. [CrossRef]
90. Anderson, H.C. Molecular biology of matrix vesicles. *Clin. Orthop. Relat. Res.* **1995**, *314*, 266–280. [CrossRef]
91. Anderson, H.C. Matrix vesicles and calcification. *Curr. Rheumatol. Rep.* **2003**, *5*, 222–226. [CrossRef]
92. Xiao, Z.; Camalier, C.E.; Nagashima, K.; Chan, K.C.; Lucas, D.A.; de la Cruz, M.J.; Gignac, M.; Lockett, S.; Issaq, H.J.; Veenstra, T.D.; et al. Analysis of the extracellular matrix vesicle proteome in mineralizing osteoblasts. *J. Cell. Physiol.* **2006**, *210*, 325–335. [CrossRef]
93. Thouverey, C.; Malinowska, A.; Balcerzak, M.; Strzelecka-Kiliszek, A.; Buchet, R.; Dadlez, M.; Pikula, S. Proteomic characterization of biogenesis and functions of matrix vesicles released from mineralizing human osteoblast-like cells. *J. Proteom.* **2011**, *74*, 1123–1134. [CrossRef]
94. Staines, K.A.; Zhu, D.; Farquharson, C.; MacRae, V.E. Identification of novel regulators of osteoblast matrix mineralization by time series transcriptional profiling. *J. Bone Miner. Metab.* **2013**, *32*, 240–251. [CrossRef]

95. Tye, C.E.; Hunter, G.K.; Goldberg, H.A. Identification of the Type I Collagen-binding Domain of Bone Sialoprotein and Characterization of the Mechanism of Interaction. *J. Biol. Chem.* **2005**, *280*, 13487–13492. [CrossRef]
96. Orimo, H. The Mechanism of Mineralization and the Role of Alkaline Phosphatase in Health and Disease. *J. Nippon Med. Sch.* **2010**, *77*, 4–12. [CrossRef]
97. Kim, H.J.; Minashima, T.; McCarthy, E.F.; A Winkles, J.; Kirsch, T. Progressive ankylosis protein (ANK) in osteoblasts and osteoclasts controls bone formation and bone remodeling. *J. Bone Miner. Res.* **2010**, *25*, 1771–1783. [CrossRef]
98. Millán, J.L. The Role of Phosphatases in the Initiation of Skeletal Mineralization. *Calcif. Tissue Int.* **2012**, *93*, 299–306. [CrossRef]
99. Woeckel, V.J.; Alves, R.D.; Swagemakers, S.M.; Eijken, M.; Chiba, H.; van der Eerden, B.C.; van Leeuwen, J.P. 1α,25-$(OH)_2D_3$ acts in the early phase of osteoblast differentiation to enhance mineralization via accelerated production of mature matrix vesicles. *J. Cell. Physiol.* **2010**, *225*, 593–600. [CrossRef]
100. Franceschi, R.T.; Romano, P.R.; Park, K.Y. Regulation of type I collagen synthesis by 1,25-dihydroxyvitamin D3 in human osteosarcoma cells. *J. Biol. Chem.* **1988**, *263*, 18938–18945. [CrossRef]
101. Hicok, K.C.; Thomas, T.; Gori, F.; Rickard, D.J.; Spelsberg, T.C.; Riggs, B.L. Development and Characterization of Conditionally Immortalized Osteoblast Precursor Cell Lines from Human Bone Marrow Stroma. *J. Bone Miner. Res.* **1998**, *13*, 205–217. [CrossRef] [PubMed]
102. Ingram, R.T.; Bonde, S.K.; Riggs, B.L.; Fitzpatrick, L.A. Effects of transforming growth factor beta (TGFβ) and 1,25 dihydroxyvitamin D3 on the function, cytochemistry and morphology of normal human osteoblast-like cells. *Differentiation* **1994**, *55*, 153–163. [CrossRef] [PubMed]
103. Siggelkow, H.; Schulz, H.; Kaesler, S.; Benzler, K.; Atkinson, M.J.; Hüfner, M. 1,25 Dihydroxyvitamin-D3 Attenuates the Confluence-Dependent Differences in the Osteoblast Characteristic Proteins Alkaline Phosphatase, Procollagen I Peptide, and Osteocalcin. *Calcif. Tissue Int.* **1999**, *64*, 414–421. [CrossRef] [PubMed]
104. Kim, H.T.; Chen, T.L. 1,25-Dihydroxyvitamin D3Interaction with Dexamethasone and Retinoic Acid: Effects on Procollagen Messenger Ribonucleic Acid Levels in Rat Osteoblast-Like Cells. *Mol. Endocrinol.* **1989**, *3*, 97–104. [CrossRef]
105. Harrison, J.R.; Petersen, D.N.; Lichtler, A.C.; Mador, A.T.; Rowe, D.W.; Kream, B.E. 1,25-Dihydroxyvitamin D_3Inhibits Transcription of Type I Collagen Genes in the Rat Osteosarcoma Cell Line ROS 17/2.8. *Endocrinology* **1989**, *125*, 327–333. [CrossRef]
106. Yang, D.; Turner, A.G.; Wijenayaka, A.R.; Anderson, P.H.; Morris, H.A.; Atkins, G.J. 1,25-Dihydroxyvitamin D3 and extracellular calcium promote mineral deposition via NPP1 activity in a mature osteoblast cell line MLO-A5. *Mol. Cell. Endocrinol.* **2015**, *412*, 140–147. [CrossRef]
107. Yajima, A.; Tsuchiya, K.; Burr, D.B.; Wallace, J.M.; Damrath, J.D.; Inaba, M.; Tominaga, Y.; Satoh, S.; Nakayama, T.; Tanizawa, T.; et al. The Importance of Biologically Active Vitamin D for Mineralization by Osteocytes After Parathyroidectomy for Renal Hyperparathyroidism. *JBMR Plus* **2019**, *3*, e10234. [CrossRef]
108. Lieben, L.; Masuyama, R.; Torrekens, S.; Van Looveren, R.; Schrooten, J.; Baatsen, P.; Lafage-Proust, M.-H.; Dresselaers, T.; Feng, J.Q.; Bonewald, L.F.; et al. Normocalcemia is maintained in mice under conditions of calcium malabsorption by vitamin D–induced inhibition of bone mineralization. *J. Clin. Investig.* **2012**, *122*, 1803–1815. [CrossRef]
109. Woeckel, V.J.; van der Eerden, B.C.; Schreuders-Koedam, M.; Eijken, M.; Van Leeuwen, J.P. 1α,25-dihydroxyvitamin D3stimulates activin A production to fine-tune osteoblast-induced mineralization. *J. Cell. Physiol.* **2013**, *228*, 2167–2174. [CrossRef]
110. Kitazawa, S.; Kajimoto, K.; Kondo, T.; Kitazawa, R. Vitamin D3 supports osteoclastogenesis via functional vitamin D response element of human RANKL gene promoter. *J. Cell. Biochem.* **2003**, *89*, 771–777. [CrossRef]
111. Shymanskyi, I.; Lisakovska, O.; Mazanova, A.; Labudzynskyi, D.; Veliky, M. Vitamin D3 Modulates Impaired Crosstalk Between RANK and Glucocorticoid Receptor Signaling in Bone Marrow Cells After Chronic Prednisolone Administration. *Front. Endocrinol.* **2018**, *9*, 303. [CrossRef]
112. Khalaf, R.M.; Almudhi, A.A. The effect of vitamin D deficiency on the RANKL/OPG ratio in rats. *J. Oral Biol. Craniofacial Res.* **2022**, *12*, 228–232. [CrossRef]
113. Kim, S.; Yamazaki, M.; Zella, L.A.; Shevde, N.K.; Pike, J.W. Activation of Receptor Activator of NF-κB Ligand Gene Expression by 1,25-Dihydroxyvitamin D3 Is Mediated through Multiple Long-Range Enhancers. *Mol. Cell. Biol.* **2006**, *26*, 6469–6486. [CrossRef]
114. Bouillon, R.; Carmeliet, G. Vitamin D and the skeleton. *Curr. Opin. Endocr. Metab. Res.* **2018**, *3*, 68–73. [CrossRef]
115. Carlberg, C.; Muñoz, A. An update on vitamin D signaling and cancer. *Semin. Cancer Biol.* **2020**, *79*, 217–230. [CrossRef]
116. Carlberg, C. Vitamin D and Its Target Genes. *Nutrients* **2022**, *14*, 1354. [CrossRef]
117. Haussler, M.R.; Livingston, S.; Sabir, Z.L.; Haussler, C.A.; Jurutka, P.W. Vitamin D Receptor Mediates a Myriad of Biological Actions Dependent on Its 1,25-Dihydroxyvitamin D Ligand: Distinct Regulatory Themes Revealed by Induction of Klotho and Fibroblast Growth Factor-23. *JBMR Plus* **2021**, *5*, e10432. [CrossRef]
118. Haussler, M.R.; Whitfield, G.K.; Kaneko, I.; Haussler, C.A.; Hsieh, D.; Hsieh, J.-C.; Jurutka, P.W. Molecular Mechanisms of Vitamin D Action. *Calcif. Tissue Int.* **2013**, *92*, 77–98. [CrossRef]
119. Owen, T.A.; Aronow, M.S.; Barone, L.M.; Bettencourt, B.; Stein, G.S.; Lian, J.B. Pleiotropic Effects of Vitamin D on Osteoblast Gene Expression Are Related to the Proliferative and Differentiated State of the Bone Cell Phenotype: Dependency upon Basal Levels of Gene Expression, Duration of Exposure, and Bone Matrix Competency in Normal Rat Osteoblast Cultures. *Endocrinology* **1991**, *128*, 1496–1504. [CrossRef]

120. Saji, F.; Shigematsu, T.; Sakaguchi, T.; Ohya, M.; Orita, H.; Maeda, Y.; Ooura, M.; Mima, T.; Negi, S. Fibroblast growth factor 23 production in bone is directly regulated by 1α,25-dihydroxyvitamin D, but not PTH. *Am. J. Physiol. Physiol.* **2010**, *299*, F1212–F1217. [CrossRef]
121. Yamamoto, R.; Minamizaki, T.; Yoshiko, Y.; Yoshioka, H.; Tanne, K.; Aubin, J.E.; Maeda, N. 1,25-dihydroxyvitamin D3 acts predominately in mature osteoblasts under conditions of high extracellular phosphate to increase fibroblast growth factor 23 production in vitro. *J. Endocrinol.* **2010**, *206*, 279–286. [CrossRef] [PubMed]
122. Quarles, L.D. Skeletal secretion of FGF-23 regulates phosphate and vitamin D metabolism. *Nat. Rev. Endocrinol.* **2012**, *8*, 276–286. [CrossRef] [PubMed]
123. Razzaque, M.S. Interactions between FGF23 and vitamin D. *Endocr. Connect.* **2022**, *11*, e220239. [CrossRef] [PubMed]
124. Lanske, B.; Densmore, M.J.; Erben, R.G. Vitamin D endocrine system and osteocytes. *BoneKEy Rep.* **2014**, *3*, 494. [CrossRef]
125. Hines, E.R.; Kolek, O.I.; Jones, M.D.; Serey, S.H.; Sirjani, N.B.; Kiela, P.R.; Jurutka, P.W.; Haussler, M.R.; Collins, J.F.; Ghishan, F.K. 1,25-Dihydroxyvitamin D3 Down-regulation of PHEX Gene Expression Is Mediated by Apparent Repression of a 110 kDa Transfactor That Binds to a Polyadenine Element in the Promoter. *J. Biol. Chem.* **2004**, *279*, 46406–46414. [CrossRef]
126. Turner, R.T.; Puzas, J.E.; Forte, M.D.; Lester, G.E.; Gray, T.K.; Howard, G.A.; Baylink, D.J. In vitro synthesis of 1 alpha,25-dihydroxycholecalciferol and 24,25-dihydroxycholecalciferol by isolated calvarial cells. *Proc. Natl. Acad. Sci. USA* **1980**, *77*, 5720–5724. [CrossRef]
127. Howard, G.A.; Turner, R.T.; Sherrard, D.J.; Baylink, D.J. Human bone cells in culture metabolize 25-hydroxyvitamin D3 to 1,25-dihydroxyvitamin D3 and 24,25-dihydroxyvitamin D3. *J. Biol. Chem.* **1981**, *256*, 7738–7740. [CrossRef]
128. Geng, S.; Zhou, S.; Glowacki, J. Effects of 25-hydroxyvitamin D3 on proliferation and osteoblast differentiation of human marrow stromal cells require CYP27B1/1α-hydroxylase. *J. Bone Miner. Res.* **2011**, *26*, 1145–1153. [CrossRef]
129. Lou, Y.-R.; Toh, T.C.; Tee, Y.H.; Yu, H. 25-Hydroxyvitamin D3 induces osteogenic differentiation of human mesenchymal stem cells. *Sci. Rep.* **2017**, *7*, srep42816. [CrossRef]
130. Geng, S.; Zhou, S.; Glowacki, J. Age-related decline in osteoblastogenesis and 1α-hydroxylase/CYP27B1 in human mesenchymal stem cells: Stimulation by parathyroid hormone. *Aging Cell* **2011**, *10*, 962–971. [CrossRef]
131. Bikle, D.D.; Patzek, S.; Wang, Y. Physiologic and pathophysiologic roles of extra renal CYP27b1: Case report and review. *Bone Rep.* **2018**, *8*, 255–267. [CrossRef]
132. Anderson, P.H.; Lam, N.N.; Turner, A.G.; Davey, R.A.; Kogawa, M.; Atkins, G.J.; Morris, H.A. The pleiotropic effects of vitamin D in bone. *J. Steroid Biochem. Mol. Biol.* **2013**, *136*, 190–194. [CrossRef]
133. Hewison, M.; Zehnder, D.; Chakraverty, R.; Adams, J.S. Vitamin D and barrier function: A novel role for extra-renal 1α-hydroxylase. *Mol. Cell. Endocrinol.* **2004**, *215*, 31–38. [CrossRef]
134. Yang, D.; Anderson, P.H.; Turner, A.G.; Morris, H.A.; Atkins, G.J. Comparison of the biological effects of exogenous and endogenous 1,25-dihydroxyvitamin D3 on the mature osteoblast cell line MLO-A5. *J. Steroid Biochem. Mol. Biol.* **2016**, *164*, 374–378. [CrossRef]
135. Meyer, M.B.; Pike, J.W. Mechanistic homeostasis of vitamin D metabolism in the kidney through reciprocal modulation of Cyp27b1 and Cyp24a1 expression. *J. Steroid Biochem. Mol. Biol.* **2020**, *196*, 105500. [CrossRef]
136. Zhou, S.; LeBoff, M.S.; Glowacki, J. Vitamin D Metabolism and Action in Human Bone Marrow Stromal Cells. *Endocrinology* **2010**, *151*, 14–22. [CrossRef]
137. St-Arnaud, R.; Jones, G. Chapter 6—CYP24A1: Structure, Function, and Physiological Role. In *Vitamin D*, 4th ed.; Feldman, D., Ed.; Academic Press: Cambridge, MA, USA, 2018; pp. 81–95.
138. Van Leeuwen, J.P.T.M.; Van Den Bemd, G.J.C.M.; Van Driel, M.; Buurman, C.J.; Pols, H.A.P. 24,25-Dihydroxyvitamin D3 and bone metabolism. *Steroids* **2001**, *66*, 375–380. [CrossRef]
139. Väisänen, S.; Dunlop, T.W.; Sinkkonen, L.; Frank, C.; Carlberg, C. Spatio-temporal Activation of Chromatin on the Human CYP24 Gene Promoter in the Presence of 1α,25-Dihydroxyvitamin D3. *J. Mol. Biol.* **2005**, *350*, 65–77. [CrossRef]
140. Henry, H.L. The 25(OH)D_3/1α,25$(OH)_2D_3$-24R-hydroxylase: A catabolic or biosynthetic enzyme? *Steroids* **2001**, *66*, 391–398. [CrossRef] [PubMed]
141. Moena, D.; Nardocci, G.; Acevedo, E.; Lian, J.; Stein, G.; Stein, J.; Montecino, M. Ezh2-dependent H3K27me3 modification dynamically regulates vitamin D3-dependent epigenetic control of CYP24A1 gene expression in osteoblastic cells. *J. Cell. Physiol.* **2020**, *235*, 5404–5412. [CrossRef]
142. Pols, H.A.; Birkenhager, J.C.; Schilte, J.P.; Visser, T.J. Evidence that the self-induced metabolism of 1,25-dihydroxyvitamin D-3 limits the homologous up-regulation of its receptor in rat osteosarcoma cells. *Biochim. et Biophys. Acta (BBA) Mol. Cell Res.* **1988**, *970*, 122–129. [CrossRef]
143. Staal, A.; vandenBemd, G.; Birkenhager, J.; Pols, H.; van Leeuwen, J. Consequences of vitamin D receptor regulation for the 1,25-dihydroxyvitamin D3-induced 24-hydroxylase activity in osteoblast-like cells: Initiation of the C24-oxidation pathway. *Bone* **1997**, *20*, 237–243. [CrossRef] [PubMed]
144. Henry, H.L.; Norman, A.W. Vitamin D: Two Dihydroxylated Metabolites Are Required for Normal Chicken Egg Hatchability. *Science* **1978**, *201*, 835–837. [CrossRef]
145. Norman, A.W.; Henry, H.L.; Malluche, H.H. 24R,25-dihydroxyvitamin D3 and 1α,25-dihydroxyvitamin D3 are both indispensable for calcium and phosphorus homeostasis. *Life Sci.* **1980**, *27*, 229–237. [CrossRef]

146. Endo, H.; Kiyoki, M.; Kawashima, K.; Naruchi, T.; Hashimoto, Y. Vitamin D3 metabolites and PTH synergistically stimulate bone formation of chick embryonic femur in vitro. *Nature* **1980**, *286*, 262–264. [CrossRef]
147. Galus, K.; Szymendera, J.; Zaleski, A.; Schreyer, K. Effects of 1α-hydroxyvitamin D3 and 24R,25-dihydroxyvitamin D3 on bone remodeling. *Calcif. Tissue Int.* **1980**, *31*, 209–213. [CrossRef]
148. Erben, R.G.; Weiser, H.; Sinowatz, F.; Rambeck, W.A.; Zucker, H. Vitamin D metabolites prevent vertebral osteopenia in ovariectomized rats. *Calcif. Tissue Int.* **1992**, *50*, 228–236. [CrossRef]
149. Matsumoto, T.; Ezawa, I.; Morita, K.; Kawanobe, Y.; Ogata, E. Effect of Vitamin D Metabolites on Bone Metabolism in a Rat Model of Postmenopausal Osteoporosis. *J. Nutr. Sci. Vitaminol.* **1985**, *31*, S61–S65. [CrossRef]
150. Kato, A.; Seo, E.G.; Einhorn, T.A.; Bishop, J.E.; Norman, A.W. Studies on 24R,25-dihydroxyvitamin D3: Evidence for a nonnuclear membrane receptor in the chick tibial fracture-healing callus. *Bone* **1998**, *23*, 141–146. [CrossRef]
151. Seo, E.-G.; Einhorn, T.A.; Norman, A.W. 24R,25-Dihydroxyvitamin D3: An Essential Vitamin D3 Metabolite for Both Normal Bone Integrity and Healing of Tibial Fracture in Chicks. *Endocrinology* **1997**, *138*, 3864–3872. [CrossRef]
152. Martineau, C.; Kaufmann, M.; Arabian, A.; Jones, G.; St-Arnaud, R. Preclinical safety and efficacy of 24R,25-dihydroxyvitamin D3 or lactosylceramide treatment to enhance fracture repair. *J. Orthop. Transl.* **2020**, *23*, 77–88. [CrossRef] [PubMed]
153. St-Arnaud, R. CYP24A1-deficient mice as a tool to uncover a biological activity for vitamin D metabolites hydroxylated at position 24. *J. Steroid Biochem. Mol. Biol.* **2010**, *121*, 254–256. [CrossRef]
154. Weisman, Y.; Salama, R.; Harell, A.; Edelstein, S. Serum 24,25-dihydroxyvitamin D and 25-hydroxyvitamin D concentrations in femoral neck fracture. *BMJ* **1978**, *2*, 1196–1197. [CrossRef]
155. Birkenhager-Frenkel, D.H.; Pols, H.A.; Zeelenberg, J.; Eijgelsheim, J.J.; Schot, R.; Nigg, A.L.; Weimar, W.; Mulder, P.G.; Birkenhager, J.C. Effects of 24r,25-dihydroxyvitamin D3 in combination with 1α-hydroxyvitamin D3 in predialysis renal insufficiency: Biochemistry and histomorphometry of cancellous bone. *J. Bone Miner. Res.* **1995**, *10*, 197–204. [CrossRef]

nutrients

MDPI

Review

Interaction of Vitamin D with Peptide Hormones with Emphasis on Parathyroid Hormone, FGF23, and the Renin-Angiotensin-Aldosterone System

Nejla Latic and Reinhold G. Erben *

Department of Biomedical Sciences, University of Veterinary Medicine, 1210 Vienna, Austria
* Correspondence: reinhold.erben@vetmeduni.ac.at; Tel.: +43-1-250-77-4550; Fax: +43-1-250-77-4599

Abstract: The seminal discoveries that parathyroid hormone (PTH) and fibroblast growth factor 23 (FGF23) are major endocrine regulators of vitamin D metabolism led to a significant improvement in our understanding of the pivotal roles of peptide hormones and small proteohormones in the crosstalk between different organs, regulating vitamin D metabolism. The interaction of vitamin D, FGF23 and PTH in the kidney is essential for maintaining mineral homeostasis. The proteohormone FGF23 is mainly secreted from osteoblasts and osteoclasts in the bone. FGF23 acts on proximal renal tubules to decrease production of the active form of vitamin D ($1,25(OH)_2D$) by downregulating transcription of 1α-hydroxylase (*CYP27B1*), and by activating transcription of the key enzyme responsible for vitamin D degradation, 24-hydroxylase (*CYP24A1*). Conversely, the peptide hormone PTH stimulates $1,25(OH)_2D$ renal production by upregulating the expression of 1α-hydroxylase and downregulating that of 24-hydroxylase. The circulating concentration of $1,25(OH)_2D$ is a positive regulator of FGF23 secretion in the bone, and a negative regulator of PTH secretion from the parathyroid gland, forming feedback loops between kidney and bone, and between kidney and parathyroid gland, respectively. In recent years, it has become clear that vitamin D signaling has important functions beyond mineral metabolism. Observation of seasonal variations in blood pressure and the subsequent identification of vitamin D receptor (VDR) and 1α-hydroxylase in non-renal tissues such as cardiomyocytes, endothelial and smooth muscle cells, suggested that vitamin D may play a role in maintaining cardiovascular health. Indeed, observational studies in humans have found an association between vitamin D deficiency and hypertension, left ventricular hypertrophy and heart failure, and experimental studies provided strong evidence for a role of vitamin D signaling in the regulation of cardiovascular function. One of the proposed mechanisms of action of vitamin D is that it functions as a negative regulator of the renin-angiotensin-aldosterone system (RAAS). This finding established a novel link between vitamin D and RAAS that was unexplored until then. During recent years, major progress has been made towards a more complete understanding of the mechanisms by which FGF23, PTH, and RAAS regulate vitamin D metabolism, especially at the genomic level. However, there are still major gaps in our knowledge that need to be filled by future research. The purpose of this review is to highlight our current understanding of the molecular mechanisms underlying the interaction between vitamin D, FGF23, PTH, and RAAS, and to discuss the role of these mechanisms in physiology and pathophysiology.

Keywords: vitamin D; vitamin D metabolism; fibroblast growth factor-23; klotho; parathyroid hormone; 1α-hydroxylase; RAAS

Citation: Latic, N.; Erben, R.G. Interaction of Vitamin D with Peptide Hormones with Emphasis on Parathyroid Hormone, FGF23, and the Renin-Angiotensin-Aldosterone System. *Nutrients* **2022**, *14*, 5186. https://doi.org/10.3390/nu14235186

Academic Editor: Carsten Carlberg

Received: 1 November 2022
Accepted: 1 December 2022
Published: 6 December 2022

1. Introduction

Vitamin D was first described in the 1920s in attempts to find the cause of rickets that had reached epidemic proportions in industrial cities in Middle and Northern Europe at that time [1–4]. However, it would take almost another century after this seminal discovery until the key factors regulating vitamin D metabolism were found, and a more complete understanding of the vitamin D hormonal system had evolved. Triggered by the finding in the 1970s that parathyroid hormone (PTH) is a major endocrine regulator of vitamin D metabolism [5–8], and by the discovery of fibroblast growth factor-23 (FGF23) in the year 2000, it became clear that peptide hormones and small proteohormones play pivotal roles in the regulation of vitamin D metabolism by participating in the crosstalk between different organs. In addition, it was discovered in the year 2002 that renin secretion in the kidney is regulated by vitamin signaling, establishing a novel link between vitamin D and the renin-angiotensin-aldosterone system (RAAS) [9].

This review aims to summarize what we currently know about the interaction of the vitamin D hormonal system with peptide hormones and small proteohormones, focussing on PTH, FGF23, and the RAAS.

2. Brief Overview of Vitamin D Metabolism

Vitamin D is a secosteroid derived from cholesterol in animals (cholecalciferol or vitamin D_3), or from ergosterol in fungi and protozoa (ergocalciferol or vitamin D_2). Both forms of vitamin D will be referred to as vitamin D in this review. Vitamin D is either taken up from dietary sources, or it is produced in the skin by UVB-mediated photochemical transformation of 7-dehydrocholesterol. Vitamin D itself is biologically inactive and needs to be metabolically activated by two hydroxylation steps occurring in the liver and the kidney. The canonical vitamin D activation pathway involves 25- and subsequent 1α-hydroxylation. The 25-hydroxylation step occurs in the liver, forming the most abundant circulating form of vitamin D, 25-hydroxyvitamin D (25(OH)D). There is only a little endocrine regulation of this step. In contrast, the 1α-hydroxylase (*CYP27B1*)-mediated synthesis of the vitamin D hormone, 1α,25-dihydroxyvitamin D (1,25$(OH)_2$D) in proximal tubules of the kidney is strictly regulated [10]. The kidney is the major source of circulating 1,25$(OH)_2$D under physiological conditions [11]. However, circulating 25(OH)D can also be converted into 1,25$(OH)_2$D locally in cells expressing 1α-hydroxylase [12,13]. 1,25$(OH)_2$D is the biologically active principle in the vitamin D hormonal system, regulating gene transcription through a nuclear receptor protein, the vitamin D receptor (VDR). It is now generally accepted that the VDR mediates all actions of the vitamin D endocrine system [14]. The VDR is ubiquitously expressed, and as a consequence vitamin D signaling has a major impact on gene regulatory networks in many cell types, regulating transcription of about 3% of the human genome [15]. Typically, the VDR-1,25$(OH)_2$D complex binds to vitamin D response elements (VDREs) in regulatory regions of target genes as a heterodimer with the retinoid X receptor (RXR) [14].

The most important physiological function of 1,25$(OH)_2$D is in the gut, stimulating intestinal absorption of calcium and phosphate. Therefore, intact vitamin D signaling is essential for bone and mineral homeostasis in most vertebrates [16]. To avoid hypercalcemia and hyperphosphatemia as a potential untoward consequence of excessive vitamin D signaling, the catabolism of 1,25$(OH)_2$D is tightly regulated. Catabolism of 25(OH)D and 1,25$(OH)_2$D in the kidney and all vitamin D target cells is mediated by 24-hydroxylase (*CYP24A1*), the key enzyme of the vitamin D inactivation pathway [17]. Several independent lines of evidence have shown that the regulation of *CYP24A1* is an important part of the homeostatic control of circulating and intracellular concentrations of 1,25$(OH)_2$D [18,19]

Although 25- and subsequent 1α-hydroxylation represent the predominant vitamin D activation pathway, other activation pathways exist, at least at a local level. It has been shown that the cytochrome p450 enzyme CYP11A1 is able to hydroxylate vitamin D_3 at the C17, C20, C22, and C23 positions, giving rise to 20(OH)D_3 and 22(OH)D_3 as the major metabolites [20,21]. Vitamin D_2 can be hydroxylated by CYP11A1 at the C17, C20, and C24 positions [20,21]. CYP11A1 is mainly found in adrenal glands, skin, and placenta. 20(OH)D_3 and 22(OH)D_3 are detectable in human serum, albeit at 30- and 15-fold lower concentrations than 25(OH)D_3 [20]. However, 1,20$(OH)_2D_3$ remained undetectable in human serum, suggesting that circulating 20(OH)D_3 is not a substrate of renal 1α-hydroxylation [20]. This is consistent with the low affinity of 1α-hydroxylase for 20(OH)D [22]. In contrast, the trihydroxylated metabolites 1,20,24$(OH)_3D_3$, 1,20,25$(OH)_3D_3$, and 1,20,26$(OH)_3D_3$ are found in adrenal extracts, suggesting that 1α-hydroxylation of CYP11A1-produced 20$(OH)D_3$ may occur locally in tissues also expressing CYP27B1 such as the adrenal glands, possibly after metabolization by CYP24A1 [20].

The main endocrine regulators of 1α– and 24-hydroxylase expression in the kidney are PTH, FGF23, and 1,25$(OH)_2$D. All these endocrine regulatory factors impact on renal 1α– and 24-hydroxylase expression in a reciprocal manner [23]. PTH stimulates the expression of 1α–hydroxylase, and suppresses that of 24-hydroxylase. On the other hand, FGF23 and 1,25$(OH)_2$D suppress 1α–hydroxylase, while at the same time stimulating 24-hydroxylase expression [17]. Hence, PTH augments renal production of 1,25$(OH)_2$D, whereas FGF23 and 1,25$(OH)_2$D suppress it.

3. The Parathyroid-Kidney Axis in the Regulation of Vitamin D Metabolism

As mentioned above, it was discovered in the 1970s that 1α,25$(OH)_2$D production in the kidney is regulated by PTH, establishing a novel feedback loop between kidney and parathyroid gland [5–8]. PTH is an 84-amino acid peptide hormone secreted by chief cells of the parathyroid gland in response to changes in the concentration of ionized blood calcium [24]. PTH acts through the parathyroid receptor 1 (PTHR1), a G-protein coupled receptor. All actions of PTH increase blood calcium: PTH signaling increases osteoclast activity, thereby promoting Ca release from the bone [25], favors calcium retention by enhancing Ca reabsorption in the thick ascending limb and in distal convoluted tubule of the kidney [26], and stimulates 1,25$(OH)_2$D synthesis in proximal renal tubules, thereby indirectly increasing Ca absorption from the intestines [27]. In turn, normalization of blood calcium levels and increased circulating 1,25$(OH)_2$D reduce PTH secretion. Hence, 1,25$(OH)_2$D and PTH form a tightly controlled feedback loop in which PTH stimulates 1,25$(OH)_2$D synthesis, whereas 1,25$(OH)_2$D inhibits PTH secretion (Figure 1). Apart from its effects on calcium metabolism, PTH has a phosphaturic action in the kidney, inhibiting phosphate reabsorption in proximal tubules by downregulating the sodium-phosphate cotransporters NaPi2a and NaPi2c [28]. Therefore, PTH also acts as a phosphate-lowering hormone.

PTH enhances renal 1,25$(OH)_2$D production by stimulating the expression of the CYP27B1 enzyme responsible for biosynthesis of active vitamin D, but also by downregulating the expression of CYP24A1, thereby inhibiting vitamin D catabolism [25]. Meyer and coworkers have provided important insights into the mechanisms of how PTH signaling controls *Cyp27b1* transcription in the kidney, using ChIP-Seq analysis [29,30]. The latter authors identified a kidney-specific regulatory region located in an intron of the adjacent *Mettl1* gene that is crucial for the regulation of *Cyp27b1* expression by PTH [29]. Mice with a deletion of the enhancer in the *Mettl1* gene showed a global *Cyp27b1* knockout phenotype with profoundly reduced basal *Cyp27b1* expression, distinctly reduced circulating 1,25$(OH)_2$D concentrations, upregulated PTH, and loss of the PTH-mediated induction of *Cyp27b1* [29]. Furthermore, PTH treatment of mice induced a rapid recruitment of the phosphorylated transcription factor CREB to the enhancer region in the *Mettl1* gene in the kidney [30]. ChIP-Seq analysis also revealed that the PTH-induced suppression of the *Cyp24a1* gene in the kidney is mediated through a kidney-specific regulatory region located downstream of the *Cyp24a1* gene [30,31]. Deletion of this regulatory region blunted the suppressive effect of PTH on *Cyp24a1* transcription in vivo [31].

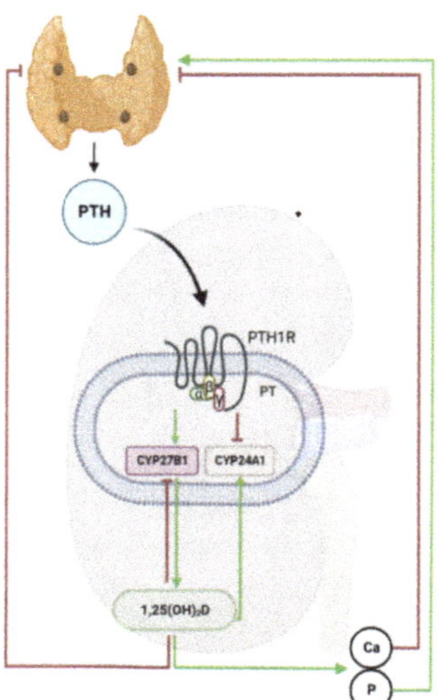

Figure 1. Parathyroid-kidney axis. Parathyroid hormone (PTH) secretion from the parathyroid gland is increased in response to a decrease in ionized plasma calcium (Ca) or to an increase in the blood concentration of inorganic phosphate (P). PTH binds to the parathyroid hormone-1 receptor (PTH1R) in proximal tubules (PT) of the kidney, promoting $1{,}25(OH)_2D$ production by upregulating 1α-hydroxylase (*Cyp27B1*) and suppressing 24-hydroxylase (*Cyp24A1*) expression. As part of a feedback inhibition, $1{,}25(OH)_2D$ downregulates its own production by stimulating *Cyp24A1* and inhibiting *Cyp27B1* expression in the kidney. Circulating $1{,}25(OH)_2D$ signals back to the parathyroid gland, where it inhibits PTH transcription and secretion. Arrows indicate the direction and nature of regulation (green, stimulation; red, inhibition). Created with BioRender.com.(Biorender 2022, Toronto, Ontario, Canada).

As part of a negative feedback mechanism, vitamin D signaling suppresses PTH transcription and secretion as evidenced by a large amount of data from in vivo and in vitro experiments as well as from clinical studies. Incubation of isolated bovine parathyroid cells with $1{,}25(OH)_2D$ for 48 h caused a reduction of PTH mRNA levels to 50% of control values [32]. This was confirmed in mouse parathyroid explants where treatment with vitamin D metabolites reduced PTH secretion [33]. In rats, injection of $1{,}25(OH)_2D$ also decreased PTH gene transcription dose- and time-dependently [34]. This reduction was associated with an upregulation of VDR mRNA levels in the parathyroid gland. Studies in mice lacking VDR have shown that $1{,}25(OH)_2D$ requires the VDR to exert its suppressive effects on *PTH* gene transcription. Administration of $1{,}25(OH)_2D$ had no effect on *PTH* transcription in global VDR knockout mice, supporting the notion that the VDR is required for modulation of PTH by vitamin D signaling. In addition, global deletion of the VDR results in elevated PTH mRNA abundance and profoundly increased PTH serum levels [35]. However, because of the ubiquitous expression of the VDR and the perturbation of mineral metabolism in these mice, it is difficult to distinguish if these effects are due to severe hypocalcemia or if they are indeed due to lack of the VDR in the parathyroid. More insights into this matter were gained when mice with a specific deletion of the VDR in the parathyroid gland were examined. In contrast to global VDR knockout mice, parathyroid-specific conditional VDR knockout mice did not show elevated *PTH* mRNA levels, but presented with moderately increased serum PTH levels [35]. The latter finding was explained by a downregulation of the calcium-sensing receptor in parathyroid-specific VDR knockout mice [35]. In agreement with the latter findings, VDR mutants on a calcium and phosphorus-enriched rescue diet were protected against secondary hyperparathyroidism (sHPT), and a switch of global VDR mutants on rescue diet to a calcium-reduced challenge diet led to severe sHPT associated with hypertrophy and hyperplasia of parathyroid glands, together with profound bone loss. Notably, sHPT was fully corrected by switching VDR mutant mice on a challenge diet back to the rescue diet, suggesting that signaling by the calcium-sensing receptor regulates chief cell function in the absence of signaling through the VDR [36].

In conclusion, there is ample evidence that active vitamin D analogs suppress PTH transcription and secretion in a VDR-dependent manner as part of a pharmacological effect. However, the suppressive effect of vitamin D signaling on *PTH* gene transcription in vivo is dispensable under physiological conditions, suggesting that VDR signaling can certainly suppress PTH secretion, but that the main regulation under normal conditions occurs via ionized blood calcium.

The exact mechanisms by which active vitamin D analogs modulate *PTH* gene transcription is still controversial. Different mechanisms have been proposed. Mackey et al. demonstrated that the VDR binds directly to a vitamin D response element in the promoter of the human *PTH* gene, independent of the RXR [37]. In the rat *PTH* gene promoter, it was found that VDR/RXR heterodimers bind to a negative, DR3-type element [38]. Another scenario suggests that the liganded VDR does not bind to the *PTH* gene promoter, but rather binds to and inactivates the vitamin D interacting receptor (VDIR), a positive effector of *PTH* gene transcription [39,40].

Interestingly, the parathyroid gland not only expresses VDR and 24-hydroxylase, but also 1α-hydroxylase [41]. Circulating 25(OH)D can indeed suppress PTH secretion, but the conversion rate is low, and 25(OH)D has much lower potency when compared to $1,25(OH)_2D$, correlating with its much lower affinity for the VDR [25]. The identification of 1α-hydroxylase in the parathyroid gland suggests that local production of $1,25(OH)_2D$ might also influence PTH secretion. However, it remains to be elucidated if local conversion of 25(OH)D into $1,25(OH)_2D$ is able to significantly modulate PTH secretion, and if expression of 1α-hydroxylase in the parathyroid gland undergoes any physiological regulation.

The importance of the vitamin D–PTH axis can be clearly seen in chronic kidney disease (CKD). In patients with CKD, renal $1,25(OH)_2D$ synthesis declines as kidney function deteriorates. As a consequence, hypocalcemia and sHPT develop [42]. Therefore, active vitamin D analogs are administered routinely with the aim to normalize PTH levels. However, although the treatment has been shown to be beneficial in terms of reducing PTH levels by 40 to 60% in different randomized control trials, development of hypercalcemia is a possible adverse effect that needs to be closely monitored [43]. It is important to note in this context that over-suppression of PTH by vitamin D analogs is a risk factor for adynamic bone disease in CKD patients [44]. Another potential pitfall of this treatment strategy is that administration of vitamin D analogs tends to become less effective over time in about 20–30% of individuals [45]. Interestingly, not all vitamin D analogs have the same effect on lowering PTH levels, and their potential to induce hypercalcemia is different [45]. Larger randomized controlled trials are necessary to determine whether one agent is superior to the other in terms of the balance between beneficial and untoward treatment effects, in order to optimize treatment strategies in these patients.

4. The Bone-Kidney Axis in the Regulation of Vitamin D Metabolism

The discovery that putative gain-of-function mutations in the *FGF23* gene are the cause of the inherited renal phosphate-wasting disease autosomal dominant hypophosphatemic rickets in the year 2000 heralded a new era in our understanding of vitamin D metabolism [46]. This discovery led to the subsequent unfolding of a previously unrecognized feedback system between bone and kidney, the bone-kidney axis, in the regulation of vitamin D metabolism. Soon after the initial description of FGF23 as a putative phosphaturic hormone [46] it became evident that FGF23 is also a powerful regulator of vitamin D metabolism. When Shimada and coworkers injected recombinant FGF23 into mice they observed not only hypophosphatemia, but also a distinct suppression of renal 1α-hydroxylase

(*CYP27B1*) mRNA expression [47]. This finding inaugurated the link between FGF23 and vitamin D metabolism.

FGF23 belongs to the group of endocrine fibroblast growth factors (FGFs), and is mainly produced by osteoblasts and osteocytes in bone [48,49]. During the secretion process, a part of the intact protein is cleaved by proteases such as FURIN into N- and C-terminal fragments at a conserved cleavage site [50]. The intact form of FGF23 is a 32 kDa glycoprotein circulating in the bloodstream. Only the intact form of FGF23 can induce signaling through a receptor complex consisting of FGF receptors (FGFR) and the co-receptor αKlotho [51,52]. FGFR1c is the main FGFR mediating FGF23 signaling under normal conditions [51,53]. The c isoform of FGFR1 is generated by alternative splicing of *Fgfr1* mRNA [53,54].

The secretion of bioactive, intact FGF23 in bone is regulated at the transcriptional and posttranscriptional level by factors such as $1,25(OH)_2D$, PTH, phosphate, iron status, and pro-inflammatory cytokines [55]. Among these factors, $1,25(OH)_2D$ is probably the most robust stimulator of FGF23 secretion as evidenced by data from in vivo and in vitro experiments. Treatment of cultured osteoblast-like cells with $1,25(OH)_2D$ results in a dose- and time-dependent increase in *Fgf23* transcription [56,57] Similarly, injection of mice with $1,25(OH)_2D$ causes a VDR-dependent rise in circulating intact Fgf23 [58,59] In addition, the blood levels of intact Fgf23 are lower in *VDR*- and 1α-hydroxylase-deficient mice than those in WT mice, showing that intact vitamin D signaling is a physiological regulator of Fgf23 secretion [58,59]. Genome-wide association studies in a large number of subjects reported that a single nucleotide polymorphism lying upstream of the *CYP24A1* gene was the strongest predictor of the blood concentrations of FGF23, confirming the importance of the vitamin D hormonal system for the regulation of circulating FGF23 levels in humans [60].

Although it is clear that $1,25(OH)_2D$ is a transcriptional regulator of the *Fgf23* gene in a VDR-dependent manner, the exact molecular mechanism of this regulation at the genomic level still remains unclear, and there may be species differences. Whereas several functional VDREs were identified in the promoter of the human *FGF23* gene [61], VDREs are absent in the murine *Fgf23* gene promoter [57], and ChIP-Seq analysis in the murine osteocyte-like cell line IDG-SW3 failed to provide evidence for VDR binding to regulatory regions in the *Fgf23* locus [62]. Therefore, at least in mice, the $1,25(OH)_2D$-induced upregulation of *Fgf23* transcription may be mediated indirectly via DNA binding of other transcription factors. The regulatory region responsible for the $1,25(OH)_2D$-induced upregulation of *Fgf23* transcription in mice has been narrowed down to an enhancer region located directly upstream of the *Fgf23* promoter [63,64]. When this enhancer region was deleted in mice, the $1,25(OH)_2D$-mediated upregulation of *Fgf23* transcription was completely lost [64]. However, the specific transcription factor(s) binding to the enhancer remain unknown.

Secretion of intact FGF23 is also regulated at the posttranslational level in the bone. Intracellular proteolysis of intact FGF23 by FURIN results in the secretion of the biologically inactive N-terminal and C-terminal fragments. Post-translational O-glycosylation at tyrosine 178 by polypeptide N-acetylgalactosaminyltransferase3 (GALNT3) inhibits FURIN-mediated cleavage, thus favoring secretion of intact FGF23. In contrast, phosphorylation at serine 180 (S180) by extracellular kinase family member 20c (FAM20) inhibits glycosylation, thus facilitating FURIN-mediated cleavage [65]. Hence, the posttranslational modification within or near the FURIN-mediated cleavage site is an important determinant of the ratio between intact and cleaved FGF23 secreted by bone cells [50]. It has been shown that PTH injection acutely alters the ratio of intact to cleaved FGF23 in mice [66]. Whether $1,25(OH)_2D$, similar to PTH, may regulate secretion of intact FGF23 not only at the transcriptional, but also at the posttranslational level remains to be elucidated.

The most important target organ of the endocrine actions of FGF23 is the kidney. FGF23 targets proximal and distal renal tubules by binding to basolateral αKlotho/FGFR1c complexes [67,68]. In proximal renal tubules, FGF23 signaling has a dual action: (i) it downregulates the apical membrane abundance of sodium-phosphate co-transporters, causing reduced transcellular phosphate uptake from the urine and, hence, increased urinary phosphate excretion [69–71]; (ii) it suppresses the transcription of 1α-hydroxylase and increases transcription of 24-hydroxylase, thereby reducing renal production of 1,25(OH)$_2$D [71]. In distal renal tubules, FGF23 signaling causes enhanced reabsorption of calcium and sodium by upregulating the apical membrane abundance of the epithelial calcium channel TRPV5 and of the sodium-chloride cotransporter NCC [67,72].

There is compelling evidence from animal experiments and rare genetic disorders in humans that intact FGF23 signaling is essential for the endocrine control of vitamin D metabolism in the kidney. In humans, loss-of-function mutations in *FGF23* or *αKLOTHO* are associated with elevated circulating 1,25(OH)$_2$D levels and subsequent hypercalcemia and hyperphosphatemia, leading to progressive soft tissue calcifications [73–75]. Similarly, knockout of the *Fgf23* of *αKlotho* genes in mice leads to early lethality caused by unleashed production of 1,25(OH)$_2$D, hypercalcemia, hyperphosphatemia, and ectopic calcifications, a phenotype that can be rescued by concomitant ablation of vitamin D signaling [76–84].

Despite the strong evidence from in vivo studies, the molecular reason for the complete failure of the physiological regulation of renal 1α- and 24-hydroxylases in the absence of FGF23 signaling is still not entirely clear. What do we know about the signaling cascade involved? Based on the similarity of the phenotypes of *αKlotho*$^{-/-}$ and *Fgf23*$^{-/-}$ mice, it is clear that the FGF23-mediated suppression of 1α-hydroxylase transcription in proximal tubular epithelium is an αKlotho-dependent process [51]. In addition, specific deletion of *Fgfr1* in proximal renal tubules has been shown to blunt the FGF23-induced suppression of circulating 1,25(OH)$_2$D levels in vivo [85]. Therefore, at least under physiological conditions, the receptor for FGF23 at the cell membrane of proximal tubules is the FGFR1c/αKlotho receptor complex. At high circulating concentrations of FGF23 such as those found in *Hyp* mice, a model of excessive endogenous FGF23 secretion, FGFR3 and 4 may also play some role in the FGF23-mediated regulation of renal vitamin D metabolism, because ablation of *Fgfr3* and *Fgfr4* increases 1,25(OH)$_2$D levels in *Hyp* mice [86]. FGFRs are tyrosine kinase receptors, and ligand binding results in dimerization and subsequent activation of intracellular phosphorylation cascades [54]. It is well known that the FGF23-mediated regulation of phosphate transport in proximal tubular epithelium involves activation of extracellular signal-regulated kinase-1/2 (ERK1/2) [68]. Activation of ERK1/2 is also one of the first steps in the control by renal 1α- and 24-hydroxylases by FGF23 signaling, as evidenced by the fact that pharmacological ERK1/2 inhibition increases serum 1,25(OH)$_2$D levels in *Hyp* mice [87,88]. However, the signaling pathway downstream of ERK1/2 remains unknown.

At the genomic level, the seminal studies of Meyer and coworkers have shed additional light on the gene regulatory networks involved in the control of 1α- and 24-hydroxylase transcription by FGF23 signaling [29,89]. ChIP-Seq experiments revealed that the kidney-specific regulatory elements leading to suppression of renal 1α-hydroxylase transcription after FGF23 injection in mice are located in introns of a neighboring gene, the *Mettl21b* gene [29]. The functional significance of these kidney specific enhancer sites were demonstrated by the fact that deletion of these sites completely ablated the FGF23-induced suppression of 1α-hydroxylase gene transcription in vivo [29]. However, the specific transcription factors binding to these regulatory regions remain to be elucidated. In addition, it is unclear why simultaneous deletion of all three FGF23-regulated enhancer sites in the *Mettl21b* gene identified by ChIP-Seq experiments did not result in uncontrolled 1,25(OH)$_2$D production and a phenotype similar to *Fgf23* or *αKlotho* deficient mice [29]. A potential explanation for this discrepancy is that FGF23 signaling has reciprocal effects on 1α- and 24-hydroxylase transcription in the kidney, at the same time suppressing 1α-hydroxylase while upregulating 24-hydroxylase transcription [71]. Therefore, the phenotype of *Fgf23* or *αKlotho* deficient mice may only develop in the absence of this dual effect on vitamin D metabolism, affecting both synthesis and degradation of 1,25(OH)$_2$D.

It was previously thought that the regulation of 24-hydroxylase transcription by FGF23 is mediated indirectly through the VDR, because treatment of global VDR knockout mice with recombinant FGF23 suppressed 1α-hydroxylase, but did not upregulate 24-hydroxylase transcription [89]. Moreover, injection of recombinant FGF23 failed to induce

24-hydroxylase transcription in kidney-specific 1α-hydroxylase knockout mice, unable to produce $1,25(OH)_2D$ in the kidney [29]. However, it is unclear whether one can infer from the latter experiments that the FGF23-mediated regulation of 24-hydroxylase is VDR-dependent, because both models are characterized by a profound downregulation of renal 24-hydroxylase expression, potentially masking acute effects of FGF23 treatment. Strong evidence in favor of a VDR independent regulation of renal 24-hydroxylase by FGF23 comes from experiments in which an enhancer region downstream of the 24-hydroxylase gene, previously identified by ChIP-Seq experiments, was deleted in mice. In contrast to the $1,25(OH)_2D$-mediated induction of 24-hydroxylase, the FGF23-mediated induction of 24-hydroxylase expression was blunted in these mice, showing that FGF23 and $1,25(OH)_2D$ have independent effects on 24-hydroxylase expression in vivo [31]. However, the exact mechanisms by which FGF23 signaling regulates renal 24-hydroxylase are still unknown.

Taken together, a paradigm has evolved during the past 20 years in which FGF23 and $1,25(OH)_2D$ represent the endocrine signal molecules linking bone and kidney in the regulation of vitamin D metabolism. In this negative feedback system, bone-derived FGF23 suppresses $1,25(OH)_2D$ production in the kidney, whereas $1,25(OH)_2D$ increases FGF23 secretion in bone (Figure 2). The pathophysiological implications of this feedback system are manifold. In diseases characterized by excessive FGF23 secretion such as X-linked hypophosphatemia or chronic kidney disease, enhanced FGF23 signaling suppresses $1,25(OH)_2D$ production in the kidney [90]. However, it is important to consider that treatment with active vitamin D analogs may increase circulating intact FGF23, which in clinical situations with already elevated levels of circulating FGF23 is increasingly recognized as an untoward side effect of the treatment due to the potential negative cardiovascular effects of FGF23 [91].

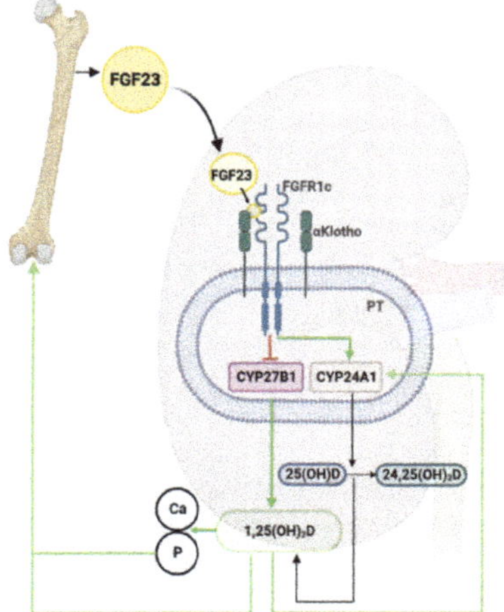

Figure 2. Bone-kidney axis. Fibroblast growth factor 23 (FGF23) is mainly produced in bone under physiological conditions. The active vitamin D hormone ($1,25(OH)_2D$) and phosphate (P) stimulate FGF23 secretion from bone. Circulating FGF23 binds to a receptor complex consisting of FGF receptor-1c (FGFR1c) and of the co-receptor αKlotho in proximal tubules (PT) of the kidney. FGF23 signaling inhibits transcription of 1α-hydroxylase (*CYP27B1*), the key enzyme for vitamin D synthesis, while stimulating transcription of 24-hydroxylase (*CYP24A1*), the main enzyme responsible for vitamin D catabolism. 25-hydroxyvitamin D (25(OH)D) produced in the liver and circulating in the bloodstream is converted in the kidney into $1,25(OH)_2D$ by CYP27B1, or into $24,25(OH)_2D$ by CYP24A1. $1,25(OH)_2D$ stimulates intestinal absorption of calcium (Ca) and P, thereby increasing the blood concentrations of both minerals. Arrows indicate the direction and nature of regulation (green, stimulation; red, inhibition). Created with BioRender.com (Biorender 2022, Toronto, Ontario, Canada).

5. Vitamin D and the Renin-Angiotensin-Aldosterone System (RAAS)

The most important physiological role of vitamin D is its function in the maintenance of mineral homeostasis. However, in the 1980s seasonal variations in the incidence of cardiovascular diseases were for the first time attributed to seasonal variations in vitamin D status [92]. Together with the findings that the VDR is expressed in cardiovascular tissues such as cardiomyocytes, endothelial and vascular smooth muscle cells, this seminal discovery suggested that vitamin D signaling may have a role in the cardiovascular system [93–96]. Indeed, observational studies in humans have found an association between vitamin D deficiency and hypertension, left ventricular hypertrophy (LVH) and heart failure [97–100], and experimental studies in mice and rats provided evidence for a role of vitamin D signaling in the regulation of cardiovascular function. One of the most important regulatory systems of hemodynamics is the RAAS. Juxtaglomerular cells in afferent arterioles in the kidneys store and secrete renin [101]. The main role of renin is to cleave the liver-derived plasma protein angiotensinogen to angiotensin I. Angiotensin I is physiologically inactive and needs to be metabolized to angiotensin II (AngII), which then binds to angiotensin II type I (AT1) and angiotensin II type II (AT2) receptors to exert its effects [101]. The cleavage of angiotensin I to angiotensin II is mediated by angiotensin-converting-enzyme (ACE1) present on vascular endothelial cells mainly in lungs and kidneys. Angiotensin II is a major regulator of peripheral vascular tone, and increases sodium reabsorption from the kidney. Furthermore, it stimulates release of aldosterone from the zona glomerulosa in the adrenal cortex. Aldosterone binds to mineralocorticoid receptors and promotes sodium and water reabsorption in the epithelial cells of the distal convoluted tubule and collecting duct, thereby increasing blood volume and subsequently elevating arterial blood pressure [101,102].

In a seminal study published in 2002, Li and coworkers reported that mice with a global VDR deletion were characterized by increased blood pressure, cardiac hypertrophy, increased renin mRNA and protein levels, as well as elevated plasma angiotensin II production [9]. The angiotensinogen levels in the liver were unchanged when compared to WT mice, suggesting that the increase in plasma AngII levels was due to renin over-activity. Although the latter study for the first time established a potential link between vitamin D signaling and the RAAS, a major caveat in that study was that global VDR knockout mice on a normal diet present with hypocalcemia and secondary hyperparathyroidism, both of which can individually influence the cardiovascular outcomes observed. The authors aimed to address this by feeding a calcium- and phosphorus-enriched rescue diet, and by examining mice at 20 days of age with the rationale that hypocalcemia in VDR knockout mice is not yet developed at this early stage. Interestingly, they found that the dietary treatment for five weeks normalized calcium levels in VDR knockout mice, but the mice still showed increased renin expression. However, the rescue diet did not normalize serum PTH, and PTH itself can stimulate the RAAS [9]. Therefore, based on the study by Li and coworkers [9], it was not entirely clear whether the VDR functions as an endocrine suppressor of renin biosynthesis. A later proposed mechanism by which 1,25$(OH)_2$D could suppress renin gene transcription is by blocking the activity of the cyclic AMP response element in the renin gene promoter [103]. A recent study in VDR knock-out mice fed the rescue diet found an increase of almost 50% in renin mRNA expression in the kidney, as well as elevated serum renin levels when compared to controls on normal or rescue diet. Interestingly, renin activity and thereby the ability to generate angiotensin I was lower in knock-out mice and there were no significant changes in any of the cardiovascular parameters measured [104]. On the other hand, Jia and coworkers found that *VDR* deficiency in 8 week-old mice increased blood pressure by elevating oxidative stress factors and RAAS activity [105]. However, the authors did not report data on mineral metabolism or PTH levels in these mice.

Zhou and coworkers utilized 1α-hydroxylase knock-out mice to investigate if the cardiovascular effect of 1,25$(OH)_2$D is dependent on calcium and/or phosphorus. 1α-hydroxylase knockout mice have a similar phenotype compared with global VDR knockout mice, making them a useful tool to investigate the role of the VDR in hypertension [106]. Similar to VDR mutant mice, 1α-hydroxylase knockout mice also presented with hypertension and cardiac hypertrophy that was associated with upregulation of the RAAS in cardiac and renal tissue [106]. Treatment of these mice with 1,25(OH)$_2$D suppressed RAAS activity and reversed the phenotype [106]. Similar to the study by Jia and coworkers, the authors did not report PTH levels, making it hard to dismiss the role of PTH in mediating cardiovascular changes.

When we analyzed aged global VDR mutant mice maintained on rescue diet since weaning, we also found increased systolic blood pressure and impaired systolic and diastolic function in 9-month-old VDR knockout mice [107]. However, these changes were due to a lower bioavailability of the vasodilator nitric oxide (NO), leading to endothelial dysfunction, increased arterial stiffness, structural remodeling of the aorta, and impaired systolic and diastolic heart function in 9-month-old mice. Interestingly, kidney renin mRNA levels, urinary aldosterone and plasma renin activity were not different between WT and VDR knockout mice fed the rescue diet, suggesting that the cardiovascular changes induced by the lack of VDR signaling were RAAS-independent. On the other hand, an increase in renin mRNA expression and serum aldosterone levels were observed in global VDR knockout mice fed the normal diet, supporting the notion that secondary hyperparathyroidism might be responsible for RAAS activation and subsequent hemodynamic changes in mice lacking VDR [107].

Mice lacking VDR specifically in the endothelium were able to shed more light on the question whether hypertension in global VDR knockout mice is caused by RAAS activation or is independent of RAAS. These animals had no changes in mineral metabolism, and thereby serve as a better model to test tissue-specific effects of vitamin D [108]. At baseline, the cardiovascular phenotype of 12-week-old endothelium-specific VDR knockout mice was comparable to that of WT mice. However, mice with VDR deletion in the endothelium were more susceptible to treatment with low doses of Ang II as evidenced by increases in systolic, diastolic, and mean arterial pressure when compared to WT mice [108]. Furthermore, in the absence of endothelial VDR, acetylcholine-induced aortic relaxation was significantly impaired, and aortic mRNA and protein abundance of eNOS was reduced, suggesting that the VDR has a protective role in regulating vascular tone and that the changes observed are independent of RAAS [108].

Interestingly, observational studies support an inverse relationship between circulating vitamin D levels and RAAS in hypertensive and non-hypertensive subjects [109–112]. This was confirmed in a small open-label study where daily treatment of vitamin D deficient, hypertensive individuals with 15,000 IU vitamin D for 4 weeks increased 25(OH)D and 1,25(OH)$_2$D, decreased PTH levels, reduced supine mean arterial pressure (MAP) and reduced kidney specific RAAS activation [113]. However, randomized controlled trials investigating the interaction between vitamin D and RAAS are scarce. In the study by McMullan and coworkers, supplementation with 50,000 IU vitamin D per week over eight weeks increased 25(OH)D levels, but neither affected blood pressure nor elements of the RAAS [114]. However, the failure to find an effect of vitamin D supplementation on RAAS in the latter study may be due to the fact that normotensive subjects were investigated, and that the subjects were vitamin D sufficient and that the dosage of vitamin D administered was lower [114]. Similarly, in recent large, randomized controlled trials such as VIDA and VITAL with over 5000 and 25,000 vitamin D sufficient subjects, respectively, no beneficial effects of vitamin D supplementation on primary and secondary cardiovascular endpoints were found [115,116]. Unfortunately, the latter trials did not include measurements of RAAS activity as endpoints. Whether the effects would be similar in vitamin D deficient individuals remains to be investigated.

ACE2 enzyme is a more recently discovered homologue of ACE1. In contrast to ACE1, ACE2 is not as ubiquitously expressed, and is found mainly in the kidneys, heart, and testis [117]. While ACE1 produces Ang II, ACE2 converts angiotensin II (Ang II) into angiotensin 1–7 (Ang 1–7), binds to Mas receptors (MasR), and has a powerful vasodilatory, antifibrotic, antiarrhythmic, and antihypertensive effect [118]. Therefore, ACE2 acts as a negative regulator of the RAAS and together with ACE1 participates in maintaining RAAS homeostasis and blood pressure. It has been reported that vitamin D signaling induces the ACE2/Ang-(1-7)/MasR axis activity, and thereby has a protective role in cardiovascular and pulmonary diseases [119]. Shedding of ACE2 by disintegrin and metalloproteinase domain 17 (ADAM17) from the membrane results in production of soluble ACE2 (sACE2). Interestingly, sACE2 appears to be a biomarker in patients with heart failure, reflecting increased ACE activity [120]. Furthermore, administration of calcitriol to rats with LPS-induced acute lung injury increased ACE2 mRNA expression, and had a positive effect on clinical manifestations and pathological changes in these animals [121,122]. Mice lacking VDR presented with a more severe acute lung injury and higher mortality when compared to WT controls, as a result of excessive induction of angiopoietin-2 and angiotensin II, suggesting that VDR signaling regulates both the angiopoietin-2 and the RAAS pathways [123]. Calcitriol and paracalcitriol decreased ACE1 concentration, ACE1/ACE2 ratio, reduced oxidative stress, and exerted renoprotective effects in diabetic rats, additionally supporting the protective effect of vitamin D [124,125]. It is tempting to speculate that some of the discrepancies observed in animal studies dealing with the effects of vitamin D signaling on the RAAS may be resolved by examining the ACE2/Ang-(1-7)/MasR axis.

While the regulation of RAAS by vitamin D has been extensively studied, the data on modulation of vitamin D metabolism by RAAS is scarce. Currently, there is no evidence that RAAS can directly influence 1α-hydroxylase activity or VDR expression. However, indirect effects of the RAAS on vitamin D metabolism are conceivable, because impacts of RAAS components on αKlotho and FGF23 have been reported. Notably, in several animal models characterized by RAAS activation there is a clear downregulation of αKlotho abundance in the kidney [126–128]. Furthermore, long-term angiotensin II infusion also downregulated renal αKlotho expression at mRNA and protein level [129]. This mechanism is thought to be AT1-dependent, because administration of losartan completely abolished the effect. Several other mechanisms of αKlotho regulation, besides the direct action on AT1 receptor, by Ang II have been proposed. It is known that oxidative stress can downregulate αKlotho expression [130]. Oral administration of a free radical scavenger in rats suppressed the downregulation of αKlotho, supporting the hypothesis that an increase in systemic oxidative stress modulates αKlotho expression [128]. There is accumulating evidence that Ang II stimulates formation of reactive oxygen species (ROS) through the NADPH oxidase system [131–133]. Indeed, Ang II receptor blockade reduced oxidative stress in vivo, suggesting that Ang II can modulate Klotho expression by influencing the formation of ROS [134]. Furthermore, Ang II can downregulate Klotho by upregulating TACE, a TNFα-converting enzyme, thereby identifying TACE inhibitors as a new potential therapeutic target in patients suffering from diseases characterized by low Klotho abundance in the kidney [135–137].

As mentioned above, FGF23 requires the presence of its coreceptor αKlotho in the kidney to exert its effects. In proximal renal tubules, FGF23 signaling inhibits $1,25(OH)_2D$ synthesis, and stimulates its catabolism by downregulating 1α-hydroxylase and upregulating 24-hydroxylase, respectively. Therefore, the effect of Ang II on αKlotho could be interpreted as an indirect modulation of vitamin D metabolism by influencing the FGF23-αKlotho axis.

Chronic kidney disease is characterized by low αKlotho and $1,25(OH)_2D$ levels and elevated circulating FGF23. High FGF23 levels are associated with left ventricular hypertrophy and heart failure [138]. Furthermore, increased circulating FGF23 additionally exacerbates $1,25(OH)_2D$ deficiency, possibly contributing to faster progression of CKD and increased mortality in these patients. Therefore, it seems plausible that targeting Klotho deficiency by blocking RAAS and thus normalizing $1,25(OH)_2D$ production may have beneficial effects in patients with CKD in regard to cardiovascular and renal outcomes. Indeed, RAAS inhibition and supplementation with vitamin D analogs are common treatment strategies in patients with CKD [139]. However, they are usually considered separately, and their potential joint effect on the FGF23-αKlotho axis is not well investigated. Adapting the existing treatment to modulate renal αKlotho levels and the FGF23-αKlotho-vitamin D axis could prove to be a more efficient therapeutic option in patients with CKD.

6. Conclusions

The purpose of this review was to highlight the current knowledge about the molecular mechanisms underlying the interaction of vitamin D, FGF23, PTH and RAAS in health and disease. FGF23 is a key regulator of vitamin D metabolism. FGF23 lowers 1,25$(OH)_2$D biosynthesis by suppressing 1α-hydroxylase transcription, and increases its degradation by stimulating 24-hydroxylase transcription in proximal tubules of the kidney, thereby down-regulating renal 1,25$(OH)_2$D production. PTH on the other hand has the opposite effect on vitamin D metabolism: it stimulates 1,25$(OH)_2$D synthesis and inhibits its catabolism. Our understanding of the genomic mechanisms involved in the FGF23- and PTH-mediated regulation of vitamin D metabolism in the kidney has increased significantly during recent years. However, some major gaps in our knowledge still remain. For example, the signaling cascades mediating the FGF23-induced regulation of 1α-hydroxylase and 24-hydroxylase transcription in the kidney are still unknown. Furthermore, studies investigating the role of vitamin D in the cardiovascular system and the relationship between vitamin D and the RAAS have provided conflicting evidence. While there are several proposed mechanisms as to how vitamin D signaling can regulate RAAS, the regulation of vitamin D metabolism by RAAS is still an area largely unexplored.

Author Contributions: N.L.: conceptualization; visualization; writing—original draft; writing—review and editing. R.G.E.: conceptualization; writing—original draft; writing—review and editing. All authors have read and agreed to the published version of the manuscript.

Funding: Open Access Funding by the University of Veterinary Medicine Vienna.

Institutional Review Board Statement: Not applicable.

Informed Consent Statement: Not applicable.

Conflicts of Interest: The authors declare no conflict of interest.

References

1. Hess, A.F.; Unger, L.J. The cure of infantile rickets by artificial light and by sunlight. *Proc. Soc. Exp. Biol. Med.* **1921**, *18*, 298. [CrossRef]
2. McCollum, E.V.; Simmonds, N.; Becker, J.E.; Shipley, P.d. Studies on experimental rickets: XXI. An experimental demonstration of the existence of a vitamin which promotes calcium deposition. *J. Biol. Chem.* **1922**, *53*, 293–312. [CrossRef]
3. Chick, H.; Dalyell, E.; Hume, M.; Smith, H.H.; Mackay, H.M.; Hams Wimberger, M.; Kinderklinik, R.T.T.U. The aetiology of rickets in infants: Prophylactic and curative observations at the Vienna University Kinderklinik. *Lancet* **1922**, *200*, 7–11. [CrossRef]
4. Wolf, G. The Discovery of Vitamin D: The Contribution of Adolf Windaus. *J. Nutr.* **2004**, *134*, 1299–1302. [CrossRef]
5. Kalra, S.; Baruah, M.P.; Sahay, R.; Sawhney, K. The history of parathyroid endocrinology. *Indian J. Endocrinol. Metab.* **2013**, *17*, 320–322. [CrossRef]
6. Garabedian, M.; Holick, M.F.; Deluca, H.F.; Boyle, I.T. Control of 25-hydroxycholecalciferol metabolism by parathyroid glands. *Proc. Natl. Acad. Sci. USA* **1972**, *69*, 1673–1676. [CrossRef]
7. Fraser, D.R.; Kodicek, E. Regulation of 25-Hydroxycholecalciferol-1-Hydroxylase Activity in Kidney by Parathyroid Hormone. *Nat. New Biol.* **1973**, *241*, 163–166. [CrossRef] [PubMed]
8. Chertow, B.S.; Baylink, D.J.; Wergedal, J.E.; Su, M.H.; Norman, A.W. Decrease in serum immunoreactive parathyroid hormone in rats and in parathyroid hormone secretion in vitro by 1,25-dihydroxycholecalciferol. *J. Clin. Investig.* **1975**, *56*, 668–678. [CrossRef] [PubMed]
9. Li, Y.C.; Kong, J.; Wei, M.; Chen, Z.-F.; Liu, S.Q.; Cao, L.-P. 1,25-Dihydroxyvitamin D3 is a negative endocrine regulator of the renin-angiotensin system. *J. Clin. Investig.* **2002**, *110*, 229–238. [CrossRef]
10. Brunette, M.G.; Chan, M.; Ferriere, C.; Roberts, K.D. Site of 1,25(OH)2 vitamin D3 synthesis in the kidney. *Nature* **1978**, *276*, 287–289. [CrossRef]
11. Fraser, D.R.; Kodicek, E. Unique Biosynthesis by Kidney of a Biologically Active Vitamin D Metabolite. *Nature* **1970**, *228*, 764–766. [CrossRef]

12. Somjen, D.; Weisman, Y.; Kohen, F.; Gayer, B.; Limor, R.; Sharon, O.; Jaccard, N.; Knoll, E.; Stern, N. 25-hydroxyvitamin D3-1alpha-hydroxylase is expressed in human vascular smooth muscle cells and is upregulated by parathyroid hormone and estrogenic compounds. *Circulation* **2005**, *111*, 1666–1671. [CrossRef]
13. Zehnder, D.; Bland, R.; Chana, R.S.; Wheeler, D.C.; Howie, A.J.; Williams, M.C.; Stewart, P.M.; Hewison, M. Synthesis of 1,25-Dihydroxyvitamin D3 by Human Endothelial Cells Is Regulated by Inflammatory Cytokines: A Novel Autocrine Determinant of Vascular Cell Adhesion. *J. Am. Soc. Nephrol.* **2002**, *13*, 621–629. [CrossRef] [PubMed]
14. Bikle, D.D. Vitamin D Metabolism, Mechanism of Action, and Clinical Applications. *Chem. Biol.* **2014**, *21*, 319–329. [CrossRef]
15. Bouillon, R.; Carmeliet, G.; Verlinden, L.; van Etten, E.; Verstuyf, A.; Luderer, H.F.; Lieben, L.; Mathieu, C.; Demay, M. Vitamin D and human health: Lessons from vitamin D receptor null mice. *Endocr. Rev.* **2008**, *29*, 726–776. [CrossRef]
16. Bouillon, R.; Suda, T. Vitamin D: Calcium and bone homeostasis during evolution. *Bonekey Rep.* **2014**, *3*, 480. [CrossRef] [PubMed]
17. Jones, G.; Prosser, D.E.; Kaufmann, M. 25-Hydroxyvitamin D-24-hydroxylase (CYP24A1): Its important role in the degradation of vitamin D. *Arch. Biochem. Biophys.* **2012**, *523*, 9–18. [CrossRef] [PubMed]
18. St-Arnaud, R.; Arabian, A.; Travers, R.; Barletta, F.; Raval-Pandya, M.; Chapin, K.; Depovere, J.; Mathieu, C.; Christakos, S.; Demay, M.B.; et al. Deficient mineralization of intramembranous bone in vitamin D-24-hydroxylase-ablated mice is due to elevated 1,25-dihydroxyvitamin D and not to the absence of 24,25-dihydroxyvitamin D. *Endocrinology* **2000**, *141*, 2658–2666. [CrossRef]
19. Nesterova, G.; Malicdan, M.C.; Yasuda, K.; Sakaki, T.; Vilboux, T.; Ciccone, C.; Horst, R.; Huang, Y.; Golas, G.; Introne, W.; et al. 1,25-(OH)2D-24 Hydroxylase (CYP24A1) Deficiency as a Cause of Nephrolithiasis. *Clin. J. Am. Soc. Nephrol.* **2013**, *8*, 649–657. [CrossRef]
20. Slominski, A.T.; Kim, T.-K.; Li, W.; Postlethwaite, A.; Tieu, E.W.; Tang, E.K.Y.; Tuckey, R.C. Detection of novel CYP11A1-derived secosteroids in the human epidermis and serum and pig adrenal gland. *Sci. Rep.* **2015**, *5*, 14875. [CrossRef] [PubMed]
21. Slominski, A.T.; Kim, T.K.; Shehabi, H.Z.; Semak, I.; Tang, E.K.; Nguyen, M.N.; Benson, H.A.; Korik, E.; Janjetovic, Z.; Chen, J.; et al. In vivo evidence for a novel pathway of vitamin D_3 metabolism initiated by P450scc and modified by CYP27B1. *FASEB J.* **2012**, *26*, 3901–3915. [CrossRef]
22. Tang, E.K.; Chen, J.; Janjetovic, Z.; Tieu, E.W.; Slominski, A.T.; Li, W.; Tuckey, R.C. Hydroxylation of CYP11A1-derived products of vitamin D3 metabolism by human and mouse CYP27B1. *Drug Metab. Dispos.* **2013**, *41*, 1112–1124. [CrossRef]
23. Verstuyf, A.; Carmeliet, G.; Bouillon, R.; Mathieu, C. Vitamin D: A pleiotropic hormone. *Kidney Int.* **2010**, *78*, 140–145. [CrossRef]
24. Gensure, R.C.; Gardella, T.J.; Jüppner, H. Parathyroid hormone and parathyroid hormone-related peptide, and their receptors. *Biochem. Biophys. Res. Commun.* **2005**, *328*, 666–678. [CrossRef]
25. Bienaime, F.; Prie, D.; Friedlander, G.; Souberbielle, J.C. Vitamin D metabolism and activity in the parathyroid gland. *Mol. Cell. Endocrinol.* **2011**, *347*, 30–41. [CrossRef]
26. Sato, T.; Courbebaisse, M.; Ide, N.; Fan, Y.; Hanai, J.-i.; Kaludjerovic, J.; Densmore, M.J.; Yuan, Q.; Toka, H.R.; Pollak, M.R.; et al. Parathyroid hormone controls paracellular Ca^{2+} transport in the thick ascending limb by regulating the tight-junction protein Claudin14. *Proc. Natl. Acad. Sci. USA* **2017**, *114*, E3344–E3353. [CrossRef] [PubMed]
27. Martin, A.; David, V.; Quarles, L.D. Regulation and function of the FGF23/klotho endocrine pathways. *Physiol. Rev.* **2012**, *92*, 131–155. [CrossRef]
28. Goretti Penido, M.; Alon, U.S. Phosphate homeostasis and its role in bone health. *Pediatr. Nephrol.* **2012**, *27*, 2039–2048. [CrossRef] [PubMed]
29. Meyer, M.B.; Benkusky, N.A.; Kaufmann, M.; Lee, S.M.; Onal, M.; Jones, G.; Pike, J.W. A kidney-specific genetic control module in mice governs endocrine regulation of the cytochrome P450 gene Cyp27b1 essential for vitamin D(3) activation. *J. Biol. Chem.* **2017**, *292*, 17541–17558. [CrossRef] [PubMed]
30. Meyer, M.B.; Benkusky, N.A.; Lee, S.M.; Yoon, S.H.; Mannstadt, M.; Wein, M.N.; Pike, J.W. Rapid genomic changes by mineralotropic hormones and kinase SIK inhibition drive coordinated renal Cyp27b1 and Cyp24a1 expression via CREB modules. *J. Biol. Chem.* **2022**, *298*, 102559. [CrossRef]
31. Meyer, M.B.; Pike, J.W. Mechanistic homeostasis of vitamin D metabolism in the kidney through reciprocal modulation of Cyp27b1 and Cyp24a1 expression. *J. Steroid. Biochem. Mol. Biol.* **2020**, *196*, 105500. [CrossRef]
32. Silver, J.; Russell, J.; Sherwood, L.M. Regulation by vitamin D metabolites of messenger ribonucleic acid for preproparathyroid hormone in isolated bovine parathyroid cells. *Proc. Natl. Acad. Sci. USA* **1985**, *82*, 4270–4273. [CrossRef]
33. Ritter, C.S.; Armbrecht, H.J.; Slatopolsky, E.; Brown, A.J. 25-Hydroxyvitamin D(3) suppresses PTH synthesis and secretion by bovine parathyroid cells. *Kidney Int.* **2006**, *70*, 654–659. [CrossRef]
34. Silver, J.; Naveh-Many, T.; Mayer, H.; Schmelzer, H.J.; Popovtzer, M.M. Regulation by vitamin D metabolites of parathyroid hormone gene transcription in vivo in the rat. *J. Clin. Investig.* **1986**, *78*, 1296–1301. [CrossRef]
35. Meir, T.; Levi, R.; Lieben, L.; Libutti, S.; Carmeliet, G.; Bouillon, R.; Silver, J.; Naveh-Many, T. Deletion of the vitamin D receptor specifically in the parathyroid demonstrates a limited role for the receptor in parathyroid physiology. *Am. J. Physiol. Renal Physiol.* **2009**, *297*, F1192–F1198. [CrossRef] [PubMed]
36. Weber, K.; Bergow, C.; Hirmer, S.; Schüler, C.; Erben, R.G. Vitamin D-independent therapeutic effects of extracellular calcium in a mouse model of adult-onset secondary hyperparathyroidism. *J. Bone Miner. Res.* **2009**, *24*, 22–32. [CrossRef] [PubMed]

37. Mackey, S.L.; Heymont, J.L.; Kronenberg, H.M.; Demay, M.B. Vitamin D receptor binding to the negative human parathyroid hormone vitamin D response element does not require the retinoid x receptor. *Mol. Endocrinol.* **1996**, *10*, 298–305. [CrossRef] [PubMed]
38. Koszewski, N.J.; Ashok, S.; Russell, J. Turning a Negative into a Positive: Vitamin D Receptor Interactions with the Avian Parathyroid Hormone Response Element. *Mol. Endocrinol.* **1999**, *13*, 455–465. [CrossRef] [PubMed]
39. Murayama, A.; Kim, M.S.; Yanagisawa, J.; Takeyama, K.; Kato, S. Transrepression by a liganded nuclear receptor via a bHLH activator through co-regulator switching. *EMBO J.* **2004**, *23*, 1598–1608. [CrossRef]
40. Ritter, C.S.; Brown, A.J. Direct suppression of Pth gene expression by the vitamin D prohormones doxercalciferol and calcidiol requires the vitamin D receptor. *J. Mol. Endocrinol.* **2011**, *46*, 63–66. [CrossRef]
41. Segersten, U.; Correa, P.; Hewison, M.; Hellman, P.; Dralle, H.; Carling, T.; Akerström, G.; Westin, G. 25-hydroxyvitamin D(3)-1alpha-hydroxylase expression in normal and pathological parathyroid glands. *J. Clin. Endocrinol. Metab.* **2002**, *87*, 2967–2972. [CrossRef] [PubMed]
42. Cozzolino, M.; Covic, A.; Martinez-Placencia, B.; Xynos, K. Treatment Failure of Active Vitamin D Therapy in Chronic Kidney Disease: Predictive Factors. *Am. J. Nephrol.* **2015**, *42*, 228–236. [CrossRef] [PubMed]
43. Coyne, D.W.; Goldberg, S.; Faber, M.; Ghossein, C.; Sprague, S.M. A randomized multicenter trial of paricalcitol versus calcitriol for secondary hyperparathyroidism in stages 3–4 CKD. *Clin. J. Am. Soc. Nephrol.* **2014**, *9*, 1620–1626. [CrossRef] [PubMed]
44. Shah, A.; Aeddula, N.R. Renal Osteodystrophy. In *StatPearls*; StatPearls Publishing LLC: Treasure Island, FL, USA, 2022.
45. Jamaluddin, E.J.; Gafor, A.H.; Yean, L.C.; Cader, R.; Mohd, R.; Kong, N.C.; Shah, S.A. Oral paricalcitol versus oral calcitriol in continuous ambulatory peritoneal dialysis patients with secondary hyperparathyroidism. *Clin. Exp. Nephrol.* **2014**, *18*, 507–514. [CrossRef]
46. White, K.E.; Evans, W.E.; O'Riordan, J.L.H.; Speer, M.C.; Econs, M.J.; Lorenz-Depiereux, B.; Grabowski, M.; Meitinger, T.; Strom, T.M. Autosomal dominant hypophosphataemic rickets is associated with mutations in *FGF23*. *Nat. Genet.* **2000**, *26*, 345–348. [CrossRef]
47. Shimada, T.; Mizutani, S.; Muto, T.; Yoneya, T.; Hino, R.; Takeda, S.; Takeuchi, Y.; Fujita, T.; Fukumoto, S.; Yamashita, T. Cloning and characterization of FGF23 as a causative factor of tumor-induced osteomalacia. *Proc. Natl. Acad. Sci. USA* **2001**, *98*, 6500–6505. [CrossRef]
48. Yoshiko, Y.; Wang, H.; Minamizaki, T.; Ijuin, C.; Yamamoto, R.; Suemune, S.; Kozai, K.; Tanne, K.; Aubin, J.E.; Maeda, N. Mineralized tissue cells are a principal source of FGF23. *Bone* **2007**, *40*, 1565–1573. [CrossRef]
49. Itoh, N.; Ohta, H.; Konishi, M. Endocrine FGFs: Evolution, Physiology, Pathophysiology, and Pharmacotherapy. *Front. Endocrinol.* **2015**, *6*, 154. [CrossRef] [PubMed]
50. Tagliabracci, V.S.; Engel, J.L.; Wiley, S.E.; Xiao, J.; Gonzalez, D.J.; Nidumanda Appaiah, H.; Koller, A.; Nizet, V.; White, K.E.; Dixon, J.E. Dynamic regulation of FGF23 by Fam20C phosphorylation, GalNAc-T3 glycosylation, and furin proteolysis. *Proc. Natl. Acad. Sci. USA* **2014**, *111*, 5520–5525. [CrossRef]
51. Urakawa, I.; Yamazaki, Y.; Shimada, T.; Iijima, K.; Hasegawa, H.; Okawa, K.; Fujita, T.; Fukumoto, S.; Yamashita, T. Klotho converts canonical FGF receptor into a specific receptor for FGF23. *Nature* **2006**, *444*, 770–774. [CrossRef] [PubMed]
52. Chen, G.; Liu, Y.; Goetz, R.; Fu, L.; Jayaraman, S.; Hu, M.-C.; Moe, O.W.; Liang, G.; Li, X.; Mohammadi, M. α-Klotho is a non-enzymatic molecular scaffold for FGF23 hormone signalling. *Nature* **2018**, *553*, 461–466. [CrossRef] [PubMed]
53. Goetz, R.; Ohnishi, M.; Ding, X.; Kurosu, H.; Wang, L.; Akiyoshi, J.; Ma, J.; Gai, W.; Sidis, Y.; Pitteloud, N.; et al. Klotho coreceptors inhibit signaling by paracrine fibroblast growth factor 8 subfamily ligands. *Mol. Cell. Biol.* **2012**, *32*, 1944–1954. [CrossRef] [PubMed]
54. Ornitz, D.M.; Itoh, N. The Fibroblast Growth Factor signaling pathway. *Wiley Interdiscip. Rev. Dev. Biol.* **2015**, *4*, 215–266. [CrossRef]
55. Ratsma, D.M.A.; Zillikens, M.C.; van der Eerden, B.C.J. Upstream Regulators of Fibroblast Growth Factor 23. *Front. Endocrinol.* **2021**, *12*, 588096. [CrossRef]
56. Kolek, O.I.; Hines, E.R.; Jones, M.D.; LeSueur, L.K.; Lipko, M.A.; Kiela, P.R.; Collins, J.F.; Haussler, M.R.; Ghishan, F.K. 1α,25-Dihydroxyvitamin D3 upregulates FGF23 gene expression in bone: The final link in a renal-gastrointestinal-skeletal axis that controls phosphate transport. *Am. J. Physiol.-Gastrointest. Liver Physiol.* **2005**, *289*, G1036–G1042. [CrossRef] [PubMed]
57. Ito, M.; Sakai, Y.; Furumoto, M.; Segawa, H.; Haito, S.; Yamanaka, S.; Nakamura, R.; Kuwahata, M.; Miyamoto, K. Vitamin D and phosphate regulate fibroblast growth factor-23 in K-562 cells. *Am. J. Physiol. Endocrinol. Metab.* **2005**, *288*, E1101–E1109. [CrossRef] [PubMed]
58. Masuyama, R.; Stockmans, I.; Torrekens, S.; Van Looveren, R.; Maes, C.; Carmeliet, P.; Bouillon, R.; Carmeliet, G. Vitamin D receptor in chondrocytes promotes osteoclastogenesis and regulates FGF23 production in osteoblasts. *J. Clin. Investig.* **2006**, *116*, 3150–3159. [CrossRef]
59. Nguyen-Yamamoto, L.; Karaplis, A.C.; St–Arnaud, R.; Goltzman, D. Fibroblast Growth Factor 23 Regulation by Systemic and Local Osteoblast-Synthesized 1,25-Dihydroxyvitamin D. *J. Am. Soc. Nephrol.* **2017**, *28*, 586–597. [CrossRef]
60. Robinson-Cohen, C.; Bartz, T.M.; Lai, D.; Ikizler, T.A.; Peacock, M.; Imel, E.A.; Michos, E.D.; Foroud, T.M.; Akesson, K.; Taylor, K.D.; et al. Genetic Variants Associated with Circulating Fibroblast Growth Factor 23. *J. Am. Soc. Nephrol.* **2018**, *29*, 2583–2592. [CrossRef]

61. Saini, R.K.; Kaneko, I.; Jurutka, P.W.; Forster, R.; Hsieh, A.; Hsieh, J.-C.; Haussler, M.R.; Whitfield, G.K. 1,25-Dihydroxyvitamin D3 Regulation of Fibroblast Growth Factor-23 Expression in Bone Cells: Evidence for Primary and Secondary Mechanisms Modulated by Leptin and Interleukin-6. *Calcif. Tissue Int.* **2013**, *92*, 339–353. [CrossRef]
62. St John, H.C.; Bishop, K.A.; Meyer, M.B.; Benkusky, N.A.; Leng, N.; Kendziorski, C.; Bonewald, L.F.; Pike, J.W. The osteoblast to osteocyte transition: Epigenetic changes and response to the vitamin D3 hormone. *Mol. Endocrinol.* **2014**, *28*, 1150–1165. [CrossRef]
63. Onal, M.; Carlson, A.H.; Thostenson, J.D.; Benkusky, N.A.; Meyer, M.B.; Lee, S.M.; Pike, J.W. A Novel Distal Enhancer Mediates Inflammation-, PTH-, and Early Onset Murine Kidney Disease-Induced Expression of the Mouse Fgf23 Gene. *JBMR Plus* **2018**, *2*, 32–47. [CrossRef] [PubMed]
64. Lee, S.M.; Carlson, A.H.; Onal, M.; Benkusky, N.A.; Meyer, M.B.; Pike, J.W. A Control Region Near the Fibroblast Growth Factor 23 Gene Mediates Response to Phosphate, 1,25(OH)2D3, and LPS In Vivo. *Endocrinology* **2019**, *160*, 2877–2891. [CrossRef]
65. Mattoo, R.L. The Roles of Fibroblast Growth Factor (FGF)-23, α-Klotho and Furin Protease in Calcium and Phosphate Homeostasis: A Mini-Review. *Indian J. Clin. Biochem.* **2014**, *29*, 8–12. [CrossRef]
66. Knab, V.M.; Corbin, B.; Andrukhova, O.; Hum, J.M.; Ni, P.; Rabadi, S.; Maeda, A.; White, K.E.; Erben, R.G.; Juppner, H.; et al. Acute Parathyroid Hormone Injection Increases C-Terminal but Not Intact Fibroblast Growth Factor 23 Levels. *Endocrinology* **2017**, *158*, 1130–1139. [CrossRef] [PubMed]
67. Andrukhova, O.; Smorodchenko, A.; Egerbacher, M.; Streicher, C.; Zeitz, U.; Goetz, R.; Shalhoub, V.; Mohammadi, M.; Pohl, E.E.; Lanske, B.; et al. FGF23 promotes renal calcium reabsorption through the TRPV5 channel. *EMBO J.* **2014**, *33*, 229–246. [CrossRef]
68. Andrukhova, O.; Zeitz, U.; Goetz, R.; Mohammadi, M.; Lanske, B.; Erben, R.G. FGF23 acts directly on renal proximal tubules to induce phosphaturia through activation of the ERK1/2-SGK1 signaling pathway. *Bone* **2012**, *51*, 621–628. [CrossRef]
69. Larsson, T.; Marsell, R.; Schipani, E.; Ohlsson, C.; Ljunggren, O.s.; Tenenhouse, H.S.; Jüppner, H.; Jonsson, K.B. Transgenic Mice Expressing Fibroblast Growth Factor 23 under the Control of the α1(I) Collagen Promoter Exhibit Growth Retardation, Osteomalacia, and Disturbed Phosphate Homeostasis. *Endocrinology* **2004**, *145*, 3087–3094. [CrossRef] [PubMed]
70. Shimada, T.; Urakawa, I.; Yamazaki, Y.; Hasegawa, H.; Hino, R.; Yoneya, T.; Takeuchi, Y.; Fujita, T.; Fukumoto, S.; Yamashita, T. FGF-23 transgenic mice demonstrate hypophosphatemic rickets with reduced expression of sodium phosphate cotransporter type IIa. *Biochem. Biophys. Res. Commun.* **2004**, *314*, 409–414. [CrossRef]
71. Shimada, T.; Hasegawa, H.; Yamazaki, Y.; Muto, T.; Hino, R.; Takeuchi, Y.; Fujita, T.; Nakahara, K.; Fukumoto, S.; Yamashita, T. FGF-23 Is a Potent Regulator of Vitamin D Metabolism and Phosphate Homeostasis. *J. Bone Miner. Res.* **2003**, *19*, 429–435. [CrossRef]
72. Andrukhova, O.; Slavic, S.; Smorodchenko, A.; Zeitz, U.; Shalhoub, V.; Lanske, B.; Pohl, E.E.; Erben, R.G. FGF23 regulates renal sodium handling and blood pressure. *EMBO Mol. Med.* **2014**, *6*, 744–759. [CrossRef] [PubMed]
73. Topaz, O.; Shurman, D.L.; Bergman, R.; Indelman, M.; Ratajczak, P.; Mizrachi, M.; Khamaysi, Z.; Behar, D.; Petronius, D.; Friedman, V.; et al. Mutations in GALNT3, encoding a protein involved in O-linked glycosylation, cause familial tumoral calcinosis. *Nat. Genet.* **2004**, *36*, 579–581. [CrossRef]
74. Araya, K.; Fukumoto, S.; Backenroth, R.; Takeuchi, Y.; Nakayama, K.; Ito, N.; Yoshii, N.; Yamazaki, Y.; Yamashita, T.; Silver, J.; et al. A novel mutation in fibroblast growth factor 23 gene as a cause of tumoral calcinosis. *J. Clin. Endocrinol. Metab.* **2005**, *90*, 5523–5527. [CrossRef] [PubMed]
75. Ichikawa, S.; Imel, E.A.; Kreiter, M.L.; Yu, X.; Mackenzie, D.S.; Sorenson, A.H.; Goetz, R.; Mohammadi, M.; White, K.E.; Econs, M.J. A homozygous missense mutation in human KLOTHO causes severe tumoral calcinosis. *J. Clin. Investig.* **2007**, *117*, 2684–2691. [CrossRef] [PubMed]
76. Shimada, T.; Kakitani, M.; Yamazaki, Y.; Hasegawa, H.; Takeuchi, Y.; Fujita, T.; Fukumoto, S.; Tomizuka, K.; Yamashita, T. Targeted ablation of Fgf23 demonstrates an essential physiological role of FGF23 in phosphate and vitamin D metabolism. *J. Clin. Investig.* **2004**, *113*, 561–568. [CrossRef] [PubMed]
77. Sitara, D.; Razzaque, M.S.; Hesse, M.; Yoganathan, S.; Taguchi, T.; Erben, R.G.; Juppner, H.; Lanske, B. Homozygous ablation of fibroblast growth factor-23 results in hyperphosphatemia and impaired skeletogenesis, and reverses hypophosphatemia in Phex-deficient mice. *Matrix Biol.* **2004**, *23*, 421–432. [CrossRef] [PubMed]
78. Anour, R.; Andrukhova, O.; Ritter, E.; Zeitz, U.; Erben, R.G. Klotho lacks a vitamin D independent physiological role in glucose homeostasis, bone turnover, and steady-state PTH secretion in vivo. *PLoS ONE* **2012**, *7*, e31376. [CrossRef]
79. Kuro-o, M.; Matsumura, Y.; Aizawa, H.; Kawaguchi, H.; Suga, T.; Utsugi, T.; Ohyama, Y.; Kurabayashi, M.; Kaname, T.; Kume, E.; et al. Mutation of the mouse klotho gene leads to a syndrome resembling ageing. *Nature* **1997**, *390*, 45–51. [CrossRef]
80. Tsujikawa, H.; Kurotaki, Y.; Fujimori, T.; Fukuda, K.; Nabeshima, Y. Klotho, a gene related to a syndrome resembling human premature aging, functions in a negative regulatory circuit of vitamin D endocrine system. *Mol. Endocrinol.* **2003**, *17*, 2393–2403. [CrossRef]
81. Yoshida, T.; Fujimori, T.; Nabeshima, Y. Mediation of unusually high concentrations of 1,25-dihydroxyvitamin D in homozygous klotho mutant mice by increased expression of renal 1alpha-hydroxylase gene. *Endocrinology* **2002**, *143*, 683–689. [CrossRef] [PubMed]
82. Murali, S.K.; Roschger, P.; Zeitz, U.; Klaushofer, K.; Andrukhova, O.; Erben, R.G. FGF23 Regulates Bone Mineralization in a 1,25(OH)2 D3 and Klotho-Independent Manner. *J. Bone Miner. Res.* **2016**, *31*, 129–142. [CrossRef]

83. Hesse, M.; Fröhlich, L.F.; Zeitz, U.; Lanske, B.; Erben, R.G. Ablation of vitamin D signaling rescues bone, mineral, and glucose homeostasis in Fgf-23 deficient mice. *Matrix Biol.* **2007**, *26*, 75–84. [CrossRef] [PubMed]
84. Streicher, C.; Zeitz, U.; Andrukhova, O.; Rupprecht, A.; Pohl, E.; Larsson, T.E.; Windisch, W.; Lanske, B.; Erben, R.G. Long-term Fgf23 deficiency does not influence aging, glucose homeostasis, or fat metabolism in mice with a nonfunctioning vitamin D receptor. *Endocrinology* **2012**, *153*, 1795–1805. [CrossRef]
85. Han, X.; Yang, J.; Li, L.; Huang, J.; King, G.; Quarles, L.D. Conditional Deletion of Fgfr1 in the Proximal and Distal Tubule Identifies Distinct Roles in Phosphate and Calcium Transport. *PLoS ONE* **2016**, *11*, e0147845. [CrossRef]
86. Li, H.; Martin, A.; David, V.; Quarles, L.D. Compound deletion of Fgfr3 and Fgfr4 partially rescues the Hyp mouse phenotype. *Am. J. Physiol. Endocrinol. Metab.* **2011**, *300*, E508–E517. [CrossRef] [PubMed]
87. Zhang, M.Y.; Ranch, D.; Pereira, R.C.; Armbrecht, H.J.; Portale, A.A.; Perwad, F. Chronic inhibition of ERK1/2 signaling improves disordered bone and mineral metabolism in hypophosphatemic (Hyp) mice. *Endocrinology* **2012**, *153*, 1806–1816. [CrossRef] [PubMed]
88. Ranch, D.; Zhang, M.Y.; Portale, A.A.; Perwad, F. Fibroblast growth factor 23 regulates renal 1,25-dihydroxyvitamin D and phosphate metabolism via the MAP kinase signaling pathway in Hyp mice. *J. Bone Miner. Res.* **2011**, *26*, 1883–1890. [CrossRef]
89. Shimada, T.; Yamazaki, Y.; Takahashi, M.; Hasegawa, H.; Urakawa, I.; Oshima, T.; Ono, K.; Kakitani, M.; Tomizuka, K.; Fujita, T.; et al. Vitamin D receptor-independent FGF23 actions in regulating phosphate and vitamin D metabolism. *Am. J. Physiol. Renal Physiol.* **2005**, *289*, F1088–F1095. [CrossRef]
90. Erben, R.G. Fibroblast growth factor-23. In *Encyclopedia of Signaling Molecules*, 2nd ed.; Choi, S., Ed.; Springer: New York, NY, USA, 2018.
91. Dorr, K.; Kammer, M.; Reindl-Schwaighofer, R.; Lorenz, M.; Prikoszovich, T.; Marculescu, R.; Beitzke, D.; Wielandner, A.; Erben, R.G.; Oberbauer, R. Randomized Trial of Etelcalcetide for Cardiac Hypertrophy in Hemodialysis. *Circ. Res.* **2021**, *128*, 1616–1625. [CrossRef]
92. Scragg, R. Seasonality of cardiovascular disease mortality and the possible protective effect of ultra-violet radiation. *Int. J. Epidemiol.* **1981**, *10*, 337–341. [CrossRef]
93. Latic, N.; Erben, R.G. Vitamin D and Cardiovascular Disease, with Emphasis on Hypertension, Atherosclerosis, and Heart Failure. *Int. J. Mol. Sci.* **2020**, *21*, 6483. [CrossRef] [PubMed]
94. Walters, M.R.; Wicker, D.C.; Riggle, P.C. 1,25-Dihydroxyvitamin D₃ receptors identified in the rat heart. *J. Mol. Cell. Cardiol.* **1986**, *18*, 67–72. [CrossRef] [PubMed]
95. Tishkoff, D.X.; Nibbelink, K.A.; Holmberg, K.H.; Dandu, L.; Simpson, R.U. Functional Vitamin D Receptor (VDR) in the T-Tubules of Cardiac Myocytes: VDR Knockout Cardiomyocyte Contractility. *Endocrinology* **2008**, *149*, 558–564. [CrossRef] [PubMed]
96. Merke, J.; Milde, P.; Lewicka, S.; Hügel, U.; Klaus, G.; Mangelsdorf, D.J.; Haussler, M.R.; Rauterberg, E.W.; Ritz, E. Identification and regulation of 1,25-dihydroxyvitamin D3 receptor activity and biosynthesis of 1,25-dihydroxyvitamin D3. Studies in cultured bovine aortic endothelial cells and human dermal capillaries. *J. Clin. Investig.* **1989**, *83*, 1903–1915. [CrossRef] [PubMed]
97. Al Mheid, I.; Patel, R.; Murrow, J.; Morris, A.; Rahman, A.; Fike, L.; Kavtaradze, N.; Uphoff, I.; Hooper, C.; Tangpricha, V.; et al. Vitamin D Status Is Associated With Arterial Stiffness and Vascular Dysfunction in Healthy Humans. *J. Am. Coll. Cardiol.* **2011**, *58*, 186–192. [CrossRef]
98. Yiu, Y.-F.; Chan, Y.-H.; Yiu, K.-H.; Siu, C.-W.; Li, S.-W.; Wong, L.-Y.; Lee, S.W.L.; Tam, S.; Wong, E.W.K.; Cheung, B.M.Y.; et al. Vitamin D Deficiency Is Associated with Depletion of Circulating Endothelial Progenitor Cells and Endothelial Dysfunction in Patients with Type 2 Diabetes. *J. Clin. Endocrinol. Metab.* **2011**, *96*, E830–E835. [CrossRef]
99. Lee, J.H.; Gadi, R.; Spertus, J.A.; Tang, F.; O'Keefe, J.H. Prevalence of Vitamin D Deficiency in Patients With Acute Myocardial Infarction. *Am. J. Cardiol.* **2011**, *107*, 1636–1638. [CrossRef]
100. Holick, M.F. Vitamin D deficiency. *N. Engl. J. Med.* **2007**, *357*, 266–281. [CrossRef]
101. Fountain, J.H.; Lappin, S.L. Physiology, Renin Angiotensin System. In *StatPearls*; StatPearls Publishing LLC: Treasure Island, FL, USA, 2022.
102. Xanthakis, V.; Vasan, R.S. Aldosterone and the risk of hypertension. *Curr. Hypertens. Rep.* **2013**, *15*, 102–107. [CrossRef]
103. Yuan, W.; Pan, W.; Kong, J.; Zheng, W.; Szeto, F.L.; Wong, K.E.; Cohen, R.; Klopot, A.; Zhang, Z.; Li, Y.C. 1,25-dihydroxyvitamin D3 suppresses renin gene transcription by blocking the activity of the cyclic AMP response element in the renin gene promoter. *J. Biol. Chem.* **2007**, *282*, 29821–29830. [CrossRef]
104. Grundmann, S.M.; Schutkowski, A.; Schreier, B.; Rabe, S.; König, B.; Gekle, M.; Stangl, G.I. Vitamin D Receptor Deficiency Does Not Affect Blood Pressure and Heart Function. *Front. Physiol.* **2019**, *10*, 1118. [CrossRef]
105. Jia, J.; Tao, X.; Tian, Z.; Liu, J.; Ye, X.; Zhan, Y. Vitamin D receptor deficiency increases systolic blood pressure by upregulating the renin-angiotensin system and autophagy. *Exp. Ther. Med.* **2022**, *23*, 314. [CrossRef] [PubMed]
106. Zhou, C.; Lu, F.; Cao, K.; Xu, D.; Goltzman, D.; Miao, D. Calcium-independent and 1,25(OH)2D3-dependent regulation of the renin-angiotensin system in 1alpha-hydroxylase knockout mice. *Kidney Int.* **2008**, *74*, 170–179. [CrossRef] [PubMed]
107. Andrukhova, O.; Slavic, S.; Zeitz, U.; Riesen, S.C.; Heppelmann, M.S.; Ambrisko, T.D.; Markovic, M.; Kuebler, W.M.; Erben, R.G. Vitamin D Is a Regulator of Endothelial Nitric Oxide Synthase and Arterial Stiffness in Mice. *Mol. Endocrinol.* **2014**, *28*, 53–64. [CrossRef] [PubMed]
108. Ni, W.; Watts, S.W.; Ng, M.; Chen, S.; Glenn, D.J.; Gardner, D.G. Elimination of vitamin D receptor in vascular endothelial cells alters vascular function. *Hypertension* **2014**, *64*, 1290–1298. [CrossRef]

109. Kota, S.K.; Kota, S.K.; Jammula, S.; Meher, L.K.; Panda, S.; Tripathy, P.R.; Modi, K.D. Renin-angiotensin system activity in vitamin D deficient, obese individuals with hypertension: An urban Indian study. *Indian J. Endocrinol. Metab.* **2011**, *15* (Suppl. 4), S395–S401. [CrossRef] [PubMed]
110. Resnick, L.M.; MÜLler, F.B.; Laragh, J.H. Calcium-Regulating Hormones in Essential Hypertension. *Ann. Intern. Med.* **1986**, *105*, 649–654. [CrossRef]
111. Forman, J.P.; Williams, J.S.; Fisher, N.D. Plasma 25-hydroxyvitamin D and regulation of the renin-angiotensin system in humans. *Hypertension* **2010**, *55*, 1283–1288. [CrossRef]
112. Vaidya, A.; Forman, J.P.; Hopkins, P.N.; Seely, E.W.; Williams, J.S. 25-Hydroxyvitamin D is associated with plasma renin activity and the pressor response to dietary sodium intake in Caucasians. *J. Renin-Angiotensin-Aldosterone Syst.* **2011**, *12*, 311–319. [CrossRef]
113. Vaidya, A.; Sun, B.; Larson, C.; Forman, J.P.; Williams, J.S. Vitamin D3 therapy corrects the tissue sensitivity to angiotensin ii akin to the action of a converting enzyme inhibitor in obese hypertensives: An interventional study. *J. Clin. Endocrinol. Metab.* **2012**, *97*, 2456–2465. [CrossRef]
114. McMullan, C.J.; Borgi, L.; Curhan, G.C.; Fisher, N.; Forman, J.P. The effect of vitamin D on renin-angiotensin system activation and blood pressure: A randomized control trial. *J. Hypertens.* **2017**, *35*, 822–829. [CrossRef] [PubMed]
115. Manson, J.E.; Cook, N.R.; Lee, I.M.; Christen, W.; Bassuk, S.S.; Mora, S.; Gibson, H.; Gordon, D.; Copeland, T.; D'Agostino, D.; et al. Vitamin D Supplements and Prevention of Cancer and Cardiovascular Disease. *N. Engl. J. Med.* **2019**, *380*, 33–44. [CrossRef] [PubMed]
116. Scragg, R.K.R. Overview of results from the Vitamin D Assessment (ViDA) study. *J. Endocrinol. Investig.* **2019**, *42*, 1391–1399. [CrossRef] [PubMed]
117. Jiang, F.; Yang, J.; Zhang, Y.; Dong, M.; Wang, S.; Zhang, Q.; Liu, F.F.; Zhang, K.; Zhang, C. Angiotensin-converting enzyme 2 and angiotensin 1–7: Novel therapeutic targets. *Nat. Rev. Cardiol.* **2014**, *11*, 413–426. [CrossRef] [PubMed]
118. Pagliaro, P.; Penna, C. ACE/ACE2 Ratio: A Key Also in 2019 Coronavirus Disease (Covid-19)? *Front. Med.* **2020**, *7*, 335. [CrossRef] [PubMed]
119. Getachew, B.; Tizabi, Y. Vitamin D and COVID-19: Role of ACE2, age, gender, and ethnicity. *J. Med. Virol.* **2021**, *93*, 5285–5294. [CrossRef] [PubMed]
120. Úri, K.; Fagyas, M.; Mányiné Siket, I.; Kertész, A.; Csanádi, Z.; Sándorfi, G.; Clemens, M.; Fedor, R.; Papp, Z.; Édes, I.; et al. New perspectives in the renin-angiotensin-aldosterone system (RAAS) IV: Circulating ACE2 as a biomarker of systolic dysfunction in human hypertension and heart failure. *PLoS ONE* **2014**, *9*, e87845. [CrossRef] [PubMed]
121. Yang, J.; XU, J.; Zhang, H. Effect of Vitamin D on ACE2 and Vitamin D receptor expression in rats with LPS-induced acute lung injury. *Chin. J. Emerg. Med.* **2016**, *25*, 1284–1289. [CrossRef]
122. Xu, J.; Yang, J.; Chen, J.; Luo, Q.; Zhang, Q.; Zhang, H. Vitamin D alleviates lipopolysaccharide-induced acute lung injury via regulation of the renin-angiotensin system. *Mol. Med. Rep.* **2017**, *16*, 7432–7438. [CrossRef] [PubMed]
123. Kong, J.; Zhu, X.; Shi, Y.; Liu, T.; Chen, Y.; Bhan, I.; Zhao, Q.; Thadhani, R.; Li, Y.C. VDR Attenuates Acute Lung Injury by Blocking Ang-2-Tie-2 Pathway and Renin-Angiotensin System. *Mol. Endocrinol.* **2013**, *27*, 2116–2125. [CrossRef]
124. Lin, M.; Gao, P.; Zhao, T.; He, L.; Li, M.; Li, Y.; Shui, H.; Wu, X. Calcitriol regulates angiotensin-converting enzyme and angiotensin converting-enzyme 2 in diabetic kidney disease. *Mol. Biol. Rep.* **2016**, *43*, 397–406. [CrossRef]
125. Riera, M.; Anguiano, L.; Clotet, S.; Roca-Ho, H.; Rebull, M.; Pascual, J.; Soler, M.J. Paricalcitol modulates ACE2 shedding and renal ADAM17 in NOD mice beyond proteinuria. *Am. J. Physiol.-Ren. Physiol.* **2016**, *310*, F534–F546. [CrossRef] [PubMed]
126. Zhou, Q.; Lin, S.; Tang, R.; Veeraragoo, P.; Peng, W.; Wu, R. Role of Fosinopril and Valsartan on Klotho Gene Expression Induced by Angiotensin II in Rat Renal Tubular Epithelial Cells. *Kidney Blood Press. Res.* **2010**, *33*, 186–192. [CrossRef] [PubMed]
127. Yoon, H.E.; Ghee, J.Y.; Piao, S.; Song, J.H.; Han, D.H.; Kim, S.; Ohashi, N.; Kobori, H.; Kuro-o, M.; Yang, C.W. Angiotensin II blockade upregulates the expression of Klotho, the anti-ageing gene, in an experimental model of chronic cyclosporine nephropathy. *Nephrol. Dial. Transplant.* **2011**, *26*, 800–813. [CrossRef] [PubMed]
128. Saito, K.; Ishizaka, N.; Mitani, H.; Ohno, M.; Nagai, R. Iron chelation and a free radical scavenger suppress angiotensin II-induced downregulation of klotho, an anti-aging gene, in rat. *FEBS Lett.* **2003**, *551*, 58–62. [CrossRef]
129. Mitani, H.; Ishizaka, N.; Aizawa, T.; Ohno, M.; Usui, S.-i.; Suzuki, T.; Amaki, T.; Mori, I.; Nakamura, Y.; Sato, M. In vivo klotho gene transfer ameliorates angiotensin II-induced renal damage. *Hypertension* **2002**, *39*, 838–843. [CrossRef]
130. Mitobe, M.; Yoshida, T.; Sugiura, H.; Shirota, S.; Tsuchiya, K.; Nihei, H. Oxidative stress decreases klotho expression in a mouse kidney cell line. *Nephron Exp. Nephrol.* **2005**, *101*, e67–e74. [CrossRef]
131. Sachse, A.; Wolf, G. Angiotensin II–induced reactive oxygen species and the kidney. *J. Am. Soc. Nephrol.* **2007**, *18*, 2439–2446. [CrossRef]
132. Zuo, Z.; Lei, H.; Wang, X.; Wang, Y.; Sonntag, W.; Sun, Z. Aging-related kidney damage is associated with a decrease in klotho expression and an increase in superoxide production. *Age* **2011**, *33*, 261–274. [CrossRef]
133. Griendling, K.K.; Minieri, C.A.; Ollerenshaw, J.D.; Alexander, R.W. Angiotensin II stimulates NADH and NADPH oxidase activity in cultured vascular smooth muscle cells. *Circ. Res.* **1994**, *74*, 1141–1148. [CrossRef]
134. Doran, D.E.; Weiss, D.; Zhang, Y.; Griendling, K.K.; Taylor, W.R. Differential effects of AT1 receptor and Ca^{2+} channel blockade on atherosclerosis, inflammatory gene expression, and production of reactive oxygen species. *Atherosclerosis* **2007**, *195*, 39–47. [CrossRef]

135. Dusso, A.; Arcidiacono, M.V.; Yang, J.; Tokumoto, M. Vitamin D inhibition of TACE and prevention of renal osteodystrophy and cardiovascular mortality. *J. Steroid Biochem. Mol. Biol.* **2010**, *121*, 193–198. [CrossRef]
136. Chen, C.-D.; Podvin, S.; Gillespie, E.; Leeman, S.E.; Abraham, C.R. Insulin stimulates the cleavage and release of the extracellular domain of Klotho by ADAM10 and ADAM17. *Proc. Natl. Acad. Sci. USA* **2007**, *104*, 19796–19801. [CrossRef]
137. Lautrette, A.; Li, S.; Alili, R.; Sunnarborg, S.W.; Burtin, M.; Lee, D.C.; Friedlander, G.; Terzi, F. Angiotensin II and EGF receptor cross-talk in chronic kidney diseases: A new therapeutic approach. *Nat. Med.* **2005**, *11*, 867–874. [CrossRef] [PubMed]
138. Grund, A.; Sinha, M.D.; Haffner, D.; Leifheit-Nestler, M. Fibroblast Growth Factor 23 and Left Ventricular Hypertrophy in Chronic Kidney Disease-A Pediatric Perspective. *Front. Pediatr.* **2021**, *9*, 702719. [CrossRef] [PubMed]
139. Christodoulou, M.; Aspray, T.J.; Schoenmakers, I. Vitamin D Supplementation for Patients with Chronic Kidney Disease: A Systematic Review and Meta-analyses of Trials Investigating the Response to Supplementation and an Overview of Guidelines. *Calcif. Tissue Int.* **2021**, *109*, 157–178. [CrossRef] [PubMed]

MDPI

Article

Month-of-Birth Effect on Muscle Mass and Strength in Community-Dwelling Older Women: The French EPIDOS Cohort

Guillaume T. Duval [1,2,3], Anne-Marie Schott [4], Dolores Sánchez-Rodríguez [5,6,7], François R. Herrmann [8] and Cédric Annweiler [1,2,3,9,10,*]

1 School of Medicine, Health Faculty, University of Angers, F-49100 Angers, France
2 Department of Geriatric Medicine and Memory Clinic, Research Center on Autonomy and Longevity, University Hospital, F-49000 Angers, France
3 UNIV ANGERS, UPRES EA 4638, F-49045 Angers, France
4 Department IMER (Information Médicale, Evaluation, Recherche), Lyon University Hospital, EA 4129, Institut National de la Santé et de la Recherche Médicale, University of Lyon, U831 Lyon, France
5 Geriatrics Department, Brugmann University Hospital, Université Libre de Bruxelles, B-1020 Brussels, Belgium
6 Geriatrics Department, Rehabilitation Research Group, Institut Hospital del Mar d'Investigations Médiques (IMIM), E-08003 Barcelona, Spain
7 Division of Public Health, Epidemiology and Health Economics, World Health Organization Collaborating Centre for Public Health Aspects of Musculo-Skeletal Health and Ageing, University of Liège, B-4000 Liège, Belgium
8 Department of Rehabilitation and Geriatrics, Geneva University Hospitals, University of Geneva, CH-1226 Thônex, Switzerland
9 Gérontopôle Autonomie Longévité des Pays de la Loire, F-44200 Nantes, France
10 Department of Medical Biophysics, Robarts Research Institute, Schulich School of Medicine and Dentistry, The University of Western Ontario, London, ON N6A 5K8, Canada
* Correspondence: cedric.annweiler@chu-angers.fr; Tel.: +33-2-41-35-47-25; Fax: +33-2-41-35-48-94

Citation: Duval, G.T.; Schott, A.-M.; Sánchez-Rodríguez, D.; Herrmann, F.R.; Annweiler, C. Month-of-Birth Effect on Muscle Mass and Strength in Community-Dwelling Older Women: The French EPIDOS Cohort. *Nutrients* **2022**, *14*, 4874. https://doi.org/10.3390/nu14224874

Academic Editor: Carsten Carlberg

Received: 13 October 2022
Accepted: 14 November 2022
Published: 18 November 2022

Abstract: Background. Vitamin D is involved in muscle health and function. This relationship may start from the earliest stages of life during pregnancy when fetal vitamin D relies on maternal vitamin D stores and sun exposure. Our objective was to determine whether there was an effect of the month of birth (MoB) on muscle mass and strength in older adults. **Methods**. Data from 7598 community-dwelling women aged $\geq$ 70 years from the French multicentric EPIDOS cohort were used in this analysis. The quadricipital strength was defined as the mean value of 3 consecutive tests of the maximal isometric voluntary contraction strength of the dominant lower limb. The muscle mass was defined as the total appendicular skeletal muscle mass measured using dual energy X-ray absorptiometry scanner. The MoB was used as a periodic function in regressions models adjusted for potential confounders including age, year of birth, latitude of recruitment center, season of testing, body mass index, number of comorbidities, IADL score, regular physical activity, sun exposure at midday, dietary protein intake, dietary vitamin D intake, use vitamin D supplements, history and current use of corticosteroids. **Results**. A total of 7133 older women had a measure of muscle strength (mean age, 80.5 $\pm$ 3.8 years; mean strength, 162.3 $\pm$ 52.1 N). Data on total ASM were available from 1321 women recruited in Toulouse, France (mean, 14.86 $\pm$ 2.04 kg). Both the sine and cosine functions of MoB were associated with the mean quadricipital strength (respectively $\beta = -2.1$, $p = 0.045$ and $\beta = -0.5$, $p = 0.025$). The sine function of MoB was associated with total ASM ($\beta = -0.2$, $p = 0.013$), but not the cosine function ($\beta = 0.1$, $p = 0.092$). Both the highest value of average quadricipital strength (mean, 163.4 $\pm$ 20.2 N) and the highest value of total ASM (15.24 $\pm$ 1.27 kg) were found among participants born in August. **Conclusions**. Summer-early fall months of birth were associated with higher muscle mass and strength in community-dwelling older women.

Keywords: vitamin D; month of birth; muscle; pregnancy; older adults

1. Introduction

Besides its classical role in the regulation of bone metabolism, vitamin D has many non-skeletal biological targets mediated by the vitamin D receptor (VDR), which is a specific vitamin D hormone receptor. Vitamin D is involved in the health and function of skeletal muscles, and serum vitamin D concentrations are positively associated with muscle mass and strength in adults [1]. The clinical relevance is that vitamin D deficiency leads to poorer muscular performance and physical deterioration [2]. Importantly, it was proposed that this relationship could start from the earliest stages of life, during pregnancy [3].

During gestation, the fetus is completely reliant on maternal vitamin D stores [4]. The maternal serum 25-hydroxyvitamin D (25OHD) concentration is highly correlated to that of the umbilical cord of the fetus [4]. Moreover, since 90% of maternal vitamin D is synthesized in the skin under the action of solar ultraviolet-B (UV-B) rays, the mother's vitamin D status is mostly influenced by the season; with higher concentrations reported during summer and early fall [5]. Thus, in the absence of supplementation, fetal vitamin D concentration largely depends on the season of pregnancy [5]. This may explain why several conditions related to hypovitaminosis D have been previously linked to the month of birth (MoB); the children born in winter being more at risk of lower height and weight, and at risk of multiple sclerosis for instance [6].

Several studies have also brought evidence that the maternal serum 25OHD concentration and UV-B rays exposure during pregnancy influence the body composition in offspring, including bone mass, degree of adiposity, and muscle mass [7,8]. However, to the best of our knowledge, the relationship between the MoB and the muscle mass and function during adulthood has not been examined yet. We hypothesized that there could be an effect of the MoB on muscles in older adults, specifically that the summer-early fall MoB would be associated with better muscle mass and function. The objective of the present study was to determine whether there was an effect of the MoB on muscle mass and strength in community-dwelling older women.

2. Materials and Methods

2.1. Participants

We used for the present analysis data from the older women included in the 'EPIDOS' study (EPIDémiologie de l'OStéoporose), a French national prospective multicentric and observational cohort study originally designed to determine the risk factors for hip fracture among community-dwelling older women. Sampling and data collection procedures have been described in detail elsewhere [9]. In summary, from 1992 to 1994, 7598 women aged 70 years and older were recruited from electoral lists in five French cities (Amiens, Lyon, Montpellier, Paris and Toulouse). All included study participants had a full medical examination, which consisted of structured questionnaires, demographical measures including the month and year of birth, and a clinical examination.

2.2. Muscle Strength Measure

The maximal isometric voluntary contraction (MVC) strength of the dominant lower limb was measured with a strain gauge fixed to a chair, while the participant was seated, leg and ankle flexed at 90° angle. The leg tested was attached to the lever arm of the strain gauge and the seat height was adjusted to the leg length of the participant. Before carrying out the test, participants were offered to practice the isometric movement in order to warm up. Verbal instructions regarding the test procedure were given by a trained evaluator. Participants pushed as hard as possible against the dynamometer. Three MVC were recorded in Newton (N), and verbal encouragement was given each time to obtain the maximal score. The average MVC strength value was calculated from a set of three consecutive contractions and used for the present analysis.

2.3. Muscle Mass Measure

The total appendicular skeletal muscle mass (ASM) was measured using a dual energy X-ray absorptiometry (DXA) scanner (QDR 4500 W Hologic, Waltham, MA, USA) at enrollment only in the center of Toulouse, France. DXA measurements were performed by a trained technician, and the DXA machine was regularly calibrated. The total ASM was defined as the sum of the two upper and lower limb muscle masses, expressed in kilograms (kg). Data on the validity of body composition parameters of the EPIDOS-Toulouse cohort have previously been published [10].

2.4. Covariates

The following covariates were included as potential confounders in the statistical models: age, body mass index (BMI), number of comorbidities, regular physical activity, instrumental activities of daily living (IADL) score (from 0 to 8, best) [11], dietary protein and vitamin D intakes, use of vitamin D supplements, history of corticosteroids use, current use of corticosteroids, sun exposure at midday, season of evaluation, and study centers.

A physical examination and a health status questionnaire were conducted to assess comorbidities (i.e., hypertension, diabetes, dyslipidemia, coronary heart disease, chronic obstructive pulmonary disease, peripheral vascular disease, cancer, stroke, Parkinson's disease and depression). Weight was measured with a beam balance scale, and height with a height gauge. BMI was calculated according to the formula: weight (kg)/height^2 (m^2). The practice of a physical activity was considered regular if the participants practiced at least one recreational physical activity (i.e., walking, gymnastics, cycling, swimming or gardening) for at least one hour per week for at least the past month. Medications taken regularly had to be brought by the participants to the clinical center during the assessment [9]. The dietary intakes of vitamin D and protein were estimated from a self-administered food frequency questionnaire, as previously published [12]. The cutaneous synthesis of vitamin D was estimated using the following standardized question: "When weather is nice, do you stay more than 15 min exposed to the sun (face and hands uncovered) between 11 a.m. and 3 p.m.?" (yes/no), as previously published [13]. Finally, the season of evaluation was recorded as follows: spring from 21 March to 20 June, summer from 21 June to 20 September, fall from 21 September to 20 December, winter from 21 December to 20 March.

2.5. Statistical Analysis

We provide here a post-hoc analysis of the EPIDOS cohort study. Firstly, the participants' characteristics were summarized using means $\pm$ standard deviations or frequencies and percentages, as appropriate. As the number of observations was higher than 40, no transform was applied. Secondly, to determine the association of the MoB (independent variable) with muscle strength and muscle mass (dependent variables), we applied multiple linear regressions. The MoB was added as a periodic function: $\beta 1 \times \sin(2\pi \times \mathrm{MoB}/12) + \beta 2 \times \cos(2\pi \times \mathrm{MoB}/12)$, where $\pi = 3.1415\ldots$ and βi are regression coefficients. p-values < 0.05 were considered significant. All statistics were performed using Stata (version 14.1; College Station, TX, USA).

2.6. Ethics

Women participating in the study were included after having given their written informed consent for research. The study was conducted in accordance with the ethical standards set forth in the Helsinki Declaration (1983). The project was approved by the local ethics committee of each city.

3. Results

Among the 7598 women recruited in the EPIDOS cohort, 7133 had a measure of quadricipital strength (mean $\pm$ standard deviation age, 80.5 $\pm$ 3.8 years), and 1321 women (mean age, 80.3 $\pm$ 3.9 years; all from Toulouse, France) had a measure of total ASM.

The mean quadricipital strength on the cohort was 162.3 ± 52.1 N, and the mean ASM 14.86 ± 2.04 kg (Table 1).

Table 1. Summary of the participants' characteristics (*n* = 7133).

Characteristics	Cohort	
	Summary Value	**(95% CI)**
Demographical measures		
Age, years	80.5 ± 3.8 (70–98)	(80.4; 80.6)
Year of birth	1912 ± 4 (1893–1922)	-
Recruitment center, n (%)		
Amiens (49°54′ N)	1488 (20.9)	(20.0; 21.8)
Paris (48°51′ N)	1457 (20.4)	(19.5; 21.3)
Lyon (45°45′ N)	1365 (19.1)	(18.2; 20.0)
Montpellier (43°36′ N)	1492 (20.9)	(20.0; 21.8)
Toulouse (43°36′ N)	1331 (18.7)	(17.8; 19.6)
Season of testing, n (%)		
Spring	1922 (26.9)	(25.9; 27.9)
Summer	1876 (26.3)	(25.3; 27.3)
Fall	1843 (25.8)	(24.8; 26.8)
Winter	1492 (20.9)	(20.0; 21.8)
Clinical measures		
Body mass index, kg/m^2	25.3 ± 4.0 (14.5–42.1)	(25.2; 25.4)
Number of comorbidities, n (%)	3.4 ± 2.0 (0–24)	(3.36; 3.44)
IADL score, /8	6.3 ± 1.3 (0–8)	(6.27; 6.33)
Regular physical activity, n (%)	3490 (48.9)	(47.7; 50.1)
Sun exposure at midday, n (%)	3609 (50.6)	(49.4; 51.8)
Dietary protein intake, g/day	69.6 ± 16.2 (8.7–162.3)	(69.2; 70.0)
Dietary vitamin D intake, µg/week	63.1 ± 31.3 (0–278.0)	(62.3; 63.9)
Use of vitamin D supplements, n (%)	985 (13.8)	(13.0; 14.6)
History of corticosteroids use, n (%)	455 (6.4)	(5.8; 7.0)
Current use of corticosteroids, n (%)	146 (2.0)	(1.7; 2.3)
Muscles measures		
Mean quadricipital strength, N	162.3 ± 52.1 (17.3–507.3)	(161.1; 163.5)
Total ASM, kg *	14.86 ± 2.04 (9.33–22.78)	(14.75; 14.97)

Data presented as mean ± standard deviation [range] where applicable. ASM: appendicular skeletal muscle mass; CI: confidence interval; SD: standard deviation; IADL: Instrumental Activities of Daily Living; *: data available for 1321 participants.

As indicated in Table 2, every MoB were represented, with a maximum of participants born in March (9.7%) and a minimum born in September (7.4%), with no significant difference.

Table 2. Months of birth among the participants (*n* = 7133).

Month of Birth	Cohort			
	n	**Percentage**	**(95% CI)**	**Cumulative Percentage**
January	630	8.83	(8.17; 9.49)	8.83
February	602	8.43	(7.79; 9.07)	17.27
March	690	9.67	(7.79; 9.07)	26.95
April	568	7.96	(7.33; 8.59)	34.91
May	603	8.45	(7.80; 9.10)	43.36
June	573	8.03	(7.40; 8.66)	51.39
July	625	8.76	(8.10; 9.42)	60.16
August	598	8.38	(7.74; 9.02)	68.54
September	524	7.35	(6.74; 7.96)	75.89
October	527	7.39	(6.78; 8.00)	83.27
November	585	8.20	(7.56; 8.84)	91.48
December	608	8.52	(7.87; 9.17)	100.00

Multiple linear regression models showed that both the sine and cosine functions of MoB were associated with the mean quadricipital strength (respectively fully adjusted β = −2.1 with p = 0.045, and fully adjusted β = −0.5 with p = 0.025), which means that the association between the MoB and mean quadricipital strength could be modeled using a combination of an inverted sine function and an inverted cosine function (Table 3, Figure 1A). Moreover, only the sine function of MoB was associated with the total ASM (fully adjusted β = −0.2, p = 0.013), but not the cosine function (fully adjusted β = 0.1, p = 0.092), which means that the association between the MoB and the total ASM was modeled by an inverted sine function (Table 3, Figure 1B). Both the highest predicted value of average quadricipital strength (mean, 163.4 ± 20.2 N) and the highest value of total ASM (15.24 ± 1.27 kg) were found among participants born in August (Table 3).

Table 3. Average values of muscles' measures predicted with the trigonometric modelling by month of birth.

Month of Birth	Mean Quadricipital Strength, N	Total ASM, kg
January	161.0 ± 20.8	15.06 ± 1.20
February	160.0 ± 19.5	15.13 ± 1.19
March	160.0 ± 19.6	15.01 ± 1.17
April	161.0 ± 19.8	14.89 ± 1.22
May	161.2 ± 19.6	14.91 ± 1.15
June	160.9 ± 19.6	15.01 ± 1.18
July	160.6 ± 20.5	14.93 ± 1.24
August	163.4 ± 20.2	15.24 ± 1.27
September	161.6 ± 20.6	15.20 ± 1.29
October	159.8 ± 20.5	15.10 ± 1.19
November	158.8 ± 19.8	15.13 ± 1.21
December	159.0 ± 19.2	14.99 ± 1.21
Beta for sin (p-value)	β = −2.1, p = 0.045	β = −0.2, p = 0.013
Beta for cos (p-value)	β = −0.5, p = 0.025	β = 0.1, p = 0.092

Data presented as mean ± standard deviation.

We also found that the BMI (β = 2.6, p = 0.001), a regular physical activity (β = 10.5, p = 0.002), and the IADL score (β = 5.9, $p < 0.001$) were positively associated with the mean quadricipital strength, although the number of comorbidities (β = −3.4, p = 0.019), the indication for vitamin D supplements (β = −5.2, p = 0.001) and the history of corticosteroids use (β = −13.0, p = 0.025) were inversely associated with the mean quadricipital strength. Finally, a more recent year of birth (β = 0.4, p = 0.001), the BMI (β = 0.3, $p < 0.001$), and the dietary protein intake (β = 0.1, $p < 0.001$) were positively associated with the total ASM, although the current use of corticosteroids (β = −0.8, p = 0.008), the need for vitamin D supplements (β = −0.5, p = 0.001) and the winter season of evaluation (β = −0.4, p = 0.016) were inversely associated with the total ASM.

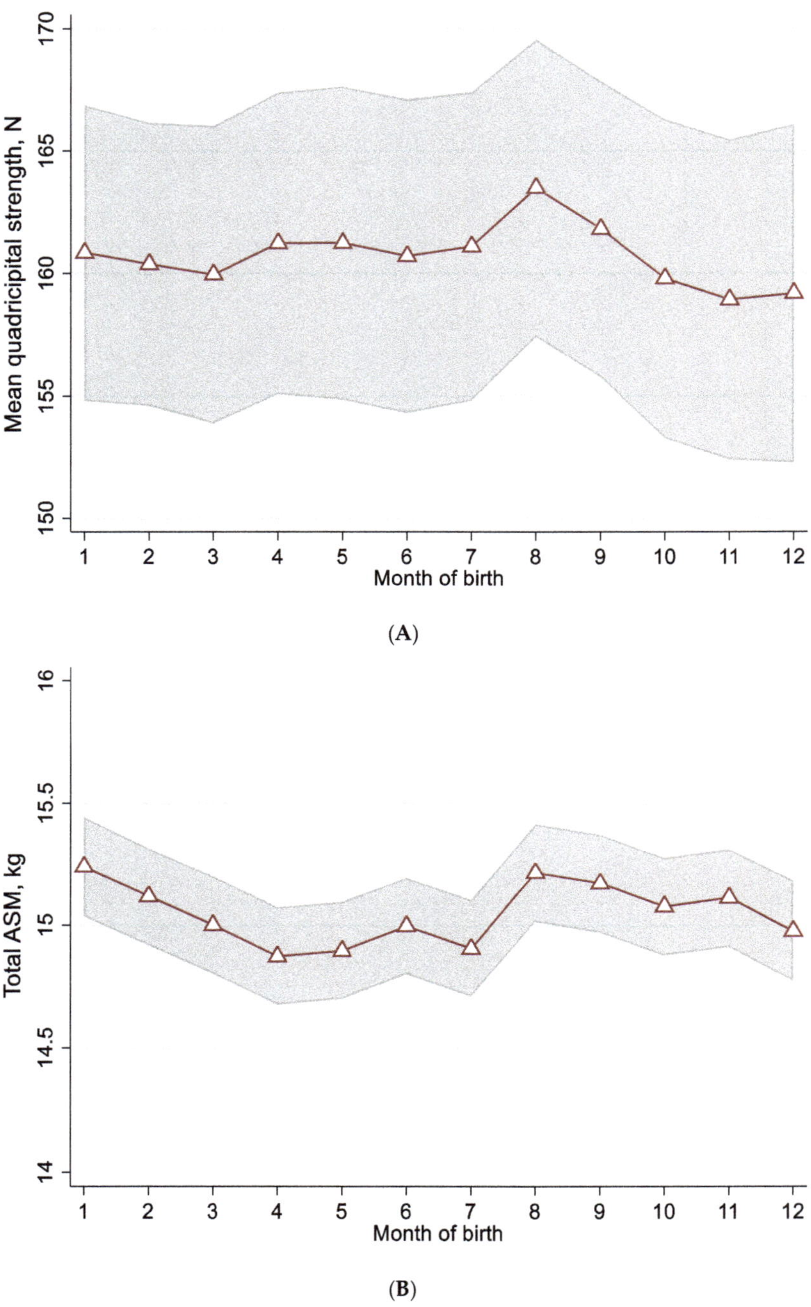

Figure 1. Trigonometric modelling by month of birth of (**A**) mean quadriceps strength (n = 7178), and (**B**) total appendicular skeletal muscle mass (ASM) (n = 1321). (**A**,**B**): Open triangles represent the average predicted values for respectively "Mean quadricipital strength" and "Total ASM" according to "Month of birth". The shaded area represents the average predicted values plus or minus the median of the standard error of the prediction. The two models are adjusted for age, year of birth, latitude of recruitment centre, season of testing, body mass index, number of comorbidities, IADL

score, regular physical activity, sun exposure at midday, dietary protein intake, dietary vitamin D intake, use of vitamin D supplements, history of corticosteroids use, and current use of corticosteroids.

4. Discussion

The main finding of this population-based study is that the month of birth was associated as a periodic function with the muscle mass and strength in a large cohort of community-dwelling older women, independently of all studied potential confounders. Summer-early fall MoBs, notably August, were associated with higher muscle mass and strength. These birthdates correspond to the participants born from pregnant women who were at the end of the second trimester and at the third trimester of pregnancy during the sunny period from May to August, i.e., the key moment for fetal muscle development.

Seasonality depends on individuals' responses to seasonal fluctuations of environmental constraints. A widely studied seasonality is the effect of MoB on human traits. For instance, previous studies reported that the MoB influences a number of organic diseases [14], mood [15], as well as some morphological traits [16]. Birthweight, a marker of prenatal supplies, depends on the MoB [17]. This effect may be sustainable as birthweight is associated with muscle mass and strength throughout the life course from childhood to older age [18]; consistent with a potential influence of early life on long-term muscle development. Precisely, an effect of MoB on height and weight was also reported during childhood, and even in later life [16]. Weber et al. [19], using a large sample of conscripts in Austria, showed that males born between February to July were taller than those born in the remaining months. Similar observations were made by Banegas et al. [20], who found that Spanish male adults born in June/July were taller than those born in December/January. However, to our knowledge, we provide here the first evidence of a MoB effect on the muscle mass and function in older adults.

Several hypotheses may explain our finding. A first explanation is that the apparent MoB effect on muscles is explained by the influence of MoB on comorbidities. On the one hand, winter MoBs are associated with greater risks of heart disease, cerebrovascular disease, malignant neoplasms, and chronic respiratory diseases, with potential adverse consequences on physical activity and muscle mass and strength in adulthood [21,22]. On the other hand, the MoB may play a role in mood disorders due to the changing length of the photoperiod, and may influence the preference of individuals to plan activities rather in the morning or in the evening. According to Caci et al., people born in March/April are eveningness with a depressive mood tendency, although those born in September/October would be morningness and with impulsivity-related personality [23]. This may have an impact on daily physical activity and thus on muscle quality. However, the number of comorbidities, the IADL score, the practice of a regular physical activity and the sun exposure habits were used as potential confounders in the present analysis, and did not alter the association of MoB with muscle mass and function. Moreover, this first set of explanation could not account for the MoB-related morphological differences reported in offspring. Thus the alternative possibility of a direct MoB effect on muscles should be considered, based on the seasonal changes of maternal exposure to UV-B radiation during pregnancy and the subsequent changes of maternal serum vitamin D concentration. In French latitudes, the sun exposure of pregnant women is deemed insufficient between October and May to allow normal vitamin D concentration [24]. As the fetus is completely reliant on maternal vitamin D stores [4], and as 90% of maternal vitamin D is synthesized under the action of solar UV-B rays, the fetus experiences seasonal changes of 25OHD concentrations throughout gestation, with potential consequences on musculature.

Emerging evidence has shown that intrauterine exposure to 25OHD during pregnancy exerts a range of effects on development of skeletal muscle [1,25], probably through the modulation of the expression of muscle transcription factors [26]. Vitamin D results in the induction of myogenesis, proliferation, differentiation and cellular apoptosis [1], and may also participate in the control of protein synthesis in muscle cells by increasing the anabolic effect of insulin and leucine on muscle cells [1]. Recently, two studies in pigs found that improving maternal vitamin D status not only increased the number of fibers in longissimus dorsi of fetuses at day 90 of gestation [3], but also induced increased weight and muscle fiber cross-sectional area in psoas major and longissimus dorsi of weaning piglets [26]. In humans, maternal serum 25OHD concentration in pregnancy was positively associated with offspring height-adjusted handgrip strength and with offspring percent lean mass [27]. Evidence is also accruing that maternal serum 25OHD concentrations during pregnancy might influence offspring body composition in childhood [7,8]. Maternal antenatal serum 25OHD concentrations have been associated positively with bone mass [8] and negatively with fat mass [7]. Consistently, a population-based mother–offspring cohort study reported that maternal vitamin D status during late pregnancy could influence muscle strength of offspring at age 4 years [27]. Findings from the Mysore Parthenon Study, a prospective mother-offspring birth cohort in India, demonstrated greater arm muscle area at 5 and 9.5 years in children born to vitamin D–replete (serum 25OHD > 50 nmol/L) compared with vitamin D–depleted (25OHD < 50 nmol/L) mothers [7]. A positive association between maternal estimated UV-B exposure in the third trimester and offspring lean mass determined by DXA at 9.9 years of age was also observed in the Avon Longitudinal Study of Parents and Children (ALSPAC) [8]. All these observations support that vitamin D promotes both prenatal and postnatal skeletal muscle development, which may account for our findings, notably that people born in summer-early fall, when vitamin D status is optimal, are more prone to exhibit higher (i.e., better) muscle mass and strength in later life.

The implications for practice and research are manifold. First, our results support the idea that muscles changes related to early life hypovitaminosis D are persistent in adult age, suggesting a trait-like association between vitamin D status and muscles; consistent with the fact that hypertrophy of skeletal muscle fibers developed prenatally is the primary mechanism by which skeletal muscle growth occurs postnatally [28]. Second, they support the fact that the vitamin D status of pregnant women is crucial for fetal development and should be closely monitored in clinical routine. Offspring exposure to high levels of vitamin D appears essential for the muscles at the end of the second trimester and during the third trimester of pregnancy. Since 70% of pregnant women have hypovitaminosis D [4], these observations are relevant to public health and call for a precautionary approach based on maternal vitamin D monitoring and eventual repletion. Third, they provide a strong rationale for conducting clinical trials in pregnant women, which is expected to reveal long-term effects of vitamin D supplements on body composition, behavioral development and physical function. We propose that future clinical trials should focus on pregnancies that give birth in winter-early spring. In this perspective, our findings participate in further elucidating the profile of ideal target populations, which is an important step to provide effective guidelines on the proper use of vitamin D supplements during pregnancy.

Our results confirmed that the use of corticosteroids is associated with reduced muscle mass and strength [29], although more frequent physical activity is associated with increased strength [30], and higher dietary intakes of proteins are associated with increased muscle mass [31]. These consensual results strengthen the consistency of our study and confirm the relevance of the MoB effect we found on muscle mass and strength.

The strengths of this study include a large sample of older adults recruited in five centers with various latitudes. Additionally, the participants recruited were all born in a large period, from 1893 to 1922, in an era without vitamin supplementation D policy for pregnant women. This could have emphasized the role of seasonality and sun exposure. We also had the opportunity to measure the muscle mass with DXA, which is more accurate and relevant than anthropometric measures such as the calf circumference. Finally, regression models were applied to measure adjusted associations. Regardless, some potential limitations of our study should be considered. Firstly, the MoB is a proxy measure of maternal vitamin D status during pregnancy, and no information was available on the actual vitamin D status of the participants' mothers during pregnancy. Secondly, the study cohort was restricted to relatively vigorous older women who may be unrepresentative of older adults in general, especially regarding the musculature. The study participants may have been also more motivated, with a greater interest in personal health issues, than the general population of older adults. Thirdly, the use of an observational design precludes inferring any causal inference. Fourthly, although we were able to control for important characteristics that could modify the association between MoB and muscles, residual potential confounders, such as the latitude of the birthplaces, might still be present. Then, the ASM was measured only in one center, Toulouse, and one latitude, 43°36′ N. Finally, no information on eventual premature birth was available, although premature delivery appears to be more frequent in mothers deficient in vitamin D [32].

5. Conclusions

In conclusion, our results show for the first time to our knowledge that the month of birth is associated as a periodic function with the muscle mass and strength in a large cohort of community-dwelling older women, with potential consequences on various health outcomes including diabetes mellitus, falls, fractures, and all-cause mortality [33]. The summer-early fall months of birth were associated with higher muscle mass and strength in late life. This suggests that enhancing maternal vitamin D status during pregnancy whether through sun exposure, diet or supplementation, might improve prenatal and postnatal muscle development. This new orientation may offer a powerful mechanism to better understand the muscular changes in older adults, and to act on their healthcare early in life by setting up vitamin D supplementation in pregnant women. However, formal testing of this hypothesis in an interventional setting should be undertaken before the development of any formal clinical recommendations.

Author Contributions: C.A. has full access to all of the data in the study, takes responsibility for the data, the analyses and interpretation, and the conduct of the research, and has the right to publish any and all data, separate and apart from the attitudes of the sponsor. Study concept and design: C.A. Acquisition of data: A.-M.S. Analysis and interpretation of data: G.T.D. and C.A. Drafting of the manuscript: G.T.D. and C.A. Critical revision of the manuscript for important intellectual content: A.-M.S., D.S.-R. and F.R.H. Statistical analysis: F.R.H. Obtaining funding: A.-M.S. Administrative, technical, or material support: A.-M.S. Supervision: C.A. All authors have read and agreed to the published version of the manuscript.

Funding: This work was supported by French Ministry of Health. The sponsor had no role in the design and conduct of the study, in the collection, management, analysis, and interpretation of the data, or in the preparation, review, or approval of the manuscript.

Institutional Review Board Statement: The study was conducted in accordance with the ethical standards set forth in the Helsinki Declaration (1983). The project was approved by the local ethics committee of each city.

Informed Consent Statement: Informed consent was obtained from all subjects involved in the study.

Data Availability Statement: Patient level data are freely available from the corresponding author at cedric.annweiler@chu-angers.fr. There is no personal identification risk within this anonymized raw data, which is available after notification and authorization of the competent authorities.

Acknowledgments: Investigators of EPIDOS study. Coordinators: Breart, Dargent-Molina, Meunier, Schott, Hans, and Delmas. Principal investigators: Baudoin and Sebert (Amiens); Chapuy and Schott (Lyon); Favier and Marcelli (Montpellier); Hausherr, Menkes and Cormier (Paris); Grandjean and Ribot (Toulouse).

Conflicts of Interest: The authors declare no conflict of interest.

Abbreviations

25OHD	25-hydroxyvitamin D
ASM	Appendicular skeletal muscle mass
BMI	body mass index
kg	kilograms
DXA	Dual Energy X-ray Absorptiometry
IADL	instrumental activities of daily living
MoB	month of birth
N	Newton
SD	standard deviations
UV-B	ultraviolet-B rays
VDR	vitamin D receptor

References

1. Ceglia, L. Vitamin D and skeletal muscle tissue and function. *Mol. Aspects Med.* **2008**, *29*, 407–414. [CrossRef]
2. Annweiler, C.; Schott, A.M.; Berrut, G.; Fantino, B.; Beauchet, O. Vitamin D-related changes in physical performance: A systematic review. *J. Nutr. Health Aging* **2009**, *13*, 893–898. [CrossRef]
3. Hines, E.A.; Coffey, J.D.; Starkey, C.W.; Chung, T.K.; Starkey, J.D. Improvement of maternal vitamin D status with 25-hydroxycholecalciferol positively impacts porcine fetal skeletal muscle development and myoblast activity. *J. Anim. Sci.* **2013**, *91*, 4116–4122. [CrossRef]
4. Við Streym, S.; Kristine Moller, U.; Rejnmark, L.; Heickendorff, L.; Mosekilde, L.; Vestergaard, P. Maternal and infant vitamin D status during the first 9 months of infant life—A cohort study. *Eur. J. Clin. Nutr.* **2013**, *67*, 1022–1028. [CrossRef]
5. Dovnik, A.; Mujezinović, F.; Treiber, M.; Pečovnik Balon, B.; Gorenjak, M.; Maver, U.; Takač, I. Seasonal variations of vitamin D concentrations in pregnant women and neonates in Slovenia. *Eur. J. Obstet. Gynecol. Reprod. Biol.* **2014**, *181*, 6–9. [CrossRef]
6. Dobson, R.; Giovannoni, G.; Ramagopalan, S. The month of birth effect in multiple sclerosis: Systematic review, meta-analysis and effect of latitude. *J. Neurol. Neurosurg. Psychiatry* **2013**, *84*, 427–432. [CrossRef]
7. Krishnaveni, G.V.; Veena, S.R.; Winder, N.R.; Hill, J.C.; Noonan, K.; Boucher, B.J.; Karat, S.C.; Fall, C.H. Maternal vitamin D status during pregnancy and body composition and cardiovascular risk markers in Indian children: The Mysore Parthenon Study. *Am. J. Clin. Nutr.* **2011**, *93*, 628–635. [CrossRef]
8. Sayers, A.; Tobias, J.H. Estimated maternal ultraviolet B exposure levels in pregnancy influence skeletal development of the child. *J. Clin. Endocrinol. Metab.* **2009**, *94*, 765–771. [CrossRef]
9. Dargent-Molina, P.; Favier, F.; Grandjean, H.; Baudoin, C.; Schott, A.M.; Hausherr, E.; Meunier, P.J.; Bréart, G. Fall-related factors and risk of hip fracture: The EPIDOS prospective study. *Lancet* **1996**, *348*, 145–149. [CrossRef]
10. Gillette-Guyonnet, S.; Nourhashemi, F.; Andrieu, S.; Cantet, C.; Albarède, J.L.; Vellas, B.; Grandjean, H. Body composition in French women 75+ years of age: The EPIDOS study. *Mech. Ageing Dev.* **2003**, *124*, 311–316. [CrossRef]
11. Lawton, M.P.; Brody, E.M. Assessment of older people: Self-maintaining and instrumental activities of daily living. *Gerontologist* **1969**, *9*, 179–186. [CrossRef]
12. Dupuy, C.; Lauwers-Cances, V.; van Kan, G.A.; Gillette, S.; Schott, A.M.; Beauchet, O.; Annweiler, C.; Vellas, B.; Rolland, Y. Dietary vitamin D intake and muscle mass in older women. Results from a cross-sectional analysis of the EPIDOS study. *J. Nutr. Health Aging* **2013**, *17*, 119–124. [CrossRef]
13. Annweiler, C.; Schott, A.M.; Beauchet, O. Proposal and validation of a quick question to rate the influence of sun exposure in geriatric epidemiological studies on vitamin D. *Int. J. Vitam. Nutr. Res.* **2012**, *82*, 412–416. [CrossRef]
14. Boland, M.R.; Shahn, Z.; Madigan, D.; Hripcsak, G.; Tatonetti, N.P. Birth month affects lifetime disease risk: A phenome-wide method. *J. Am. Med. Inform. Assoc.* **2015**, *22*, 1042–1053. [CrossRef]
15. Tonetti, L.; Milfont, T.L.; Tilyard, B.A.; Natale, V. Month of birth and mood seasonality: A comparison between countries in the northern and southern hemispheres. *Psychiatry Clin. Neurosci.* **2013**, *67*, 133–138. [CrossRef]
16. Henneberg, M.; Louw, G.J. Height and weight differencies among South African urban school children born in various months of the year. *Am. J. Hum. Biol.* **1990**, *2*, 227–233. [CrossRef]
17. Adair, L.S.; Pollitt, E. Seasonal variation in pre- and postpartum maternal body measurement and infant birthweights. *Am. J. Phys. Anthropol.* **1983**, *62*, 325–331. [CrossRef]
18. Yliharsila, H.; Kajantie, E.; Osmond, C.; Forsen, T.; Barker, D.J.; Eriksson, J.G. Birth size, adult body composition and muscle strength in later life. *Int. J. Obes.* **2007**, *31*, 1392–1399. [CrossRef]
19. Weber, G.W.; Prossinger, H.; Seidler, H. Height depends on month of birth. *Nature* **1998**, *391*, 754–755. [CrossRef]
20. Banegas, J.R.; Rodriguez-Artalejo, F.; Graciani, A.; De La Cruz, J.J.; Gutierrez-Fisac, J.L. Month of birth and height of Spanish middle-aged men. *Ann. Hum. Biol.* **2001**, *28*, 15–20.
21. Deng, J.; Wang, J.; Xiao, C.; Xu, S.; Gao, X.; Pan, F. The influence of birth month on total and cardiovascular mortality: A population-based surveillance study. *Chronobiol. Int.* **2020**, *37*, 1772–1777. [CrossRef]

22. Vaiserman, A. Season-of-birth phenomenon in health and longevity: Epidemiologic evidence and mechanistic considerations. *J. Dev. Orig. Health Dis.* **2021**, *12*, 849–858. [CrossRef]
23. Caci, H.; Robert, P.; Dossios, C.; Boyer, P. Morningness-Eveningness for Children Scale: Psychometric properties and month of birth effect. *Encephale* **2005**, *31*, 56–64. [CrossRef]
24. Annweiler, C.; Legrand, E.; Souberbielle, J.C. Vitamin D in adults: Update on testing and supplementation. *Geriatr. Psychol. Neuropsychiatr. Vieil.* **2018**, *16*, 7–22. [CrossRef]
25. Girgis, C.M.; Clifton-Bligh, R.J.; Hamrick, M.W.; Holick, M.F.; Gunton, J.E. The roles of vitamin D in skeletal muscle: Form, function, and metabolism. *Endocr. Rev.* **2012**, *34*, 33–83. [CrossRef]
26. Zhou, H.; Chen, Y.; Lv, G.; Zhuo, Y.; Lin, Y.; Feng, B.; Fang, Z.; Che, L.; Li, J.; Xu, S.; et al. Improving maternal vitamin D status promotes prenatal and postnatal skeletal muscle development of pig offspring. *Nutrition* **2016**, *32*, 1144–1152. [CrossRef]
27. Harvey, N.C.; Moon, R.J.; Sayer, A.A.; Ntani, G.; Davies, J.H.; Javaid, M.K.; Robinson, S.M.; Godfrey, K.M.; Inskip, H.M.; Cooper, C.; et al. Maternal antenatal vitamin D status and offspring muscle development: Findings from the Southampton Women's Survey. *J. Clin. Endocrinol. Metab.* **2014**, *99*, 330–337. [CrossRef]
28. Lauridsen, C. Triennial Growth Symposiumdestablishment of the 2012 vitamin D requirements in swine with focus on dietary forms and levels of vitamin D. *J. Anim. Sci.* **2014**, *92*, 910–916. [CrossRef]
29. Fardet, L.; Flahault, A.; Kettaneh, A.; Tiev, K.P.; Généreau, T.; Tolédano, C.; Lebbé, C.; Cabane, J. Corticosteroid-induced clinical adverse events: Frequency, risk factors and patient's opinion. *Br. J. Dermatol.* **2007**, *157*, 142–148. [CrossRef]
30. Felman, A.L. Immediate effects of exercise on apparent limb mass and circumference. *Int. Z. Angew. Physiol.* **1963**, *20*, 38–44. [CrossRef]
31. Tipton, K.D.; Elliott, T.A.; Cree, M.G.; Wolf, S.E.; Sanford, A.P.; Wolfe, R.R. Ingestion of casein and whey proteins result in muscle anabolism after resistance exercise. *Med. Sci. Sports Exerc.* **2004**, *36*, 2073–2081. [CrossRef] [PubMed]
32. Zhou, S.S.; Tao, Y.H.; Huang, K.; Zhu, B.B.; Tao, F.B. Vitamin D and risk of preterm birth: Up-to-date meta-analysis of randomized controlled trials and observational studies. *J. Obstet. Gynaecol. Res.* **2017**, *43*, 247–256. [CrossRef] [PubMed]
33. Cooper, R.; Kuh, D.; Hardy, R.; Mortality Review Group; FALCon and HALCyon Study Teams. Objectively measured physical capability levels and mortality: Systematic review and meta-analysis. *BMJ* **2010**, *341*, c4467. [CrossRef] [PubMed]

Review

Novel CYP11A1-Derived Vitamin D and Lumisterol Biometabolites for the Management of COVID-19

Shariq Qayyum [1,2,3], Radomir M. Slominski [4], Chander Raman [1] and Andrzej T. Slominski [1,5,*]

1 Department of Dermatology, University of Alabama at Birmingham, Birmingham, AL 35294, USA
2 Research Program in Men's Health: Aging and Metabolism, The Center for Clinical Investigation, Brigham and Women's Hospital, Harvard Medical School, 221 Longwood Avenue, Boston, MA 02115, USA
3 Department of Biological Chemistry and Molecular Pharmacology, Harvard Medical School, Boston, MA 02115, USA
4 Department of Genetics, Informatics Institute, University of Alabama at Birmingham, Birmingham, AL 35294, USA
5 Pathology and Laboratory Medicine Service, VA Medical Center, Birmingham, AL 35294, USA
* Correspondence: aslominski@uabmc.edu

Abstract: Vitamin D deficiency is associated with a higher risk of SARS-CoV-2 infection and poor outcomes of the COVID-19 disease. However, a satisfactory mechanism explaining the vitamin D protective effects is missing. Based on the anti-inflammatory and anti-oxidative properties of classical and novel (CYP11A1-derived) vitamin D and lumisterol hydroxymetabolites, we have proposed that they would attenuate the self-amplifying damage in lungs and other organs through mechanisms initiated by interactions with corresponding nuclear receptors. These include the VDR mediated inhibition of NFκβ, inverse agonism on RORγ and the inhibition of ROS through activation of NRF2-dependent pathways. In addition, the non-receptor mediated actions of vitamin D and related lumisterol hydroxymetabolites would include interactions with the active sites of SARS-CoV-2 transcription machinery enzymes (M^{pro};main protease and RdRp;RNA dependent RNA polymerase). Furthermore, these metabolites could interfere with the binding of SARS-CoV-2 RBD with ACE2 by interacting with ACE2 and TMPRSS2. These interactions can cause the conformational and dynamical motion changes in TMPRSS2, which would affect TMPRSS2 to prime SARS-CoV-2 spike proteins. Therefore, novel, CYP11A1-derived, active forms of vitamin D and lumisterol can restrain COVID-19 through both nuclear receptor-dependent and independent mechanisms, which identify them as excellent candidates for antiviral drug research and for the educated use of their precursors as nutrients or supplements in the prevention and attenuation of the COVID-19 disease.

Keywords: vitamin D; lumisterol; SARS-CoV-2; anti-inflammatory; ACE2; M^{pro}; RdRp

Citation: Qayyum, S.; Slominski, R.M.; Raman, C.; Slominski, A.T. Novel CYP11A1-Derived Vitamin D and Lumisterol Biometabolites for the Management of COVID-19. *Nutrients* **2022**, *14*, 4779. https://doi.org/10.3390/nu14224779

Academic Editor: Federica I. Wolf

Received: 15 August 2022
Accepted: 8 November 2022
Published: 11 November 2022

1. Introduction

COVID-19 is still the top health issue in the world. Vaccines approved for COVID-19 provide protection against some strains of SARS-CoV-2; however, new mutant strains develop in continuity and some of them escape immunity provided by current vaccines [1]. The infection from SARS-CoV-2 has severe adverse outcomes with a significantly higher mortality rate than influenza. The major cause of death in COVID-19 is acute respiratory distress syndrome (ARDS) caused by cytokine storm [2,3]. This enhanced hyperactivated innate immune response against the virus causes severe damage to the patient's body/organs which might be fatal (Figure 1). In this mini-review we will discuss how vitamin D and its derivatives can be helpful against the infection caused by SARS-CoV-2.

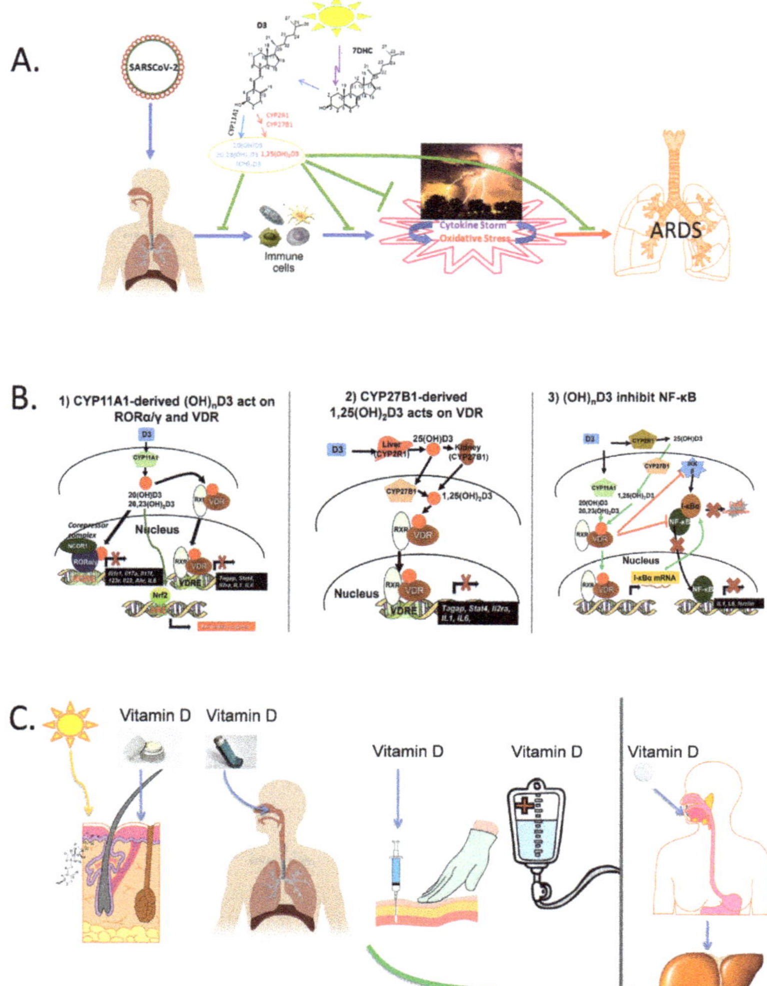

Figure 1. Possible mechanisms by which vitamin D can counteract the COVID-19 illness. In panel (**A**) it is proposed that the novel hydroxyderivatives of vitamin D3, in similar manner as 1,25(OH)$_2$D3, inhibit cytokine storm and oxidative stress, with net attenuating effect on ARDS and multiorgan failure induced by COVID-19. Panel (**B**) proposes a mechanism of action of canonical and non-canonical vitamin D-hydroxyderivatives. Vitamin D signaling in mononuclear cells involves the activation of the VDR or inverse agonism on RORγ with downstream inhibition of inflammatory genes and the suppression of oxidative stress through the activation of NRF2. VDR, vitamin D receptor; RXR, retinoid X receptor; ROR, retinoic acid orphan receptor, RORE, ROR response element; ARE, antioxidant response element; VDRE, vitamin D response element; NRF2, transcription factor NF-E2-related factor 2. Panel (**C**) shows how different routes of vitamin D delivery impact vitamin D hydroxylation/activation patterns. Reprinted with permission from the publisher [86].

Vitamin D, a prohormone, is a fat soluble secosteroid, which is formed in the skin after the absorption of UVB energy by the B ring of 7-dehydrocholesterol (7DHC) [4–6]. The prolonged exposure of 7-DHC to UVB leads to the phototransformation of pre-vitamin D3 to tachysterol and lumisterol [4–6]. It can also be ingested from the diet and supplements. As a prohormone, it must be activated to exert the biological activity. In the classical activation pathway vitamin D3 is metabolized by several cytochromes P450 (CYPs) enzymes before being transformed to its known active form 1,25-dihydroxyvitamin D3 (1,25(OH)$_2$D3) [7–10]. 1,25(OH)$_2$D3 activates the nuclear vitamin D (VDR) receptor, which controls not only body calcium metabolism [9,11–13] but also many important physiological functions, including the regulation of the innate and adaptive immunity [6,9,14–17]. Vitamin D can be activated by two pathways known as a canonical, with sequential hydroxylation at C25 and C1α [18], and a non-canonical, activated by CYP11A1 [19,20]. The canonical pathway includes the metabolism of vitamin D3 to 25-hydroxyvitamin D3 (25(OH)D3) by CYP2R1 and CYP27A1 in the liver and final C1α hydroxylation in the kidney to the biologically active form 1,25(OH)$_2$D3 by CYP27B1 [7–9]. This pathway also operates in the peripheral tissues, including skin [8,18,21–24].

The phenotypic effect of 1,25(OH)$_2$D3 is predominantly mediated through an interaction with the VDR leading to the transcriptional activation of more than 3000 genes [4,6,8–10]. Non-genomic regulatory actions for 1,25(OH)$_2$D3 were also described [6,8,10]. 1,25(OH)$_2$D3, in addition of regulating body calcium metabolism, also regulates diverse functions on the systemic, tissue, and cellular levels [4,6,8–15,18–20]. Of general interest are the anti-oxidative and anti-inflammatory properties of 1,25(OH)$_2$D3, which have been appreciated for almost two decades [4,6,8–15,18–20]. The latter includes the downregulation of pro-inflammatory cytokines production through the inhibition of NFkB [6,12–14].

Recently discovered non-canonical pathways of vitamin D activation are initiated by an obligatory enzyme of steroidogenesis, CYP11A1, in a complex process involving several CYPs and producing more than a dozen of hydroxyderivatives [8,20,25–27]. CYP11A1 is not only expressed in adrenals, placenta, and gonads [28] but also in immune cells [29] and other peripheral organs including skin [19,30]. The CYP11A1-derived hydroxyderivatives are non-calcemic or low calcemic [31–34] and can therefore be used at high concentrations for therapeutic purposes [35–39]. They are also detectable in natural products including honey [40] and human serum [25,41,42]. Similarly, to 1,25(OH)$_2$D3, they can alter gene expression by binding on the genomic site of the VDR [33,43–46]. CYP11A1-derived vitamin D3 hydroxyderivatives also bind to other nuclear receptors, including aryl hydrocarbon receptor (AhR) [47–49], retinoic acid orphan receptors (ROR)α and γ [44,50], liver X receptors (LXR)α and β [51] and can change their expression and activities [27]. The CYP11A1-derived hydroxymetabolites of vitamin D3, including 20(OH)D3 and 20,23(OH)$_2$D3, have demonstrated anti-inflammatory and anti-oxidative effects [27,36,42,46,52–56], which are similar to the effects described for the classical active form of vitamin D3, 1,25(OH)$_2$D3 [6,12–14,18–20].

Novel pathways of 7DHC transformation by CYP11A1 [30,57,58], with the further phototransformation of the 5.7-dienal products to corresponding secosteroids [27,59–64] and the hydroxylations of lumisterol by CYP11A1 and CYP27A1, were also discovered [65–67], with their products being biologically active and detectable in human body and acting on RORα and γ as inverse agonists and as agonists on LXR α and β and on the non-genomic site of the VDR [51,66,68]. Most recently, an enzymatic activation of tachysterol was reported with the metabolic products acting on the VDR, AhR, LXRs, and RORs [63]. These metabolites exert similar biological effects as the classical active form of vitamin D [1,25(OH)$_2$D3] and they have their unique activity pattern towards various nuclear receptors, aside of the classical VDR/RXR complex [9].

Although SARS-CoV-2 infection in human cells involves multiple factors, in this review we are focusing on two main interactions listed below. The spike protein (S) of SARS-CoV-2 facilitates the entry of the virus into human cells by engaging angiotensin-converting enzyme 2 (ACE2) as their entry receptor [69] and further cellular serine protease TMPRSS2 is used for priming of S protein [70–72]. The association between ACE2 and Spike protein is critically important and current vaccines (mRNA) are developed to inhibit this interaction [73]. The actions of CYP11A-derived vitamin D3-hydroxymetabolites, canonical 1,25(OH)$_2$D3 and lumisterol hydroxymetabolites with SARS-CoV-2 replication machinery enzymes were previously explored [74], which included molecular modeling on classical vitamin D compounds [75]. The significance of these studies is further discussed in this review.

The SARS-CoV-2 virus replicates within host cells using its cellular and enzymatic components. M^{pro}, also termed 3CL protease, is a 33.8-kDa cysteine protease which helps in the maturation of functional polypeptides involved in the assembly of replication-transcription machinery [76–78]. M^{pro} digests the polyprotein at no less than 11 conserved sites, starting with the autolytic cleavage of this enzyme itself from pp1a and pp1ab, which are individual nonstructural proteins essential for viral genome replication [76]. Another enzyme important to the life cycle of SARS-CoV-2 is RdRp (RNA-dependent RNA polymerase), which catalyzes the replication [79] of RNA from an RNA template. SARS-CoV-2 use an RdRp complex for the replication of their genome and for the transcription of their genes [79]. M^{pro} and RdRp are enzymes required for viral replication and are not homologous to any gene in the human genome. Hence, they are very attractive targets for the development of anti-viral drugs against COVID-19. Therefore, we will further discuss how the hydroxymetabolites of vitamin D and of lumisterol can counter SARS-CoV-2 infection at different stages of this process.

2. CYP11A1-Derived Vitamin D and Lumisterol Hydroxymetabolites Exert Anti-Inflammatory and Antioxidant Effects

The cytokine storm is a response to viral infection, which causes immune cells to release several pro-inflammatory cytokines/chemokines (interferons, interleukins 1, 6 and 17, chemokines, colony-stimulating factors, and tumor necrosis factors (TNF)), leading to hyper inflammation and organ damage [80–82]. This process in the lung leads to acute lung injury and ARDS. Along with = ARDS, another factor which plays a role in damage to the tissue and cells is oxidative stress. It is secondary to the production of reactive oxygen species (ROS) and reactive nitrogen species (RNS) [83–85]. CYP11A1-derived vitamin D3 and lumisterol hydroxymetabolites exhibit potent anti-inflammatory activities through the inhibition of IL-1, IL-6, IL-17, TNFα and INFγ production and/or other pro-inflammatory pathways [25,36,37,43,46,50,52,54–56,68], which are similar to those mediated by classical 1,25(OH)$_2$D3 [4,6,8–15,18–20]. The anti-inflammatory effects of active forms of vitamin D can be mediated through the downregulation of NF-κB, involving action on VDR and inverse agonism on RORγ leading to the attenuation of Th17 responses (Figure 1B). These compounds also induce anti-oxidative and reparative responses with mechanism of action involving the activation of NRF2 and p53 signaling pathways [27,42,52,53,68,86]. Interestingly, the anti-viral role of NRF2 is also recognized [87]. The use of 1,25(OH)$_2$D3 has its limitations because of the toxicity that includes hypercalcemia [7,88]. However, CYP11A1-derived 20(OH)D3, 20(OH)D2, and 20,23(OH)$_2$D3 are not calcemic even at very high doses [31–34]. Hence, vitamin D and its metabolites can be used as economic nutritional supplements to counter the effects of the SARS-CoV-2 infection like cytokine storm [86,89].

3. Inhibition of the Interaction between ACE2 and SARS-CoV-2 Spike RBD

The ACE2 interacts with the receptor-binding domain (RBD) region of the spike protein [90,91]. SARS-CoV-2 and SARS-CoV-2 RBD are typically in standing up state and resting state, respectively. SARS-COV-2 RBD has higher binding affinity to ACE2, but its lying-down state makes it less accessible to ACE2 and other inhibitory or neutralizing agents [90,91]. The use of host protease (furin, TMPRSS2, cathepsins etc.) for its activation helps as a strategy to overcome the lying down state and maintaining its high binging affinity to ACE2 [90–92]. This interaction is critical and important for drug development against COVID-19 infection. Molecular modeling and simulation [93] evaluated the binding of vitamin D3 and its hydroxyderivatives to SARS-CoV-2 RBD, and their potential to inhibit its interaction with ACE2. The study showed that vitamin D3 and its hydroxyderivatives can function as inhibitors of TMPRSS2 and inhibit the SARS-CoV-2 receptor binding domain (RBD) binding to the ACE2 [93]. Molecular dynamics (MD) simulations for the interactions of 1,25(OH)$_2$D3 have shown the favorable binding free energy of ACE2, SARS-CoV-2 RBD, and TMPRSS2 with 1,25(OH)$_2$D3 [93]. The binding free energy of ACE2, SARS-CoV-2, and TMPRSS2 with 1,25(OH)$_2$D3 were -18.55 ± 4.16, -16.97 ± 1.69, and -21.04 ± 1.53 kcal/mol separately, further indicating that ACE2, SARS-CoV-2 RBD, and TMPRSS2 show favorable binding with 1,25(OH)$_2$D3 [93]. The predicted interaction of 1,25(OH)$_2$D3 with SARS-CoV-2 RBD and ACE2 could result in the conformation and dynamical motion changes of the binding surfaces between SARS-CoV-2 RBD and ACE2, leading to the interruption of the binding of SARS-CoV-2 RBD with ACE2 [93]. The interaction of 1,25(OH)$_2$D3 with TMPRSS2 also caused the conformational and dynamical motion changes of TMPRSS2, which could affect TMPRSS2 to prime SARS-CoV-2 spike proteins [93]. These studies [93] have indicated that vitamin D3 and its biologically active hydroxymetabolites have the theoretical potential to prevent the cellular entry of SARS-CoV-2 by serving as the inhibitor of TMPRSS2 and blocking the binding of SARS-CoV-2 RBD with ACE2.

Molecular modeling was also used for the virtual screening of antiviral compounds to SARS-CoV-2 non-structural proteins [94]. The described interactions between spike protein and ACE2 can be also disrupted by other compounds, including vitamin D as described by other research groups [95–97]. These included interactions with vitamins, retinoids, steroids, vitamin D derivatives, and dihydrotachysterol as examples. The detailed mechanisms of action were discussed in [95–97]. In addition, vitamin D was identified as a potential inhibitor of COVID-19 Nsp15 endoribonuclease binding sites [98].

To confirm our predictions on novel vitamin D3 and lumisterol hydroxymetabolites, we have used an inhibitor screening kit (SARS-CoV-2 inhibitor screening kit, Acro Biosytems) (Table 1). The kit uses a colorimetric ELISA platform, which measures the binding of immobilized SARS-CoV-2 S protein and biotinylated human ACE2. Top compounds with best binding energy were theoretically predicted previously [93], selected for this assay, showing the inhibition of the interaction between ACE2 and RBD (Table 1). 20(OH)L3 showed the highest level of inhibition of the RBD-ACE2 interaction followed by 25*S*27(OH)L3 and 20(OH)D3 at concentrations of 2×10^{-7} M. The compounds were observed to be effective in μM concentrations, which is promising for future clinical and preclinical testing.

Table 1. Inhibition of ACE2 and RBD interaction by the hydroxymetabolites. Inhibition by the selected metabolites was observed concentration of 2×10^{-7} M using a SARS-CoV-2 inhibitor screening kit from Acro Biosytems. The assay followed the manufacture's protocol of the SARS-CoV-2 (B.1.617.2) Inhibitor Screening Kit (Spike RBD) (1 Innovation Way, Newark, DE 19711, USA). Data were analyzed by one way ANOVA using GraphPad Prism statistical software.

No.	Name of the Ligand	Inhibition in Enzyme Activity (%)	*p*-Value
1.	20(OH)D3	46.057	0.013
2.	1,20(OH)$_2$D3	29.222	NS *
3.	1,25(OH)$_2$D3	36.876	0.034
4.	20,23(OH)$_2$D3	36.152	0.018
5.	24(OH)L3	32.343	0.0265
6.	20(OH)L3	74.552	0.001
7.	25*S*27(OH)L	51.722	0.005
8.	20,22(OH)$_2$L3	13.074	NS *

NS * Not significant.

In addition, the treatment of HaCaT keratinocytes with these hydoxymetabolites changed the expression of ACE2 and TMPRSS2 in a metabolite-specific manner. 1,20(OH)$_2$D3, 1,25(OH)$_2$D3, and 24(OH)L3 suppressed the expression of ACE2, and TMPRSS2 expression was inhibited by 20(OH)D3, 1,20(OH)$_2$D3, 1,25(OH)$_2$D3, 20(OH)L3, and 24(OH)L3 (Supplementary Figure S1). This suggests that these compounds can inhibit the bonding of ACE2 and RBD not only by directly blocking the binding but also by altering the expression of these receptors' genes.

4. Inhibition of the Activity of the Replication Enzymes of SARS-CoV-2

SARS-CoV-2 replicates inside the host cell after its entry. The viral particles utilize host resources, but viral replication machinery plays crucial role in its replication. These viral specific factors do not share homologies with human proteins and are therefore targets for drug development against COVID-19 [74]. We selected two SARS-CoV-2 replication enzymes, RdRp or nsp12 and 3C-like protease (3CLpro or M^{pro}), based on their recognized importance for drug development [76]. Although there are reports that vitamin D and its metabolites have potential to inhibit other viral protein, we selected these proteins as we have had experimental potential to confirm our results [74]. 3CL-Chymotrypsin, such as Protease or Main protease (M^{pro}), is one of the two proteolytic enzymes that helps in cleaving the replicase polyprotein 1ab in SARS-CoV-2 at 11 specific sites, the recognition sites being Leu-Gln (Ser, Ala, Gly) in order to release 12 nsps (nsp4, nsp6-16) that are essential for viral replication as well as viral assembly [3,99]. This enzyme shares no common cleavage site with any human protease and its functional importance in the life cycle of the virus makes it an attractive target for drug development. Similarly, RNA-dependent RNA Polymerase (RdRp) is an enzyme that is responsible for the replication of RNA from an RNA template [100]. RdRp is another conserved protein of retroviruses and is also a proven target for the development of antiviral drugs [79]. We performed molecular docking on the active sites of these two enzymes and found that the hydroxyderivatives of vitamin D3 and lumisterol were binding efficiently on the active sites of these enzymes with similar affinities to known therapeutics danoprevir, lopinavir, and ritonavir serving as positive controls [74]. We further confirmed the inhibition of the enzyme activity in the presence of the compounds.

Danoprevir, lopinavir, and ritonavir were used as a standard for the comparison of predicted energies of the top 10 selected compounds [74,94]. A virtuous complementarity to the M^{pro} binding pocket was observed for top compounds, which indicates a possibility that these metabolites have ability to hinder the substrate accessibility, inhibiting the enzymatic activity in process. Significant interactions between the selected metabolites and the critically important residues of the M^{pro} substrate-binding pocket were observed (Figure 2) and predicting a block to the substrate-binding pocket of COVID-19 M^{pro} (Figure 2) [74]. The detailed analysis of the residues interaction with these metabolites was reported previously [74]. The inhibition of M^{pro} was confirmed using 3CL Protease, MBP-tagged (SARS-CoV-2) Assay (BPS Biosciences). The 25(OH)L3, 24(OH)L3, and 20S(OH)7DHC being most effective at inhibiting M^{pro} activity by 10–19% at a concentration of 2×10^{-7} M (Figure 3A). Similarly, selected metabolites showed interactions with critically essential residues of SARS-CoV-2 RdRp (Figure 4) [74]. These sterols and secosteroids presented a similar binding pattern to inhibitor remdesivir on RdRp active sites [74,101]. The binding prototype of the compounds showed a virtuous complementarity to the SARS-CoV-2 RdRp binding pocket predicting that they can inhibit the enzymatic activity (Figure 3B). These metabolites showed inhibitory activity ranging from 40–60% at a concentration of 10^{-7} M (Figure 3B) with 25(OH)L3 with an IC_{50} of 0.5 μM followed by 1,25(OH)$_2$D3 and 20S(OH)L3, which had an IC_{50} of 1 μM. Thus, our published work [74] has demonstrated that novel 7DHC, lumisterol, and vitamin D3 hydroxymetabolites have the potential to inhibit SARS-CoV-2 infection by restricting its replication cycle. Interestingly, unbiased retrospective analyses of microarray data obtained with epithelial cells indicated the antiviral effects of 20,23(OH)$_2$D3 with similar effects for 1,25(OH)$_2$D3 [86]. A potential role of hydroxylumisterols appears to be strengthened by recent findings, showing their similarity in structure to 25(OH)L3, where 25-hydroxycholesterol can act as a potent SARS-CoV-2 inhibitor [102], and that cholesterol 25-hydroxylase inhibits SARS-CoV-2 [103] and oxysterols show anti-viral activity [104].

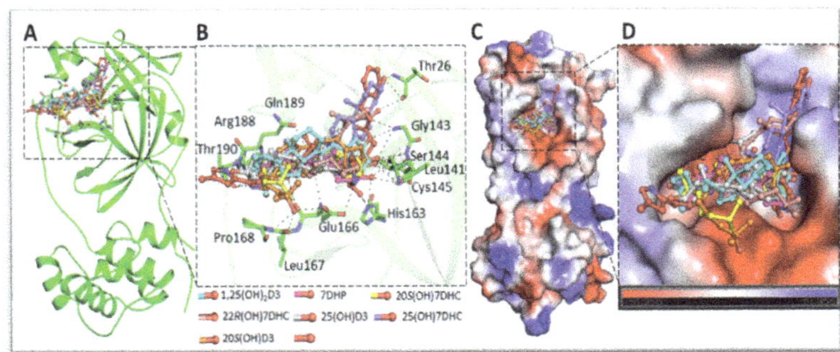

Figure 2. The binding pattern of identified compounds with SARS-CoV-2 M^{pro}. (**A**) structural representation of the protein in complex with selected sterols and secosteroids. (**B**) selected compounds blocking the binding pocket and making significant interactions with the functionally important residues of SARS-CoV-2 M^{pro}. (**C**) surface representation of conserved substrate-binding pocket of SARS-CoV-2 M^{pro} in complex with selected compounds. (**D**) zoomed view of the substrate-binding pocket of SARS-CoV-2 M^{pro} in complex with selected compounds. Reprinted with permission from the publisher [74].

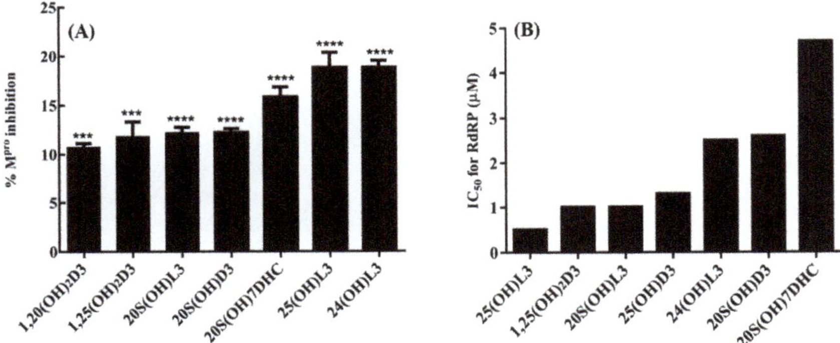

Figure 3. Enzyme inhibition by the selected sterols and secosteroids. (**A**) the M^{pro} enzyme inhibition by the selected metabolites at concentration of 2×10^{-7} M. The inhibition percentages were calculated using the formula: % inhibition = 100 × [1(X Minimum)/(Maximum–Minimum)]. Minimum = negative control without any enzyme (0% enzyme activity); Maximum = positive control with enzyme and substrate (100% enzyme activity). The test sets included enzymes, substrates, and the test compounds, and excitation at a wavelength of 360 nm and the detection of emission at a wavelength of 460 nm was observed for change in enzyme activity. The statistical significance of differences was evaluated by one-way ANOVA; *** $p< 0.001$ and **** $p < 0.0001$ for all conditions relative to ethanol blank, $n = 3$. (**B**) the RdRp enzyme activity inhibition by selected sterols and secosteroids. The inhibition percentages were calculated using the formula: % inhibition = 100 × [1 − (X-Minimum)/(Maximum–Minimum)]. Minimum = negative control without any enzyme (0% enzyme activity); Maximum = positive control with enzyme and substrate (100% enzyme activity). The statistical significance of differences was evaluated by one-way ANOVA; *** $p < 0.001$ and **** $p < 0.0001$ for all conditions relative to the ethanol blank, $n = 3$. RdRp, RNA-dependent RNA polymerase. Reprinted with permission from the publisher [74].

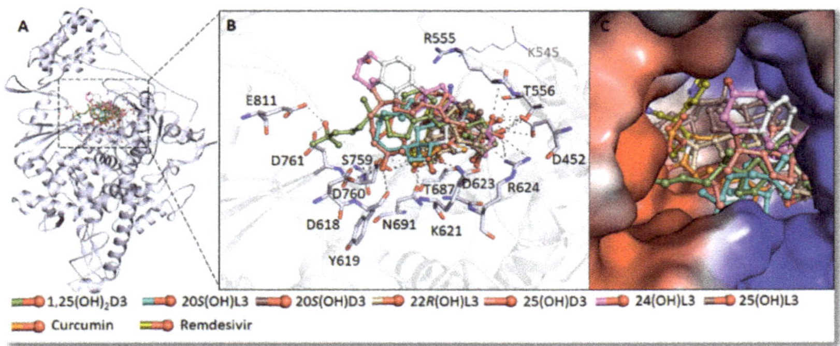

Figure 4. The binding pattern of identified sterols and secosteroids with SARS-CoV-2 RdRp. (**A**) structural representation of the protein in complex with selected compounds. (**B**) active site residues of the RdRp-binding pocket making significant interactions with each of the identified compounds. (**C**) surface view of the RdRp active site with the electrostatic potential from red (negative) to blue (positive) in complex with selected compounds. RdRp, RNAdependent RNA polymerase. Reprinted with permission from the publisher [74].

5. Hypothesis

Overall, the above considerations provide strong support for the ability of D3, L3, and 7DHC hydroxymetabolites to counter the different stages of SARS-CoV-2 infection and most of them are non-calcemic. These metabolites can attenuate cytokine storm and have an ability to inhibit the viral replication enzymes. A deficiency of these hydroxymetabolites may contribute to the transition of SARS-CoV-2 patients from asymptomatic to symptomatic. Vitamin D deficiency in the body will lead to reduced levels of the vitamin D hydroxymetabolites and consequently diminished capacity to attenuate cytokine storm. Anti-inflammatory effects by vitamin D and lumisterol hydroxymetabolites were observed at 0.1 μM concentration. Similarly, 20(OH)L3, 25S27(OH)L3, and 20(OH)D3 were able to inhibit RBD-ACE2 interaction, which is necessary for cellular entry of the virus. For M^{pro}, a significant inhibition has been observed at 0.1 μM of 25(OH)D3, which is close to its plasma concentration [8]. Of note, low pre-infection 25(OH)D3 levels are associated with a higher severity of the COVID-19 illness [105]. For RdRp, the IC_{50} for 25(OH)D3 was 1.3 μM, approximately one order of magnitude above its plasma concentration. For the hydroxylumisterols tested, plasma concentrations are unknown except for 20(OH)L3 where a value of 0.25 μM has been reported [66]. Also based on the enzymology, 25(OH)L3, which had the lowest IC_{50} (0.5 μM) for the inhibition of RdRp, is likely to have substantially higher concentrations. Interestingly, similar in structure 25(OH)cholesterol has been shown to have anti-SARS-CoV-2 activities [102,103], while cholesterol 25-hydroxylase generated anti-inflammatory environments [106] and oxysterols are recognized for anti-viral activities [104].

Therefore, defects in vitamin D or lumisterol delivery either orally as nutrients or supplements or their production in the skin after UVB exposure can lead to the deficiency of corresponding hydroxymetabolites; which have demonstrated anti-viral potential [74,86,93]. This family of compounds contains dozens of molecules [27], which can influence the different stages of SARS-CoV-2 infection. This will require further validation, as has been carried out for other molecules, including classical vitamin D3 derivatives [48,75,94–98]. Vaccines against SARS-CoV-2 are clearly a major advance in controlling COVID-19; however, new viral variants emphasize the need for alternative therapeutic or nutritional approaches. Therefore, consideration for novel vitamin D and L3 metabolites for anti-viral drugs that could attenuate COVID-19 is warranted.

6. Concluding Remarks

There are reports demonstrating a strong association for the pre-infection deficiency of vitamin D in hospitalized COVID-19 patients [100] and increased disease severity and mortality [102,103]. The oral supplementation of vitamin D may affect SARS-CoV-2 infection outcomes. Several clinical trials are ongoing, assessing the ability of vitamin D to prevent COVID-19 infection and disease severity [104–110], showing its importance in managing COVID-19. However, the precise mechanism of vitamin D action against COVID-19 is still unresolved. Here, we have formulated a hypothesis for the mechanism of action of vitamin D and lumisterol hydroxymetabolites against COVID-19. It includes the enzymatic activation of vitamin D3 or sterol precursors with receptor-independent [74,93] or nuclear receptor-dependent activities downstream of VDR, LXR, and AhR activation or inverse agonism on RORs [27,51,86]. Similarly, tachysterol and its metabolites represent additional candidates for at least receptor-mediated activities [63]. We also acknowledge other mechanism-oriented work on classical vitamin D hydroxyderivatives in the prevention or therapy of COVID-19 that has been reported or reviewed recently [75,96,111–117]. Therefore, further clinical testing for their therapeutic use would represent an important step in understanding the beneficial actions of vitamin D3, lumisterol, and possibly tachysterol derivatives [63,97] in COVID-19 with significant implications for the nutritional approach. Vitamin D, lumisterol, or tachysterol ingested through the diet or as supplements would serve as prohormones for further activation in the body towards biologically active forms.

7. Patent

Patent application pending (WO2022006446A1), which includes the experimental portion of this work.

Supplementary Materials: The following supporting information can be downloaded at: https://www.mdpi.com/article/10.3390/nu1422477/s1, Figure S1: The change in the expression of *ACE2* and *TMPRSS2* in keratinocyte HaCaT cell line after treatment with listed vitamin D and lumisterol metabolites at 2×10^{-7} M [39].

Author Contributions: S.Q., R.M.S., C.R. and A.T.S.—conceptualization, formal analysis, writing, figure preparation, review, and editing. All authors have read and agreed to the published version of the manuscript.

Funding: NIH (grants nos. 1R01AR073004, R01AR071189, and R21AI149267-01A1) and VA merit (grant no. 1I01BX004293-01A1).

Institutional Review Board Statement: Not applicable.

Informed Consent Statement: Not applicable.

Data Availability Statement: Not applicable. The discussed papers were selected through a PubMed search and current reading of articles on COVID-19 and the role of vitamin D and sterol compounds in the therapy and prevention of the disease.

Conflicts of Interest: The authors declare no conflict of interest.

References

1. Coronaviridae Study Group of the International Committee on Taxonomy of Viruses. The species Severe acute respiratory syndrome-related coronavirus: Classifying 2019-nCoV and naming it SARS-CoV-2. *Nat. Microbiol.* **2020**, *5*, 536–544. [CrossRef] [PubMed]
2. Mehta, P.; McAuley, D.F.; Brown, M.; Sanchez, E.; Tattersall, R.S.; Manson, J.J.; on behalf of the Hlh Across Speciality Collaboration, U.K. COVID-19: Consider cytokine storm syndromes and immunosuppression. *Lancet* **2020**, *395*, 1033–1034. [CrossRef]
3. Zhang, C.; Shi, L.; Wang, F.S. Liver injury in COVID-19: Management and challenges. *Lancet Gastroenterol. Hepatol.* **2020**, *5*, 428–430. [CrossRef]
4. Holick, M.F.; Frommer, J.E.; McNeill, S.C.; Richtand, N.M.; Henley, J.W.; Potts, J.T., Jr. Photometabolism of 7-dehydrocholesterol to previtamin D3 in skin. *Biochem. Biophys. Res. Commun.* **1977**, *76*, 107–114. [CrossRef]
5. Holick, M.F.; Clark, M.B. The photobiogenesis and metabolism of vitamin D. *Fed. Proc.* **1978**, *37*, 2567–2574. [PubMed]
6. Wacker, M.; Holick, M.F. Sunlight and Vitamin D: A global perspective for health. *Dermato-Endocrinol.* **2013**, *5*, 51–108. [CrossRef]
7. Holick, M.F. Vitamin D deficiency. *N. Engl. J. Med.* **2007**, *357*, 266–281. [CrossRef]
8. Tuckey, R.C.; Cheng, C.Y.S.; Slominski, A.T. The serum vitamin D metabolome: What we know and what is still to discover. *J. Steroid Biochem. Mol. Biol.* **2019**, *186*, 4–21. [CrossRef]

9. Bikle, D.D. Vitamin D: Newer Concepts of Its Metabolism and Function at the Basic and Clinical Level. *J. Endocr. Soc.* **2020**, *4*, bvz038. [CrossRef]
10. Jenkinson, C. The vitamin D metabolome: An update on analysis and function. *Cell Biochem. Funct.* **2019**, *37*, 408–423. [CrossRef]
11. Zmijewski, M.A.; Carlberg, C. Vitamin D receptor(s): In the nucleus but also at membranes? *Exp. Dermatol.* **2020**, *29*, 876–884. [CrossRef] [PubMed]
12. Carlberg, C. Vitamin D Genomics: From In Vitro to In Vivo. *Front. Endocrinol.* **2018**, *9*, 250. [CrossRef] [PubMed]
13. Haussler, M.R.; Jurutka, P.W.; Mizwicki, M.; Norman, A.W. Vitamin D receptor (VDR)-mediated actions of 1alpha,25(OH)(2)vitamin D(3): Genomic and non-genomic mechanisms. *Best Pr. Res. Clin. Endocrinol. Metab.* **2011**, *25*, 543–559. [CrossRef] [PubMed]
14. Abhimanyu; Coussens, A.K. The role of UV radiation and vitamin D in the seasonality and outcomes of infectious disease. *Photochem. Photobiol. Sci.* **2017**, *16*, 314–338. [CrossRef] [PubMed]
15. Dankers, W.; Colin, E.M.; van Hamburg, J.P.; Lubberts, E. Vitamin D in Autoimmunity: Molecular Mechanisms and Therapeutic Potential. *Front. Immunol.* **2016**, *7*, 697. [CrossRef] [PubMed]
16. Chun, R.F.; Liu, P.T.; Modlin, R.L.; Adams, J.S.; Hewison, M. Impact of vitamin D on immune function: Lessons learned from genome-wide analysis. *Front. Physiol.* **2014**, *5*, 151. [CrossRef]
17. Bouillon, R.; Marcocci, C.; Carmeliet, G.; Bikle, D.; White, J.H.; Dawson-Hughes, B.; Lips, P.; Munns, C.F.; Lazaretti-Castro, M.; Giustina, A.; et al. Skeletal and Extraskeletal Actions of Vitamin D: Current Evidence and Outstanding Questions. *Endocr. Rev.* **2019**, *40*, 1109–1151. [CrossRef]
18. Bikle, D.D. Vitamin D metabolism, mechanism of action, and clinical applications. *Chem. Biol.* **2014**, *21*, 319–329. [CrossRef]
19. Slominski, R.M.; Raman, C.; Elmets, C.; Jetten, A.M.; Slominski, A.T.; Tuckey, R.C. The significance of CYP11A1 expression in skin physiology and pathology. *Mol. Cell Endocrinol.* **2021**, *530*, 111238. [CrossRef]
20. Slominski, A.T.; Li, W.; Kim, T.K.; Semak, I.; Wang, J.; Zjawiony, J.K.; Tuckey, R.C. Novel activities of CYP11A1 and their potential physiological significance. *J. Steroid Biochem. Mol. Biol.* **2015**, *151*, 25–37. [CrossRef]
21. Holick, M.F.; Smith, E.; Pincus, S. Skin as the site of vitamin D synthesis and target tissue for 1,25-dihydroxyvitamin D3. Use of calcitriol (1,25-dihydroxyvitamin D3) for treatment of psoriasis. *Arch. Dermatol.* **1987**, *123*, 1677–1683a. [CrossRef] [PubMed]
22. Reichrath, J.; Saternus, R.; Vogt, T. Endocrine actions of vitamin D in skin: Relevance for photocarcinogenesis of non-melanoma skin cancer, and beyond. *Mol. Cell Endocrinol.* **2017**, *453*, 96–102. [CrossRef] [PubMed]
23. Bikle, D.D. Vitamin D: An ancient hormone. *Exp. Dermatol.* **2011**, *20*, 7–13. [CrossRef]
24. Slominski, A.T.; Kim, T.K.; Li, W.; Tuckey, R.C. Classical and non-classical metabolic transformation of vitamin D in dermal fibroblasts. *Exp. Dermatol.* **2016**, *25*, 231–232. [CrossRef]
25. Slominski, A.T.; Kim, T.K.; Li, W.; Postlethwaite, A.; Tieu, E.W.; Tang, E.K.Y.; Tuckey, R.C. Detection of novel CYP11A1-derived secosteroids in the human epidermis and serum and pig adrenal gland. *Sci. Rep.* **2015**, *5*, 14875. [CrossRef]
26. Slominski, A.T.; Kim, T.K.; Shehabi, H.Z.; Semak, I.; Tang, E.K.; Nguyen, M.N.; Benson, H.A.; Korik, E.; Janjetovic, Z.; Chen, J.; et al. In vivo evidence for a novel pathway of vitamin D(3) metabolism initiated by P450scc and modified by CYP27B1. *FASEB J.* **2012**, *26*, 3901–3915. [CrossRef]
27. Slominski, A.T.; Chaiprasongsuk, A.; Janjetovic, Z.; Kim, T.K.; Stefan, J.; Slominski, R.M.; Hanumanthu, V.S.; Raman, C.; Qayyum, S.; Song, Y.; et al. Photoprotective properties of vitamin D and lumisterol hydroxyderivatives. *Cell Biochem. Biophys.* **2020**, *78*, 165–180. [CrossRef] [PubMed]
28. Miller, W.L.; Auchus, R.J. The molecular biology, biochemistry, and physiology of human steroidogenesis and its disorders. *Endocr. Rev.* **2011**, *32*, 81–151. [CrossRef] [PubMed]
29. Slominski, R.M.; Tuckey, R.C.; Manna, P.R.; Jetten, A.M.; Postlethwaite, A.; Raman, C.; Slominski, A.T. Extra-adrenal glucocorticoid biosynthesis: Implications for autoimmune and inflammatory disorders. *Genes Immun.* **2020**, *21*, 150–168. [CrossRef]
30. Slominski, A.; Zjawiony, J.; Wortsman, J.; Semak, I.; Stewart, J.; Pisarchik, A.; Sweatman, T.; Marcos, J.; Dunbar, C.; Tuckey, R.C. A novel pathway for sequential transformation of 7-dehydrocholesterol and expression of the P450scc system in mammalian skin. *Europ. J. Biochem.* **2004**, *271*, 4178–4188. [CrossRef]
31. Slominski, A.T.; Janjetovic, Z.; Fuller, B.E.; Zmijewski, M.A.; Tuckey, R.C.; Nguyen, M.N.; Sweatman, T.; Li, W.; Zjawiony, J.; Miller, D.; et al. Products of vitamin D3 or 7-dehydrocholesterol metabolism by cytochrome P450scc show anti-leukemia effects, having low or absent calcemic activity. *PLoS ONE* **2010**, *5*, e9907. [CrossRef] [PubMed]
32. Wang, J.; Slominski, A.; Tuckey, R.C.; Janjetovic, Z.; Kulkarni, A.; Chen, J.; Postlethwaite, A.E.; Miller, D.; Li, W. 20-hydroxyvitamin D inhibits proliferation of cancer cells with high efficacy while being non-toxic. *Anticancer Res.* **2012**, *32*, 739–746. [PubMed]
33. Slominski, A.T.; Kim, T.K.; Janjetovic, Z.; Tuckey, R.C.; Bieniek, R.; Yue, J.; Li, W.; Chen, J.; Nguyen, M.N.; Tang, E.K.; et al. 20-Hydroxyvitamin D2 is a noncalcemic analog of vitamin D with potent antiproliferative and prodifferentiation activities in normal and malignant cells. *Am. J. Physiol. Cell Physiol.* **2011**, *300*, C526–C541. [CrossRef]
34. Chen, J.; Wang, J.; Kim, T.K.; Tieu, E.W.; Tang, E.K.; Lin, Z.; Kovacic, D.; Miller, D.D.; Postlethwaite, A.; Tuckey, R.C.; et al. Novel vitamin D analogs as potential therapeutics: Metabolism, toxicity profiling, and antiproliferative activity. *Anticancer Res.* **2014**, *34*, 2153–2163. [PubMed]
35. Slominski, A.; Janjetovic, Z.; Tuckey, R.C.; Nguyen, M.N.; Bhattacharya, K.G.; Wang, J.; Li, W.; Jiao, Y.; Gu, W.; Brown, M.; et al. 20S-hydroxyvitamin D3, noncalcemic product of CYP11A1 action on vitamin D3, exhibits potent antifibrogenic activity in vivo. *J. Clin. Endocrinol. Metab.* **2013**, *98*, E298–E303. [CrossRef] [PubMed]

36. Postlethwaite, A.E.; Tuckey, R.C.; Kim, T.K.; Li, W.; Bhattacharya, S.K.; Myers, L.K.; Brand, D.D.; Slominski, A.T. 20S-Hydroxyvitamin D3, a secosteroid produced in humans, is anti-inflammatory and inhibits murine autoimmune arthritis. *Front. Immunol.* **2021**, *12*, 678487. [CrossRef]
37. Myers, L.K.; Winstead, M.; Kee, J.D.; Park, J.J.; Zhang, S.; Li, W.; Yi, A.K.; Stuart, J.M.; Rosloniec, E.F.; Brand, D.D.; et al. 1,25-Dihydroxyvitamin D3 and 20-Hydroxyvitamin D3 Upregulate LAIR-1 and Attenuate Collagen Induced Arthritis. *Int. J. Mol. Sci.* **2021**, *22*, 13342. [CrossRef]
38. Skobowiat, C.; Oak, A.S.; Kim, T.K.; Yang, C.H.; Pfeffer, L.M.; Tuckey, R.C.; Slominski, A.T. Noncalcemic 20-hydroxyvitamin D3 inhibits human melanoma growth in in vitro and in vivo models. *Oncotarget* **2017**, *8*, 9823–9834. [CrossRef]
39. Janjetovic, Z.; Postlethwaite, A.; Kang, H.S.; Kim, T.K.; Tuckey, R.C.; Crossman, D.K.; Qayyum, S.; Jetten, A.M.; Slominski, A.T. Antifibrogenic Activities of CYP11A1-derived Vitamin D3-hydroxyderivatives Are Dependent on RORgamma. *Endocrinology* **2021**, *162*, bqaa198. [CrossRef]
40. Kim, T.K.; Atigadda, V.; Brzeminski, P.; Fabisiak, A.; Tang, E.K.Y.; Tuckey, R.C.; Slominski, A.T. Detection of 7-dehydrocholesterol and vitamin D3 derivatives in honey. *Molecules* **2020**, *25*, 2583. [CrossRef]
41. Jenkinson, C.; Desai, R.; Slominski, A.T.; Tuckey, R.C.; Hewison, M.; Handelsman, D.J. Simultaneous measurement of 13 circulating vitamin D3 and D2 mono and dihydroxy metabolites using liquid chromatography mass spectrometry. *Clin. Chem. Lab. Med.* **2021**, *59*, 1642–1652. [CrossRef] [PubMed]
42. Slominski, A.T.; Janjetovic, Z.; Kim, T.K.; Wasilewski, P.; Rosas, S.; Hanna, S.; Sayre, R.M.; Dowdy, J.C.; Li, W.; Tuckey, R.C. Novel non-calcemic secosteroids that are produced by human epidermal keratinocytes protect against solar radiation. *J. Steroid Biochem. Mol. Biol.* **2015**, *148*, 52–63. [CrossRef] [PubMed]
43. Kim, T.K.; Wang, J.; Janjetovic, Z.; Chen, J.; Tuckey, R.C.; Nguyen, M.N.; Tang, E.K.; Miller, D.; Li, W.; Slominski, A.T. Correlation between secosteroid-induced vitamin D receptor activity in melanoma cells and computer-modeled receptor binding strength. *Mol. Cell Endocrinol.* **2012**, *361*, 143–152. [CrossRef] [PubMed]
44. Slominski, A.T.; Kim, T.K.; Hobrath, J.V.; Oak, A.S.W.; Tang, E.K.Y.; Tieu, E.W.; Li, W.; Tuckey, R.C.; Jetten, A.M. Endogenously produced nonclassical vitamin D hydroxy-metabolites act as "biased" agonists on VDR and inverse agonists on RORalpha and RORgamma. *J. Steroid Biochem. Mol. Biol.* **2017**, *173*, 42–56. [CrossRef] [PubMed]
45. Lin, Z.; Chen, H.; Belorusova, A.Y.; Bollinger, J.C.; Tang, E.K.Y.; Janjetovic, Z.; Kim, T.K.; Wu, Z.; Miller, D.D.; Slominski, A.T.; et al. 1alpha,20S-dihydroxyvitamin D3 interacts with vitamin D receptor: Crystal structure and route of chemical synthesis. *Sci. Rep.* **2017**, *7*, 10193. [CrossRef]
46. Lin, Z.; Marepally, S.R.; Goh, E.S.Y.; Cheng, C.Y.S.; Janjetovic, Z.; Kim, T.K.; Miller, D.D.; Postlethwaite, A.E.; Slominski, A.T.; Tuckey, R.C.; et al. Investigation of 20S-hydroxyvitamin D3 analogs and their 1alpha-OH derivatives as potent vitamin D receptor agonists with anti-inflammatory activities. *Sci. Rep.* **2018**, *8*, 1478. [CrossRef]
47. Slominski, A.T.; Kim, T.K.; Janjetovic, Z.; Brozyna, A.A.; Zmijewski, M.A.; Xu, H.; Sutter, T.R.; Tuckey, R.C.; Jetten, A.M.; Crossman, D.K. Differential and overlapping effects of 20,23$(OH)_2$D3 and 1,25$(OH)_2$D3 on gene expression in human epidermal keratinocytes: Identification of AhR as an alternative receptor for 20,23$(OH)_2$D3. *Int. J. Mol. Sci.* **2018**, *19*, 3072. [CrossRef]
48. Song, Y.; Slominski, R.M.; Qayyum, S.; Kim, T.K.; Janjetovic, Z.; Raman, C.; Tuckey, R.C.; Song, Y.; Slominski, A.T. Molecular and structural basis of interactions of vitamin D3 hydroxyderivatives with aryl hydrocarbon receptor (AhR): An integrated experimental and computational study. *Int. J. Biol. Macromol.* **2022**, *209*, 1111–1123. [CrossRef]
49. Brzeminski, P.; Fabisiak, A.; Slominski, R.M.; Kim, T.K.; Janjetovic, Z.; Podgorska, E.; Song, Y.; Saleem, M.; Reddy, S.B.; Qayyum, S.; et al. Chemical synthesis, biological activities and action on nuclear receptors of 20S(OH)D3, 20S,25$(OH)_2$D3, 20S,23S$(OH)_2$D3 and 20S,23R$(OH)_2$D3. *Bioorg. Chem.* **2022**, *121*, 105660. [CrossRef]
50. Slominski, A.T.; Kim, T.K.; Takeda, Y.; Janjetovic, Z.; Brozyna, A.A.; Skobowiat, C.; Wang, J.; Postlethwaite, A.; Li, W.; Tuckey, R.C.; et al. RORalpha and ROR gamma are expressed in human skin and serve as receptors for endogenously produced noncalcemic 20-hydroxy- and 20,23-dihydroxyvitamin D. *FASEB J.* **2014**, *28*, 2775–2789. [CrossRef]
51. Slominski, A.T.; Kim, T.K.; Qayyum, S.; Song, Y.; Janjetovic, Z.; Oak, A.S.W.; Slominski, R.M.; Raman, C.; Stefan, J.; Mier-Aguilar, C.A.; et al. Vitamin D and lumisterol derivatives can act on liver X receptors (LXRs). *Sci. Rep.* **2021**, *11*, 8002. [CrossRef] [PubMed]
52. Chaiprasongsuk, A.; Janjetovic, Z.; Kim, T.K.; Tuckey, R.C.; Li, W.; Raman, C.; Panich, U.; Slominski, A.T. CYP11A1-derived vitamin D3 products protect against UVB-induced inflammation and promote keratinocytes differentiation. *Free Radic. Biol. Med.* **2020**, *155*, 87–98. [CrossRef] [PubMed]
53. Chaiprasongsuk, A.; Janjetovic, Z.; Kim, T.K.; Jarrett, S.G.; D'Orazio, J.A.; Holick, M.F.; Tang, E.K.Y.; Tuckey, R.C.; Panich, U.; Li, W.; et al. Protective effects of novel derivatives of vitamin D3 and lumisterol against UVB-induced damage in human keratinocytes involve activation of Nrf2 and p53 defense mechanisms. *Redox Biol.* **2019**, *24*, 101206. [CrossRef] [PubMed]
54. Slominski, A.T.; Kim, T.K.; Li, W.; Yi, A.K.; Postlethwaite, A.; Tuckey, R.C. The role of CYP11A1 in the production of vitamin D metabolites and their role in the regulation of epidermal functions. *J. Steroid Biochem. Mol. Biol.* **2014**, *144PA*, 28–39. [CrossRef] [PubMed]
55. Janjetovic, Z.; Tuckey, R.C.; Nguyen, M.N.; Thorpe, E.M., Jr.; Slominski, A.T. 20,23-dihydroxyvitamin D3, novel P450scc product, stimulates differentiation and inhibits proliferation and NF-kappaB activity in human keratinocytes. *J. Cell Physiol.* **2010**, *223*, 36–48. [PubMed]

56. Janjetovic, Z.; Zmijewski, M.A.; Tuckey, R.C.; DeLeon, D.A.; Nguyen, M.N.; Pfeffer, L.M.; Slominski, A.T. 20-Hydroxycholecalciferol, product of vitamin D3 hydroxylation by P450scc, decreases NF-kappaB activity by increasing IkappaB alpha levels in human keratinocytes. *PLoS ONE* **2009**, *4*, e5988. [CrossRef]
57. Slominski, A.T.; Kim, T.K.; Chen, J.; Nguyen, M.N.; Li, W.; Yates, C.R.; Sweatman, T.; Janjetovic, Z.; Tuckey, R.C. Cytochrome P450scc-dependent metabolism of 7-dehydrocholesterol in placenta and epidermal keratinocytes. *Int. J. Biochem. Cell Biol.* **2012**, *44*, 2003–2018. [CrossRef]
58. Guryev, O.; Carvalho, R.A.; Usanov, S.; Gilep, A.; Estabrook, R.W. A pathway for the metabolism of vitamin D3: Unique hydroxylated metabolites formed during catalysis with cytochrome P450scc (CYP11A1). *Proc. Natl. Acad. Sci. USA* **2003**, *100*, 14754–14759. [CrossRef]
59. Zmijewski, M.A.; Li, W.; Zjawiony, J.K.; Sweatman, T.W.; Chen, J.; Miller, D.D.; Slominski, A.T. Photo-conversion of two epimers (20R and 20S) of pregna-5,7-diene-3beta, 17alpha, 20-triol and their bioactivity in melanoma cells. *Steroids* **2009**, *74*, 218–228. [CrossRef]
60. Zmijewski, M.A.; Li, W.; Chen, J.; Kim, T.K.; Zjawiony, J.K.; Sweatman, T.W.; Miller, D.D.; Slominski, A.T. Synthesis and photochemical transformation of 3beta,21-dihydroxypregna-5,7-dien-20-one to novel secosteroids that show anti-melanoma activity. *Steroids* **2011**, *76*, 193–203. [CrossRef]
61. Slominski, A.; Kim, T.K.; Zmijewski, M.A.; Janjetovic, Z.; Li, W.; Chen, J.; Kusniatsova, E.I.; Semak, I.; Postlethwaite, A.; Miller, D.D.; et al. Novel vitamin D photoproducts and their precursors in the skin. *Dermato-Endocrinology* **2013**, *5*, 7–19. [CrossRef] [PubMed]
62. Zmijewski, M.A.; Li, W.; Zjawiony, J.K.; Sweatman, T.W.; Chen, J.; Miller, D.D.; Slominski, A.T. Synthesis and photo-conversion of androsta- and pregna-5,7-dienes to vitamin D3-like derivatives. *Photochem. Photobiol. Sci.* **2008**, *7*, 1570–1576. [CrossRef] [PubMed]
63. Slominski, A.T.; Kim, T.K.; Slominski, R.M.; Song, Y.; Janjetovic, Z.; Podgorska, E.; Reddy, S.B.; Song, Y.; Raman, C.; Tang, E.K.Y.; et al. Metabolic activation of tachysterol3 to biologically active hydroxyderivatives that act on VDR, AhR, LXRs, and PPARgamma receptors. *FASEB J.* **2022**, *36*, e22451. [CrossRef] [PubMed]
64. Li, W.; Chen, J.; Janjetovic, Z.; Kim, T.K.; Sweatman, T.; Lu, Y.; Zjawiony, J.; Tuckey, R.C.; Miller, D.; Slominski, A. Chemical synthesis of 20S-hydroxyvitamin D3, which shows antiproliferative activity. *Steroids* **2010**, *75*, 926–935. [CrossRef]
65. Tuckey, R.C.; Slominski, A.T.; Cheng, C.Y.; Chen, J.; Kim, T.K.; Xiao, M.; Li, W. Lumisterol is metabolized by CYP11A1: Discovery of a new pathway. *Int. J. Biochem. Cell Biol.* **2014**, *55*, 24–34. [CrossRef]
66. Slominski, A.T.; Kim, T.K.; Hobrath, J.V.; Janjetovic, Z.; Oak, A.S.W.; Postlethwaite, A.; Lin, Z.; Li, W.; Takeda, Y.; Jetten, A.M.; et al. Characterization of a new pathway that activates lumisterol in vivo to biologically active hydroxylumisterols. *Sci. Rep.* **2017**, *7*, 11434. [CrossRef]
67. Tuckey, R.C.; Li, W.; Ma, D.; Cheng, C.Y.S.; Wang, K.M.; Kim, T.K.; Jeayeng, S.; Slominski, A.T. CYP27A1 acts on the pre-vitamin D3 photoproduct, lumisterol, producing biologically active hydroxy-metabolites. *J. Steroid Biochem. Mol. Biol.* **2018**, *181*, 1–10. [CrossRef]
68. Chaiprasongsuk, A.; Janjetovic, Z.; Kim, T.K.; Schwartz, C.J.; Tuckey, R.C.; Tang, E.K.Y.; Raman, C.; Panich, U.; Slominski, A.T. Hydroxylumisterols, photoproducts of pre-vitamin D3, protect human keratinocytes against UVB-induced damage. *Int. J. Mol. Sci.* **2020**, *21*, 9374. [CrossRef]
69. Li, W.; Moore, M.J.; Vasilieva, N.; Sui, J.; Wong, S.K.; Berne, M.A.; Somasundaran, M.; Sullivan, J.L.; Luzuriaga, K.; Greenough, T.C.; et al. Angiotensin-converting enzyme 2 is a functional receptor for the SARS coronavirus. *Nature* **2003**, *426*, 450–454. [CrossRef]
70. Matsuyama, S.; Nagata, N.; Shirato, K.; Kawase, M.; Takeda, M.; Taguchi, F. Efficient activation of the severe acute respiratory syndrome coronavirus spike protein by the transmembrane protease TMPRSS2. *J. Virol.* **2010**, *84*, 12658–12664. [CrossRef]
71. Glowacka, I.; Bertram, S.; Müller, M.A.; Allen, P.; Soilleux, E.; Pfefferle, S.; Steffen, I.; Tsegaye, T.S.; He, Y.; Gnirss, K.; et al. Evidence that TMPRSS2 activates the severe acute respiratory syndrome coronavirus spike protein for membrane fusion and reduces viral control by the humoral immune response. *J. Virol.* **2011**, *85*, 4122–4134. [CrossRef]
72. Shulla, A.; Heald-Sargent, T.; Subramanya, G.; Zhao, J.; Perlman, S.; Gallagher, T. A transmembrane serine protease is linked to the severe acute respiratory syndrome coronavirus receptor and activates virus entry. *J. Virol.* **2011**, *85*, 873–882. [CrossRef] [PubMed]
73. Hasegawa, H. Development of Corona-virus-disease-19 Vaccines. *JMA J.* **2021**, *4*, 187–190. [CrossRef] [PubMed]
74. Qayyum, S.; Mohammad, T.; Slominski, R.M.; Hassan, M.I.; Tuckey, R.C.; Raman, C.; Slominski, A.T. Vitamin D and lumisterol novel metabolites can inhibit SARS-CoV-2 replication machinery enzymes. *Am. J. Physiol. Endocrinol. Metab.* **2021**, *321*, E246–E251. [CrossRef] [PubMed]
75. Oristrell, J.; Oliva, J.C.; Subirana, I.; Casado, E.; Domínguez, D.; Toloba, A.; Aguilera, P.; Esplugues, J.; Fafián, P.; Grau, M. Association of Calcitriol Supplementation with Reduced COVID-19 Mortality in Patients with Chronic Kidney Disease: A Population-Based Study. *Biomedicines* **2021**, *9*, 509. [CrossRef] [PubMed]
76. Jin, Z.; Du, X.; Xu, Y.; Deng, Y.; Liu, M.; Zhao, Y.; Zhang, B.; Li, X.; Zhang, L.; Peng, C.; et al. Structure of M(pro) from SARS-CoV-2 and discovery of its inhibitors. *Nature* **2020**, *582*, 289–293. [CrossRef]
77. Amin, S.A.; Banerjee, S.; Singh, S.; Qureshi, I.A.; Gayen, S.; Jha, T. First structure-activity relationship analysis of SARS-CoV-2 virus main protease (Mpro) inhibitors: An endeavor on COVID-19 drug discovery. *Mol. Divers.* **2021**, *25*, 1827–1838. [CrossRef]

78. Yang, J.; Lin, X.; Xing, N.; Zhang, Z.; Zhang, H.; Wu, H.; Xue, W. Structure-Based Discovery of Novel Nonpeptide Inhibitors Targeting SARS-CoV-2 M(pro). *J. Chem. Inf. Model.* **2021**, *61*, 3917–3926. [CrossRef]
79. Mishra, A.; Rathore, A.S. RNA dependent RNA polymerase (RdRp) as a drug target for SARS-CoV2. *J. Biomol. Struct. Dyn.* **2021**, *40*, 6039–6051. [CrossRef]
80. Pugin, J.; Ricou, B.; Steinberg, K.P.; Suter, P.M.; Martin, T.R. Proinflammatory activity in bronchoalveolar lavage fluids from patients with ARDS, a prominent role for interleukin-1. *Am. J. Respir. Crit. Care Med.* **1996**, *153*, 1850–1856. [CrossRef]
81. Tisoncik, J.R.; Korth, M.J.; Simmons, C.P.; Farrar, J.; Martin, T.R.; Katze, M.G. Into the eye of the cytokine storm. *Microbiol. Mol. Biol. Rev.* **2012**, *76*, 16–32. [CrossRef] [PubMed]
82. Ragab, D.; Salah Eldin, H.; Taeimah, M.; Khattab, R.; Salem, R. The COVID-19 Cytokine Storm; What We Know So Far. *Front. Immunol.* **2020**, *11*, 1446. [CrossRef] [PubMed]
83. Schieber, M.; Chandel, N.S. ROS function in redox signaling and oxidative stress. *Curr. Biol.* **2014**, *24*, R453–R462. [CrossRef]
84. Ye, S.; Lowther, S.; Stambas, J. Inhibition of reactive oxygen species production ameliorates inflammation induced by influenza A viruses via upregulation of SOCS1 and SOCS3. *J. Virol.* **2015**, *89*, 2672–2683. [CrossRef] [PubMed]
85. Vlahos, R.; Stambas, J.; Bozinovski, S.; Broughton, B.R.; Drummond, G.R.; Selemidis, S. Inhibition of Nox2 oxidase activity ameliorates influenza A virus-induced lung inflammation. *PLoS Pathog.* **2011**, *7*, e1001271. [CrossRef]
86. Slominski, R.M.; Stefan, J.; Athar, M.; Holick, M.F.; Jetten, A.M.; Raman, C.; Slominski, A.T. COVID-19 and Vitamin D: A lesson from the skin. *Exp. Dermatol.* **2020**, *29*, 885–890. [CrossRef] [PubMed]
87. Herengt, A.; Thyrsted, J.; Holm, C.K. NRF2 in Viral Infection. *Antioxidants* **2021**, *10*, 1491. [CrossRef]
88. Podgorska, E.; Sniegocka, M.; Mycinska, M.; Trybus, W.; Trybus, E.; Kopacz-Bednarska, A.; Wiechec, O.; Krzykawska-Serda, M.; Elas, M.; Krol, T.; et al. Acute hepatologic and nephrologic effects of calcitriol in Syrian golden hamster (*Mesocricetus auratus*). *Acta Biochim. Pol.* **2018**, *65*, 351–358. [CrossRef]
89. Slominski, A.T.; Slominski, R.M.; Goepfert, P.A.; Kim, T.K.; Holick, M.F.; Jetten, A.M.; Raman, C. Reply to Jakovac and to Rocha et al.: Can vitamin D prevent or manage COVID-19 illness? *Am. J. Physiol. Endocrinol. Metab.* **2020**, *319*, E455–E457. [CrossRef]
90. Hoffmann, M.; Kleine-Weber, H.; Schroeder, S.; Kruger, N.; Herrler, T.; Erichsen, S.; Schiergens, T.S.; Herrler, G.; Wu, N.H.; Nitsche, A.; et al. SARS-CoV-2 Cell Entry Depends on ACE2 and TMPRSS2 and Is Blocked by a Clinically Proven Protease Inhibitor. *Cell* **2020**, *181*, 271–280 e278. [CrossRef]
91. Perrotta, F.; Matera, M.G.; Cazzola, M.; Bianco, A. Severe respiratory SARS-CoV2 infection: Does ACE2 receptor matter? *Respir. Med.* **2020**, *168*, 105996. [CrossRef] [PubMed]
92. Shang, J.; Wan, Y.; Luo, C.; Ye, G.; Geng, Q.; Auerbach, A.; Li, F. Cell entry mechanisms of SARS-CoV-2. *Proc. Natl. Acad. Sci. USA* **2020**, *117*, 11727–11734. [CrossRef] [PubMed]
93. Song, Y.; Qayyum, S.; Greer, R.A.; Slominski, R.M.; Raman, C.; Slominski, A.T.; Song, Y. Vitamin D3 and its hydroxyderivatives as promising drugs against COVID-19: A computational study. *J. Biomol. Struct. Dyn.* **2021**, *17*, 11727–11734. [CrossRef] [PubMed]
94. Singh, A.; Dhar, R. A large-scale computational screen identifies strong potential inhibitors for disrupting SARS-CoV-2 S-protein and human ACE2 interaction. *J. Biomol. Struct. Dyn.* **2021**, *17*, 1–14. [CrossRef] [PubMed]
95. Shoemark, D.K.; Colenso, C.K.; Toelzer, C.; Gupta, K.; Sessions, R.B.; Davidson, A.D.; Berger, I.; Schaffitzel, C.; Spencer, J.; Mulholland, A.J. Molecular simulations suggest vitamins, retinoids and steroids as ligands of the free fatty acid pocket of the SARS-CoV-2 spike protein. *Angew. Chem. Int. Ed.* **2021**, *60*, 7098–7110. [CrossRef]
96. Mansouri, A.; Kowsar, R.; Zakariazadeh, M.; Hakimi, H.; Miyamoto, A. The impact of calcitriol and estradiol on the SARS-CoV-2 biological activity: A molecular modeling approach. *Sci. Rep.* **2022**, *12*, 717. [CrossRef]
97. Fidan, O.; Mujwar, S.; Kciuk, M. Discovery of adapalene and dihydrotachysterol as antiviral agents for the Omicron variant of SARS-CoV-2 through computational drug repurposing. *Mol. Divers.* **2022**, 1–3. [CrossRef]
98. Shalayel, M.H.; Al-Mazaideh, G.M.; Aladaileh, S.H.; Al-Swailmi, F.K.; Al-Thiabat, M.G. Vitamin D is a potential inhibitor of COVID-19: In silico molecular docking to the binding site of SARS-CoV-2 endoribonuclease Nsp15. *Pak. J. Pharm. Sci.* **2020**, *33*, 2179–2186. [CrossRef]
99. Dai, W.; Zhang, B.; Jiang, X.M.; Su, H.; Li, J.; Zhao, Y.; Xie, X.; Jin, Z.; Peng, J.; Liu, F.; et al. Structure-based design of antiviral drug candidates targeting the SARS-CoV-2 main protease. *Science* **2020**, *368*, 1331–1335. [CrossRef]
100. Lung, J.; Lin, Y.S.; Yang, Y.H.; Chou, Y.L.; Shu, L.H.; Cheng, Y.C.; Liu, H.T.; Wu, C.Y. The potential chemical structure of anti-SARS-CoV-2 RNA-dependent RNA polymerase. *J. Med Virol.* **2020**, *92*, 693–697. [CrossRef]
101. Yin, W.; Mao, C.; Luan, X.; Shen, D.D.; Shen, Q.; Su, H.; Wang, X.; Zhou, F.; Zhao, W.; Gao, M.; et al. Structural basis for inhibition of the RNA-dependent RNA polymerase from SARS-CoV-2 by remdesivir. *Science* **2020**, *368*, 1499–1504. [CrossRef] [PubMed]
102. Zu, S.; Deng, Y.-Q.; Zhou, C.; Li, J.; Li, L.; Chen, Q.; Li, X.-F.; Zhao, H.; Gold, S.; He, J.; et al. 25-Hydroxycholesterol is a potent SARS-CoV-2 inhibitor. *Cell Res.* **2020**, *30*, 1043–1045. [CrossRef] [PubMed]
103. Wang, S.; Li, W.; Hui, H.; Tiwari, S.K.; Zhang, Q.; Croker, B.A.; Rawlings, S.; Smith, D.; Carlin, A.F.; Rana, T.M. Cholesterol 25-Hydroxylase inhibits SARS-CoV-2 and other coronaviruses by depleting membrane cholesterol. *EMBO J.* **2020**, *39*, e106057. [CrossRef] [PubMed]
104. Wnętrzak, A.; Chachaj-Brekiesz, A.; Kuś, K.; Lipiec, E.; Dynarowicz-Latka, P. Oxysterols can act antiviral through modification of lipid membrane properties–The Langmuir monolayer study. *J. Steroid Biochem. Mol. Biol.* **2022**, *220*, 106092. [CrossRef] [PubMed]

105. Dror, A.A.; Morozov, N.; Daoud, A.; Namir, Y.; Yakir, O.; Shachar, Y.; Lifshitz, M.; Segal, E.; Fisher, L.; Mizrachi, M.; et al. Pre-infection 25-hydroxyvitamin D3 levels and association with severity of COVID-19 illness. *PLoS ONE* **2022**, *17*, e0263069. [CrossRef]
106. Takahashi, H.; Nomura, H.; Iriki, H.; Kubo, A.; Isami, K.; Mikami, Y.; Mukai, M.; Sasaki, T.; Yamagami, J.; Kudoh, J.; et al. Cholesterol 25-hydroxylase is a metabolic switch to constrain T cell-mediated inflammation in the skin. *Sci. Immunol.* **2021**, *6*, eabb6444. [CrossRef]
107. Chiodini, I.; Gatti, D.; Soranna, D.; Merlotti, D.; Mingiano, C.; Fassio, A.; Adami, G.; Falchetti, A.; Eller-Vainicher, C.; Rossini, M.; et al. Vitamin D Status and SARS-CoV-2 Infection and COVID-19 Clinical Outcomes. *Front. Public Health* **2021**, *9*, 736665. [CrossRef]
108. Kaya, M.O.; Pamukçu, E.; Yakar, B. The role of vitamin D deficiency on COVID-19: A systematic review and meta-analysis of observational studies. *Epidemiol. Health* **2021**, *43*, e2021074. [CrossRef]
109. Leaf, D.E.; Ginde, A.A. Vitamin D3 to Treat COVID-19: Different Disease, Same Answer. *JAMA* **2021**, *325*, 1047–1048. [CrossRef]
110. Kazemi, A.; Mohammadi, V.; Aghababaee, S.K.; Golzarand, M.; Clark, C.C.T.; Babajafari, S. Association of Vitamin D Status with SARS-CoV-2 Infection or COVID-19 Severity: A Systematic Review and Meta-analysis. *Adv. Nutr.* **2021**, *12*, 1636–1658. [CrossRef]
111. Bae, J.H.; Choe, H.J.; Holick, M.F.; Lim, S. Association of vitamin D status with COVID-19 and its severity: Vitamin D and COVID-19: A narrative review. *Rev. Endocr. Metab. Disord.* **2022**, *23*, 579–599. [CrossRef] [PubMed]
112. Rastogi, A.; Bhansali, A.; Khare, N.; Suri, V.; Yaddanapudi, N.; Sachdeva, N.; Puri, G.D.; Malhotra, P. Short term, high-dose vitamin D supplementation for COVID-19 disease: A randomised, placebo-controlled, study (SHADE study). *Postgrad. Med. J.* **2022**, *98*, 87–90. [CrossRef] [PubMed]
113. Martineau, A.R.; Forouhi, N.G. Vitamin D for COVID-19: A case to answer? *Lancet Diabetes Endocrinol.* **2020**, *8*, 735–736. [CrossRef]
114. Lau, F.H.; Majumder, R.; Torabi, R.; Saeg, F.; Hoffman, R.; Cirillo, J.D.; Greiffenstein, P. Vitamin D insufficiency is prevalent in severe COVID-19. *medRxiv* **2020**. [CrossRef]
115. Luo, X.; Liao, Q.; Shen, Y.; Li, H.; Cheng, L. Vitamin D Deficiency Is Associated with COVID-19 Incidence and Disease Severity in Chinese People. *J. Nutr.* **2021**, *151*, 98–103. [CrossRef]
116. Chauss, D.; Freiwald, T.; McGregor, R.; Yan, B.; Wang, L.; Nova-Lamperti, E.; Kumar, D.; Zhang, Z.; Teague, H.; West, E.E.; et al. Autocrine vitamin D signaling switches off pro-inflammatory programs of TH1 cells. *Nat. Immunol.* **2022**, *23*, 62–74. [CrossRef]
117. Quesada-Gomez, J.M.; Lopez-Miranda, J.; Entrenas-Castillo, M.; Casado-Díaz, A.; Nogues, Y.S.X.; Mansur, J.L.; Bouillon, R. Vitamin D Endocrine System and COVID-19: Treatment with Calcifediol. *Nutrients* **2022**, *14*, 2716. [CrossRef]

MDPI

Article

Vitamin D and the Ability to Produce 1,25(OH)$_2$D Are Critical for Protection from Viral Infection of the Lungs

Juhi Arora [1], Devanshi R. Patel [1,†], McKayla J. Nicol [1,†], Cassandra J. Field [1,†], Katherine H. Restori [1], Jinpeng Wang [1], Nicole E. Froelich [1], Bhuvana Katkere [1], Josey A. Terwilliger [1], Veronika Weaver [1], Erin Luley [2], Kathleen Kelly [2], Girish S. Kirimanjeswara [1], Troy C. Sutton [1,*] and Margherita T. Cantorna [1,*]

[1] Department of Veterinary and Biomedical Sciences, The Pennsylvania State University, University Park, PA 16802, USA; jua268@psu.edu (J.A.); drp5323@psu.edu (D.R.P.); mjn5181@psu.edu (M.J.N.); cif5202@psu.edu (C.J.F.); khr114@psu.edu (K.H.R.); jinpeng.wong1990@gmail.com (J.W.); nef5148@psu.edu (N.E.F.); bxk33@psu.edu (B.K.); josey.terwilliger@gmail.com (J.A.T.); vcw100@psu.edu (V.W.); gsk125@psu.edu (G.S.K.)

[2] Animal Diagnostic Laboratory, The Pennsylvania State University, University Park, PA 16802, USA; ehl5008@psu.edu (E.L.); kmk6898@psu.edu (K.K.)

* Correspondence: tcs38@psu.edu (T.C.S.); mxc69@psu.edu (M.T.C.)

† These authors contributed equally to this work.

Abstract: Vitamin D supplementation is linked to improved outcomes from respiratory virus infection, and the COVID-19 pandemic renewed interest in understanding the potential role of vitamin D in protecting the lung from viral infections. Therefore, we evaluated the role of vitamin D using animal models of pandemic H1N1 influenza and severe acute respiratory syndrome coronavirus-2 (SARS-CoV-2) infection. In mice, dietary-induced vitamin D deficiency resulted in lung inflammation that was present prior to infection. Vitamin D sufficient (D+) and deficient (D−) wildtype (WT) and D+ and D− Cyp27B1 (Cyp) knockout (KO, cannot produce 1,25(OH)$_2$D) mice were infected with pandemic H1N1. D− WT, D+ Cyp KO, and D− Cyp KO mice all exhibited significantly reduced survival compared to D+ WT mice. Importantly, survival was not the result of reduced viral replication, as influenza M gene expression in the lungs was similar for all animals. Based on these findings, additional experiments were performed using the mouse and hamster models of SARS-CoV-2 infection. In these studies, high dose vitamin D supplementation reduced lung inflammation in mice but not hamsters. A trend to faster weight recovery was observed in 1,25(OH)$_2$D treated mice that survived SARS-CoV-2 infection. There was no effect of vitamin D on SARS-CoV-2 N gene expression in the lung of either mice or hamsters. Therefore, vitamin D deficiency enhanced disease severity, while vitamin D sufficiency/supplementation reduced inflammation following infections with H1N1 influenza and SARS-CoV-2.

Keywords: vitamin D; influenza; SARS-CoV-2; lung; inflammation

Citation: Arora, J.; Patel, D.R.; Nicol, M.J.; Field, C.J.; Restori, K.H.; Wang, J.; Froelich, N.E.; Katkere, B.; Terwilliger, J.A.; Weaver, V.; et al. Vitamin D and the Ability to Produce 1,25(OH)$_2$D Are Critical for Protection from Viral Infection of the Lungs. *Nutrients* **2022**, *14*, 3061. https://doi.org/10.3390/nu14153061

Academic Editor: Carsten Carlberg

Received: 29 June 2022
Accepted: 20 July 2022
Published: 26 July 2022

1. Introduction

Low vitamin D status is associated with poorer outcomes following acute respiratory diseases including influenza [1]. Vitamin D supplements are touted as being useful in high doses for reducing the severity of seasonal influenza [2–4]. The recent emergence of severe acute respiratory syndrome (SARS)-coronavirus (CoV)-2 and the ongoing pandemic led to a renewed interest in high-dose vitamin D supplements to prevent and treat severe SARS-CoV-2 disease (i.e., COVID-19) [4]. Infection with SARS-CoV-2 results in local and systemic inflammation that when controlled may enhance survival and clinical outcomes of COVID-19 [5]. Severe respiratory illness can also be caused by influenza viruses or co-infection with influenza and coronaviruses. An association between low vitamin D status and severe COVID-19 was postulated, and accordingly, vitamin D supplementation was proposed to be beneficial to treat COVID-19 [6]. A recent systemic review concluded that low circulating levels of vitamin D (serum 25(OH)D, 25D) were associated with more severe symptoms and higher mortality in patients with COVID-19 [7]. Interventions that control the lung inflammatory response to viruses have the potential to benefit the global population.

Vitamin D, and the active form of vitamin D (1,25$(OH)_2$D, 1,25D), has been implicated to play a role in the anti-viral response; however, this effect may be specific to different viruses. For example, 1,25D treatment of T cells from human immunodeficiency virus (HIV)-infected individuals in vitro resulted in a decrease in viral RNA transcription by a direct reduction in NF-κB, which reactivates proviral HIV [8]. Conversely, 1,25D treatment of respiratory syncytial virus (RSV)-infected human epithelial cells in vitro did not affect viral replication [9]. Production of antimicrobial peptides such as cathelicidin, β-defensin etc. and production of cytokines such as TNFα, IL-5, IL-1β, IL-6, IL-10, and type I interferons at the mucosal surface are important parts of the innate immune response against viruses [10,11]. Several studies show that 1,25D and other vitamin D analogs induce cathelicidin production in response to virus infection [12]. Cathelicidin LL-37 was shown to bind and kill viruses including influenza viruses in vitro [13–17]. Therefore, it is possible that vitamin D through the induction of cathelicidin could directly target SARS-CoV-2 and influenza. 1,25D also limits inflammatory responses by decreasing IFNγ, IL-6, and TNFα: however, the effects of 1,25D to reduce inflammation could be detrimental for the ability of the host to clear some viruses [18,19].

Lung epithelial cells are vitamin D targets since they express the vitamin D receptor (VDR) and are regulated by 1,25D treatments. In mice, 1,25D treatments reduced inflammation following lipopolysaccharide-induced lung injury and regulated angiotensin converting enzyme (ACE) expression in the rat lung epithelium [20]. In mice, infection with an influenza H9N2 virus induced mRNA for the VDR in the lung and 1,25D-treated animals had reduced lung inflammation [21]. Treating mice with the 1,25D precursor, 25hydroxyvitamin D (25D), also protected mice from subsequent H1N1 influenza infection [22]. 1,25D treatment of human bronchial epithelial cells suppressed IL-6 and protected the cells from oxidative damage [23]. VDR knockout (KO) mice had reduced expression of tight junction proteins such as zonula occludens-1, occludin, and claudins (2,4, and 12), suggesting an important role for the VDR in maintaining the integrity of the lung [24]. Vitamin D has direct effects on the lung epithelium, and 1,25D suppresses inflammation in the lung.

Extensive association studies in humans led to the proposal that high dose vitamin D supplements could be beneficial for protection from severe influenza and SARS-CoV-2 infections. Since cause and effect are extremely difficult to determine in human studies, we sought to evaluate the effects of vitamin D on the lung anti-viral response in animal models. The data from mice and hamsters suggest that vitamin D supplementation reduces inflammation in the lung following pandemic H1N1 and SARS-CoV-2 infection. D− mice had lung inflammation even without infection. We showed previously that feeding D− diets to mice that cannot produce 1,25D (Cyp KO) resulted in severe vitamin D deficiency [25]. Influenza disease was greatest in D− Cyp KO, and the least amount of disease was in D+ WT mice. The survival of D+ Cyp KO mice was significantly less than D+ WT mice following an influenza infection. Vitamin D supplementation reduced lung inflammation and *Ifnb* expression in the lung of mice following SARS-CoV-2 infection. Vitamin D treatments had no effect on the expression of viral RNA for either SARS-CoV-2 or H1N1 influenza in the lungs of hamsters or mice. Instead, the data support an important role for vitamin D and 1,25D in controlling the host inflammatory response to viruses in the lungs.

2. Materials and Methods

2.1. Animal Models

C57BL/6 WT, K18hACE2 (hACE, Jackson Laboratories, Bar Harbor, ME, USA), and Cyp KO (gift from Dr. Hector DeLuca, University of Wisconsin, Madison, WI, USA) mice were bred (mice) and housed (mice and hamsters) according to approved IACUC protocols at the Pennsylvania State University (University Park, PA, USA). For the experiments, age-

and sex-matched mice were fed: chow diets (D+) (Lab diets #5053, Arden Hills, MN, USA) or purified diets with (D+) and without (D−) vitamin D (Envigo, T.D. 89123, Madison, WI, USA). For some experiments, D+ mice were fed corn oil alone or corn oil with 1,25D. For some experiments, D− mice were fed corn oil alone or corn oil with one of two doses of vitamin D3 (Sigma-Aldrich, C9756, St. Louis, MO, USA). Age- and sex-matched Golden Syrian Hamsters were purchased from Envigo (Indianapolis, IN, USA) and maintained on the chow (D+) or D− diet (Envigo, T.D.120008) and orally fed corn oil or corn oil with vitamin D3. Serum was collected to monitor the vitamin D status of mice and hamsters.

2.2. Serum 25 Hydroxy Vitamin D (25D) Measurements

Serum 25D levels were measured using an ELISA kit and standards as per the manufacturer's instructions (25-OH D, Eagle Biosciences, Amherst, NH, USA). The limits of detection were 1.6 ng/mL 25D.

2.3. SARS-CoV-2 Infection

SARS-CoV-2 USA-WA-1/2020 (Centers for Disease Control and Prevention, BEI Resources, NIAID, NIH: NR-52281) was used for infecting both mice and hamsters. hACE2 mice (*n* = 7–8/group) were anesthetized using isofluorane and infected with 100−1000 $TCID_{50}$ units of SARS-CoV-2 in 50 µL phosphate buffered saline. Hamsters were sedated with ketamine (150 mg/kg), atropine (0.015 mg/kg), and xylazine (7.5 mg/kg) via intraperitoneal injection and intranasally inoculated with 10,000 $TCID_{50}$ units of SARS-CoV-2 in 100 µL Dulbecco's Modified Eagle Media. Hamsters were given atipamezole (1 mg/kg) subcutaneously to reverse the sedation. Equal numbers of males and females were used for all experiments, and body weights and symptoms were monitored daily until the endpoint criteria were reached or day 14 post-infection.

One series of experiments in hACE2 mice used standard D+ rodent chow (Lab diets #5053, Arden Hills, MN, USA) diets with oral dosing of corn oil or 10 ng/day of 1,25D diluted in 10 µL of corn oil beginning the day before infection and continuing until sacrifice. Additional experiments used mice or hamsters fed D− diets with or without oral vitamin D3 (*n* = 8–16/group). D− hACE2 mice were dosed orally with corn oil (D−), 0.125 µg vitamin D3/day (D+), or 2.5 µg vitamin D3/day (D++) beginning 8 weeks prior to infection and continuing throughout the experiment (*n* = 12–18/group). Hamsters were placed on D− diets with corn oil or 8 µg vitamin D3 (D+) in corn oil/day starting 11 days before infection and continuing throughout the experiment (*n* = 5–6/group).

2.4. H1N1 Infection

D+ and D− WT and D+ and D− Cyp KO littermates (*n* = 12–18/group) were fed identical diets with and without vitamin D. Mice were anesthetized with isofluorane to inoculate them intranasally with 10–30 $TCID_{50}$ mouse-adapted A/H1N1/California/04/2009 influenza in 50 µL Gibco's reduced-serum minimum essential media (ThermoFisher, 31985-070, Waltham, MA, USA) [26]. Body weight and clinical signs were monitored daily until they either reached end-point criteria or at day 14 post-infection.

2.5. Biocontainment and Animal Care and Use

All studies with SARS-CoV-2 were conducted in a biosafety level 3 enhanced (BSL3+) laboratory. This facility is approved for BSL3+ respiratory pathogen studies by the U.S. Department of Agriculture and Centers for Disease Control. Studies with pandemic H1N1 influenza were conducted under biosafety level 2 enhanced conditions. All animal studies were conducted in compliance with the Animal Care and Use Committee under protocol numbers: 202001693, 202001516, 202001440, and 202001638.

2.6. RNA Isolation and Quantitative PCR

Tissues were homogenized in TRIzol reagent (Sigma-Aldrich, St. Louis, MO, USA). Total RNA was extracted from the tissues using chloroform-isopropanol precipitation and quantified using NanoDrop (ThermoFisher, Waltham, MA, USA). A total of 1–2 µg RNA was reverse transcribed using AMV Reverse Transcriptase (Promega, Madison, WI, USA). Primers were purchased from Integrated DNA Technologies (IDT, Coralville, IA, USA) and are listed in Supplementary Table S1. Gene expression was quantitated using the SYBR green mix (Azura Genomics, Raynham, MA, USA) and StepOne Plus system (Applied Biosystems, Carlsbad, CA, USA). Primers for the SARS-CoV-2 N gene were a commercial kit from IDT (Cat #10006713). Gene expression was calculated using the delta-delta C_T method using GAPDH and uninfected control tissues. Gene expression was normalized to day 0 uninfected controls.

2.7. Histology

Lung tissues were collected from hamsters and mice and fixed in 10% formalin. These tissues were embedded in paraffin, sectioned, and H & E stained by the PSU Animal Diagnostic Lab Histology lab. Scoring of lung pathology associated with SARS-CoV-2 infection was conducted by veterinary pathologists certified by the American College of Veterinary Pathologists that were blinded to vitamin D3 treatment status. The criteria for the histopathologic evaluation of the tissue sections were conducted as described for H1N1 infected mouse lungs (n = 2–5/group) [27–29], SARS-CoV-2 infected mouse lung sections (n = 4–6/group) [30,31], and hamster lung sections (n = 4/group) [32–34]. The individual parameters evaluated the extent of pneumonia, damage to alveoli, and lymphocytic infiltration by various specific parameters described in Tables S2–S4.

2.8. Statistical Analysis

Results are represented as the mean ± SEM. Statistical analysis was performed using Prism ver. 10 (GraphPad, San Diego, CA, USA) using multiple t-tests, one-way ANOVA, two-way ANOVA, mixed-effects analysis with Bonferroni multiple comparisons tests, the Kruskal–Wallis test with Dunn's multiple comparison, the unpaired t-test, and the log-rank survival test as applicable. Data were checked for normal distribution and outliers. $p < 0.05$ was the cutoff to determine significance.

3. Results

3.1. The Effect of Vitamin D on H1N1 Infection

To evaluate the role of vitamin D deficiency during respiratory virus infection, WT and Cyp KO mice were fed a D+ or D− diet. Regardless of the genotype, serum 25D levels were significantly different between mice fed the D+ or D− diet (Figure 1A). As expected, D+ Cyp KO mice accumulated 25D and had higher levels of 25D than D+ WT mice [25]. Following infection with pandemic H1N1 influenza, respiratory distress was evident in all mice by 7 days post-infection (Figure 1B). D+ WT mice showed the least amount of respiratory distress, and after day 10, D+ WT and D+ Cyp KO mice no longer exhibited symptoms of respiratory distress (Figure 1B). D− WT and D− Cyp KO mice had greater symptoms of respiratory distress that were not completely resolved by day 14 post-infection in the D− Cyp KO mice (Figure 1B). Only 1 of 18 D+ WT mice died following H1N1 infection (Figure 1C). D− WT mice had lower survival compared to D+ WT mice (Figure 1C). Conversely, both Cyp KO groups showed decreased survival; 62% from D+ Cyp KO and 57% from D− Cyp KO mice, respectively. The expression of the influenza M gene at day 4 post-infection was not different in the four groups of influenza-infected mice (data not shown). Lung sections from D− WT and D− Cyp KO mice (d0) but not D+ WT mice showed signs of inflammation even before infection (Figure 1D). The amount of alveolar hemorrhage was significantly more at day 4 post-infection in the D− Cyp KO compared to D+ WT or D− WT (Figure 1D). At day 14 post-infection, the lung sections of D+ Cyp KO, D− WT, and D− Cyp KO mice appeared more severe than D+ WT (Figure 1E). Therefore, vitamin D deficiency and the inability to produce 1,25D increased the susceptibility of mice to H1N1 influenza infection.

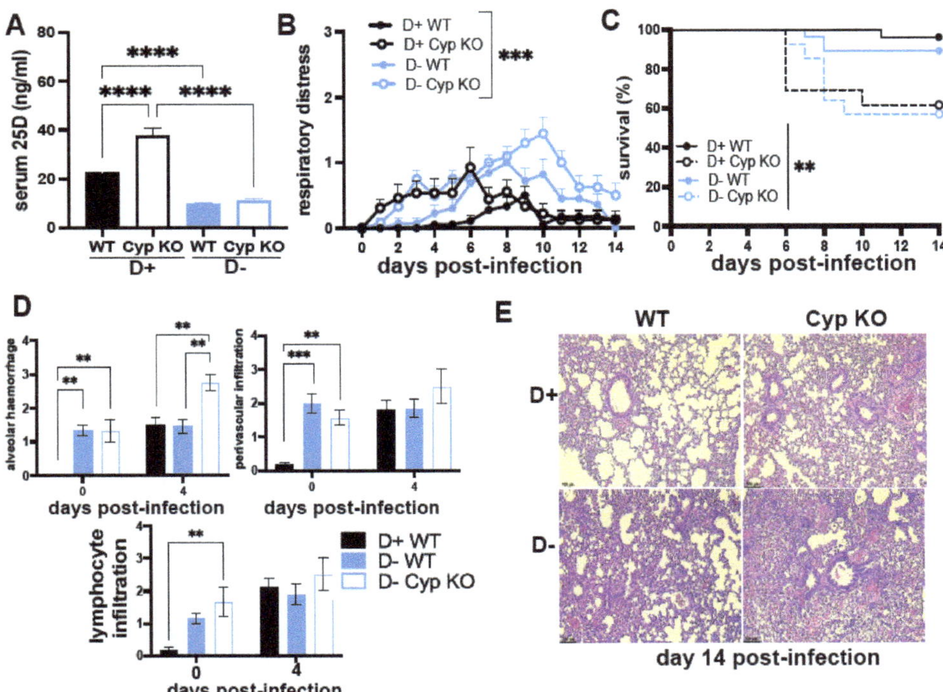

Figure 1. Vitamin D deficiency and Cyp27B1 KO increases susceptibility to H1N1 infection. D+ and D− WT and D+ and D− Cyp KO littermates were infected with H1N1 influenza (n = 12–18 per group). (**A**) Serum was collected at day 14 post-infection to measure 25D. Mice were monitored for (**B**) respiratory distress symptoms and (**C**) survival of D+ WT (n = 18), D+ Cyp KO (n = 13), D− WT (n = 13), and D− Cyp KO (n = 12) mice until day 14 post-infection. Lung tissues were collected for histology from uninfected and day 4 post-infected mice for histology. H & E-stained sections were scored for (**D**) lung alveolar hemorrhage, perivascular infiltration, and lymphocyte infiltration in D+ WT (n = 4) and D− WT (n = 4–5) and D− Cyp KO (n = 2–3) mice at day 0 and day 4 post-infection. (**E**) Representative histology of the lung of D+ WT (score = 3), D+ Cyp KO (score = 4), D− WT (score = 5.5), and D− Cyp KO (score = 6) at day 14 post-infection. Sections from D− WT, D+ Cyp KO, and D− Cyp KO show increased lymphocyte infiltration, alveolar hemorrhage and constricted bronchiolar spaces compared to D+ WT. Values are the mean ± SEM. Statistical significance was assessed using one-way ANOVA with Bonferroni multiple comparison test for (**A**) two-way ANOVA with Bonferroni multiple comparison test for (**B**,**D**) and log rank (Mantel–Cox) survival analysis for (**C**). ** $p < 0.01$, *** $p < 0.001$, and **** $p < 0.0001$.

3.2. *Mouse and Hamster Models of SARS-CoV-2 Infection*

SARS-CoV-2 does not effectively infect WT mice. The transgenic expression of human (h)ACE-2 was shown to allow SARS-CoV-2 infection in mice [31]. hACE-2 mice were infected with 100, 200, and 315 $TCID_{50}$ SARS-CoV-2, and the mice were evaluated for pre-determined euthanasia endpoints for up to 14 days post-infection. The dose of SARS-CoV-2 that resulted in the sacrifice of 50% of the mice was 200 $TCID_{50}$ (Figure 2A). The mice infected with 100 $TCID_{50}$ failed to reach the euthanasia endpoints, were sacrificed at day 14 post-infection and had only minimal weight loss (Figure 2A). The viral gene copy number for the nucleocapsid (N) protein was measured in the lungs of mice at 100 and 200 $TCID_{50}$. High copy numbers of the N gene were detected as early as day 2 post-infection and remained high until day 6 post infection (Figure 2B). The N gene copy number significantly decreased by day 14 post-infection (Figure 2B). Interestingly, the mice that were infected with 100 $TCID_{50}$ had high amounts of the N gene in the lung even though they showed only mild symptoms of infection, and there were no differences between the N gene expression in the lung between the 100 and 200 $TCID_{50}$ inoculum (Figure 2A,B). Lungs from SARS-CoV-2-infected mice were used to measure mRNA for the 1alpha hydroxylase that produces 1,25D (*Cyp27B1*), the vitamin D receptor (*Vdr*), and the 24 hydroxlase that degrades vitamin D (*Cyp24A1*). There was no change in *Vdr* expression in the lung at day 2 or day 4 post-infection (Figure 2C). Conversely, *Cyp27B1* and *Cyp24A1* expression was higher at day 2 and day 4 post-infection than in the uninfected lung (Figure 2C). *Cyp27B1* expression was significantly lower in the day 4 than the day 2 post-infection lung (Figure 2C). Infection of the hACE2 mice with SARS-CoV-2 induced the expression of two genes that regulate vitamin D metabolism in the lungs.

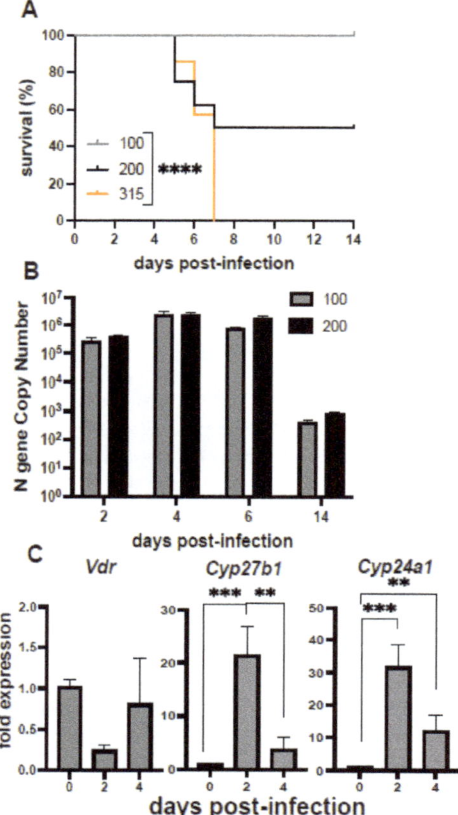

Figure 2. SARS-CoV-2 infection. K18-hACE2 mice were intranasally inoculated with 100 ($n = 8$), 200 ($n = 8$), and 315 ($n = 7$) $TCID_{50}$ of SARS-CoV-2 virus. Mice were sacrificed when they met predetermined endpoints (worsening conjunctivitis, lethargy, labored breathing, and/or dehydration) or at day 14 post-infection. (**A**) Mice were monitored for survival until day 14 post-infection. Mice were euthanized, and lung tissue was collected for (**B**) SARS-CoV-2 N gene expression at days 2–14 post-infection, (**C**) *Vdr, Cyp24A1*, and *Cyp27B1* mRNA gene expression (n = 4/group/timepoint) at day 0, day 2, and day 4 post-infection in the lung. Samples were normalized to uninfected control tissue. Values are the mean ± SEM. Statistical significance was assessed using log-rank test for trend (**A**), two-way ANOVA model for main effects only for (**B**), and one-way ANOVA with Bonferroni multiple comparison test on log-transformed expression values for (**C**). ** $p < 0.01$, *** $p < 0.001$, and **** $p < 0.0001$.

Hamsters infected with SARS-CoV-2 are reflective of human SARS-CoV-2 infection in that the virus infects the lower respiratory tract, causing similar respiratory sequelae [35]. We previously showed that hamsters infected with 10^5 $TCID_{50}$ SARS-CoV-2 lost weight shortly after infection, and none of the hamsters died following infection [33]. The weight loss peaked by day 6 post-infection, and the hamsters recovered completely by day 10 post-infection [33].

3.3. The Effect of Vitamin D on SARS-CoV-2 Infection

To establish differing vitamin D statuses, hACE2 mice were given a vitamin D deficient chow and were orally dosed with the vehicle (D−), 0.125 µg/day (D+), or 2.5 µg/day (D++) for 8 weeks prior to infection. Serum 25D levels were higher in D+ mice and significantly higher in D++ mice before and at 14 days post-infection (Figure 3A). The D+ dose was inadequate to raise serum 25D levels significantly over the D− values (Figure 3A). The survival of the SARS-CoV-2-infected mice was not affected by the D+ or the D++ treatments (Figure 3B). At day 6 post-infection the amount of N gene expression was the same in D− and D++ mice (Figure 3C). The expression of the *Vdr, Cyp27B1, Cyp24A1* was not different in D− and D++ mice at day 6 post-infection (Figure 4A). *Ifnb* and *Ifng* were induced by SARS-CoV-2 infection, while *Ifna* was not (uninfected control set at 1, Figure 4B). Expression of *Ifnb* was significantly lower in the D++ lung as compared to the D− lung at day 6 post-infection (Figure 4B). Mice surviving until day 14 post-infection showed no difference in lung histopathology scores between D− and D+ mice (Figure 3D,E). The D++ lung histopathology scores showed significantly reduced type II hyperplasia, significantly reduced alveolar remodeling, and lower (not significant) total histopathology scores (Figure 3D,E). The final series of experiments tested whether the active form of vitamin D (1,25D) could prevent the lethality of SARS-CoV-2 infection. There was no effect of 1,25D on survival from a lethal dose of SARS-CoV-2 (1,000 $TCID_{50}$) (Figure 3F). There was a trend for faster weight recovery in the surviving 1,25D-treated mice that did not reach significance (Figure 3F). There was an effect of D++ treatment to decrease lung histopathology and a trend towards faster recovery in 1,25D-treated mice infected with SARS-CoV-2.

Serum 25D levels in hamsters fed on the chow diet were not significantly different than serum 25D levels in hamsters fed D− diets for 4 weeks (Figure 5A). Therefore, to control for the diet, hamsters were fed D− diets and then fed orally with the vehicle (D−) or with 8 µg/day of vitamin D3 (D+) beginning 14 days before the SARS-CoV-2 infection and continuing throughout the experiment. Confirming the effectiveness of the dietary intervention, the serum 25D levels were significantly higher in D+ hamsters as compared to D− hamsters before infection on day 0, and this effect was maintained until day 14 post-infection (Figure 5A). The SARS-CoV-2 N gene was detected on day 3 but not at day 6 post-infection in the lungs (Figure 5B), and there was no difference in N gene expression between the D+ versus D− lungs (Figure 5B). Surprisingly, SARS-CoV-2 N gene expression was also detected in the colon tissues of hamsters at both day 3 and day 6 post-infection compared to uninfected control tissues (Figure 5B). Expression of the N gene in the colon was 1000-fold less than in the infected lung (Figure 5). N gene expression was significantly higher in the D− colon compared to baseline values from uninfected tissue controls but not different from uninfected controls in the D+ colon at day 3 post-infection (Figure 5B). The histopathology of the lungs showed significantly more damage at day 6 than day 3 post-infection (Figure 5C,D). There was no effect of vitamin D on the weight loss or histopathology scores following SARS-CoV-2 infection of the hamsters (Figure 5D,E).

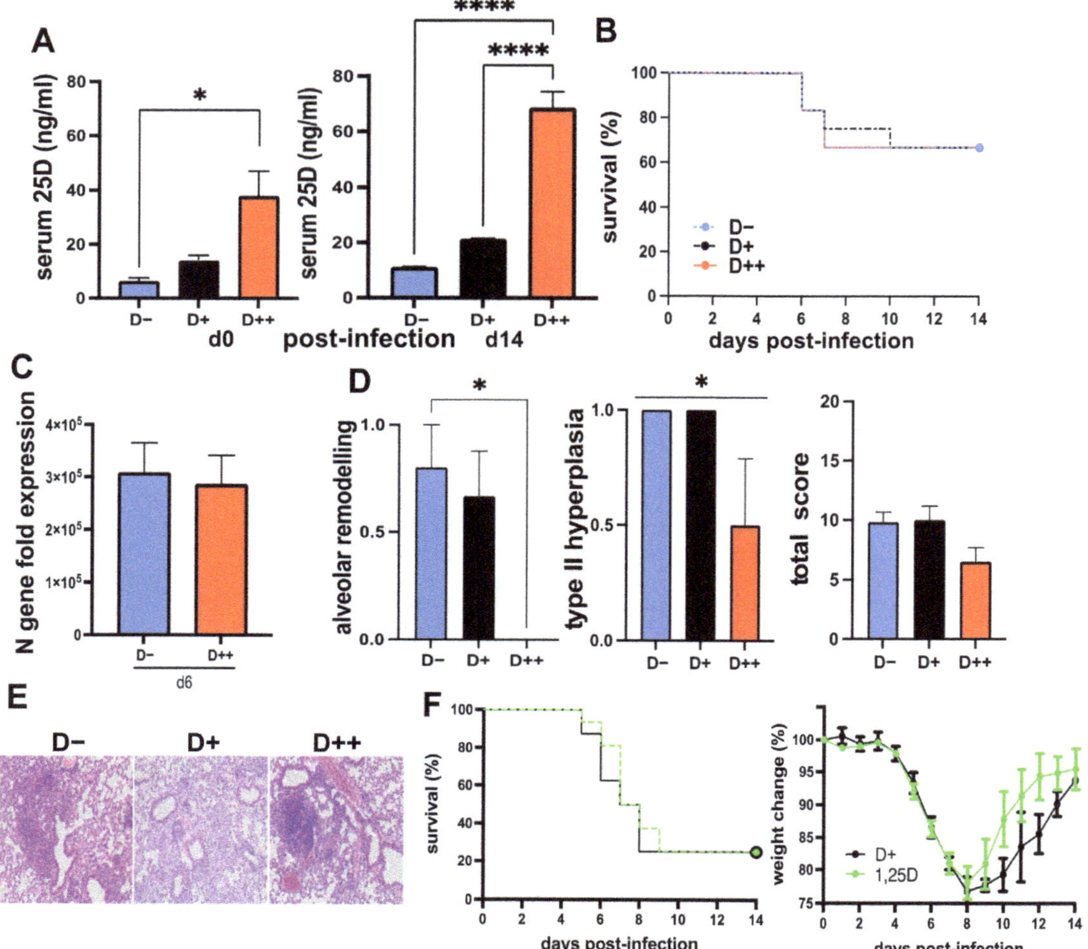

Figure 3. High dose vitamin D reduces lung inflammation following SARS-CoV-2 infection. hACE2 mice were fed vitamin D deficient (D−, $n = 18$), vitamin D sufficient (D+, $n = 12$), or vitamin D supplemented (D++, $n = 18$) diet and infected with 200 $TCID_{50}$ SARS-CoV-2. (**A**) Serum 25D was measured before and at day 14 post-infection, and (**B**) the survival of D−, D+, and D++ mice. (**C**) SARS-CoV-2 N gene expression in the lungs ($n = 6$ mice/group) at day 6 post-infection. Gene expression relative to uninfected D+ controls. (**D**) Alveolar remodeling, type II pneumocyte hyperplasia, and total histology score in D−, D+, and D++ mice ($n = 4$–6 mice/group) at day 14 post-infection. (**E**) Representative histology images for D− (score = 9), D+ (score = 11), and D++ (score = 5). D+ hACE2 ($n = 8$) mice at day 14 post-infection. 1,25$(OH)_2$D (1,25D)-treated D+ hACE 2 ($n = 16$) mice were infected with 1000 $TCID_{50}$ SARS-CoV-2. (**F**) Survival and body weight change over the course of infection. Values are the mean ± SEM. Statistical significance was assessed using one-way ANOVA with Bonferroni multiple comparison test for (**A**,**D**), log rank (Mantel–Cox) test for each of the groups for (**B**,**F**), unpaired *t*-test on log-transformed expression values for (**C**), and two-way ANOVA with Bonferroni multiple comparison test for (**F**). * $p < 0.05$ and **** $p < 0.0001$.

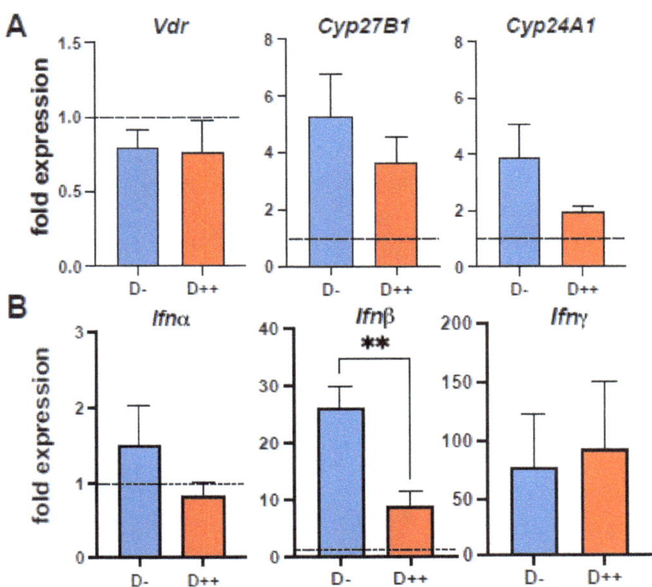

Figure 4. D++ mice have lower *Ifnb* at day 6 post-SARS-CoV-2 infection in the lung. Lung mRNA from day 6 SARS-CoV-2 infected D− ($n = 5$) and D++ ($n = 5$) mice. (**A**) *Vdr*, *Cyp27B1*, *Cyp24A1*, and (**B**) *Ifna*, *Ifnb*, and *Ifng* relative to *Gapdh* and uninfected controls set at 1 (dashed line). Values are the mean ± SEM. Statistical significance was assessed using unpaired *t*-test on log-transformed values. ** $p < 0.01$.

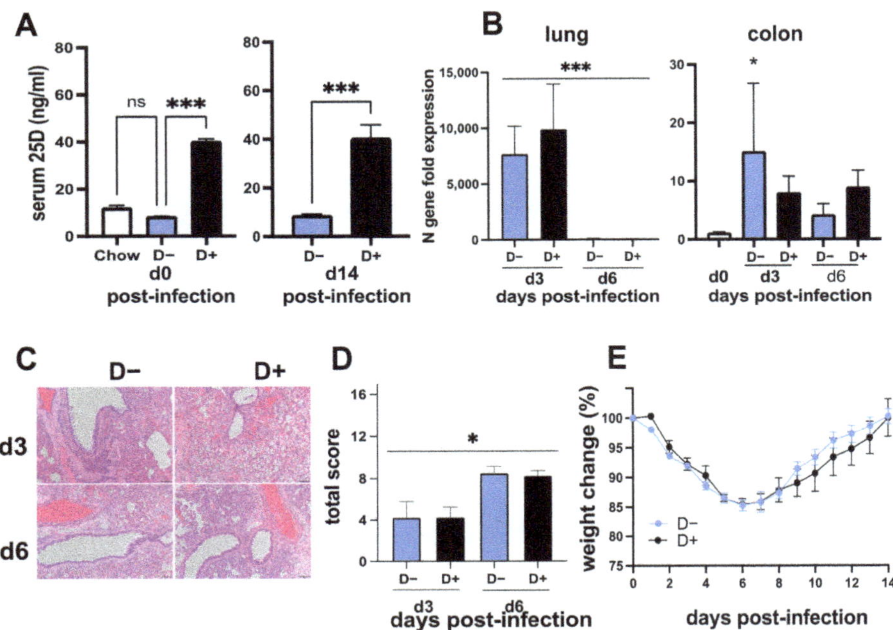

Figure 5. SARS-CoV-2 infection of hamsters. Hamsters were fed D− diets with corn oil (D−) or 8 µg D3/day dissolved in corn oil (D+) and then infected with SARS-CoV-2 virus. (**A**) Serum levels of 25D

in chow-fed, D+, and D− fed hamsters before (n = 2–10 hamsters/group) or from D− and D+ hamsters 14 days after (n = 5–6 hamsters/group) SARS-CoV-2 infection. Hamsters were euthanized at day 3 and day 6 post-infection, and tissues were collected for gene expression as well as histology. (**B**) SARS-CoV-2 N gene expression in the lung and colon relative to uninfected control (n = 4 hamsters per group and timepoint). (**C**) Representative histopathology sections of the lungs at day 3 (D− score = 6, D+ score = 4) and day 6 (D− score = 10, D+ score = 9) post-infection showed increased lymphocyte infiltration and lesions in D− and D+ hamsters at day 6 post-infection and (**D**) total histopathology scores (n = 4 hamsters/group) were determined based on the scoring criteria outlined in Table S4. (**E**) Change in body weight following infection with SARS-CoV-2 (n= 5–6 hamsters/group). Values are the mean ± SEM. Statistical significance was assessed using Kruskal–Wallis test with Dunn's multiple comparison (day 0) and unpaired t-test (day 14) for (**A**), one-way ANOVA with Bonferroni multiple comparison test on log-transformed expression values for (**B**), one-way ANOVA for (**D**), and two-way ANOVA with Bonferroni multiple comparison test for (**E**). * $p < 0.05$ and *** $p < 0.001$.

4. Discussion

Vitamin D deficiency resulted in lung inflammation in the absence of infection. Infected D− mice had more severe lung inflammation and respiratory symptoms than D+ or D++ mice when infected with either H1N1 influenza or SARS-CoV-2. The data point to shared effects of vitamin D to control inflammation in the lung following influenza or coronavirus infection. D− mice had significantly more inflammation than D+ mice following H1N1 influenza infection (Figure 1). High-dose vitamin D3 treatment (D++) resulted in some protection of mice from SARS-CoV-2 (Figure 3). In addition, 1,25D treatment showed a trend towards faster recovery of surviving mice from SARS-CoV-2 (Figure 3). Others have shown that 1,25D-treated mice had reduced lung inflammation [20,21], and treating D+ mice with 25D had a small protective effect on weight loss and lethality following H1N1 infection [22]. A recent clinical trial that used 25D in humans showed reduced mortality in hospitalized patients with COVID-19 [36,37]. However, it is unclear whether 25D treatment would be effective in mouse or hamster models of SARS-CoV-2. This is the first study that investigated the effects of vitamin D in animal models of SARS-CoV-2. The data suggest that vitamin D and 1,25D may be effective to protect the lung from SARS-CoV-2. SARS-CoV-2 infection induced *Cyp27B1* and *Cyp24A1* in the lung of mice suggesting a role for vitamin D metabolites in the lung response to SARS-CoV-2 infection (Figure 2). There are likely shared and unique mechanisms by which vitamin D regulates the host response to influenza versus SARS-CoV-2. A better understanding of the mechanisms by which vitamin D regulates the anti-viral response in the lung to both influenza and coronaviruses is needed to inform clinical studies.

Cyp27B1 KO mice cannot produce 1,25D, which induces Cyp24A1 and degrades 25D and 1,25D. Feeding Cyp27B1 KO mice D+ diets results in the accumulation of 25D (Figure 1A, [25,38]). 25D is a low-affinity ligand for the VDR, and it was shown that high amounts of 25D can replace the need for 1,25D for the regulation of calcium homeostasis and osteomalacia [39]. Previous experiments showed that D+ Cyp KO and D+ WT mice cleared a bacterial infection in the gut with similar kinetics [25]. Conversely, D+ Cyp KO mice had higher lethality and more severe inflammation than D+ WT mice when infected with H1N1 influenza (Figure 1). The effects of Cyp27B1 expression on host resistance to a bacteria could be different than the effects on host resistance to a virus. It would be interesting to determine the effect of the Cyp27B1 deletion on host resistance to other respiratory viruses including SARS-CoV-2. Conversely, the differential effect of Cyp27B1 could be due to the location of the infection in the gut versus the lung. Regardless, it seems that the ability of the host to produce Cyp27B1 is important for the mice to survive a H1N1 lung infection.

The dietary interventions to generate D−, D+, and D++ hACE2 mice resulted in D++, but not D+, mice having higher serum 25D than D− mice (Figure 4). Interestingly, the hamster studies suggest that the commercially available chow may not be adequate to raise serum 25D levels (Figure 5A). The data suggest that the adequacy of vitamin D should be considered in evaluating studies that infect chow-fed hamsters with SARS-CoV-2. At day 6 post-SARS-CoV-2 infection, the D++ hACE2 mice had less inflammation and lower IFN-β in the lung than the D− mice (Figure 3). The results are consistent with the anti-inflammatory effects of vitamin D. Suppression of type-1 inflammatory cytokines by vitamin D underlies the effects of vitamin D and 1,25D to suppress immune-mediated diseases [40–42]. The benefits of 25D from influenza infection were associated with a reduction in IFN-γ in the lung [22]. Recently, Chauss et.al showed that 1,25D promotes anti-inflammatory responses by switching off IFN-γ production from Th1 cells and upregulating IL-10 [43]. IFN-γ and IFN-β production is essential for effective viral clearance; viruses have mechanisms to evade the IFN responses, and severe COVID-19 is associated with dysregulation of IFN responses [44–46]. Down-regulation of IFN-β by vitamin D is associated with protection from inflammation in the lung following a virus infection with either influenza or SARS-CoV-2.

Importantly, there was no effect of vitamin D on the SARS-CoV-2 N gene or H1N1 M gene expression in the lungs of mice or hamsters. This indicates that vitamin D did not reduce or inhibit viral replication in the lung. We found SARS-CoV-2 N gene expression in the hamster colons. Interestingly, D− colons had relatively more SARS-CoV-2 N gene expression than D+ colons. The implications of having SARS-CoV-2 in the colon but not the lung would need to be determined, and it would be important to quantitate live virus in the tissues. Unfortunately, we did not save colons from our SARS-CoV-2 mouse studies. Vitamin D is shown to be a strong inducer of cathelicidin LL-37 in human cells [47]. There is some evidence that LL-37 can directly kill some viruses including influenza viruses [13–17]. Treating mice with a high dose (500 μg/day) of human LL-37 peptide protected from lethal influenza infection and significantly reduced viral titers at day 3 post-H1N1 infection in the lung [48]. LL-37 inhibited binding of the SARS-CoV-2 spike protein containing pseudo viruses both in vivo and in vitro blocking entry via ACE2 [49]. There were no effects of vitamin D in vivo on the expression of viral genes for SARS-CoV-2 or H1N1 influenza in the lung. The cathelicidin peptides found in mice are not the same as the LL-37 in humans, and the mouse cathelicidin is not regulated by vitamin D [50]. The lack of a vitamin D effect on SARS-CoV-2 was shown in mice and hamster lungs. It is unclear whether the hamster cathelicidin gene has vitamin D response elements. Furthermore, no studies have been conducted to test the effect of vitamin D on SARS-CoV-2 in vitro. Therefore, an effect of vitamin D through the induction of anti-bacterial peptides, such as LL-37, that reduces viral titers cannot be ruled out.

The data from mouse H1N1 and mouse and hamster SARS-CoV-2 infection point towards a protective role of vitamin D against acute viral infection in the lung. The data are in line with several meta-analyses and observational studies suggesting a beneficial effect of vitamin D in the lung [51–54]. Vitamin D supplementation was reported to have a moderately protective effect during acute respiratory tract infections, such as influenza and COVID-19; however, this outcome was affected by a number of variables such as dose frequency, season of supplementation, and pre-existing conditions [51–54]. Patients with existing vitamin D deficiency, especially patients >80 years of age, are at a higher risk of being infected with SARS-CoV-2 and developing severe disease [51–54]. High-dose vitamin D supplementation with either calcifediol or cholecalciferol was shown to reduce overall mortality and severe outcomes such as intensive care admission [55–59]. Vitamin D deficiency is associated with acute respiratory distress syndrome (ARDS), and several clinical trials are underway to investigate whether vitamin D supplementation can reduce the development of ARDS [60]. However, a recent randomized controlled study (reported in medRxiv, [61]) that identified vitamin D deficiency and then treated it showed that vitamin D supplementation did not reduce the risk of acute respiratory infection or the risk of COVID-19 infection. With the limitations of performing a clinical study for a nutrient such as vitamin D, our study in animals suggests that vitamin D does control the host response to H1N1 and SARS-CoV-2 infection in the lung. The data point to an effect of vitamin D to control the cytokine response and resolve inflammation in the lung following a viral infection. Understanding the mechanisms and timing of the vitamin D effects in animal models would inform future clinical trials.

There are several limitations of the current animal studies. There were technical limitations due to the cost, training, and restrictions needed to use the BSL3 facility for SARS-CoV-2 infections safely. Experiments in the BSL-3 facility had small sample sizes. In addition, viruses including H1N1 and SARS-CoV-2 do not naturally infect mice, and so the viruses used were passaged (H1N1), or a transgenic mouse was needed to allow infection (hACE) with SARS-CoV-2. In addition, vitamin D metabolism is not identical in animals and humans. To our surprise, hamsters fed on standard chow diets replete in vitamin D were found to have low serum 25D levels. It is possible that vitamin D metabolism in hamsters is different than in the mouse and humans, which might impact our results. Mechanistic studies in humans are difficult, so it is important to use animals to determine pathways and processes that are regulated by vitamin D. The data in humans and animals do support a role of vitamin D to control the inflammatory responses in the lung providing protection following a viral infection with H1N1 or SARS-CoV-2.

5. Conclusions

Together, the data support an important role for vitamin D and Cyp27B1 in the regulation of the host response to H1N1 and SARS-CoV-2 viruses. The role of vitamin D includes the restraining of the IFN response shortly after infection. Vitamin D-deficient hosts had pre-existing inflammation in the lungs that contributed to susceptibility to viral infection. Future experiments should continue to determine the mechanisms by which vitamin D regulates the anti-viral response in the lungs and whether there are differences in the effect of vitamin D on host resistance to H1N1 influenza and SARS-CoV-2.

Supplementary Materials: The following supporting information can be downloaded at: https://www.mdpi.com/article/10.3390/nu14153061/s1. Table S1: Primer sequences for qPCR, Table S2: Histological evaluation of SARS-CoV-2-infected mouse lung, Table S3: Histological evaluation of SARS-CoV-2-infected hamster lung, and Table S4: Histological evaluation of H1N1-infected mouse lung.

Author Contributions: Conceptualization, N.E.F., G.S.K., T.C.S. and M.T.C.; Formal analysis, J.A., J.W., E.L. and K.K.; Funding acquisition, M.T.C.; Investigation, J.A., D.R.P., M.J.N., C.J.F., K.H.R., J.W., N.E.F., B.K., J.A.T., V.W., E.L., K.K. and M.T.C.; Methodology, J.A., D.R.P., M.J.N., C.J.F., K.H.R., J.W., N.E.F., B.K., J.A.T., V.W., E.L. and K.K.; Project administration, J.A., D.R.P., M.J.N., C.J.F., B.K., V.W., G.S.K., T.C.S. and M.T.C.; Resources, G.S.K., T.C.S. and M.T.C.; Supervision, K.H.R., B.K., G.S.K. and T.C.S.; Visualization, J.A., M.J.N., J.W. and N.E.F.; Writing—original draft, J.A. and M.T.C.; Writing—review and editing, J.A., D.R.P., M.J.N., C.J.F., K.H.R., J.A.T., V.W., K.K., G.S.K., T.C.S. and M.T.C. All authors have read and agreed to the published version of the manuscript.

Funding: Funding of the work NIH R01AT005378 to M.T.C., R01AI123521 to G.S.K., T32DK120509 to N.E.F. and T32GM108563 to J.A., USDA NIFA award PEN04771 to M.T.C., G.S.K. and T.C.S.

Institutional Review Board Statement: All studies with SARS-CoV-2 were conducted in a biosafety level 3 enhanced (BSL3+) laboratory. This facility is approved for BSL3+ respiratory pathogen studies by the US Department of Agriculture and Centers for Disease Control. Studies with pandemic H1N1 influenza were conducted under biosafety level 2 enhanced conditions. All animal studies were conducted in compliance with the Animal Care and Use Committee under protocol numbers: 202001693, 202001516, 202001440, and 202001638.

Informed Consent Statement: Not applicable.

Data Availability Statement: Not applicable.

Conflicts of Interest: The authors declare no conflict of interest.

References

1. Zhu, Z.; Zhu, X.; Gu, L.; Zhan, Y.; Chen, L.; Li, X. Association Between Vitamin D and Influenza: Meta-Analysis and Systematic Review of Randomized Controlled Trials. *Front. Nutr.* **2021**, *8*, 799709. [CrossRef] [PubMed]
2. Cannell, J.J.; Vieth, R.; Umhau, J.C.; Holick, M.F.; Grant, W.B.; Madronich, S.; Garland, C.F.; Giovannucci, E. Epidemic influenza and vitamin D. *Epidemiol Infect.* **2006**, *134*, 1129–1140. [CrossRef] [PubMed]
3. Aglipay, M.; Birken, C.S.; Parkin, P.C.; Loeb, M.B.; Thorpe, K.; Chen, Y.; Laupacis, A.; Mamdani, M.; Macarthur, C.; Hoch, J.S.; et al. Effect of High-Dose vs Standard-Dose Wintertime Vitamin D Supplementation on Viral Upper Respiratory Tract Infections in Young Healthy Children. *JAMA* **2017**, *318*, 245–254. [CrossRef] [PubMed]
4. Grant, W.B.; Lahore, H.; McDonnell, S.L.; Baggerly, C.A.; French, C.B.; Aliano, J.L.; Bhattoa, H.P. Evidence that Vitamin D Supplementation Could Reduce Risk of Influenza and COVID-19 Infections and Deaths. *Nutrients* **2020**, *12*, 988. [CrossRef]
5. Chen, G.; Wu, D.; Guo, W.; Cao, Y.; Huang, D.; Wang, H.; Wang, T.; Zhang, X.; Chen, H.; Yu, H.; et al. Clinical and immunological features of severe and moderate coronavirus disease 2019. *J. Clin. Investig.* **2020**, *130*, 2620–2629. [CrossRef]
6. Mohan, M.; Cherian, J.J.; Sharma, A. Exploring links between vitamin D deficiency and COVID-19. *PLoS Pathog.* **2020**, *16*, e1008874. [CrossRef]
7. Akbar, M.R.; Wibowo, A.; Pranata, R.; Setiabudiawan, B. Low Serum 25-hydroxyvitamin D (Vitamin D) Level Is Associated With Susceptibility to COVID-19, Severity, and Mortality: A Systematic Review and Meta-Analysis. *Front. Nutr.* **2021**, *8*, 660420. [CrossRef]
8. Nunnari, G.; Fagone, P.; Lazzara, F.; Longo, A.; Cambria, D.; Di Stefano, G.; Palumbo, M.; Malaguarnera, L.; Di Rosa, M. Vitamin D3 inhibits TNFα-induced latent HIV reactivation in J-LAT cells. *Mol. Cell. Biochem.* **2016**, *418*, 49–57. [CrossRef]
9. Hansdottir, S.; Monick, M.M.; Lovan, N.; Powers, L.; Gerke, A.; Hunninghake, G.W. Vitamin D decreases respiratory syncytial virus induction of NF-kappaB-linked chemokines and cytokines in airway epithelium while maintaining the antiviral state. *J. Immunol.* **2010**, *184*, 965–974. [CrossRef]
10. Gruber-Bzura, B.M. Vitamin D and Influenza-Prevention or Therapy? *Int. J. Mol. Sci.* **2018**, *19*, 2419. [CrossRef]
11. Tripathi, S.; Wang, G.; White, M.; Qi, L.; Taubenberger, J.; Hartshorn, K.L. Antiviral Activity of the Human Cathelicidin, LL-37, and Derived Peptides on Seasonal and Pandemic Influenza A Viruses. *PLoS ONE* **2015**, *10*, e0124706. [CrossRef] [PubMed]
12. Saleh, M.; Welsch, C.; Cai, C.; Döring, C.; Gouttenoire, J.; Friedrich, J.; Haselow, K.; Sarrazin, C.; Badenhoop, K.; Moradpour, D.; et al. Differential modulation of hepatitis C virus replication and innate immune pathways by synthetic calcitriol-analogs. *J. Steroid Biochem. Mol. Biol.* **2018**, *183*, 142–151. [CrossRef] [PubMed]
13. Tripathi, S.; Tecle, T.; Verma, A.; Crouch, E.; White, M.; Hartshorn, K.L. The human cathelicidin LL-37 inhibits influenza A viruses through a mechanism distinct from that of surfactant protein D or defensins. *J. Gen. Virol.* **2013**, *94*, 40–49. [CrossRef] [PubMed]
14. Gordon, Y.J.; Huang, L.C.; Romanowski, E.G.; Yates, K.A.; Proske, R.J.; McDermott, A.M. Human cathelicidin (LL-37), a multifunctional peptide, is expressed by ocular surface epithelia and has potent antibacterial and antiviral activity. *Curr. Eye Res.* **2005**, *30*, 385–394. [CrossRef]
15. Howell, M.D.; Jones, J.F.; Kisich, K.O.; Streib, J.E.; Gallo, R.L.; Leung, D.Y.M. Selective Killing of Vaccinia Virus by LL-37: Implications for Eczema Vaccinatum. *J. Immunol.* **2004**, *172*, 1763. [CrossRef]
16. Currie, S.M.; Findlay, E.G.; McHugh, B.J.; Mackellar, A.; Man, T.; Macmillan, D.; Wang, H.; Fitch, P.M.; Schwarze, J.; Davidson, D.J. The Human Cathelicidin LL-37 Has Antiviral Activity against Respiratory Syncytial Virus. *PLoS ONE* **2013**, *8*, e73659. [CrossRef]
17. Currie, S.M.; Gwyer Findlay, E.; McFarlane, A.J.; Fitch, P.M.; Böttcher, B.; Colegrave, N.; Paras, A.; Jozwik, A.; Chiu, C.; Schwarze, J.; et al. Cathelicidins Have Direct Antiviral Activity against Respiratory Syncytial Virus In Vitro and Protective Function In Vivo in Mice and Humans. *J. Immunol.* **2016**, *196*, 2699. [CrossRef]
18. Helming, L.; Böse, J.; Ehrchen, J.; Schiebe, S.; Frahm, T.; Geffers, R.; Probst-Kepper, M.; Balling, R.; Lengeling, A. 1α,25-dihydroxyvitamin D3 is a potent suppressor of interferon γ–mediated macrophage activation. *Blood* **2005**, *106*, 4351–4358. [CrossRef] [PubMed]
19. Sundaram, M.E.; Coleman, L.A. Vitamin D and influenza. *Adv. Nutr.* **2012**, *3*, 517–525. [CrossRef]
20. Xu, J.; Yang, J.; Chen, J.; Luo, Q.; Zhang, Q.; Zhang, H. Vitamin D alleviates lipopolysaccharideinduced acute lung injury via regulation of the reninangiotensin system. *Mol. Med. Rep.* **2017**, *16*, 7432–7438. [CrossRef]
21. Gui, B.; Chen, Q.; Hu, C.; Zhu, C.; He, G. Effects of calcitriol (1, 25-dihydroxy-vitamin D3) on the inflammatory response induced by H9N2 influenza virus infection in human lung A549 epithelial cells and in mice. *Virol. J.* **2017**, *14*, 10. [CrossRef] [PubMed]
22. Hayashi, H.; Okamatsu, M.; Ogasawara, H.; Tsugawa, N.; Isoda, N.; Matsuno, K.; Sakoda, Y. Oral Supplementation of the Vitamin D Metabolite 25(OH)D3 Against Influenza Virus Infection in Mice. *Nutrients* **2020**, *12*, 2000. [CrossRef] [PubMed]
23. Pfeffer, P.E.; Lu, H.; Mann, E.H.; Chen, Y.H.; Ho, T.R.; Cousins, D.J.; Corrigan, C.; Kelly, F.J.; Mudway, I.S.; Hawrylowicz, C.M. Effects of vitamin D on inflammatory and oxidative stress responses of human bronchial epithelial cells exposed to particulate matter. *PLoS ONE* **2018**, *13*, e0200040. [CrossRef] [PubMed]
24. Chen, H.; Lu, R.; Zhang, Y.G.; Sun, J. Vitamin D Receptor Deletion Leads to the Destruction of Tight and Adherens Junctions in Lungs. *Tissue Barriers* **2018**, *6*, 1–13. [CrossRef] [PubMed]
25. Lin, Y.D.; Arora, J.; Diehl, K.; Bora, S.A.; Cantorna, M.T. Vitamin D Is Required for ILC3 Derived IL-22 and Protection From Citrobacter rodentium Infection. *Front. Immunol.* **2019**, *10*, 1. [CrossRef]

26. Ye, J.; Sorrell, E.M.; Cai, Y.; Shao, H.; Xu, K.; Pena, L.; Hickman, D.; Song, H.; Angel, M.; Medina, R.A.; et al. Variations in the hemagglutinin of the 2009 H1N1 pandemic virus: Potential for strains with altered virulence phenotype? *PLoS Pathog.* **2010**, *6*, e1001145. [CrossRef]
27. Meyerholz, D.K.; Beck, A.P. Histopathologic Evaluation and Scoring of Viral Lung Infection. *Methods Mol. Biol.* **2020**, *2099*, 205–220.
28. Rodriguez, A.E.; Bogart, C.; Gilbert, C.M.; McCullers, J.A.; Smith, A.M.; Kanneganti, T.-D.; Lupfer, C.R. Enhanced IL-1β production is mediated by a TLR2-MYD88-NLRP3 signaling axis during coinfection with influenza A virus and Streptococcus pneumoniae. *PLoS ONE* **2019**, *14*, e0212236. [CrossRef]
29. Zablockienė, B.; Kačergius, T.; Ambrozaitis, A.; Žurauskas, E.; Bratchikov, M.; Jurgauskienė, L.; Zablockis, R.; Gravenstein, S. Zanamivir Diminishes Lung Damage in Influenza A Virus-infected Mice by Inhibiting Nitric Oxide Production. *In Vivo* **2018**, *32*, 473–478.
30. Oladunni, F.S.; Park, J.-G.; Pino, P.A.; Gonzalez, O.; Akhter, A.; Allué-Guardia, A.; Olmo-Fontánez, A.; Gautam, S.; Garcia-Vilanova, A.; Ye, C.; et al. Lethality of SARS-CoV-2 infection in K18 human angiotensin-converting enzyme 2 transgenic mice. *Nat. Commun.* **2020**, *11*, 6122. [CrossRef]
31. Winkler, E.S.; Bailey, A.L.; Kafai, N.M.; Nair, S.; McCune, B.T.; Yu, J.; Fox, J.M.; Chen, R.E.; Earnest, J.T.; Keeler, S.P.; et al. SARS-CoV-2 infection of human ACE2-transgenic mice causes severe lung inflammation and impaired function. *Nat. Immunol.* **2020**, *21*, 1327–1335. [CrossRef] [PubMed]
32. Gerhards, N.M.; Cornelissen, J.B.W.J.; van Keulen, L.J.M.; Harders-Westerveen, J.; Vloet, R.; Smid, B.; Vastenhouw, S.; van Oort, S.; Hakze-van der Honing, R.W.; Gonzales, J.L.; et al. Predictive Value of Precision-Cut Lung Slices for the Susceptibility of Three Animal Species for SARS-CoV-2 and Validation in a Refined Hamster Model. *Pathogens* **2021**, *10*, 824. [CrossRef]
33. Field, C.J.; Heinly, T.A.; Patel, D.R.; Sim, D.G.; Luley, E.; Gupta, S.L.; Vanderford, T.H.; Wrammert, J.; Sutton, T.C. Immune durability and protection against SARS-CoV-2 re-infection in Syrian hamsters. *Emerg. Microbes Infect.* **2022**, *11*, 1103–1114. [CrossRef] [PubMed]
34. Gruber, A.D.; Firsching, T.C.; Trimpert, J.; Dietert, K. Hamster models of COVID-19 pneumonia reviewed: How human can they be? *Vet. Pathol.* **2021**, *59*, 528–545. [CrossRef] [PubMed]
35. Chan, J.F.; Zhang, A.J.; Yuan, S.; Poon, V.K.; Chan, C.C.; Lee, A.C.; Chan, W.M.; Fan, Z.; Tsoi, H.W.; Wen, L.; et al. Simulation of the clinical and pathological manifestations of Coronavirus Disease 2019 (COVID-19) in golden Syrian hamster model: Implications for disease pathogenesis and transmissibility. *Clin. Infect. Dis.* **2020**, *71*, 2428–2446. [CrossRef]
36. Nogues, X.; Ovejero, D.; Pineda-Moncusi, M.; Bouillon, R.; Arenas, D.; Pascual, J.; Ribes, A.; Guerri-Fernandez, R.; Villar-Garcia, J.; Rial, A.; et al. Calcifediol treatment and COVID-19-related outcomes. *J. Clin. Endocrinol. Metab.* **2021**, *106*, e4017–e4027. [CrossRef]
37. Carpagnano, G.E.; Di Lecce, V.; Quaranta, V.N.; Zito, A.; Buonamico, E.; Capozza, E.; Palumbo, A.; Di Gioia, G.; Valerio, V.N.; Resta, O. Vitamin D deficiency as a predictor of poor prognosis in patients with acute respiratory failure due to COVID-19. *J. Endocrinol. Investig.* **2020**, *44*, 765–771. [CrossRef]
38. Bora, S.A.; Kennett, M.J.; Smith, P.B.; Patterson, A.D.; Cantorna, M.T. Regulation of vitamin D metabolism following disruption of the microbiota using broad spectrum antibiotics. *J. Nutr. Biochem.* **2018**, *56*, 65–73. [CrossRef]
39. Rowling, M.J.; Gliniak, C.; Welsh, J.; Fleet, J.C. High dietary vitamin D prevents hypocalcemia and osteomalacia in CYP27B1 knockout mice. *J. Nutr.* **2007**, *137*, 2608–2615. [CrossRef]
40. Cantorna, M.T.; Rogers, C.J.; Arora, J. Aligning the Paradoxical Role of Vitamin D in Gastrointestinal Immunity. *Trends Endocrinol. Metab.* **2019**, *30*, 459–466. [CrossRef]
41. Cantorna, M.T.; Yu, S.; Bruce, D. The paradoxical effects of vitamin D on type 1 mediated immunity. *Mol. Asp. Med.* **2008**, *29*, 369–375. [CrossRef] [PubMed]
42. Ooi, J.H.; Chen, J.; Cantorna, M.T. Vitamin D regulation of immune function in the gut: Why do T cells have vitamin D receptors? *Mol. Asp. Med.* **2012**, *33*, 77–82. [CrossRef]
43. Chauss, D.; Freiwald, T.; McGregor, R.; Yan, B.; Wang, L.; Nova-Lamperti, E.; Kumar, D.; Zhang, Z.; Teague, H.; West, E.E.; et al. Autocrine vitamin D signaling switches off pro-inflammatory programs of TH1 cells. *Nat. Immunol.* **2022**, *23*, 62–74. [CrossRef] [PubMed]
44. Stehle, C.; Hernandez, D.C.; Romagnani, C. Innate lymphoid cells in lung infection and immunity. *Immunol. Rev.* **2018**, *286*, 102–119. [CrossRef]
45. Eskandarian Boroujeni, M.; Sekrecka, A.; Antonczyk, A.; Hassani, S.; Sekrecki, M.; Nowicka, H.; Lopacinska, N.; Olya, A.; Kluzek, K.; Wesoly, J.; et al. Dysregulated Interferon Response and Immune Hyperactivation in Severe COVID-19: Targeting STATs as a Novel Therapeutic Strategy. *Front. Immunol.* **2022**, *13*, 888897. [CrossRef] [PubMed]
46. Wang, H.; Li, W.; Zheng, S.J. Advances on Innate Immune Evasion by Avian Immunosuppressive Viruses. *Front. Immunol.* **2022**, *13*, 901913. [CrossRef] [PubMed]
47. White, J.H. Vitamin D as an inducer of cathelicidin antimicrobial peptide expression: Past, present and future. *J. Steroid Biochem. Mol. Biol.* **2010**, *121*, 234–238. [CrossRef]
48. Barlow, P.G.; Svoboda, P.; Mackellar, A.; Nash, A.A.; York, I.A.; Pohl, J.; Davidson, D.J.; Donis, R.O. Antiviral activity and increased host defense against influenza infection elicited by the human cathelicidin LL-37. *PLoS ONE* **2011**, *6*, e25333. [CrossRef]
49. Wang, C.; Wang, S.; Li, D.; Chen, P.; Han, S.; Zhao, G.; Chen, Y.; Zhao, J.; Xiong, J.; Qiu, J.; et al. Human Cathelicidin Inhibits SARS-CoV-2 Infection: Killing Two Birds with One Stone. *ACS Infect. Dis.* **2021**, *7*, 1545–1554. [CrossRef]

50. Dimitrov, V.; White, J.H. Species-specific regulation of innate immunity by vitamin D signaling. *J. Steroid Biochem. Mol. Biol.* **2016**, *164*, 246–253. [CrossRef]
51. Baktash, V.; Hosack, T.; Patel, N.; Shah, S.; Kandiah, P.; Van den Abbeele, K.; Mandal, A.K.J.; Missouris, C.G. Vitamin D status and outcomes for hospitalised older patients with COVID-19. *Postgrad Med. J.* **2021**, *97*, 442–447. [CrossRef] [PubMed]
52. Katz, J.; Yue, S.; Xue, W. Increased risk for COVID-19 in patients with vitamin D deficiency. *Nutrition* **2021**, *84*, 111106. [CrossRef]
53. D'Avolio, A.; Avataneo, V.; Manca, A.; Cusato, J.; De Nicolò, A.; Lucchini, R.; Keller, F.; Cantù, M. 25-Hydroxyvitamin D Concentrations Are Lower in Patients with Positive PCR for SARS-CoV-2. *Nutrients* **2020**, *12*, 1359. [CrossRef]
54. Panagiotou, G.; Tee, S.A.; Ihsan, Y.; Athar, W.; Marchitelli, G.; Kelly, D.; Boot, C.S.; Stock, N.; Macfarlane, J.; Martineau, A.R.; et al. Low serum 25-hydroxyvitamin D (25[OH]D) levels in patients hospitalized with COVID-19 are associated with greater disease severity. *Clin. Endocrinol.* **2020**, *93*, 508–511. [CrossRef] [PubMed]
55. Oristrell, J.; Oliva, J.C.; Casado, E.; Subirana, I.; Domínguez, D.; Toloba, A.; Balado, A.; Grau, M. Vitamin D supplementation and COVID-19 risk: A population-based, cohort study. *J. Endocrinol. Investig.* **2022**, *45*, 167–179. [CrossRef]
56. Sluyter, J.D.; Camargo, C.A.; Waayer, D.; Lawes, C.M.M.; Toop, L.; Khaw, K.T.; Scragg, R. Effect of Monthly, High-Dose, Long-Term Vitamin D on Lung Function: A Randomized Controlled Trial. *Nutrients* **2017**, *9*, 1353. [CrossRef] [PubMed]
57. Annweiler, G.; Corvaisier, M.; Gautier, J.; Dubée, V.; Legrand, E.; Sacco, G.; Annweiler, C. Vitamin D Supplementation Associated to Better Survival in Hospitalized Frail Elderly COVID-19 Patients: The GERIA-COVID Quasi-Experimental Study. *Nutrients* **2020**, *12*, 3377. [CrossRef]
58. Annweiler, C.; Beaudenon, M.; Gautier, J.; Gonsard, J.; Boucher, S.; Chapelet, G.; Darsonval, A.; Fougère, B.; Guérin, O.; Houvet, M.; et al. High-dose versus standard-dose vitamin D supplementation in older adults with COVID-19 (COVIT-TRIAL): A multicenter, open-label, randomized controlled superiority trial. *PLoS Med.* **2022**, *19*, e1003999. [CrossRef]
59. Entrenas Castillo, M.; Entrenas Costa, L.M.; Vaquero Barrios, J.M.; Alcalá Díaz, J.F.; López Miranda, J.; Bouillon, R.; Quesada Gomez, J.M. "Effect of calcifediol treatment and best available therapy versus best available therapy on intensive care unit admission and mortality among patients hospitalized for COVID-19: A pilot randomized clinical study". *J. Steroid Biochem. Mol. Biol.* **2020**, *203*, 105751. [CrossRef]
60. Quesada-Gomez, J.M.; Entrenas-Castillo, M.; Bouillon, R. Vitamin D receptor stimulation to reduce acute respiratory distress syndrome (ARDS) in patients with coronavirus SARS-CoV-2 infections: Revised Ms SBMB 2020_166. *J. Steroid Biochem. Mol. Biol.* **2020**, *202*, 105719. [CrossRef]
61. Jolliffe, D.A.; Holt, H.; Greenig, M.; Talaei, M.; Perdek, N.; Pfeffer, P.; Maltby, S.; Symons, J.; Barlow, N.L.; Normandale, A.; et al. Vitamin D Supplements for Prevention of COVID-19 or other Acute Respiratory Infections: A Phase 3 Randomized Controlled Trial (CORONAVIT). *medRxiv* **2022**. [CrossRef]

MDPI

Article

A Single Vitamin D_3 Bolus Supplementation Improves Vitamin D Status and Reduces Proinflammatory Cytokines in Healthy Females

Hadeil M. Alsufiani [1,2,3,*], Shareefa A. AlGhamdi [1,2,3], Huda F. AlShaibi [1,4], Sawsan O. Khoja [1,2,3], Safa F. Saif [1] and Carsten Carlberg [5,6,*]

1 Department of Biochemistry, Faculty of Sciences, King Abdulaziz University, Jeddah 21589, Saudi Arabia
2 Vitamin D Pharmacogenomics Research Group, King Abdulaziz University, Jeddah 21589, Saudi Arabia
3 Experimental Biochemistry Unit, King Fahd Medical Research Center, King Abdulaziz University, Jeddah 21589, Saudi Arabia
4 Embryonic Stem Cell Unit, King Fahd Medical Research Center, King Abdulaziz University, Jeddah 21589, Saudi Arabia
5 Institute of Animal Reproduction and Food Research, Polish Academy of Sciences, PL 10-748 Olsztyn, Poland
6 Institute of Biomedicine, School of Medicine, University of Eastern Finland, FI 70211 Kuopio, Finland
* Correspondence: halsufiani@kau.edu.sa (H.M.A.); c.carlberg@pan.olsztyn.pl (C.C.)

Abstract: Vitamin D deficiency is a global health problem that not only leads to metabolic bone disease but also to many other illnesses, most of which are associated with chronic inflammation. Thus, our aim was to investigate the safety and effectiveness of a single high dose of vitamin D_3 (80,000 IU) on vitamin D status and proinflammatory cytokines such as interleukin (IL)6, IL8 and tumor necrosis factor (TNF) in healthy Saudi females. Fifty healthy females were recruited and orally supplemented with a single vitamin D_3 bolus (80,000 IU). All participants donated fasting blood samples at baseline, one day and thirty days after supplementation. Serum 25-hydroxyvitamin D_3 (25(OH)D_3), IL6, IL8, TNF, calcium, phosphate, parathyroid hormone (PTH) and blood lipid levels were determined. Serum 25(OH)D_3 significantly increased one and thirty days after supplementation when compared with baseline without causing elevation in calcium or phosphate or a decrease in PTH to abnormal levels. In contrast, the concentrations of the three representative proinflammatory cytokines decreased gradually until the end of the study period. In conclusion, a single high dose (80,000 IU) is effective in improving serum vitamin D status and reducing the concentration of the proinflammatory cytokines in a rapid and safe way in healthy females.

Keywords: vitamin D deficiency; single high dose; vitamin D_3 supplementation; proinflammatory cytokines; IL6; IL8; TNF; 25(OH)D_3

Citation: Alsufiani, H.M.; AlGhamdi, S.A.; AlShaibi, H.F.; Khoja, S.O.; Saif, S.F.; Carlberg, C. A Single Vitamin D_3 Bolus Supplementation Improves Vitamin D Status and Reduces Proinflammatory Cytokines in Healthy Females. *Nutrients* **2022**, *14*, 3963. https://doi.org/10.3390/nu14193963

Academic Editor: Andrea Fabbri

Received: 9 September 2022
Accepted: 23 September 2022
Published: 24 September 2022

1. Introduction

Vitamin D_3 is a micronutrient that can be synthesized in human skin from the cholesterol precursor 7-dehydrocholesterol through energy provided by the ultraviolet-B (UVB) component of sunlight [1]. Recent lifestyle and work–life changes towards indoor activities as well as the use of clothing and sunscreen for sunburn protection outdoors have reduced the chances of filling up vitamin D_3 stores. This results in far lower average vitamin D status in today's modern societies than in more traditionally living populations [2–5]. Even in sunny Saudi Arabia, a substantial proportion of the population is considered vitamin D-deficient [6]. This increases the risk not only of muscle weakness (sarcopenia) and early onset of osteoporosis but also leads to an increase in autoimmune diseases, such as type 1 diabetes, arthritis, multiple sclerosis, cancer, cardiovascular diseases and Alzheimer's disease [7,8]. Therefore, vitamin D deficiency is a global health problem that not only leads to musculoskeletal problems but also to many other illnesses, most of which are associated with chronic inflammation [9,10].

In the liver, vitamin D_3 is hydroxylated to 25(OH)D_3, which is the most stable vitamin D_3 metabolite circulating in the blood. Therefore, 25(OH)D_3 serum levels serve as a biomarker for the vitamin D status. In the kidneys, 25(OH)D_3 is further metabolized to the physiologically most active vitamin D metabolite, 1,25-dihydroxyvitamin D_3 (1,25(OH)$_2$$D_3$) [11]. The lipophilic nature of 1,25(OH)$_2$$D_3$ allows the molecule to pass through cellular and nuclear membranes and to act in the nucleus as a high-affinity ligand to the transcription factor vitamin D receptor (VDR), i.e., 1,25(OH)$_2$$D_3$ has a direct effect on gene regulation [12,13]. Besides the kidneys, 1,25(OH)$_2$$D_3$ is also synthesized locally in a number of tissues and cell types expressing VDR. Taking all presently investigated tissues and cell types together, there are more than 20,000 VDR binding sites in the human genome, and significant changes in the transcriptome profile occur in over 1000 human genes [14].

Examples of vitamin D target tissues include immune cells such as T cells, B cells and monocytes, which are the major components of peripheral blood mononuclear cells (PBMCs) [15–17]. One hallmark of vitamin D's effects is the regulation of genes involved in the regulation of inflammatory processes. Accordingly, there is an interplay between vitamin D signaling and other signaling cascades involved in inflammation [18,19].

The impact of 1,25(OH)$_2$$D_3$ treatment on the expression of the proinflammatory cytokines IL6, IL8 and TNF was extensively studied in PBMCs from healthy donors, primary monocytes/macrophages as well as in monocytic cell lines, indicating that the VDR ligand causes their down-regulation on an mRNA and protein level [20–27]. Importantly, not only does the treatment of cell culture models with 1,25(OH)$_2$$D_3$ promote changes in gene expression, but also the supplementation of individuals with vitamin D_3 leads to the same results. Most of these studies were conducted on patients with diverse inflammatory diseases, such as COVID-19, colorectal cancer, irritable bowel syndrome, obesity and diabetes [28–34], and only a few studies were performed with healthy individuals [35–37].

Vitamin D intervention studies usually use different doses of vitamin D_3 supplementation either daily or weekly for several weeks or months, and only a few of them used a single high dose. The pharmacology of vitamin D shows that the proper half-life for dose periods is longer than daily supplementation, and many dosing regimens suggest that high vitamin D_3 doses at less frequent periods are more suitable and have become a broad practice [38]. Moreover, from an experimental point of view, the use of a single high vitamin D_3 dose is more suitable for observing the direct effects of vitamin D on the expression of its target genes, such as multiple cytokines, both on the mRNA and protein level. Accordingly, the aim of this study was to investigate the safety and effectiveness of a single high dose of vitamin D_3 (80,000 IU) on the vitamin D status and the serum levels of representative proinflammatory cytokines IL6, IL8 and TNF in healthy Saudi females.

2. Materials and Methods

2.1. Study Design and Participants

Fifty healthy Saudi females aged between 18 and 60 were recruited from King Abdul Aziz University and King Fahad Medical Research Center's staff and their families from January to December 2019. The total sample size was calculated based on a power analysis (using G*Power software, version 3.1.9.7, Düsseldorf, Germany) that indicated a 95% chance of a 0.5 effect size between the tested groups at the 5% level (two-tailed). Exclusion criteria included the presence of cancer, liver or kidney diseases, the intake of vitamin D supplements during the last three months, and non-Saudis. All participants received a single high dose of vitamin D_3 (80,000 IU) orally administered (Figure 1). This dose was chosen since previous experience in the vitamin D intervention studies VitDbol (https://clinicaltrials.gov/ct2/show/NCT02063334) (accessed on 19 September 2022) and VitDHiD (https://clinicaltrials.gov/ct2/show/NCT03537027) (accessed on 19 September 2022) indicated that 80,000 IU vitamin D_3 is a safe monthly dose in healthy individuals. This study was approved by the ethical committee of the Faculty of Medicine, King Abdulaziz University (reference number 30-18), and all participants provided written informed consent.

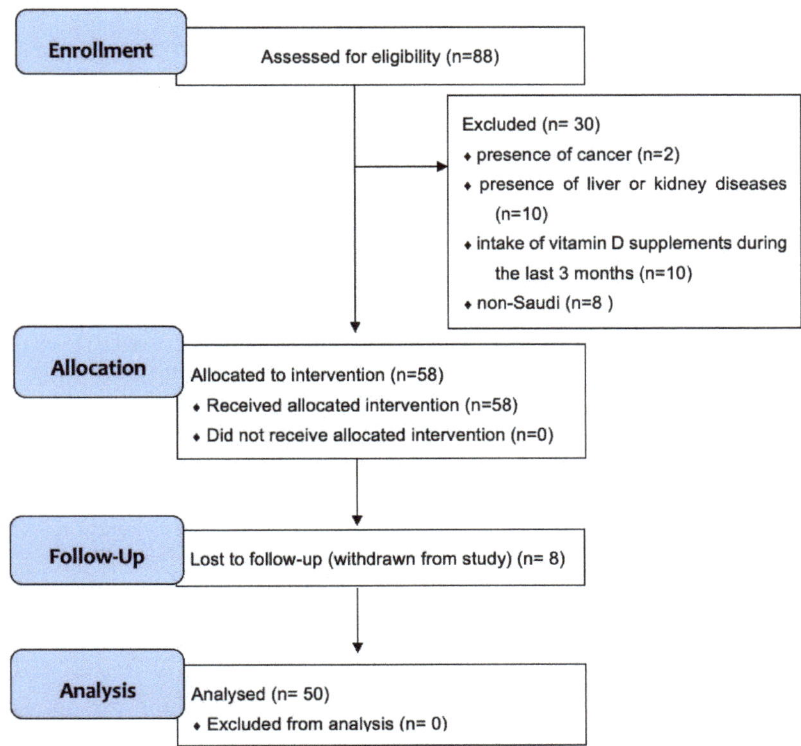

Figure 1. Flow chart showing the flow of the participants throughout the study. n = number of individuals.

2.2. Anthropometric Measurements

Height and weight were measured by using an electronic scale and a portable stadiometer from Seca (Hamburg, Germany), respectively, and the body mass index (BMI) was calculated for all participants. In addition, waist and hip circumference were measured using Seca tape, and the waist-to-hip ratio (WHR) was then calculated.

2.3. Biochemical Measurements

All participants donated fasting blood samples at three different time points; at baseline (day 0), after one day (day 1) and after thirty days (day 30) of oral administration of a single high dose of vitamin D_3 (80,000 IU). Serum was isolated and stored at −80 °C for later measurements of biochemical parameters including lipid profile, phosphorus (PHOS), calcium (CAL), parathyroid hormone (PTH), 25(OH)D_3 and proinflammatory cytokines.

Quantitative determination of serum cholesterol (CHOL), low-density lipoproteins (LDL) and triglycerides (TAG) was performed using a Siemens Dimension Vista instrument. Serum CAL and PHOS were measured using a kit from Siemens Healthcare Diagnostic Limited, Dimension Vista System UK (Cat. No K1023 and Cat. No K1061, respectively). Serum PTH was measured using a chemiluminescent microparticle immunoassay (CMIA) technique kit from Abbott (Cat. No 8K25). Serum vitamin D status was determined by measuring 25(OH)D_3 via the Abbott Architect 25-OH Vitamin D assay kit. Finally, the proinflammatory cytokines IL6, IL8 and TNF were measured using Human Interleukin 6 ELISA Kit by Bioassay Technology Laboratory (Cat. No E0089Hu), Human Interleukin 8 ELISA Kit by Bioassay Technology Laboratory (Cat. No E0089Hu) and Human Tumor Necrosis Factor Alpha ELISA Kit by Bioassay Technology Laboratory (Cat. No E0082Hu), respectively.

2.4. Statistical Analysis

All statistical analyses were performed using IBM SPSS software version 24 (SPSS Inc., Chicago, IL, USA) and graphs were represented using GraphPad prism 7. Data were presented as mean ± standard error of mean (SEM). Repeated measures one-way analysis of variance (ANOVA) followed by Bonferroni's multiple comparison test was used to determine the significant differences in mean serum levels of 25(OH)D_3, IL6, IL8 and TNF, CHOL, TAG, LDL, PHOS, CAL and PTH between days 0, 1 and 30 of vitamin D_3 supplementation. The statistical significance threshold was taken as $p < 0.05$.

3. Results

Fifty females with a mean age of 29 years participated in this study. At baseline, their mean BMI was 23.6 kg/m^2 and their mean WHR was 0.77. All biochemical parameters including CHOL, LDL, TAG, PHOS, CAL, and PTH were in the normal range intervals indicating a good health status of all participants (Table 1). After supplementation with vitamin D_3, no changes were found in most biochemical parameters except in CHOL, PHOS and PTH levels. The changes in these parameters were minor and did not reach abnormal levels.

Table 1. Demographic and clinical characteristics of study participants at baseline, day 1 and day 30 following a single high dose of vitamin D_3 supplementation (n = 50).

	Baseline	Day 1	Day 30
Age (years)	28.9 ± 0.9		
Height (cm)	158.9 ± 0.7	NA	NA
Weight (kg)	59.9 ± 1.8	NA	NA
BMI (kg/m^2)	23.6 ± 0.7	NA	NA
Waist circumference (cm)	74.5 ± 2.1	NA	NA
Hip circumference (cm)	97.7 ± 2.5	NA	NA
WHR	0.77 ± 0.02	NA	NA
CHOL (mM)	4.34 ± 0.12	4.26 ± 0.12	4.13 ± 0.11 *
TAG (mM)	1.05 ± 0.07	2.39 ± 1.34	1.07 ± 0.08
LDL (mM)	2.78 ± 0.11	2.34 ± 0.18	2.53 ± 0.09
PHOS (mM)	1.24 ± 0.03	1.18 ± 0.03 *	1.24 ± 0.03
CAL (mM)	2.29 ± 0.02	2.24 ± 0.01	2.22 ± 0.02
PTH (pM)	5.25 ± 0.44	4.44 ± 0.30 *	4.17 ± 0.28 **

Data are presented as mean ± SEM. * $p < 0.05$, ** $p < 0.01$ when compared with baseline. NA: Data are not available.

The mean serum 25(OH)D_3 concentration at baseline was 41.9 ± 4.1 nM, and 72% of study participants had an insufficient vitamin D status of less than 50 nM (Table 2). The average vitamin D status significantly increased to 66.3 ± 3.5 nM at day 1 and 68.9 ± 2.5 nM at day 30 (Figure 2). This represents an average increase by 24.4 and 26.9 nM and a shift from deficiency and insufficiency to sufficiency for 76% and 94% of the study participants, respectively, at days 1 and 30 after vitamin D_3 bolus supplementation (Table 2).

Table 2. Prevalence of vitamin D deficiency among study participants at baseline, day 1 and day 30 following a single high dose of vitamin D_3 supplementation (n = 50).

Serum Vitamin D Status *	Baseline N (%)	Day 1 N (%)	Day 30 N (%)
Deficiency $25(OH)D_3 < 30$ nM	24 (48%)	0 (0%)	0 (0%)
Insufficiency $25(OH)D_3$ of 30–50 nM	12 (24%)	12 (24%)	3 (6%)
Sufficiency $25(OH)D_3 > 50$ nM	14 (28%)	38 (76%)	47 (94%)

* classification was based on US Institute of Medicine (IOM).

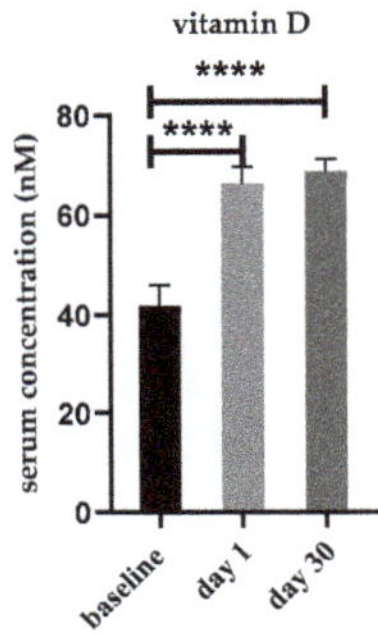

Figure 2. Mean serum $25(OH)D_3$ concentrations at baseline, day 1 and day 30 following a single high dose of vitamin D_3 supplementation (n = 50). Error bars show SEM. **** $p < 0.0001$.

Mean serum IL6 concentrations significantly decreased from 405 ± 30 ng/L at baseline to 350 ± 30 ng/L at day 1 and even 137 ± 20 ng/L at day 30 (Figure 3). This represents an average decrease by 55 and 269 ng/L, respectively. Similar trends were also found for serum IL8 concentrations, where baseline levels gradually decreased from 506 ± 40 ng/L to 455 ± 35 ng/L at day 1 and 192 ± 10 ng/L at day 30 (Figure 3) and for serum TNF levels, which significantly decreased from 165 ± 8 ng/L at baseline to 156 ± 7 ng/L at day 1 and 63 ± 3 ng/L at day 30 (Figure 3). Interestingly, neither Pearson nor Spearman correlation analysis provided any significant correlation between the vitamin D status and the expression level of the proinflammatory cytokines.

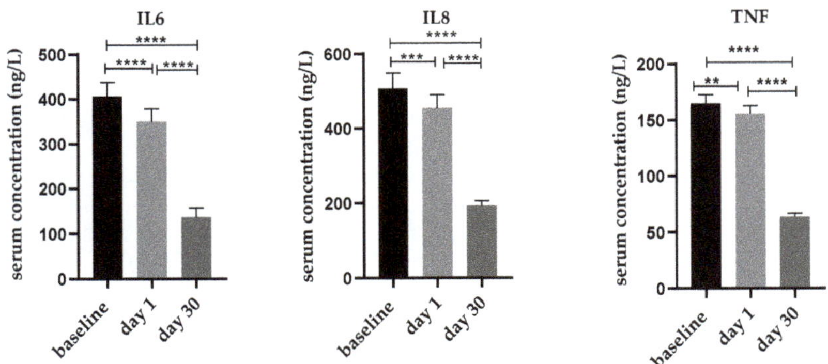

Figure 3. Mean serum levels of the proinflammatory cytokines IL6, IL8 and TNF at baseline, day 1 and day 30 following a single high dose of vitamin D_3 supplementation (n = 50). Error bars show SEM. ** $p < 0.01$, *** $p < 0.001$, **** $p < 0.0001$.

4. Discussion

The purpose of this study was to investigate the effectiveness of a single high dose of vitamin D_3 (80,000 IU) on the vitamin D status and the representative proinflammatory cytokines IL6, IL8 and TNF in healthy Saudi females. The vitamin D_3 bolus increased the vitamin D status within a month by nearly 27 nM and achieved a shift in the study participants from vitamin D deficiency and insufficiency to sufficiency. In fact, the approximately 60% increase in vitamin D status was already visible within one day. This result is comparable to a previous study conducted in female adults supplemented with a single high dose of vitamin D_3 (100,000 IU) [39]. For comparison, when a lower dose was used (50,000 IU), the percent increase in serum 25(OH)D_3 concentrations was only 30% [40]. Other previous studies conducted on adults supplemented daily with different doses of vitamin D_3 ranging from 200 to 600 IU for 2 to 5 months showed a similar or lower percent increase in serum 25(OH)D_3 levels [41–44].

A potential chronic toxicity of vitamin D would result from the administration of doses far above the maximally recommended daily dose of 4000 IU vitamin D_3 for months or years that will increase serum 25(OH)D_3 concentrations to 250 nM or more. In addition to elevated serum 25(OH)D_3 concentrations, vitamin D toxicity can be diagnosed by severe hypercalcemia and by very low or undetectable PTH activity [45]. Accordingly, oral supplementation with a single high dose (80,000 IU) is sufficient to increase the level of serum 25(OH)D_3 in a rapid, suitable and safe way, as none of our study participants reached a vitamin D status of more than 125 nM. Moreover, no abnormal changes were found in either serum calcium or PTH levels after supplementation.

An association between high serum 25(OH)D_3 concentrations and low concentrations of the proinflammatory cytokines IL6, IL8 and TNF was reported previously [46–48]. In the present study, low serum 25(OH)D_3 concentrations at baseline were observed in concordance with the high concentration of the proinflammatory cytokines, but these correlations did not reach statistical significance. Importantly, a single vitamin D_3 bolus was sufficient to significantly increase the vitamin D status within one month and in parallel resulted in the reduction in protein levels of IL6, IL8 and TNF by 67, 62 and 61%, respectively, at the end of the study. The downregulation of the expression of the proinflammatory cytokines may be explained by the increased activation of VDR by an elevated vitamin D status. The latter may have caused a raise in 1,25(OH)$_2D_3$ levels in the nuclei of VDR-expressing PBMCs. Although the genes *IL6* and *TNF* are not known as primary vitamin D target genes, a network of secondary and indirect effects of VDR activation can lead to changes in their expression [49]. However, the *IL8* gene is known as a primary vitamin D target [50].

In contrast to our results, Smith et al. (2017) reported that a single high dose of vitamin D_3 (250,000 IU) did not change serum IL6 and IL8 levels in healthy adults, which could be due to the small sample size of their study [36]. Moreover, daily supplementation with low doses of vitamin D_3 (4000 IU) did not affect serum IL6 concentrations in healthy adults [35]. However, serum TNF concentrations were reported to decrease after supplementation of healthy male and female adults with 4000 IU vitamin D_3 for 20 days [37]. Studies conducted on patients with inflammation-related diseases showed that daily supplementation with different doses of vitamin D_3 ranging from 1000 to 50,000 IU for several weeks or months decreased not only serum TNF concentrations but also serum IL6 and IL8 levels [28–32,51–55]. Thus, a single high dose (80,000 IU) of vitamin D_3 is as effective in reducing proinflammatory cytokines as daily doses.

5. Conclusions

An important finding of this study was that oral supplementation with a single high dose (80,000 IU) is effective in improving the serum's vitamin D status and decreasing the concentration of the proinflammatory cytokines in a rapid, suitable and safe way in healthy females. This will help in preventing and reducing vitamin D deficiency, as well as related inflammatory diseases, in the general population. Further research needs to be performed in order investigate the effectiveness of this single high dose on pro- and anti-inflammatory markers in various inflammatory diseases.

Author Contributions: H.M.A., S.A.A., S.O.K. and C.C. conceived the idea and designed the study. S.F.S. carried out the experiments. H.M.A., S.A.A. and S.F.S. analyzed the data. H.M.A., S.A.A. and H.F.A. performed the literature search and wrote the manuscript. S.O.K. and C.C. reviewed the original manuscript. All authors have read and agreed to the published version of the manuscript.

Funding: The authors thank the deanship of scientific research (DSR). This work was supported by the deanship of scientific research (DSR), King Abdulaziz University, Saudi Arabia under Grant [RG: 21-130-38]. C.C. receives funding from the European Union's Horizon2020 research and innovation program under grant agreement no. 952601.

Institutional Review Board Statement: The study was conducted in accordance with the Declaration of Helsinki and approved by the ethical committee of the Faculty of Medicine, King Abdulaziz University (reference number 30-18).

Informed Consent Statement: Informed consent was obtained from all subjects involved in the study.

Data Availability Statement: Data are available when requested.

Acknowledgments: The authors thank the deanship of scientific research (DSR). This work was supported by the deanship of scientific research (DSR), King Abdulaziz University, Saudi Arabia under Grant [RG: 21-130-38].

Conflicts of Interest: The authors declare no conflict of interest.

References

1. Wacker, M.; Holick, M.F. Sunlight and Vitamin D: A global perspective for health. *Dermatoendocrinol* **2013**, *5*, 51–108. [CrossRef]
2. Carlberg, C. Chapter Ten—Molecular approaches for optimizing vitamin D supplementation. In *Vitamins & Hormones*; Litwack, G., Ed.; Academic Press: New York, NY, USA, 2016; Volume 100, pp. 255–271.
3. Carlberg, C.; Haq, A. The concept of the personal vitamin D response index. *J. Steroid Biochem. Mol. Biol.* **2018**, *175*, 12–17. [CrossRef]
4. Luxwolda, M.F.; Kuipers, R.S.; Kema, I.P.; van der Veer, E.; Dijck-Brouwer, D.A.; Muskiet, F.A. Vitamin D status indicators in indigenous populations in East Africa. *Eur. J. Nutr.* **2013**, *52*, 1115–1125. [CrossRef]
5. Luxwolda, M.F.; Kuipers, R.S.; Kema, I.P.; Dijck-Brouwer, D.A.; Muskiet, F.A. Traditionally living populations in East Africa have a mean serum 25-hydroxyvitamin D concentration of 115 nmol/l. *Br. J. Nutr.* **2012**, *108*, 1557–1561. [CrossRef]
6. Al-Alyani, H.; Al-Turki, H.A.; Al-Essa, O.N.; Alani, F.M.; Sadat-Ali, M. Vitamin D deficiency in Saudi Arabians: A reality or simply hype: A meta-analysis (2008–2015). *J. Family Community Med.* **2018**, *25*, 1. [CrossRef]
7. Holick, M.F. Vitamin D deficiency. *N. Engl. J. Med.* **2007**, *357*, 266–281. [CrossRef] [PubMed]
8. Grant, W.B.; Boucher, B.J.; Al Anouti, F.; Pilz, S. Comparing the evidence from observational studies and randomized controlled trials for nonskeletal health effects of vitamin D. *Nutrients* **2022**, *14*, 3811. [CrossRef]
9. Chun, R.F.; Liu, P.T.; Modlin, R.L.; Adams, J.S.; Hewison, M. Impact of vitamin D on immune function: Lessons learned from genome-wide analysis. *Front. Physiol.* **2014**, *5*, 151. [CrossRef]
10. Di Filippo, L.; De Lorenzo, R.; Giustina, A.; Rovere-Querini, P.; Conte, C. Vitamin D in osteosarcopenic obesity. *Nutrients* **2022**, *14*, 1816. [CrossRef]
11. Carlberg, C.; Polly, P. Gene regulation by vitamin D_3. *Crit. Rev. Eukaryot. Gene Expr.* **1998**, *8*, 19–42. [CrossRef] [PubMed]
12. Haussler, M.R.; Haussler, C.A.; Jurutka, P.W.; Thompson, P.D.; Hsieh, J.C.; Remus, L.S.; Selznick, S.H.; Whitfield, G.K. The vitamin D hormone and its nuclear receptor: Molecular actions and disease states. *J. Endocrinol.* **1997**, *154*, S57–S73. [PubMed]
13. Carlberg, C.; Dunlop, T.W. An integrated biological approach to nuclear receptor signaling in physiological control and disease. *Crit. Rev. Eukaryot. Gene Expr.* **2006**, *16*, 1–22. [CrossRef] [PubMed]
14. Tuoresmäki, P.; Väisänen, S.; Neme, A.; Heikkinen, S.; Carlberg, C. Patterns of genome-wide VDR locations. *PLoS ONE* **2014**, *9*, e96105. [CrossRef] [PubMed]
15. Seuter, S.; Neme, A.; Carlberg, C. Epigenome-wide effects of vitamin D and their impact on the transcriptome of human monocytes involve CTCF. *Nucleic Acids Res.* **2016**, *44*, 4090–4104. [CrossRef] [PubMed]
16. Ramagopalan, S.V.; Heger, A.; Berlanga, A.J.; Maugeri, N.J.; Lincoln, M.R.; Burrell, A.; Handunnetthi, L.; Handel, A.E.; Disanto, G.; Orton, S.M.; et al. A ChIP-seq defined genome-wide map of vitamin D receptor binding: Associations with disease and evolution. *Genome Res.* **2010**, *20*, 1352–1360. [CrossRef] [PubMed]
17. Handel, A.E.; Sandve, G.K.; Disanto, G.; Berlanga-Taylor, A.J.; Gallone, G.; Hanwell, H.; Drabløs, F.; Giovannoni, G.; Ebers, G.C.; Ramagopalan, S.V. Vitamin D receptor ChIP-seq in primary CD4+ cells: Relationship to serum 25-hydroxyvitamin D levels and autoimmune disease. *BMC Med.* **2013**, *11*, 163. [CrossRef] [PubMed]

18. Wöbke, T.K.; Sorg, B.L.; Steinhilber, D. Vitamin D in inflammatory diseases. *Front. Physiol.* **2014**, *5*, 244. [CrossRef]
19. Malmberg, H.R.; Hanel, A.; Taipale, M.; Heikkinen, S.; Carlberg, C. Vitamin D treatment sequence is critical for transcriptome modulation of immune challenged primary human cells. *Front. Immunol.* **2021**, *12*, 754056. [CrossRef]
20. Zhang, Y.; Leung, D.Y.; Richers, B.N.; Liu, Y.; Remigio, L.K.; Riches, D.W.; Goleva, E. Vitamin D inhibits monocyte/macrophage proinflammatory cytokine production by targeting MAPK phosphatase-1. *J. Immunol.* **2012**, *188*, 2127–2135. [CrossRef]
21. Willheim, M.; Thien, R.; Schrattbauer, K.; Bajna, E.; Holub, M.; Gruber, R.; Baier, K.; Pietschmann, P.; Reinisch, W.; Scheiner, O.; et al. Regulatory effects of 1α,25-dihydroxyvitamin D_3 on the cytokine production of human peripheral blood lymphocytes. *J. Clin. Endocrinol. Metab.* **1999**, *84*, 3739–3744. [CrossRef]
22. Joshi, S.; Pantalena, L.C.; Liu, X.K.; Gaffen, S.L.; Liu, H.; Rohowsky-Kochan, C.; Ichiyama, K.; Yoshimura, A.; Steinman, L.; Christakos, S.; et al. 1,25-dihydroxyvitamin D_3 ameliorates Th17 autoimmunity via transcriptional modulation of interleukin-17A. *Mol. Cell Biol.* **2011**, *31*, 3653–3669. [CrossRef]
23. Di Rosa, M.; Malaguarnera, G.; De Gregorio, C.; Palumbo, M.; Nunnari, G.; Malaguarnera, L. Immuno-modulatory effects of vitamin D3 in human monocyte and macrophages. *Cell Immunol.* **2012**, *280*, 36–43. [CrossRef] [PubMed]
24. Müller, K.; Bendtzen, K. Inhibition of human T lymphocyte proliferation and cytokine production by 1,25-dihydroxyvitamin D_3. Differential effects on CD45RA+ and CD45R0+ cells. *Autoimmunity* **1992**, *14*, 37–43. [CrossRef]
25. Panichi, V.; De Pietro, S.; Andreini, B.; Bianchi, A.M.; Migliori, M.; Taccola, D.; Giovannini, L.; Tetta, C.; Palla, R. Calcitriol modulates in vivo and in vitro cytokine production: A role for intracellular calcium. *Kidney Int.* **1998**, *54*, 1463–1469. [CrossRef]
26. Rausch-Fan, X.; Leutmezer, F.; Willheim, M.; Spittler, A.; Bohle, B.; Ebner, C.; Jensen-Jarolim, E.; Boltz-Nitulescu, G. Regulation of cytokine production in human peripheral blood mononuclear cells and allergen-specific th cell clones by 1α,25-dihydroxyvitamin D_3. *Int. Arch. Allergy Immunol.* **2002**, *128*, 33–41. [CrossRef]
27. Prabhu Anand, S.; Selvaraj, P.; Narayanan, P.R. Effect of 1,25 dihydroxyvitamin D_3 on intracellular IFN-gamma and TNF-α positive T cell subsets in pulmonary tuberculosis. *Cytokine* **2009**, *45*, 105–110. [CrossRef]
28. Sabico, S.; Enani, M.A.; Sheshah, E.; Aljohani, N.J.; Aldisi, D.A.; Alotaibi, N.H.; Alshingetti, N.; Alomar, S.Y.; Alnaami, A.M.; Amer, O.E.; et al. Effects of a 2-week 5000 IU versus 1000 IU vitamin D_3 supplementation on recovery of symptoms in patients with mild to moderate Covid-19: A randomized clinical trial. *Nutrients* **2021**, *13*, 2170. [CrossRef]
29. Haidari, F.; Abiri, B.; Iravani, M.; Ahmadi-Angali, K.; Vafa, M. Randomized study of the effect of vitamin D and omega-3 fatty acids cosupplementation as adjuvant chemotherapy on inflammation and nutritional status in colorectal cancer patients. *J. Diet. Suppl.* **2020**, *17*, 384–400. [CrossRef]
30. Khalighi Sikaroudi, M.; Mokhtare, M.; Janani, L.; Faghihi Kashani, A.H.; Masoodi, M.; Agah, S.; Abbaspour, N.; Dehnad, A.; Shidfar, F. Vitamin D_3 supplementation in diarrhea-predominant irritable bowel syndrome patients: The effects on symptoms improvement, serum corticotropin-releasing hormone, and interleukin-6—a randomized clinical trial. *Complement. Med. Res.* **2020**, *27*, 302–309. [CrossRef]
31. Mirzaei, K.; Hossein-Nezhad, A.; Keshavarz, S.A.; Eshaghi, S.M.; Koohdani, F.; Saboor-Yaraghi, A.A.; Hosseini, S.; Tootee, A.; Djalali, M. Insulin resistance via modification of PGC1α function identifying a possible preventive role of vitamin D analogues in chronic inflammatory state of obesity. A double blind clinical trial study. *Minerva Med.* **2014**, *105*, 63–78.
32. Imanparast, F.; Javaheri, J.; Kamankesh, F.; Rafiei, F.; Salehi, A.; Mollaaliakbari, Z.; Rezaei, F.; Rahimi, A.; Abbasi, E. The effects of chromium and vitamin D_3 co-supplementation on insulin resistance and tumor necrosis factor-alpha in type 2 diabetes: A randomized placebo-controlled trial. *Appl. Physiol. Nutr. Metab.* **2020**, *45*, 471–477. [CrossRef] [PubMed]
33. Wimalawansa, S.J. Rapidly increasing serum 25(OH)D boosts the immune system, against infections-sepsis and COVID-19. *Nutrients* **2022**, *14*, 2997. [CrossRef] [PubMed]
34. Hopefl, R.; Ben-Eltriki, M.; Deb, S. Association between vitamin D levels and inflammatory markers in COVID-19 patients: A meta-analysis of observational studies. *J. Pharm. Pharm. Sci.* **2022**, *25*, 124–136. [CrossRef]
35. Chandler, P.D.; Scott, J.B.; Drake, B.F.; Ng, K.; Manson, J.E.; Rifai, N.; Chan, A.T.; Bennett, G.G.; Hollis, B.W.; Giovannucci, E.L.; et al. Impact of vitamin D supplementation on inflammatory markers in African Americans: Results of a four-arm, randomized, placebo-controlled trial. *Cancer Prev. Res.* **2014**, *7*, 218–225. [CrossRef] [PubMed]
36. Smith, E.M.; Alvarez, J.A.; Kearns, M.D.; Hao, L.; Sloan, J.H.; Konrad, R.J.; Ziegler, T.R.; Zughaier, S.M.; Tangpricha, V. High-dose vitamin D_3 reduces circulating hepcidin concentrations: A pilot, randomized, double-blind, placebo-controlled trial in healthy adults. *Clin. Nutr.* **2017**, *36*, 980–985. [CrossRef] [PubMed]
37. Alkhedaide, A.Q.H.; Alshehri, Z.S.; Soliman, M.M.; Althumali, K.W.; Abu-Elzahab, H.S.; Baiomy, A.A.A. Vitamin D_3 supplementation improves immune and inflammatory response in vitamin D deficient adults in Taif, Saudi Arabia. *Biomed. Res.* **2016**, *27*, 1049–1053.
38. Vieth, R. Chapter 57—The pharmacology of vitamin D. In *Vitamin D*, 3rd ed.; Feldman, D., Pike, J.W., Adams, J.S., Eds.; Academic Press: San Diego, CA, USA, 2011; pp. 1041–1066.
39. Witham, M.D.; Adams, F.; Kabir, G.; Kennedy, G.; Belch, J.J.; Khan, F. Effect of short-term vitamin D supplementation on markers of vascular health in South Asian women living in the UK—A randomised controlled trial. *Atherosclerosis* **2013**, *230*, 293–299. [CrossRef]
40. Raimundo, F.V.; Lang, M.A.; Scopel, L.; Marcondes, N.A.; Araújo, M.G.; Faulhaber, G.A.; Furlanetto, T.W. Effect of fat on serum 25-hydroxyvitamin D levels after a single oral dose of vitamin D in young healthy adults: A double-blind randomized placebo-controlled study. *Eur. J. Nutr.* **2015**, *54*, 391–396. [CrossRef] [PubMed]

41. Yao, P.; Lu, L.; Hu, Y.; Liu, G.; Chen, X.; Sun, L.; Ye, X.; Zheng, H.; Chen, Y.; Hu, F.B.; et al. A dose-response study of vitamin D_3 supplementation in healthy Chinese: A 5-arm randomized, placebo-controlled trial. *Eur. J. Nutr.* **2016**, *55*, 383–392. [CrossRef]
42. Pilz, S.; Hahn, A.; Schön, C.; Wilhelm, M.; Obeid, R. Effect of two different multimicronutrient supplements on vitamin D status in women of childbearing age: A randomized trial. *Nutrients* **2017**, *9*, 30. [CrossRef]
43. Shirvani, A.; Kalajian, T.A.; Song, A.; Allen, R.; Charoenngam, N.; Lewanczuk, R.; Holick, M.F. Variable genomic and metabolomic responses to varying doses of vitamin D supplementation. *Anticancer Res.* **2020**, *40*, 535–543. [CrossRef]
44. Pettersen, J.A. Does high dose vitamin D supplementation enhance cognition?: A randomized trial in healthy adults. *Exp. Gerontol.* **2017**, *90*, 90–97. [CrossRef] [PubMed]
45. Marcinowska-Suchowierska, E.; Kupisz-Urbańska, M.; Łukaszkiewicz, J.; Płudowski, P.; Jones, G. Vitamin D toxicity-a clinical perspective. *Front. Endocrinol.* **2018**, *9*, 550. [CrossRef]
46. De Vita, F.; Lauretani, F.; Bauer, J.; Bautmans, I.; Shardell, M.; Cherubini, A.; Bondi, G.; Zuliani, G.; Bandinelli, S.; Pedrazzoni, M.; et al. Relationship between vitamin D and inflammatory markers in older individuals. *Age* **2014**, *36*, 9694. [CrossRef] [PubMed]
47. Laird, E.; McNulty, H.; Ward, M.; Hoey, L.; McSorley, E.; Wallace, J.M.; Carson, E.; Molloy, A.M.; Healy, M.; Casey, M.C.; et al. Vitamin D deficiency is associated with inflammation in older Irish adults. *J. Clin. Endocrinol. Metab.* **2014**, *99*, 1807–1815. [CrossRef]
48. Liefaard, M.C.; Ligthart, S.; Vitezova, A.; Hofman, A.; Uitterlinden, A.G.; Kiefte-de Jong, J.C.; Franco, O.H.; Zillikens, M.C.; Dehghan, A. Vitamin D and C-reactive protein: A Mendelian randomization study. *PLoS ONE* **2015**, *10*, e0131740. [CrossRef]
49. Hanel, A.; Carlberg, C. Time-resolved gene expression analysis monitors the regulation of inflammatory mediators and attenuation of adaptive immune response by vitamin D. *Int. J. Mol. Sci.* **2022**, *23*, 911. [CrossRef] [PubMed]
50. Ryynänen, J.; Carlberg, C. Primary 1,25-dihydroxyvitamin D_3 response of the interleukin 8 gene cluster in human monocyte- and macrophage-like cells. *PLoS ONE* **2013**, *8*, e78170. [CrossRef] [PubMed]
51. Ghorbani, Z.; Togha, M.; Rafiee, P.; Ahmadi, Z.S.; Rasekh Magham, R.; Djalali, M.; Shahemi, S.; Martami, F.; Zareei, M.; Razeghi Jahromi, S.; et al. Vitamin D_3 might improve headache characteristics and protect against inflammation in migraine: A randomized clinical trial. *Neurol. Sci.* **2020**, *41*, 1183–1192. [CrossRef]
52. Esfandiari, A.; Pourghassem Gargari, B.; Noshad, H.; Sarbakhsh, P.; Mobasseri, M.; Barzegari, M.; Arzhang, P. The effects of vitamin D_3 supplementation on some metabolic and inflammatory markers in diabetic nephropathy patients with marginal status of vitamin D: A randomized double blind placebo controlled clinical trial. *Diabetes Metab. Syndr.* **2019**, *13*, 278–283. [CrossRef]
53. Gagnon, C.; Daly, R.M.; Carpentier, A.; Lu, Z.X.; Shore-Lorenti, C.; Sikaris, K.; Jean, S.; Ebeling, P.R. Effects of combined calcium and vitamin D supplementation on insulin secretion, insulin sensitivity and β-cell function in multi-ethnic vitamin D-deficient adults at risk for type 2 diabetes: A pilot randomized, placebo-controlled trial. *PLoS ONE* **2014**, *9*, e109607. [CrossRef]
54. Beilfuss, J.; Berg, V.; Sneve, M.; Jorde, R.; Kamycheva, E. Effects of a 1-year supplementation with cholecalciferol on interleukin-6, tumor necrosis factor-alpha and insulin resistance in overweight and obese subjects. *Cytokine* **2012**, *60*, 870–874. [CrossRef] [PubMed]
55. Pincikova, T.; Paquin-Proulx, D.; Sandberg, J.K.; Flodström-Tullberg, M.; Hjelte, L. Clinical impact of vitamin D treatment in cystic fibrosis: A pilot randomized, controlled trial. *Eur. J. Clin. Nutr.* **2017**, *71*, 203–205. [CrossRef]

MDPI

Review

Vitamin D and Cancer: An Historical Overview of the Epidemiology and Mechanisms

Alberto Muñoz [1] and William B. Grant [2,*]

1 Instituto de Investigaciones Biomédicas "Alberto Sols", Consejo Superior de Investigaciones Científicas, Universidad Autónoma de Madrid, CIBERONC and IdiPAZ, 28029 Madrid, Spain; amunoz@iib.uam.es

2 Sunlight, Nutrition and Health Research Center, P.O. Box 641603, San Francisco, CA 94164-1603, USA

* Correspondence: wbgrant@infionline.net; Tel.: +14-15-409-1980

Abstract: This is a narrative review of the evidence supporting vitamin D's anticancer actions. The first section reviews the findings from ecological studies of cancer with respect to indices of solar radiation, which found a reduced risk of incidence and mortality for approximately 23 types of cancer. Meta-analyses of observational studies reported the inverse correlations of serum 25-hydroxyvitamin D [25(OH)D] with the incidence of 12 types of cancer. Case-control studies with a 25(OH)D concentration measured near the time of cancer diagnosis are stronger than nested case-control and cohort studies as long follow-up times reduce the correlations due to changes in 25(OH)D with time. There is no evidence that undiagnosed cancer reduces 25(OH)D concentrations unless the cancer is at a very advanced stage. Meta-analyses of cancer incidence with respect to dietary intake have had limited success due to the low amount of vitamin D in most diets. An analysis of 25(OH)D-cancer incidence rates suggests that achieving 80 ng/mL vs. 10 ng/mL would reduce cancer incidence rates by 70 ± 10%. Clinical trials have provided limited support for the UVB-vitamin D-cancer hypothesis due to poor design and execution. In recent decades, many experimental studies in cultured cells and animal models have described a wide range of anticancer effects of vitamin D compounds. This paper will review studies showing the inhibition of tumor cell proliferation, dedifferentiation, and invasion together with the sensitization to proapoptotic agents. Moreover, 1,25-$(OH)_2D_3$ and other vitamin D receptor agonists modulate the biology of several types of stromal cells such as fibroblasts, endothelial and immune cells in a way that interferes the apparition of metastases. In sum, the available mechanistic data support the global protective action of vitamin D against several important types of cancer.

Keywords: 25-hydroxyvitamin D; 1,25-$(OH)_2D_3$; antitumor action; breast cancer; case-control studies; colorectal cancer; cohort studies; ecological studies; epidemiological studies; randomized controlled trials; UVB; vitamin D

Citation: Muñoz, A.; Grant, W.B. Vitamin D and Cancer: An Historical Overview of the Epidemiology and Mechanisms. *Nutrients* **2022**, *14*, 1448. https://doi.org/10.3390/nu14071448

Academic Editor: Andrea Fabbri

Received: 14 March 2022
Accepted: 28 March 2022
Published: 30 March 2022

Publisher's Note: MDPI stays neutral with regard to jurisdictional claims in published maps and institutional affiliations.

1. Introduction

The role of vitamin D in reducing the risk of cancer incidence and death has been studied for years. A search of PubMed on 10 March 2022 searching for "cancer" and "vitamin D" or "vitamin D_3" in the title or abstract found 6732 publications starting in 1949. Of these, 523 were published prior to 2000; 1630 were published from 2000 through 2009; 1797 were published from 2010 through 2014; and 2782 were published in or after 2015. Publications with vitamin D and cancer in the title or abstract rose from 13 in 1990, 34 in 1995, 75 in 2000, 170 in 2005, 338 in 2010, 401 in 2012, and between 400 and 500 per year since then.

The earliest studies were ecological studies of cancer mortality rates with respect to indices of solar total or UVB radiation or laboratory studies of mechanisms of vitamin D metabolites on cancer cells. As time progressed, observational studies of cancer incidence with respect to serum 25-hydroxyvitamin D [25(OH)D] took place, and studies of the mechanisms of vitamin on cancer incidence, progression, and metastasis were conducted. Later, randomized controlled trials (RCTs) of cancer risk with respect to vitamin D supplementation were conducted, and as more observational studies accrued, meta-analyses were conducted. Along the way, research approaches built on previous studies. However, since there are many sources of vitamin D, UVB exposure, diet, and supplements, and since 25(OH)D concentrations vary with time, both seasonally and over long periods, and since quantifying 25(OH)D concentrations can be uncertain and is not always conducted in studies, all such human studies of vitamin D and cancer are subject to error. There are also methodological issues, such as how to adjust for when 25(OH)D was measured. In addition, what was found in one group of people may not apply to other groups, such as those with different diets, geographical location, clothing, occupation, age, genetics, and BMI. Thus, all the epidemiological studies and RCTs have inherent limitations. However, by taking a comprehensive look at the findings from many types of studies and trying to identify those that are most reliable, a reasonable picture can emerge. What has emerged is that 25(OH)D concentrations play very important roles in the incidence, progression, and death for many types of cancer. While the roles of vitamin D in cancer are not fully understood, there is enough information for clinical and public health decisions to be made.

The epidemiology of vitamin D and cancer can be examined through the prisms of ecological studies, observational studies, and clinical trials. This review looks at findings from ecological studies of cancer risk with respect to indices of solar ultraviolet-B (UVB) doses, observational studies of cancer risk with respect to serum 25(OH)D concentration and oral vitamin D intake, and randomized controlled trials (RCTs) of cancer risk with respect to vitamin D supplementation.

Epidemiological data prompted the study of the putative anticancer action of vitamin D in the laboratory. Two important considerations in the study of the action of 1,25-$(OH)_2D_3$ and analogues in experimental cancer systems are the expression of vitamin D receptor (VDR), which is frequently low or absent, and the high doses of its ligands that are usually required to observe effects. A lack of VDR is linked to transcriptional (by silencing by DNA methylation or repression by SNAIL1/2), posttranscriptional (by several microRNAs) or posttranslational (phosphorylation, alteration of subcellular localization) inhibitory mechanisms, and low cell responsiveness to VDR ligands is often associated with upregulation of the 1,25-$(OH)_2D_3$ degrading enzyme CYP24A1 in tumor cells. These are two reasons for the absence of the 1,25-$(OH)_2D_3$ effects in some studies. An additional consideration is that, though fully convinced of the value of animal models, we will almost exclusively review studies performed in human systems in this paper.

2. Epidemiological Studies

2.1. Ecological Studies

Ecological studies treat defined populations as entities and compare health outcomes with respect to risk-modifying factors averaged for each population. The groups are usually defined by geographical location but also can be defined by other factors such as occupation. For vitamin D, various indices related to solar UVB dose can be used—for example, annual solar radiation, summertime solar UVB dose, and latitude. Other risk-modifying factors can be added to adjust for confounding factors. Ecological studies offer some advantages: the data required are generally readily available, often with large datasets, and the analyses are easy to do.

Thus, it is not surprising that the first epidemiological study linking vitamin D to a reduced risk of cancer, albeit indirectly, was an ecological study. In 1936, Peller reported that people who developed skin cancer from light exposure, such as from their occupation, had lower rates of internal cancers [1]. In 1937, he showed that sailors in the U.S. Navy, who had extremely high sun exposure, had eight times the expected rate of skin cancer but only 40% of the expected rate of internal cancers [2]. In 1941, Apperly showed that skin cancer mortality rates increased directly in a non-linear fashion with respect to a solar radiation index in the U.S., while total cancer mortality rates decreased in a linear fashion [3]. Evidently, the fact that these three articles were related to vitamin D production went unnoticed until they were cited in a review published in 1993 by Ainsleigh [4].

In 1974, the brothers Cedric and Frank Garland were beginning graduate school at the Johns Hopkins School of Public Health. They attended a lecture by Robert N. Hoover, one author of the *Atlas of Cancer Mortality for U.S. Counties, 1950–1969* [5]. They were struck by the map for mortality, by county, for cancer of the large intestine except the rectum in white males. It showed low rates in three southwest states and high rates in approximately 15 northeast states. The Garlands reasoned that because vitamin D production is the most important health effect of sun exposure, vitamin D must reduce the risk of cancer in the large intestine (colon). They submitted manuscripts to several journals before one was finally accepted and published in the UK in 1980 [6]. They next found support for their hypothesis in terms of the reduced risk of colorectal cancer with respect to dietary vitamin D and calcium [7], prediagnostic serum 25(OH)D concentration, and risk of colon cancer [8]. They later published early ecological studies on solar radiation and the risk of breast cancer [9] and ovarian cancer [10]. Cedric Garland described their discovery and later work in an online posting at Grassrootshealth.net [11].

In 1999, the National Cancer Institute published the *Atlas of Cancer Mortality in the United States, 1950–1994* [12]. That revised edition used 10 colors (five shades each of blue and red) to show mortality rates for 38 cancers (see the breast cancer map in Garland's web post [11] as well as for other cancers at www.sunarc.org, both accessed on 24 February 2022) rather than only five in the earlier version [5]. Data were also displayed for 3053 counties and 506 state economic areas (totals of data for contiguous counties), and showed results for white people (including Hispanics) and black people separately. Through the previous work of one author (W.B.G.) at NASA in Virginia at the time, a map was available of surface-level solar UVB doses in the United States for July 1992 [www.sunarc.org (accessed 24 on February 2022)]. Solar UVB decreases with increasing latitude, albeit with higher doses at any latitude west of the Rocky Mountains than to the east. That effect is due to a combination of higher surface elevation in the west as well as a thinner stratospheric ozone layer owing to the prevailing westerly winds pushing the tropopause up as the air masses cross the Rocky Mountains. Inverse correlations were found for 11 cancers with respect to solar UVB doses for white Americans and several types of cancer for black Americans [13]. A new set of analyses, this time by state, included several risk-modifying factors: alcohol consumption, Hispanic heritage, lung cancer as an index of smoking, poverty status, and urban/rural residence [14]. However, the attribution to solar UVB did not change much between the two articles.

Later, a separate analysis regarding cancer mortality rates for black Americans was published [15]. Significant inverse correlations were found for lung cancer for males and breast cancer for females. The results for colon, esophageal, gastric, and rectal cancer suggested an inverse correlation with respect to solar UVB, but alcohol consumption rates and lung cancer mortality rates also had similar regression coefficients. As a result, UVB did not have a low enough p-value to satisfy the Bonferroni criteria. The results were weak because of the lower numbers of black participants in addition to having lower 25(OH)D concentrations [16].

Several ecological studies of UVB and cancer incidence or mortality rates have been published, particularly between 2002 and 2012 [17]. They helped encourage observational studies, mechanism studies, and clinical trials to explore the relationship between vitamin D and cancer. Single-country studies are preferred because people in individual countries tend to have many similarities, such as clothing preferences, diet, and religion, as well as differences, such as smoking, socioeconomic status, and urban/rural residences. Those comparisons can often be modeled. In addition, variations in solar UVB doses tend to be significant [18,19].

Table 1 outlines the more important solar single-country UVB–cancer ecological studies starting in 2002. Most are from mid-latitude countries, but one is from a subtropical country (Iran) and two encompass the Arctic Circle. Most studies used UVB data from NASA's

Total Ozone Mapping Spectrometer (TOMS) satellite instrument [20], but other indices were used as well, including latitude and global solar radiation.

Table 1. Characteristics of large single-country ecological studies of cancer incidence or mortality rates with respect to solar UVB doses.

Country(ies)	Solar UVB Index	Latitude (°N)	Incidence or Mortality; Years of Data	No. of Cases	Confounding Factors	Ref.
U.S.	Surface UVB, July 1992, TOMS	25–45	Mortality, 1950–1994	9.5 million, 1970–1994	None	[13]
Japan	Annual hours of solar radiation	30–45	Mortality, 2000	180,000	Fat intake for colon, rectum, and prostate; salt intake for stomach cancer	[21]
U.S. (white pop.)	Surface UVB, July 1992	25–45	Mortality, 1950–1994	9.5 million, 1970–1994	Alcohol consumption, Hispanic heritage, lung cancer (index for smoking), poverty, urban/rural residence	[14]
U.S.	300–320 nm, TOMS, north vs. south	25–45	Incidence, 1998–2002; mortality, 1993–2002	Incidence, 3.4 million; mortality, 3.5 million	Age, air quality, alcohol, exercise, income, outdoor occupation, poverty, smoking, urban/rural residence	[22]
Japan	Global solar radiation	30–45	Mortality, 1998–2002	~900,000	Dietary factors, smoking, socioeconomic conditions	[23]
China	TOMS, 305 nm	22–50	Incidence, 1998–2002; mortality, 1990–1992		Urban/rural residence	[18]
Russia	Latitude	43–69	Incidence, mortality, 2008	incidence, ~250,000; deaths, ~140,000	None	[24]
Nordic countries	Lip cancer less lung cancer incidence	55–70	Incidence, 1961–2005	2.8 million	Lung cancer	[25]

Pop., population; TOMS, NASA's Total Ozone Mapping Spectrometer satellite instrument.

One ecological study was based on data by occupation from a study involving 2.8 million cancer incidence cases from 15 million inhabitants of the five Nordic countries aged 30–64 years in the 10-year censuses from 1960 to 1990 [26]. The study included 53 occupational categories. A novel index, lip cancer less lung cancer, was used for long-term UVB exposure [25]. A suspected important risk factor for lip cancer was solar UVB exposure [27]. A study conducted in Denmark reported that outdoor workers employed for more than 10 years had twice the rate of lip cancer than nonmelanoma skin cancer [28]. Smoking also is a well-known risk factor for lip cancer. As expected, people in occupational categories associated with outdoor work, such as farmers, forestry workers, and gardeners, had the lowest cancer incidence rates.

Table 2 presents findings regarding the incidence of specific cancers for males and females with respect to the UVB indices used. Cancers are listed in descending order of incidence rates in the United States in 2009 to show that as the number of cases decreases, so does the likelihood of finding significant correlations with solar UVB. Note that the results from the United States [22], Russia [24], and the Nordic countries [25] are in good agreement.

Table 2. Ecological studies of cancer incidence rates with respect to indices of solar UVB doses.

Incidence [29] (×1000)	Cancer	USA [22]	China [18]	Russia [24]	Nordic [25]
219.4	Lung		–M, FNS, –R, –U		M, FNS
194.3	Breast	F	–F, –R, –U		M, F
192.3	Prostate	M		–M	MNS
147.0	Colorectal		M, F, R		
106.1	Colon	M, F			M, F
71.0	Bladder, urinary	M, F	–M, –F, –R, –U		M, F
68.7	Melanoma	–M, –F		M + F	M
66.0	Non-Hodgkin lymphoma	M, F			NS
57.8	Kidney	M, F		M + F	M, FNS
44.8	Leukemia	M, F	MNS, FNS, R, –U		
42.5	Pancreas	M, F		M + F	M, FNS
42.2	Uterus, corpus	F			FNS
40.9	Rectum	M, F			M, FNS
37.2	Thyroid	MNS, F			
35.7	Oral cavity and pharynx	–M, –F			
23.1	Oral				M
22.6	Myeloma	M, F		M + F	
22.6	Liver		–M, –F, –R, –U		M, FNS
22.1	Brain				M
21.6	Ovary	FNS			
21.1	Stomach (gastric)	M, F	M, F, R, –U	M + F	M?, FNS
16.5	Esophagus	M	M, F, R, –U	M + F	MNS
12.6	Pharynx		–M, –F, –R, –U	–(M + F)	
12.3	Larynx				M
11.3	Cervix	–F	F, R,–U		
9.8	Gallbladder	F			M
9.8	Biliary, other	M, F		M + F	
8.5	Hodgkin lymphoma	M, F			
8.4	Testis				NS
6.2	Small intestine	M, F			M
5.9	Skin, other	–M, –F		–(M + F)	–M
5.3	Anus, etc.	–M, –F			
3.6	Vulva	F			

F, female; FNS, female nonsignificant; M, male; MNS, male nonsignificant; R, rural residence; U, urban residence, –, direct correlation; ?, uncertain.

Table 3 is similar to Table 2 except for showing mortality rates, not incidence rates, and cancers are listed in descending order with respect to cancer mortality rates in the United States in 2009. Note the good general agreement between the findings for mortality rates in Table 3 with incidence rates in Table 2. The main exception is that solar UVB dose was inversely correlated with mortality rates for several cancers in China, for which it was directly correlated with incidence rates.

Table 3. Ecological studies of cancer mortality rates with respect to indices of solar UVB doses.

Mortality [29] (×1000)	Cancer	Japan [23]	USA [14]	USA [22]	China [18]	Russia [24]
159.4	Lung	M, F			M, F, R, U	
69.1	Colorectal	M			M, F, R	
49.9	Colon		M, F	M, F		M + F
40.6	Breast	FNS	M, F	F	F, R	–(M + F)
35.2	Pancreas	M, F	M, FNS	M, F		M + F
27.4	Prostate	MNS	MNS	M		M
21.9	Leukemia			M, F	MNS, FNS	
19.5	Non-Hodgkin lymphoma		M, F	M, F		

Table 3. *Cont.*

Mortality [29] (×1000)	Cancer	Japan [23]	USA [14]	USA [22]	China [18]	Russia [24]
19.2	Rectum		M, F	M, F		M + F
18.2	Liver	M		–M, –F	M, F, R	
14.6	Ovary		F	F		F
14.5	Esophagus	M	M, F	M	M, F, R	M + F
14.3	Bladder, urinary		M, F	M, F	M, F, R	M + F
13.9	Kidney		M, F	M, F		M + F
12.9	Brain			–M, –F		
10.6	Myeloma			M, F		M + F
10.6	Stomach (gastric)	M, FNS	M, F	M, F	M, F, U	M + F
8.7	Melanoma			–M, –F		M + F
7.8	Uterus, corpus		F	F		
7.6	Oral cavity and pharynx			–M, –F		
5.4	Oral		MNS, FNS			
4.1	Cervix		F	–F	–F, –R, –U	
3.7	Larynx		M, F?	MNS, FNS		M + F
3.4	Gallbladder	MNS, F	M, F	M, F		
3.4	Biliary, other			M. F		
2.9	Skin, other			–M, –F		–(M + F)
2.2	Pharynx				–M. –F, –R, –U	
1.6	Thyroid			MNS, F		
1.5	Bone and joint			–M, –F		
1.3	Hodgkin lymphoma		M, F	M, F		
1.1	Small intestine			MNS, F		
0.9	Vulva			F		F
0.7	Anus, etc.			–M, –F		M + F

F, female; FNS, female nonsignificant; M, male; MNS, male nonsignificant; R, rural residence; U, urban residence, –, direct correlation; ?, uncertain.

2.2. *Observational Studies Based on Residential UVB Doses*

Related to ecological studies of solar UVB and cancer risk are observational studies of ambient solar UVB doses and cancer risk. Cancer incidence data from the prospective National Institutes of Health—AARP Diet and Health Study were used with solar UVB dose data at residential locations to assess the relationship between UVB and cancer risk [30]. The study was limited to participants living in California, Florida, Georgia (Atlanta), Louisiana, Michigan (Detroit), Pennsylvania, and North Carolina. During the 9 years of follow-up, 75,917 participants developed cancer. Erythemal UV data for July from TOMS for 1978–1993 and 1996–2005 were used. Data were adjusted for age; sex; body mass index (BMI); caloric intake; intake of fruit, vegetables, and red and white meat; alcohol consumption; tobacco smoking; education; physical activity; and median household income. Over 9 years of follow-up, UV exposure was inversely associated with total cancer risk (highest vs. lowest quartile) and decreased risk of non-Hodgkin lymphoma and colon, squamous-cell lung, pleural, prostate, kidney, and bladder cancers (all $p_{\text{trend}} < 0.05$). UV exposure was associated with increased melanoma risk.

Another example is a nested case–control (NCC) study using 373 esophageal and 249 gastric cancer cases from the UK Biobank with respect to UVB doses at the residential location [31]. Annual solar UVB doses ranged from ~500 kJ/m^2 in the south to ~750 kJ/m^2 in the north. Five controls were matched to each case. Data were available for many cancer risk-modifying factors. Significant reductions were found for adjusted esophageal cancer, adjusted lower-third esophageal cancer, and adjusted gastric cancer, in agreement with ecological studies noted previously.

A further discussion of observational studies of cancer incidence and death with respect to solar UVB is in progress.

2.3. Observational Studies Based on Serum 25(OH)D Concentrations

Observational studies examine correlations between risk-modifying factors and health outcomes such as cancer incidence, survival, and mortality rates. Observational studies include cohort studies, both prospective and retrospective; case–control (CC) studies; and cross-sectional studies. Each type has advantages and disadvantages. For example, most observational studies regarding vitamin D use serum 25(OH)D concentrations as the index of vitamin D status, but assays used to measure 25(OH)D concentrations vary in quality [32]. Furthermore, serum 25(OH)D concentrations change with the seasons and over long periods [33]. Some studies use dietary vitamin D, i.e., oral vitamin D, including dietary sources and supplements. However, using dietary sources to assess vitamin D intake is problematic because diet generally accounts for less than 300 IU/d in the United States. Although meat is an important source of vitamin D as 25(OH)D [34], most food frequency tables do not include data on meat [35]. Some studies use personal or geographical solar UVB doses. This review emphasizes those that use serum 25(OH)D concentrations but will also include a few that used solar UVB doses.

Generally, CC studies of cancer risk report a stronger reduction with respect to serum 25(OH)D concentrations than do other observational studies. However, observational studies using serum 25(OH)D concentration from blood drawn before cancer diagnosis are generally considered more accurate than those in which blood is drawn near the time of cancer diagnosis.

Researchers have hypothesized that because RCTs have generally not been able to confirm findings from observational studies for many health outcomes, including cancer, having the disease may reduce 25(OH)D concentrations; that is, "reverse causation" [36,37]. However, that effect has been shown only for acute inflammatory diseases such as acute respiratory tract infections [38].

Although systemic inflammation may play a role in cancer risk, the inflammation does not rise as high as in, say, COVID-19. Reports on levels of C-reactive protein levels, an index of systemic inflammation, at the time of diagnosis show that for COVID-19, values can range from 1 to 120 mg/L as severity increases [39], whereas for cancer, they are between 1 and 4 mg/L [40]. Thus, systemic inflammation is not high at the time of cancer diagnosis. We are not aware of any other factor that could result in reverse causality regarding 25(OH)D concentrations for undiagnosed cancer. As will be discussed, the main reason for discrepancies between observational studies and RCTs of vitamin D and cancer is that the RCTs have not been properly designed and conducted.

Two articles reported that the longer the follow-up time in observational studies of 25(OH)D concentration and cancer risk, the lower the effect of 25(OH)D concentration [41,42]. The same effect has been found for all-cause mortality rates [43]. The reasons include that serum 25(OH)D concentrations change for several reasons and that 25(OH)D concentration near the time of diagnosis is more important than earlier concentrations, even though cancer may develop over a long period. Figure 1 in Grant's 2012 report [43] shows that the correlation coefficient between serum 25(OH)D concentrations repeated in the same group of participants drops to approximately 0.4 after 14 years.

Most observational studies of 25(OH)D concentration and cancer incidence are prospective cohort or NCC studies. An NCC study of 25(OH)D concentration and incidence of colorectal cancer (CRC) based on two Harvard cohorts [44] is reviewed here to show the complexity of such studies. The Health Professionals Follow-up Study (HPFS), with 18,225 male participants who supplied a blood sample, had 179 cases of CRC during follow-up periods up to 8 years. The analysis of results from the cohort was combined with results from the Nurses' Health Study (NHS) of women, of whom 32,826 gave blood samples, and 193 developed CRC during 11 years of follow-up [45]. In the HPFS, values for many factors were recorded at baseline in 1994, including season of blood donation, BMI, physical activity, aspirin use, smoking, alcohol intake, intake of vitamin D, calcium and retinol, and meat intake. Analyses were made for colon, rectal, and CRC with respect to quantiles of 25(OH)D, showing that though the trend in 25(OH)D concentrations was not significant for HPFS alone, it was significant when combined with results from NHS. The pooled odds ratio (OR) for CRC for high versus low quintile of 25(OH)D was 0.66 (95% confidence interval [95% CI], 0.42–1.05; $p_{trend} = 0.01$). The risk of rectal cancer increased with respect to 25(OH)D in the HPFS but decreased in the NHS. Interesting findings also were shown for lifestyle characteristics, including BMI, physical activity, calcium intake, retinol intake, and effect of 25(OH)D measured in winter or summer. Thus, with 372 CRC cases, it was possible to find support for 25(OH)D concentrations reducing the risk of colon cancer and CRC.

A meta-analysis published in 2007 based on five NCC studies found a predicted 50 ± 20% reduction in CRC for 34 ng/mL vs. 6 ng/mL [46].

A pooled analysis of 12 NCC studies for CRC for men showed a relative risk (RR) of 0.93 (95% CI, 0.86–1.00), whereas the pooled analysis for 13 studies for women reported an RR of 0.81 (95% CI, 0.75–0.87) [47]. For men and women combined, the RR was 0.87 (95% CI, 0.75–0.87). A significant reduction in RR was shown for women between approximately 25 and 45 ng/mL, but no significant reduction was evident for men at any range. This analysis did not adjust for follow-up time between blood draw and cancer diagnosis. To examine the effect of follow-up time, plots were made of the ORs or RRs from the meta-analysis by McCullough and colleagues [47]. Table 4 shows the data used. Information regarding the relative weight for each study was not available, so plots were made of OR against follow-up time. Figure 1 shows the results. The RR for zero follow-up time should be approximately 0.75 for men and 0.77 for women. The regression fit to the data for men is OR = 0.74 + 0.031x years, $r = 0.79$, adjusted $r^2 = 0.59$, $p = 0.002$; the regression fit to the data for women is OR = 0.77 + 0.008x years, $r = 0.25$, adjusted $r^2 = 0$, $p = 0.42$. Thus, the lower effect of 25(OH)D on men versus that of women shown in Figure 1 in McCullough and colleagues [47] is due to not accounting for the degradation of the 25(OH)D effect with a longer follow-up time. Providing evidence that the results for men and women should be similar is supported by ecological studies in the United States [14].

Table 4. Data related to Figure 2 in McCullough and colleagues [47].

Study	Follow-Up (Years)	RR	Ref.
Men			
ATBC2	12.5	1.17	[48]
PHS	9.50	1.06	[49]
CLUE II	3.20	0.99	[50]
HPFS	6.30	0.99	[51]
JANUS	5.10	0.93	[52]
EPIC	3.60	0.86	[53]
MEC	1.50	0.86	[54]
CPS-II	3.20	0.83	[55]
JPHC	5.10	0.83	[56]
CARET	4.90	0.82	[57]
PLCO	5.40	0.81	[58]
ABCT1	3.50	0.77	[59]
Women			
ORDET	10.8	1.03	[60]
JPHC	5.10	0.94	[56]
JANUS	5.10	0.90	[52]
BGS	2.30	0.90	[61]
CLUE-II	9.00	0.87	[50]
WHI	3.20	0.87	[62]
NHS	9.60	0.84	[51]
CPS-II	3.20	0.77	[55]
WHS	8.00	0.77	[63]
EPIC	3.60	0.73	[53]

Table 4. *Cont.*

Study	Follow-Up (Years)	RR	Ref.
NYUWHS	12.3	0.72	[64]
PLCO	5.40	0.67	[58]
MEC	1.50	0.63	[54]

ATBC, Alpha-Tocopherol, Beta-Carotene Cancer Prevention Study; BGS, Breakthrough Generations Study; CARET, Carotene and Retinol Efficacy Trial; CLUE II, Cancer Prevention Study II Nutrition Cohort; CPS-II, Cancer Prevention Study II; EPIC, European Prospective Investigation into Cancer and Nutrition; HPFS, Health Professionals Follow-up Study; JANUS, JANUS Serum Bank, Norway; JPHC, Japan Public Health Center-Based Prospective Study; MEC, multiethnic cohort study'; NYUWHS; New York University, Women's Health Study; ORDET, Hormones and Diet in the Etiology of Breast Cancer Risk; PHS, Physicians' Health Study; PLCO, Prostate, Lung, Colorectal and Ovarian Cancer Screening Trial; RR, relative risk; WHI, Women's Health Initiative.

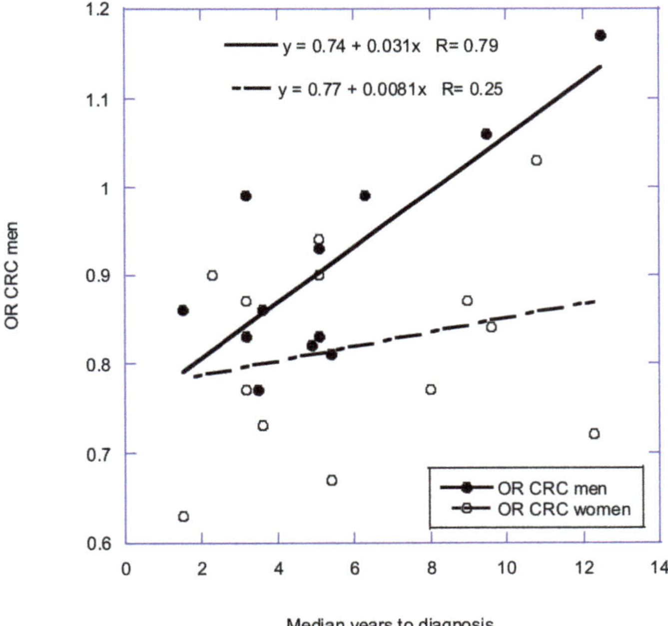

Figure 1. Plot of odds ratio (OR) for CRC against median years to diagnosis for data for men and women used in McCullough and colleagues [47].

In contrast to CRC, prospective and NCC studies with follow-up times greater than 4 years seldom show a significant inverse correlation between serum 25(OH)D concentration and incidence of breast cancer. Breast cancer can develop rapidly, with progression strongly affected by 25(OH)D concentration. Breast cancer is one of the few cancers that have a seasonality in diagnosis, with the highest diagnosis rates in spring and fall [65]. The authors of that study suggested that solar UVB, through producing vitamin D, lowers the risk of breast cancer in summer, whereas higher concentrations of melatonin reduce risk in winter. As a result, many more CC studies of breast cancer with 25(OH)D measured at the time of diagnosis exist than that for CRC.

CC studies of breast cancer incidence with respect to serum 25(OH)D concentrations in pre- and postmenopausal women are discussed first [66,67]. The premenopausal study included 289 cases and 595 matched controls; the postmenopausal study included 1394 cases and 1365 controls. In the premenopausal study, the adjusted OR (aOR) for 25(OH)D >24 ng/mL versus <12 ng/mL was 0.48 (95% CI, 0.29–0.70) and the p_{trend} value for the quantiles was 0.0006. In the postmenopausal study, the aOR for 25(OH)D >30 ng/mL versus <12 ng/mL was 0.31 (95% CI, 0.24–0.42) and the p_{trend} value of the quintiles was <0.0001. In both studies, the risk increased more rapidly as 25(OH)D concentrations decreased below 12 ng/mL. Those two studies show that several individual factors affect cancer risk but, in general, have little impact on the role of 25(OH)D concentration.

The present study incorporated a search at Google Scholar and the National Library of Medicine's PubMed database for meta-analyses of cancer incidence or mortality rate with respect to serum 25(OH)D concentration. The most recent meta-analyses were favored. For several cancers, Table 5 includes more than one meta-analysis. Of the 44 studies listed as CC in the meta-analysis of breast cancer by Song and colleagues [68], 26 were true CC studies in which serum 25(OH)D concentration was measured near the time of cancer diagnosis for both cases and controls, with 14,851 cases and 30,979 controls. The remaining 18 studies were NCC studies or, in one case, a cross-sectional study. The number of breast cancer cases was 17,871, whereas the number of controls was 21,753. The analysis for cohort studies of breast cancer incidence in that study included the observational study of breast cancer incidence for participants in either two vitamin D plus calcium RCTs or the Grassrootshealth.net community-based cohort [69]. Because those participants generally had serum 25(OH)D measured every 6 months to 1–2 years, that study should have been combined with the CC studies. It reported an 82% lower risk of breast cancer for 25(OH)D concentration >60 ng/mL versus <20 ng/mL (rate ratio = 0.18 [95% CI, 0.04–0.62]).

Table 5. Meta-analyses of observational studies of incidence risk of individual cancer sites related to serum 25(OH)D concentration.

Cancer Site	*N* Studies, Cases, Controls	Type of Study	Follow-Up (Years)	RR (95% CI), High vs. Low	Ref.
All	8, —, —	Prospective, incidence	5–28	0.86 (0.73–1.02)	[70]
All	17, —, —	Prospective, mortality	5–28	0.81 (0.71–0.93)	[70]
Bladder	5, 1251, 1332	CC and NCC, incidence	0 (4), 12, 13	0.70 (0.56–0.88)	[71]
Bladder	2, 2264, 2258	Cohort, incidence	14, 28	0.80 (0.67–0.94)	[71]
Breast	44, 29,095, 53,060	CC and NCC, incidence		0.57 (0.48–0.66)	[68]
Breast	6, 2257, —	Cohort, incidence		1.17 (0.92–1.48)	[68]
Colorectal	11, —, —	1 CC, 9 NCC, 1 meta-analysis, incidence	0–20	0.60 (0.53–0.68)	[72]
Colorectal	6, 1252, —	Cohort, incidence	8–20	0.80 (0.66–0.97)	[72]
Colorectal	15, 6691, —	NCC, incidence		0.67 (0.59–0.76)	[73]
Head and neck	5, —, —	Cohort, incidence	7, 15	0.68 (0.59–0.78)	[74]
Liver	8, 992, —	Cohort, incidence	6–28	0.78 (0.63–0.95)	[75]
Liver	6, 776, —	Cohort, incidence	(0.75), 16–22	0.53 (0.41–0.68)	[76]
Lung	8, 1386, —	Cohort, incidence	7–26	0.72 (0.61–0.85)	[77]
Lung	9, —, —	7 Cohort, 2 CC, incidence		0.84 (0.74–0.95)	[78]
Lung	3, —, —	1 Cohort, 2 CC, mortality		0.76 (0.61–0.94)	[78]
Lung	12, —, —	7 Cohort, 5 CC		1.05 (0.95–1.16)	[79]
Ovarian	8, —, —	CC, cohort, NCC		0.86 (0.56–1.33)	[80]
Pancreatic	5, 1068, —	2 Cohort, 3 NCC, incidence	6.5–21	1.02 (0.66–1.57)	[81]
Pancreatic	5, 2003, —	Cohort, mortality	6.5–21	0.81 (0.68–0.96)	[81]
Prostate	19, 12,786	16 NCC, 3 cohort, incidence		1.15 (1.06–1.24)	[82]
Renal	5, —, —	4 Cohort (+1 CC, 3.5% weighting), incidence	(0), 7–22	0.76 (0.64–0.89)	[83]
Renal	1, —, —	CC, incidence	0	0.30 (0.13–0.72)	[83]
Thyroid	6, 387, 457	CC, incidence	0	Deficiency, 1.30 (1.00–1.69), $p = 0.05$	[84]

95% CI, 95% confidence interval; CC, case–control study; NCC, nested case–control study; parentheses for follow-up years indicate numbers for a very small percentage of the total; RR, relative risk; —, no data.

From the data in Table 5, it is apparent that CC and NCC studies report greater reductions in cancer risk for high versus low 25(OH)D concentration. The reason may be that cohort studies are conducted for longer than CC or NCC studies. That difference lowers the benefit due to 25(OH)D concentrations as a result of changes in 25(OH)D concentration, as discussed previously. Another finding is that studies of mortality rates show greater reductions than studies of incidence rates. That finding is similar to findings in RCTs of cancer as reported, for example, in the VITAL study [85] as well as in a meta-analysis of results from vitamin D–cancer RCTs [86]. The reason for that finding is probably the presence of many risk-modifying factors that affect cancer incidence but few factors other than vitamin D that affect angiogenesis around tumors, cancer progression, and metastasis into stromal tissue.

Table 6 presents findings from a few meta-analyses of observational studies of vitamin D intake, both from diet and from supplements, and cancer risk. The reductions in cancer risk from oral intake are generally much lower than what is found with respect to serum 25(OH)D concentration studies, largely because differences in oral intakes did not have an observable effect on serum 25(OH)D concentrations. In addition, results with respect to serum 25(OH)D concentrations were not given.

Table 6. Meta-analyses of observational studies of the risk of incidence of individual cancer sites related to vitamin D intake.

Cancer Site	*N* Studies	Type of Study	RR (95% CI), High vs. Low Vitamin D Intake	Ref.
Breast	17	8 CC, 9 cohorts	0.97 (0.92–1.07), per 400 IU/d	[68]
Colorectal	12	CC	0.75 (0.67–0.81)	[72]
Colorectal	6	Cohort	0.89 (0.80–1.02)	[72]
Head and neck	3		0.75 (0.58–0.97)	[74]
Lung	6	Cohort	0.89 (0.83–0.97)	[77]
Lung	5	Cohort	0.85 (0.74–0.98)	[79]
Renal	4	CC	0.80 (0.67–0.95)	[83]
Renal	4	Cohort	0.97 (0.77–1.22)	[83]
Overall cancer death			0.84 (0.74–0.95)	[87]

CC, case–control study; NCC, nested case–control study.

Table 7 presents estimates of the OR for maximum 25(OH)D concentration compared with minimum concentration for several cancers. The reviews obtained from these values did not give numerical values, so they were estimated by inspecting the graphs.

Table 7. Estimates of odds ratio for maximum 25(OH)D concentration compared with minimum concentration for several cancers.

Cancer	Min 25(OH)D (ng/mL)	Max 25(OH)D (ng/mL)	OR (95% CI)	Ref.
All, inc	2	25	~0.6	[70]
Bladder, inc	3	30	~0.55 (0.35–0.70)	[71]
Breast, inc (Song et al.)	5	85	~0.2 (0.1–0.3)	[68]
Breast, inc	15	70	0.18 (0.04–0.62)	[69]
Colorectal, inc	4	55	~0.4 (0.3–0.5)	[73]
Colorectal, inc	10	50	~0.7 (0.4–1.0)	[88]
Liver, inc	4	30	0.35 (0.21–0.48)	[76]
Liver, inc	5	30	~0.6 (0.5–0.7)	[75]
Lung, inc	6	21	0.87 (0.76–0.97)	[89]
Lung, inc	10	24	0.80 (0.61–0.98)	[78]
Lung, mort	10	42	0.37 (0.25–0.53)	[78]
Prostate, inc	0	60	~1.3 (1.1–1.8)	[82]
Prostate, mort	4	43	~0.55 (0.2–1.1)	[90]

Inc, incidence; mort, mortality; OR, odds ratio.

Figure 2 shows the plot of OR for cancer incidence against the difference between minimum and maximum 25(OH)D concentration. The plot indicates a nearly linear relationship between serum 25(OH)D concentration and OR. The linearity between OR and 25(OH)D concentration is supported by results in the breast cancer study by McDonnell and colleagues [69]. Many studies have few participants with 25(OH)D concentrations above 40 ng/mL, thereby limiting the ability to investigate the effects of higher 25(OH)D concentrations.

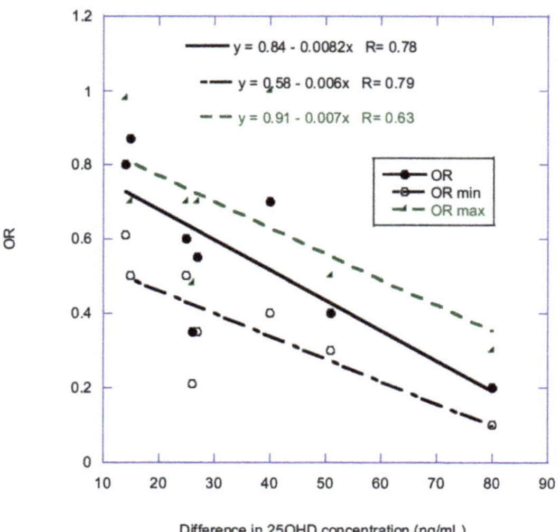

Figure 2. Plot of OR for cancer incidence versus the difference between minimum and maximum 25(OH)D concentration, using data from Table 7, omitting data for all cancer, breast cancer in McDonnell and colleagues [69], and data for prostate cancer.

2.4. *RCTs of Vitamin D and Cancer Risk*

According to a review published in 2019 [86], nine RCTs have studied how vitamin D supplementation affects cancer incidence, of which five also studied the effect on cancer mortality rate. The relative risk of vitamin D supplementation in the treatment versus placebo groups for cancer incidence was 0.98 (95% CI, 0.93–1.03), whereas for cancer, the mortality rate was 0.87 (95% CI 0.79–0.96). Results did not change significantly if they were analyzed by daily intake versus nondaily intake in a large bolus or attained 25(OH)D concentration >40 ng/mL. However, as pointed out in a recent review by Pilz and colleagues, RCTs rarely found a significant benefit from vitamin D supplementation [91].

The information on most of the trials discussed in [86] plus another published thereafter are presented in Tables 8 and 9. As can be seen in Table 8, none of the trials were well designed based on what is now known. Not all trials measured baseline 25(OH)D concentration and when they did, the concentrations were almost always above mean population values. Only five reported achieving 25(OH)D concentrations, and both baseline and achieved concentrations were generally based on a fraction of all participants. Four trials used infrequent bolus doses, which were done to improve compliance but resulted in large variations in 25(OH)D concentration between doses since the half-life of 25(OH)D is approximately two weeks. Some of the trials also gave calcium to the treatment arm but not the control arm. In all cases, participants were permitted to take modest vitamin D supplement doses and solar UVB exposure was not controlled. The mean BMI was generally high in the trials, which is a problem since those with higher BMI do not have the same response for a similar change in 25(OH)D concentration as those with lower BMI. For example, the VITAL study [86] reported that participants with BMI <25 kg/m^2 of body surface area had a significantly reduced risk of cancer from vitamin D supplementation (hazard ratio = 0.76 [95% CI, 0.63–0.90]) but not for higher BMI categories, even though the change in 25(OH)D was near 12 ng/mL for all three BMI categories. The apparent reason is that obesity is an important risk factor for cancer and vitamin D has a limited ability to overcome the mechanisms whereby obesity increases risk of cancer [92]. Finally, only a few of the trials were explicitly designed with cancer incidence a primary outcome.

Table 8. Characteristics of ten RCTs that investigated the effect of vitamin D supplementation on risk of cancer incidence and/or mortality rate.

Location	Mean Baseline and Achieved 25(OH)D (ng/mL), Treatment Arm	Vitamin D Dose (IU) Frequency in Treatment Arm	Duration (Years)	Mean BMI (kg/m^2)	Original Purpose	Reference
UK		100,000/ (4 months)	5.5	24 ± 3	fracture incidence, cause of death	[93]
USA		400/day + 1 g/day Ca	7	28?	colorectal cancer incidence, mortality	[94]
Nebraska, USA	29, 38	1100/day + 1.5 g/day Ca; 1.5 g/day Ca	4	29 ± 6	fracture incidence	[95]
Australia	21, 24–48	500,000/year			falls and fractures	[96]
England, Scotland		800/day; 1 g/d Ca; 800/day + 1 g/day Ca	3			[97]
Nebraska, USA	33, 44	2000/day + 1500 mg/day Ca	4	30 ± 7	cancer	[98]
New Zealand	26, –	100,000/mo	3.3 ± 0.8	28 ± 5	disease incidence with respect to bolus dose of vitamin D	[99]
USA	30, 41	2000/day	5.3	31	cancer and cardiovascular disease risk	[85]
Australia	31 ± 10, 46 ± 12	60,000/ month	5	27?	mortality by disease	[100]

Table 9. Outcomes of ten RCTs that investigated the effect of vitamin D supplementation on risk of cancer incidence and/or mortality rate with respect to intention to treat.

Location	Number of Participants, Cancer Cases, Deaths, Treatment Arm	Number of Participants, Cancer Cases, Deaths, Non-Vitamin D Arm	RR, Incidence (95% CI)	RR, Mortality (95% CI)	Reference
UK	1345, 163, 63	1341, 147, 72	1.11 (0.86–1.42)	0.86 (0.61–1.20)	[93]
USA	18,176, 1634, 344	18,106, 1655, 382	0.98 (0.91–1.05)	0.89 (0.77–1.03)	[94]
Nebraska, USA	446, 13, –	733, 37, –	0.76 (0.38–1.55)		[95]
Australia	1131, 7	1125, 10	0.70 (0.27–1.82)		[96]
England, Scotland	1306, 182, 78; 1311, 189, 95	1343, 187, 73; 1332, 165, 83	1.24 (0.80–2.28)	1.26 (0.73–3.26)	[97]
Nebraska, USA	1156, 45, –	1147, 64, –	0.70 (0.47–1.02)		[98]
New Zealand	2558, 302, –	2550, 293, –	1.01 (0.81–1.25)		[99]
USA	12,927, 793, 154	12,946, 824, 187	0.96 (0.88–1.06)	0.83 (0.67–1.02)	[85]
Meta-analysis for ten incidence trials and five mortality rate trials			0.98 (0.93–1.03)	0.87 (0.79–0.96)	[86]
Australia	21,315, –, 221	10,662, –, 189		1.15 (0.96–1.39)	[100]

Only one outcome based on intention to treat was significantly reduced, that of cancer mortality rate in the VITAL trial [85]. Nonetheless, a meta-analysis of five trials found a significant reduction in the cancer mortality rate [86].

The main problem with vitamin D RCTs seems to be that they are generally designed and conducted by following guidelines for pharmaceutical drugs rather than nutrients. For drugs, the only source of the agent is assumed to be what is given to participants in the treatment arm, and a linear dose–response relationship is presumed. Neither assumption is valid for vitamin D. As a result, participants generally have mean 25(OH)D concentrations above the population's mean values, participants are given small doses of vitamin D, and participants in both the treatment and control arms are permitted to take additional vitamin D supplements as well as produce vitamin D through solar UVB exposure.

Robert Heaney outlined the guidelines for nutrient RCTs in 2014 [101], which were updated in 2018 [102]. The principal guidelines adapted for vitamin D are that:

- Baseline 25(OH)D concentrations should be measured and used as a criterion for inclusion in the study;
- The vitamin D dose should be large enough to increase 25(OH)D concentration to the point at which it would have an observable effect on health outcomes;
- Achieved 25(OH)D concentrations should be measured;
- Conutrient status must be optimized to ensure that vitamin D is the only nutrient-related limiting factor in the response.

No RCT investigating the role of vitamin D in reducing risk of cancer appears to have followed those guidelines.

Some secondary results of the vitamin D–cancer RCTs have yielded useful information. The VITAL study also reported that African American participants had a trend for reduced risk of cancer incidence (hazard ratio = 0.77 [95% CI, 0.59–1.01]). According to the report's supplementary material for African Americans who supplied 25(OH)D concentration values, the baseline 25(OH)D was 25.0 ng/mL, and the achieved 25(OH)D concentration was 39.7 ng/mL. Those values are in contrast to 31.4 and 42.4 ng/mL, respectively, for non-Hispanic white participants.

In addition, two RCTs showed some effect of vitamin D plus calcium supplementation on risk of cancer [95,98]. When those data were pooled with data from the Grassroots Health volunteer cohort and analyzed by achieved 25(OH)D concentration, the incidence rate of breast cancer for women with 25(OH)D concentrations ≥60 versus <20 ng/mL had a rate ratio of 0.18 (95% CI, 0.04–0.62; $p = 0.006$).

3. Perspectives on Epidemiological Studies

3.1. Ecological Studies

As would be generally expected, incidence and mortality rates are generally inversely correlated with solar UVB indices unless UVB exposure is linked to increased risk, such as that for melanoma and other skin cancer. The direct correlation with oral cavities and the pharynx in the United States is consistent with UVB exposure's being a risk factor for lip cancer. Solar UVB exposure increases human papillomavirus (HPV) concentrations, as evidenced by peak rates of positive Pap smears for cervical cancer in Denmark in August [103]. HPV is a risk factor for head and neck cancer [104]. HPV is also hypothesized to be an important risk factor for melanoma [105].

The finding that the incidence rates for several cancers are directly correlated with solar UV in China, whereas most of the cancer mortality rates are inversely correlated, is probably owing to the fact that air pollution levels are much higher in northern than in southern China [106]. In addition, vitamin D generally reduces the risk of cancer mortality rates rather than incidence rates. The reasons may include that although many factors affect cancer incidence, few factors affect cancer progression and metastasis.

Because the countries included are different in many respects, including diet, ethnicity, latitude, and pollution level, ecological studies offer strong evidence that UVB irradiance affects cancers similarly regardless of many other factors.

An important reason why ecological studies have shown robust relationships between indices of solar UVB doses is that they included many cases of cancer. Researchers conducting earlier ecological studies were more likely than researchers of more recent studies to find significant correlations with UVB doses because people back then spent more time in the sun without concern for skin cancer or photoaging, and obesity rates were lower.

3.2. Observational Studies

Several findings are important from the analyses presented regarding observational studies.

First, the inverse relationships between serum 25(OH)D concentration and cancer incidence or mortality rates are similar to those between solar UVB and cancer reported in ecological studies. The primary exception is for head and neck cancer; serum risk was inversely correlated with both serum 25(OH)D concentration and vitamin D intake. However, ecological studies showed direct correlations between solar UVB and both incidence and mortality rates for oral cavity/pharynx and pharynx cancers, although one study reported an inverse relationship for laryngeal cancer [25].

Secondly, a long follow-up time was again found to significantly decrease the observed beneficial effect of 25(OH)D concentration. For example, the meta-analysis of CRC risk with respect to 25(OH)D concentration by Hernandez-Alonso and colleagues [72] had 11 studies (one CC, nine NCC, and one meta-analysis) and six prospective cohort studies. The OR for the CC study was 0.45 (95% CI, 0.36–0.57). For the nine NCC studies, the mean follow-up time was near 8 years, and the OR was 0.63, whereas for the prospective cohort studies, the mean follow-up time was 13 years, and the OR was 0.80 (95% CI, 0.66–0.97).

Some parties have argued that CC studies with 25(OH)D concentration measured near the time of diagnosis would be the best type of observational study due to possible reverse causality [53]. There is no evidence to indicate that having undiagnosed cancer reduces 25(OH)D concentration other than perhaps decreasing with the progression cancer stage. Thus, CC studies, which are easier to conduct than prospective studies, are preferred.

The epidemiological and mechanical evidence regarding solar UVB exposure and vitamin D presented here generally satisfy Hill's criteria for causality in a biological system (based on Kosh's postulates) [107–109]. The only weakness is that RCTs have not yielded strong support, largely because they were poorly designed and conducted. However, as argued by Dr. Thomas R. Frieden, former head of the U.S. Centers for Disease Control and Prevention, in a review in *The New England Journal of Medicine*, RCTs have substantial limitations [110]. The review tabulates the strength and limitations of 11 study designs, including RCTs, prospective cohort, retrospective cohort, case-control, and ecological studies. It concludes by stating that there is no single, best approach to the study of health interventions, and clinical and public health decisions are almost always made with imperfect data.

3.3. Historical Overview

Many of the articles reviewed regarding epidemiological studies of solar UVB dose or exposure and vitamin D played important roles in developing the understanding of the role of vitamin D in reducing risk of cancer incidence and mortality rates. Table 10 lists a few of them in chronological order. Note that the importance of some of the articles, notably those reported prior to 1980, was not recognized until many years later.

Table 10. List of epidemiological studies that had important findings in the history of solar UVB exposure and/or vitamin D and cancer.

Year	Finding	Reference
1936	Sun exposure can cause skin cancer but reduce risk of internal cancer.	[1]
1937	US Navy personnel highly exposed to sun had high skin cancer rates but low internal cancer rates.	[2]
1941	Cancer mortality rates for whites in the U.S. found inversely related to a solar radiation index while skin cancer (melanoma) mortality rates were directly related.	[3]
1980	Annual solar radiation dose inversely correlated with colon cancer mortality rate, USA, vitamin D production suggested.	[6]
1985	Dietary vitamin D and calcium inversely correlated with colorectal cancer incidence.	[7]
1989	Serum 25(OH)D concentration inversely correlated with colon cancer incidence.	[8]

Table 10. *Cont.*

Year	Finding	Reference
1990	Annual solar radiation dose inversely correlated with breast cancer mortality rate in the U.S.	[9]
2002	Mortality rates for thirteen types of cancer are inversely correlated with solar UVB doses in the U.S., 1970–1994.	[13]
2006	A Harvard cohort study finding that incidence of several types of cancer were inversely correlated with predicted 25(OH)D concentration.	[111]
2006	An ecological study in the U.S. finding that incidence and mortality rates for many types of cancer were inversely correlated with solar UVB doses.	[22]
2007	A meta-analysis presenting a 25(OH)D concentration-colorectal cancer incidence relationship.	[46]
2007	An RCT conducted in the U.S. finding that vitamin D supplementation significantly reduced risk of all-cancer incidence rate.	[95]

4. Mechanisms Introduction

The first experimental studies supporting this effect of 1,25-$(OH)_2D_3$ were reported in 1981. They addressed the inhibition of human melanoma cell proliferation and the induction of the differentiation of mouse myeloid leukemia cells and were by D. Feldman's and T. Suda's groups, respectively [112,113]. Since then, many laboratories have described a high number of antitumoral effects of 1,25-$(OH)_2D_3$ on a variety of molecular mechanisms and cellular processes during carcinogenesis. Previous reviews have discussed some of these mechanisms in particular cancer types [114–119]. In this review, we update the current knowledge on 1,25-$(OH)_2D_3$ antitumor mechanisms.

4.1. Inhibition of Tumor Cell Proliferation

1,25-$(OH)_2D_3$ exerts an antiproliferative action on tumor cells by direct and indirect mechanisms that are partially redundant and sometimes function simultaneously in target cells. Of note, this action is mostly independent of *TP53* tumor suppressor gene status.

Direct mechanisms. In many cancer cell types, 1,25-$(OH)_2D_3$ directly arrests the cell cycle in the G_0/G_1 phase by downregulating cyclin-dependent kinases (CDKs: CDK4, CDK6) and repressing the genes that encode cyclins D1 and C (*CCND1, CCNC*) and CDK inhibitors $p21^{CIP1/WAF1}$ (*CDKN1A*), $p27^{KIP1}$ (*CDKN1B*) and p19 (*CDKN2D*) [116,119]. The induction of $p27^{KIP1}$ expression takes place at the promoter/transcriptional level and posttranslationally by the inhibition of its degradation [120–122]. These effects hamper retinoblastoma (Rb) protein phosphorylation and thus the activation of the E2F family of transcription factors, which trigger a series of target genes that are critical to entering the cell cycle from the quiescent state. In addition, an Rb-independent G_1 arrest has been described that is probably a consequence of the repression of the *MYC* oncogene [123]. Thus, 1,25-$(OH)_2D_3$ represses *MYC* expression via direct [124] or indirect transcriptional inhibition by antagonism of the Wnt/β-catenin pathway [125,126], the induction of cystatin D [127] or the MYC antagonist MAD/MXD1 [128], by repressing long non-coding *(lnc)RNA CCAT2* [129] or by promoting MYC protein degradation [130] in several carcinoma cell types.

In some systems (colon and gastric cancer cells), 1,25-$(OH)_2D_3$ downregulates other proliferative genes such as *FOS, JUN, JUNB,* and *JUND* proto-oncogenes, *G0S2* (G_0/G_1 switch 2), and *CD44*, while it upregulates *GADD45A* (growth arrest and DNA damage 45a), *MEG3* (Maternally expressed gene 3, a lncRNA) and *NAT2* (N-acetyltransferase 2) [131–134]. Additionally, 1,25-$(OH)_2D_3$ induces antiproliferative genes such as *CEBPA* (CCAAT-enhancer-binding protein-α) and *IGFBP3* (insulin-like growth factor binding protein-3) in breast, prostate, or colon carcinoma cells, respectively [131,135,136]. IGFBP3 mediates the induction of $p21^{CIP1/WAF1}$ by 1,25-$(OH)_2D_3$ in prostate carcinoma cells [136], and microRNA *miR-145* the repression of *CDK2, CDK6, CCNA2,* and *E2F3* genes and the antiproliferative effect of 1,25-(OH)$_2$D$_3$ in gastric cancer cells [137]. In breast carcinoma and anaplastic thyroid cancer cells, 1,25-(OH)$_2$D$_3$ causes G$_2$/M phase arrest probably as a consequence of the downregulation of CDK2 activity due to the E2F blockade by non-phosphorylated Rb protein [138]. vitamin D analogues also inhibit proliferation through induction of G$_1$ phase arrest of some hematological cancer cells (lymphoma, myeloma, B-cell acute lymphoblastic leukemia and acute myeloid leukemia) [139].

Indirect mechanisms. 1,25-(OH)$_2$D$_3$ interferes with several mitogen signaling pathways in a context-dependent fashion. Thus, 1,25-(OH)$_2$D$_3$ decreases the expression of epidermal growth factor receptor (EGFR) and promotes its ligand-induced internalization in colon carcinoma cells [140,141]. Additionally, it diminishes EGFR signaling through the induction of E-cadherin and the repression of SPROUTY-2 and the renin-angiotensin system [125,142–144]. 1,25-(OH)$_2$D$_3$ and certain analogues interfere with the insulin-like growth factor (IGF)-I/II pathway by inhibiting IGF-II secretion and increasing IGFBP3 and IGFBP6 levels, and by inducing type II IGF receptor (IGFR-II), which accelerates IGF-II degradation and downregulates this pathway [145,146]. In oral squamous cell carcinoma cells, the 1,25-(OH)$_2$D$_3$ analogue Eldecalcitol antagonizes the mitogenic action of fibroblast growth factor (FGF)1/2 by repressing nuclear factor *kappa* B (NF-*k*B) and inducing *miR6887-5p*, which targets 3′UTR mRNA of heparin-binding protein 17/FGF-binding protein-1 (HBp17/FGFBP-1), a FGF2 chaperone [147,148]. In addition, 1,25-(OH)$_2$D$_3$ inhibits the mitogenic action of platelet-derived growth factor (PDGF)-BB in prostate cancer cells by downregulating PDGF receptor β [149]. The effect of 1,25-(OH)$_2$D$_3$ on hepatocyte growth factor (HGF) signaling is cell-type dependent. It is inhibitory in hepatocellular cells by reducing the expression of c-Met, the tyrosine kinase HGF receptor [150] and in promyelocytic leukemia cells by downregulating HGF RNA [151], but activating in some non-tumoral cell types [152].

1,25-(OH)$_2$D$_3$ also diminishes the proliferation of breast cancer cells by inhibiting estrogen synthesis and signaling through estrogen receptor (ER)α [153] and by downregulating RAS expression and the phosphorylation of its downstream effectors MEK and ERK1/2 [154]. The inhibition of pituitary transcription factor (Pit)-1 is another antiproliferative effect of 1,25-(OH)$_2$D$_3$ in breast cancer cells. Pit-1 expression is higher in tumors than in normal breast. It regulates growth hormone (GH) and prolactin (PRL) secretion and leads to increased cell proliferation, invasiveness, and metastasis [155]. 1,25-(OH)$_2$D$_3$ reduces Pit-1 expression and the increase in cell proliferation either directly or indirectly through GH and/or PRL [156].

Another indirect mechanism of the antiproliferative effect of 1,25-(OH)$_2$D$_3$ is the regulation of miRs. Thus, *miR-22* is induced by 1,25-(OH)$_2$D$_3$ and contributes to its antiproliferative effect on colon carcinoma cells 1,25-(OH)$_2$D$_3$ [157] and has antitumor effects in other carcinomas. Additionally, a recent study indicates that *miR-1278* sensitizes cells to 1,25-(OH)$_2$D$_3$ by suppressing the expression of CYP24A1 [158].

Transforming growth factor (TGF)-β is a strong inhibitor of epithelial cell proliferation in normal cells and at early steps in the tumorigenic process. 1,25-(OH)$_2$D$_3$ activates latent TGF-β and induces the expression of type I TGF-β receptor, which sensitizes breast and colon carcinoma cells to the growth inhibitory action of TGF-β [159,160]. Of note, TGF-β signaling is blocked in around 30% of colon cancers due to mutation of the genes encoding TGF-β receptor type II, SMAD2, or SMAD4. In contrast, TGF-β promotes at late stages epithelial-to-mesenchymal transition (EMT), migration, invasion, immunosuppression, and metastasis. As discussed in the following sections, these protumorigenic effects of TGF-β on tumor and stromal cells later in carcinogenesis are counteracted by 1,25(OH)$_2$D$_3$.

Concordantly with the association between low vitamin D status and poorer overall survival and progression-free survival in myeloid and lymphoid malignancies [161], in several types of leukemic cells, 1,25-(OH)$_2$D$_3$ regulates essential pathways for survival and proliferation such as TLR, STAT1/3 or PI3K/AKT that are induced by immune cell–cell or cytokine activation [162,163].

4.2. Sensitization to Apoptosis, Combined Action with Chemotherapy and Radiotherapy

Obviously, 1,25-$(OH)_2D_3$ per se does not induce apoptosis or any other type of cell death. However, it controls the expression of genes involved in apoptosis in cell systems in a way that is compatible with sensitization to the induction of apoptosis by other agents. Thus, in colon, prostate, and breast carcinoma cells, 1,25-$(OH)_2D_3$ upregulates several pro-apoptotic proteins (BAX, BAK, BAG, BAD, G0S2) and suppresses survival and anti-apoptotic proteins (thymidylate synthase, survivin, BCL-2, BCL-XL). In this way, it favors the release of cytochrome C from mitochondria and the activation of caspases 3 and 9 that lead to apoptosis promoted by a variety of signals [116,117]. Moreover, 1,25-$(OH)_2D_3$ induces apoptosis in ovarian carcinoma cells by caspase 9 activation [164] and by downregulation of telomerase reverse transcriptase (hTERT) via the induction of *miR-498* [165,166]. Intriguingly, while the aforementioned effects seem to be independent of the *TP53* gene, a study has proposed that mutant p53 protein interacts physically with VDR in breast cancer cells, converting the ligand into an anti-apoptotic agent by mechanisms that remain unclear [167].

In addition, 1,25-$(OH)_2D_3$ and metformin have additive/synergistic antiproliferative and proapoptotic effects in colon carcinoma and other types of cells, which are modulated but not hampered by *TP53* status [168]. Moreover, in an in vitro model developed to evaluate the crosstalk between tumor-associated macrophages and colon carcinoma cells, 1,25-$(OH)_2D_3$ restored the sensitivity of these cells to TRAIL-induced apoptosis by interfering with the release of interleukin (IL)-1β by macrophages [169]. Interestingly, the *TP53* mutation and suppression of *miR-17~92* polycistron are highly toxic in non-small lung cancer cell lines due to the upregulation of VDR signaling [170].

Based on these data, many completed and ongoing studies investigate the antitumor action of the combination of 1,25-$(OH)_2D_3$ and a variety of chemotherapeutic agents (5-fluorouracil, gemcitabine, paclitaxel, imatinib, and cisplatin, among others), inhibitors (of EGFR, HER2, HER4, JAK1/2 tyrosine kinases, estrogen or aromatase) and apoptosis inducers (dexamethasone, trichostatin A and 5-aza-2′-deoxycytidine, among others) in cells and animal models of several types of cancers see [116,119] and references therein. The definitive results of these studies are expected to constitute the foundation for clinical trials.

4.3. Regulation of Autophagy

Autophagy is a process of elimination of cytoplasmic waste materials and dysfunctional organelles that serves as a cytoprotective mechanism but that, when excessive, leads to cell death. vitamin D activates autophagy in many organs in healthy conditions to preserve homeostasis. It can also induce autophagy as protection against cell damage caused by intracellular microbial infection, oxidative stress, inflammation, aging, and cancer [171].

In cancer, VDR ligands trigger autophagic death by inducing crucial genes in several cancer cell types. Thus, 1,25-$(OH)_2D_3$ and its analogues de-repress the key autophagic MAP1LC3B (LC3B) gene and activate 5′-AMP-activated protein kinase (AMPK) via increased cytosolic Ca^{2+} and activation of Ca^{2+}/calmodulin-dependent protein kinase β in breast carcinoma cells [172]. In Kaposi's sarcoma cells [173] and myeloid leukemia cells [174], vitamin D compounds inhibit PI3K/AKT/mTOR signaling and activate Beclin-1-dependent autophagy. 1,25-$(OH)_2D_3$ also induces autophagy through the mTOR pathway in Pfeiffer diffuse large B lymphoma cells [175] and is mediated by activation of DNA damage-inducible transcript 4 (DDIT4), in cutaneous squamous cell carcinoma cells [176]. In addition, a recent study has shown that 1,25-$(OH)_2D_3$ promotes autophagy in acute myeloid leukemia cells by inhibiting miR-17-5p-induced Beclin-1 overexpression [177].

Moreover, 1,25-$(OH)_2D_3$ or EB1089 increase radiation efficiency via promotion of autophagic cell death in a VDR- and p53-dependent fashion in non-small cell lung cancer and breast cancer cells [178–181]. Additionally, synergy between 1,25-$(OH)_2D_3$ and temozolomide in tumor reduction and prolonged survival time has been reported in rat-cultured glioblastoma cells and in an orthotopic xenograft model [182].

4.4. Induction of Cell Differentiation, Inhibition of Epithelial-to-Mesenchymal Transition

Cell differentiation is usually, but not necessarily, linked to an arrest in proliferation, and both processes put a brake on tumorigenesis. Carcinoma is the most frequent type of solid cancer. Carcinomas originate from the transformation of epithelial cells in a process that involves the early loss of two key features of their differentiated phenotype: apical-basal polarity and adhesiveness (cell–cell and cell–extracellular matrix, ECM). Loss of epithelial differentiation results from the acquisition of a cellular program called epithelial-mesenchymal transition (EMT), which implies changes in gene expression, triggered by a group of transcription factors (EMT-TFs: mainly SNAIL1, SNAIL2, ZEB1, ZEB2 and TWIST1). EMT provides tumor cells with features of malignancy such as migratory capacity, stemness and diminished apoptosis that facilitate invasion and metastasis and possibly cause resistance to cytotoxic chemotherapy and radiotherapy, and to immunotherapy [183]. The EMT process is activated by a variety of agents and signals that induce or activate the EMT-TFs, such as TGF-β, Wnt, Notch, and ligands of several receptors with tyrosine kinase activity and cytokine receptors.

1,25-$(OH)_2D_3$ has a prodifferentiation effect on several types of carcinoma cells either by direct upregulation of epithelial genes and/or the repression of key EMT-TFs, as shown in [184,185]. In breast cancer cells, 1,25-$(OH)_2D_3$ promotes the formation of focal adhesion contacts, structures of binding to the ECM, by increasing the expression of several integrins, paxillin and focal adhesion kinase. Additionally, 1,25-$(OH)_2D_3$ reduces the expression of the mesenchymal marker N-cadherin and the myoepithelial proteins P-cadherin, integrins α_6 and β_4, and α-smooth muscle actin, which are associated with more aggressive and lethal forms of human breast cancer [186]. In colon carcinoma cells, 1,25-$(OH)_2D_3$ upregulates an array of intercellular adhesion molecules that are constituents of adherens junctions and tight junctions, including E-cadherin, occludin, claudin-2 and -12, and ZO-1 and -2 [125,131]. As mentioned by JoEllen Welsh in an excellent recent review [187], breast cancer heterogeneity is reflected in available model systems of this disease, including human breast cancer cell lines. These differ in the expression of VDR and other hormone receptors and in their global gene expression profile and phenotype. Consequently, results vary widely in laboratory studies of 1,25-$(OH)_2D_3$ and other VDR ligands, which show a heterogeneous, usually multilevel protective action that affects a variety of pathways (ERBB2/NEU-ERK-AKT, WNT/β-catenin, JAK-STAT, NF-κB, ERα). These studies have rendered only a few genes that are commonly regulated: *CYP24A1*, *CLMN*, *EFTD1* and *SERPINB1*.

Remarkably, the induction of E-cadherin by 1,25-$(OH)_2D_3$ in colon carcinoma cells has been reproduced in tumor cell lines derived from breast, prostate, non-small cell lung, and squamous cell carcinomas, usually associated with an increase in epithelial differentiation [184]. The mechanism of E-cadherin induction by 1,25$(OH)_2D_3$ in human colon cancer cells is transcriptional indirect. It requires transient activation of the RhoA-ROCK-p38MAPK-MSK1 signaling pathway [126]. Phosphatidylinositol 5-phosphate 4-kinase type II β is also needed for E-cadherin induction by 1,25-$(OH)_2D_3$ in these cells [188]. In agreement with the transcriptional regulation, 1,25-$(OH)_2D_3$ treatment causes partial demethylation of CpG sites of *CDH1* promoter in MDA-MB-231 triple-negative breast cancer cells [189]. In addition, 1,25-$(OH)_2D_3$ induces and/or redistributes several cytokeratins, F-actin, vinculin, plectin, filamin A and paxillin that modulate the actin cytoskeleton and the intermediate filament network, changing stress fibers and the ECM binding structures (focal adhesion contacts and hemidesmosomes) [125,126]. In summary, 1,25$(OH)_2D_3$ increases cell–cell and cell-ECM adhesion.

1,25-$(OH)_2D_3$ inhibits SNAIL1 and ZEB1 expression in non-small cell lung carcinoma cells, accompanied by an increase in E-cadherin expression, vimentin downregulation, and maintenance of epithelial morphology [190]. The 1,25-$(OH)_2D_3$ analogue MART-10 inhibits EMT in breast and pancreatic cancer cells through the downregulation of SNAIL1, SNAIL2 and TWIST1 in breast cancer cells [191,192]. 1,25-$(OH)_2D_3$ causes the downregulation of SNAIL1 and SNAIL2 in colon and ovarian carcinoma cells [193,194].

In addition, 1,25-$(OH)_2D_3$ induces several modulators of the epithelial phenotype that can influence the expression of these EMT-TF. Thus, it increases by a transcriptional indirect mechanism the expression of *KDM6B*, a histone H3 lysine 27 demethylase that mediates the induction of a highly adhesive epithelial phenotype in human colon cancer cells [195]. *KDM6B* depletion upregulates SNAIL1, ZEB1, and ZEB2 and increases the expression of mesenchymal markers fibronectin and LEF-1, and claudin-7. Accordingly, *KDM6B* and *SNAI1* RNA expression correlate inversely in samples from human colon cancer patients [195]. Furthermore, 1,25-$(OH)_2D_3$ directly upregulates the expression of cystatin D, which represses SNAIL1, SNAIL2, ZEB1, and ZEB2, and induces the expression of E-cadherin and other adhesion proteins such as occludin and p120-catenin. Accordingly, cystatin D and E-cadherin protein expression directly correlate in colon cancer, and loss of cystatin D is associated with poor tumor differentiation [127]. The *SPRY2* gene encodes SPROUTY-2, a modulator of tyrosine kinase receptor signaling that is strongly repressed by 1,25$(OH)_2D_3$ in colon carcinoma cells [143]. SPROUTY-2 promotes EMT through upregulation of ZEB1 and downregulation of the epithelial splicing regulator ESRP1. Consequently, SPROUTY-2 represses genes that encode E-cadherin, claudin-7, and occludin and the important regulators of the polarized epithelial phenotype LLGL2, PATJ, and ST14 [143,196].

The induction of differentiation seems to be a less important protective mechanism of 1,25-$(OH)_2D_3$ in hematological malignancies than in solid cancers. 1,25-$(OH)_2D_3$ induces differentiation almost exclusively of acute myeloid leukemia cells [197–199]. Thus, 1,25$(OH)_2D_3$ increases the expression of markers of the monocyte-macrophage phenotype such as CD14 and some proteins involved in phagocytosis and adherence to substratum, including CD11b [139,200]. A number of genes and proteins have been proposed as mediators of this prodifferentiation action of 1,25-$(OH)_2D_3$, such as PI3K, *CEBPB*, and *CDKN1A* [201–203]. Differentiation of acute myeloid leukemia cells was also described by the combination of 1,25-$(OH)_2D_3$ with l-asparaginase [204]. Interestingly, a recent study reports that liganded VDR has a strong prodifferentiation effect in acute myeloid leukemia cells harboring mutations in *IDH* gene encoding isocitrate dehydrogenase. This is the case because the oncometabolite 2-hydroxyglutarate that is produced by mutant IDH potentiates VDR signaling in a CEBPα-dependent manner [205]. In addition, prodifferentiation effects of VDR agonists have been reported in follicular non-Hodgkin's lymphoma cells, with increased expression of mature B-cell markers [206].

4.5. Antagonism of Wnt/β-Catenin Signaling Pathway

The Wnt/β-catenin signaling pathway is activated by several members of the Wnt family of secreted proteins (19 in humans) during ontogenesis and adult life, which play important roles in the development and homeostasis of many tissues and organs. The binding of these Wnt factors to plasma membrane co-receptor (Frizzled-LRP) complexes inhibits the degradation of β-catenin protein in the cytoplasm that is promoted by the products of tumor suppressor genes APC and AXIN, which leads to β-catenin accumulation and partial translocation into the cell nucleus. Nuclear β-catenin acts as a transcriptional co-activator of genes bound by the T-cell factor (TCF) family of transcriptional repressors [207]. The long list of β-catenin/TCF target genes includes some that are crucial for cell survival and proliferation (MYC, CCND1), EMT, migration/invasion, and other tumoral processes (Stanford University Wnt homepage: https://web.stanford.edu/group/nusselab/cgi-bin/wnt/) (accessed on 19 March 2022). These genes are active during ontogenesis but remain mostly silent in adult life except in some situations such as wound healing. Recent data suggest that Wnt factors only prime β-catenin signaling. This causes basal activation of the pathway that only becomes fully activated in the presence of R-spondin (RSPO)1–4. Upon binding to their membrane LGR4–6 receptors, the secreted RSPO family members inactivate two E3 ubiquitin ligases (RNF43, ZNRF3) that mediate Frizzled degradation. In this way, RSPOs extend Frizzled half-life at the cell surface and so potentiate Wnt signaling.

The Wnt/β-catenin pathway is an important player in cancer as it is aberrantly activated by mutation (APC, AXIN, CTNNB1/β-catenin, RSPO2/3, and RNF43 genes), overexpression of Wnt factors/receptors or silencing of Wnt signaling inhibitors (DICKKOPF/DKKs, SFRPs) leading to the activation or potentiation of carcinogenesis [208]. This is particularly important in colorectal cancer, as massive sequencing efforts have revealed that the mutation of at least one Wnt/β-catenin pathway gene is present in over 94% of primary tumors and metastases [209,210], while a variable proportion of other cancers (liver, breast, lung and leukemia, among others) also show abnormal pathway activation. Despite its clinical relevance, no inhibitors of the Wnt/β-catenin pathway have been approved up to now.

The first description of the antagonism of the Wnt/β-catenin pathway by 1,25-$(OH)_2D_3$ was reported in colon carcinoma cells by a double mechanism: (a) liganded VDR binds nuclear β-catenin, which hampers the formation of transcriptionally active β-catenin/TCF complexes, and (b) induction E-cadherin expression that attracts newly synthesized β-catenin protein to the plasma membrane adherens junctions. In that way, it decreases β-catenin nuclear accumulation [125]. Other mechanisms of interference of the Wnt/β-catenin signaling pathway by 1,25-$(OH)_2D_3$ have been subsequently described in colon, breast, ovarian, hepatocellular, renal, head, and neck carcinomas, and in Kaposi's sarcoma, see [211]. These mechanisms include the increase in AXIN, TCF4 or DKK1 level, modulation of TLR7, reduction of total or nuclear β-catenin, and enhancement of LRP6 degradation [212–217]. In addition, a paracrine mechanism of Wnt/β-catenin signaling has been proposed based on interruption by 1,25-$(OH)_2D_3$ of the secretion of the Wnt stimulator IL-β by environmental macrophages [218].

4.6. Inhibition of Angiogenesis

1,25-$(OH)_2D_3$ inhibits cancer angiogenesis by acting at two levels: tumor cells and endothelial cells. In diverse types of carcinoma cells (colon, prostate, and breast), the anti-angiogenic action of 1,25-$(OH)_2D_3$ relies to a great extent on its ability to inhibit two major angiogenesis promoters: it suppresses the expression and activity of hypoxia-inducible factor (HIF)-1α, a key transcription factor in hypoxia-induced angiogenesis, and of vascular endothelial growth factor (VEGF)-A. Additionally, 1,25-$(OH)_2D_3$ induces the angiogenesis inhibitor thrombospondin-1 [219,220]. In colon tumor cells, modulation of the angiogenic phenotype is also mediated by the control of genes encoding inhibitors of differentiation (ID)-1/2 and by the repression of DKK4, a weak Wnt antagonist that promotes angiogenesis and invasion and is upregulated in colon tumors [219,221]. 1,25-$(OH)_2D_3$ alone and more strongly in combination with cisplatin suppresses VEGF activity in ovarian cancer cells [222]. By modulating VEGF receptor (VEGFR) 2, 1,25-$(OH)_2D_3$ or calcipotriol, it enhances the efficacy of the VEGFR inhibitor Cediranib in malignant melanoma cells [223]. Another antiangiogenic mechanism of 1,25-$(OH)_2D_3$ is the reduction of IL-8 secretion by prostate cancer cells through the inhibition of NF-κB [224]. Intriguingly, variable and sometimes opposite effects of 1,25-$(OH)_2D_3$ on angiogenesis have been reported, as in a xenograft breast cancer model, where it inhibits TSP-1 and increases VEGF expression [225]. Likewise, 1,25-$(OH)_2D_3$ induces VEGF synthesis and action in some non-tumoral cell systems, see [152].

1,25-$(OH)_2D_3$ also has inhibitory effects on tumor-derived endothelial cells. It reduces their proliferation and sprouting in vitro and diminishes the blood vessel density in xenograft tumors in breast, squamous cell carcinoma, bladder and prostate cancer models [226–230].

4.7. Inhibition of Cancer Cell Migration, Invasion and Metastasis

1,25-$(OH)_2D_3$ inhibits the migratory and invasive phenotype of cancer cells as a result of its effects on the cytoskeleton and adhesive properties and on the expression of proteases, protease inhibitors and ECM proteins. To a variable extent, these effects are linked to inhibition of EMT and the TGF-β and Wnt/β-catenin signaling pathways.

As mentioned above, in carcinoma cells, 1,25-$(OH)_2D_3$ induces E-cadherin and other proteins of adhesion structures and modulates actin and intermediate filament networks, which results in increased cell–cell and cell–ECM adhesion [125,186,194,217,231–233]. By promoting intercellular adhesion via upregulation of E-cadherin, 1,25-$(OH)_2D_3$ suppresses prostate cancer cell rolling and adhesion to microvascular endothelial cells, which is a step in extravasation that precedes metastasis [234]. In addition, vitamin D deficiency increases breast cancer metastasis to the lung by enhancing EMT and the CXCL12/CXCR4 chemokine axis [235].

1,25-$(OH)_2D_3$ reduces breast, renal, and prostate carcinoma cell migration and invasion by downregulating the expression and/or activity of N-cadherin, the ECM components tenascin C and periostin, several integrins and metalloproteases (MMP-1, -2, and -9) and serine proteases (plasminogen activator), while it upregulates protease inhibitors and the pro-adhesive actin cytoskeleton adaptor protein PDLIM2 [236–240]. In triple-negative breast cancer cells, 1,25-$(OH)_2D_3$ decreases hyaluronic acid synthesis [241], and inhibits bladder cancer cell migration partially via the induction of miR-101-3p [242]. In pancreatic adenocarcinoma cells, 1,25-$(OH)_2D_3$ ameliorates the pro-invasive action of tumor necrosis factor (TNF)-α by decreasing the expression of miR-221 and increasing that of the tissue inhibitor of metalloproteinase (TIMP)-3 [243].

4.8. Stromal Effects: Cancer-Associated Fibroblasts

Today, the critical role of stroma in the carcinogenic process is clear. Fibroblasts are the main cellular component of tumor stroma (Cancer-Associated Fibroblasts, CAF). This is a heterogeneous cell population of multiple origins (tissue-resident fibroblasts, myeloid precursors, pericytes and adipocytes, among others) and features that is acquired via the change to an "activation phenotype". It is thought to promote cancer invasion, angiogenesis and metastasis; inhibit the immune response; and reduce intratumoral delivery and the activity of chemotherapeutic agents [244,245]. However, the protective effects of CAF have also been described in some systems, and reprogramming their phenotype is accepted as a more advisable strategy than their elimination [246,247]. Early studies showed that VDR agonists have antifibrotic and antitumoral effects by antagonizing TGF-β in the intestine, liver, and pancreas [248–252].

1,25-$(OH)_2D_3$ regulated over one hundred genes in human CAF isolated from tumor biopsies of five breast cancer patients [253]. The induced gene signature reflects an antiproliferative and anti-inflammatory effect of 1,25$(OH)_2D_3$. Importantly, 1,25-$(OH)_2D_3$ inhibits the protumoral action of human colon CAF by reprograming them to a less activated phenotype. Thus, 1,25$(OH)_2D_3$ reduces the capacity of CAF to alter the ECM and their ability to promote the migration of colon carcinoma cells [254]. 1,25-$(OH)_2D_3$ regulates over one thousand genes in colon CAF that are involved in cell adhesion, differentiation and migration, tissue remodeling, blood vessel development, and the inflammatory response. Remarkably, 1,25$(OH)_2D_3$ imposes a gene signature that correlates with a better prognosis for colon cancer patients [254]. Curiously, in contrast to the antagonism reported in colon carcinoma cells, 1,25-$(OH)_2D_3$ and Wnt3A have an additive, partially overlapping effect in colon fibroblasts [255,256]. In line with the results in colon CAF, 1,25$(OH)_2D_3$ decreases the amount of miR-10a-5p found in the exosomes secreted by human pancreatic CAF, which attenuates the promigratory and pro-invasive effects that these CAF exert on pancreatic carcinoma cells [257]. Of note, a recent study reported that calcipotriol promotes an antitumorigenic phenotype of pancreatic CAF by reducing the release of prostaglandin (PG) E_2, IL-6, periostin, and other factors. However, it reduces T-cell-mediated immunity [258]. Clearly, the action of VDR agonists on fibroblasts associated with distinct human cancers is a highly interesting, open line of research.

4.9. Effects on Cancer Stem Cells

Cancer stem cells (CSC) are supposedly a small population of cells present in tumors that are responsible for tumor initiation, growth, malignization, metastasis, and resistance to therapies. They originate from the mutational and epigenetic alteration of normal stem cells that maintain the homeostasis of tissues in adult life and behave as a source of new functional differentiated cells following injuries or in aging. The characterization and study of CSC present two unresolved problems: (a) the lack of confirmed universal or even tissue-specific markers, and (b) the existence of cell plasticity in tumors that implies differentiation/dedifferentiation processes during tumorigenesis and thus the lack of a stable stem phenotype but, instead, interconversion of stem and non-stem cells.

At present, there are two systems to study CSC: organoid cultures generated by CSC present in patient-derived tumor biopsies and subcultures of established, immortal tumor cell lines enriched in populations of cells expressing putative CSC markers and/or selected by their capacity to grow in suspension. Clearly, fresh, primary organoids are a more valuable system. They are three-dimensional (3D), self-organized multicellular structures generated by normal stem cells or CSC (that allow matched normal and tumor organoids to be obtained from a patient) that grow embedded in an ECM covered by a complex, tissue-specific, usually serum-free medium [259,260]. Organoids recapitulate some of the features of a particular organ or tumor of origin and are quite stable genetically, and thus are considered a better system to study cancer processes than 2D cell lines grown for decades on plastic dishes [261]. 1,25(OH)$_2$D$_3$ profoundly and differentially regulates the gene expression profile of colon cancer patient-derived normal and tumor organoid cultures. 1,25(OH)$_2$D$_3$ induced stemness-related genes (*LGR5, SMOC2, LRIG1*, and others) in normal but not tumor organoids [262]. In both normal and tumor organoids, 1,25(OH)$_2$D$_3$ reduced cell proliferation and the expression of proliferation and tumorigenesis genes that affected only a few Wnt/β-catenin target genes (*MYC, DKK4*). Importantly, 1,25(OH)$_2$D$_3$ induced some features of epithelial differentiation in tumor organoids cultured in proliferation medium, such as microvilli, adhesion structures, partial chromatin condensation, and increased cytoplasmic organelles. These effects were also observed in rectal tumor organoids [263].

Concordantly, 1,25(OH)$_2$D$_3$-regulated genes were involved in cell proliferation, differentiation, adhesion, and migration in another study using patient-derived colon organoids [264]. Moreover, MDL-811, an allosteric activator of the sirtuin (SIRT)6 deacetylase, reduced cell proliferation in colon carcinoma cell lines and patient-derived organoids and has a synergistic antitumoral effect in combination with vitamin D in *Apc*$^{min/+}$ mice [265]. However, conflicting data have been found in normal, nontumoral organoids: whereas 1,25-(OH)$_2$D$_3$ increased the stemness genes and the undifferentiated associated cell phenotype in organoids from healthy colon and rectum tissues of a dozen individuals [262,263], it enhanced the differentiation of organoids established from a benign region of a radical prostatectomy from a single patient [266].

A series of studies have examined the action of VDR agonists on putative breast cancer stem or progenitor cells identified by some markers (CD44^{hi}/CD24^{low} and/or ADH1$^+$) that can grow as floating, nonadherent spheres (mammospheres). In these systems, 1,25(OH)$_2$D$_3$ or the BXL1024 analogue reduced the population of putative CSC and the formation of mammospheres and the expression of pluripotency markers (OCT4, KL-4), Notch ligands and target genes, and genes involved in proliferation, EMT, invasion, metastasis, and chemoresistance 32,467,291 [267–269].

Organoids formed by cells isolated from patient-derived xenografts (not obtained directly from human biopsies but on injection and growth in mice) that acquired resistance in vitro to Trastuzumab-emtansine (T-DM1; composed of the humanized monoclonal anti-HER2 antibody Trastuzumab covalently linked to the microtubule-inhibitory agent DMI) constitute an intermediate system to the two discussed above. In this system, two vitamin D analogues (UVB1 and EM1) reduce the formation and growth of organoids [270].

4.10. *Effects on the Immune System*

1,25-(OH)$_2$D$_3$ is an important modulator of the immune system, as reflected by the expression of VDR by almost all types of immune cells [271–273]. 1,25-(OH)$_2$D$_3$ is an enhancer of innate immune reactions against infections and tumor cells by activating the responsive cells (macrophages, natural killer (NK) cells, and neutrophils). Conversely, and in line with its accepted anti-inflammatory action (that may contribute to the inhibition of cancers associated with chronic inflammation), 1,25-$(OH)_2D_3$ is commonly presented as a repressor of the adaptive immune reactions by deactivating antigen-presenting cells (induction of tolerogenic dendritic cells) and $CD4^+$ type-1 helper T (Th1) response (production of interferon-γ, IL-1, IL-6, IL-12...), and by promoting the suppressive Th2 and Treg responses (production of IL-10, IL-4, IL-5, IL-13...) [273,274]. Moreover, in macrophages, 1,25-$(OH)_2D_3$ has been proposed to promote a switch from the pro-inflammatory M1 phenotype (producing IL-1β, IL-6, TNF-α, RANKL, COX) towards the anti-inflammatory protumoral M2 phenotype and to reduce the T-cell stimulatory capacity of macrophages [275,276]. This is somehow counterintuitive as it would represent a potential protumoral effect that cannot be easily attributed to a conserved evolutionary agent such as vitamin D. Some other studies discussed below have introduced putative explanations.

Since naïve T-cells express VDR at a very low level that increases only after activation of the T-cell receptor [277], the role of 1,25-$(OH)_2D_3$ may conceivably be related to the late downregulation of the activated adaptive response. This view agrees with the usual description of repressive 1,25-$(OH)_2D_3$ action in experimental settings following overstimulation of the cells, and it may constitute a safety mechanism to prevent undesirable long-lasting immune activation, potentially leading to inflammation or autoimmunity [278,279]. Concordant with this idea and the anticancer action of 1,25-$(OH)_2D_3$, a series of studies have revealed antitumor effects at the level of several types of immune cells.

Interestingly, a study in mice orthotopically implanted with breast tumors has revealed that vitamin D decreases tumor growth and increases the amount of tumor-infiltrating cytolytic CD8+ T-cells, a usual marker of antitumor response. This effect is lost in high-fat diet conditions [280]. Moreover, in pancreatic cancer, 1,25-$(OH)_2D_3$ inhibits the T-cell suppressive function of myeloid-derived suppressor cells [281].

An important mechanism of 1,25-$(OH)_2D_3$ is the inhibition of the NF-κB pathway. In turn, this causes the downregulation of multiple cytokines and their effects [282]. 1,25-$(OH)_2D_3$ inhibits NF–κB at different levels: by inactivating the p65 subunit of the NF-κB complex and upregulating the inhibitor subunit IκB. In addition, 1,25-$(OH)_2D_3$ inhibits the PG-endoperoxide synthase (PTGS-2, also known as COX-2) [283–285]. 1,25$(OH)_2D_3$ reduces the protumorigenic effect of PG E_2 in prostate cancer cells by inhibiting COX-2 and so decreasing the levels of PG E_2 and two PG receptors (EP2 and FP) [286]. Importantly, vitamin D and calcium favorably modulate the balance of expression of COX-2 and 15-hydroxyPG dehydrogenase, its physiological antagonist, in the normal-appearing colorectal mucosa of patients with colorectal adenoma [287]. vitamin D enhances the tumoricidal activity of NK cells and macrophages [288,289]. 1,25-$(OH)_2D_3$ probably has a dual effect of stimulating the differentiation from monocytes to macrophages and their cell killing activity, including antibody-dependent cell cytotoxicity (ADCC). It may later balance these effects by promoting the M1 to M2 phenotypic switch ([279] and references therein). In addition, 1,25-$(OH)_2D_3$ enhances the susceptibility of hematological and solid cancer cells to NK cell cytotoxicity through downregulation of *miR-302c* and *miR-520c* [289].

The potentiation of ADCC of macrophages and NK cells may be a relevant antitumor action of 1,25-$(OH)_2D_3$ in clinical cases, particularly in patients treated with antibodies, of which the major mechanism of action is ADCC. Thus, several studies have shown that vitamin D deficiency impairs the macrophage and/or NK cell-mediated cytotoxicity of Rituximab (anti-CD20) in diffuse large B-cell, follicular, and Burkitt lymphoma patients [288,290,291], and of Cetuximab (anti-EGFR) in colon cancer cell lines [292]. In addition, some evidence of benefit has been observed in breast cancer patients treated with Trastuzumab (anti-HER2) and in melanoma patients treated with Bevacizumab (anti-VEGF) [290,293].

Agents that target programmed death (PD)-1 or its ligand PD-L1 immune checkpoint inhibitors (ICI) have attracted great attention in cancer therapy. Interestingly, 1,25-$(OH)_2D_3$

upregulates PD-L1 in human (but not mouse)-cultured epithelial and immune cells [294], while vitamin D treatment increases PD-1 expression in $CD24^{+}CD25^{+int}$ T-cells in Crohn's disease patients [295] and PD-L1 in epithelial and immune cells in melanoma patients [296]. These data suggest the possibility of combined treatments with VDR agonists and these ICIs, and perhaps others in development.

In conclusion, it is conceivable that 1,25-$(OH)_2D_3$ works as a general homeostatic regulator of the immune system, ensuring an appropriate global defense against challenges like tumors and infections.

4.11. *Animal Models*

Many studies on animal diet, chemical, genetic, and xenograft models (mainly for colon and breast cancer) have shown the antitumor actions of vitamin D compounds. This in vivo action is difficult to dissect and probably results from a variable combination of mechanisms in the distinct systems that were assayed, including the inhibition of tumor cell growth, EMT, invasiveness, angiogenesis, and metastasis. Importantly, as occurs in cultured cancer cells, vitamin D antitumor action is mostly independent of *TP53* gene status [119,187].

4.12. *Systemic Effects: Detoxification and Microbiome*

4.12.1. Detoxification

The elimination of xenobiotics or the detoxification process involves chemical modification (phase I reactions: oxidation, hydrolysis, etc.) and subsequent conjugations to water-soluble molecules (phase II reactions) carried out by a large number of enzymes. 1,25-$(OH)_2D_3$ regulates some of these enzymes in the intestine and liver [297]. This may have a positive effect on the prevention of tumorigenesis and perhaps another more controversial impact on the inactivation of chemotherapeutic drugs [298].

1,25-$(OH)_2D_3$ induces CYP3A4, a major human drug-metabolizing enzyme, SULT2A, a phase II sulfotransferase, and members of the multidrug resistance-associated protein (MRP) family in colon carcinoma cells [299,300]. CYP3A4, SULT2A1, and MRP3 are involved in the elimination of lithocholic acid (LCA), a secondary bile acid LCA that induces DNA damage and inhibits DNA repair enzymes in colonic cells. Accordingly, LCA promotes colon cancer in experimental animals, and high levels of LCA have been found in colon cancer patients [301,302]. Interestingly, LCA binds weakly and activates VDR, and so it activates its own degradation [303]. Another example is enhancement by 1,25$(OH)_2D_3$ of the benzo[a]pyrene metabolism via CYP1A1 in macrophages [304].

4.12.2. Microbiome

Alteration of the intestinal microbiome (dysbiosis) is connected to colon cancer and possibly other neoplasias [305]. Many experimental studies in mice have shown that vitamin D deficiency promotes gut permeability, colon mucosa bacterial infiltration, and translocation of intestinal pathogens. These effects lead to changes in immune cell populations and gut inflammation, and cancer—an overall condition that is improved after vitamin D supplementation [306,307]. As bacteria lack VDR, the effect of vitamin D is mediated by the host. Importantly, genome-wide association analysis of the gut microbiome in two large cohorts of individuals identified VDR as a factor that influences the gut microbiota [308]. A conditioned medium from probiotic lactic acid bacteria showed increased expression of VDR and of its target *CAMP* gene encoding cathelicidin in cultured colon carcinoma cells and organoids. It protected against the inflammatory response induced by TNF-α [309]. The protective action against dysbiosis and the intestinal tumorigenesis of liganded VDR have been proposed to be at least partially mediated by the inhibition of the JAK/STAT pathway [310].

4.13. Discussion of Mechanistic Studies

The vast array of effects that 1,25-$(OH)_2D_3$ has in a wide variety of experimental systems of a high number of cancer types agrees with a selected evolutionary role in protection against tumoral processes. The underlying mechanisms include the control of tumor cell survival (autophagy, apoptosis) and phenotype (differentiation), and the inhibition of their proliferation, invasiveness, and metastasis; attenuation of the proliferation and phenotypic features of some CSC; modulation of the physiology of diverse non-tumoral stromal cells (fibroblasts, endothelial cells); and the regulation of several types of immune cells and responses. Table 11 summarizes the references corresponding to key studies focused on the most relevant topics of the anticancer action of vitamin D.

Table 11. vitamin D anticancer mechanisms in experimental model systems. List of key representative references.

Mechanism	Cancer Type Model	References
Inhibition of cell proliferation	Breast, prostate, colon, ovarian, gastric thyroid, hepatocellular, leukemias, lymphomas	[111,119–150,152–156,158,159,161,162]
Induction of differentiation	Leukemia, colon, breast	[112,124–126,138,176,185,187,196–205]
EMT inhibition	Colon, ovarian, breast, pancreas	[126,142,189–195]
Sensitization of autophagy	Colon, prostate, breast, ovarian, lung	[115,116,118,163–165,168,169]
Induction of autophagy	Breast, Kaposi's sarcoma, lymphoma, cutaneous squamous cell carcinoma, leukemia	[171–181]
Wnt/β-catenin antagonism	Colon, breast, ovarian, hepatocellular, renal, head and neck, Kaposi's sarcoma	[124,210–217]
Invasion, angiogenesis, metastasis	Colon, prostate, breast, ovarian, renal, pancreas	[193–216,218–223,230–242]
Cancer-associated fibroblasts	Breast, colon, pancreas, liver	[248,250,252–257]
Normal/cancer stem cells	Breast, colon, pancreas, liver	[261–269]
Detoxification and microbiome	Colon, perhaps other cancer types	[296–303,305–309]
Immune system regulation	Many	[272–288]
Combination with immunotherapy	Lymphoma, melanoma, colon, breast	[289–295]

Together, these effects reflect a multilevel anticancer action of vitamin D. Therefore, an appropriate vitamin D status of the organism should be maintained to minimize the risk and severe consequences of many neoplasias. Further supporting this, the toxicity of vitamin D supplementation is limited, acceptable, and clearly lower than that of current anticancer drugs and therapies. We are not aware of any other natural or synthetic compound that has such an array of antitumor activities combined with low toxicity. Doubtless, the available experimental results meet Koch's postulate for biological causality regarding the existence of a global mechanism of action behind the association between vitamin D deficiency and high incidence and, especially, the mortality of several major cancer types found in observational and epidemiological studies. Hopefully, the further development of current and possibly, novel studies on the wide range of mechanisms of VDR agonists in a variety of biological systems will allow us to elucidate the anticancer action of vitamin D (Figure 3).

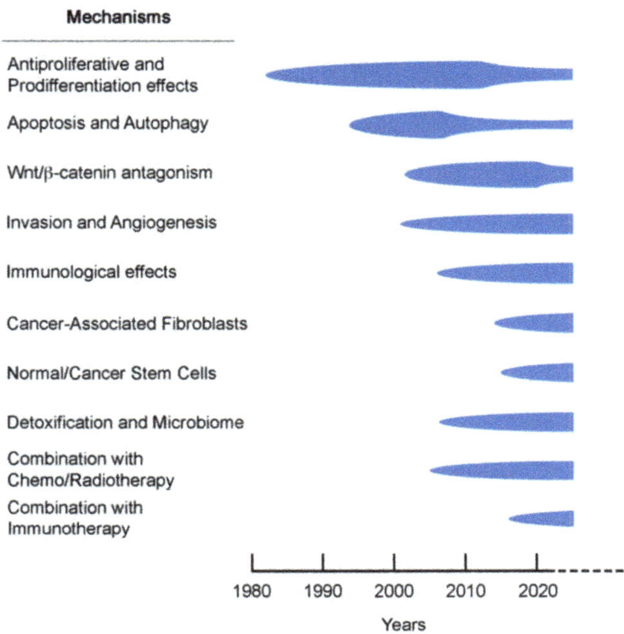

Figure 3. Time flow-chart of studies on the anticancer mechanisms of vitamin D compounds with some key references that are discussed in the text.

5. Outlook

On the basis of this review of ecological and observational studies, it seems that an efficient way to strengthen the links between vitamin D and cancer is to conduct more CC studies of cancer incidence. Such studies would measure 25(OH)D concentration, C-reactive protein, and other relevant factors, as well as obtain the history of UVB exposure, vitamin D supplementation, and dietary sources of vitamin D. The next step is to then find appropriate controls using, perhaps, the propensity score analysis, as done in a study of breast cancer survival with respect to de novo vitamin D supplementation [311]. In addition, care should be taken to investigate the effect of vitamin D supplementation and 25(OH)D concentration on cancer risk for various subgroups based on such factors as age, BMI, diet, ethnicity, geographical location, etc.

Future laboratory research on the anticancer action of vitamin D is desirable to develop a deeper understanding of the individual response to treatment with VDR agonists. To this end, *omics* studies using genomic, epigenomic, transcriptomic, proteomic, and metabolomic approaches must be integrated to understand and foresee personal susceptibility/sensitivity to each compound, which has been defined as "the personal vitamin D response index" [312]. Clearly, the characterization of biomarkers of compound activity and patient response in different cancer types will be important. Since 1,25-$(OH)_2D_3$ regulates the same pathways but distinct genes of them in mice and humans [313], studies should preferentially be carried out in human systems. Among them, it seems that primary cell cultures and organoids should be used instead of classical, long-term established cell lines.

Given the increasingly important role attributed to the stroma in tumorigenesis, the effects of vitamin D compounds on CAF, endothelial cells, and specific types of immune cells require attention. Likewise, the association of chronic inflammation with several types of cancer and the pro-inflammatory action of adipocytes suggest the interest in studying the effects of vitamin D in this context.

Another open field for research is combination therapies. Up until now, experimental studies have focused on the combination of VDR agonists and chemotherapeutic agents, sometimes with radiotherapy. Obviously, this should be continued and extended to the exponentially growing field of cancer immunotherapies.

Author Contributions: Writing, review, and editing: A.M. and W.B.G. All authors have read and agreed to the published version of the manuscript.

Funding: The work in A.M. laboratory is funded by the Agencia Estatal de Investigación (PID2019-104867RB-I00/AEI/10.13039/501100011033) and the Instituto de Salud Carlos III—Fondo Europeo de Desarrollo Regional (CIBERONC/CB16/12/00273).

Conflicts of Interest: W.B.G.'s nonprofit organization, Sunlight, Nutrition and Health Research Center, receives funding from Bio-Tech Pharmacal, Inc. (Fayetteville, AR, USA). A.M. has no conflict of interest to declare.

References

1. Peller, S. Carcinogenesis as a means of reducing cancer mortlity. *Lancet* **1936**, *228*, 552–556. [CrossRef]
2. Peller, S.; Stephenson, C.S. Skin ititition and cancer in the United States Navy. *Am. J. Med. Sci.* **1937**, *194*, 326–333. [CrossRef]
3. Apperly, F.L. The Relation of Solar Radiation to Cancer Mortality in North America. *Cancer Res.* **1941**, *1*, 191–195.
4. Ainsleigh, H.G. Beneficial effects of sun exposure on cancer mortality. *Prev. Med.* **1993**, *22*, 132–140. [CrossRef] [PubMed]
5. Mason, T.J.; McKay, F.W.; Hoover, R.; Blot, W.J.; Fraumeni, J.F., Jr. *Atlas of Cancer Mortality for U.S. Counties: 1950–1969*; U.S. Department of Health, Education, and Welfare: Washington, DC, USA, 1975.
6. Garland, C.F.; Garland, F.C. Do sunlight and vitamin D reduce the likelihood of colon cancer? *Int. J. Epidemiol.* **1980**, *9*, 227–231. [CrossRef] [PubMed]
7. Garland, C.; Shekelle, R.B.; Barrett-Connor, E.; Criqui, M.H.; Rossof, A.H.; Paul, O. Dietary vitamin D and calcium and risk of colorectal cancer: A 19-year prospective study in men. *Lancet* **1985**, *1*, 307–309. [CrossRef]
8. Garland, C.F.; Comstock, G.W.; Garland, F.C.; Helsing, K.J.; Shaw, E.K.; Gorham, E.D. Serum 25-hydroxyvitamin D and colon cancer: Eight-year prospective study. *Lancet* **1989**, *2*, 1176–1178. [CrossRef]
9. Garland, F.C.; Garland, C.F.; Gorham, E.D.; Young, J.F. Geographic variation in breast cancer mortality in the United States: A hypothesis involving exposure to solar radiation. *Prev. Med.* **1990**, *19*, 614–622. [CrossRef]
10. Lefkowitz, E.S.; Garland, C.F. Sunlight, vitamin D, and ovarian cancer mortality rates in US women. *Int. J. Epidemiol.* **1994**, *23*, 1133–1136. [CrossRef]
11. Garland, C. The Summer of 1974, Or...How I Found My Life's Mission. Available online: https://www.grassrootshealth.net/?s=The+Summer+of+1974%2C+Or...How+I+Found+My+Life%27s+Mission (accessed on 23 February 2022).
12. Devesa, S.S.; Grauman, D.J.; Blot, W.J.; Pennello, G.A.; Hoover, R.N.; Fraumeni, J.F., Jr. *Atlas of Cancer Mortality in the United States, 1950–1994*; National Institutes of Health; National Cancer Institue: Bethesda, MD, USA, 1999.
13. Grant, W.B. An estimate of premature cancer mortality in the U.S. due to inadequate doses of solar ultraviolet-B radiation. *Cancer* **2002**, *94*, 1867–1875. [CrossRef] [PubMed]
14. Grant, W.B.; Garland, C.F. The association of solar ultraviolet B (UVB) with reducing risk of cancer: Multifactorial ecologic analysis of geographic variation in age-adjusted cancer mortality rates. *Anticancer Res.* **2006**, *26*, 2687–2699. [PubMed]
15. Grant, W.B. Lower vitamin-D production from solar ultraviolet-B irradiance may explain some differences in cancer survival rates. *J. Natl. Med. Assoc.* **2006**, *98*, 357–364. [PubMed]
16. Ames, B.N.; Grant, W.B.; Willett, W.C. Does the High Prevalence of vitamin D Deficiency in African Americans Contribute to Health Disparities? *Nutrients* **2021**, *13*, 499. [CrossRef] [PubMed]
17. Moukayed, M.; Grant, W.B. Molecular link between vitamin D and cancer prevention. *Nutrients* **2013**, *5*, 3993–4021. [CrossRef]
18. Chen, W.; Clements, M.; Rahman, B.; Zhang, S.; Qiao, Y.; Armstrong, B.K. Relationship between cancer mortality/incidence and ambient ultraviolet B irradiance in China. *Cancer Causes Control* **2010**, *21*, 1701–1709. [CrossRef]
19. Fioletov, V.E.; McArthur, L.J.; Mathews, T.W.; Marrett, L. Estimated ultraviolet exposure levels for a sufficient vitamin D status in North America. *J. Photochem. Photobiol. B* **2010**, *100*, 57–66. [CrossRef]
20. Herman, J.R.; Krotkov, N.; Celarier, E.; Larko, D.; Lebow, G. Distribution of UV radiation at the Earth's surface from TOMSmeasured UV-backscattered radiances. *J. Geophys. Res.* **1999**, *104*, 12059–12076. [CrossRef]
21. Mizoue, T. Ecological study of solar radiation and cancer mortality in Japan. *Health Phys.* **2004**, *87*, 532–538. [CrossRef]
22. Boscoe, F.P.; Schymura, M.J. Solar ultraviolet-B exposure and cancer incidence and mortality in the United States, 1993–2002. *BMC Cancer* **2006**, *6*, 264. [CrossRef]
23. Fukuda, Y.; Nakaya, T.; Nakao, H.; Yahata, Y.; Imai, H. Multilevel analysis of solar radiation and cancer mortality using ecological data in Japan. *Biosci. Trends* **2008**, *2*, 235–240.
24. Borisenkov, M.F. Latitude of residence and position in time zone are predictors of cancer incidence, cancer mortality, and life expectancy at birth. *Chronobiol. Int.* **2011**, *28*, 155–162. [CrossRef] [PubMed]
25. Grant, W.B. Role of solar UVB irradiance and smoking in cancer as inferred from cancer incidence rates by occupation in Nordic countries. *Dermatoendocrinology* **2012**, *4*, 203–211. [CrossRef] [PubMed]

26. Pukkala, E.; Martinsen, J.I.; Lynge, E.; Gunnarsdottir, H.K.; Sparen, P.; Tryggvadottir, L.; Weiderpass, E.; Kjaerheim, K. Occupation and cancer—Follow-up of 15 million people in five Nordic countries. *Acta Oncol.* **2009**, *48*, 646–790. [CrossRef]
27. Grant, W.B. A meta-analysis of second cancers after a diagnosis of nonmelanoma skin cancer: Additional evidence that solar ultraviolet-B irradiance reduces the risk of internal cancers. *J. Steroid Biochem. Mol. Biol.* **2007**, *103*, 668–674. [CrossRef] [PubMed]
28. Kenborg, L.; Jorgensen, A.D.; Budtz-Jorgensen, E.; Knudsen, L.E.; Hansen, J. Occupational exposure to the sun and risk of skin and lip cancer among male wage earners in Denmark: A population-based case-control study. *Cancer Causes Control* **2010**, *21*, 1347–1355. [CrossRef] [PubMed]
29. Jemal, A.; Siegel, R.; Ward, E.; Hao, Y.; Xu, J.; Thun, M.J. Cancer statistics, 2009. *CA Cancer J. Clin.* **2009**, *59*, 225–249. [CrossRef] [PubMed]
30. Lin, S.W.; Wheeler, D.C.; Park, Y.; Cahoon, E.K.; Hollenbeck, A.R.; Freedman, D.M.; Abnet, C.C. Prospective study of ultraviolet radiation exposure and risk of cancer in the United States. *Int. J. Cancer* **2012**, *131*, E1015–E1023. [CrossRef] [PubMed]
31. O'Sullivan, F.; van Geffen, J.; van Weele, M.; Zgaga, L. Annual Ambient UVB at Wavelengths that Induce vitamin D Synthesis is Associated with Reduced Esophageal and Gastric Cancer Risk: A Nested Case-Control Study. *Photochem. Photobiol.* **2018**, *94*, 797–806. [CrossRef]
32. Sempos, C.T.; Durazo-Arvizu, R.A.; Binkley, N.; Jones, J.; Merkel, J.M.; Carter, G.D. Developing vitamin D dietary guidelines and the lack of 25-hydroxyvitamin D assay standardization: The ever-present past. *J. Steroid Biochem. Mol. Biol.* **2016**, *164*, 115–119. [CrossRef]
33. Ginde, A.A.; Liu, M.C.; Camargo, C.A., Jr. Demographic differences and trends of vitamin D insufficiency in the US population, 1988–2004. *Arch. Intern. Med.* **2009**, *169*, 626–632. [CrossRef]
34. Crowe, F.L.; Steur, M.; Allen, N.E.; Appleby, P.N.; Travis, R.C.; Key, T.J. Plasma concentrations of 25-hydroxyvitamin D in meat eaters, fish eaters, vegetarians and vegans: Results from the EPIC-Oxford study. *Public Health Nutr.* **2011**, *14*, 340–346. [CrossRef] [PubMed]
35. Cashman, K.D.; O'Sullivan, S.M.; Galvin, K.; Ryan, M. Contribution of vitamin D2 and D3 and Their Respective 25-Hydroxy Metabolites to the Total vitamin D Content of Beef and Lamb. *Curr. Dev. Nutr.* **2020**, *4*, nzaa112. [CrossRef] [PubMed]
36. Autier, P.; Boniol, M.; Pizot, C.; Mullie, P. vitamin D status and ill health: A systematic review. *Lancet Diabetes Endocrinol.* **2014**, *2*, 76–89. [CrossRef]
37. Autier, P.; Mullie, P.; Macacu, A.; Dragomir, M.; Boniol, M.; Coppens, K.; Pizot, C.; Boniol, M. Effect of vitamin D supplementation on non-skeletal disorders: A systematic review of meta-analyses and randomised trials. *Lancet Diabetes Endocrinol.* **2017**, *5*, 986–1004. [CrossRef]
38. Smolders, J.; van den Ouweland, J.; Geven, C.; Pickkers, P.; Kox, M. Letter to the Editor: vitamin D deficiency in COVID-19: Mixing up cause and consequence. *Metabolism* **2021**, *115*, 154434. [CrossRef] [PubMed]
39. Wang, L. C-reactive protein levels in the early stage of COVID-19. *Med. Mal. Infect.* **2020**, *50*, 332–334. [CrossRef] [PubMed]
40. Allin, K.H.; Bojesen, S.E.; Nordestgaard, B.G. Baseline C-reactive protein is associated with incident cancer and survival in patients with cancer. *J. Clin. Oncol.* **2009**, *27*, 2217–2224. [CrossRef]
41. Grant, W.B. Effect of interval between serum draw and follow-up period on relative risk of cancer incidence with respect to 25-hydroxyvitamin D level: Implications for meta-analyses and setting vitamin D guidelines. *Dermatoendocrinology* **2011**, *3*, 199–204. [CrossRef]
42. Grant, W.B. 25-hydroxyvitamin D and breast cancer, colorectal cancer, and colorectal adenomas: Case-control versus nested case-control studies. *Anticancer Res.* **2015**, *35*, 1153–1160.
43. Grant, W.B. Effect of follow-up time on the relation between prediagnostic serum 25-hydroxyvitamin D and all-cause mortality rate. *Dermatoendocrinology* **2012**, *4*, 198–202. [CrossRef]
44. Wu, K.; Feskanich, D.; Fuchs, C.S.; Willett, W.C.; Hollis, B.W.; Giovannucci, E.L. A nested case control study of plasma 25-hydroxyvitamin D concentrations and risk of colorectal cancer. *J. Natl. Cancer Inst.* **2007**, *99*, 1120–1129. [CrossRef] [PubMed]
45. Feskanich, D.; Ma, J.; Fuchs, C.S.; Kirkner, G.J.; Hankinson, S.E.; Hollis, B.W.; Giovannucci, E.L. Plasma vitamin D metabolites and risk of colorectal cancer in women. *Cancer Epidemiol. Biomark. Prev.* **2004**, *13*, 1502–1508.
46. Gorham, E.D.; Garland, C.F.; Garland, F.C.; Grant, W.B.; Mohr, S.B.; Lipkin, M.; Newmark, H.L.; Giovannucci, E.; Wei, M.; Holick, M.F. Optimal vitamin D status for colorectal cancer prevention: A quantitative meta analysis. *Am. J. Prev. Med.* **2007**, *32*, 210–216. [CrossRef] [PubMed]
47. McCullough, M.L.; Zoltick, E.S.; Weinstein, S.J.; Fedirko, V.; Wang, M.; Cook, N.R.; Eliassen, A.H.; Zeleniuch-Jacquotte, A.; Agnoli, C.; Albanes, D.; et al. Circulating vitamin D and Colorectal Cancer Risk: An International Pooling Project of 17 Cohorts. *J. Natl. Cancer Inst.* **2019**, *111*, 158–169. [CrossRef] [PubMed]
48. Weinstein, S.J.; Yu, K.; Horst, R.L.; Ashby, J.; Virtamo, J.; Albanes, D. Serum 25-hydroxyvitamin D and risks of colon and rectal cancer in Finnish men. *Am. J. Epidemiol.* **2011**, *173*, 499–508. [CrossRef]
49. Lee, J.E.; Li, H.; Chan, A.T.; Hollis, B.W.; Lee, I.M.; Stampfer, M.J.; Wu, K.; Giovannucci, E.; Ma, J. Circulating levels of vitamin D and colon and rectal cancer: The Physicians' Health Study and a meta-analysis of prospective studies. *Cancer Prev. Res.* **2011**, *4*, 735–743. [CrossRef] [PubMed]
50. Kakourou, A.; Koutsioumpa, C.; Lopez, D.S.; Hoffman-Bolton, J.; Bradwin, G.; Rifai, N.; Helzlsouer, K.J.; Platz, E.A.; Tsilidis, K.K. Interleukin-6 and risk of colorectal cancer: Results from the CLUE II cohort and a meta-analysis of prospective studies. *Cancer Causes Control* **2015**, *26*, 1449–1460. [CrossRef] [PubMed]

51. Song, M.; Wu, K.; Chan, A.T.; Fuchs, C.S.; Giovannucci, E.L. Plasma 25-hydroxyvitamin D and risk of colorectal cancer after adjusting for inflammatory markers. *Cancer Epidemiol. Biomark. Prev.* **2014**, *23*, 2175–2180. [CrossRef] [PubMed]
52. Langseth, H.; Gislefoss, R.E.; Martinsen, J.I.; Dillner, J.; Ursin, G. Cohort Profile: The Janus Serum Bank Cohort in Norway. *Int. J. Epidemiol.* **2017**, *46*, 403–404. [CrossRef] [PubMed]
53. Jenab, M.; Bueno-de-Mesquita, H.B.; Ferrari, P.; van Duijnhoven, F.J.; Norat, T.; Pischon, T.; Jansen, E.H.; Slimani, N.; Byrnes, G.; Rinaldi, S.; et al. Association between pre-diagnostic circulating vitamin D concentration and risk of colorectal cancer in European populations:a nested case-control study. *BMJ* **2010**, *340*, b5500. [CrossRef]
54. Woolcott, C.G.; Wilkens, L.R.; Nomura, A.M.; Horst, R.L.; Goodman, M.T.; Murphy, S.P.; Henderson, B.E.; Kolonel, L.N.; Le Marchand, L. Plasma 25-hydroxyvitamin D levels and the risk of colorectal cancer: The multiethnic cohort study. *Cancer Epidemiol. Biomark. Prev.* **2010**, *19*, 130–134. [CrossRef]
55. McCullough, M.L.; Robertson, A.S.; Rodriguez, C.; Jacobs, E.J.; Chao, A.; Carolyn, J.; Calle, E.E.; Willett, W.C.; Thun, M.J. Calcium, vitamin D, dairy products, and risk of colorectal cancer in the Cancer Prevention Study II Nutrition Cohort (United States). *Cancer Causes Control* **2003**, *14*, 1–12. [CrossRef] [PubMed]
56. Otani, T.; Iwasaki, M.; Sasazuki, S.; Inoue, M.; Tsugane, S.; Japan Public Health Center-Based Prospective Study, G. Plasma vitamin D and risk of colorectal cancer: The Japan Public Health Center-Based Prospective Study. *Br. J. Cancer* **2007**, *97*, 446–451. [CrossRef] [PubMed]
57. Cheng, T.Y.; Goodman, G.E.; Thornquist, M.D.; Barnett, M.J.; Beresford, S.A.; LaCroix, A.Z.; Zheng, Y.; Neuhouser, M.L. Estimated intake of vitamin D and its interaction with vitamin A on lung cancer risk among smokers. *Int. J. Cancer* **2014**, *135*, 2135–2145. [CrossRef] [PubMed]
58. Weinstein, S.J.; Purdue, M.P.; Smith-Warner, S.A.; Mondul, A.M.; Black, A.; Ahn, J.; Huang, W.Y.; Horst, R.L.; Kopp, W.; Rager, H.; et al. Serum 25-hydroxyvitamin D, vitamin D binding protein and risk of colorectal cancer in the Prostate, Lung, Colorectal and Ovarian Cancer Screening Trial. *Int. J. Cancer* **2015**, *136*, E654–E664. [CrossRef]
59. Tangrea, J.; Helzlsouer, K.; Pietinen, P.; Taylor, P.; Hollis, B.; Virtamo, J.; Albanes, D. Serum levels of vitamin D metabolites and the subsequent risk of colon and rectal cancer in Finnish men. *Cancer Causes Control* **1997**, *8*, 615–625. [CrossRef] [PubMed]
60. Schernhammer, E.S.; Sperati, F.; Razavi, P.; Agnoli, C.; Sieri, S.; Berrino, F.; Krogh, V.; Abbagnato, C.; Grioni, S.; Blandino, G.; et al. Endogenous sex steroids in premenopausal women and risk of breast cancer: The ORDET cohort. *Breast Cancer Res.* **2013**, *15*, R46. [CrossRef] [PubMed]
61. Swerdlow, A.J.; Jones, M.E.; Schoemaker, M.J.; Hemming, J.; Thomas, D.; Williamson, J.; Ashworth, A. The Breakthrough Generations Study: Design of a long-term UK cohort study to investigate breast cancer aetiology. *Br. J. Cancer* **2011**, *105*, 911–917. [CrossRef]
62. Neuhouser, M.L.; Manson, J.E.; Millen, A.; Pettinger, M.; Margolis, K.; Jacobs, E.T.; Shikany, J.M.; Vitolins, M.; Adams-Campbell, L.; Liu, S.; et al. The influence of health and lifestyle characteristics on the relation of serum 25-hydroxyvitamin D with risk of colorectal and breast cancer in postmenopausal women. *Am. J. Epidemiol.* **2012**, *175*, 673–684. [CrossRef] [PubMed]
63. Chandler, P.D.; Buring, J.E.; Manson, J.E.; Giovannucci, E.L.; Moorthy, M.V.; Zhang, S.; Lee, I.M.; Lin, J.H. Circulating vitamin D Levels and Risk of Colorectal Cancer in Women. *Cancer Prev. Res.* **2015**, *8*, 675–682. [CrossRef] [PubMed]
64. Scarmo, S.; Afanasyeva, Y.; Lenner, P.; Koenig, K.L.; Horst, R.L.; Clendenen, T.V.; Arslan, A.A.; Chen, Y.; Hallmans, G.; Lundin, E.; et al. Circulating levels of 25-hydroxyvitamin D and risk of breast cancer: A nested case-control study. *Breast Cancer Res.* **2013**, *15*, R15. [CrossRef]
65. Oh, E.Y.; Ansell, C.; Nawaz, H.; Yang, C.H.; Wood, P.A.; Hrushesky, W.J. Global breast cancer seasonality. *Breast Cancer Res. Treat.* **2010**, *123*, 233–243. [CrossRef] [PubMed]
66. Abbas, S.; Linseisen, J.; Slanger, T.; Kropp, S.; Mutschelknauss, E.J.; Flesch-Janys, D.; Chang-Claude, J. Serum 25-hydroxyvitamin D and risk of post-menopausal breast cancer–results of a large case-control study. *Carcinogenesis* **2008**, *29*, 93–99. [CrossRef] [PubMed]
67. Abbas, S.; Chang-Claude, J.; Linseisen, J. Plasma 25-hydroxyvitamin D and premenopausal breast cancer risk in a German case-control study. *Int. J. Cancer* **2009**, *124*, 250–255. [CrossRef] [PubMed]
68. Song, D.; Deng, Y.; Liu, K.; Zhou, L.; Li, N.; Zheng, Y.; Hao, Q.; Yang, S.; Wu, Y.; Zhai, Z.; et al. vitamin D intake, blood vitamin D levels, and the risk of breast cancer: A dose-response meta-analysis of observational studies. *Aging* **2019**, *11*, 12708–12732. [CrossRef]
69. McDonnell, S.L.; Baggerly, C.A.; French, C.B.; Baggerly, L.L.; Garland, C.F.; Gorham, E.D.; Hollis, B.W.; Trump, D.L.; Lappe, J.M. Breast cancer risk markedly lower with serum 25-hydroxyvitamin D concentrations >/=60 vs >20 ng/mL (150 vs 50 nmol/L): Pooled analysis of two randomized trials and a prospective cohort. *PLoS ONE* **2018**, *13*, e0199265. [CrossRef]
70. Han, J.; Guo, X.; Yu, X.; Liu, S.; Cui, X.; Zhang, B.; Liang, H. 25-Hydroxyvitamin D and Total Cancer Incidence and Mortality: A Meta-Analysis of Prospective Cohort Studies. *Nutrients* **2019**, *11*, 2295. [CrossRef] [PubMed]
71. Zhao, Y.; Chen, C.; Pan, W.; Gao, M.; He, W.; Mao, R.; Lin, T.; Huang, J. Comparative efficacy of vitamin D status in reducing the risk of bladder cancer: A systematic review and network meta-analysis. *Nutrition* **2016**, *32*, 515–523. [CrossRef] [PubMed]
72. Hernandez-Alonso, P.; Boughanem, H.; Canudas, S.; Becerra-Tomas, N.; Fernandez de la Puente, M.; Babio, N.; Macias-Gonzalez, M.; Salas-Salvado, J. Circulating vitamin D levels and colorectal cancer risk: A meta-analysis and systematic review of case-control and prospective cohort studies. *Crit. Rev. Food Sci. Nutr.* **2021**, *61*, 1–17. [CrossRef] [PubMed]
73. Garland, C.F.; Gorham, E.D. Dose-response of serum 25-hydroxyvitamin D in association with risk of colorectal cancer: A meta-analysis. *J. Steroid Biochem. Mol. Biol.* **2017**, *168*, 1–8. [CrossRef] [PubMed]

74. Pu, Y.; Zhu, G.; Xu, Y.; Zheng, S.; Tang, B.; Huang, H.; Wu, I.X.Y.; Huang, D.; Liu, Y.; Zhang, X. Association between vitamin D Exposure and Head and Neck Cancer: A Systematic Review with Meta-Analysis. *Front. Immunol.* **2021**, *12*, 627226. [CrossRef] [PubMed]
75. Guo, X.F.; Zhao, T.; Han, J.M.; Li, S.; Li, D. vitamin D and liver cancer risk: A meta-analysis of prospective studies. *Asia Pac. J. Clin. Nutr.* **2020**, *29*, 175–182. [CrossRef] [PubMed]
76. Zhang, Y.; Jiang, X.; Li, X.; Gaman, M.A.; Kord-Varkaneh, H.; Rahmani, J.; Salehi-Sahlabadi, A.; Day, A.S.; Xu, Y. Serum vitamin D Levels and Risk of Liver Cancer: A Systematic Review and Dose-Response Meta-Analysis of Cohort Studies. *Nutr. Cancer* **2021**, *73*, 1–9. [CrossRef] [PubMed]
77. Liu, J.; Dong, Y.; Lu, C.; Wang, Y.; Peng, L.; Jiang, M.; Tang, Y.; Zhao, Q. Meta-analysis of the correlation between vitamin D and lung cancer risk and outcomes. *Oncotarget* **2017**, *8*, 81040–81051. [CrossRef]
78. Feng, J.; Shan, L.; Du, L.; Wang, B.; Li, H.; Wang, W.; Wang, T.; Dong, H.; Yue, X.; Xu, Z.; et al. Clinical improvement following vitamin D3 supplementation in Autism Spectrum Disorder. *Nutr. Neurosci.* **2017**, *20*, 284–290. [CrossRef]
79. Wei, H.; Jing, H.; Wei, Q.; Wei, G.; Heng, Z. Associations of the risk of lung cancer with serum 25-hydroxyvitamin D level and dietary vitamin D intake: A dose-response PRISMA meta-analysis. *Medicine* **2018**, *97*, e12282. [CrossRef]
80. Xu, J.; Chen, K.; Zhao, F.; Huang, D.; Zhang, H.; Fu, Z.; Xu, J.; Wu, Y.; Lin, H.; Zhou, Y.; et al. Association between vitamin D/calcium intake and 25-hydroxyvitamin D and risk of ovarian cancer: A dose-response relationship meta-analysis. *Eur. J. Clin. Nutr.* **2021**, *75*, 417–429. [CrossRef]
81. Zhang, X.; Huang, X.Z.; Chen, W.J.; Wu, J.; Chen, Y.; Wu, C.C.; Wang, Z.N. Plasma 25-hydroxyvitamin D levels, vitamin D intake, and pancreatic cancer risk or mortality: A meta-analysis. *Oncotarget* **2017**, *8*, 64395–64406. [CrossRef]
82. Gao, J.; Wei, W.; Wang, G.; Zhou, H.; Fu, Y.; Liu, N. Circulating vitamin D concentration and risk of prostate cancer: A dose-response meta-analysis of prospective studies. *Ther. Clin. Risk Manag.* **2018**, *14*, 95–104. [CrossRef]
83. Wu, J.; Yang, N.; Youan, M. Dietary and circulating vitamin D and risk of renal cell carcinoma: A meta-analysis of observational studies. *Int. Braz. J. Urol.* **2021**, *47*, 733–744. [CrossRef]
84. Zhao, J.; Wang, H.; Zhang, Z.; Zhou, X.; Yao, J.; Zhang, R.; Liao, L.; Dong, J. vitamin D deficiency as a risk factor for thyroid cancer: A meta-analysis of case-control studies. *Nutrition* **2019**, *57*, 5–11. [CrossRef] [PubMed]
85. Manson, J.E.; Cook, N.R.; Lee, I.M.; Christen, W.; Bassuk, S.S.; Mora, S.; Gibson, H.; Gordon, D.; Copeland, T.; D'Agostino, D.; et al. vitamin D Supplements and Prevention of Canc.cer and Cardiovascular Disease. *N. Engl. J. Med.* **2019**, *380*, 33–44. [CrossRef] [PubMed]
86. Keum, N.; Lee, D.H.; Greenwood, D.C.; Manson, J.E.; Giovannucci, E. vitamin D supplementation and total cancer incidence and mortality: A meta-analysis of randomized controlled trials. *Ann. Oncol.* **2019**, *30*, 733–743. [CrossRef]
87. Zhang, X.; Niu, W. Meta-analysis of randomized controlled trials on vitamin D supplement and cancer incidence and mortality. *Biosci. Rep.* **2019**, *39*, BSR20190396. [CrossRef] [PubMed]
88. Ekmekcioglu, C.; Haluza, D.; Kundi, M. 25-Hydroxyvitamin D Status and Risk for Colorectal Cancer and Type 2 Diabetes Mellitus: A Systematic Review and Meta-Analysis of Epidemiological Studies. *Int. J. Environ. Res. Public Health* **2017**, *14*, 20127. [CrossRef] [PubMed]
89. Chen, G.C.; Zhang, Z.L.; Wan, Z.; Wang, L.; Weber, P.; Eggersdorfer, M.; Qin, L.Q.; Zhang, W. Circulating 25-hydroxyvitamin D and risk of lung cancer: A dose-response meta-analysis. *Cancer Causes Control* **2015**, *26*, 1719–1728. [CrossRef]
90. Song, Z.Y.; Yao, Q.; Zhuo, Z.; Ma, Z.; Chen, G. Circulating vitamin D level and mortality in prostate cancer patients: A dose-response meta-analysis. *Endocr. Connect.* **2018**, *7*, R294–R303. [CrossRef]
91. Pilz, S.; Trummer, C.; Theiler-Schwetz, V.; Grubler, M.R.; Verheyen, N.D.; Odler, B.; Karras, S.N.; Zittermann, A.; Marz, W. Critical Appraisal of Large vitamin D Randomized Controlled Trials. *Nutrients* **2022**, *14*, 303. [CrossRef]
92. De Pergola, G.; Silvestris, F. Obesity as a major risk factor for cancer. *J. Obes.* **2013**, *2013*, 291546. [CrossRef]
93. Trivedi, D.P.; Doll, R.; Khaw, K.T. Effect of four monthly oral vitamin D3 (cholecalciferol) supplementation on fractures and mortality in men and women living in the community: Randomised double blind controlled trial. *BMJ* **2003**, *326*, 469. [CrossRef]
94. Wactawski-Wende, J.; Kotchen, J.M.; Anderson, G.L.; Assaf, A.R.; Brunner, R.L.; O'Sullivan, M.J.; Margolis, K.L.; Ockene, J.K.; Phillips, L.; Pottern, L.; et al. Calcium plus vitamin D supplementation and the risk of colorectal cancer. *N. Engl. J. Med.* **2006**, *354*, 684–696. [CrossRef] [PubMed]
95. Lappe, J.M.; Travers-Gustafson, D.; Davies, K.M.; Recker, R.R.; Heaney, R.P. vitamin D and calcium supplementation reduces cancer risk: Results of a randomized trial. *Am. J. Clin. Nutr.* **2007**, *85*, 1586–1591. [CrossRef] [PubMed]
96. Sanders, K.M.; Stuart, A.L.; Williamson, E.J.; Simpson, J.A.; Kotowicz, M.A.; Young, D.; Nicholson, G.C. Annual high-dose oral vitamin D and falls and fractures in older women: A randomized controlled trial. *JAMA* **2010**, *303*, 1815–1822. [CrossRef] [PubMed]
97. Avenell, A.; MacLennan, G.S.; Jenkinson, D.J.; McPherson, G.C.; McDonald, A.M.; Pant, P.R.; Grant, A.M.; Campbell, M.K.; Anderson, F.H.; Cooper, C.; et al. Long-term follow-up for mortality and cancer in a randomized placebo-controlled trial of vitamin D(3) and/or calcium (RECORD trial). *J. Clin. Endocrinol. Metab.* **2012**, *97*, 614–622. [CrossRef] [PubMed]
98. Lappe, J.; Garland, C.; Gorham, E. vitamin D Supplementation and Cancer Risk. *JAMA* **2017**, *318*, 299–300. [CrossRef]
99. Scragg, R.; Khaw, K.T.; Toop, L.; Sluyter, J.; Lawes, C.M.M.; Waayer, D.; Giovannucci, E.; Camargo, C.A., Jr. Monthly High-Dose vitamin D Supplementation and Cancer Risk: A Post Hoc Analysis of the vitamin D Assessment Randomized Clinical Trial. *JAMA Oncol.* **2018**, *4*, e182178. [CrossRef]

100. Neale, R.E.; Baxter, C.; Romero, B.D.; McLeod, D.S.A.; English, D.R.; Armstrong, B.K.; Ebeling, P.R.; Hartel, G.; Kimlin, M.G.; O'Connell, R.; et al. The D-Health Trial: A randomised controlled trial of the effect of vitamin D on mortality. *Lancet Diabetes Endocrinol.* **2022**, *10*, 120–128. [CrossRef]
101. Heaney, R.P. Guidelines for optimizing design and analysis of clinical studies of nutrient effects. *Nutr. Rev.* **2014**, *72*, 48–54. [CrossRef]
102. Grant, W.B.; Boucher, B.J.; Bhattoa, H.P.; Lahore, H. Why vitamin D clinical trials should be based on 25-hydroxyvitamin D concentrations. *J. Steroid Biochem. Mol. Biol.* **2018**, *177*, 266–269. [CrossRef]
103. Hrushesky, W.J.; Sothern, R.B.; Rietveld, W.J.; Du Quiton, J.; Boon, M.E. Season, sun, sex, and cervical cancer. *Cancer Epidemiol. Biomark. Prev.* **2005**, *14*, 1940–1947. [CrossRef]
104. Marur, S.; D'Souza, G.; Westra, W.H.; Forastiere, A.A. HPV-associated head and neck cancer: A virus-related cancer epidemic. *Lancet Oncol.* **2010**, *11*, 781–789. [CrossRef]
105. Merrill, S.J.; Subramanian, M.; Godar, D.E. Worldwide cutaneous malignant melanoma incidences analyzed by sex, age, and skin type over time (1955–2007): Is HPV infection of androgenic hair follicular melanocytes a risk factor for developing melanoma exclusively in people of European-ancestry? *Dermatoendocrinology* **2016**, *8*, e1215391. [CrossRef] [PubMed]
106. Loomis, D.; Huang, W.; Chen, G. The International Agency for Research on Cancer (IARC) evaluation of the carcinogenicity of outdoor air pollution: Focus on China. *Chin. J. Cancer* **2014**, *33*, 189–196. [CrossRef] [PubMed]
107. Hill, A.B. The Environment and Disease: Association or Causation? *Proc. R Soc. Med.* **1965**, *58*, 295–300. [CrossRef]
108. Grant, W.B. How strong is the evidence that solar ultraviolet B and vitamin D reduce the risk of cancer?: An examination using Hill's criteria for causality. *Dermatoendocrinology* **2009**, *1*, 17–24. [CrossRef] [PubMed]
109. Mohr, S.B.; Gorham, E.D.; Alcaraz, J.E.; Kane, C.I.; Macera, C.A.; Parsons, J.K.; Wingard, D.L.; Garland, C.F. Does the evidence for an inverse relationship between serum vitamin D status and breast cancer risk satisfy the Hill criteria? *Dermatoendocrinology* **2012**, *4*, 152–157. [CrossRef]
110. Frieden, T.R. Evidence for Health Decision Making—Beyond Randomized, Controlled Trials. *N. Engl. J. Med.* **2017**, *377*, 465–475. [CrossRef]
111. Giovannucci, E.; Liu, Y.; Rimm, E.B.; Hollis, B.W.; Fuchs, C.S.; Stampfer, M.J.; Willett, W.C. Prospective study of predictors of vitamin D status and cancer incidence and mortality in men. *J. Natl. Cancer Inst.* **2006**, *98*, 451–459. [CrossRef]
112. Colston, K.; Colston, M.J.; Feldman, D. 1,25-dihydroxyvitamin D3 and malignant melanoma: The presence of receptors and inhibition of cell growth in culture. *Endocrinology* **1981**, *108*, 1083–1086. [CrossRef]
113. Abe, E.; Miyaura, C.; Sakagami, H.; Takeda, M.; Konno, K.; Yamazaki, T.; Yoshiki, S.; Suda, T. Differentiation of mouse myeloid leukemia cells induced by 1alpha,25-dihydroxyvitamin D3. *Proc. Natl. Acad. Sci. USA* **1981**, *78*, 4990–4994. [CrossRef]
114. Feldman, D.; Krishnan, A.V.; Swami, S.; Giovannucci, E.; Feldman, B.J. The role of vitamin D in reducing cancer risk and progression. *Nat. Rev. Cancer* **2014**, *14*, 342–357. [CrossRef] [PubMed]
115. Ferrer-Mayorga, G.; Larriba, M.J.; Crespo, P.; Muñoz, A. Mechanisms of action of vitamin D in colon cancer. *J. Steroid Biochem. Mol. Biol.* **2019**, *185*, 1–6. [CrossRef]
116. Wu, X.; Hu, W.; Lu, L.; Zhao, Y.; Zhou, Y.; Xiao, Z.; Zhang, L.; Zhang, H.; Li, X.; Li, W.; et al. Repurposing vitamin D for treatment of human malignancies via targeting tumor microenvironment. *Acta Pharm. Sinica B* **2019**, *9*, 203–219. [CrossRef] [PubMed]
117. Markowska, A.; Antoszczak, M.; Kojs, Z.; Bednarek, W.; Markowska, J.; Huczynski, A. Role of vitamin D3 in selected malignant neoplasms. *Nutrition* **2020**, *79*, 110964. [CrossRef]
118. Carlberg, C.; Velleuer, E. vitamin D and the risk for cancer: A molecular analysis. *Biochem. Pharmacol.* **2022**, *196*, 114735. [CrossRef] [PubMed]
119. Vanhevel, J.; Verlinden, L.; Doms, S.; Wildiers, H.; Verstuyf, A. The role of vitamin D in breast cancer risk and progression. *Endocr. Relat. Cancer* **2022**, *29*, R33–R55. [CrossRef] [PubMed]
120. Huang, Y.-C.; Chen, J.-Y.; Hung, W.-C. vitamin D_3 receptor/Sp1 complex is required for the induction of $p27^{KIP1}$ expression by vitamin D_3. *Oncogene* **2004**, *23*, 4856–4861. [CrossRef] [PubMed]
121. Yang, E.S.; Burnstein, K.L. vitamin D inhibits G1 to S progression in LNCaP prostate cancer cells through p27Kip1 stabilization and Cdk2 mislocalization to the cytoplasm. *J. Biol. Chem.* **2003**, *278*, 46862–46868. [CrossRef]
122. Li, P.; Li, C.; Zhao, X.; Zhang, X.; Nicosia, S.V.; Bai, W. $p27^{Kip1}$ stabilization and G_1 arrest by 1,25-dihydroxyvitamin D_3 in ovarian cancer cells mediated through down-regulation of cyclin E/cyclin-dependent kinase 2 and Skp1-Cullin-F-box protein/Skp2 ubiquitin ligase. *J. Biol. Chem.* **2004**, *279*, 25260–25267. [CrossRef]
123. Washington, M.N.; Kim, J.S.; Weigel, N.L. 1alpha,25-dihydroxyvitamin D3 inhibits C4-2 prostate cancer cell growth via a retinoblastoma protein (Rb)-independent G1 arrest. *Prostate* **2011**, *71*, 98–110. [CrossRef]
124. Toropainen, S.; Väisänen, S.; Heikkinen, S.; Carlberg, C. The down-regulation of the human MYC gene by the nuclear hormone 1alpha,25-dihydroxyvitamin D3 is associated with cycling of corepressors and histone deacetylases. *J. Mol. Biol.* **2010**, *400*, 284–294. [CrossRef]
125. Pálmer, H.G.; González-Sancho, J.M.; Espada, J.; Berciano, M.T.; Puig, I.; Baulida, J.; Quintanilla, M.; Cano, A.; García de Herreros, A.; Lafarga, M.; et al. vitamin D_3 promotes the differentiation of colon carcinoma cells by the induction of E-cadherin and the inhibition of b-catenin signaling. *J. Cell Biol.* **2001**, *154*, 369–387. [CrossRef] [PubMed]

126. Ordóñez-Morán, P.; Larriba, M.J.; Pálmer, H.G.; Valero, R.A.; Barbáchano, A.; Duñach, M.; García de Herreros, A.; Villalobos, C.; Berciano, M.T.; Lafarga, M.; et al. RhoA-ROCK and p38MAPK-MSK1 mediate vitamin D effects on gene expression, phenotype, and Wnt pathway in colon cancer cells. *J. Cell Biol.* **2008**, *183*, 697–710. [CrossRef]
127. Álvarez-Díaz, S.; Valle, N.; García, J.M.; Peña, C.; Freije, J.M.; Quesada, V.; Astudillo, A.; Bonilla, F.; López-Otín, C.; Muñoz, A. Cystatin D is a candidate tumor suppressor gene induced by vitamin D in human colon cancer cells. *J. Clin. Investig.* **2009**, *119*, 2343–2358. [CrossRef] [PubMed]
128. Salehi-Tabar, R.; Nguyen-Yamamoto, L.; Tavera-Mendoza, L.E.; Quail, T.; Dimitrov, V.; An, B.S.; Glass, L.; Goltzman, D.; White, J.H. vitamin D receptor as a master regulator of the c-MYC/MXD1 network. *Proc. Natl. Acad. Sci. USA* **2012**, *109*, 18827–18832. [CrossRef] [PubMed]
129. Wang, L.; Zhou, S.; Guo, B. vitamin D Suppresses Ovarian Cancer Growth and Invasion by Targeting Long Non-Coding RNA CCAT2. *Int. J. Mol. Sci.* **2020**, *21*, 72334. [CrossRef] [PubMed]
130. Salehi-Tabar, R.; Memari, B.; Wong, H.; Dimitrov, V.; Rochel, N.; White, J.H. The Tumor Suppressor FBW7 and the vitamin D Receptor Are Mutual Cofactors in Protein Turnover and Transcriptional Regulation. *Mol. Cancer Res. MCR* **2019**, *17*, 709–719. [CrossRef] [PubMed]
131. Pálmer, H.G.; Sánchez-Carbayo, M.; Ordóñez-Morán, P.; Larriba, M.J.; Cordón-Cardó, C.; Muñoz, A. Genetic signatures of differentiation induced by 1a,25-dihydroxyvitamin D_3 in human colon cancer cells. *Cancer Res.* **2003**, *63*, 7799–7806. [PubMed]
132. Zhu, Y.; Chen, P.; Gao, Y.; Ta, N.; Zhang, Y.; Cai, J.; Zhao, Y.; Liu, S.; Zheng, J. MEG3 Activated by vitamin D Inhibits Colorectal Cancer Cells Proliferation and Migration via Regulating Clusterin. *EBioMedicine* **2018**, *30*, 148–157. [CrossRef] [PubMed]
133. Zhu, C.; Wang, Z.; Cai, J.; Pan, C.; Lin, S.; Zhang, Y.; Chen, Y.; Leng, M.; He, C.; Zhou, P.; et al. VDR Signaling via the Enzyme NAT2 Inhibits Colorectal Cancer Progression. *Front. Pharmacol.* **2021**, *12*, 727704. [CrossRef] [PubMed]
134. Li, Q.; Li, Y.; Jiang, H.; Xiao, Z.; Wu, X.; Zhang, H.; Zhao, Y.; Du, F.; Chen, Y.; Wu, Z.; et al. vitamin D suppressed gastric cancer cell growth through downregulating CD44 expression in vitro and in vivo. *Nutrition* **2021**, *91*, 111413. [CrossRef] [PubMed]
135. Dhawan, P.; Weider, R.; Christakos, S. CCAAT enhancer-binding protein alpha is a molecular target of 1,25-dihydroxyvitamin D3 in MCF-7 breast cancer cells. *J. Biol. Chem.* **2009**, *284*, 3086–3095. [CrossRef]
136. Boyle, B.J.; Zhao, X.Y.; Cohen, P.; Feldman, D. Insulin-like growth factor binding protein-3 mediates 1 alpha,25-dihydroxyvitamin D(3) growth inhibition in the LNCaP prostate cancer cell line through p21/WAF1. *J. Urol.* **2001**, *165*, 1319–1324. [CrossRef]
137. Chang, S.; Gao, L.; Yang, Y.; Tong, D.; Guo, B.; Liu, L.; Li, Z.; Song, T.; Huang, C. miR-145 mediates the antiproliferative and gene regulatory effects of vitamin D3 by directly targeting E2F3 in gastric cancer cells. *Oncotarget* **2015**, *6*, 7675–7685. [CrossRef] [PubMed]
138. Peng, W.; Wang, K.; Zheng, R.; Derwahl, M. 1,25 dihydroxyvitamin D3 inhibits the proliferation of thyroid cancer stem-like cells via cell cycle arrest. *Endocr. Res.* **2016**, *41*, 71–80. [CrossRef]
139. Kulling, P.M.; Olson, K.C.; Olson, T.L.; Feith, D.J.; Loughran, T.P., Jr. vitamin D in hematological disorders and malignancies. *Eur. J. Haematol.* **2017**, *98*, 187–197. [CrossRef]
140. Tong, W.-M.; Kállay, E.; Hofer, H.; Hulla, W.; Manhardt, T.; Peterlik, M.; Cross, H.S. Growth regulation of human colon cancer cells by epidermal growth factor and 1,25-dihydroxyvitamin D_3 is mediated by mutual modulation of receptor expression. *Eur. J. Cancer* **1998**, *34*, 2119–2125. [CrossRef]
141. Tong, W.-M.; Hofer, H.; Ellinger, A.; Peterlik, M.; Cross, H.S. Mechanism of antimitogenic action of vitamin D in human colon carcinoma cells: Relevance for suppression of epidermal growth factor-stimulated cell growth. *Oncol. Res.* **1999**, *11*, 77–84. [PubMed]
142. Andl, C.D.; Rustgi, A.K. No one-way street: Cross-talk between e-cadherin and receptor tyrosine kinase (RTK) signaling: A mechanism to regulate RTK activity. *Cancer Biol. Ther.* **2005**, *4*, 28–31. [CrossRef] [PubMed]
143. Barbáchano, A.; Ordóñez-Morán, P.; García, J.M.; Sánchez, A.; Pereira, F.; Larriba, M.J.; Martínez, N.; Hernández, J.; Landolfi, S.; Bonilla, F.; et al. SPROUTY-2 and E-cadherin regulate reciprocally and dictate colon cancer cell tumourigenicity. *Oncogene* **2010**, *29*, 4800–4813. [CrossRef] [PubMed]
144. Dougherty, U.; Mustafi, R.; Sadiq, F.; Almoghrabi, A.; Mustafi, D.; Kreisheh, M.; Sundaramurthy, S.; Liu, W.; Konda, V.J.; Pekow, J.; et al. The renin-angiotensin system mediates EGF receptor-vitamin D receptor cross-talk in colitis-associated colon cancer. *Clin. Cancer Res. Off. J. Am. Assoc. Cancer Res.* **2014**, *20*, 5848–5859. [CrossRef]
145. Oh, Y.S.; Kim, E.J.; Schaffer, B.S.; Kang, Y.H.; Binderup, L.; MacDonald, R.G.; Park, J.H.Y. Synthetic low-calcaemic vitamin D_3 analogues inhibit secretion of insulin-like growth factor II and stimulate production of insulin-like growth factor-binding protein-6 in conjunction with growth suppression of HT-29 colon cancer cells. *Mol. Cell. Endocrinol.* **2001**, *183*, 141–149. [CrossRef]
146. Leng, S.L.; Leeding, K.S.; Whitehead, R.H.; Bach, L.A. Insulin-like growth factor (IGF)-binding protein-6 inhibits IGF-II-induced but not basal proliferation and adhesion of LIM 1215 colon cancer cells. *Mol. Cell. Endocrinol.* **2001**, *174*, 121–127. [CrossRef]
147. Rosli, S.N.; Shintani, T.; Toratani, S.; Usui, E.; Okamoto, T. 1alpha,25(OH)(2)D(3) inhibits FGF-2 release from oral squamous cell carcinoma cells through down-regulation of HBp17/FGFBP-1. *In Vitro Cell. Dev. Biol. Anim.* **2014**, *50*, 802–806. [CrossRef]
148. Higaki, M.; Shintani, T.; Hamada, A.; Rosli, S.N.Z.; Okamoto, T. Eldecalcitol (ED-71)-induced exosomal miR-6887-5p suppresses squamous cell carcinoma cell growth by targeting heparin-binding protein 17/fibroblast growth factor-binding protein-1 (HBp17/FGFBP-1). *In Vitro Cell. Dev. Biol. Anim.* **2020**, *56*, 222–233. [CrossRef]
149. Nazarova, N.; Golovko, O.; Blauer, M.; Tuohimaa, P. Calcitriol inhibits growth response to Platelet-Derived Growth Factor-BB in human prostate cells. *J. Steroid Biochem. Mol. Biol.* **2005**, *94*, 189–196. [CrossRef]

150. Wu, F.S.; Zheng, S.S.; Wu, L.J.; Teng, L.S.; Ma, Z.M.; Zhao, W.H.; Wu, W. Calcitriol inhibits the growth of MHCC97 heptocellular cell lines by down-modulating c-met and ERK expressions. *Liver Int.* **2007**, *27*, 700–707. [CrossRef]
151. Inaba, M.; Koyama, H.; Hino, M.; Okuno, S.; Terada, M.; Nishizawa, Y.; Nishino, T.; Morii, H. Regulation of release of hepatocyte growth factor from human promyelocytic leukemia cells, HL-60, by 1,25-dihydroxyvitamin D3, 12-O-tetradecanoylphorbol 13-acetate, and dibutyryl cyclic adenosine monophosphate. *Blood* **1993**, *82*, 53–59. [CrossRef]
152. Larriba, M.J.; González-Sancho, J.M.; Bonilla, F.; Muñoz, A. Interaction of vitamin D with membrane-based signaling pathways. *Front. Physiol.* **2014**, *5*, 60. [CrossRef]
153. Krishnan, A.V.; Swami, S.; Feldman, D. vitamin D and breast cancer: Inhibition of estrogen synthesis and signaling. *J. Steroid Biochem. Mol. Biol.* **2010**, *121*, 343–348. [CrossRef]
154. Zheng, W.; Cao, L.; Ouyang, L.; Zhang, Q.; Duan, B.; Zhou, W.; Chen, S.; Peng, W.; Xie, Y.; Fan, Q.; et al. Anticancer activity of 1,25-(OH)2D3 against human breast cancer cell lines by targeting Ras/MEK/ERK pathway. *OncoTargets Ther.* **2019**, *12*, 721–732. [CrossRef] [PubMed]
155. Ben-Batalla, I.; Seoane, S.; García-Caballero, T.; Gallego, R.; Macia, M.; González, L.O.; Vizoso, F.; Pérez-Fernández, R. Deregulation of the Pit-1 transcription factor in human breast cancer cells promotes tumor growth and metastasis. *J. Clin. Investig.* **2010**, *120*, 4289–4302. [CrossRef]
156. Perez-Fernandez, R.; Seoane, S.; Garcia-Caballero, T.; Segura, C.; Macia, M. vitamin D, Pit-1, GH, and PRL: Possible roles in breast cancer development. *Curr. Med. Chem.* **2007**, *14*, 3051–3058. [CrossRef] [PubMed]
157. Álvarez-Díaz, S.; Valle, N.; Ferrer-Mayorga, G.; Lombardía, L.; Herrera, M.; Domínguez, O.; Segura, M.F.; Bonilla, F.; Hernando, E.; Muñoz, A. MicroRNA-22 is induced by vitamin D and contributes to its antiproliferative, antimigratory and gene regulatory effects in colon cancer cells. *Hum. Mol. Genet.* **2012**, *21*, 2157–2165. [CrossRef] [PubMed]
158. Lin, W.; Zou, H.; Mo, J.; Jin, C.; Jiang, H.; Yu, C.; Jiang, Z.; Yang, Y.; He, B.; Wang, K. Micro1278 Leads to Tumor Growth Arrest, Enhanced Sensitivity to Oxaliplatin and vitamin D and Inhibits Metastasis via KIF5B, CYP24A1, and BTG2, Respectively. *Front. Oncol.* **2021**, *11*, 637878. [CrossRef] [PubMed]
159. Yang, L.; Yang, J.; Venkateswarlu, S.; Ko, T.; Brattain, M.G. Autocrine TGFbeta signaling mediates vitamin D3 analog-induced growth inhibition in breast cells. *J. Cell. Physiol.* **2001**, *188*, 383–393. [CrossRef] [PubMed]
160. Chen, A.; Davis, B.H.; Sitrin, M.D.; Brasitus, T.A.; Bissonnette, M. Transforming growth factor-b 1 signaling contributes to Caco-2 cell growth inhibition induced by $1,25(OH)_2D_3$. *Am. J. Physiol. Gastrointest. Liver Physiol.* **2002**, *283*, G864–G874. [CrossRef] [PubMed]
161. Ito, Y.; Honda, A.; Kurokawa, M. Impact of vitamin D level at diagnosis and transplantation on the prognosis of hematological malignancy: A meta-analysis. *Blood Adv.* **2021**, *6*, 1499–1511. [CrossRef]
162. Gerousi, M.; Psomopoulos, F.; Kotta, K.; Tsagiopoulou, M.; Stavroyianni, N.; Anagnostopoulos, A.; Anastasiadis, A.; Gkanidou, M.; Kotsianidis, I.; Ntoufa, S.; et al. The Calcitriol/vitamin D Receptor System Regulates Key Immune Signaling Pathways in Chronic Lymphocytic Leukemia. *Cancers* **2021**, *13*, 285. [CrossRef]
163. Olson, K.C.; Kulling, P.M.; Olson, T.L.; Tan, S.F.; Rainbow, R.J.; Feith, D.J.; Loughran, T.P., Jr. vitamin D decreases STAT phosphorylation and inflammatory cytokine output in T-LGL leukemia. *Cancer Biol. Ther.* **2017**, *18*, 290–303. [CrossRef]
164. McGlorthan, L.; Paucarmayta, A.; Casablanca, Y.; Maxwell, G.L.; Syed, V. Progesterone induces apoptosis by activation of caspase-8 and calcitriol via activation of caspase-9 pathways in ovarian and endometrial cancer cells in vitro. *Apoptosis Int. J. Program. Cell Death* **2021**, *26*, 184–194. [CrossRef] [PubMed]
165. Jiang, F.; Bao, J.; Li, P.; Nicosia, S.V.; Bai, W. Induction of ovarian cancer cell apoptosis by 1,25-dihydroxyvitamin D3 through the down-regulation of telomerase. *J. Biol. Chem.* **2004**, *279*, 53213–53221. [CrossRef] [PubMed]
166. Kasiappan, R.; Shen, Z.; Tse, A.K.; Jinwal, U.; Tang, J.; Lungchukiet, P.; Sun, Y.; Kruk, P.; Nicosia, S.V.; Zhang, X.; et al. 1,25-Dihydroxyvitamin D3 suppresses telomerase expression and human cancer growth through microRNA-498. *J. Biol. Chem.* **2012**, *287*, 41297–41309. [CrossRef]
167. Stambolsky, P.; Tabach, Y.; Fontemaggi, G.; Weisz, L.; Maor-Aloni, R.; Siegfried, Z.; Shiff, I.; Kogan, I.; Shay, M.; Kalo, E.; et al. Modulation of the vitamin D3 response by cancer-associated mutant p53. *Cancer Cell* **2010**, *17*, 273–285. [CrossRef] [PubMed]
168. Abu El Maaty, M.A.; Wölfl, S. Effects of 1,25(OH)(2)D(3) on Cancer Cells and Potential Applications in Combination with Established and Putative Anti-Cancer Agents. *Nutrients* **2017**, *9*, 87. [CrossRef]
169. Kaler, P.; Galea, V.; Augenlicht, L.; Klampfer, L. Tumor associated macrophages protect colon cancer cells from TRAIL-induced apoptosis through IL-1beta-dependent stabilization of Snail in tumor cells. *PLoS ONE* **2010**, *5*, e11700. [CrossRef] [PubMed]
170. Borkowski, R.; Du, L.; Zhao, Z.; McMillan, E.; Kosti, A.; Yang, C.R.; Suraokar, M.; Wistuba, I.I.; Gazdar, A.F.; Minna, J.D.; et al. Genetic mutation of p53 and suppression of the miR-17 approximately 92 cluster are synthetic lethal in non-small cell lung cancer due to upregulation of vitamin D Signaling. *Cancer Res.* **2015**, *75*, 666–675. [CrossRef] [PubMed]
171. Bhutia, S.K. vitamin D in autophagy signaling for health and diseases: Insights on potential mechanisms and future perspectives. *J. Nutr. Biochem.* **2022**, *99*, 108841. [CrossRef] [PubMed]
172. Hoyer-Hansen, M.; Bastholm, L.; Szyniarowski, P.; Campanella, M.; Szabadkai, G.; Farkas, T.; Bianchi, K.; Fehrenbacher, N.; Elling, F.; Rizzuto, R.; et al. Control of macroautophagy by calcium, calmodulin-dependent kinase kinase-beta, and Bcl-2. *Mol. Cell* **2007**, *25*, 193–205. [CrossRef]
173. Suares, A.; Tapia, C.; Gonzalez-Pardo, V. VDR agonists down regulate PI3K/Akt/mTOR axis and trigger autophagy in Kaposi's sarcoma cells. *Heliyon* **2019**, *5*, e02367. [CrossRef] [PubMed]

174. Wang, J.; Lian, H.; Zhao, Y.; Kauss, M.A.; Spindel, S. vitamin D3 induces autophagy of human myeloid leukemia cells. *J. Biol. Chem.* **2008**, *283*, 25596–25605. [CrossRef] [PubMed]
175. Han, J.; Tang, Y.; Zhong, M.; Wu, W. Antitumor effects and mechanisms of 1,25(OH)2D3 in the Pfeiffer diffuse large B lymphoma cell line. *Mol. Med. Rep.* **2019**, *20*, 5064–5074. [CrossRef] [PubMed]
176. Zhang, X.; Luo, F.; Li, J.; Wan, J.; Zhang, L.; Li, H.; Chen, A.; Chen, J.; Cai, T.; He, X.; et al. DNA damage-inducible transcript 4 is an innate guardian for human squamous cell carcinoma and an molecular vector for anti-carcinoma effect of 1,25(OH)2 D3. *Exp. Dermatol.* **2019**, *28*, 45–52. [CrossRef] [PubMed]
177. Wang, W.; Liu, J.; Chen, K.; Wang, J.; Dong, Q.; Xie, J.; Yuan, Y. vitamin D promotes autophagy in AML cells by inhibiting miR-17-5p-induced Beclin-1 overexpression. *Mol. Cell. Biochem.* **2021**, *476*, 3951–3962. [CrossRef] [PubMed]
178. Demasters, G.; Di, X.; Newsham, I.; Shiu, R.; Gewirtz, D.A. Potentiation of radiation sensitivity in breast tumor cells by the vitamin D3 analogue, EB 1089, through promotion of autophagy and interference with proliferative recovery. *Mol. Cancer Ther.* **2006**, *5*, 2786–2797. [CrossRef] [PubMed]
179. Wilson, E.N.; Bristol, M.L.; Di, X.; Maltese, W.A.; Koterba, K.; Beckman, M.J.; Gewirtz, D.A. A switch between cytoprotective and cytotoxic autophagy in the radiosensitization of breast tumor cells by chloroquine and vitamin D. *Horm. Cancer* **2011**, *2*, 272–285. [CrossRef] [PubMed]
180. Bristol, M.L.; Di, X.; Beckman, M.J.; Wilson, E.N.; Henderson, S.C.; Maiti, A.; Fan, Z.; Gewirtz, D.A. Dual functions of autophagy in the response of breast tumor cells to radiation: Cytoprotective autophagy with radiation alone and cytotoxic autophagy in radiosensitization by vitamin D 3. *Autophagy* **2012**, *8*, 739–753. [CrossRef] [PubMed]
181. Sharma, K.; Goehe, R.W.; Di, X.; Hicks, M.A., 2nd; Torti, S.V.; Torti, F.M.; Harada, H.; Gewirtz, D.A. A novel cytostatic form of autophagy in sensitization of non-small cell lung cancer cells to radiation by vitamin D and the vitamin D analog, EB 1089. *Autophagy* **2014**, *10*, 2346–2361. [CrossRef] [PubMed]
182. Bak, D.H.; Kang, S.H.; Choi, D.R.; Gil, M.N.; Yu, K.S.; Jeong, J.H.; Lee, N.S.; Lee, J.H.; Jeong, Y.G.; Kim, D.K.; et al. Autophagy enhancement contributes to the synergistic effect of vitamin D in temozolomide-based glioblastoma chemotherapy. *Exp. Ther. Med.* **2016**, *11*, 2153–2162. [CrossRef] [PubMed]
183. Dongre, A.; Weinberg, R.A. New insights into the mechanisms of epithelial-mesenchymal transition and implications for cancer. *Nat. Rev. Mol. Cell Biol.* **2019**, *20*, 69–84. [CrossRef] [PubMed]
184. Larriba, M.J.; Garcia de Herreros, A.; Muñoz, A. vitamin D and the Epithelial to Mesenchymal Transition. *Stem Cells Int.* **2016**, *2016*, 6213872. [CrossRef] [PubMed]
185. Fernández-Barral, A.; Bustamante-Madrid, P.; Ferrer-Mayorga, G.; Barbáchano, A.; Larriba, M.J.; Muñoz, A. vitamin D Effects on Cell Differentiation and Stemness in Cancer. *Cancers* **2020**, *12*, 2413. [CrossRef] [PubMed]
186. Pendás-Franco, N.; González-Sancho, J.M.; Suarez, Y.; Aguilera, O.; Steinmeyer, A.; Gamallo, C.; Berciano, M.T.; Lafarga, M.; Muñoz, A. vitamin D regulates the phenotype of human breast cancer cells. *Differ. Res. Biol. Divers.* **2007**, *75*, 193–207. [CrossRef] [PubMed]
187. Welsh, J. vitamin D and Breast Cancer: Mechanistic Update. *J. Bone Miner. Res. Plus* **2021**, *5*, e10582. [CrossRef] [PubMed]
188. Kouchi, Z.; Fujiwara, Y.; Yamaguchi, H.; Nakamura, Y.; Fukami, K. Phosphatidylinositol 5-phosphate 4-kinase type II beta is required for vitamin D receptor-dependent E-cadherin expression in SW480 cells. *Biochem. Biophys. Res. Commun.* **2011**, *408*, 523–529. [CrossRef] [PubMed]
189. Lopes, N.; Carvalho, J.; Duraes, C.; Sousa, B.; Gomes, M.; Costa, J.L.; Oliveira, C.; Paredes, J.; Schmitt, F. 1Alpha,25-dihydroxyvitamin D3 induces de novo E-cadherin expression in triple-negative breast cancer cells by CDH1-promoter demethylation. *Anticancer Res.* **2012**, *32*, 249–257. [PubMed]
190. Upadhyay, S.K.; Verone, A.; Shoemaker, S.; Qin, M.; Liu, S.; Campbell, M.; Hershberger, P.A. 1,25-Dihydroxyvitamin D3 (1,25(OH)2D3) Signaling Capacity and the Epithelial-Mesenchymal Transition in Non-Small Cell Lung Cancer (NSCLC): Implications for Use of 1,25(OH)2D3 in NSCLC Treatment. *Cancers* **2013**, *5*, 1504–1521. [CrossRef] [PubMed]
191. Chiang, K.C.; Chen, S.C.; Yeh, C.N.; Pang, J.H.; Shen, S.C.; Hsu, J.T.; Liu, Y.Y.; Chen, L.W.; Kuo, S.F.; Takano, M.; et al. MART-10, a less calcemic vitamin D analog, is more potent than 1alpha,25-dihydroxyvitamin D3 in inhibiting the metastatic potential of MCF-7 breast cancer cells in vitro. *J. Steroid Biochem. Mol. Biol.* **2014**, *139*, 54–60. [CrossRef] [PubMed]
192. Chiang, K.C.; Yeh, C.N.; Hsu, J.T.; Jan, Y.Y.; Chen, L.W.; Kuo, S.F.; Takano, M.; Kittaka, A.; Chen, T.C.; Chen, W.T.; et al. The vitamin D analog, MART-10, represses metastasis potential via downregulation of epithelial-mesenchymal transition in pancreatic cancer cells. *Cancer Lett.* **2014**, *354*, 235–244. [CrossRef] [PubMed]
193. Findlay, V.J.; Moretz, R.E.; Wang, C.; Vaena, S.G.; Bandurraga, S.G.; Ashenafi, M.; Marshall, D.T.; Watson, D.K.; Camp, E.R. Slug expression inhibits calcitriol-mediated sensitivity to radiation in colorectal cancer. *Mol. Carcinog.* **2014**, *53*, E130–E139. [CrossRef] [PubMed]
194. Hou, Y.F.; Gao, S.H.; Wang, P.; Zhang, H.M.; Liu, L.Z.; Ye, M.X.; Zhou, G.M.; Zhang, Z.L.; Li, B.Y. 1alpha,25(OH)(2)D(3) Suppresses the Migration of Ovarian Cancer SKOV-3 Cells through the Inhibition of Epithelial-Mesenchymal Transition. *Int. J. Mol. Sci.* **2016**, *17*, 1285. [CrossRef] [PubMed]
195. Pereira, F.; Barbáchano, A.; Silva, J.; Bonilla, F.; Campbell, M.J.; Muñoz, A.; Larriba, M.J. KDM6B/JMJD3 histone demethylase is induced by vitamin D and modulates its effects in colon cancer cells. *Hum. Mol. Genet.* **2011**, *20*, 4655–4665. [CrossRef] [PubMed]

196. Barbáchano, A.; Fernández-Barral, A.; Pereira, F.; Segura, M.F.; Ordóñez-Morán, P.; Carrillo-de Santa Pau, E.; González-Sancho, J.M.; Hanniford, D.; Martinez, N.; Costales-Carrera, A.; et al. SPROUTY-2 represses the epithelial phenotype of colon carcinoma cells via upregulation of ZEB1 mediated by ETS1 and miR-200/miR-150. *Oncogene* **2016**, *35*, 2991–3003. [CrossRef] [PubMed]
197. Koeffler, H.P.; Amatruda, T.; Ikekawa, N.; Kobayashi, Y.; DeLuca, H.F. Induction of macrophage differentiation of human normal and leukemic myeloid stem cells by 1,25-dihydroxyvitamin D3 and its fluorinated analogues. *Cancer Res.* **1984**, *44*, 5624–5628. [PubMed]
198. Tanaka, H.; Abe, E.; Miyaura, C.; Shiina, Y.; Suda, T. 1 alpha,25-dihydroxyvitamin D3 induces differentiation of human promyelocytic leukemia cells (HL-60) into monocyte-macrophages, but not into granulocytes. *Biochem. Biophys. Res. Commun.* **1983**, *117*, 86–92. [CrossRef]
199. Abe, J.; Moriya, Y.; Saito, M.; Sugawara, Y.; Suda, T.; Nishii, Y. Modulation of cell growth, differentiation, and production of interleukin-3 by 1 alpha,25-dihydroxyvitamin D3 in the murine myelomonocytic leukemia cell line WEHI-3. *Cancer Res.* **1986**, *46*, 6316–6321.
200. Gocek, E.; Studzinski, G.P. vitamin D and differentiation in cancer. *Crit. Rev. Clin. Lab. Sci.* **2009**, *46*, 190–209. [CrossRef]
201. Hmama, Z.; Nandan, D.; Sly, L.; Knutson, K.L.; Herrera-Velit, P.; Reiner, N.E. 1alpha,25-dihydroxyvitamin D(3)-induced myeloid cell differentiation is regulated by a vitamin D receptor-phosphatidylinositol 3-kinase signaling complex. *J. Exp. Med.* **1999**, *190*, 1583–1594. [CrossRef]
202. Ji, Y.; Studzinski, G.P. Retinoblastoma protein and CCAAT/enhancer-binding protein beta are required for 1,25-dihydroxyvitamin D3-induced monocytic differentiation of HL60 cells. *Cancer Res.* **2004**, *64*, 370–377. [CrossRef]
203. Marchwicka, A.; Marcinkowska, E. Regulation of Expression of CEBP Genes by Variably Expressed vitamin D Receptor and Retinoic Acid Receptor alpha in Human Acute Myeloid Leukemia Cell Lines. *Int. J. Mol. Sci.* **2018**, *19*, 1918. [CrossRef]
204. Song, J.H.; Park, E.; Kim, M.S.; Cho, K.M.; Park, S.H.; Lee, A.; Song, J.; Kim, H.J.; Koh, J.T.; Kim, T.S. l-Asparaginase-mediated downregulation of c-Myc promotes 1,25(OH)2 D3-induced myeloid differentiation in acute myeloid leukemia cells. *Int. J. Cancer* **2017**, *140*, 2364–2374. [CrossRef]
205. Sabatier, M.; Boet, E.; Zaghdoudi, S.; Guiraud, N.; Hucteau, A.; Polley, N.; Cognet, G.; Saland, E.; Lauture, L.; Farge, T.; et al. Activation of vitamin D Receptor Pathway Enhances Differentiating Capacity in Acute Myeloid Leukemia with Isocitrate Dehydrogenase Mutations. *Cancers* **2021**, *13*, 5243. [CrossRef]
206. Hickish, T.; Cunningham, D.; Colston, K.; Millar, B.C.; Sandle, J.; Mackay, A.G.; Soukop, M.; Sloane, J. The effect of 1,25-dihydroxyvitamin D3 on lymphoma cell lines and expression of vitamin D receptor in lymphoma. *Br. J. Cancer* **1993**, *68*, 668–672. [CrossRef]
207. Nusse, R.; Clevers, H. Wnt/beta-Catenin Signaling, Disease, and Emerging Therapeutic Modalities. *Cell* **2017**, *169*, 985–999. [CrossRef] [PubMed]
208. Polakis, P. Wnt signaling in cancer. *Cold Spring Harb. Perspect. Biol.* **2012**, *4*, a008052. [CrossRef] [PubMed]
209. The_Cancer_Genome_Atlas_Network. Comprehensive molecular characterization of human colon and rectal cancer. *Nature* **2012**, *487*, 330–337. [CrossRef]
210. Yaeger, R.; Chatila, W.K.; Lipsyc, M.D.; Hechtman, J.F.; Cercek, A.; Sanchez-Vega, F.; Jayakumaran, G.; Middha, S.; Zehir, A.; Donoghue, M.T.A.; et al. Clinical Sequencing Defines the Genomic Landscape of Metastatic Colorectal Cancer. *Cancer Cell* **2018**, *33*, 125–136. [CrossRef]
211. González-Sancho, J.M.; Larriba, M.J.; Muñoz, A. Wnt and vitamin D at the Crossroads in Solid Cancer. *Cancers* **2020**, *12*, 3434. [CrossRef]
212. Aguilera, O.; Peña, C.; García, J.M.; Larriba, M.J.; Ordóñez-Morán, P.; Navarro, D.; Barbáchano, A.; López de Silanes, I.; Ballestar, E.; Fraga, M.F.; et al. The Wnt antagonist DICKKOPF-1 gene is induced by 1alpha,25-dihydroxyvitamin D_3 associated to the differentiation of human colon cancer cells. *Carcinogenesis* **2007**, *28*, 1877–1884. [CrossRef] [PubMed]
213. Beildeck, M.E.; Islam, M.; Shah, S.; Welsh, J.; Byers, S.W. Control of TCF-4 expression by VDR and vitamin D in the mouse mammary gland and colorectal cancer cell lines. *PLoS ONE* **2009**, *4*, e7872. [CrossRef] [PubMed]
214. Jin, D.; Zhang, Y.G.; Wu, S.; Lu, R.; Lin, Z.; Zheng, Y.; Chen, H.; Cs-Szabo, G.; Sun, J. vitamin D receptor is a novel transcriptional regulator for Axin1. *J. Steroid Biochem. Mol. Biol.* **2017**, *165*, 430–437. [CrossRef] [PubMed]
215. Arensman, M.D.; Nguyen, P.; Kershaw, K.M.; Lay, A.R.; Ostertag-Hill, C.A.; Sherman, M.H.; Downes, M.; Liddle, C.; Evans, R.M.; Dawson, D.W. Calcipotriol Targets LRP6 to Inhibit Wnt Signaling in Pancreatic Cancer. *Mol. Cancer Res. MCR* **2015**, *13*, 1509–1519. [CrossRef] [PubMed]
216. Chen, J.; Katz, L.H.; Munoz, N.M.; Gu, S.; Shin, J.H.; Jogunoori, W.S.; Lee, M.H.; Belkin, M.D.; Kim, S.B.; White, J.C.; et al. vitamin D Deficiency Promotes Liver Tumor Growth in Transforming Growth Factor-beta/Smad3-Deficient Mice Through Wnt and Toll-like Receptor 7 Pathway Modulation. *Sci. Rep.* **2016**, *6*, 30217. [CrossRef]
217. Xu, S.; Zhang, Z.H.; Fu, L.; Song, J.; Xie, D.D.; Yu, D.X.; Xu, D.X.; Sun, G.P. Calcitriol inhibits migration and invasion of renal cell carcinoma cells by suppressing Smad2/3-, STAT3- and beta-catenin-mediated epithelial-mesenchymal transition. *Cancer Sci.* **2020**, *111*, 59–71. [CrossRef] [PubMed]
218. Kaler, P.; Augenlicht, L.; Klampfer, L. Macrophage-derived IL-1beta stimulates Wnt signaling and growth of colon cancer cells: A crosstalk interrupted by vitamin D3. *Oncogene* **2009**, *28*, 3892–3902. [CrossRef] [PubMed]

219. Fernández-García, N.I.; Pálmer, H.G.; García, M.; González-Martín, A.; del Rio, M.; Barettino, D.; Volpert, O.; Muñoz, A.; Jiménez, B. 1a,25-Dihydroxyvitamin D_3 regulates the expression of *Id1* and *Id2* genes and the angiogenic phenotype of human colon carcinoma cells. *Oncogene* **2005**, *24*, 6533–6544. [CrossRef] [PubMed]
220. Ben-Shoshan, M.; Amir, S.; Dang, D.T.; Dang, L.H.; Weisman, Y.; Mabjeesh, N.J. 1alpha,25-dihydroxyvitamin D3 (Calcitriol) inhibits hypoxia-inducible factor-1/vascular endothelial growth factor pathway in human cancer cells. *Mol. Cancer Ther.* **2007**, *6*, 1433–1439. [CrossRef] [PubMed]
221. Pendás-Franco, N.; García, J.M.; Peña, C.; Valle, N.; Pálmer, H.G.; Heinaniemi, M.; Carlberg, C.; Jiménez, B.; Bonilla, F.; Muñoz, A.; et al. DICKKOPF-4 is induced by TCF/beta-catenin and upregulated in human colon cancer, promotes tumour cell invasion and angiogenesis and is repressed by 1alpha,25-dihydroxyvitamin D_3. *Oncogene* **2008**, *27*, 4467–4477. [CrossRef] [PubMed]
222. Kim, J.H.; Park, W.H.; Suh, D.H.; Kim, K.; No, J.H.; Kim, Y.B. Calcitriol Combined with Platinum-based Chemotherapy Suppresses Growth and Expression of Vascular Endothelial Growth Factor of SKOV-3 Ovarian Cancer Cells. *Anticancer Res.* **2021**, *41*, 2945–2952. [CrossRef]
223. Piotrowska, A.; Beserra, F.P.; Wierzbicka, J.M.; Nowak, J.I.; Zmijewski, M.A. vitamin D Enhances Anticancer Properties of Cediranib, a VEGFR Inhibitor, by Modulation of VEGFR2 Expression in Melanoma Cells. *Front. Oncol.* **2021**, *11*, 763895. [CrossRef] [PubMed]
224. Bao, B.Y.; Yao, J.; Lee, Y.F. 1alpha, 25-dihydroxyvitamin D3 suppresses interleukin-8-mediated prostate cancer cell angiogenesis. *Carcinogenesis* **2006**, *27*, 1883–1893. [CrossRef] [PubMed]
225. García-Quiroz, J.; Rivas-Suárez, M.; Garcia-Becerra, R.; Barrera, D.; Martínez-Reza, I.; Ordaz-Rosado, D.; Santos-Martinez, N.; Villanueva, O.; Santos-Cuevas, C.L.; Avila, E.; et al. Calcitriol reduces thrombospondin-1 and increases vascular endothelial growth factor in breast cancer cells: Implications for tumor angiogenesis. *J. Steroid Biochem. Mol. Biol.* **2014**, *144*, 215–222. [CrossRef] [PubMed]
226. Mantell, D.J.; Owens, P.E.; Bundred, N.J.; Mawer, E.B.; Canfield, A.E. 1 alpha,25-dihydroxyvitamin D(3) inhibits angiogenesis in vitro and in vivo. *Circ. Res.* **2000**, *87*, 214–220. [CrossRef] [PubMed]
227. Bernardi, R.J.; Johnson, C.S.; Modzelewski, R.A.; Trump, D.L. Antiproliferative effects of 1alpha,25-dihydroxyvitamin D(3) and vitamin D analogs on tumor-derived endothelial cells. *Endocrinology* **2002**, *143*, 2508–2514. [CrossRef] [PubMed]
228. Chung, I.; Wong, M.K.; Flynn, G.; Yu, W.D.; Johnson, C.S.; Trump, D.L. Differential antiproliferative effects of calcitriol on tumor-derived and matrigel-derived endothelial cells. *Cancer Res.* **2006**, *66*, 8565–8573. [CrossRef] [PubMed]
229. Chung, I.; Han, G.; Seshadri, M.; Gillard, B.M.; Yu, W.D.; Foster, B.A.; Trump, D.L.; Johnson, C.S. Role of vitamin D receptor in the antiproliferative effects of calcitriol in tumor-derived endothelial cells and tumor angiogenesis in vivo. *Cancer Res.* **2009**, *69*, 967–975. [CrossRef] [PubMed]
230. Flynn, G.; Chung, I.; Yu, W.D.; Romano, M.; Modzelewski, R.A.; Johnson, C.S.; Trump, D.L. Calcitriol (1,25-dihydroxycholecalciferol) selectively inhibits proliferation of freshly isolated tumor-derived endothelial cells and induces apoptosis. *Oncology* **2006**, *70*, 447–457. [CrossRef] [PubMed]
231. Sung, V.; Feldman, D. 1,25-Dihydroxyvitamin D3 decreases human prostate cancer cell adhesion and migration. *Mol. Cell. Endocrinol.* **2000**, *164*, 133–143. [CrossRef]
232. Tokar, E.J.; Webber, M.M. Cholecalciferol (vitamin D3) inhibits growth and invasion by up-regulating nuclear receptors and 25-hydroxylase (CYP27A1) in human prostate cancer cells. *Clin. Exp. Metastasis* **2005**, *22*, 275–284. [CrossRef] [PubMed]
233. Chen, S.; Zhu, J.; Zuo, S.; Ma, J.; Zhang, J.; Chen, G.; Wang, X.; Pan, Y.; Liu, Y.; Wang, P. 1,25(OH)2D3 attenuates TGF-beta1/beta2-induced increased migration and invasion via inhibiting epithelial-mesenchymal transition in colon cancer cells. *Biochem. Biophys. Res. Commun.* **2015**, *468*, 130–135. [CrossRef]
234. Hsu, J.W.; Yasmin-Karim, S.; King, M.R.; Wojciechowski, J.C.; Mickelsen, D.; Blair, M.L.; Ting, H.J.; Ma, W.L.; Lee, Y.F. Suppression of prostate cancer cell rolling and adhesion to endothelium by 1alpha,25-dihydroxyvitamin D3. *Am. J. Pathol.* **2011**, *178*, 872–880. [CrossRef] [PubMed]
235. Li, J.; Luco, A.L.; Camirand, A.; St-Arnaud, R.; Kremer, R. vitamin D Regulates CXCL12/CXCR4 and Epithelial-to-Mesenchymal Transition in a Model of Breast Cancer Metastasis to Lung. *Endocrinology* **2021**, *162*, bqab049. [CrossRef]
236. González-Sancho, J.M.; Alvarez-Dolado, M.; Muñoz, A. 1,25-Dihydroxyvitamin D3 inhibits tenascin-C expression in mammary epithelial cells. *FEBS Lett.* **1998**, *426*, 225–228. [CrossRef]
237. Koli, K.; Keski-Oja, J. 1alpha,25-dihydroxyvitamin D3 and its analogues down-regulate cell invasion-associated proteases in cultured malignant cells. *Cell Growth Differ. Mol. Biol. J. Am. Assoc. Cancer Res.* **2000**, *11*, 221–229.
238. Bao, B.Y.; Yeh, S.D.; Lee, Y.F. 1alpha,25-dihydroxyvitamin D3 inhibits prostate cancer cell invasion via modulation of selective proteases. *Carcinogenesis* **2006**, *27*, 32–42. [CrossRef] [PubMed]
239. Wilmanski, T.; Barnard, A.; Parikh, M.R.; Kirshner, J.; Buhman, K.; Burgess, J.; Teegarden, D. 1alpha,25-Dihydroxyvitamin D Inhibits the Metastatic Capability of MCF10CA1a and MDA-MB-231 Cells in an In Vitro Model of Breast to Bone Metastasis. *Nutr. Cancer* **2016**, *68*, 1202–1209. [CrossRef] [PubMed]
240. Vanoirbeek, E.; Eelen, G.; Verlinden, L.; Carmeliet, G.; Mathieu, C.; Bouillon, R.; O'Connor, R.; Xiao, G.; Verstuyf, A. PDLIM2 expression is driven by vitamin D and is involved in the pro-adhesion, and anti-migration and -invasion activity of vitamin D. *Oncogene* **2014**, *33*, 1904–1911. [CrossRef] [PubMed]
241. Narvaez, C.J.; Grebenc, D.; Balinth, S.; Welsh, J.E. vitamin D regulation of HAS2, hyaluronan synthesis and metabolism in triple negative breast cancer cells. *J. Steroid Biochem. Mol. Biol.* **2020**, *201*, 105688. [CrossRef]

242. Ma, Y.; Luo, W.; Bunch, B.L.; Pratt, R.N.; Trump, D.L.; Johnson, C.S. 1,25D3 differentially suppresses bladder cancer cell migration and invasion through the induction of miR-101-3p. *Oncotarget* **2017**, *8*, 60080–60093. [CrossRef] [PubMed]
243. Cheng, Y.H.; Chiang, E.I.; Syu, J.N.; Chao, C.Y.; Lin, H.Y.; Lin, C.C.; Yang, M.D.; Tsai, S.Y.; Tang, F.Y. Treatment of 13-cis retinoic acid and 1,25-dihydroxyvitamin D3 inhibits TNF-alpha-mediated expression of MMP-9 protein and cell invasion through the suppression of JNK pathway and microRNA 221 in human pancreatic adenocarcinoma cancer cells. *PLoS ONE* **2021**, *16*, e0247550. [CrossRef] [PubMed]
244. Ohlund, D.; Elyada, E.; Tuveson, D. Fibroblast heterogeneity in the cancer wound. *J. Exp. Med.* **2014**, *211*, 1503–1523. [CrossRef] [PubMed]
245. Barrett, R.L.; Pure, E. Cancer-associated fibroblasts and their influence on tumor immunity and immunotherapy. *eLife* **2020**, *9*, e57243. [CrossRef] [PubMed]
246. Rhim, A.D.; Oberstein, P.E.; Thomas, D.H.; Mirek, E.T.; Palermo, C.F.; Sastra, S.A.; Dekleva, E.N.; Saunders, T.; Becerra, C.P.; Tattersall, I.W.; et al. Stromal elements act to restrain, rather than support, pancreatic ductal adenocarcinoma. *Cancer Cell* **2014**, *25*, 735–747. [CrossRef] [PubMed]
247. Ozdemir, B.C.; Pentcheva-Hoang, T.; Carstens, J.L.; Zheng, X.; Wu, C.C.; Simpson, T.R.; Laklai, H.; Sugimoto, H.; Kahlert, C.; Novitskiy, S.V.; et al. Depletion of carcinoma-associated fibroblasts and fibrosis induces immunosuppression and accelerates pancreas cancer with reduced survival. *Cancer Cell* **2014**, *25*, 719–734. [CrossRef]
248. Abramovitch, S.; Dahan-Bachar, L.; Sharvit, E.; Weisman, Y.; Ben Tov, A.; Brazowski, E.; Reif, S. vitamin D inhibits proliferation and profibrotic marker expression in hepatic stellate cells and decreases thioacetamide-induced liver fibrosis in rats. *Gut* **2011**, *60*, 1728–1737. [CrossRef] [PubMed]
249. Sherman, M.H.; Yu, R.T.; Engle, D.D.; Ding, N.; Atkins, A.R.; Tiriac, H.; Collisson, E.A.; Connor, F.; Van Dyke, T.; Kozlov, S.; et al. vitamin D receptor-mediated stromal reprogramming suppresses pancreatitis and enhances pancreatic cancer therapy. *Cell* **2014**, *159*, 80–93. [CrossRef] [PubMed]
250. Ding, N.; Yu, R.T.; Subramaniam, N.; Sherman, M.H.; Wilson, C.; Rao, R.; Leblanc, M.; Coulter, S.; He, M.; Scott, C.; et al. A vitamin D receptor/SMAD genomic circuit gates hepatic fibrotic response. *Cell* **2013**, *153*, 601–613. [CrossRef] [PubMed]
251. Durán, A.; Hernández, E.D.; Reina-Campos, M.; Castilla, E.A.; Subramaniam, S.; Raghunandan, S.; Roberts, L.R.; Kisseleva, T.; Karin, M.; Diaz-Meco, M.T.; et al. p62/SQSTM1 by Binding to vitamin D Receptor Inhibits Hepatic Stellate Cell Activity, Fibrosis, and Liver Cancer. *Cancer Cell* **2016**, *30*, 595–609. [CrossRef] [PubMed]
252. Tao, Q.; Wang, B.; Zheng, Y.; Jiang, X.; Pan, Z.; Ren, J. vitamin D prevents the intestinal fibrosis via induction of vitamin D receptor and inhibition of transforming growth factor-beta1/Smad3 pathway. *Dig. Dis. Sci.* **2015**, *60*, 868–875. [CrossRef] [PubMed]
253. Campos, L.T.; Brentani, H.; Roela, R.A.; Katayama, M.L.; Lima, L.; Rolim, C.F.; Milani, C.; Folgueira, M.A.; Brentani, M.M. Differences in transcriptional effects of 1alpha,25 dihydroxyvitamin D3 on fibroblasts associated to breast carcinomas and from paired normal breast tissues. *J. Steroid Biochem. Mol. Biol.* **2013**, *133*, 12–24. [CrossRef]
254. Ferrer-Mayorga, G.; Gómez-López, G.; Barbáchano, A.; Fernández-Barral, A.; Peña, C.; Pisano, D.G.; Cantero, R.; Rojo, F.; Muñoz, A.; Larriba, M.J. vitamin D receptor expression and associated gene signature in tumour stromal fibroblasts predict clinical outcome in colorectal cancer. *Gut* **2017**, *66*, 1449–1462. [CrossRef] [PubMed]
255. Niell, N.; Larriba, M.J.; Ferrer-Mayorga, G.; Sánchez-Pérez, I.; Cantero, R.; Real, F.X.; Del Peso, L.; Muñoz, A.; González-Sancho, J.M. The human PKP2/plakophilin-2 gene is induced by Wnt/beta-catenin in normal and colon cancer-associated fibroblasts. *Int. J. Cancer* **2018**, *142*, 792–804. [CrossRef] [PubMed]
256. Ferrer-Mayorga, G.; Niell, N.; Cantero, R.; González-Sancho, J.M.; Del Peso, L.; Muñoz, A.; Larriba, M.J. vitamin D and Wnt3A have additive and partially overlapping modulatory effects on gene expression and phenotype in human colon fibroblasts. *Sci. Rep.* **2019**, *9*, 8085. [CrossRef] [PubMed]
257. Kong, F.; Li, L.; Wang, G.; Deng, X.; Li, Z.; Kong, X. VDR signaling inhibits cancer-associated-fibroblasts' release of exosomal miR-10a-5p and limits their supportive effects on pancreatic cancer cells. *Gut* **2019**, *68*, 950–951. [CrossRef] [PubMed]
258. Gorchs, L.; Ahmed, S.; Mayer, C.; Knauf, A.; Fernandez Moro, C.; Svensson, M.; Heuchel, R.; Rangelova, E.; Bergman, P.; Kaipe, H. The vitamin D analogue calcipotriol promotes an anti-tumorigenic phenotype of human pancreatic CAFs but reduces T cell mediated immunity. *Sci. Rep.* **2020**, *10*, 17444. [CrossRef]
259. Fujii, M.; Sato, T. Somatic cell-derived organoids as prototypes of human epithelial tissues and diseases. *Nat. Mater.* **2021**, *20*, 156–169. [CrossRef]
260. Schutgens, F.; Clevers, H. Human Organoids: Tools for Understanding Biology and Treating Diseases. *Annu. Rev. Pathol.* **2020**, *15*, 211–234. [CrossRef]
261. Barbachano, A.; Fernández-Barral, A.; Bustamante-Madrid, P.; Prieto, I.; Rodriguez-Salas, N.; Larriba, M.J.; Muñoz, A. Organoids and Colorectal Cancer. *Cancers* **2021**, *13*, 2657. [CrossRef]
262. Fernández-Barral, A.; Costales-Carrera, A.; Buira, S.P.; Jung, P.; Ferrer-Mayorga, G.; Larriba, M.J.; Bustamante-Madrid, P.; Dominguez, O.; Real, F.X.; Guerra-Pastrián, L.; et al. vitamin D differentially regulates colon stem cells in patient-derived normal and tumor organoids. *FEBS J.* **2020**, *287*, 53–72. [CrossRef]
263. Costales-Carrera, A.; Fernández-Barral, A.; Bustamante-Madrid, P.; Dominguez, O.; Guerra-Pastrián, L.; Cantero, R.; Del Peso, L.; Burgos, A.; Barbáchano, A.; Muñoz, A. Comparative Study of Organoids from Patient-Derived Normal and Tumor Colon and Rectal Tissue. *Cancers* **2020**, *12*, 2302. [CrossRef]

264. Vaughan-Shaw, P.G.; Blackmur, J.P.; Grimes, G.; Ooi, L.Y.; Ochocka-Fox, A.M.; Dunbar, K.; von Kriegsheim, A.; Rajasekaran, V.; Timofeeva, M.; Walker, M.; et al. vitamin D treatment induces in vitro and ex vivo transcriptomic changes indicating anti-tumor effects. *FASEB J.* **2022**, *36*, e22082. [CrossRef] [PubMed]
265. Shang, J.; Zhu, Z.; Chen, Y.; Song, J.; Huang, Y.; Song, K.; Zhong, J.; Xu, X.; Wei, J.; Wang, C.; et al. Small-molecule activating SIRT6 elicits therapeutic effects and synergistically promotes anti-tumor activity of vitamin D3 in colorectal cancer. *Theranostics* **2020**, *10*, 5845–5864. [CrossRef] [PubMed]
266. McCray, T.; Pacheco, J.V.; Loitz, C.C.; Garcia, J.; Baumann, B.; Schlicht, M.J.; Valyi-Nagy, K.; Abern, M.R.; Nonn, L. vitamin D sufficiency enhances differentiation of patient-derived prostate epithelial organoids. *iScience* **2021**, *24*, 101974. [CrossRef] [PubMed]
267. Shan, N.L.; Minden, A.; Furmanski, P.; Bak, M.J.; Cai, L.; Wernyj, R.; Sargsyan, D.; Cheng, D.; Wu, R.; Kuo, H.D.; et al. Analysis of the Transcriptome: Regulation of Cancer Stemness in Breast Ductal Carcinoma In Situ by vitamin D Compounds. *Cancer Prev. Res.* **2020**, *13*, 673–686. [CrossRef] [PubMed]
268. So, J.Y.; Wahler, J.; Das Gupta, S.; Salerno, D.M.; Maehr, H.; Uskokovic, M.; Suh, N. HES1-mediated inhibition of Notch1 signaling by a Gemini vitamin D analog leads to decreased CD44(+)/CD24(-/low) tumor-initiating subpopulation in basal-like breast cancer. *J. Steroid Biochem. Mol. Biol.* **2015**, *148*, 111–121. [CrossRef]
269. Wahler, J.; So, J.Y.; Cheng, L.C.; Maehr, H.; Uskokovic, M.; Suh, N. vitamin D compounds reduce mammosphere formation and decrease expression of putative stem cell markers in breast cancer. *J. Steroid Biochem. Mol. Biol.* **2015**, *148*, 148–155. [CrossRef]
270. Ferronato, M.J.; Nadal Serrano, M.; Arenas Lahuerta, E.J.; Bernado Morales, C.; Paolillo, G.; Martinez-Sabadell Aliguer, A.; Santalla, H.; Mascaro, M.; Vitale, C.; Fall, Y.; et al. vitamin D analogues exhibit antineoplastic activity in breast cancer patient-derived xenograft cells. *J. Steroid Biochem. Mol. Biol.* **2021**, *208*, 105735. [CrossRef] [PubMed]
271. Ao, T.; Kikuta, J.; Ishii, M. The Effects of vitamin D on Immune System and Inflammatory Diseases. *Biomolecules* **2021**, *11*, 1624. [CrossRef] [PubMed]
272. Hanel, A.; Neme, A.; Malinen, M.; Hamalainen, E.; Malmberg, H.R.; Etheve, S.; Tuomainen, T.P.; Virtanen, J.K.; Bendik, I.; Carlberg, C. Common and personal target genes of the micronutrient vitamin D in primary immune cells from human peripheral blood. *Sci. Rep.* **2020**, *10*, 21051. [CrossRef]
273. Chun, R.F.; Liu, P.T.; Modlin, R.L.; Adams, J.S.; Hewison, M. Impact of vitamin D on immune function: Lessons learned from genome-wide analysis. *Front. Physiol.* **2014**, *5*, 151. [CrossRef] [PubMed]
274. Catala-Moll, F.; Ferrete-Bonastre, A.G.; Godoy-Tena, G.; Morante-Palacios, O.; Ciudad, L.; Barbera, L.; Fondelli, F.; Martínez-Cáceres, E.M.; Rodriguez-Ubreva, J.; Li, T.; et al. vitamin D receptor, STAT3, and TET2 cooperate to establish tolerogenesis. *Cell Rep.* **2022**, *38*, 110244. [CrossRef] [PubMed]
275. Korf, H.; Wenes, M.; Stijlemans, B.; Takiishi, T.; Robert, S.; Miani, M.; Eizirik, D.L.; Gysemans, C.; Mathieu, C. 1,25-Dihydroxyvitamin D3 curtails the inflammatory and T cell stimulatory capacity of macrophages through an IL-10-dependent mechanism. *Immunobiology* **2012**, *217*, 1292–1300. [CrossRef]
276. Zhang, X.; Zhou, M.; Guo, Y.; Song, Z.; Liu, B. 1,25-Dihydroxyvitamin D(3) Promotes High Glucose-Induced M1 Macrophage Switching to M2 via the VDR-PPARgamma Signaling Pathway. *BioMed Res. Int.* **2015**, *2015*, 157834. [CrossRef]
277. Von Essen, M.R.; Kongsbak, M.; Schjerling, P.; Olgaard, K.; Odum, N.; Geisler, C. vitamin D controls T cell antigen receptor signaling and activation of human T cells. *Nat. Immunol.* **2010**, *11*, 344–349. [CrossRef] [PubMed]
278. El-Sharkawy, A.; Malki, A. vitamin D Signaling in Inflammation and Cancer: Molecular Mechanisms and Therapeutic Implications. *Molecules* **2020**, *25*, 3219. [CrossRef]
279. Dankers, W.; Colin, E.M.; van Hamburg, J.P.; Lubberts, E. vitamin D in Autoimmunity: Molecular Mechanisms and Therapeutic Potential. *Front. Immunol.* **2016**, *7*, 697. [CrossRef] [PubMed]
280. Karkeni, E.; Morin, S.O.; Bou Tayeh, B.; Goubard, A.; Josselin, E.; Castellano, R.; Fauriat, C.; Guittard, G.; Olive, D.; Nunes, J.A. vitamin D Controls Tumor Growth and CD8+ T Cell Infiltration in Breast Cancer. *Front. Immunol.* **2019**, *10*, 1307. [CrossRef] [PubMed]
281. Fleet, J.C.; Burcham, G.N.; Calvert, R.D.; Elzey, B.D.; Ratliff, T.L. 1alpha, 25 Dihydroxyvitamin D (1,25(OH)2D) inhibits the T cell suppressive function of myeloid derived suppressor cells (MDSC). *J. Steroid Biochem. Mol. Biol.* **2020**, *198*, 105557. [CrossRef] [PubMed]
282. Sun, D.; Luo, F.; Xing, J.C.; Zhang, F.; Xu, J.Z.; Zhang, Z.H. 1,25(OH)2 D3 inhibited Th17 cells differentiation via regulating the NF-kappaB activity and expression of IL-17. *Cell Prolif.* **2018**, *51*, e12461. [CrossRef] [PubMed]
283. Cohen-Lahav, M.; Shany, S.; Tobvin, D.; Chaimovitz, C.; Douvdevani, A. vitamin D decreases NFkappaB activity by increasing IkappaBalpha levels. *Nephrol. Dial. Transpl.* **2006**, *21*, 889–897. [CrossRef]
284. Tse, A.K.; Zhu, G.Y.; Wan, C.K.; Shen, X.L.; Yu, Z.L.; Fong, W.F. 1alpha,25-Dihydroxyvitamin D3 inhibits transcriptional potential of nuclear factor kappa B in breast cancer cells. *Mol. Immunol.* **2010**, *47*, 1728–1738. [CrossRef] [PubMed]
285. Krishnan, A.V.; Feldman, D. Mechanisms of the anti-cancer and anti-inflammatory actions of vitamin D. *Annu. Rev. Pharmacol. Toxicol.* **2011**, *51*, 311–336. [CrossRef]
286. Moreno, J.; Krishnan, A.V.; Swami, S.; Nonn, L.; Peehl, D.M.; Feldman, D. Regulation of prostaglandin metabolism by calcitriol attenuates growth stimulation in prostate cancer cells. *Cancer Res.* **2005**, *65*, 7917–7925. [CrossRef] [PubMed]

287. Gibbs, D.C.; Fedirko, V.; Baron, J.A.; Barry, E.L.; Flanders, W.D.; McCullough, M.L.; Yacoub, R.; Raavi, T.; Rutherford, R.E.; Seabrook, M.E.; et al. Inflammation Modulation by vitamin D and Calcium in the Morphologically Normal Colorectal Mucosa of Patients with Colorectal Adenoma in a Clinical Trial. *Cancer Prev. Res.* **2021**, *14*, 65–76. [CrossRef]
288. Bruns, H.; Buttner, M.; Fabri, M.; Mougiakakos, D.; Bittenbring, J.T.; Hoffmann, M.H.; Beier, F.; Pasemann, S.; Jitschin, R.; Hofmann, A.D.; et al. vitamin D-dependent induction of cathelicidin in human macrophages results in cytotoxicity against high-grade B cell lymphoma. *Sci. Transl. Med.* **2015**, *7*, 282ra247. [CrossRef] [PubMed]
289. Min, D.; Lv, X.B.; Wang, X.; Zhang, B.; Meng, W.; Yu, F.; Hu, H. Downregulation of miR-302c and miR-520c by 1,25(OH)2D3 treatment enhances the susceptibility of tumour cells to natural killer cell-mediated cytotoxicity. *Br. J. Cancer* **2013**, *109*, 723–730. [CrossRef]
290. Neumann, F.; Acker, F.; Schormann, C.; Pfreundschuh, M.; Bittenbring, J.T. Determination of optimum vitamin D3 levels for NK cell-mediated rituximab- and obinutuzumab-dependent cellular cytotoxicity. *Cancer Immunol. Immunother.* **2018**, *67*, 1709–1718. [CrossRef] [PubMed]
291. Bittenbring, J.T.; Neumann, F.; Altmann, B.; Achenbach, M.; Reichrath, J.; Ziepert, M.; Geisel, J.; Regitz, E.; Held, G.; Pfreundschuh, M. vitamin D deficiency impairs rituximab-mediated cellular cytotoxicity and outcome of patients with diffuse large B-cell lymphoma treated with but not without rituximab. *J. Clin. Oncol.* **2014**, *32*, 3242–3248. [CrossRef] [PubMed]
292. Mortara, L.; Gariboldi, M.B.; Bosi, A.; Bregni, M.; Pinotti, G.; Guasti, L.; Squizzato, A.; Noonan, D.M.; Monti, E.; Campiotti, L. vitamin D Deficiency has a Negative Impact on Cetuximab-Mediated Cellular Cytotoxicity against Human Colon Carcinoma Cells. *Target. Oncol.* **2018**, *13*, 657–665. [CrossRef] [PubMed]
293. Lipplaa, A.; Fernandes, R.; Marshall, A.; Lorigan, P.; Dunn, J.; Myers, K.A.; Barker, E.; Newton-Bishop, J.; Middleton, M.R.; Corrie, P.G. 25-hydroxyvitamin D serum levels in patients with high risk resected melanoma treated in an adjuvant bevacizumab trial. *Br. J. Cancer* **2018**, *119*, 793–800. [CrossRef] [PubMed]
294. Dimitrov, V.; Bouttier, M.; Boukhaled, G.; Salehi-Tabar, R.; Avramescu, R.G.; Memari, B.; Hasaj, B.; Lukacs, G.L.; Krawczyk, C.M.; White, J.H. Hormonal vitamin D up-regulates tissue-specific PD-L1 and PD-L2 surface glycoprotein expression in humans but not mice. *J. Biol. Chem.* **2017**, *292*, 20657–20668. [CrossRef] [PubMed]
295. Bendix, M.; Greisen, S.; Dige, A.; Hvas, C.L.; Bak, N.; Jorgensen, S.P.; Dahlerup, J.F.; Deleuran, B.; Agnholt, J. vitamin D increases programmed death receptor-1 expression in Crohn's disease. *Oncotarget* **2017**, *8*, 24177–24186. [CrossRef]
296. Stucci, L.S.; D'Oronzo, S.; Tucci, M.; Macerollo, A.; Ribero, S.; Spagnolo, F.; Marra, E.; Picasso, V.; Orgiano, L.; Marconcini, R.; et al. vitamin D in melanoma: Controversies and potential role in combination with immune check-point inhibitors. *Cancer Treat. Rev.* **2018**, *69*, 21–28. [CrossRef] [PubMed]
297. Kutuzova, G.D.; DeLuca, H.F. 1,25-Dihydroxyvitamin D3 regulates genes responsible for detoxification in intestine. *Toxicol. Appl. Pharmacol.* **2007**, *218*, 37–44. [CrossRef]
298. Lindh, J.D.; Bjorkhem-Bergman, L.; Eliasson, E. Vitamin D and drug-metabolising enzymes. *Photochem. Photobiol. Sci.* **2012**, *11*, 1797–1801. [CrossRef] [PubMed]
299. Chatterjee, B.; Echchgadda, I.; Song, C.S. Vitamin D receptor regulation of the steroid/bile acid sulfotransferase SULT2A1. *Methods Enzymol.* **2005**, *400*, 165–191. [CrossRef] [PubMed]
300. Wang, Z.; Schuetz, E.G.; Xu, Y.; Thummel, K.E. Interplay between Vitamin D and the drug metabolizing enzyme CYP3A4. *J. Steroid Biochem. Mol. Biol.* **2013**, *136*, 54–58. [CrossRef] [PubMed]
301. Ajouz, H.; Mukherji, D.; Shamseddine, A. Secondary bile acids: An underrecognized cause of colon cancer. *World J. Surg. Oncol.* **2014**, *12*, 164. [CrossRef]
302. Peterlik, M. Role of bile acid secretion in human colorectal cancer. *Wien. Med. Wochenschr.* **2008**, *158*, 539–541. [CrossRef] [PubMed]
303. Makishima, M.; Lu, T.T.; Xie, W.; Whitfield, G.K.; Domoto, H.; Evans, R.M.; Haussler, M.R.; Mangelsdorf, D.J. vitamin D receptor as an intestinal bile acid sensor. *Science* **2002**, *296*, 1313–1316. [CrossRef]
304. Matsunawa, M.; Akagi, D.; Uno, S.; Endo-Umeda, K.; Yamada, S.; Ikeda, K.; Makishima, M. vitamin D receptor activation enhances benzo[a]pyrene metabolism via CYP1A1 expression in macrophages. *Drug Metab. Dispos.* **2012**, *40*, 2059–2066. [CrossRef] [PubMed]
305. Chen, G.Y. The Role of the Gut Microbiome in Colorectal Cancer. *Clin. Colon Rectal Surg.* **2018**, *31*, 192–198. [CrossRef] [PubMed]
306. Zhou, X.; Chen, C.; Zhong, Y.N.; Zhao, F.; Hao, Z.; Xu, Y.; Lai, R.; Shen, G.; Yin, X. Effect and mechanism of vitamin D on the development of colorectal cancer based on intestinal flora disorder. *J. Gastroenterol. Hepatol.* **2020**, *35*, 1023–1031. [CrossRef] [PubMed]
307. Malaguarnera, L. vitamin D and microbiota: Two sides of the same coin in the immunomodulatory aspects. *Int. Immunopharmacol.* **2020**, *79*, 106112. [CrossRef]
308. Wang, J.; Thingholm, L.B.; Skieceviciene, J.; Rausch, P.; Kummen, M.; Hov, J.R.; Degenhardt, F.; Heinsen, F.A.; Ruhlemann, M.C.; Szymczak, S.; et al. Genome-wide association analysis identifies variation in vitamin D receptor and other host factors influencing the gut microbiota. *Nat. Genet.* **2016**, *48*, 1396–1406. [CrossRef]
309. Lu, R.; Shang, M.; Zhang, Y.G.; Jiao, Y.; Xia, Y.; Garrett, S.; Bakke, D.; Bauerl, C.; Martinez, G.P.; Kim, C.H.; et al. Lactic Acid Bacteria Isolated From Korean Kimchi Activate the vitamin D Receptor-autophagy Signaling Pathways. *Inflamm. Bowel Dis.* **2020**, *26*, 1199–1211. [CrossRef]

310. Zhang, Y.G.; Lu, R.; Wu, S.; Chatterjee, I.; Zhou, D.; Xia, Y.; Sun, J. vitamin D Receptor Protects Against Dysbiosis and Tumorigenesis via the JAK/STAT Pathway in Intestine. *Cell. Mol. Gastroenterol. Hepatol.* **2020**, *10*, 729–746. [CrossRef] [PubMed]
311. Madden, J.M.; Murphy, L.; Zgaga, L.; Bennett, K. De novo vitamin D supplement use post-diagnosis is associated with breast cancer survival. *Breast Cancer Res. Treat.* **2018**, *172*, 179–190. [CrossRef]
312. Carlberg, C.; Haq, A. The concept of the personal vitamin D response index. *J. Steroid Biochem. Mol. Biol.* **2018**, *175*, 12–17. [CrossRef]
313. Dimitrov, V.; Barbier, C.; Ismailova, A.; Wang, Y.; Dmowski, K.; Salehi-Tabar, R.; Memari, B.; Groulx-Boivin, E.; White, J.H. vitamin D-regulated Gene Expression Profiles: Species-specificity and Cell-specific Effects on Metabolism and Immunity. *Endocrinology* **2021**, *162*, bqaa218. [CrossRef]

MDPI

Review

The Effect of Vitamin D and Its Analogs in Ovarian Cancer

Karina Piatek, Martin Schepelmann and Enikö Kallay *

Center for Pathophysiology, Infectiology and Immunology, Institute for Pathophysiology and Allergy Research, Medical University of Vienna, Waehringer Guertel 18-20, 1090 Vienna, Austria
* Correspondence: enikoe.kallay@meduniwien.ac.at; Tel.: +43-1-40400-51230

Abstract: Ovarian cancer is one of the deadliest cancers in women, due to its heterogeneity and usually late diagnosis. The current first-line therapies of debulking surgery and intensive chemotherapy cause debilitating side effects. Therefore, there is an unmet medical need to find new and effective therapies with fewer side effects, or adjuvant therapies, which could reduce the necessary doses of chemotherapeutics. Vitamin D is one of the main regulators of serum calcium and phosphorus homeostasis, but it has also anticancer effects. It induces differentiation and apoptosis, reduces proliferation and metastatic potential of cancer cells. However, doses that would be effective against cancer cause hypercalcemia. For this reason, synthetic and less calcemic analogs have been developed and tested in terms of their anticancer effect. The anticancer role of vitamin D is best understood in colorectal, breast, and prostate cancer and much less research has been done in ovarian cancer. In this review, we thus summarize the studies on the role of vitamin D and its analogs in vitro and in vivo in ovarian cancer models.

Keywords: ovarian cancer; vitamin D; vitamin D analogs

1. Ovarian Cancer

Ovarian cancer (OC) is the seventh deadliest and the eighth most common cancer in women, affecting 313,000 women and causing 207,000 deaths in 2020 (International Agency for Research on Cancer). OC is also called the "silent killer" because it is usually diagnosed at a late stage when the chances of a cure are already very low. Most OCs are diagnosed at stage III (51%) or IV (29%), where the 5-year survival is only 42% or 26%, respectively [1]. See Figure 1 for the most important facts of ovarian cancer. Ovarian tumors arise not only, as previously thought, in tissues of the ovary, but recent data shows that in some cases they can also start in the distal fallopian tube [2]. OC is a highly heterogeneous disease. Heterogeneity is high not only among the different types of ovarian tumors, but also within a single tumor [3]. Based on its origin, OC has seven histological types: epithelial tumors, mesenchymal tumors, mixed epithelial and mesenchymal tumors, sex cord stromal tumors, germ cell tumors miscellaneous tumors, and tumor-like lesions [4]. Around 90% of all ovarian tumors are of epithelial origin [5]. Because of their high heterogeneity, epithelial ovarian carcinomas (EOC) are divided into subgroups by the World Health Organization according to cell type: serous tumors, mucinous tumors, endometroid tumors, clear cell tumors, seromucinous tumors, Brenner tumors, and other carcinomas [4]. According to the International Federation of Gynecology and Obstetrics (FIGO), OC has four stages based on macroscopic and microscopic examination before and after surgery as well as cytology [6]. At stage I and II, the tumor is present mainly in ovaries and fallopian tubes, whereas at stage III it has spread already to local lymph nodes and peritoneum outside the pelvis. At the highest stage IV, distant metastases are present [7].

Treatment of OC can be local or systemic. Selection of the therapy depends on the type and stage of the disease. First-line therapy is usually debulking surgery and chemotherapy with platinum-based compounds and taxanes. There are also other, less common treatments such as radiation therapy, hormone therapy, but also targeted therapy, which is

Citation: Piatek, K.; Schepelmann, M.; Kallay, E. The Effect of Vitamin D and Its Analogs in Ovarian Cancer. *Nutrients* **2022**, *14*, 3867. https://doi.org/10.3390/nu14183867

Academic Editor: Carsten Carlberg

Received: 29 July 2022
Accepted: 14 September 2022
Published: 18 September 2022

directed at genes and proteins specific for the cancers. The most common targeted therapy uses Poly(ADP-Ribose)-Polymerase (PARP) inhibitors. Unfortunately, current first-line chemotherapeutic treatments cause serious side effects such as dizziness, fatigue, nausea, vomiting, and diarrhea. That is why there is an urgent need to develop new adjuvant or curative therapeutic approaches that would decrease the required dose or duration of the classical chemotherapy, thus reducing side effect severity.

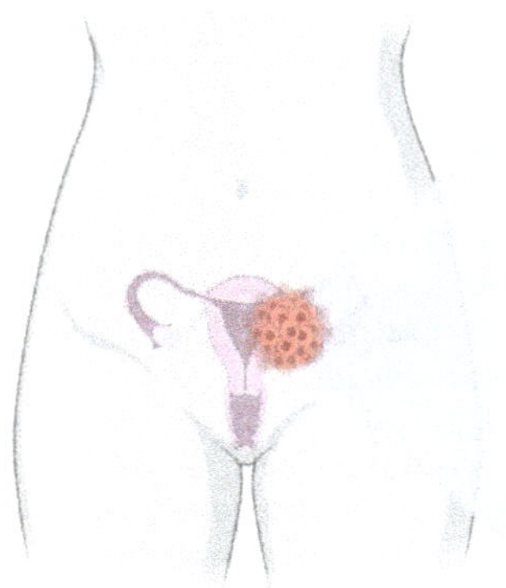

Figure 1. Ovarian cancer facts.

2. Vitamin D and Vitamin D Analogs

Ultraviolet rays transform 7-dehydrocholesterol in the skin to cholecalciferol, or vitamin D3 [8]. Vitamin D is then transported to the liver where it is hydroxylated on position 25 to the main circulating metabolite 25-hydroxyvitamin D3 (25D3) by 25-hydroxylase encoded by the gene *CYP2R1*. 25D3, bound to the vitamin D binding protein, is transported to the target tissues, where it is hydroxylated on the position 1α to the main active hormone 1,25-dihydroxyvitamin D3, or calcitriol (1,25D3), by the rate-limiting enzyme 1α-hydroxylase, coded by the *CYP27B1* gene [8]. The kidneys are the main site for this hydroxylation step, but many other tissues express *CYP27B1* and thus are able to synthetize the active hormone 1,25D3. Both 25D3 and 1,25D3 are degraded by the enzyme 24-hydroxylase, encoded by *CYP24A1* [8]. The main role of 1,25D3, bound to its receptor, the transcription factor vitamin D receptor (VDR), is the regulation of serum calcium and phosphorus homeostasis, but it also regulates the expression of many genes that might play a role in the development of cancer [9]. The main indicator of the organism's vitamin D status is serum 25D3, although the optimal status is still a subject of debate. Levels <12 ng/mL are considered as severe vitamin D deficiency, while sufficiency varies between 20-30 ng/mL, depending on the expert bodies or societies [10–12].

The main function of 1,25D3 is to maintain the proper levels of calcium and phosphorus in serum. Thus, higher doses of 1,25D3, which would be effective against cancer, can cause hypercalcemia as a side effect. For this reason, various synthetic vitamin D analogs have been developed (representative structures of parent compounds and analogs are shown in Figure 2). MT19c is a vitamin D2 (the plant-derived ergocalciferol) derivate, created by Diels–Alder cycloaddition of *N*-methyl,1,2,4-triazolinedione and esterification with bromoacetic acid [13]. This last modification is shared with the analog B3Cd, which is a 3-bromoacetoxy derivative of 25D3 [14]. EB1089 was one of the first developed vitamin D analogs. It is based on the 1,25D3 structure by elongating the side chain by one carbon and the introduction of terminal ethyl groups and introduction of double bonds at positions 22 and 24 [15]. The PRI-1906 and PRI-1907 analogs were designed based on the 1,25-hydroxylated form (1,25D2) of the plant-derived vitamin D2 (ergocalciferol), with an extended side chain and introduced double bonds. PRI-1906 carries terminal methyl and

PRI-1907 ethyl groups [16]. The analogs PRI-5201 and PRI-5202 are based on PRI-1906 and PRI-1907 by removing the methylidene group of the A-ring [17].

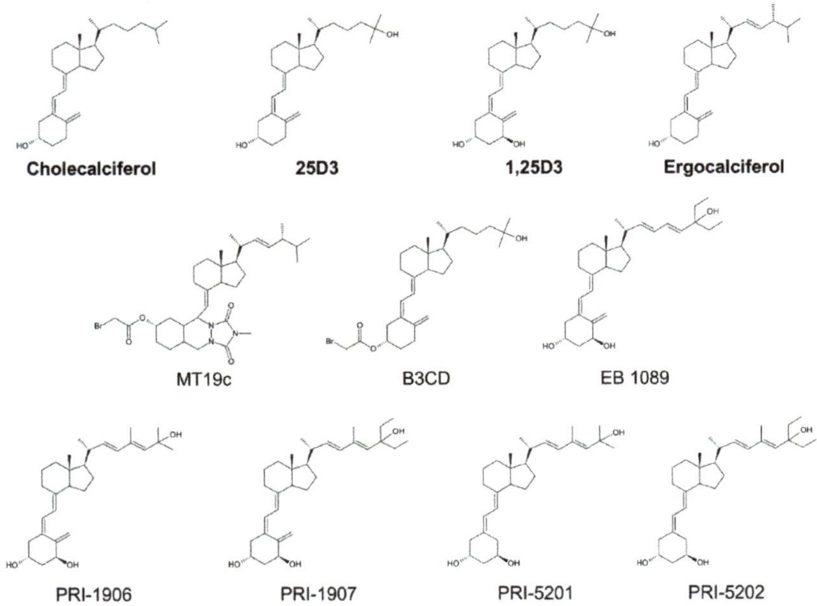

Figure 2. Chemical structures of the most important parent compounds (in bold) and the vitamin D analogs mentioned in the review.

3. Effect of Vitamin D on Ovarian Cancer Epidemiology

Several recent studies reviewed the anticancer effects of different vitamins on selected female malignancies [18–20]; in our review, we focus only on the effect of vitamin D in ovarian cancer. The role of vitamin D levels in epidemiology of OC is still unclear. One of the first studies on the correlation between vitamin D and cancer found that sunlight can be a protective factor against OC-associated mortality [21]. A report from Australia concluded that exposure to ambient ultraviolet radiation may reduce the risk of EOC [22]. In another cohort study from Australia, researchers have shown that higher 25D3 serum levels at the stage of diagnosis correlated significantly with longer survival of women with diagnosed invasive OC [23]. In a European population, a Mendelian randomized study found that genetically lower 25D3 levels correlated inversely with higher susceptibility to OC [24]. Additionally, predicted higher concentrations of 25D3 (based on GWAS studies) were associated with reduced risk of EOC [25]. However, current data on this topic are inconsistent. In another Mendelian randomized study, researchers concluded that vitamin D levels had no effect in seven cancers, including OC [26]. A further study found that genetically low plasma 25D3 concentrations were not associated with increased cancer risk and mortality rates [27]. Similarly, a study conducted in African women from Nigeria found no significant correlation between serum 25D3 levels and the risk of EOC [28]. A meta-analysis of 21 articles with almost one million participants concluded that vitamin D intake could not decrease the risk of OC [29].

4. Mechanism of Anticancer Activity in Ovarian Cancer Models In Vitro

The heterogeneity of OC is mirrored by the diversity of the existing cell lines. Even those cell lines that were obtained from the same type of tumors show high diversity in their sensitivity to vitamin D and its analogs [30]. This might be due to differences in their mutational landscape, but also in the expression of the components of the vitamin D system.

While VDR is present in most of the known cell lines, the expression is highly variable, as is also the expression of *CYP27B1* and *CYP24A1*. The results comparing the expression level of VDR in normal and malignant ovarian tissues are inconsistent. One study reported higher levels in ovarian tumors compared with healthy tissue [31], while others found that the VDR level was lower in tumors than in normal ovaries [32]. Treatment of OC cell lines with 1,25D3 or its analogs has no effect on VDR mRNA levels, while it might increase protein levels in a cell line-dependent manner [30,33]. Interestingly, the effect of 1,25D3 or its analogs on the expression of *CYP24A1* in different OC cell lines does not predict their anticancer effect [30].

4.1. Effect of 1,25D3 and Its Analogs on the Hallmarks of Cancer

Besides its major role to maintain calcium-phosphate homeostasis, 1,25D3 regulates most hallmarks of cancer. It inhibits proliferation, angiogenesis, and metastasis, induces differentiation and apoptosis, and regulates the immune system [34,35]. The anticancer effects of 1,25D3 were best documented for colorectal, breast, and prostate cancer [36–39]. Much less is known about the effect of 1,25D3 and its analogs in OC cells (Figure 3).

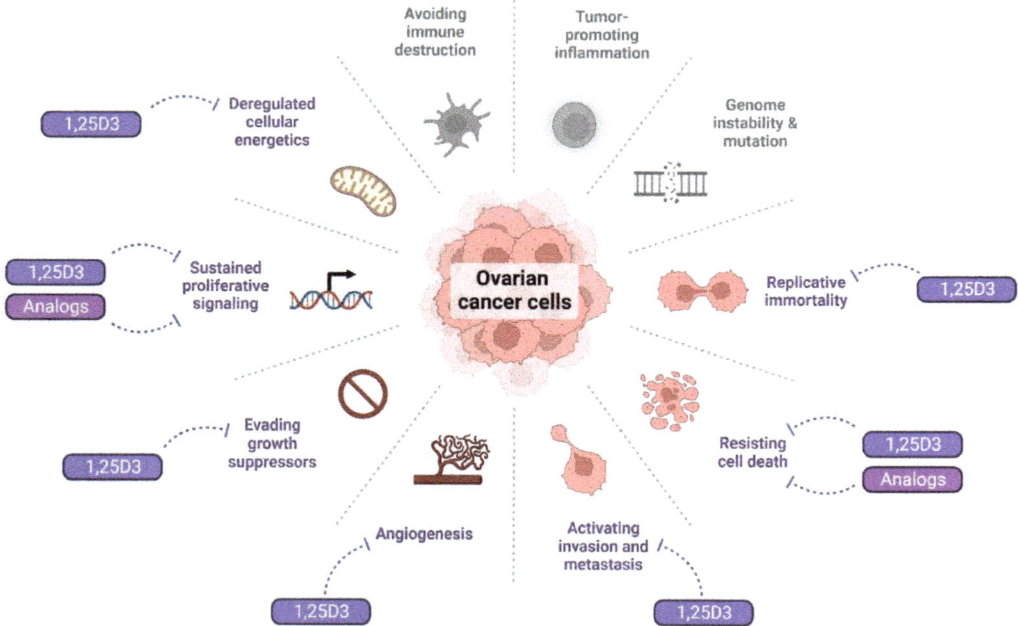

Figure 3. Impact of 1,25D3 and its analogs on the hallmarks of cancer in OC cells. In color, the hallmarks affected by 1,25D3 (see main text below), in gray, those where no relevant studies were found. The dotted blunt arrows indicate inhibition.

4.2. Effect of 1,25D3 and Its Analogs on Proliferative Signals

1,25D3 inhibited proliferation by reducing the cell number of a patient-derived high grade serous ovarian cancer (HGSOC) cell line, while no effect was seen in a further HGSOC line [30]. In OVCAR3 cells, 1,25D3 induced cell cycle arrest either at the G1/S or G2/M checkpoint [40,41]. The expression of several genes involved in the cell cycle was also inhibited in some (e.g., OVCAR3, CAOV3, OV2008) but not in other cell lines (e.g., OVCAR5, SKOV3). Interestingly, another study found that SKOV3 cells were sensitive to 1,25D3, which reduced both their proliferation and viability [42]. In OVCAR3 cells, 1,25D3 prevented cell cycle progression through the inhibition of the CKD2-Rb-E2F axis. In these cells, 1,25D3 increased p27 and decreased cyclin E and A expression [40]. Proliferation of OVCAR3 and SKOV3 cells was inhibited also by the 25D3 analog B3CD [14].

Epidermal growth factor receptor (EGFR) is often upregulated in OC, conveying a proliferative advantage to these tumors [43,44]. In vitamin D-sensitive OC cells, 1,25D3 downregulated EGFR expression, reducing their proliferative potential [41]. The OVCAR3 cells were also sensitive to the 1,25D3 analog, EB1089, which was more active in reducing EGFR expression than 1,25D3 [41]. The relevance of the impact of vitamin D on EGFR is underlined also by the fact that EGFR seems to be one of the key genes associated with resistance to platinum therapy of OC [44].

We also observed that different HGSOC cell lines responded differently to various analogs of 1,25D2. While all tested analogs reduced cell number and viability of 13,781 cells, in the 14,433 and 8714 cells, none of the analogs affected cell viability significantly ([30] and unpublished data).

4.3. *Effect of 1,25D3 and Its Analogs on Cell Death*

Very often, p53 is mutated in ovarian tumors, reducing the apoptotic ability of these cells. Therefore, finding compounds that would induce apoptosis even in the presence of a mutated p53 is of utmost importance. Interestingly, 1,25D3 and the EB1089 analog were able to stimulate apoptosis through p53-independent ways, by upregulating Growth Arrest and DNA Damage-inducible 45 (GADD45), or the cyclin-dependent kinases p21 and p27 [45]. The pro-apoptotic role of GADD45 proteins is well documented and they regulate many cellular functions, e.g., DNA repair, cell cycle, and senescence [46]. In cell lines from clear cell ovarian carcinoma (ES-2, TOV-21G), papillary serous adenocarcinoma (OV-90) and endometrioid carcinoma (TOV-112D), 1,25D3 activated the intrinsic apoptotic pathway by reducing the membrane potential of the mitochondria, increasing cytochrome C release, and activating caspase 9 [47]. In the cell lines that expressed the progesterone receptor, the effect of 1,25D3 was significantly higher when given together with progesterone [48].

1,25D3 increased sensitivity of the ovarian epithelial adenocarcinoma cell line SKOV3 to radiation-induced apoptosis by supporting the formation of reactive oxygen species (ROS) [49]. Another study has shown that, in SKOV3 cells, 1,25D3 induced apoptosis and potentiated the cytotoxic effect of cisplatin, increasing the activity of caspase 3/7. It also increased expression of the pro-apoptotic protein Bax and the cleaved PARP in a dose-dependent manner [42]. B3CD induced apoptosis by activating the p38 MAPK pathway [14].

4.4. *Effect of 1,25D3 and Its Analogs on Metastatic Potential*

Epithelial to mesenchymal transition (EMT) is considered as the driver of invasion and metastasis. 1,25D3 inhibited EMT in SCOV3 cancer cells [50]. One of the first organs OC disseminates to is the peritoneum [51]. Vitamin D prevented the TGF-β-induced mesenchymal transition and thus the transformation of the peritoneal mesothelial cells into cancer-associated mesothelial cells (CAM) by maintaining high e-cadherin levels and blocking the upregulation of the EMT-associated markers α-smooth muscle actin, slug and the matrix metalloproteinases (MMP) 9 and 2, and that of thrombospondin-1, a gene involved in both the TGF-β and focal adhesion pathways [52]. Treatment of the germline-derived immortalized ovarian cancer cell line A2780 with 1,25D3 inhibited migration of the cells and their adhesion to fibronectin. Pre-treatment of the cells with 1,25D3 also reduced their metastatic potential when injected in immunodeficient mice [53].

Several long noncoding RNAs, e.g., lnc-BCAS1-4_1, play an important role in the regulation of EMT by 1,25D3 [54]. OC patients with high levels of the lncRNA *TOPORS Antisense RNA 1* (*TOPORS-AS1*) in their tumors had favorable overall survival compared with those expressing low levels. As *TOPORS-AS1* is a target for VDR, it has been suggested that the inhibitory effect of VDR in ovarian cancer cells could be mediated through *TOPORS-AS1* [55].

In a recent study, 1,25D3 inhibited the self-renewal capacity of ovarian cancer stem cells (CSC) by reduction of their sphere formation rate and inhibition of the expression of stem cell markers, such as CD44, SOX2, or OCT4 [56].

4.5. Effect of 1,25D3 and Its Analogs on Replicative Immortality and Angiogenesis

In the OVCAR3 cells, 1,25D3 inhibited a further hallmark of cancer, the replicative immortality, by downregulating activity and expression of the telomerase [57]. One mechanism by which 1,25D3 regulated telomerase expression in these cells was the upregulation of miR-498, which then degrades the telomerase mRNA, leading to their apoptotic cell death [58].

In SKOV3, 1,25D3 inhibited VEGF expression and activity and enhanced the anti-angiogenic effect of cisplatin [42].

There is not enough information on the role that vitamin D could play in regulating other hallmarks. Although it is known that vitamin D affects the immune system and plays an important role in inflammation, little is known about its effect on these hallmarks in OC. One study has shown that 1,25D3 was able to reduce the tumor promoting effect of M2 macrophages in ovarian cancer cells [59]. Another interesting finding was that the vitamin D target human cathelicidin, well known as an effector molecule of the innate immune system, promotes OC progression. It seems that OC cells induce the expression of cathelicidin in macrophages in a VDR-dependent manner [60]. This would suggest a detrimental effect of vitamin D on OC development.

5. Mechanism of Anticancer Activity in Ovarian Cancer Models In Vivo

The anticancer activity of 1,25D3 and its analogs in OC models has already been studied broadly in vitro but there are only a few studies about their effects in vivo. Figure 4 summarizes the observed effects found in in vivo studies.

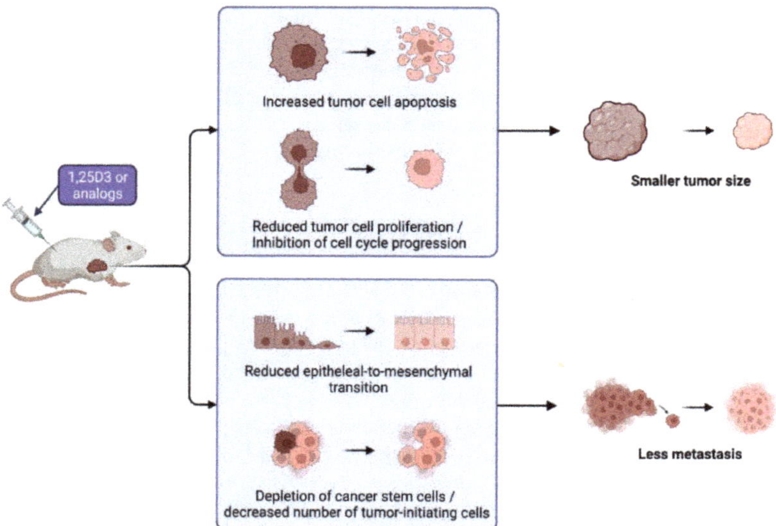

Figure 4. Effects of 1,25D3 and its analogs in in vivo models of ovarian cancer.

In a mouse model of peritoneal metastasis, vitamin D3 protected the microvilli on the peritoneum, thus preventing the interaction of CAMs with cancer cells by inhibiting Smad-dependent TGF-β signaling, thus inhibiting peritoneal dissemination of the ES-2 OC cells [52].

Vitamin D affects at different stages of the carcinogenesis process. In a mouse model of OC, induced by 7,12-dimethylbenz[a]anthracene (DMBA), 1,25D3 administration reduced tumor size significantly at the stage of initiation, promotion and entire period of the experiment. The general condition of the mice in the treated groups was significantly better than in the untreated controls [61]. Mice treated with vitamin 1,25D3 also had lower levels of CA125, which is considered a potential ovarian tumor marker [61,62].

1,25D3 as an anticancer agent is known to also regulate CSCs. Srivastava et al. studied the effect of 1,25D3 on OC stem cells in vivo in a mouse xenograft tumor model using the 2008 cell line. 1,25D3 delayed tumor growth and depleted the ovarian CSCs. The effect on the CSCs was studied in isolated xenograft tumor cells by measuring CD44 and CD117 positive cells, which are markers for stem cells. 1,25D3 significantly reduced the CSC population in ovarian xenografts in vivo [63].

Only a few analogs of vitamin D3 or 25D3 were tested in vivo in ovarian cancer models. MT19c reduced tumor growth in the SKOV3 xenograft model in nude mice and in a syngeneic rat ovarian cancer models and decreased the expression of genes involved in energy metabolism. The treatment reduced the expression of EGFR, and inhibited PI-3 kinase [13]. The 25D3 derivative B3CD was tested on the SKOV-3 xenograft model. The compound delayed tumor growth in the majority of the mice, and in some cases even led to full regression. However, in some of the mice, B3CD accelerated tumor growth [14]. The study with the 1,25D3 analog EB 1089 found that EB 1089 suppressed the growth of OVCAR3 tumor xenografts in mice. Histological analysis of tumor sections showed that EB 1089 induced apoptosis and decreased proliferation in the tumor [45].

More in vivo studies are needed to understand if vitamin D analogs should be carried further into clinical trials.

6. Future Perspectives

The impact of vitamin D and its analogs on OC is still unclear, although the in vitro studies are promising. Further to the research we summarized, a few studies reported that 1,25D3 is able to potentiate the effect of some of the chemotherapeutics used in the treatment of OC. 1,25D3 increased the cytotoxic effect of carboplatin and paclitaxel in serous-, mucinous-, and endometrioid-type OC cell lines [33]. Although it has been shown that 1,25D3 is a PARP inhibitor [64], the studies to test if it would potentiate the effect of the PARP inhibitors used in OC treatment are still missing. PARP inhibitors are used for the therapy of recurrent OC, however, the majority of the patients develop resistance, mediated by cancer stem cells [65]. As 1,25D3 was shown to reduce the number of CSCs in OC [63], this suggests a high clinical potential in preventing resistance to this class of drugs. Therefore, it would be of utmost importance to understand what makes OC cells responsive or resistant to the anticancer effects of vitamin D and its analogs. More needs to be done especially in vivo to gain a clear picture about the impact and potential application of vitamin D and its less calcemic analogs in OC.

Author Contributions: Conceptualization, K.P. and E.K.; investigation, K.P. and E.K.; resources, M.S. and E.K.; writing—original draft preparation, K.P. and E.K.; writing—review and editing, K.P., E.K. and M.S.; visualization, M.S.; supervision, E.K. and M.S.; project administration, E.K. and M.S.; funding acquisition, M.S. and E.K. All authors have read and agreed to the published version of the manuscript.

Funding: This research was funded by the FWF Austrian Science Fund, grant number P 29948-B28 (to E.K.) and FWF Austrian Science Fund and Herzfelder Family Fund, grant number P 32840-B (to M.S.).

Institutional Review Board Statement: Not applicable.

Informed Consent Statement: Not applicable.

Data Availability Statement: Not applicable.

Acknowledgments: The authors wish to thank Andrzej Kutner for helpful comments on the chemical structures. All figures were created with BioRender.com. Open Access Funding by the Austrian Science Fund (FWF).

Conflicts of Interest: The authors declare no conflict of interest. The funders had no role in the design of the study; in the collection, analyses, or interpretation of data; in the writing of the manuscript; or in the decision to publish the results.

References

1. Torre, L.A.; Trabert, B.; DeSantis, C.E.; Miller, K.D.; Samimi, G.; Runowicz, C.D.; Gaudet, M.M.; Jemal, A.; Siegel, R.L. Ovarian cancer statistics, 2018. *CA Cancer J. Clin.* **2018**, *68*, 284–296. [CrossRef] [PubMed]
2. Ducie, J.; Dao, F.; Considine, M.; Olvera, N.; Shaw, P.A.; Kurman, R.J.; Shih, I.M.; Soslow, R.A.; Cope, L.; Levine, D.A. Molecular analysis of high-grade serous ovarian carcinoma with and without associated serous tubal intra-epithelial carcinoma. *Nat. Commun.* **2017**, *8*, 990. [CrossRef] [PubMed]
3. Kossai, M.; Leary, A.; Scoazec, J.Y.; Genestie, C. Ovarian Cancer: A Heterogeneous Disease. *Pathobiology* **2018**, *85*, 41–49. [CrossRef]
4. Adhikari, L.H.L. Ovarian Neoplasms WHO Classification Review. Available online: https://www.pathologyoutlines.com/topic/ovarytumorwhoclassif.html (accessed on 17 May 2022).
5. Reid, B.M.; Permuth, J.B.; Sellers, T.A. Epidemiology of ovarian cancer: A review. *Cancer Biol. Med.* **2017**, *14*, 9–32. [CrossRef]
6. Matulonis, U.A.; Sood, A.K.; Fallowfield, L.; Howitt, B.E.; Sehouli, J.; Karlan, B.Y. Ovarian cancer. *Nat. Rev. Dis. Primers* **2016**, *2*, 16061. [CrossRef] [PubMed]
7. Berek, J.S.; Crum, C.; Friedlander, M. Cancer of the ovary, fallopian tube, and peritoneum. *Int. J. Gynaecol. Obstet.* **2012**, *119* (Suppl. 2), S118–S129. [CrossRef]
8. Bouillon, R.; Carmeliet, G.; Verlinden, L.; van Etten, E.; Verstuyf, A.; Luderer, H.F.; Lieben, L.; Mathieu, C.; Demay, M. Vitamin D and human health: Lessons from vitamin D receptor null mice. *Endocr. Rev.* **2008**, *29*, 726–776. [CrossRef]
9. Carlberg, C.; Munoz, A. An update on vitamin D signaling and cancer. *Semin. Cancer Biol.* **2022**, *79*, 217–230. [CrossRef]
10. Amrein, K.; Scherkl, M.; Hoffmann, M.; Neuwersch-Sommeregger, S.; Kostenberger, M.; Tmava Berisha, A.; Martucci, G.; Pilz, S.; Malle, O. Vitamin D deficiency 2.0: An update on the current status worldwide. *Eur. J. Clin. Nutr.* **2020**, *74*, 1498–1513. [CrossRef]
11. Holick, M.F.; Binkley, N.C.; Bischoff-Ferrari, H.A.; Gordon, C.M.; Hanley, D.A.; Heaney, R.P.; Murad, M.H.; Weaver, C.M.; Endocrine, S. Evaluation, treatment, and prevention of vitamin D deficiency: An Endocrine Society clinical practice guideline. *J. Clin. Endocrinol. Metab.* **2011**, *96*, 1911–1930. [CrossRef]
12. Bresson, J.L.; Burlingame, B.; Dean, T.; Fairweather-Tait, S.; Heinonen, M.; Hirsch-Ernst, K.I.; Mangelsdorf, I.; McArdle, H.; Naska, A.; Neuhauser-Berthold, M.; et al. Dietary reference values for vitamin D. *Efsa J.* **2016**, *14*, e04547. [CrossRef]
13. Moore, R.G.; Lange, T.S.; Robinson, K.; Kim, K.K.; Uzun, A.; Horan, T.C.; Kawar, N.; Yano, N.; Chu, S.R.; Mao, Q.; et al. Efficacy of a non-hypercalcemic vitamin-D2 derived anti-cancer agent (MT19c) and inhibition of fatty acid synthesis in an ovarian cancer xenograft model. *PLoS ONE* **2012**, *7*, e34443. [CrossRef]
14. Lange, T.S.; Stuckey, A.R.; Robison, K.; Kim, K.K.; Singh, R.K.; Raker, C.A.; Brard, L. Effect of a vitamin D(3) derivative (B3CD) with postulated anti-cancer activity in an ovarian cancer animal model. *Invest. New Drugs* **2010**, *28*, 543–553. [CrossRef]
15. Davicco, M.J.; Coxam, V.; Gaumet, N.; Lebecque, P.; Barlet, J.P. EB 1089, a calcitriol analogue, decreases fetal calcium content when injected into pregnant rats. *Exp. Physiol.* **1995**, *80*, 449–456. [CrossRef]
16. Baurska, H.; Klopot, A.; Kielbinski, M.; Chrobak, A.; Wijas, E.; Kutner, A.; Marcinkowska, E. Structure-function analysis of vitamin D(2) analogs as potential inducers of leukemia differentiation and inhibitors of prostate cancer proliferation. *J. Steroid Biochem. Mol. Biol.* **2011**, *126*, 46–54. [CrossRef]
17. Pietraszek, A.; Malinska, M.; Chodynski, M.; Krupa, M.; Krajewski, K.; Cmoch, P.; Wozniak, K.; Kutner, A. Synthesis and crystallographic study of 1,25-dihydroxyergocalciferol analogs. *Steroids* **2013**, *78*, 1003–1014. [CrossRef] [PubMed]
18. Markowska, A.; Antoszczak, M.; Markowska, J.; Huczynski, A. Role of Vitamin K in Selected Malignant Neoplasms in Women. *Nutrients* **2022**, *14*, 3401. [CrossRef] [PubMed]
19. Markowska, A.; Antoszczak, M.; Markowska, J.; Huczynski, A. Role of Vitamin E in Selected Malignant Neoplasms in Women. *Nutr. Cancer* **2022**, *74*, 1163–1170. [CrossRef]
20. Markowska, A.; Antoszczak, M.; Markowska, J.; Huczynski, A. Role of Vitamin C in Selected Malignant Neoplasms in Women. *Nutrients* **2022**, *14*, 882. [CrossRef]
21. Lefkowitz, E.S.; Garland, C.F. Sunlight, vitamin D, and ovarian cancer mortality rates in US women. *Int. J. Epidemiol.* **1994**, *23*, 1133–1136. [CrossRef]
22. Tran, B.; Jordan, S.J.; Lucas, R.; Webb, P.M.; Neale, R.; Australian Ovarian Cancer Study Group. Association between ambient ultraviolet radiation and risk of epithelial ovarian cancer. *Cancer Prev. Res. (Phila)* **2012**, *5*, 1330–1336. [CrossRef] [PubMed]
23. Webb, P.M.; de Fazio, A.; Protani, M.M.; Ibiebele, T.I.; Nagle, C.M.; Brand, A.H.; Blomfield, P.I.; Grant, P.; Perrin, L.C.; Neale, R.E.; et al. Circulating 25-hydroxyvitamin D and survival in women with ovarian cancer. *Am. J. Clin. Nutr.* **2015**, *102*, 109–114. [CrossRef] [PubMed]
24. Ong, J.S.; Cuellar-Partida, G.; Lu, Y.; Fasching, P.A.; Hein, A.; Burghaus, S.; Beckmann, M.W.; Lambrechts, D.; Van Nieuwenhuysen, E.; Vergote, I.; et al. Association of vitamin D levels and risk of ovarian cancer: A Mendelian randomization study. *Int. J. Epidemiol.* **2016**, *45*, 1619–1630. [CrossRef] [PubMed]
25. Ong, J.S.; Dixon-Suen, S.C.; Han, X.; An, J.; Liyanage, U.; Me Research, T.; Dusingize, J.C.; Schumacher, J.; Gockel, I.; Böhmer, A.; et al. A comprehensive re-assessment of the association between vitamin D and cancer susceptibility using Mendelian randomization. *Nat. Commun.* **2021**, *12*, 246. [CrossRef] [PubMed]

26. Dimitrakopoulou, V.I.; Tsilidis, K.K.; Haycock, P.C.; Dimou, N.L.; Al-Dabhani, K.; Martin, R.M.; Lewis, S.J.; Gunter, M.J.; Mondul, A.; Shui, I.M.; et al. Circulating vitamin D concentration and risk of seven cancers: Mendelian randomisation study. *BMJ* **2017**, *359*, j4761. [CrossRef]
27. Ong, J.S.; Gharahkhani, P.; An, J.; Law, M.H.; Whiteman, D.C.; Neale, R.E.; MacGregor, S. Vitamin D and overall cancer risk and cancer mortality: A Mendelian randomization study. *Hum. Mol. Genet.* **2018**, *27*, 4315–4322. [CrossRef]
28. Sajo, E.A.; Okunade, K.S.; Olorunfemi, G.; Rabiu, K.A.; Anorlu, R.I. Serum vitamin D deficiency and risk of epithelial ovarian cancer in Lagos, Nigeria. *Ecancermedicalscience* **2020**, *14*, 1078. [CrossRef]
29. Xu, J.; Chen, K.; Zhao, F.; Huang, D.; Zhang, H.; Fu, Z.; Xu, J.; Wu, Y.; Lin, H.; Zhou, Y.; et al. Association between vitamin D/calcium intake and 25-hydroxyvitamin D and risk of ovarian cancer: A dose-response relationship meta-analysis. *Eur. J. Clin. Nutr.* **2021**, *75*, 417–429. [CrossRef]
30. Piatek, K.; Kutner, A.; Cacsire Castillo-Tong, D.; Manhardt, T.; Kupper, N.; Nowak, U.; Chodynski, M.; Marcinkowska, E.; Kallay, E.; Schepelmann, M. Vitamin D Analogs Regulate the Vitamin D System and Cell Viability in Ovarian Cancer Cells. *Int. J. Mol. Sci.* **2021**, *23*, 172. [CrossRef]
31. Friedrich, M.; Rafi, L.; Mitschele, T.; Tilgen, W.; Schmidt, W.; Reichrath, J. Analysis of the vitamin D system in cervical carcinomas, breast cancer and ovarian cancer. *Recent. Results Cancer Res.* **2003**, *164*, 239–246. [CrossRef]
32. Brozyna, A.A.; Kim, T.K.; Zablocka, M.; Jozwicki, W.; Yue, J.; Tuckey, R.C.; Jetten, A.M.; Slominski, A.T. Association among Vitamin D, Retinoic Acid-Related Orphan Receptors, and Vitamin D Hydroxyderivatives in Ovarian Cancer. *Nutrients* **2020**, *12*, 3541. [CrossRef] [PubMed]
33. Kuittinen, T.; Rovio, P.; Luukkaala, T.; Laurila, M.; Grenman, S.; Kallioniemi, A.; Maenpaa, J. Paclitaxel, Carboplatin and 1,25-D3 Inhibit Proliferation of Ovarian Cancer Cells In Vitro. *Anticancer Res.* **2020**, *40*, 3129–3138. [CrossRef]
34. Wacker, M.; Holick, M.F. Vitamin D—Effects on skeletal and extraskeletal health and the need for supplementation. *Nutrients* **2013**, *5*, 111–148. [CrossRef]
35. Feldman, D.; Krishnan, A.V.; Swami, S.; Giovannucci, E.; Feldman, B.J. The role of vitamin D in reducing cancer risk and progression. *Nat. Rev. Cancer* **2014**, *14*, 342–357. [CrossRef]
36. Ferrer-Mayorga, G.; Larriba, M.J.; Crespo, P.; Munoz, A. Mechanisms of action of vitamin D in colon cancer. *J. Steroid Biochem Mol. Biol.* **2019**, *185*, 1–6. [CrossRef] [PubMed]
37. Mahendra, A.; Choudhury, B.K.; Sharma, T.; Bansal, N.; Bansal, R.; Gupta, S. Vitamin D and gastrointestinal cancer. *J. Lab. Physicians* **2018**, *10*, 1–5. [CrossRef] [PubMed]
38. Vanhevel, J.; Verlinden, L.; Doms, S.; Wildiers, H.; Verstuyf, A. The role of vitamin D in breast cancer risk and progression. *Endocr. Relat. Cancer* **2022**, *29*, R33–R55. [CrossRef]
39. Swami, S.; Krishnan, A.V.; Feldman, D. Vitamin D metabolism and action in the prostate: Implications for health and disease. *Mol. Cell Endocrinol.* **2011**, *347*, 61–69. [CrossRef]
40. Li, P.; Li, C.; Zhao, X.; Zhang, X.; Nicosia, S.V.; Bai, W. p27(Kip1) stabilization and G(1) arrest by 1,25-dihydroxyvitamin D(3) in ovarian cancer cells mediated through down-regulation of cyclin E/cyclin-dependent kinase 2 and Skp1-Cullin-F-box protein/Skp2 ubiquitin ligase. *J. Biol. Chem.* **2004**, *279*, 25260–25267. [CrossRef]
41. Shen, Z.; Zhang, X.; Tang, J.; Kasiappan, R.; Jinwal, U.; Li, P.; Hann, S.; Nicosia, S.V.; Wu, J.; Zhang, X.; et al. The coupling of epidermal growth factor receptor down regulation by 1alpha,25-dihydroxyvitamin D3 to the hormone-induced cell cycle arrest at the G1-S checkpoint in ovarian cancer cells. *Mol. Cell Endocrinol.* **2011**, *338*, 58–67. [CrossRef]
42. Kim, J.H.; Park, W.H.; Suh, D.H.; Kim, K.; No, J.H.; Kim, Y.B. Calcitriol Combined With Platinum-based Chemotherapy Suppresses Growth and Expression of Vascular Endothelial Growth Factor of SKOV-3 Ovarian Cancer Cells. *Anticancer Res.* **2021**, *41*, 2945–2952. [CrossRef] [PubMed]
43. Olbromski, P.J.; Pawlik, P.; Bogacz, A.; Sajdak, S. Identification of New Molecular Biomarkers in Ovarian Cancer Using the Gene Expression Profile. *J. Clin. Med.* **2022**, *11*, 3888. [CrossRef] [PubMed]
44. Han, L.; Guo, X.; Du, R.; Guo, K.; Qi, P.; Bian, H. Identification of key genes and pathways related to cancer-associated fibroblasts in chemoresistance of ovarian cancer cells based on GEO and TCGA databases. *J. Ovarian Res.* **2022**, *15*, 75. [CrossRef] [PubMed]
45. Zhang, X.; Jiang, F.; Li, P.; Li, C.; Ma, Q.; Nicosia, S.V.; Bai, W. Growth suppression of ovarian cancer xenografts in nude mice by vitamin D analogue EB1089. *Clin. Cancer Res.* **2005**, *11*, 323–328. [CrossRef]
46. Tamura, R.E.; de Vasconcellos, J.F.; Sarkar, D.; Libermann, T.A.; Fisher, P.B.; Zerbini, L.F. GADD45 proteins: Central players in tumorigenesis. *Curr. Mol. Med.* **2012**, *12*, 634–651. [CrossRef]
47. McGlorthan, L.; Paucarmayta, A.; Casablanca, Y.; Maxwell, G.L.; Syed, V. Progesterone induces apoptosis by activation of caspase-8 and calcitriol via activation of caspase-9 pathways in ovarian and endometrial cancer cells in vitro. *Apoptosis* **2021**, *26*, 184–194. [CrossRef]
48. Rodriguez, G.C.; Turbov, J.; Rosales, R.; Yoo, J.; Hunn, J.; Zappia, K.J.; Lund, K.; Barry, C.P.; Rodriguez, I.V.; Pike, J.W.; et al. Progestins inhibit calcitriol-induced CYP24A1 and synergistically inhibit ovarian cancer cell viability: An opportunity for chemoprevention. *Gynecol. Oncol.* **2016**, *143*, 159–167. [CrossRef]
49. Ji, M.T.; Nie, J.; Nie, X.F.; Hu, W.T.; Pei, H.L.; Wan, J.M.; Wang, A.Q.; Zhou, G.M.; Zhang, Z.L.; Chang, L.; et al. 1alpha,25(OH)2D3 Radiosensitizes Cancer Cells by Activating the NADPH/ROS Pathway. *Front. Pharmacol.* **2020**, *11*, 945. [CrossRef]

50. Hou, Y.F.; Gao, S.H.; Wang, P.; Zhang, H.M.; Liu, L.Z.; Ye, M.X.; Zhou, G.M.; Zhang, Z.L.; Li, B.Y. 1alpha,25(OH)(2)D(3) Suppresses the Migration of Ovarian Cancer SKOV-3 Cells through the Inhibition of Epithelial-Mesenchymal Transition. *Int. J. Mol. Sci.* **2016**, *17*, 1285. [CrossRef]
51. Siegel, R.L.; Miller, K.D.; Fuchs, H.E.; Jemal, A. Cancer Statistics, 2021. *CA Cancer J. Clin.* **2021**, *71*, 7–33. [CrossRef]
52. Kitami, K.; Yoshihara, M.; Tamauchi, S.; Sugiyama, M.; Koya, Y.; Yamakita, Y.; Fujimoto, H.; Iyoshi, S.; Uno, K.; Mogi, K.; et al. Peritoneal Restoration by Repurposing Vitamin D Inhibits Ovarian Cancer Dissemination via Blockade of the TGF-beta1/Thrombospondin-1 Axis. *Matrix Biol.* **2022**, *109*, 70–90. [CrossRef] [PubMed]
53. Abdelbaset-Ismail, A.; Pedziwiatr, D.; Suszynska, E.; Sluczanowska-Glabowska, S.; Schneider, G.; Kakar, S.S.; Ratajczak, M.Z. Vitamin D3 stimulates embryonic stem cells but inhibits migration and growth of ovarian cancer and teratocarcinoma cell lines. *J. Ovarian Res.* **2016**, *9*, 26. [CrossRef] [PubMed]
54. Xue, Y.; Wang, P.; Jiang, F.; Yu, J.; Ding, H.; Zhang, Z.; Pei, H.; Li, B. A Newly Identified lncBCAS1-4_1 Associated With Vitamin D Signaling and EMT in Ovarian Cancer Cells. *Front. Oncol.* **2021**, *11*, 691500. [CrossRef] [PubMed]
55. Fu, Y.; Katsaros, D.; Biglia, N.; Wang, Z.; Pagano, I.; Tius, M.; Tiirikainen, M.; Rosser, C.; Yang, H.; Yu, H. Vitamin D receptor upregulates lncRNA TOPORS-AS1 which inhibits the Wnt/beta-catenin pathway and associates with favorable prognosis of ovarian cancer. *Sci. Rep.* **2021**, *11*, 7484. [CrossRef] [PubMed]
56. Ji, M.; Liu, L.; Hou, Y.; Li, B. 1alpha,25Dihydroxyvitamin D3 restrains stem celllike properties of ovarian cancer cells by enhancing vitamin D receptor and suppressing CD44. *Oncol. Rep.* **2019**, *41*, 3393–3403. [CrossRef] [PubMed]
57. Jiang, F.; Bao, J.; Li, P.; Nicosia, S.V.; Bai, W. Induction of ovarian cancer cell apoptosis by 1,25-dihydroxyvitamin D3 through the down-regulation of telomerase. *J. Biol. Chem.* **2004**, *279*, 53213–53221. [CrossRef] [PubMed]
58. Kasiappan, R.; Shen, Z.; Tse, A.K.; Jinwal, U.; Tang, J.; Lungchukiet, P.; Sun, Y.; Kruk, P.; Nicosia, S.V.; Zhang, X.; et al. 1,25-Dihydroxyvitamin D3 suppresses telomerase expression and human cancer growth through microRNA-498. *J. Biol. Chem.* **2012**, *287*, 41297–41309. [CrossRef]
59. Guo, Y.; Jiang, F.; Yang, W.; Shi, W.; Wan, J.; Li, J.; Pan, J.; Wang, P.; Qiu, J.; Zhang, Z.; et al. Effect of 1alpha,25(OH)2D3-Treated M1 and M2 Macrophages on Cell Proliferation and Migration Ability in Ovarian Cancer. *Nutr. Cancer* **2022**, *74*, 2632–2643. [CrossRef]
60. Li, D.; Wang, X.; Wu, J.L.; Quan, W.Q.; Ma, L.; Yang, F.; Wu, K.Y.; Wan, H.Y. Tumor-produced versican V1 enhances hCAP18/LL-37 expression in macrophages through activation of TLR2 and vitamin D3 signaling to promote ovarian cancer progression in vitro. *PLoS ONE* **2013**, *8*, e56616. [CrossRef]
61. Liu, L.; Hu, Z.; Zhang, H.; Hou, Y.; Zhang, Z.; Zhou, G.; Li, B. Vitamin D postpones the progression of epithelial ovarian cancer induced by 7, 12-dimethylbenz [a] anthracene both in vitro and in vivo. *Onco Targets Ther.* **2016**, *9*, 2365–2375. [CrossRef]
62. Gandhi, T.; Bhatt, H. *Cancer Antigen 125;* StatPearls: Treasure Island, FL, USA, 2022.
63. Srivastava, A.K.; Rizvi, A.; Cui, T.; Han, C.; Banerjee, A.; Naseem, I.; Zheng, Y.; Wani, A.A.; Wang, Q.E. Depleting ovarian cancer stem cells with calcitriol. *Oncotarget* **2018**, *9*, 14481–14491. [CrossRef] [PubMed]
64. Rizvi, A.; Naseem, I. Causing DNA damage and stopping DNA repair—Vitamin D supplementation with Poly(ADP-ribose) polymerase 1 (PARP1) inhibitors may cause selective cell death of cancer cells: A novel therapeutic paradigm utilizing elevated copper levels within the tumour. *Med. Hypotheses* **2020**, *144*, 110278. [CrossRef] [PubMed]
65. Bellio, C.; DiGloria, C.; Foster, R.; James, K.; Konstantinopoulos, P.A.; Growdon, W.B.; Rueda, B.R. PARP Inhibition Induces Enrichment of DNA Repair-Proficient CD133 and CD117 Positive Ovarian Cancer Stem Cells. *Mol. Cancer Res.* **2019**, *17*, 431–445. [CrossRef] [PubMed]

Review

Vitamin D Derivatives in Acute Myeloid Leukemia: The Matter of Selecting the Right Targets

Ewa Marcinkowska

Department of Biotechnology, University of Wroclaw, Joliot-Curie 14a, 50-383 Wroclaw, Poland; ema@cs.uni.wroc.pl; Tel.: +48-71-375-2929

Abstract: Acute myeloid leukemia (AML) is an aggressive and often fatal hematopoietic malignancy. A very attractive way to treat myeloid leukemia, called "differentiation therapy", was proposed when in vitro studies showed that some compounds are capable of inducing differentiation of AML cell lines. One of the differentiation-inducing agents, all-*trans*-retinoic acid (ATRA), which can induce granulocytic differentiation in AML cell lines, has been introduced into clinics to treat patients with acute promyelocytic leukemia (APL) in which a PML-RARA fusion protein is generated by a chromosomal translocation. ATRA has greatly improved the treatment of APL. Since 1,25-dihydroxyvitamin D (1,25D) is capable of inducing monocytic differentiation of leukemic cells, the idea of treating other AMLs with vitamin D analogs was widely accepted. However, early clinical trials in which cancer patients were treated either with 1,25D or with analogs did not lead to conclusive results. Recent results have shown that AML types with certain mutations, such as isocitrate dehydrogenase (IDH) mutations, may be the right targets for differentiation therapy using 1,25D, due to upregulation of vitamin D receptor (VDR) pathway.

Keywords: acute myeloid leukemia; blast; 1,25-dihydroxyvitamin D; analogs; all-*trans*-retinoic acid; differentiation; immunomodulation

Citation: Marcinkowska, E. Vitamin D Derivatives in Acute Myeloid Leukemia: The Matter of Selecting the Right Targets. *Nutrients* **2022**, *14*, 2851. https://doi.org/10.3390/nu14142851

Academic Editor: Carsten Carlberg

Received: 17 June 2022
Accepted: 9 July 2022
Published: 12 July 2022

1. Introduction

Acute myeloid leukemia (AML) is a malignancy of the myeloid blood lineage, characterized by the rapid growth of abnormal cells (blasts) in the bone marrow. The blast cells overgrow bone marrow, preventing normal blood cell production, and expanding to circulation, where they are unable to function properly. Since leukocytes produced in bone marrow belong to the immune system, every AML is accompanied by an immune deficiency resulting in vulnerability to infections. In addition, inability to produce appropriate amounts of red blood cells and platelets results in anemia and bleeding [1].

The primary goal in the treatment of AML is an elimination of leukemic blasts. However, chemotherapy blocks not only the proliferation of blasts, but also the proliferation of immune cells, an essential step in immune cells' activation. Therefore, chemotherapy-induced immunodeficiency adds to leukemia-induced immunodeficiency [2].

AML is a relatively rare disease which constitutes about 1% of all malignancies. It is a disease common in elderly people and very rare in children, with about 25% of cases diagnosed among adults aged 65–74 years and 34% among these aged 75 and older [3]. AML is the most heterogeneous hematologic malignancy with about 200 known underlying mutations [4]. For more than 40 years, all AML patients have been treated using standard intensive chemotherapy, combining anthracycline and cytarabine. For patients who responded with complete remission after intensive chemotherapy, stem cell transplantation was their treatment of choice [3]. However, it should be remembered that most AML patients are elderly and not fit for either intensive chemotherapy or stem cell transplantation. Understanding disease heterogeneity has allowed for the development of lower-intensity and more targeted treatments for elderly patients who are unfit for intensive treatments [3].

Leukemic blasts are inhibited in their differentiation by either genetic abnormalities or by gene-expression anomalies. These cells do not express the proteins important for the function of their normal counterparts. Therefore, finding a method of forced differentiation of leukemic blasts seemed to be a particularly attractive solution for AML patients. Differentiation therapy is based on forced transcription of the genes that are crucial for the function of normal counterparts to leukemic blasts. This concept has been based on the findings concerning normal hematopoiesis, where the eventual cell fate is governed by spatiotemporal fluctuations in transcription factor concentrations, which either cooperate or compete in driving target-gene expression [5]. Some of these transcription factors have critical roles in lineage selection [6], while others govern cell cycle exit and expression of lineage-specific genes [7]. There are several reasons why transcription factors in leukemic blasts do not operate properly: one of them may be epigenetic silencing of the gene, while the others are mutations [8,9]. The general idea of this type of therapy is presented in Figure 1.

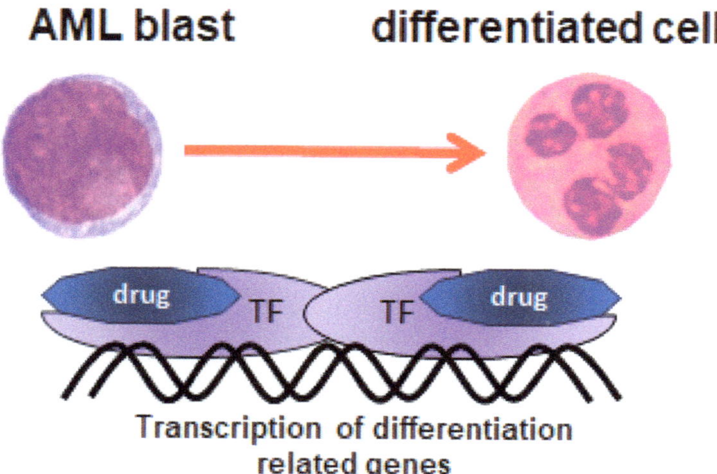

Figure 1. The general idea of differentiation therapy. AML—acute myeloid leukemia; TF—transcription factor.

2. All-*trans*-Retinoic Acid (ATRA)

Acute promyelocytic leukemia (APL) is a subtype of AML characterized by uncontrolled expansion of blasts, which are blocked at the promyelocytic stage of hematopoiesis. Cytogenetically, APL is characterized by a translocation between the long arms of chromosomes 15 and 17 [t(15;17)]. This aberration leads to the fusion between the promyelocytic leukemia gene (*PML*) located on chromosome 15q21, and the retinoic acid receptor α gene (*RARA*) from chromosome 17q21, forming the chimeric oncogene *PML-RARA* [10]. In its first description in 1957, APL was considered to be the most malignant form of AML, accompanied by severe bleeding and very short survival time [11]. Retinoic acid receptor α (RARα) is a nuclear receptor activated by two metabolites of retinoic acid (RA): all-*trans*-RA (ATRA) or 9-*cis*-RA. When dimerized with a retinoid X receptor α (RXRα), it binds to response elements located in the promoters of target genes, activating their transcription. In the absence of the ligand, RARα/RXRα induces chromatin condensation and repression of transcription [12]. Activated RARα/RXRα regulates many genes crucial for myeloid differentiation, for example these encoding transcription factors PU.1 and CCAAT/enhancer-binding proteins α and ε (C/EBPα and C/EBPε) [13–15].

Fusion protein in APL contains the *N*-terminal part of PML protein and the C-terminal part of RARα, and in terms of function it influences transcription. ATRA at physiological concentrations is unable to release complexes of co-repressors from PML-RARα, leading to transcription blockade [16]. It has been noticed, however, that supra-physiological concentrations of ATRA are able to cause the exchange of co-repressors to co-activators, activating the transcription of genes responsible for granulocytic differentiation [17]. Importantly, the blasts lose their immortality following differentiation processes, and start to die by apoptosis [18]. In fact, surprisingly, the very first demonstration that ATRA is capable of inducing granulocytic differentiation was in using HL60 cell line, which is not an APL subtype [19]. However, in clinical situations only patients who have the t(15;17) mutation respond to ATRA treatment, which was reported for the first time in 1988 [20]. Despite experiencing rapid remission when treated with ATRA alone, the patients suffered from relapse within 6 months. Arsenic trioxide (ATO) used in the patients who relapsed after initial treatment with ATRA had significantly improved results [21,22]. The mechanisms of beneficial action of ATO in APL are SUMOylation, ubiquitination, and eventual degradation of the PML part of the fusion protein [23]. Most of the current protocols combine ATRA, ATO, and cytostatics, such as cytarabine or idarubicin. Using these protocols, complete remission (CR) can be achieved in 90–100% of patients, while overall survival (OS) rates can be achieved in 86–97% of patients [24]. This highlights the great success of differentiation therapy, indicating that the proper combinations of drugs with complementing mechanisms of action are needed.

There were many attempts to widen the success of ATRA therapy beyond APL subtypes of AML. There were some clinical trials in which ATRA was added to chemotherapy [25]. Analysis of one trial suggested that the beneficial effects of ATRA were restricted to the subgroup of patients with a mutated nucleophosmin 1 (NPM1) gene, and without fms-like tyrosine kinase 3 (FLT3)-internal tandem duplication (ITD) [26]. Unfortunately, in other trials this beneficial effect was not observed [27,28]. In fact, in some cases ATRA may even worsen the patient's situation, as it was in the case of the patient with t(4;15)(q31;q22) translocation, resulting in the expression of the TMEM154-RASGRF1 fusion protein. This patient was treated with ATRA and died from rapid disease progression, which was related to ATRA-induced activation of RARγ, a RAR isoform responsible for hematopoietic stem cell renewal and proliferation [29].

3. 1,25-Dihydroxyvitamin D_3 (1,25D)

The possibility to use 1,25-dihydroxyvitamin D_3 (1,25D) in differentiation therapy originated from a study published in 1981, where mouse myeloid leukemia cells exposed in culture to 1,25D were induced to differentiate into functional macrophages [30]. This discovery was extended to human HL60 cells soon after [31,32]. The beneficial actions of 1,25D against AML were also presented in mouse models of this disease [33,34].

The idea to use 1,25D against cancers originated from epidemiological studies. These studies indicated an association between an increased risk of developing colorectal cancer and a low level of 25D in the blood [35,36], as well as an increased risk of developing breast cancer and a low blood level of 25D [37,38]. The role of 1,25D in solid cancers has been discussed in a detailed manner in another paper from this Special Issue [39].

1,25D is an active metabolite of vitamin D, which, despite being named a "vitamin", is a steroid hormone [40]. It is produced by the human body from cholesterol and, similarly to other steroid hormones, its effective concentration is strictly regulated by feedback mechanisms. Vitamin D is produced from 7-dehydrocholesterol in human skin when exposed to UV light. Activation of vitamin D is controlled by cytochrome P450 mixed-function oxidases (CYPs) and occurs in two steps: 25-hydroxylation followed by 1α-hydroxylation [41]. The first stage of activation occurs in the liver, where vitamin D undergoes enzymatic hydroxylation by 25-hydroxylase (CYP2R1/CYP27A1), converting it to 25-hydroxyvitamin D (25D). Then, 25D is transported to the kidneys, where it undergoes further hydroxylation at C-1 by 1α-hydroxylase (CY27B1) and results in the formation of the active metabolite, 1,25D. Hydroxylation of 1,25D at carbon atom C-24, catalyzed by 24-hydroxylase of 1,25D (CYP24A1), is the first step of its inactivation. Since the gene encoding CYP24A1 is the most strongly upregulated 1,25D target, it provides negative feedback to the activity of 1,25D and controls the effective concentration of this highly active compound [42]. The metabolism of vitamin D is presented in Figure 2.

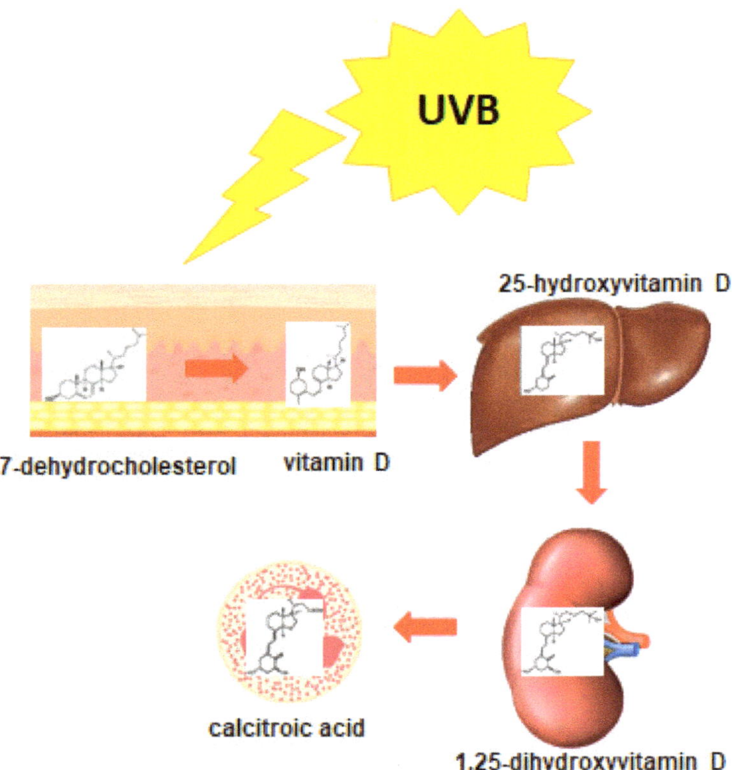

Figure 2. Vitamin D metabolism. Vitamin D is produced in human skin from 7-dehydrocholesterol following exposure to UVB. Then, vitamin D undergoes two hydroxylations: at C-25 in the liver by 25-hydroxylase, and at C-1 in the kidneys by 1α-hydroxylase. Degradation of 1,25-dihydroxyvitamin D (1,25D) into inactive metabolite (calcitroic acid) occurs by hydroxylation at C-24 by 24-hydroxylase in all cells which express vitamin D receptor (VDR).

The major and most well known role of 1,25D is to maintain the calcium phosphate homeostasis of the organism [43], but it is well-documented that 1,25D regulates other vital processes, such as differentiation and proliferation of the cells [40]. The vitamin D receptor (VDR), similarly to RARα, is the nuclear receptor which after binding its ligand translocates to the cell nucleus, where it acts as a ligand-activated transcription factor. VDR, after binding 1,25D, heterodimerizes with RXRα in order to regulate transcription of target genes [44]. There are hundreds of VDR-regulated genes [45], many of them responsible for maintaining calcium phosphate homeostasis [43]; however, there are also many genes involved in immune functions, exemplified by CD14, encoding a macrophage co-receptor for bacterial LPS [46]. The overview of 1,25D/VDR intracellular pathway is presented in Figure 3.

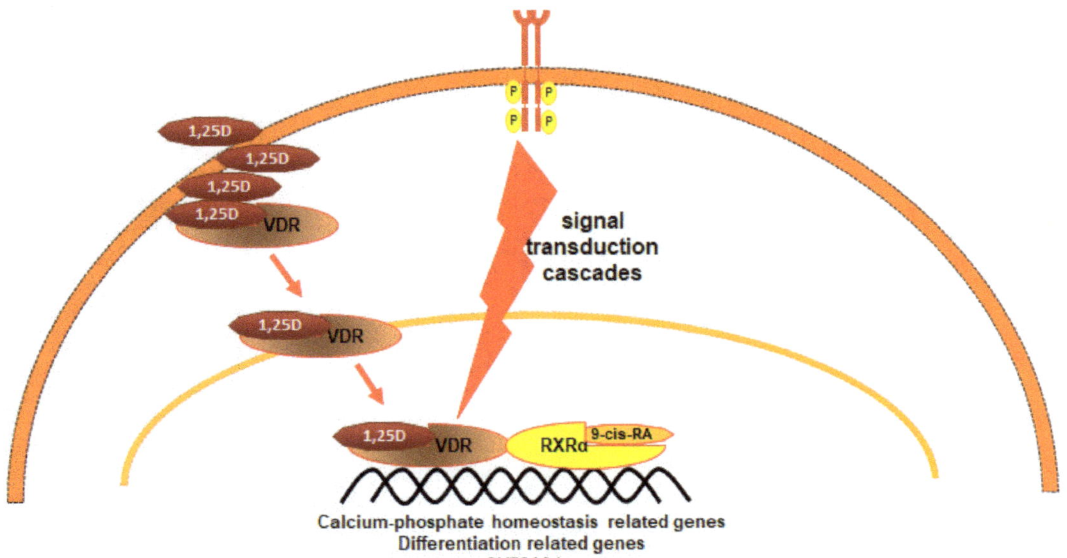

Figure 3. Vitamin D receptor (VDR) pathway. 1,25D translocates through the plasma membrane and binds to its receptor in the cytosol. Ligated VDR is transported to the cell nucleus, where it dimerizes with RXRα. VDR/RXRα complex binds to response elements in the DNA to regulate transcription of target genes. Signal transduction from membrane receptors participates in the activity and stability of VDR [47–49].

Encouraging results of in vitro and murine studies prompted some clinical trials conducted with small groups of patients with myelodysplastic syndrome (MDS) and AML [50,51]. In these trials either 1,25D or its precursor 25D were used, but results were variable and inconclusive. In general, combination treatments resulted in better outcomes than 1,25D alone [52,53]. For example, the combination of 1,25D, AraC, and hydroxyurea resulted in complete or partial responses in 79% of patients with AML [54].

4. Low-Calcemic Analogs of 1,25D

One of the problems with therapeutic uses of 1,25D is its calcemic action and possible consequences of hypercalcemia [55]. In fact, in some of the very few clinical trials in which 1,25D was used against MDS, patients suffered from hypercalcemia [56,57]. The symptoms of hypercalcemia might vary from mild to severe, such as nausea, fatigue, loss of appetite, arrhythmia, kidney failure, calcification of soft tissues, and decalcification of bones [58]. This problem may be overcome by use of low-calcemic analogs which are available from many laboratories [59].

Many analogs of 1,25D have been synthesized with intention to split its activities. The idea was to reduce calcemic actions and retain pro-differentiating activities. Despite the fact that numerous analogs have been available for over 30 years, it is still not clear how the split of these activities is obtained [60]. The most puzzling is the fact that there is only one VDR which mediates calcemic and pro-differentiating actions. It is possible, then, that different analogs activate different intracellular signaling pathways, but it is still not clear how this would be achieved [61].

Analogs of 1,25D have been modified in one or more sites of the structure of the parental compound [59]. Some modifications are minor, but some change the structure substantially [62]. It is noteworthy that not only analogs of 1,25D can be used as agonists of VDR: lithocholic acid (LCA) is a natural ligand, and a very weak agonist of VDR. Modifications of LCA structure can substantially increase the pro-differentiation potency of

LCA, without affecting calcium phosphate homeostasis [63–65]. Unfortunately, the clinical trials using analogs of 1,25D were also far from these for ATRA in APL [66].

5. The Heterogeneity of AML

The most likely source of failure in differentiation therapy using 1,25D and analogs lies in the heterogeneity of AML. There are two systems of AML classifications, the French–American–British (FAB) system from 1976 [67], and the World Health Organization (WHO) system from 2008 [68]. In the FAB system, all AMLs are divided into 8 groups, based predominantly on the cell morphology and cytochemical staining [69]. The later WHO system divided AMLs into 7 groups. This system is much more complicated because it is based on a combination of clinical characteristics, morphology, immunophenotype, cytogenetics, and molecular genetics of the blasts. It takes prognostic factors known to affect the treatment and the outcome of the leukemia into consideration [68,70]. Neither of these classifications is ideal; therefore, there are some attempts to make amendments [3]. APL is an M3 subtype according to FAB, and belongs to group 1 according to WHO (AML with recurrent genetic abnormalities). In addition to variability of driver mutations in AML, there is also intrinsic heterogeneity in each patient resulting from clonal diversification of blasts [71]. The most frequent mutations in AML have been identified and are used to guide treatment and predict outcome. These are *NPM1* mutations, DNA methyltansferase 3A (*DNMT3A*) mutations, *FLT3* mutations, isocitrate dehydrogenase (*IDH*) mutations, ten-eleven translocation 2 (*TET2*) mutations, runt-related transcription factor (*RUNX1*) mutations, CCAAT enhancer binding protein α (*CEBPA*) mutations, additional sex comb-like 1 (*ASXL1*) mutations, mixed lineage leukemia (*MLL*) mutations, protein p53 (*TP53*) mutations, c-Kit mutations, or *PML-RARA* translocation t(15,17)(q22;q12). Out of these examples, only the M3 subtype, characterized by *PML-RARA*, is susceptible to ATRA-based differentiation therapy.

6. AMLs Resistant to 1,25D

The lessons learnt from ATRA therapies prompted studies focused on identification of AML subtypes sensitive and resistant to 1,25D-induced differentiation. In one study, the majority of patient's blasts did not respond to 1,25D or to the analogs with monocytic differentiation [72]. Figure 4 shows that only about 25% of the blasts were responsive. The correlation study performed using blasts isolated from AML patients indicated that blasts carrying FLT3 mutations are resistant to 1,25D and to its analogs [73]. Surprisingly, available cell lines which carry FLT3 mutations, MV-11 and MOLM-13, are responsive in vitro to 1,25D and to analogs [74]. There are some possible explanations for this phenomenon, including that the correlation observed was not due to a causal implication, or that the cell lines grown in vitro for many years had changed their phenotype due to epigenetic changes.

The data from ALM patients indicate that *VDR* expression levels positively correlate with patients' survival. VDR controls the stemness of blast cells and promotes their differentiation [75].

The cell line which was found to be completely resistant to 1,25D-induced cell differentiation is KG1 [76]. This cell line has very low expression of *VDR* gene as compared to other AML cell lines, very low levels of VDR protein, and almost no response of VDR target *CYP24A1* [74]. KG1 cells originated from 8p11 myeloproliferative syndrome, a blood disease which rapidly develops into AML [77]. KG1 cells are characterized by a chromosomal translocation where FGFR1 oncogene partner 2 (*FOP2*)—the fibroblast growth factor receptor 1 (*FGFR1*) fusion gene—encodes a constitutively active fusion protein FOP2–FGFR1. This fusion protein constitutively activates signal transducer and activator of transcription (STAT) 1 and STAT5 [78,79]. Disruption of this fusion gene restored expression of VDR gene, and sensitivity to 1,25D-induced monocytic differentiation [80]. Whether or not a similar situation exists in patients with 8p11 myeloproliferative syndrome remains to be elucidated. The obstacle to study this is that the mutations observed in this syndrome are not routinely tested in patients with AML [81,82].

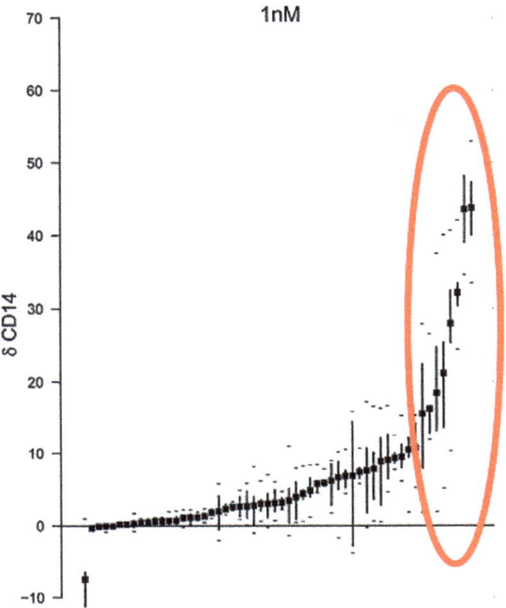

Figure 4. The monocytic differentiation of blasts from AML patients in response to 1 nM 1,25D and 1 nM analogs. The blasts of AML patients were isolated from peripheral blood and exposed to either 1 nM 1,25D or to one of the eight low-calcemic analogs at 1 nM concentration. Mean gain in expression of CD14 cell surface antigen for each patient is presented as a dot (•). Quartiles of response are marked by vertical lines, while minimum and maximum values are marked by dashes (–). Red oval surrounds the data from patients whose blasts were susceptible to 1,25D and to analogs. Adapted from [72].

7. AMLs Sensitive to 1,25D

It seems obvious that in order to benefit from immuno-stimulating activity of 1,25D in patients with AML, it is necessary to define the subtypes of the disease which are sensitive to 1,25D-induced differentiation.

An interesting observation was made about AML cases with IDH mutations. These mutations result in the production of the (R)-2-hydroxyglutarate (2-HG), which causes a hypermethylation, and dysregulates hematopoietic differentiation. One specific mutation in IDH is a R132H substitution. AML blasts with this specific mutation have been shown to have certain transcription factor genes upregulated when compared to the cells without this mutation. *CEBPA* gene and resulting protein C/EBPα were enriched in mutated cells. Interestingly, AML blasts harboring this particular mutation were more responsive to ATRA than blasts with wild-type (wt) IDH. Moreover, a cell-permeable form of 2-HG sensitized wt-IDH1 AML cells to ATRA-induced myeloid differentiation [83]. AML cells with IDH-R132H mutation also have higher levels of VDR and RXRα proteins than the cells with wt-IDH. Consequently, these cells respond better to 1,25D than wt-IDH cells, and even better to the combination of 1,25D and ATRA [84].

In fact, combination therapy using 1,25D and ATRA was postulated long ago, when VDR protein was found to be upregulated in ATRA-treated Kasumi-1 cells [85]. However, the regulation of *VDR* gene by ATRA is quite complex, and depends on the cell context [74]. This is because an abundant and unligated RARα acts as a suppressor of *VDR* transcription, while following ligation with ATRA or with RARα agonists, starts to act as an activator [86]. This shows that patient-tailored combination therapy should be advised.

Another recent observation about the sensitivity of AML cells to 1,25D concerns the cells with overexpression of *FGFRs*. In addition to chromosomal translocations, *FGFR* genes may be affected by other mutations. Gene amplification of *FGFR1* was discovered in squamous cell lung cancers and estrogen-receptor-positive breast cancers, while *FGFR2* in some gastric cancers and in some triple-negative breast cancers [87,88]. There are data that indicate that the *FGFR1* gene is amplified in some cases of AML also [25]. In AML cell lines, overexpression of *FGFR1-3* caused enhanced sensitivity to 1,25D-induced differentiation, due to enhanced expression of *VDR* gene (Figure 5) [89]. Whether a similar regulation exists in the AML blasts of patients remains to be studied.

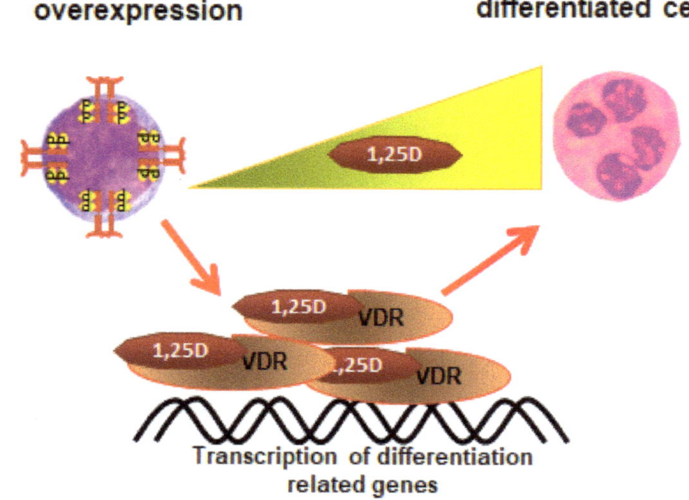

Figure 5. Differentiation of AML blasts with *FGFR 1-3* overexpression. The AML blasts with overexpression of *FGFR 1-3* produce more VDR protein than wild-type cells, and therefore are more susceptible to 1,25D-induced differentiation.

The *FGFR* family contains five genes, out of which four encode transmembrane tyrosine kinase receptors that exist in multiple splicing variants. Binding of the ligand to FGFRs results in a dimerization of these receptors and transphosphorylation of their tyrosine kinase domains [90]. As a result, FGFRs activate different signaling cascades including mitogen-activated protein kinase (MAPK), phosphatidylinositol 3-kinase (PI3K), and phospholipase Cγ (PLCγ) [91]. It has been shown in the past that activating some of the MAPK pathways, namely JNK and Erk-1,2 pathways, enhances 1,25D-induced cell differentiation [92,93]. In contrast, constitutively active FGFRs, such as in FOP2–FGFR1 fusion kinase, cause downstream activation of signal transducer and activator of transcription (STAT) pathways [94]. Our unpublished data indicate that activation of STAT1 is responsible for low *VDR* expression.

8. Conclusions

AML is a disease of the elderly, and the proportion of older people is increasing steadily in modern societies. The current estimate of the probability of developing cancer is one in two for people born after 1960 [95], and despite the fact that AML is a relatively rare malignancy, its numbers will grow in the near future. For more than 40 years, all AML patients have been treated using standard intensive chemotherapy, but intensive chemotherapy cannot be used for elderly people. When chemotherapy is given to elderly patients, they are often unable to tolerate it. Consequently, there is a need for gentler drugs for use alone or in a combined treatment. Differentiation therapy provides a much milder approach to treating malignancy, and should be advanced. However, the great success of ATRA-based differentiation therapy against APL has shown that this type of therapy must be targeted to molecular lesions susceptible to differentiation-inducing drugs. Recent data indicate that similarly to ATRA, 1,25D, or its analogs should be applied only to these patients who are likely to respond. Recent advances in next-generation sequencing, transcriptome analysis, immunophenotyping, and multiparameter flow cytometry will provide the means to delivering patient-tailored and tolerable differentiation therapies in the near future.

Funding: This research described in this paper was funded by the National Science Centre in Poland (grant OPUS No 2016/23/B/NZ5/00065).

Institutional Review Board Statement: Not applicable.

Informed Consent Statement: Not applicable.

Conflicts of Interest: The author declares no conflict of interest.

References

1. Pelcovits, A.; Niroula, R. Acute Myeloid Leukemia: A Review. *Rhode Isl. Med. J.* **2013**, *103*, 38–40.
2. Weycker, D.; Barron, R.; Kartashov, A.; Legg, J.; Lyman, G. Incidence, treatment, and consequences of chemotherapy-induced febrile neutropenia in the inpatient and outpatient settings. *J. Oncol. Pharm. Pract.* **2014**, *20*, 190–198. [CrossRef] [PubMed]
3. Newell, L.F.; Cook, R.J. Advances in acute myeloid leukemia. *BMJ* **2021**, *375*, n2026. [CrossRef] [PubMed]
4. Padmakumar, D.; Chandraprabha, V.R.; Gopinath, P.; Vimala Devi, A.R.T.; Anitha, G.R.J.; Sreelatha, M.M.; Padmakumar, A.; Sreedharan, H. A concise review on the molecular genetics of acute myeloid leukemia. *Leuk. Res.* **2021**, *111*, 16. [CrossRef]
5. Friedman, A. Transcriptional control of granulocyte and monocyte development. *Oncogene* **2007**, *26*, 6816–6828. [CrossRef]
6. Brown, G.; Hughes, P.; Michell, R.; Rolink, A.; Ceredig, R. The sequential determination model of hematopoiesis. *Trends Immunol.* **2007**, *28*, 442–448. [CrossRef]
7. Theilgaard-Mönch, K.; Pundhir, S.; Reckzeh, K.; Su, J.; Tapia, M.; Furtwängler, B.; Jendholm, J.; Jakobsen, J.S.; Hasemann, M.S.; Knudsen, K.J.; et al. Transcription factor-driven coordination of cell cycle exit and lineage-specification in vivo during granulocytic differentiation: In memoriam Professor Niels Borregaard. *Nat. Commun.* **2022**, *13*, 022–31332. [CrossRef]
8. Heidari, N.; Abroun, S.; Bertacchini, J.; Vosoughi, T.; Rahim, F.; Saki, N. Significance of Inactivated Genes in Leukemia: Pathogenesis and Prognosis. *Cell J.* **2017**, *19*, 9–26.
9. Nie, Y.; Su, L.; Li, W.; Gao, S. Novel insights of acute myeloid leukemia with CEBPA deregulation: Heterogeneity dissection and re-stratification. *Crit. Rev. Oncol. Hematol.* **2021**, *163*, 1. [CrossRef]
10. Rowley, J.; Golomb, H.; Dougherty, C. 15/17 translocation, a consistent chromosomal change in acute promyelocytic leukaemia. *Lancet* **1977**, *1*, 549–550. [CrossRef]
11. Hillestad, L. Acute promyelocytic leukemia. *Acta Med. Scand.* **1957**, *159*, 189–194. [CrossRef]
12. Mark, M.; Chambon, P. Functions of RARs and RXRs in vivo: Genetic dissection of the retinoid signaling pathway. *Pure Appl. Chem.* **2003**, *75*, 1709–1732. [CrossRef]
13. Iwasaki, H.; Somoza, C.; Shigematsu, H.; Duprez, E.; Iwasaki-Arai, J.; Mizuno, S.; Arinobu, Y.; Geary, K.; Zhang, P.; Dayaram, T.; et al. Distinctive and indispensable roles of PU.1 in maintenance of hematopoietic stem cells and their differentiation. *Blood* **2005**, *106*, 1590–1600. [CrossRef] [PubMed]
14. Duprez, E.; Wagner, K.; Koch, H.; Tenen, D. C/EBPbeta: A major PML-RARA-responsive gene in retinoic acid-induced differentiation of APL cells. *EMBO J.* **2003**, *22*, 5806–5816. [CrossRef] [PubMed]
15. Morosetti, R.; Park, D.; Chumakov, A.; Grillier, I.; Shiohara, M.; Gombart, A.; Nakamaki, T.; Weinberg, K.; Koeffler, H. A novel, myeloid transcription factor, C/EBPepsilon, is upregulated during granulocytic, but not monocytic, differentiation. *Blood* **1997**, *90*, 2591–2600. [CrossRef]
16. de Thé, H.; Lavau, C.; Marchio, A.; Chomienne, C.; Degos, L.; Dejean, A. The PML-RAR alpha fusion mRNA generated by the t(15;17) translocation in acute promyelocytic leukemia encodes a functionally altered RAR. *Cell* **1991**, *66*, 675–684. [CrossRef]
17. De Braekeleer, E.; Douet-Guilbert, N.; De Braekeleer, M. RARA fusion genes in acute promyelocytic leukemia: A review. *Expert Rev. Hematol.* **2014**, *7*, 347–357. [CrossRef]
18. Gianni, M.; Ponzanelli, I.; Mologni, L.; Reichert, U.; Rambaldi, A.; Terao, M.; Garattini, E. Retinoid-dependent growth inhibition, differentiation and apoptosis in acute promyelocytic leukemia cells. Expression and activation of caspases. *Cell Death Differ.* **2000**, *7*, 447–460. [CrossRef]
19. Breitman, T.; Selonick, S.; Collins, S. Induction of differentiation of the human promyelocytic leukemia cell line (HL-60) by retinoic acid. *Proc. Natl. Acad. Sci. USA* **1980**, *77*, 2936–2940. [CrossRef]

20. Huang, M.; Ye, Y.; Chen, S.; Chai, J.; Lu, J.; Zhoa, L.; Gu, L.; Wang, Z. Use of all-trans retinoic acid in the treatment of acute promyelocytic leukemia. *Blood* **1988**, *72*, 567–572. [CrossRef]
21. Shen, Y.; Shen, Z.; Yan, H.; Chen, J.; Zeng, X.; Li, J.; Li, X.; Wu, W.; Xiong, S.; Zhao, W.; et al. Studies on the clinical efficacy and pharmacokinetics of low-dose arsenic trioxide in the treatment of relapsed acute promyelocytic leukemia: A comparison with conventional dosage. *Leukemia* **2001**, *15*, 735–741. [CrossRef] [PubMed]
22. Shen, Z.; Shi, Z.; Fang, J.; Gu, B.; Li, J.; Zhu, Y.; Shi, J.; Zheng, P.; Yan, H.; Liu, Y.; et al. All-trans retinoic acid/As_2O_3 combination yields a high quality remission and survival in newly diagnosed acute promyelocytic leukemia. *Proc. Natl. Acad. Sci. USA* **2004**, *101*, 5328–5335. [CrossRef] [PubMed]
23. de Thé, H.; Le Bras, M.; Lallemand-Breitenbach, V. The cell biology of disease: Acute promyelocytic leukemia, arsenic, and PML bodies. *J. Cell Biol.* **2012**, *198*, 11–21. [CrossRef]
24. McCulloch, D.; Brown, C.; Iland, H. Retinoic acid and arsenic trioxide in the treatment of acute promyelocytic leukemia: Current perspectives. *Onco Targets Ther.* **2017**, *10*, 1585–1601. [CrossRef] [PubMed]
25. Wu, Q.; Bhole, A.; Qin, H.; Karp, J.; Malek, S.; Cowell, J.; Ren, M. SCLLTargeting FGFR1 to suppress leukemogenesis in syndromic and de novo AML in murine models. *Oncotarget* **2016**, *7*, 49733–49742. [CrossRef]
26. Schlenk, R.; Dohner, K.; Kneba, M.; Götze, K.; Hartmann, F.; del Valle, F.; Kirchen, H.; Koller, E.; Fischer, J.T.; Bullinger, L.; et al. Gene mutations and response to treatment with all-trans retinoic acid in elderly patients with acute myeloid leukemia. Results from the AMLSG Trial AML HD98B. *Haematologica* **2009**, *94*, 54–60. [CrossRef] [PubMed]
27. Burnett, A.K.; Hills, R.K.; Green, C.; Jenkinson, S.; Koo, K.; Patel, Y.; Guy, C.; Gilkes, A.; Milligan, D.W.; Goldstone, A.H.; et al. The impact on outcome of the addition of all-trans retinoic acid to intensive chemotherapy in younger patients with nonacute promyelocytic acute myeloid leukemia: Overall results and results in genotypic subgroups defined by mutations in NPM1, FLT3, and CEBPA. *Blood* **2010**, *115*, 948–956.
28. Schlenk, R.F.; Lübbert, M.; Benner, A.; Lamparter, A.; Krauter, J.; Herr, W.; Martin, H.; Salih, H.R.; Kündgen, A.; Horst, H.A.; et al. All-trans retinoic acid as adjunct to intensive treatment in younger adult patients with acute myeloid leukemia: Results of the randomized AMLSG 07-04 study. *Ann. Hematol.* **2016**, *95*, 1931–1942. [CrossRef]
29. Watts, J.; Perez, A.; Pereira, L.; Fan, Y.-S.; Brown, G.; Vega, F.; Petrie, K.; Swords, R.; Zelent, A. A case of AML characterized by a novel t(4;15)(q31;q22) translocation that confers a growth-stimulatory response to retinoid-based therapy. *Int. J. Mol. Sci.* **2017**, *18*, 1492. [CrossRef]
30. Abe, E.; Miamura, C.; Sakagami, H.; Takeda, M.; Konno, K.; Yamazaki, T.; Yoshiki, S.; Suda, T. Differentiation of mouse myeloid leukemia cells induced by 1-alpha,25-dihydroxyvitamin D_3. *Proc. Natl. Acad. Sci. USA* **1981**, *78*, 4990–4994. [CrossRef]
31. Studzinski, G.; Bhandal, A.; Brelvi, Z. Cell cycle sensitivity of HL-60 cells to the differentiation-inducing effects of 1-alpha,25-dihydroxyvitamin D_3. *Cancer Res.* **1985**, *45*, 3898–3905.
32. Studzinski, G.P.; Bhandal, A.K.; Brelvi, Z.S. A system for monocytic differentiation of leukemic cells HL60 by a short exposure to 1,25-dihydroxycholecalciferol. *Proc. Soc. Exp. Biol. Med.* **1985**, *179*, 288–295. [CrossRef] [PubMed]
33. Sharabani, H.; Izumchenko, E.; Wang, Q.; Kreinin, R.; Steiner, M.; Barvish, Z.; Kafka, M.; Sharoni, Y.; Levy, J.; Uskokovic, M.; et al. Cooperative antitumor effects of vitamin D_3 derivatives and rosemary preparations in a mouse model of myeloid leukemia. *Int. J. Cancer* **2006**, *118*, 3012–3021. [CrossRef] [PubMed]
34. Nachliely, M.; Sharony, E.; Bolla, N.; Kutner, A.; Danilenko, M. Prodifferentiation Activity of Novel Vitamin D_2 Analogs PRI-1916 and PRI-1917 and Their Combinations with a Plant Polyphenol in Acute Myeloid Leukemia Cells. *Int. J. Mol. Sci.* **2016**, *17*, 1068. [CrossRef] [PubMed]
35. Chandler, P.; Buring, J.; Manson, J.; Giovannucci, E.; Moorthy, M.; Zhang, S.; Lee, I.; Lin, J. Circulating Vitamin D Levels and Risk of Colorectal Cancer in Women. *Cancer Prev. Res.* **2015**, *8*, 675–682. [CrossRef]
36. Mohr, S.; Gorham, E.; Kim, J.; Hofflich, H.; Cuomo, R.; Garland, C. Could vitamin D sufficiency improve the survival of colorectal cancer patients? *J. Steroid Biochem. Mol. Biol.* **2015**, *148*, 239–244. [CrossRef]
37. Mohr, S.; Gorham, E.; Kim, J.; Hofflich, H.; Garland, C. Meta-analysis of vitamin D sufficiency for improving survival of patients with breast cancer. *Anticancer Res.* **2014**, *34*, 1163–1166.
38. Maalmi, H.; Ordóñez-Mena, J.; Schöttker, B.; Brenner, H. Serum 25-hydroxyvitamin D levels and survival in colorectal and breast cancer patients: Systematic review and meta-analysis of prospective cohort studies. *Eur. J. Cancer* **2014**, *50*, 1510–1521. [CrossRef]
39. Muñoz, A.; Grant, W.B. Vitamin D and Cancer: An Historical Overview of the Epidemiology and Mechanisms. *Nutrients* **2022**, *14*, 1448. [CrossRef]
40. Carlberg, C. The physiology of vitamin D-far more than calcium and bone. *Front. Physiol.* **2014**, *5*, 335. [CrossRef]
41. Prosser, D.; Jones, G. Enzymes involved in the activation and inactivation of vitamin D. *Trends Biochem. Sci.* **2004**, *29*, 664–673. [CrossRef]
42. Vaisanen, S.; Dunlop, T.; Sinkkonen, L.; Frank, C.; Carlberg, C. Spatio-temporal activation of chromatin on the human CYP24 gene promoter in the presence of 1alpha,25-dihydroxyvitamin D_3. *J. Mol. Biol.* **2005**, *350*, 65–77. [CrossRef] [PubMed]
43. Holick, M. Vitamin D and bone health. *J. Nutr.* **1996**, *126*, 1159S–1164S. [CrossRef] [PubMed]
44. Aranda, A.; Pascual, A. Nuclear hormone receptors and gene expression. *Physiol. Rev.* **2001**, *81*, 1269–1304. [CrossRef] [PubMed]
45. Pike, J.; Meyer, M. Fundamentals of vitamin D hormone-regulated gene expression. *J. Steroid Biochem. Mol. Biol.* **2014**, *144 Pt A*, 5–11. [CrossRef]

46. Carlberg, C.; Seuter, S.; de Mello, V.; Schwab, U.; Voutilainen, S.; Pulkki, K.; Nurmi, T.; Virtanen, J.; Tuomainen, T.; Uusitupa, M. Primary vitamin D target genes allow a categorization of possible benefits of vitamin D_3 supplementation. *PLoS ONE* **2013**, *8*, e71042. [CrossRef]
47. Gocek, E.; Kielbinski, M.; Marcinkowska, E. Activation of intracellular signaling pathways is necessary for an increase in VDR expression and its nuclear translocation. *FEBS Lett.* **2007**, *581*, 1751–1757. [CrossRef]
48. Hsieh, J.; Dang, H.; Galligan, M.; Whitfield, G.; Haussler, C.; Jurutka, P.; Haussler, M. Phosphorylation of human vitamin D receptor serine-182 by PKA suppresses 1,25$(OH)_2D_3$-dependent transactivation. *Biochem. Biophys. Res. Commun.* **2004**, *324*, 801–809. [CrossRef]
49. Hsieh, J.; Jurutka, P.; Galligan, M.; Terpening, C.; Haussler, C.; Samuels, D.; Shimizu, Y.; Shimizu, N.; Haussler, M. Human vitamin D receptor is selectively phosphorylated by protein kinase C on serine 51, a residue crucial to its trans-activation function. *Proc. Natl. Acad. Sci. USA* **1991**, *88*, 9315–9319. [CrossRef]
50. Irino, S.; Taoka, T. Treatment of myelodysplastic syndrome and acute myelogenous leukemia with vitamin D_3 [1 alpha$(OH)D_3$]. *Gan Kagaku Ryoho Cancer Chemother.* **1988**, *15*, 1183–1190.
51. Nakayama, S.; Ishikawa, T.; Yabe, H.; Nagai, K.; Kasakura, S.; Uchino, H. Successful treatment of a patient with acute myeloid leukemia with 1 alpha$(OH)D_3$. *Nihon Ketsueki Gakkai Zasshi* **1988**, *51*, 1026–1030.
52. Hellström, E.; Robèrt, K.; Gahrton, G.; Mellstedt, H.; Lindemalm, C.; Einhorn, S.; Björkholm, M.; Grimfors, G.; Udén, A.; Samuelsson, J. Therapeutic effects of low-dose cytosine arabinoside, alpha-interferon, 1 alpha-hydroxyvitamin D_3 and retinoic acid in acute leukemia and myelodysplastic syndromes. *Eur. J. Haematol.* **1988**, *40*, 449–459. [CrossRef] [PubMed]
53. Hellström, E.; Robèrt, K.; Samuelsson, J.; Lindemalm, C.; Grimfors, G.; Kimby, E.; Oberg, G.; Winqvist, I.; Billström, R.; Carneskog, J. Treatment of myelodysplastic syndromes with retinoic acid and 1 alpha-hydroxy-vitamin D_3 in combination with low-dose ara-C is not superior to ara-C alone. Results from a randomized study. The Scandinavian Myelodysplasia Group (SMG). *Eur. J. Haematol.* **1990**, *45*, 255–261. [CrossRef]
54. Ferrero, D.; Campa, E.; Dellacasa, C.; Campana, S.; Foli, C.; Boccadoro, M. Differentiating agents + low-dose chemotherapy in the management of old/poor prognosis patients with acute myeloid leukemia or myelodysplastic syndrome. *Haematologica* **2004**, *89*, 619–620. [PubMed]
55. Donovan, P.J.; Sundac, L.; Pretorius, C.J.; d'Emden, M.C.; McLeod, D.S.A. Calcitriol-Mediated Hypercalcemia: Causes and Course in 101 Patients. *J. Clin. Endocrinol. Metab.* **2013**, *98*, 4023–4029. [CrossRef] [PubMed]
56. Motomura, S.; Kanamori, H.; Maruta, A.; Kodama, F.; Ohkubo, T. The effect of 1-hydroxyvitamin D_3 for prolongation of leukemic transformation-free survival in myelodysplastic syndromes. *Am. J. Hematol.* **1991**, *38*, 67–68. [CrossRef]
57. Mellibovsky, L.; Díez, A.; Pérez-Vila, E.; Serrano, S.; Nacher, M.; Aubía, J.; Supervía, A.; Recker, R. Vitamin D treatment in myelodysplastic syndromes. *Br. J. Haematol.* **1998**, *100*, 516–520. [CrossRef]
58. Hathcock, J.N.; Shao, A.; Vieth, R.; Heaney, R. Risk assessment for vitamin D. *Am. J. Clin. Nutr.* **2007**, *85*, 6–18. [CrossRef]
59. Nadkarni, S.; Chodynski, M.; Corcoran, A.; Marcinkowska, E.; Brown, G.; Kutner, A. Double point modified analogs of vitamin D as potent activators of vitamin D receptor. *Curr. Pharm. Des.* **2015**, *21*, 1741–1763. [CrossRef]
60. Norman, A.W.; Zhou, J.Y.; Henry, H.L.; Uskokovic, M.R.; Koeffler, H.P. Structure-function studies on analogues of 1 alpha,25-dihydroxyvitamin D_3: Differential effects on leukemic cell growth, differentiation, and intestinal calcium absorption. *Cancer Res.* **1990**, *50*, 6857–6864.
61. Zmijewski, M.A.; Carlberg, C. Vitamin D receptor(s): In the nucleus but also at membranes? *Exp. Dermatol.* **2020**, *29*, 876–884. [CrossRef] [PubMed]
62. Maestro, M.A.; Molnár, F.; Carlberg, C. Vitamin D and Its Synthetic Analogs. *J. Med. Chem.* **2019**, *62*, 6854–6875. [CrossRef] [PubMed]
63. Gaikwad, S.; González, C.M.; Vilariño, D.; Lasanta, G.; Villaverde, C.; Mouriño, A.; Verlinden, L.; Verstuyf, A.; Peluso-Iltis, C.; Rochel, N.; et al. Lithocholic acid-based design of noncalcemic vitamin D receptor agonists. *Bioorg. Chem.* **2021**, *111*, 30. [CrossRef] [PubMed]
64. González, C.M.; Gaikwad, S.; Lasanta, G.; Loureiro, J.; Nilsson, N.; Peluso-Iltis, C.; Rochel, N.; Mouriño, A. Design, synthesis and evaluation of side-chain hydroxylated derivatives of lithocholic acid as potent agonists of the vitamin D receptor (VDR). *Bioorg. Chem.* **2021**, *115*, 22. [CrossRef] [PubMed]
65. Sasaki, H.; Masuno, H.; Kawasaki, H.; Yoshihara, A.; Numoto, N.; Ito, N.; Ishida, H.; Yamamoto, K.; Hirata, N.; Kanda, Y.; et al. Lithocholic Acid Derivatives as Potent Vitamin D Receptor Agonists. *J. Med. Chem.* **2020**, *64*, 516–526. [CrossRef] [PubMed]
66. Harrison, J.; Bershadskiy, A. Clinical experience using vitamin D and analogs in the treatment of myelodysplasia and acute myeloid leukemia: A review of the literature. *Leuk. Res. Treat.* **2012**, *125814*, 8. [CrossRef] [PubMed]
67. Bennett, J.; Catovsky, D.; Daniel, M.; Flandrin, G.; Galton, D.; Gralnick, H.; Sultan, C. Proposals for the classification of the acute leukaemias. French-American-British (FAB) co-operative group. *Br. J. Haematol.* **1976**, *33*, 451–458. [CrossRef]
68. Vardiman, J.W.; Thiele, J.; Arber, D.A.; Brunning, R.D.; Borowitz, M.J.; Porwit, A.; Harris, N.L.; Le Beau, M.M.; Hellström-Lindberg, E.; Tefferi, A.; et al. The 2008 revision of the World Health Organization (WHO) classification of myeloid neoplasms and acute leukemia: Rationale and important changes. *Blood* **2009**, *114*, 937–951. [CrossRef]
69. Gralnick, H.R.; Galton, D.A.G.; Catovsky, D.; Sultan, C.; Bennett, J.M. Classification of Acute Leukemia. *Ann. Intern. Med.* **1977**, *87*, 740–753. [CrossRef]

70. Vardiman, J.; Harris, N.; Brunning, R. The World Health Organization (WHO) classification of the myeloid neoplasms. *Blood* **2002**, *100*, 2292–2302. [CrossRef]
71. Li, S.; Mason, C.E.; Melnick, A. Genetic and epigenetic heterogeneity in acute myeloid leukemia. *Curr. Opin. Genet. Dev.* **2016**, *36*, 100–106. [CrossRef] [PubMed]
72. Baurska, H.; Kiełbiński, M.; Biecek, P.; Haus, O.; Jaźwiec, B.; Kutner, A.; Marcinkowska, E. Monocytic differentiation induced by side-chain modified analogs of vitamin D in ex vivo cells from patients with acute myeloid leukemia. *Leuk. Res.* **2014**, *38*, 638–647. [CrossRef] [PubMed]
73. Gocek, E.; Kielbinski, M.; Baurska, H.; Haus, O.; Kutner, A.; Marcinkowska, E. Different susceptibilities to 1,25-dihydroxyvitamin D_3-induced differentiation of AML cells carrying various mutations. *Leuk. Res.* **2010**, *34*, 649–657. [CrossRef] [PubMed]
74. Gocek, E.; Marchwicka, A.; Baurska, H.; Chrobak, A.; Marcinkowska, E. Opposite regulation of vitamin D receptor by ATRA in AML cells susceptible and resistant to vitamin D-induced differentiation. *J. Steroid Biochem. Mol. Biol.* **2012**, *132*, 220–226. [CrossRef]
75. Paubelle, E.; Zylbersztejn, F.; Maciel, T.; Carvalho, C.; Mupo, A.; Cheok, M.; Lieben, L.; Sujobert, P.; Decroocq, J.; Yokoyama, A.; et al. Vitamin D Receptor Controls Cell Stemness in Acute Myeloid Leukemia and in Normal Bone Marrow. *Cell Rep.* **2020**, *30*, 739–754. [CrossRef] [PubMed]
76. Munker, R.; Norman, A.; Koeffler, H. Vitamin D compounds. Effect on clonal proliferation and differentiation of human myeloid cells. *J. Clin. Investig.* **1986**, *78*, 424–430. [CrossRef]
77. Sohal, J.; Chase, A.; Mould, S.; Corcoran, M.; Oscier, D.; Iqbal, S.; Parker, S.; Welborn, J.; Harris, R.; Martinelli, G.; et al. Identification of four new translocations involving FGFR1 in myeloid disorders. *Genes Chromosomes Cancer* **2001**, *32*, 155–163. [CrossRef]
78. Gu, T.; Goss, V.; Reeves, C.; Popova, L.; Nardone, J.; Macneill, J.; Walters, D.; Wang, Y.; Rush, J.; Comb, M.; et al. Phosphotyrosine profiling identifies the KG-1 cell line as a model for the study of FGFR1 fusions in acute myeloid leukemia. *Blood* **2006**, *108*, 4202–4204. [CrossRef]
79. Jin, Y.; Zhen, Y.; Haugsten, E.; Wiedlocha, A. The driver of malignancy in KG-1a leukemic cells, FGFR1OP2–FGFR1, encodes an HSP90 addicted oncoprotein. *Cell. Signal.* **2011**, *23*, 1758–1766. [CrossRef]
80. Marchwicka, A.; Corcoran, A.; Berkowska, K.; Marcinkowska, E. Restored expression of vitamin D receptor and sensitivity to 1,25-dihydroxyvitamin D_3 in response to disrupted fusion FOP2-FGFR1 gene in acute myeloid leukemia cells. *Cell Biosci.* **2016**, *6*, 7. [CrossRef]
81. Jackson, C.; Medeiros, L.; Miranda, R. 8p11 myeloproliferative syndrome: A review. *Hum. Pathol.* **2010**, *41*, 461–476. [CrossRef] [PubMed]
82. Liu, J.J.; Meng, L. 8p11 Myeloproliferative syndrome with t(8;22)(p11;q11): A case report. *Exp. Ther. Med.* **2018**, *16*, 1449–1453. [CrossRef] [PubMed]
83. Boutzen, H.; Saland, E.; Larrue, C.; de Toni, F.; Gales, L.; Castelli, F.A.; Cathebas, M.; Zaghdoudi, S.; Stuani, L.; Kaoma, T.; et al. Isocitrate dehydrogenase 1 mutations prime the all-trans retinoic acid myeloid differentiation pathway in acute myeloid leukemia. *J. Exp. Med.* **2016**, *213*, 483–497. [CrossRef]
84. Sabatier, M.; Boet, E.; Zaghdoudi, S.; Guiraud, N.; Hucteau, A.; Polley, N.; Cognet, G.; Saland, E.; Lauture, L.; Farge, T.; et al. Activation of Vitamin D Receptor Pathway Enhances Differentiating Capacity in Acute Myeloid Leukemia with Isocitrate Dehydrogenase Mutations. *Cancers* **2021**, *13*, 5243. [CrossRef]
85. Manfredini, R.; Trevisan, F.; Grande, A.; Tagliafico, E.; Montanari, M.; Lemoli, R.; Visani, G.; Tura, S.; Ferrari, S.; Ferrari, S. Induction of a functional vitamin D receptor in all-trans-retinoic acid-induced monocytic differentiation of M2-type leukemic blast cells. *Cancer Res.* **1999**, *59*, 3803–3811.
86. Marchwicka, A.; Cebrat, M.; Łaszkiewicz, A.; Śnieżewski, Ł.; Brown, G.; Marcinkowska, E. Regulation of vitamin D receptor expression by retinoic acid receptor alpha in acute myeloid leukemia cells. *J. Steroid Biochem. Mol. Biol.* **2016**, *159*, 121–130. [CrossRef]
87. Katoh, M.; Nakagama, H. FGF receptors: Cancer biology and therapeutics. *Med. Res. Rev.* **2014**, *34*, 280–300. [CrossRef] [PubMed]
88. Katoh, M. FGFR inhibitors: Effects on cancer cells, tumor microenvironment and whole-body homeostasis (Review). *Int. J. Mol. Med.* **2016**, *38*, 3–15. [CrossRef]
89. Marchwicka, A.; Jakuszak, A.; Grembowska, A.; Kumari, P.; Marcinkowska, E. The Influence of Overexpressed Fibroblast Growth Factor Receptors Towards Vitamin D Receptor Expression and Activity. *Preprints* **2021**, *2021100304*. [CrossRef]
90. Xie, Y.; Su, N.; Yang, J.; Tan, Q.; Huang, S.; Jin, M.; Ni, Z.; Zhang, B.; Zhang, D.; Luo, F.; et al. FGF/FGFR signaling in health and disease. *Signal. Transduct. Target. Ther.* **2020**, *5*, 181.
91. Eswarakumar, V.P.; Lax, I.; Schlessinger, J. Cellular signaling by fibroblast growth factor receptors. *Cytokine Growth Factor Rev.* **2005**, *16*, 139–149. [CrossRef] [PubMed]
92. Wang, X.; Rao, J.; Studzinski, G.P. Inhibition of p38 MAP kinase activity up-regulates multiple MAP kinase pathways and potentiates 1,25-dihydroxyvitamin D_3-induced differentiation of human leukemia HL60 cells. *Exp. Cell Res.* **2000**, *258*, 425–437. [CrossRef] [PubMed]
93. Wang, X.; Studzinski, G. Inhibition of p38 MAP kinase potentiates the JNK/SAPK pathway and AP-1 activity in monocytic but not in macrophage or granulocytic differentiation of HL60 cells. *J. Cell Biochem.* **2001**, *82*, 68–77. [CrossRef] [PubMed]

94. Hart, K.; Robertson, S.; Kanemitsu, M.; Meyer, A.; Tynan, J.; Donoghue, D. Transformation and Stat activation by derivatives of FGFR1, FGFR3, and FGFR4. *Oncogene* **2000**, *19*, 3309–3320. [CrossRef]
95. Sasieni, P.D.; Shelton, J.; Ormiston-Smith, N.; Thomson, C.S.; Silcocks, P.B. What is the lifetime risk of developing cancer? The effect of adjusting for multiple primaries. *Br. J. Cancer* **2011**, *105*, 460–465. [CrossRef]

nutrients

MDPI

Review

Vitamin D and the Central Nervous System: Causative and Preventative Mechanisms in Brain Disorders

Xiaoying Cui [1,2] and Darryl W. Eyles [1,2,*]

1 Queensland Centre for Mental Health Research, The Park Centre for Mental Health, Wacol Q4076, Australia
2 Queensland Brain Institute, University of Queensland, St Lucia Q4076, Australia
* Correspondence: eyles@uq.edu.au

Abstract: Twenty of the last one hundred years of vitamin D research have involved investigations of the brain as a target organ for this hormone. Our group was one of the first to investigate brain outcomes resulting from primarily restricting dietary vitamin D during brain development. With the advent of new molecular and neurochemical techniques in neuroscience, there has been increasing interest in the potential neuroprotective actions of vitamin D in response to a variety of adverse exposures and how this hormone could affect brain development and function. Rather than provide an exhaustive summary of this data and a listing of neurological or psychiatric conditions that vitamin D deficiency has been associated with, here, we provide an update on the actions of this vitamin in the brain and cellular processes vitamin D may be targeting in psychiatry and neurology.

Keywords: brain; development; vitamin D deficiency; neuroprotection; disease mechanisms

Citation: Cui, X.; Eyles, D.W. Vitamin D and the Central Nervous System: Causative and Preventative Mechanisms in Brain Disorders. *Nutrients* **2022**, *14*, 4353. https://doi.org/10.3390/nu14204353

Academic Editor: Carsten Carlberg

Received: 5 August 2022
Accepted: 13 September 2022
Published: 17 October 2022

1. Introduction

The vitamin D receptor is part of the nuclear receptor super family containing members such as testosterone, estradiol, cortisol, progesterone, vitamin A derivative all-trans retinoic acid and the thyroid hormones [1]. These factors have important roles in the differentiation of all organs including the brain. It is now 21 years since we first suggested that vitamin D was a "possible" neurosteroid [2]. At that time, all we knew from existing research was that immunohistochemical evidences for the vitamin D receptor (VDR) had been shown in non-neuronal glial cells and that vitamin D may affect neurotrophic factor expression in vitro. There was very little knowledge about how vitamin D could exert its genomic effects in brain cells and a complete absence of any understanding about whether, like other neurosteroids, vitamin D would also have more rapid non-genomic actions. Over the last two decades, research on neurons, non-neuronal brain cells, in both human and animal brains, has confirmed not only does the brain possess the molecular machinery for vitamin D's actions, but also that VDRs are functional, that liganded VDRs directly regulate the expression of target genes and that vitamin D in the form of 1,25-hydroxyvitamin D_3, $1,25(OH)_2D_3$ can rapidly alter ion channel function [3]. Moreover, only very recently have new epigenetic mechanisms been revealed as gene regulatory pathways that vitamin D targets to affect gene expression in the developing brain [4,5]. As a result of all this research, vitamin D can now be considered an important steroid in brain development [6]. More recently the neuroprotective functions of vitamin D have been highlighted in models of adult brain disorders. Here we integrate these data and outline the molecules and processes vitamin D appears to target in both developing and mature brains and how such actions shape behaviour and brain function.

We are now aware of the large number of non-skeletal targets for vitamin D. In brain cells, changing vitamin D status alters cytokine regulation and has been shown to affect cell differentiation, neurotrophin expression, intracellular calcium signalling, neurotransmitter release, anti-oxidant activity, anti-inflammatory actions, stress responsivity and the expression of genes/proteins important to neuron physiology [7,8]. The purpose of this short update is not to exhaustively summarise all such associations, but to bring the reader up to speed with recent findings using the very latest techniques in molecular manipulation/quantitation and cell visualisation.

With respect to brain disorders, there is now reliable epidemiology linking embryonic or neonatal vitamin D deficiency with an increased risk of neurodevelopmental disorders, such as schizophrenia [9,10], autism [11–14] and, more recently, attention deficit hyperactivity disorder (ADHD) [15–17]. There are also numerous studies suggesting adult vitamin D deficiency can be correlated with certain neurodegenerative conditions. This clinical epidemiological literature has been reviewed elsewhere [18,19], and we will return to this in the final section of this article. Our purpose here will be to focus on the latest preclinical studies modelling these epidemiological links and to discuss plausible biological mechanisms behind disease-relevant phenotypes.

Although we were pioneers in this area, works from a number of laboratories have firmly established the biological plausibility for how low levels of vitamin D could adversely affect how brains form and how this could lead to subtle changes in brain function and behaviour. Our task now is to discover exactly how low levels of vitamin D impair the function of specific brain cells/circuits leading to adult brain disorders and whether correcting vitamin levels and/or the processes affected by impaired vitamin D signalling can diminish phenotype/symptom severity.

2. Vitamin D Signalling in the Brain, the Basics and Controversies

Early studies reporting levels of vitamin D metabolite levels in the brain produced widely varied results. This has been to a large extent due to technical difficulties in extraction and quantification methods. The major circulatory form of vitamin D, 25-hydroxyvitamin D_3, (25(OH)D_3), and its active hormonal form, 1,25-hydroxyvitamin D_3, 1,25(OH)$_2D_3$, have been reported to be present in brain [20,21]. Whilst the exact concentrations are debatable, they are routinely reported as far lower than blood levels. The later use of LC/MS/MS with isotope dilution provides absolute chemical identification due to its selectivity and greater sensitivity. Using this technology, one group shows 25(OH)D_3 to be around (4 ng/g tissue) in rodent brain [22,23]. This group used atmospheric pressure, photoionisation which may avoid ion-suppression artifacts which may have affected earlier methods. Very recently, Fu and colleagues developed a method to measure vitamin D metabolites in the human brain and found 25(OH)D_3 was measurable, but levels were far lower than those found in the blood around 0.2–0.3 ng/g tissue [24].

With respect to the VDR, despite immunohistochemical studies confirming its presence in human, chick, rat, mouse and zebrafish brains, [25–30] claims that some of the antibodies used may have been less-than-specific and have invited dispute [31,32]. Now unambiguous evidence has been provided using mass-spectrophotometry to electrophoretically resolve proteins from adult rodent brains to identify five unique VDR peptides with a confidence interval greater than 99% [33]. There is also close anatomical overlap between brain regions in rats and humans, with respect to VDR location [25–27,29,34,35]. Consistent close cross-species overlap in VDR distribution validates the use of rodents in modelling vitamin D-related brain outcomes.

Immunohistochemistry, mRNA and protein analyses all reveal a gradual expression of the VDR in the developing brain [33,36–38]. In the developing brain, the VDR is concentrated in differentiating fields, such as the ependymal surface of the lateral ventricles [39] (the greatest source of cell division in the brain), consistent with vitamin D actions as a differentiation agent. In particular, VDR expression in the nucleus of dopamine neurons which appear very early in brain development correlates tightly with the ontogeny of these neurons [35]. New genomic technologies have now allowed VDR location and activity to be assessed in animal brains. Using clustered regularly interspaced short palindromic repeats (CRISPR), Liu et al. inserted a Cre-expression sequence driven by the endogenous VDR promotor. When such animals were mated with a tdTomato reporter mouse ((Ai14; B6.Cg-Gt(ROSA)26Sor tm14 (CAG-tdTomato)Hze/J), intense tdTomato fluorescence was detected in multiple brain regions, including caudate putamen, amygdala, reticular thalamic nucleus, cortex and less intense tdTomato fluorescence in hippocampus, hypothalamus, paraventricular thalamic nucleus (PVH), dorsal raphe and bed nucleus of the stria terminalis (BNST). RNAscope confirmed the presence of VDR RNA in these same tdTomato positive neurons. However, perhaps most importantly, $1,25(OH)_2D_3$ (5 μM) selectively depolarised the tdTomato positive but not negative neurons in the PVH, indicating vitamin D directly regulates neuronal activity, but only in VDR-expressing neurons [40].

Immunohistochemical evidences for the enzymes that convert $25(OH)D_3$ to the active hormonal form of vitamin D, $1,25(OH)_2D_3$, *CYP*27B1, and enzymes responsible for its breakdown, *CYP*24A1, have been provided in the foetal [41] and adult human brain [25,42], suggesting the active hormone can be made or removed locally in human brains. The most thorough investigation of expression of the major vitamin D metabolising enzymes *CYP*24A1, *CYP*271B and *CYP*271A in brain cell types was undertaken by Landel and colleagues [43]. Using primary cultures of neurons, astrocytes, microglia and oligodendrocytes (encompassing all major brain cell types), this group showed a broad distribution across all cell types but at a lower level than the kidney and liver. Strikingly, however, the addition of $1,25(OH)_2D_3$ induced a profound upregulation of *CYP*24A1 only in astrocytes and microglia, a finding also previously observed [44]. Again, genomic technologies have been employed to study the activity of these enzymes. Hendrix et al. created a transgenic mouse that expresses the luciferase reporter under the control of a full-length human CYP27B1 promoter. The organ distribution of CYP27B1 luciferase activity is similar to endogenous CYP27B1 with highest activity in the kidney, testis, brain, skin and bone [45]. Unfortunately, this useful model apparently has not yet been utilised to study what factors upregulate CYP27B1 brain activity.

3. Vitamin D and Normal Brain Development

A wealth of experimental evidence exists regarding the plausibility for how altering vitamin D signalling could affect critical events in brain development, such as axonal elongation, neurotransmitter synthesis, neurotrophin production and later brain function. Although there are some comprehensive reviews in this space [46,47], here, we bring this up to date.

3.1. Vitamin D and Neurite Growth

The addition of $1,25(OH)_2D_3$ to embryonic hippocampal neurons in culture increases neurite outgrowth possibly via an increased nerve growth factor (NGF) [48,49]. We also now have new unpublished data replicating the neurite-promoting potential of $1,25(OH)_2D_3$ in developing dopamine neurons differentiated from (a) a neuroblastoma cell line and (b) dopamine neurons in explant mesencephalic cultures. This action is in line with most other neurosteroids. Now, we are exploring the molecular mechanisms behind these effects. Others have chosen to lesion peripheral neurons and to show that vitamin D enhances axonal repair, myelination and functional recovery [50,51]. The $1,25(OH)_2D_3$ treatment also restores neurite outgrowth in models where it is impeded. The knockdown of the important neuronal epigenetic regulator, MeCP2, in cortical neurons blunts neurite extension potentially through reducing the activity of NFκB pathways. The $1,25(OH)_2D_3$ restores normal outgrowth in this model [52]. Vitamin D deficiency has also been shown to reduce peripheral nerve fibre density [53], and vitamin D has been considered a potential therapeutic in spinal cord repair [54].

3.2. Vitamin D and Neurotrophic Factors

The first evidence that vitamin D had any action on brain cells came from the early studies from Didier Wion's group in glioblastoma cells. Early studies showed that vitamin D promoted the expression of NT-3, NT-4 and nerve growth factor (NGF) [48,49,55–57]. Vitamin D-mediated increases in NGF were shown to be highly relevant to neuronal survival in vitro [48,58–60].

Given its prominent role in dopaminergic neuron differentiation (see below) and survival in earlier studies, there was a strong focus on vitamin D and neurotrophic factors important for dopaminergic neurons, such as glial cell line-derived neurotrophic factor (GDNF) [61,62] and now brain-derived neurotrophic factor (BDNF) [63]. Blocking vitamin D-mediated increase in GDNF synthesis prevents vitamin D's trophic effects on these neurons [64]. We have described the genomic proof that vitamin D regulates the transcription of both receptors for GDNF. The 1,25(OH)$_2$D$_3$ suppresses GDNF family receptor alpha 1 (GFRa1), but ligand bound VDR binds to the promoter of the other major receptor for this neurotrophin, the proto-oncogene tyrosine-protein kinase receptor Ret (C-Ret), to upregulate C-Ret expression. Accordingly, the maternal absence of vitamin D decreases C-Ret expression in the developing rat mesencephalon [65].

The effect of vitamin D on neurotrophic factors in the developing brain has been far less explored. One study showed DVD-deficiency in rats reduced NGF and GDNF protein in neonatal brains [66]. Reductions in BDNF and transforming growth factor-β1 (Tgf-β1) in DVD-deficient embryonic mouse brains have also been described [67].

Most recently, there have been numerous studies showing a convincing link between hippocampal BDNF levels and vitamin D status in adult animals. Consistent with findings in development where the absence of vitamin D leads to a BDNF deficit, very recent studies show chronic treatment with 1,25(OH)$_2$D$_3$ induces a profound upregulation in BDNF in rat hippocampus [68]. Others show stress- or drug-induced memory deficits correspond with decreased hippocampal BDNF expression that can be restored via chronic supplementation with high doses of cholecalciferol [69,70]. Another study revealed age-induced memory deficits could be reversed with cholecalciferol supplementation with corresponding increases in hippocampal BDNF and NGF expression [71]. In a model of type 2 diabetes in mice, BDNF and phosphorylation of its downstream effector CREB are reduced in the brain, and cholecalciferol restores these and associated behavioural deficits [72].

In summarising this data, 1,25(OH)$_2$D$_3$ *increases*, and the developmental dietary absence of vitamin D *reduces* the expression of these crucial neurotrophic factors in neurons [73] and glia in developing and adult brains. Vitamin D's regulation of neurotrophic factors remains a central feature in brain ontogeny.

3.3. Vitamin D Regulates the Development of Dopamine Neurons

Since we first reported the VDR within the human substantia nigra [25], we have gone on to confirm the VDR is prominent in neuromelanin containing Tyrosine Hydroxylase (TH) positive human nigral neurons [35]. Vitamin D deficiency during development leads to a reduction in specification factors crucial for dopamine neurons [74,75], alters dopamine turnover [76] and leads to decreased lateral positioning of dopamine neurons that will form the substantia nigra [75]. Importantly, similar developmental anomalies in dopamine neuron ontogeny created by other adverse developmental exposures are rescued by maternal treatment with 1,25(OH)$_2$D$_3$ [77]. To better understand the molecular processes involved, we have used neuroblast cells in which the VDR has been over-expressed. In such a model, we have shown 1,25(OH)$_2$D$_3$ increases TH mRNA [78], drives cells down a dopaminergic lineage [79] and directly targets the promoter of Catechol-o-methyl transferase (COMT), an important enzyme in brain dopamine turnover [79].

One very recent study has examined microRNAs in developing dopamine neurons as epigenetic factors in neuronal maturation. This study showed that in flow cytometry-sorted E14 dopamine neurons from DVD-deficient rat embryos, the expression of a number of microRNAs was increased and that gene-ontology analysis indicated these microRNAs were involved in neuronal differentiation. These authors then examined the functional consequences of this in developing dopamine neurons and showed that some of these microRNAs decreased neurite outgrowth [4]. This represents the first evidence in the brain indicating how vitamin D may epigenetically differentiate dopamine neurons.

Taken together, the decreased expression of dopamine-promoting neurotrophic and specification agents and delayed dopamine neuron maturation in DVD-deficient embryonic brains along with the abundant in vitro evidences for 1,25(OH)$_2$D$_3$ as a maturation agent for dopamine neurons, all confirm a central role for vitamin D in dopamine ontogeny. Given the links between DVD-deficiency and schizophrenia [9,10] and that dopamine abnormalities are perhaps causal in this disease, further studies examining this possible mechanistic link are needed.

4. Developmental Vitamin D Deficiency Effect on Brain Function and Behaviour

The long-term effects of DVD-deficiency on offspring behaviour have also been extensively studied in both rats and mice. Outcomes vary based on the species used/strain and duration and degree of vitamin D deficiency. Notably, there are alterations in critical early dam/pup interactions, such as maternal licking and grooming [80] and pup ultrasonic vocalisations [5,80], along with signs of delayed motor development [5]. As juveniles or adults, offspring have social behavioural deficits and alterations in stereotyped behavioural phenotypes of relevance to autism [5,80]. Interestingly, vitamin D deficiency leads to subtle increases in testosterone levels in maternal blood [81], a finding which has long-been considered a risk-modifying factor in autism [82]. Additionally, foetal male brains from vitamin D deficient dams have increased testosterone compared to control males. This study went on to show silencing of the major enzyme involved in testosterone breakdown, aromatase, by hypermethylation of its promoter in the brain may be the mechanism [81].

DVD-deficiency also leads to long-term changes in adult behaviours. Novelty [83], or exposure to psychomimetics, such as the N-methyl-D-aspartic acid receptor (NMDA-R) antagonist, MK-801 [84–86] or amphetamine [87], all increase locomotor activity in the adult offspring of DVD-deficient dams. Along with sensitivity to these exposures/agents that release dopamine, DVD-deficient rats have a greater response to antipsychotics that block dopamine 2 receptors [84,88]. Brain functional measures are also abnormal in these animals. Long-term potentiation, a cellular correlate of learning and memory and latent inhibition, a measure of attentional processing are also abnormal in DVD-deficient adults [89,90]. With respect to cognition, DVD-deficient offspring have impaired response inhibition, a deficit normalised by the antipsychotic clozapine [91]. Associative learning is also impaired in DVD-deficient adults [92].

Only recently have investigators begun to assess the impact of vitamin D treatment in dams exposed to either DVD-deficiency or another developmental animal model that induces brain changes, maternal immune activation. Such interventions reverse behavioural phenotypes produced by these models related to anxiety, depression cognition, stereotyped and social behaviours [93,94].

As discussed previously, given the stark differences in the effects of DVD-deficiency on brain development between rats and mice, it is not surprising that the behavioural phenotypes in adult offspring also vary widely [95]. For instance, although there are some similarities in novelty-induced hyperlocomotion [84] and perseverative responses [96], outcomes vary widely in tests of exploratory behaviour [88,97] and locomotor response to psychomimetics [98].

Whether DVD-deficiency affects cognitive performance in children is not clear. Two studies show DVD-deficient children have delayed cognitive development [99,100]. However, these findings were not replicated in much larger studies [15]. A recent meta-analysis that summarised 31 studies could find no conclusive evidence for the association between maternal or child vitamin D status and behavioural or cognitive outcomes in children and adolescents. A summary of twelve mother–child studies (n = 17,136) and five studies just in children (n = 1091) showed low maternal or child 25(OH)D$_3$ levels led to impaired behavioural outcomes in children. In contrast, fifteen mother–child studies (n = 20,778) and eight studies in children (n = 7496) showed no association [101]. A large Finnish nested case-control study (1607 children with learning difficulties with matched controls) also showed no link between maternal 25(OH)D$_3$ levels and behaviour [102].

One small randomised clinical trial (55 infants) in healthy term infants showed a low dose of vitamin D supplementation might benefit gross motor function compared to a higher dose of vitamin D supplementation, but this study lacked a placebo control [103]. There is also a not-insubstantial number of studies that link maternal or childhood vitamin D deficiency to neurodevelopmental psychiatric disorders such as autism, ADHD or schizophrenia [9–11,13,14,100,104,105]. This association is not universally replicated [106,107]; however, the mean population 25(OH)D_3 levels in these later studies were much higher, making the association difficult to test. Randomised controlled trials of cholecalciferol supplementation in children with ASD or ADHD show diverse outcomes, some showed no beneficial effects and others reported some symptom relief [108–114]. In the future, trials examining vitamin D levels in mothers/children prior to supplementation are required as supplementing vitamin D-sufficient mothers or children is likely to have minimal effect, as shown in other clinical trials with vitamin D [115].

5. Vitamin D Is Neuroprotective in Neurons and Adult Brain

There are now sufficient epidemiological studies also linking low levels of 25OHD_3 with various neurodegenerative conditions such as Alzheimer's disease, Parkinson's disease and multiple sclerosis. Though many such studies suffer from reverse causality (i.e., low levels of vitamin D are a consequence of disease-induced behavioural changes rather than causal), there has been sufficient interest to launch numerous clinical supplementation trials [116–118]. There have also been studies linking exposure to various toxins with low 25OHD_3 levels. Such links have stimulated researchers to initiate mechanistic studies in neuronal cell systems and animal models. Along with the wealth of studies showing vitamin D is neuroprotective via neurotrophin production, here, we have chosen not to focus on the clinical epidemiology, but rather on vitamin D's regulatory actions on calcium in the brain and its neuroprotective actions against reactive oxygen species (ROS) and inflammation as well as its actions in mitigating stress as plausible prophylactic/therapeutic mechanisms. We summarised the potential mechanisms underlying the action of vitamin D (Figure 1). Much of this work is very recent.

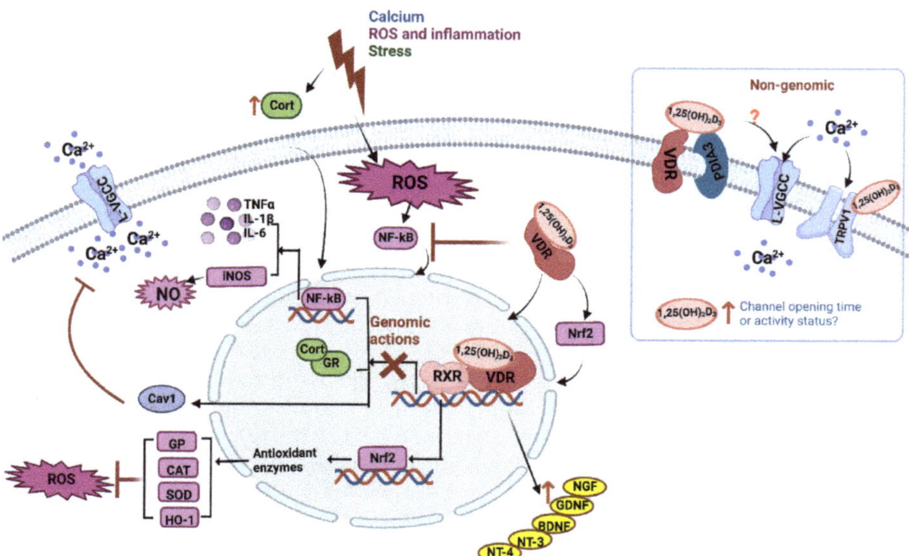

Figure 1. Neuroprotective actions of vitamin D in brain.

Calcium-related mechanisms (in blue). Non-genomic actions. The 1,25$(OH)_2D_3$ acutely facilitates calcium influx via L-VGCC. The 1,25$(OH)_2D_3$; can also directly bind to TRPV1 to induce calcium influx. Both actions facilitate neuronal function. *Genomic actions.* Chronic vitamin D treatment decreases Cav1 (a L-VGCC subunit) expression which will reduce calcium influx in response to ROS/inflammation or stress.

ROS/inflammation-related mechanisms (in purple). ROS enhances NF-κB nuclear translocation to promote proinflammatory cytokine production. The $1,25(OH)_2D_3$ acts to inhibit this nuclear translocation. By also inhibiting NF-kB expression, $1,25(OH)_2D_3$ reduces iNOS expression and thus, reduces NO production. Finally, $1,25(OH)_2D_3$ increases Nrf2 to enhance the expression of anti-oxidant enzymes including GP, CAT, SOD and HO-1, thus countering ROS toxicity. The $1,25(OH)_2D_3$ also directly facilitates Nrf2 nuclear translocation.

Stress-related mechanisms (in green). Stress increases corticosterone synthesis. In the brain, corticosterone increases inflammatory cytokine production via its receptor (GR). The $1,25(OH)_2D_3$ is neuroprotective by antagonising GR expression.

Neurotrophin-related mechanisms (in yellow). A variety of neurotrophic factors have been shown to counter the effects of stress- or toxin-induced neuronal damage. Historically, $1,25(OH)_2D_3$ has been shown to increase NGF, GDNF, BDNF, NT-3 and NT-4 under such conditions.

5.1. Calcium Regulation

Calcium transients are required for normal neuronal function, but if calcium is unbuffered, it is toxic to brain cells. For more than 30 years, we have known how vitamin D regulates calcium uptake in bone cells [119,120] and similar mechanisms appear to act in neurons and the brain. In cultured neurons, $1,25(OH)_2D_3$ retards calcium influx via the downregulation of L-type voltage-sensitive calcium channels, thus potentially preventing toxic outcomes [59,121–123]. In contrast, the rapid non-genomic actions of vitamin D produce the opposite effect, increasing calcium influx in cortical slices, a process that is again dependent on L-type calcium channels [124]. In a seminal study, the non-genomic rapid actions of $1,25(OH)_2D_3$ were investigated in cortical neurons using calcium imaging, electrophysiology and molecular biological techniques. This study confirmed that physiological concentrations of $1,25(OH)_2D_3$ lead to rapid calcium influx, but only in some neurons. Somatic nucleated patch recordings revealed a rapid, $1,25(OH)_2D_3$-evoked increase in high-voltage-activated calcium currents mediated by L-type voltage-gated calcium channels [125]. Whether any of these actions are caused by the putative membrane VDR, Protein disulfide-isomerase A3 (PDIA3), in the brain remains unknown. Genetic variants in L-type voltage-gated calcium channels continue to be implicated in schizophrenia [126]. Given the epidemiological links between DVD-deficiency and schizophrenia [9,10], continued studies in this area are warranted.

Recent research also discovered that $25OHD_3$ and $1,25(OH)_2D_3$ directly bind to the transient receptor potential vanilloid subfamily member 1 (TRPV1) channel [127]. Binding to the same region as the TRPV1 agonist capsaicin, $25OHD_3$ can weakly activate TRPV1 and inhibit capsaicin-induced TRPV1 activity. TRVP1 activity modulates immune cell activation and cytokine production through regulating intracellular calcium to mediate nociceptive signals. This, therefore, may be one mechanism for how vitamin D may modulate nociceptive pain pathways, as vitamin D deficiency has been linked to chronic pain [128], although oxidative mechanisms have also been proposed [129].

5.2. ROS and Inflammation

Vitamin D increases anti-oxidants, such as glutathione and cytochrome c, to mediate anti-oxidant actions in cultured neurons [121,130] and the brain [131,132]. Along with extracellular calcium, vitamin D deficiency in the adult brain increases ROS along with producing impairments in gamma-aminobutyric acid (GABA) and glutamate release. Importantly, reintroducing dietary vitamin D normalises all deficits [133].

In the brain, microglia are the immunologically responsive cells responsible for the production of inflammatory regulators such as nitric oxide (NO). Early studies showed $1,25(OH)_2D_3$ blocks inducible nitric oxide synthetase in the rat brain in response to autoimmune or inflammatory factors [134–136]. Later studies suggested that oxidative stress may upregulate *CYP*27B1 in microglia to induce 1,25(OH)$_2$D$_3$ production locally at the site of NO or ROS production to mediate vitamin D's anti-oxidant effects in the brain [133,137,138].

In primary cultured neurons, hypoxia induces apoptotic cell death and interferes with normal calcium signalling. One study, that chose to use cholecalciferol rather than calcitriol, showed that when added to cultured primary neurons exposed to hypoxia, cholecalciferol (10 nM) was anti-apoptotic and preserved calcium signalling though this effect was reversed at high doses. Upregulation of hypoxia-induced factor (HIF)-1α and/or BDNF were considered as the possible protective mechanisms [139], though conversion to calcitriol was not assessed.

DVD-deficiency would also appear to render the foetal environment more prone to oxidative stress. Ali and colleagues reported that DVD-deficient rat placenta produces more of the inflammatory cytokines IL-6 and 1L-1β upon challenge with a viral inflammatory agent [140]. Separately, when microglia were cultured from DVD-deficient mouse brains, they were increased in number, were hyperproliferative and had increased ROS production. Culturing these cells in the presence of calcitriol reversed these changes [141].

In recent years, there appears to have been intense interest in vitamin D's protective actions against ROS and inflammation in the brain induced by numerous disease models. For instance, in hypothyroid juvenile rats, cholecalciferol supplementation (100 or 500 IU/kg/day) prevented hypothyroidism-induced cognitive and learning memory impairments [142]. Plausible mechanisms included elevations in the anti-oxidant enzyme superoxide dismutase (SOD) and thiol content in hippocampus and cortical tissue and reductions in malondialdehyde (MDA), a marker of oxidative stress. Similar restorative outcomes on cognition have been achieved in models of acute inflammation and again these same anti-oxidant processes in the brain were invoked by vitamin D supplementation [143].

In spontaneously hypertensive (SH) rats, infusion of 1,25(OH)$_2$D$_3$ into the hypothalamic paraventricular nucleus (PVN), a brain region maintaining baroreflex and autonomic function, prevented the elevation of ROS stress-related proteins such as NOX2, NOX4 and p22^{phox}. Chronic calcitriol infusion also attenuated microglial activation and reduced tumour necrosis factor (TNF)-α, IL-1β and IL-6 inflammatory cytokine production. Likely, mechanisms involved the inhibition of the high-mobility group box-1(HMGB1)-receptor for advanced glycation end products (RAGE)/Toll-like receptor (TLR4) and NF-κB in the PVN of SH rats [144].

The effect of vitamin D on NO is also detected in a rat model of cerebral ischemia reperfusion model. Seven days of calcitriol administration prior to ischemic surgery reduced stroke-induced elevation of MDA and NO and increased total anti-oxidant capacity. These effects could be attributed to vitamin D-increasing nuclear factor erythroid 2-related factor 2(Nrf2), a transcription factor that decreases oxidative stress following stroke and/or heme oxygenase (HO-1), a major cytoprotective enzyme with anti-oxidative, and anti-inflammatory properties [145,146].

In a traumatic brain injury model, calcitriol treatment reduced MDA production and promoted autophagic flux and activated Nrf2 pathways. Autophagy reduces oxidative stress by a timely removal of damaged substances [147]. Nrf2 is a redox-sensitive transcription factor that binds to anti-oxidant response elements to promote the expression of enzymes/proteins for detoxication and anti-oxidation [148]. Confirmation that these two processes were central to calcitriols protective actions in traumatic brain injury was shown by inhibiting the autophagy using chloroquine or deleting Nrf2 as either action blocked calcitriol's protective effects [149].

In a model of lead-induced neurotoxicity and oxidative stress, decreases in anti-oxidant molecules GSH, SOD and catalase and increased ROS production are observed in rat cortex. Cholecalciferol reversed these changes possibly via Nrf2 and/or NF-κB mechanism [150]. In a model of experimentally induced epilepsy in young male rats, cholecalciferol not only reduced seizure severity, but corrected associated memory deficits, reduced extracellular calcium, restored anti-oxidative enzymes SOD and glutathione-related enzymes and reduced inflammatory cytokine production in the hippocampus [151].

One neurological disorder commonly associated with ROS-mediated brain damage is Parkinsons's disease (PD). The motor deficits produced by PD are believed to be caused by dopamine neurons dying selectively in the substantia nigra and the associated reduced dopamine release in the dorsal striatum. This is modelled in animals via the intracranial delivery of relatively selective dopaminergic terminal toxins, such as 6-hydroxy dopamine (6-OHDA), or nigral toxins such as 1-methyl-4-phenyltetrahydropyrine (MPTP). In a 6-OHDA-treated mouse, cholecalciferol treatment two weeks after surgical lesion attenuated 6-OHDA-induced increases in the microglia marker CD11b, IL-1β and the oxidative stress marker p47phox, a primary modulator of NADPH oxidase activity and restored some motor functional deficits [152]. Another 6-OHDA study showed that cholecalciferol prevented characteristic losses in dopamine synthetic enzymes and transporters, preserved motor function as well as reduced lipid peroxidation [153]. Another study using MPTP showed co-administration with calcitriol improved motor deficits, reduced dopamine neuron toxicity and reduced ROS production [154]. The mechanism proposed was via an interaction between the liganded VDR and poly(ADP-ribose) polymerase-1(PARP1) to reduce its contribution to ROS-induced cell death. PARP1 pathways have been proposed as one mechanism for dopamine cell death in PD.

ROS production is also linked with other degenerative processes such as Alzheimer's disease. Animal models for Alzheimer's disease frequently employ genetic models that over-express the Tau or amyloid (Aβ) proteins that are closely linked with disease pathology. In an amyloid presenilin model, 13 weeks of vitamin D deficiency exacerbated the ROS production in this model by downregulating superoxide dismutase 1 (SOD1), glutathione peroxidase 4 and enhanced the expression of IL-1β, IL-6 and TNFα, along with increased Aβ production and Tau phosphorylation [155].

In another amyloid model where rats are injected with Aβ1-40, cholecalciferol supplementation reduced Aβ-induced MDA levels, increased SOD activity and improved hippocampal neuronal survival [156]. Prolonged vitamin D hypovitaminosis in mice (from 6-weeks-old to 6-months-old) altered the expression of genes involved in amyloid precursor protein homeostasis (Snca, Nep, Psmb5), oxidative stress (Park7), inflammation (Casp4), lipid metabolism (Abca1), signal transduction (Gnb5) and neurogenesis (Plat) [157]. In another distinct amyloid protein mutant mouse model (3xtg-AD), vitamin D levels are dramatically reduced at 9 and 12 months of age. Vitamin D supplementation in this model improves memory possibly via the suppression of collapsin response mediator protein-2 (CRMP2) phosphorylation [158]. An in vitro study has also shown $1,25(OH)_2D_3$ alleviates Tau hyperphosphorylation and reduces ROS in a neuronal cell model treated with Aβ possibly via vitamin D's role in enhancing GDNF [159].

Experimentally induced autoimmune encephalitis (EAE) is a mouse model mimicking the autoimmune reaction to myelin proteins inducing multiple sclerosis-like pathology. In this model, calcitriol administration decreased the severity of EAE by attenuating inflammation and demyelination at the spinal cord [160]. This same study also showed calcitriol treatment reduces lymphocyte, macrophage and activated microglia infiltration into the brain. EAE mice have increased blood brain barrier permeability thought to be due at least in part to a reduction in endothelial tight-junction proteins such as ZO-1. This same study showed calcitriol increases this proteins expression. Calcitriol also reduced EAE-induced lipid hydroxylation and enhanced the anti-oxidant enzymes glutathione peroxidase, catalase and SOD.

The $1,25(OH)_2D_3$ not only suppresses inflammation in models of demyelination, but also enhances the differentiation or survival of oligodendrocyte progenitor cells in the spinal cord of a MOG35–55-induced EAE mouse model [161]. The promotion of oligodendrocyte maturation by $1,25(OH)_2D_3$ is also observed in a cuprizone-induced EAE mouse model [162]. Krabbe disease is an inherited leukodystrophy. This demyelinating condition is caused by a galactocerebrosidase (GALC) deficit that results in loss of oligodendrocytes and demyelination. In an animal model of this disorder ($GALC^{twi/twi}$; twitcher mouse), supplementing the heterozygous $GALC^{+/-}$ dam from birth to weaning with cholecalciferol delayed onset of disease-induced locomotor deficits and tremors and extend the life span of offspring [163].

5.3. Glucocorticoids and Stress

Glucocorticoid release is the classic endocrine response to stress, and protracted exposure induces neuronal shrinkage then cell death [164]. The effects of $1,25(OH)_2D_3$ and glucocorticoids in the body can be considered antagonistic [165–169]. Similarly, in the brain, $1,25(OH)_2D_3$ antagonises the effects of dexamethasone (a corticosterone agonist) on hippocampal neuron differentiation and glucocorticoid receptor function [170]. Interestingly, dexamethasone can decrease $1,25(OH)_2D_3$ synthesis in the hippocampus and prefrontal cortex, indicating this process is reversible [171].

Behaviourally, vitamin D antagonises the depression-like phenotypes induced by chronic cortisol administration in animals [172–175]. Possible mechanisms include regulation of hippocampal glucocorticoid receptors or the restoration of dopamine levels in the reward centres in the brain [175]. Chronic mild stress in rats leads to increased corticosteroids, inflammatory markers and decreased anti-oxidant enzymes SOD and glutathione peroxidase and catalase in the hippocampus and prefrontal cortex. Simultaneous cholecalciferol treatment reverses these stress-mediated effects [69,176].

Chronic unpredictable stress increases immobility in a widely used test of behavioural despair (tail suspension test). Cholecalciferol treatment reduces this stress-induced immobility. Inhibiting the synthesis of the neurotransmitter serotonin abolishes cholecalciferol's actions in this test, suggesting vitamin D might act via increased serotonin synthesis [177]. Support for this idea comes from studies in glioblastoma cells showing functional VDREs are localised at −7 kb and −10 kb upstream of the serotonin, synthesising enzyme tryptophan hydroxylase [178]. Other studies chronically administering corticosterone replicate the cholecalciferol's reversal of these stress-induced behaviours and suggest either alterations to glucocorticoid signalling in the brain or reductions in stress-associated ROS production in the brain are the protective mechanisms [172,179]. Stress can be measured in animals using other behavioural paradigms such as immobility in a forced swim test. Chronic unpredictable stress increases immobility in this test. Cholecalciferol supplementation reverses this immobility as well as reduces serum corticosterone/ACTH levels and increases BDNF and NT-3/NT-4 levels in the hippocampus [174].

DVD-deficiency may also alter maternal response to stress in rats [180] and mice [181] and can adversely affect maternal care [80], which is well-known to induce permanent changes in offspring stress-response [182]. The translational potential of these later findings is enhanced given that a randomised clinical trial showed that 50,000 IU of cholecalciferol alone every 2 weeks, or in combination with Omega 3 fatty acids (1 g/day), for 8 weeks significantly reduced anxiety and improved sleep quality in women of reproductive age with pre-diabetes and hypovitaminosis D [183].

6. Hypervitaminosis D and Adverse CNS Outcomes

While the role of hypovitaminosis D in brain function has been extensively investigated, the effect of hypervitaminosis D has received less attention. This is despite an inverted U-shaped cellular response to $1,25(OH)_2D_3$ being frequently reported, i.e., both high and low levels of $1,25(OH)_2D_3$ induce adverse outcomes. Hypervitaminosis D is rare in humans, generally resulting from excess vitamin D supplementation or diseases such as sarcoidosis that produces excess $1,25(OH)_2D_3$ due to activated macrophages [184]. Though not focused on neurological conditions, an older review concluded "*There is accumulating evidence that both high and low serum calcidiol concentrations are associated with an increased risk of chronic diseases*" [185]. Hypervitaminosis D always results in hypercalcaemia, which may be toxic to brain function. Animal studies showed that high cholecalciferol intake, 25,000 IU/kg, for four consecutive days reduces brainwave activity [186]. Senescence-accelerated-mouse-phenotype (SAMP) strain-8 mice, an animal model of accelerated human ageing, showed a progressive increase in serum $25OHD_3$, which coexists with reduced cognitive function and increased capillary permeability [187]. Deleting Fibroblast growth factor (FGF-23) in mice increased *CYP*27B1 levels, leading to increased $1,25(OH)_2D_3$ synthesis. FGF-23 null mice display early ageing, and deleting *CYP*27B1 delayed premature ageing [188]. Considering the risk of hypervitaminosis D, serum vitamin D levels should be monitored in clinical trials of vitamin D supplementation.

7. Conclusions and Future Challenges

Here we have provided an up-to-date summary of research detailing vitamin D turnover, synthesis genomic and non-genomic actions in the brain, neurons and non-neuronal cells. We advise caution when considering much of the prior literature employing either dietary restrictions which often produced hypo-calcaemia or the use of constitutive knock out models permanently reducing $1,25(OH)_2D_3$ synthesis or impairing VDR signalling, as these models produce too many non-CNS effects to make brain-related outcomes interpretable. Although there are still strain and species differences when choosing a model organism, contemporary DVD- or AVD-deficient models in rodents do not produce hypocalcaemic offspring and continue to produce findings of apparent relevance to the fields of psychiatry and neurology.

Increasingly, epidemiological studies associate low levels of vitamin D either prenatally or at birth with psychiatric conditions such as autism and schizophrenia. The epidemiology for degenerative diseases as diverse as Alzheimer's disease, Parkinson's disease and multiple sclerosis all continue to indicate a role of optimal vitamin D status throughout life. This raises the possibility of simple dietary supplementation as an adjunct to current therapies. Although not practicably possible for developmental conditions with adult onset such as schizophrenia, this could eventually be considered in early onset psychiatric disorders, such as autism or ADHD, or for maintaining neural integrity in degenerative conditions via large placebo-controlled, randomised clinical supplementation trials.

We would also like to take this opportunity to highlight design issues in many observational studies opportunistically linking early-life vitamin D deficiency with psychiatric disorders. Most of the published observational epidemiological studies linking early-life vitamin D status with a psychiatric diagnosis never address reverse causality (the condition changes behaviours that lead to less sun exposure). By way of illustration, a very high-profile recent report in the New England Journal of Medicine showed all mental illnesses were associated with an increased risk in a general medical condition [189]. In other words, patients with psychiatric conditions are generally suffering from other conditions that will curtail behaviour perhaps altering diet, exercise and exposure to sunshine. We urge all future epidemiological studies that seek to examine the relationship between vitamin D and psychiatric or neurological conditions to rigorously control for the often-poor general health of patients. This, of course, is less of an issue for gestational (DVD-deficiency) exposures in otherwise healthy mothers.

Important data have emerged from a recent mendelian randomisation study examining gene pathways related to $25(OH)D_3$ blood concentrations. This study found no evidence that genetic factors involved in the synthesis of $25(OH)D_3$ were causal for psychiatric disorders [190]. We have interpreted this to mean any link between $25(OH)D_3$ levels and brain-related outcomes are likely to be solely driven by environmental factors.

Autism is perhaps the developmental brain disorder most regularly linked with DVD-deficiency. However, well-conducted studies refuting this link are now emerging. For the studies describing an inverse relationship between maternal vitamin D levels and autism, they all had mean $25(OH)D_3$ levels of <50 nM which is considered by some authors to represent a cut off for vitamin D deficiency [11–13,191]. Two recent studies have failed to find this inverse association [106,192]. So, at face value, this appears a failure to replicate previous studies. However, it is crucial to note that in these last two studies, the mean levels of $25(OH)D_3$ were actually very high (>70–80 nM) and there were very few individuals that were actually vitamin D-deficient, meaning the association could not be properly tested. These same six studies all used the same laboratory to analyse samples, so technical bias (so common amongst vitamin D studies in different populations) could be ruled out. This suggests a threshold effect rather than any continuous relationship between DVD-deficiency and autism. We highlight this particular relationship to illustrate some of the confusion regarding statements regarding potential causality between vitamin D and various brain-related clinical disorders.

The incidence of hypovitaminosis D in both pregnant women and their newborns and the general population remains concerning [193]. Clearly, more rigorous study design is required taking into consideration such issues to bring clarity to the future epidemiological studies. Better quality studies in the future are needed given the substantial emotional and financial burden psychiatric and neurological disorders place on the patient and community. The opportunity to use such a simple, safe and inexpensive intervention as vitamin D supplementation as treatment or as an adjunct to existing therapies in such disorders remains extremely attractive from a public health perspective.

Author Contributions: D.W.E. and X.C. conceived and wrote the manuscript. All authors have read and agreed to the published version of the manuscript.

Funding: This research was funded by the National Health and Medical Research Council (APPS 1124721 and 1141699) and the Queensland State Government.

Acknowledgments: We thank Suzy Alexander for assistance creating the figure.

Conflicts of Interest: The authors have no conflicts of interest.

Abbreviations

VDR, vitamin D receptor; *PDIA3*, Protein disulfide-isomerase A3; *RXR*, retinoid X receptor; *L-VGCC*, L-type voltage-gated calcium channel; *Cav1*, L-type voltage-gated calcium subunit 1; *TRPV1*, transient receptor potential vanilloid subfamily member 1; *ROS*, reactive oxygen species; *iNOS*, inducible nitric oxide synthetase; *NO*, nitric oxide; *NF-kB*, nuclear factor kappa-B; *Nrf2*, nuclear factor erythroid 2-related factor 2; *GP*, glutathione peroxidase; *CAT*, catalase; *SOD*, superoxide dismutase; *HO-1*, heme oxygenase; *NGF*, nerve growth factor; *BDNF*, brain-derived neurotrophic factor; *GDNF*, glial cell line-derived neurotrophic factor; *NT-3*, neurotrophin 3; *NT-4*, neurotrophin 4; *GR*, glucocorticoid receptor; *CORT*, corticosterone; *TNFα*, tumour necrosis factor α; *IL-6*, interlukin-6; *IL-1B*, interlukin-1 Beta. (Figure created from BioRender.com).

References

1. Evans, R.M. The nuclear receptor superfamily: A rosetta stone for physiology. *Mol. Endocrinol.* **2005**, *19*, 1429–1438. [CrossRef] [PubMed]
2. McGrath, J.; Feron, F.; Eyles, D. Vitamin D: The neglected neurosteroid? *Trends Neurosci.* **2001**, *24*, 570–572. [CrossRef]
3. Cui, X.; Gooch, H.; Petty, A.; McGrath, J.J.; Eyles, D. Vitamin D and the brain: Genomic and non-genomic actions. *Mol. Cell. Endocrinol.* **2017**, *453*, 131–143. [CrossRef] [PubMed]
4. Pertile, R.; Kiltschewskij, D.G.; Geaghan, M.; Barnett, M.; Cui, X.; Cairns, M.J.; Eyles, D.W. Developmental vitamin D-deficiency increases the expression of microRNAs involved in dopamine neuron development. *Brain Res.* **2022**, *1789*, 147953. [CrossRef] [PubMed]
5. Ali, A.; Vasileva, S.; Langguth, M.; Alexander, S.; Cui, X.; Whitehouse, A.; McGrath, J.J.; Eyles, D. Deveopmental vitamin D deficiency produces behavioral phenotypes of relevance to autism in an animal model. *Nutrients* **2019**, *11*, 1187. [CrossRef] [PubMed]
6. Melcangi, R.C.; Panzica, G. Neuroactive steroids: An update of their roles in central and peripheral nervous system. *Psychoneuroendocrinology* **2009**, *34*, S1–S8. [CrossRef]
7. Eyles, D.W.; Burne, T.H.; McGrath, J.J. Vitamin D, effects on brain development, adult brain function and the links between low levels of vitamin D and neuropsychiatric disease. *Front. Neuroendocrinol.* **2013**, *34*, 47–64. [CrossRef]
8. McCann, J.C.; Ames, B.N. Is there convincing biological or behavioral evidence linking vitamin D deficiency to brain dysfunction? *FASEB J.* **2008**, *22*, 982–1001. [CrossRef]

9. McGrath, J.J.; Eyles, D.W.; Pedersen, C.B.; Anderson, C.; Ko, P.; Burne, T.H.; Norgaard-Pedersen, B.; Hougaard, D.M.; Mortensen, P.B. Neonatal vitamin D status and risk of schizophrenia: A population-based case-control study. *Arch. Gen. Psychiatry* **2010**, *67*, 889–894. [CrossRef]
10. Eyles, D.W.; Trzaskowski, M.; Vinkhuyzen, A.A.E.; Mattheisen, M.; Meier, S.; Gooch, H.; Anggono, V.; Cui, X.; Tan, M.C.; Burne, T.H.J.; et al. The association between neonatal vitamin D status and risk of schizophrenia. *Sci. Rep.* **2018**, *8*, 17692. [CrossRef]
11. Vinkhuyzen, A.A.E.; Eyles, D.W.; Burne, T.H.J.; Blanken, L.M.E.; Kruithof, C.J.; Verhulst, F.; Jaddoe, V.W.; Tiemeier, H.; McGrath, J.J. Gestational vitamin D deficiency and autism-related traits: The Generation R study. *Mol. Psychiatry* **2016**, *23*, 240–246. [CrossRef] [PubMed]
12. Vinkhuyzen, A.; Eyles, D.; Burne, T.; Blanken, L.; Kruithof, C.; Verhulst, F.; White, T.; Jaddoe, V.W.; Tiemeier, H.; McGrath, J. Gestational Vitamin D Deficiency and Autism Spectrum Disorder. *Br. J. Psychiatry Open* **2017**, *3*, 85–90. [CrossRef] [PubMed]
13. Lee, B.K.; Eyles, D.W.; Magnusson, C.; Newschaffer, C.J.; McGrath, J.J.; Kvaskoff, D.; Ko, P.; Dalman, C.; Karlsson, H.; Gardner, R.M. Developmental vitamin D and autism spectrum disorders: Findings from the Stockholm Youth Cohort. *Mol. Psychiatry* **2019**, *26*, 1578–1588. [CrossRef] [PubMed]
14. Sourander, A.; Upadhyaya, S.; Surcel, H.M.; Hinkka-Yli-Salomaki, S.; Cheslack-Postava, K.; Silwal, S.; Sucksdorff, M.; McKeague, I.W.; Brown, A.S. Maternal Vitamin D Levels During Pregnancy and Offspring Autism Spectrum Disorder. *Biol. Psychiatry* **2021**, *90*, 790–797. [CrossRef] [PubMed]
15. Strøm, M.; Halldorsson, T.; Hansen, S.; Granström, C.; Maslova, E.; Petersen, S.B.; Cohen, A.S.; Olsen, S.F. Vitamin D measured in maternal serum and offspring neurodevelopmental outcomes: A prospective study with long-term follow-up. *Ann. Nutr. Metab.* **2014**, *64*, 254–261. [CrossRef]
16. Morales, E.; Julvez, J.; Torrent, M.; Ballester, F.; Rodríguez-Bernal, C.L.; Andiarena, A.; Vegas, O.; Castilla, A.M.; Rodriguez-Dehli, C.; Tardón, A.; et al. Vitamin D in Pregnancy and Attention Deficit Hyperactivity Disorder-like Symptoms in Childhood. *Epidemiology* **2015**, *26*, 458–465. [CrossRef]
17. Arns, M.; van der Heijden, K.B.; Arnold, L.E.; Kenemans, J.L. Geographic variation in the prevalence of ateention-deficit/hyperactivity disorder: The sunny perspective. *Biol. Psychiatry* **2013**, *74*, 585–590. [CrossRef]
18. Groves, N.; McGrath, J.; Burne, T. Adult vitamin D deficiency and adverse Brain Outcomes. In *Vitamin D Vol 2 Health, Disease and Therapeutics*, 4th ed.; Feldman, D., Ed.; Elsevier: London, UK, 2018; Volume 1, pp. 1147–1158.
19. Cui, X.; McGrath, J.J.; Burne, T.H.J.; Eyles, D.W. Vitamin D and schizophrenia: 20 years on. *Mol. Psychiatry* **2021**, *26*, 2708–2720. [CrossRef]
20. Holmoy, T.; Moen, S.M.; Gundersen, T.A.; Holick, M.F.; Fainardi, E.; Castellazzi, M.; Casetta, I. 25-hydroxyvitamin D in cerebrospinal fluid during relapse and remission of multiple sclerosis. *Mult. Scler.* **2009**, *15*, 1280–1285. [CrossRef]
21. Balabanova, S.; Richter, H.P.; Antoniadis, G.; Homoki, J.; Kremmer, N.; Hanle, J.; Teller, W.M. 25-Hydroxyvitamin D, 24, 25-dihydroxyvitamin D and 1,25-dihydroxyvitamin D in human cerebrospinal fluid. *Klin. Wochenschr.* **1984**, *62*, 1086–1090. [CrossRef]
22. Ahonena, L.; Maireb, F.B.R.; Savolainenc, S.; Koprac, J.; Vreekenb, R.J.; Hankemeierb, T.; Myöhänenc, T.; Kyllia, P.; Kostiainena, R. Analysis of oxysterols and vitamin D metabolites in mouse brain and cell line samples by ultra-high-performance liquid chromatography-atmospheric pressure photoionization–mass spectrometry. *J. Chromatogr. A* **2014**, *1364*, 214–222. [CrossRef] [PubMed]
23. Xue, Y.; He, X.; Li, H.-D.; Deng, Y.; Yan, M.; Cai, H.-L.; Tang, M.-M.; Dang, R.-L.; Jiang, P. Simultaneous quantification of 25-Hydroxyvitamin D3 and 24,25-dihydroxyvitamin D3 in rats shows strong correlations between serum and brain tissue levels. *Int. J. Endocrinol.* **2015**, *2015*, 296531. [CrossRef] [PubMed]
24. Fu, X.; Dolnikowski, G.G.; Patterson, W.B.; Dawson-Hughes, B.; Zheng, T.; Morris, M.C.; Holland, T.M.; Booth, S.L. Determination of Vitamin D and Its Metabolites in Human Brain Using an Ultra-Pressure LC-Tandem Mass Spectra Method. *Curr. Dev. Nutr.* **2019**, *3*, nzz074. [CrossRef] [PubMed]
25. Eyles, D.W.; Smith, S.; Kinobe, R.; Hewison, M.; McGrath, J.J. Distribution of the vitamin D receptor and 1 alpha-hydroxylase in human brain. *J. Chem. Neuroanat.* **2005**, *29*, 21–30. [CrossRef]
26. Prufer, K.; Jirikowski, G.F. 1.25-Dihydroxyvitamin D3 receptor is partly colocalized with oxytocin immunoreactivity in neurons of the male rat hypothalamus. *Cell. Mol. Biol.* **1997**, *43*, 543–548.
27. Clemens, T.L.; McGlade, S.A.; Garrett, K.P.; Horiuchi, N.; Hendy, G.N. Tissue-specific regulation of avian vitamin D-dependent calcium-binding protein 28-kDa mRNA by 1,25-dihydroxyvitamin D3. *J. Biol. Chem.* **1988**, *263*, 13112–13116. [CrossRef]
28. Prufer, K.; Veenstra, T.D.; Jirikowski, G.F.; Kumar, R. Distribution of 1,25-dihydroxyvitamin D3 receptor immunoreactivity in the rat brain and spinal cord. *J. Chem. Neuroanat.* **1999**, *16*, 135–145. [CrossRef]
29. Walbert, T.; Jirikowski, G.F.; Prufer, K. Distribution of 1,25-dihydroxyvitamin D3 receptor immunoreactivity in the limbic system of the rat. *Horm. Metab. Res.* **2001**, *33*, 525–531. [CrossRef]
30. Craig, T.A.; Sommer, S.; Sussman, C.R.; Grande, J.P.; Kumar, R. Expression and regulation of the vitamin D receptor in the zebrafish, Danio rerio. *J. Bone Miner. Res.* **2008**, *23*, 1486–1496. [CrossRef]
31. Wang, Y.; Becklund, B.R.; DeLuca, H.F. Identification of a highly specific and versatile vitamin D receptor antibody. *Arch. Biochem. Biophys.* **2010**, *494*, 166–177. [CrossRef]
32. Wang, Y.; DeLuca, H.F. Is the vitamin D receptor found in muscle? *Endocrinology* **2011**, *152*, 354–363. [CrossRef] [PubMed]

33. Eyles, D.W.; Liu, P.Y.; Josh, P.; Cui, X. Intracellular distribution of the vitamin D receptor in the brain: Comparison with classic target tissues and redistribution with development. *Neuroscience* **2014**, *268*, 1–9. [CrossRef] [PubMed]
34. Stumpf, W.E.; O'Brien, L.P. 1,25 (OH)2 vitamin D3 sites of action in the brain. An autoradiographic study. *Histochemistry* **1987**, *87*, 393–406. [CrossRef] [PubMed]
35. Cui, X.; Pelekanos, M.; Liu, P.Y.; Burne, T.H.J.; McGrath, J.J.; Eyles, D. The vitamin D receptor in dopamine neurons; its presence in human substantia nigra and its ortogenesis in rat midbrain. *Neuroscience* **2013**, *236*, 77–87. [CrossRef]
36. Veenstra, T.D.; Prufer, K.; Koenigsberger, C.; Brimijoin, S.W.; Grande, J.P.; Kumar, R. 1,25-Dihydroxyvitamin D3 receptors in the central nervous system of the rat embryo. *Brain Res.* **1998**, *804*, 193–205. [CrossRef]
37. Burkert, R.; McGrath, J.; Eyles, D. Vitamin D receptor expression in the embryonic rat brain. *Neurosci. Res. Commun.* **2003**, *33*, 63–71. [CrossRef]
38. Erben, R.G.; Soegiarto, D.W.; Weber, K.; Zeitz, U.; Lieberherr, M.; Gniadecki, R.; Moller, G.; Adamski, J.; Balling, R. Deletion of deoxyribonucleic acid binding domain of the vitamin D receptor abrogates genomic and nongenomic functions of vitamin D. *Mol. Endocrinol.* **2002**, *16*, 1524–1537. [CrossRef]
39. Cui, X.; McGrath, J.J.; Burne, T.H.; Mackay-Sim, A.; Eyles, D.W. Maternal vitamin D depletion alters neurogenesis in the developing rat brain. *Int. J. Dev. Neurosci.* **2007**, *25*, 227–232. [CrossRef]
40. Liu, H.; He, Y.; Beck, J.; da Silva Teixeira, S.; Harrison, K.; Xu, Y.; Sisley, S. Defining vitamin D receptor expression in the brain using a novel VDR(Cre) mouse. *J. Comp. Neurol.* **2021**, *529*, 2362–2375. [CrossRef]
41. Fu, G.K.; Lin, D.; Zhang, M.Y.; Bikle, D.D.; Shackleton, C.H.; Miller, W.L.; Portale, A.A. Cloning of human 25-hydroxyvitamin D-1 alpha-hydroxylase and mutations causing vitamin D-dependent rickets type 1. *Mol. Endocrinol.* **1997**, *11*, 1961–1970.
42. Zehnder, D.; Bland, R.; Williams, M.C.; McNinch, R.W.; Howie, A.J.; Stewart, P.M.; Hewison, M. Extrarenal expression of 25-hydroxyvitamin d(3)-1 alpha-hydroxylase. *J. Clin. Endocrinol. Metab.* **2001**, *86*, 888–894. [PubMed]
43. Landel, V.; Stephan, D.; Cui, X.; Eyles, D.; Feron, F. Differential expression of vitamin D-associated enzymes and receptors in brain cell subtypes. *J. Steroid Biochem. Mol. Biol.* **2018**, *177*, 129–134. [CrossRef] [PubMed]
44. Naveilhan, P.; Neveu, I.; Baudet, C.; Ohyama, K.Y.; Brachet, P.; Wion, D. Expression of 25(OH) vitamin D3 24-hydroxylase gene in glial cells. *Neuroreport* **1993**, *5*, 255–257. [CrossRef] [PubMed]
45. Hendrix, I.; Anderson, P.; May, B.; Morris, H. Regulation of gene expression by the CYP27B1 promoter-study of a transgenic mouse model. *J. Steroid Biochem. Mol. Biol.* **2004**, *89*, 139–142. [CrossRef] [PubMed]
46. Eyles, D.W.; McGrath, J.J. Vitamin D brain development and function. In *Vitamin D Vol 1 Biochemistry, Physiology and Diagnostics*; Feldman, D., Ed.; Elsevier: London, UK, 2018; Volume 1, pp. 563–581.
47. Eyles, D.W. Vitamin D: Brain and Behavior. *JBMR Plus* **2021**, *5*, e10419. [CrossRef]
48. Brown, J.; Bianco, J.I.; McGrath, J.J.; Eyles, D.W. 1,25-dihydroxyvitamin D3 induces nerve growth factor, promotes neurite outgrowth and inhibits mitosis in embryonic rat hippocampal neurons. *Neurosci. Lett.* **2003**, *343*, 139–143. [CrossRef]
49. Marini, F.; Bartoccini, E.; Cascianelli, G.; Voccoli, V.; Baviglia, M.G.; Magni, M.V.; Garcia-Gil, M.; Albi, E. Effect of 1alpha,25-dihydroxyvitamin D3 in embryonic hippocampal cells. *Hippocampus* **2010**, *20*, 696–705. [CrossRef]
50. Horst, R.L.; Napoli, J.L.; Littledike, E.T. Discrimination in the metabolism of orally dosed ergocalciferol and cholecalciferol by the pig, rat and chick. *Biochem. J.* **1982**, *204*, 185–189. [CrossRef]
51. Chabas, J.F.; Alluin, O.; Rao, G.; Garcia, S.; Lavaut, M.N.; Risso, J.J.; Legre, R.; Magalon, G.; Khrestchatisky, M.; Marqueste, T.; et al. Vitamin D2 potentiates axon regeneration. *J. Neurotrauma* **2008**, *25*, 1247–1256. [CrossRef]
52. Ribeiro, M.C.; Moore, S.M.; Kishi, N.; Macklis, J.D.; MacDonald, J.L. Vitamin D Supplementation Rescues Aberrant NF-kappaB Pathway Activation and Partially Ameliorates Rett Syndrome Phenotypes in Mecp2 Mutant Mice. *eNeuro* **2020**, *7*. [CrossRef]
53. Tague, S.E.; Smith, P.G. Vitamin D deficiency leads to sensory and sympathetic denervation of the rat synovium. *Neuroscience* **2014**, *279*, 77–93. [CrossRef] [PubMed]
54. Feron, F.; Marqueste, T.; Bianco, J.; Gueye, Y.; Chabas, J.F.; Decherchi, P. Repairing the spinal cord with vitamin D: A promising strategy. *Biol. Aujourdhui* **2014**, *208*, 69–75. [CrossRef] [PubMed]
55. Neveu, I.; Naveilhan, P.; Jehan, F.; Baudet, C.; Wion, D.; De Luca, H.F.; Brachet, P. 1,25-dihydroxyvitamin D3 regulates the synthesis of nerve growth factor in primary cultures of glial cells. *Brain Res. Mol. Brain Res.* **1994**, *24*, 70–76. [CrossRef]
56. Wion, D.; MacGrogan, D.; Neveu, I.; Jehan, F.; Houlgatte, R.; Brachet, P. 1,25-Dihydroxyvitamin D3 is a potent inducer of nerve growth factor synthesis. *J. Neurosci. Res.* **1991**, *28*, 110–114. [CrossRef]
57. Neveu, I.; Naveilhan, P.; Baudet, C.; Brachet, P.; Metsis, M. 1,25-dihydroxyvitamin D3 regulates NT-3, NT-4 but not BDNF mRNA in astrocytes. *Neuroreport* **1994**, *6*, 124–126. [CrossRef]
58. Dursun, E.; Gezen-Ak, D.; Yilmazer, S. A novel perspective for Alzheimer's disease: Vitamin D receptor suppression by amyloid-beta and preventing the amyloid-beta induced alterations by vitamin D in cortical neurons. *J. Alzheimers. Dis.* **2011**, *23*, 207–219. [CrossRef]
59. Gezen-Ak, D.; Dursun, E.; Yilmazer, S. The Effects of Vitamin D Receptor Silencing on the Expression of LVSCC-A1C and LVSCC-A1D and the Release of NGF in Cortical Neurons. *PLoS ONE* **2011**, *6*, e17553. [CrossRef]
60. Saporito, M.S.; Wilcox, H.M.; Hartpence, K.C.; Lewis, M.E.; Vaught, J.L.; Carswell, S. Pharmacological induction of nerve growth factor mRNA in adult rat brain. *Exp. Neurol.* **1993**, *123*, 295–302. [CrossRef]

61. Granholm, A.C.; Reyland, M.; Albeck, D.; Sanders, L.; Gerhardt, G.; Hoernig, G.; Shen, L.; Westphal, H.; Hoffer, B. Glial cell line-derived neurotrophic factor is essential for postnatal survival of midbrain dopamine neurons. *J. Neurosci.* **2000**, *20*, 3182–3190. [CrossRef]
62. Oo, T.F.; Burke, R.E. The time course of developmental cell death in phenotypically defined dopaminergic neurons of the substantia nigra. *Brain Res. Dev. Brain Res.* **1997**, *98*, 191–196. [CrossRef]
63. Shirazi, H.A.; Rasouli, J.; Ciric, B.; Rostami, A.; Zhang, G.X. 1,25-Dihydroxyvitamin D3 enhances neural stem cell proliferation and oligodendrocyte differentiation. *Exp. Mol. Pathol.* **2015**, *98*, 240–245. [CrossRef] [PubMed]
64. Orme, R.P.; Bhangal, M.S.; Fricker, R.A. Calcitriol imparts neuroprotection in vitro to midbrain dopaminergic neurons by upregulating GDNF expression. *PLoS ONE* **2013**, *23*, e62040. [CrossRef] [PubMed]
65. Pertile, R.; Cui, X.; Hammond, L.A.; Eyles, D.W. Vitamin D regulation of GDNF/Ret signaling in dopaminergic neurons. *FASEB J.* **2018**, *32*, 819–828. [CrossRef] [PubMed]
66. Eyles, D.; Brown, J.; Mackay-Sim, A.; McGrath, J.; Feron, F. Vitamin D3 and brain development. *Neuroscience* **2003**, *118*, 641–653. [CrossRef]
67. Hawes, J.E.; Tesic, D.; Whitehouse, A.J.; Zosky, G.R.; Smith, J.T.; Wyrwoll, C.S. Maternal vitamin D deficiency alters fetal brain development in the BALB/c mouse. *Behav. Brain Res.* **2015**, *286*, 192–200. [CrossRef]
68. Abdollahzadeh, M.; Panahpour, H.; Ghaheri, S.; Saadati, H. Calcitriol supplementation attenuates cisplatin-induced behavioral and cognitive impairments through up-regulation of BDNF in male rats. *Brain Res. Bull.* **2022**, *181*, 21–29. [CrossRef]
69. Bakhtiari-Dovvombaygi, H.; Izadi, S.; Zare, M.; Asgari Hassanlouei, E.; Dinpanah, H.; Ahmadi-Soleimani, S.M.; Beheshti, F. Vitamin D3 administration prevents memory deficit and alteration of biochemical parameters induced by unpredictable chronic mild stress in rats. *Sci. Rep.* **2021**, *11*, 16271. [CrossRef]
70. Mansouri, F.; Ghanbari, H.; Marefati, N.; Arab, Z.; Salmani, H.; Beheshti, F.; Hosseini, M. Protective effects of vitamin D on learning and memory deficit induced by scopolamine in male rats: The roles of brain-derived neurotrophic factor and oxidative stress. *Naunyn-Schmiedeberg's Arch. Pharmacol.* **2021**, *394*, 1451–1466. [CrossRef]
71. Bayat, M.; Kohlmeier, K.A.; Haghani, M.; Haghighi, A.B.; Khalili, A.; Bayat, G.; Hooshmandi, E.; Shabani, M. Co-treatment of vitamin D supplementation with enriched environment improves synaptic plasticity and spatial learning and memory in aged rats. *Psychopharmacology* **2021**, *238*, 2297–2312. [CrossRef]
72. Tan, X.; Gao, L.; Cai, X.; Zhang, M.; Huang, D.; Dang, Q.; Bao, L. Vitamin D3 alleviates cognitive impairment through regulating inflammatory stress in db/db mice. *Food Sci. Nutr.* **2021**, *9*, 4803–4814. [CrossRef]
73. Manjari, S.K.V.; Maity, S.; Poornima, R.; Yau, S.Y.; Vaishali, K.; Stellwagen, D.; Komal, P. Restorative Action of Vitamin D3 on Motor Dysfunction Through Enhancement of Neurotrophins and Antioxidant Expression in the Striatum. *Neuroscience* **2022**, *492*, 67–81. [CrossRef] [PubMed]
74. Cui, X.; Pelekanos, M.; Burne, T.H.; McGrath, J.J.; Eyles, D.W. Maternal vitamin D deficiency alters the expression of genes involved in dopamine specification in the developing rat mesencephalon. *Neurosci. Lett.* **2010**, *486*, 220–223. [CrossRef] [PubMed]
75. Luan, W.; Hammond, L.A.; Cotter, E.; Osborne, G.W.; Alexander, S.A.; Nink, V.; Cui, X.; Eyles, D.W. Developmental Vitamin D (DVD) Deficiency Reduces Nurr1 and TH Expression in Post-mitotic Dopamine Neurons in Rat Mesencephalon. *Mol. Neurobiol.* **2018**, *55*, 2243–2453. [CrossRef] [PubMed]
76. Kesby, J.P.; Cui, X.; Ko, P.; McGrath, J.J.; Burne, T.H.; Eyles, D.W. Developmental vitamin D deficiency alters dopamine turnover in neonatal rat forebrain. *Neurosci. Lett.* **2009**, *461*, 155–158. [CrossRef] [PubMed]
77. Luan, W.; Hammond, L.A.; Vuillermot, S.; Meyer, U.; Eyles, D.W. Maternal vitamin D prevents abnormal dopaminergic development and function in a mouse model of prenatal immune activation. *Sci. Rep.* **2018**, *8*, 9741. [CrossRef] [PubMed]
78. Cui, X.; Pertile, R.; Liu, P.; Eyles, D.W. Vitamin D regulates tyrosine hydroxylase expression: N-cadherin a possible mediator. *Neuroscience* **2015**, *304*, 90–100. [CrossRef]
79. Pertile, R.A.N.; Cui, X.; Eyles, D.W. Vitamin D signalling and the differentiation of developing dopamine systems. *Neuroscience* **2016**, *333*, 193–203. [CrossRef]
80. Yates, N.J.; Tesic, D.; Feindel, K.W.; Smith, J.T.; Clarke, M.W.; Wale, C.; Crew, R.C.; Wharfe, M.; Whitehouse, A.J.; Wyrwoll, C.S. Vitamin D is crucial for maternal care and offspring social behaviour in rats. *J. Endocrinol.* **2018**, *237*, 73–85. [CrossRef]
81. Ali, A.A.; Cui, X.; Pertile, R.A.N.; Li, X.; Medley, G.; Alexander, S.A.; Whitehouse, A.J.O.; McGrath, J.J.; Eyles, D.W. Developmental vitamin D deficiency increases foetal exposure to testosterone. *Mol. Autism.* **2020**, *11*, 96. [CrossRef]
82. Xu, X.J.; Shou, X.J.; Li, J.; Jia, M.X.; Zhang, J.S.; Guo, Y.; Wei, Q.Y.; Zhang, X.T.; Han, S.P.; Zhang, R.; et al. Mothers of autistic children: Lower plasma levels of oxytocin and Arg-vasopressin and a higher level of testosterone. *PLoS ONE* **2013**, *8*, e74849. [CrossRef]
83. Burne, T.H.; Becker, A.; Brown, J.; Eyles, D.W.; Mackay-Sim, A.; McGrath, J.J. Transient prenatal Vitamin D deficiency is associated with hyperlocomotion in adult rats. *Behav. Brain Res.* **2004**, *154*, 549–555. [CrossRef]
84. Kesby, J.P.; Burne, T.H.; McGrath, J.J.; Eyles, D.W. Developmental vitamin D deficiency alters MK 801-induced hyperlocomotion in the adult rat: An animal model of schizophrenia. *Biol. Psychiatry* **2006**, *60*, 591–596. [CrossRef]
85. Kesby, J.P.; O'Loan, J.C.; Alexander, S.; Deng, C.; Huang, X.F.; McGrath, J.J.; Eyles, D.W.; Burne, T.H. Developmental vitamin D deficiency alters MK-801-induced behaviours in adult offspring. *Psychopharmacology* **2012**, *220*, 455–463. [CrossRef]
86. O'Loan, J.; Eyles, D.W.; Kesby, J.; Ko, P.; McGrath, J.J.; Burne, T.H. Vitamin D deficiency during various stages of pregnancy in the rat; its impact on development and behaviour in adult offspring. *Psychoneuroendocrinology* **2007**, *32*, 227–234. [CrossRef]

87. Kesby, J.P.; Cui, X.; O'Loan, J.; McGrath, J.J.; Burne, T.H.; Eyles, D.W. Developmental vitamin D deficiency alters dopamine-mediated behaviors and dopamine transporter function in adult female rats. *Psychopharmacology* **2010**, *208*, 159–168. [CrossRef]
88. Becker, A.; Grecksch, G. Pharmacological treatment to augment hole board habituation in prenatal Vitamin D-deficient rats. *Behav. Brain Res.* **2006**, *166*, 177–183. [CrossRef]
89. Grecksch, G.; Ruthrich, H.; Hollt, V.; Becker, A. Transient prenatal vitamin D deficiency is associated with changes of synaptic plasticity in the dentate gyrus in adult rats. *Psychoneuroendocrinology* **2009**, *34*, S258–S264. [CrossRef]
90. Becker, A.; Eyles, D.W.; McGrath, J.J.; Grecksch, G. Transient prenatal vitamin D deficiency is associated with subtle alterations in learning and memory functions in adult rats. *Behav. Brain Res.* **2005**, *161*, 306–312. [CrossRef]
91. Turner, K.M.; Young, J.W.; McGrath, J.J.; Eyles, D.W.; Burne, T.H.J. Cognitive performance and response inhibition in developmentally vitamin D (DVD)-deficient rats. *Behav. Brain Res.* **2013**, *242*, 47–53. [CrossRef]
92. Overeem, K.; Alexander, S.; Burne, T.H.J.; Ko, P.; Eyles, D.W. Developmental Vitamin D Deficiency in the Rat Impairs Recognition Memory, but Has No Effect on Social Approach or Hedonia. *Nutrients* **2019**, *11*, 2713. [CrossRef]
93. Vuillermot, S.; Luan, W.; Meyer, U.; Eyles, D. Vitamin D treatment during pregnancy prevents autism-related phenotypes in a mouse model of maternal immune activation. *Mol. Autism.* **2017**, *8*, 9. [CrossRef]
94. Kazemi, F.; Babri, S.; Keyhanmehr, P.; Farid-Habibi, M.; Rad, S.N.; Farajdokht, F. Maternal vitamin D supplementation and treadmill exercise attenuated vitamin D deficiency-induced anxiety-and depressive-like behaviors in adult male offspring rats. *Nutr. Neurosci.* **2022**, 1–13. [CrossRef] [PubMed]
95. Schoenrock, S.A.; Tarantino, L.M. Deveopmental vitamin D deficiency and schizophrenia: The role of animal models. *Genes Brain Behav.* **2016**, *15*, 45–61. [CrossRef] [PubMed]
96. Harms, L.H.; Turner, K.M.; Eyles, D.W.; Young, J.W.; McGrath, J.J.; BUrne, T.H.J. Attentional processing in C57BL/6J mice exposed to developmental vitamin D deficiency. *PLoS ONE* **2012**, *7*, e35896.
97. Harms, L.R.; Eyles, D.W.; McGrath, J.J.; Mackay-Sim, A.; Burne, T.H. Developmental vitamin D deficiency alters adult behaviour in 129/SvJ and C57BL/6J mice. *Behav. Brain Res.* **2008**, *187*, 343–350. [CrossRef] [PubMed]
98. Harms, L.H.; Cowin, G.; Eyles, D.W.; Kurniawan, N.; McGrath, J.J.; Burne, T.H.J. Neuroanatomy and psychomimetic-induced locomotion in C57BL/6J and 129/X1SvJ mice exposed to developmental vitamin D deficiency. *Behav. Brain Res.* **2012**, *230*, 125–131. [CrossRef]
99. Hart, P.H.; Lucas, R.M.; Walsh, J.P.; Zosky, G.R.; Whitehouse, A.J.; Zhu, K.; Allen, K.L.; Kusel, M.M.; Anderson, D.; Mountain, J.A. Vitamin D in fetal development: Findings from a birth cohort study. *Pediatrics* **2015**, *135*, e167–e173. [CrossRef]
100. Voltas, N.; Canals, J.; Hernandez-Martinez, C.; Serrat, N.; Basora, J.; Arija, V. Effect of Vitamin D Status during Pregnancy on Infant Neurodevelopment: The ECLIPSES Study. *Nutrients* **2020**, *12*, 3196. [CrossRef]
101. Mutua, A.M.; Mogire, R.M.; Elliott, A.M.; Williams, T.N.; Webb, E.L.; Abubakar, A.; Atkinson, S.H. Effects of vitamin D deficiency on neurobehavioural outcomes in children: A systematic review. *Wellcome Open Res.* **2020**, *5*, 28. [CrossRef]
102. Arrhenius, B.; Upadhyaya, S.; Hinkka-Yli-Salomaki, S.; Brown, A.S.; Cheslack-Postava, K.; Ohman, H.; Sourander, A. Prenatal Vitamin D Levels in Maternal Sera and Offspring Specific Learning Disorders. *Nutrients* **2021**, *13*, 3321. [CrossRef]
103. Wicklow, B.; Gallo, S.; Majnemer, A.; Vanstone, C.; Comeau, K.; Jones, G.; L'Abbe, M.; Khamessan, A.; Sharma, A.; Weiler, H.; et al. Impact of Vitamin D Supplementation on Gross Motor Development of Healthy Term Infants: A Randomized Dose-Response Trial. *Phys. Occup. Ther. Pediatr.* **2016**, *36*, 330–342. [CrossRef] [PubMed]
104. Whitehouse, A.J.; Holt, B.J.; Serralha, M.; Holt, P.G.; Kusel, M.M.; Hart, P.H. Maternal serum vitamin D levels during pregnancy and offspring neurocognitive development. *Pediatrics* **2012**, *129*, 485–493. [CrossRef] [PubMed]
105. Hanieh, S.; Ha, T.T.; Simpson, J.A.; Thuy, T.T.; Khuong, N.C.; Thoang, D.D.; Tran, T.D.; Tuan, T.; Fisher, J.; Biggs, B.A. Maternal vitamin D status and infant outcomes in rural Vietnam: A prospective cohort study. *PLoS ONE* **2014**, *9*, e99005. [CrossRef] [PubMed]
106. Schmidt, R.J.; Niu, Q.; Eyles, D.W.; Hansen, R.L.; Iosif, A.M. Neonatal vitamin D status in relation to autism spectrum disorder and developmental delay in the CHARGE case-control study. *Autism. Res.* **2019**, *12*, 976–988. [CrossRef] [PubMed]
107. Lopez-Vicente, M.; Sunyer, J.; Lertxundi, N.; Gonzalez, L.; Rodriguez-Dehli, C.; Espada Saenz-Torre, M.; Vrijheid, M.; Tardon, A.; Llop, S.; Torrent, M.; et al. Maternal circulating Vitamin D3 levels during pregnancy and behaviour across childhood. *Sci. Rep.* **2019**, *9*, 14792. [CrossRef]
108. Sass, L.; Vinding, R.K.; Stokholm, J.; Bjarnadottir, E.; Noergaard, S.; Thorsen, J.; Sunde, R.B.; McGrath, J.; Bonnelykke, K.; Chawes, B.; et al. High-Dose Vitamin D Supplementation in Pregnancy and Neurodevelopment in Childhood: A Prespecified Secondary Analysis of a Randomized Clinical Trial. *JAMA Netw. Open* **2020**, *3*, e2026018. [CrossRef]
109. Li, B.; Xu, Y.; Zhang, X.; Zhang, L.; Wu, Y.; Wang, X.; Zhu, C. The effect of vitamin D supplementation in treatment of children with autism spectrum disorder: A systematic review and meta-analysis of randomized controlled trials. *Nutr. Neurosci.* **2022**, *25*, 835–845. [CrossRef]
110. Dehbokri, N.; Noorazar, G.; Ghaffari, A.; Mehdizadeh, G.; Sarbakhsh, P.; Ghaffary, S. Effect of vitamin D treatment in children with attention-deficit hyperactivity disorder. *World J. Pediatr.* **2019**, *15*, 78–84. [CrossRef]
111. Elshorbagy, H.H.; Barseem, N.F.; Abdelghani, W.E.; Suliman, H.A.I.; Al-Shokary, A.H.; Abdulsamea, S.E.; Elsadek, A.E.; Abdel Maksoud, Y.H.; Nour El Din, D. Impact of Vitamin D Supplementation on Attention-Deficit Hyperactivity Disorder in Children. *Ann. Pharmacother.* **2018**, *52*, 623–631. [CrossRef]

112. Miller, M.C.; Pan, X.; Eugene Arnold, L.; Mulligan, A.; Connor, S.; Bergman, R.; deBeus, R.; Roley-Roberts, M.E. Vitamin D levels in children with attention deficit hyperactivity disorder: Association with seasonal and geographical variation, supplementation, inattention severity, and theta:beta ratio. *Biol. Psychol.* **2021**, *162*, 108099. [CrossRef]
113. Mohammadpour, N.; Jazayeri, S.; Tehrani-Doost, M.; Djalali, M.; Hosseini, M.; Effatpanah, M.; Davari-Ashtiani, R.; Karami, E. Effect of vitamin D supplementation as adjunctive therapy to methylphenidate on ADHD symptoms: A randomized, double blind, placebo-controlled trial. *Nutr. Neurosci.* **2018**, *21*, 202–209. [CrossRef] [PubMed]
114. Zhou, P.; Wolraich, M.L.; Cao, A.H.; Jia, F.Y.; Liu, B.; Zhu, L.; Liu, Y.; Li, X.; Li, C.; Peng, B.; et al. Adjuvant effects of vitamin A and vitamin D supplementation on treatment of children with attention-deficit/hyperactivity disorder: A study protocol for a randomised, double-blinded, placebo-controlled, multicentric trial in China. *BMJ Open* **2021**, *11*, e050541. [CrossRef] [PubMed]
115. Pilz, S.; Trummer, C.; Theiler-Schwetz, V.; Grubler, M.R.; Verheyen, N.D.; Odler, B.; Karras, S.N.; Zittermann, A.; Marz, W. Critical Appraisal of Large Vitamin D Randomized Controlled Trials. *Nutrients* **2022**, *14*, 303. [CrossRef] [PubMed]
116. Suzuki, M.; Yoshioka, M.; Hashimoto, M.; Murakami, M.; Noya, M.; Takahashi, D.; Urashima, M. Randomized, double-blind, placebo-controlled trial of vitamin D supplementation in Parkinson disease. *Am. J. Clin. Nutr.* **2013**, *97*, 1004–1013. [CrossRef]
117. Kang, J.H.; Vyas, C.M.; Okereke, O.I.; Ogata, S.; Albert, M.; Lee, I.M.; D'Agostino, D.; Buring, J.E.; Cook, N.R.; Grodstein, F.; et al. Effect of vitamin D on cognitive decline: Results from two ancillary studies of the VITAL randomized trial. *Sci. Rep.* **2021**, *11*, 23253. [CrossRef]
118. Feige, J.; Moser, T.; Bieler, L.; Schwenker, K.; Hauer, L.; Sellner, J. Vitamin D Supplementation in Multiple Sclerosis: A Critical Analysis of Potentials and Threats. *Nutrients* **2020**, *12*, 783. [CrossRef]
119. Lieberherr, M. Effects of vitamin D3 metabolites on cytosolic free calcium in confluent mouse osteoblasts. *J. Biol. Chem.* **1987**, *262*, 13168–13173. [CrossRef]
120. Caffrey, J.M.; Farach-Carson, M.C. Vitamin D3 metabolites modulate dihydropyridine-sensitive calcium currents in clonal rat osteosarcoma cells. *J. Biol. Chem.* **1989**, *264*, 20265–20274. [CrossRef]
121. Ibi, M.; Sawada, H.; Nakanishi, M.; Kume, T.; Katsuki, H.; Kaneko, S.; Shimohama, S.; Akaike, A. Protective effects of 1 alpha,25-(OH)(2)D-3 against the neurotoxicity of glutamate and reactive oxygen species in mesencephalic culture. *Neuropharmacology* **2001**, *40*, 761–771. [CrossRef]
122. Brewer, L.D.; Thibault, V.; Chen, K.C.; Langub, M.C.; Landfield, P.W.; Porter, N.M. Vitamin D hormone confers neuroprotection in parallel with downregulation of L-type calcium channel expression in hippocampal neurons. *J. Neurosci.* **2001**, *21*, 98–108. [CrossRef]
123. Gezen-Ak, D.; Dursun, E.; Yilmazer, S. Vitamin D inquiry in hippocampal neurons: Consequences of vitamin D-VDR pathway disruption on calcium channel and the vitamin D requirement. *Neurol. Sci.* **2013**, *34*, 1453–1458. [CrossRef] [PubMed]
124. Zanatta, L.; Goulart, P.B.; Gonçalves, R.; Pierozan, P.; Winkelmann-Duarte, E.C.; Woehl, V.M.; Pessoa-Pureur, R.; Silva, F.R.; Zamoner, A. 1α,25-dihydroxyvitamin D(3) mechanism of action: Modulation of L-type calcium channels leading to calcium uptake and intermediate filament phosphorylation in cerebral cortex of young rats. *Biochim. Biophys. Acta* **2012**, *1823*, 1708–1719. [CrossRef] [PubMed]
125. Gooch, H.; Cui, X.; Anggono, V.; Trzaskowski, M.; Tan, M.C.; Eyles, D.W.; Burne, T.H.J.; Jang, S.E.; Mattheisen, M.; Hougaard, D.M.; et al. 1,25-Dihydroxyvitamin D modulates L-type voltage-gated calcium channels in a subset of neurons in the developing mouse prefrontal cortex. *Transl. Psychiatry* **2019**, *9*, 281. [CrossRef]
126. Schizophrenia Working Group of the Psychiatric Genomics Consortium. Biological insights from 108 schizophrenia-associated genetic loci. *Nature* **2014**, *511*, 421–427. [CrossRef] [PubMed]
127. Long, W.; Fatehi, M.; Soni, S.; Panigrahi, R.; Philippaert, K.; Yu, Y.; Kelly, R.; Boonen, B.; Barr, A.; Golec, D.; et al. Vitamin D is an endogenous partial agonist of the transient receptor potential vanilloid 1 channel. *J. Physiol.* **2020**, *598*, 4321–4338. [CrossRef] [PubMed]
128. Holick, M.F. Vitamin D deficiency. *N. Engl. J. Med.* **2007**, *357*, 266–281. [CrossRef]
129. Almeida Moreira Leal, L.K.; Lima, L.A.; Alexandre de Aquino, P.E.; Costa de Sousa, J.A.; Jatai Gadelha, C.V.; Felicio Calou, I.B.; Pereira Lopes, M.J.; Viana Lima, F.A.; Tavares Neves, K.R.; Matos de Andrade, G.; et al. Vitamin D (VD3) antioxidative and anti-inflammatory activities: Peripheral and central effects. *Eur. J. Pharmacol.* **2020**, *879*, 173099. [CrossRef]
130. Uberti, F.; Morsanuto, V.; Bardelli, C.; Molinari, C. Protective effects of 1α,25-Dihydroxyvitamin D3 on cultured neural cells exposed to catalytic iron. *Physiol. Rep.* **2016**, *4*, e12769. [CrossRef]
131. Chen, K.B.; Lin, A.M.; Chiu, T.H. Systemic vitamin D3 attenuated oxidative injuries in the locus coeruleus of rat brain. *Ann. N. Y. Acad. Sci.* **2003**, *993*, 313–324; discussion 345–349. [CrossRef]
132. Lin, A.M.; Fan, S.F.; Yang, D.M.; Hsu, L.L.; Yang, C.H. Zinc-induced apoptosis in substantia nigra of rat brain: Neuroprotection by vitamin D3. *Free Radic. Biol. Med.* **2003**, *34*, 1416–1425. [CrossRef]
133. Kasatkina, L.A.; Tarasenko, A.S.; Krupko, O.O.; Kuchmerovska, T.M.; Lisakovska, O.O.; Trikash, I.O. Vitamin D deficiency induces the excitation/inhibition brain imbalance and the proinflammatory shift. *Int. J. Biochem. Cell Biol.* **2020**, *119*, 105665. [CrossRef] [PubMed]
134. Garcion, E.; Nataf, S.; Berod, A.; Darcy, F.; Brachet, P. 1,25-Dihydroxyvitamin D3 inhibits the expression of inducible nitric oxide synthase in rat central nervous system during experimental allergic encephalomyelitis. *Brain Res. Mol. Brain Res.* **1997**, *45*, 255–267. [CrossRef]

135. Garcion, E.; Sindji, L.; Montero-Menei, C.; Andre, C.; Brachet, P.; Darcy, F. Expression of inducible nitric oxide synthase during rat brain inflammation: Regulation by 1,25-dihydroxyvitamin D3. *Glia* **1998**, *22*, 282–294. [CrossRef]
136. Lefebvre d'Hellencourt, C.; Montero-Menei, C.N.; Bernard, R.; Couez, D. Vitamin D3 inhibits proinflammatory cytokines and nitric oxide production by the EOC13 microglial cell line. *J. Neurosci. Res.* **2003**, *71*, 575–582. [CrossRef] [PubMed]
137. Hur, J.; Lee, P.H.; Kim, M.J.; Cho, Y.-W. Regulatory effect of 25-hydroxyvitamin D3 on nitric oxide production in activated microglia. *Korean J. Physiol. Pharmacol.* **2014**, *18*, 397–402. [CrossRef]
138. Huang, Y.; Ho, Y.; Lai, C.; Chiu, C.; Wang, Y. 1,25-dihydroxyvitamin D3 attenuates endotoxin-induced production of inflammatory mediators by inhibiting MAPK activation in primary cortical neuron-glia cultures. *J. Neuroinflammation* **2015**, *12*, 147. [CrossRef]
139. Loginova, M.; Mishchenko, T.; Savyuk, M.; Guseva, S.; Gavrish, M.; Krivonosov, M.; Ivanchenko, M.; Fedotova, J.; Vedunova, M. Double-Edged Sword of Vitamin D3 Effects on Primary Neuronal Cultures in Hypoxic States. *Int. J. Mol. Sci.* **2021**, *22*, 5417. [CrossRef]
140. Ali, A.; Cui, X.; Alexander, S.; Eyles, D. The placental immune response is dysregulated developmentally vitamin D deficient rats: Relevance to autism. *J. Steroid. Biochem. Mol. Biol.* **2018**, *180*, 73–80. [CrossRef]
141. Alessio, N.; Belardo, C.; Trotta, M.C.; Paino, S.; Boccella, S.; Gargano, F.; Pieretti, G.; Ricciardi, F.; Marabese, I.; Luongo, L.; et al. Vitamin D Deficiency Induces Chronic Pain and Microglial Phenotypic Changes in Mice. *Int. J. Mol. Sci.* **2021**, *22*, 3604. [CrossRef]
142. Rastegar-Moghaddam, S.H.; Hosseini, M.; Alipour, F.; Rajabian, A.; Ebrahimzadeh Bideskan, A. The effects of vitamin D on learning and memory of hypothyroid juvenile rats and brain tissue acetylcholinesterase activity and oxidative stress indicators. *Naunyn-Schmiedeberg's Arch. Pharmacol.* **2022**, *395*, 337–351. [CrossRef]
143. Mokhtari-Zaer, A.; Hosseini, M.; Salmani, H.; Arab, Z.; Zareian, P. Vitamin D3 attenuates lipopolysaccharide-induced cognitive impairment in rats by inhibiting inflammation and oxidative stress. *Life Sci.* **2020**, *253*, 117703. [CrossRef] [PubMed]
144. Xu, M.L.; Yu, X.J.; Zhao, J.Q.; Du, Y.; Xia, W.J.; Su, Q.; Du, M.M.; Yang, Q.; Qi, J.; Li, Y.; et al. Calcitriol ameliorated autonomic dysfunction and hypertension by down-regulating inflammation and oxidative stress in the paraventricular nucleus of SHR. *Toxicol. Appl. Pharmacol.* **2020**, *394*, 114950. [CrossRef] [PubMed]
145. Khassafi, N.; Zahraei, Z.; Vahidinia, Z.; Karimian, M.; Azami Tameh, A. Calcitriol Pretreatment Attenuates Glutamate Neurotoxicity by Regulating NMDAR and CYP46A1 Gene Expression in Rats Subjected to Transient Middle Cerebral Artery Occlusion. *J. Neuropathol. Exp. Neurol.* **2022**, *81*, 252–259. [CrossRef]
146. Vahidinia, Z.; Khassafi, N.; Tameh, A.A.; Karimian, M.; Zare-Dehghanani, Z.; Moradi, F.; Joghataei, M.T. Calcitriol Ameliorates Brain Injury in the Rat Model of Cerebral Ischemia-Reperfusion Through Nrf2/HO-1 Signalling Axis: An in Silico and in Vivo Study. *J. Stroke Cerebrovasc. Dis.* **2022**, *31*, 106331. [CrossRef] [PubMed]
147. Filomeni, G.; De Zio, D.; Cecconi, F. Oxidative stress and autophagy: The clash between damage and metabolic needs. *Cell Death Differ.* **2015**, *22*, 377–388. [CrossRef] [PubMed]
148. Bhowmick, S.; D'Mello, V.; Caruso, D.; Abdul-Muneer, P.M. Traumatic brain injury-induced downregulation of Nrf2 activates inflammatory response and apoptotic cell death. *J. Mol. Med.* **2019**, *97*, 1627–1641. [CrossRef] [PubMed]
149. Cui, C.; Wang, C.; Jin, F.; Yang, M.; Kong, L.; Han, W.; Jiang, P. Calcitriol confers neuroprotective effects in traumatic brain injury by activating Nrf2 signaling through an autophagy-mediated mechanism. *Mol. Med.* **2021**, *27*, 118. [CrossRef]
150. Hosseinirad, H.; Shahrestanaki, J.K.; Moosazadeh Moghaddam, M.; Mousazadeh, A.; Yadegari, P.; Afsharzadeh, N. Protective Effect of Vitamin D3 Against Pb-Induced Neurotoxicity by Regulating the Nrf2 and NF-kappaB Pathways. *Neurotox. Res.* **2021**, *39*, 687–696. [CrossRef]
151. Jiang, H.; Zhang, S. Therapeutic effect of acute and chronic use of different doses of vitamin D3 on seizure responses and cognitive impairments induced by pentylenetetrazole in immature male rats. *Iran. J. Basic Med. Sci.* **2022**, *25*, 84–95. [CrossRef]
152. Bayo-Olugbami, A.; Nafiu, A.B.; Amin, A.; Ogundele, O.M.; Lee, C.C.; Owoyele, B.V. Vitamin D attenuated 6-OHDA-induced behavioural deficits, dopamine dysmetabolism, oxidative stress, and neuro-inflammation in mice. *Nutr. Neurosci.* **2022**, *25*, 823–834. [CrossRef]
153. Lima, L.A.R.; Lopes, M.J.P.; Costa, R.O.; Lima, F.A.V.; Neves, K.R.T.; Calou, I.B.F.; Andrade, G.M.; Viana, G.S.B. Vitamin D protects dopaminergic neurons against neuroinflammation and oxidative stress in hemiparkinsonian rats. *J. Neuroinflammation* **2018**, *15*, 249. [CrossRef] [PubMed]
154. Hu, J.; Wu, J.; Wan, F.; Kou, L.; Yin, S.; Sun, Y.; Li, Y.; Zhou, Q.; Wang, T. Calcitriol Alleviates MPP(+)- and MPTP-Induced Parthanatos Through the VDR/PARP1 Pathway in the Model of Parkinson's Disease. *Front. Aging Neurosci.* **2021**, *13*, 657095. [CrossRef] [PubMed]
155. Fan, Y.G.; Pang, Z.Q.; Wu, T.Y.; Zhang, Y.H.; Xuan, W.Q.; Wang, Z.; Yu, X.; Li, Y.C.; Guo, C.; Wang, Z.Y. Vitamin D deficiency exacerbates Alzheimer-like pathologies by reducing antioxidant capacity. *Free Radic. Biol. Med.* **2020**, *161*, 139–149. [CrossRef] [PubMed]
156. Mehrabadi, S.; Sadr, S.S. Administration of Vitamin D3 and E supplements reduces neuronal loss and oxidative stress in a model of rats with Alzheimer's disease. *Neurol. Res.* **2020**, *42*, 862–868. [CrossRef] [PubMed]
157. Grimm, M.O.W.; Lauer, A.A.; Grosgen, S.; Thiel, A.; Lehmann, J.; Winkler, J.; Janitschke, D.; Herr, C.; Beisswenger, C.; Bals, R.; et al. Profiling of Alzheimer's disease related genes in mild to moderate vitamin D hypovitaminosis. *J. Nutr. Biochem.* **2019**, *67*, 123–137. [CrossRef]

158. Lin, F.Y.; Lin, Y.F.; Lin, Y.S.; Yang, C.M.; Wang, C.C.; Hsiao, Y.H. Relative D3 vitamin deficiency and consequent cognitive impairment in an animal model of Alzheimer's disease: Potential involvement of collapsin response mediator protein-2. *Neuropharmacology* **2020**, *164*, 107910. [CrossRef]
159. Lin, C.I.; Chang, Y.C.; Kao, N.J.; Lee, W.J.; Cross, T.W.; Lin, S.H. 1,25(OH)2D3 Alleviates Abeta(25-35)-Induced Tau Hyperphosphorylation, Excessive Reactive Oxygen Species, and Apoptosis Through Interplay with Glial Cell Line-Derived Neurotrophic Factor Signaling in SH-SY5Y Cells. *Int. J. Mol. Sci.* **2020**, *21*, 4215. [CrossRef] [PubMed]
160. De Oliveira, L.R.C.; Mimura, L.A.N.; Fraga-Silva, T.F.C.; Ishikawa, L.L.W.; Fernandes, A.A.H.; Zorzella-Pezavento, S.F.G.; Sartori, A. Calcitriol Prevents Neuroinflammation and Reduces Blood-Brain Barrier Disruption and Local Macrophage/Microglia Activation. *Front. Pharmacol.* **2020**, *11*, 161. [CrossRef]
161. Shirazi, H.A.; Rasouli, J.; Ciric, B.; Wei, D.; Rostami, A.; Zhang, G.X. 1,25-Dihydroxyvitamin D3 suppressed experimental autoimmune encephalomyelitis through both immunomodulation and oligodendrocyte maturation. *Exp. Mol. Pathol.* **2017**, *102*, 515–521. [CrossRef]
162. Nystad, A.E.; Wergeland, S.; Aksnes, L.; Myhr, K.M.; Bo, L.; Torkildsen, O. Effect of high-dose 1.25 dihydroxyvitamin D3 on remyelination in the cuprizone model. *APMIS* **2014**, *122*, 1178–1186. [CrossRef]
163. Paintlia, M.K.; Singh, I.; Singh, A.K. Effect of vitamin D3 intake on the onset of disease in a murine model of human Krabbe disease. *J. Neurosci. Res.* **2015**, *93*, 28–42. [CrossRef] [PubMed]
164. Sapolsky, R.M. Stress, Glucocorticoids, and Damage to the Nervous System: The Current State of Confusion. *Stress* **1996**, *1*, 1–19. [CrossRef] [PubMed]
165. Chen, T.L.; Cone, C.M.; Morey-Holton, E.; Feldman, D. Glucocorticoid regulation of 1,25(OH)2-vitamin D3 receptors in cultured mouse bone cells. *J. Biol. Chem.* **1982**, *257*, 13564–13569. [CrossRef]
166. Chen, T.L.; Cone, C.M.; Morey-Holton, E.; Feldman, D. 1 alpha,25-dihydroxyvitamin D3 receptors in cultured rat osteoblast-like cells. Glucocorticoid treatment increases receptor content. *J. Biol. Chem.* **1983**, *258*, 4350–4355. [CrossRef]
167. Neveu, I.; Barbot, N.; Jehan, F.; Wion, D.; Brachet, P. Antagonistic effects of dexamethasone and 1,25-dihydroxyvitamin D3 on the synthesis of nerve growth factor. *Mol. Cell. Endocrinol.* **1991**, *78*, R1–R6. [CrossRef]
168. Neveu, I.; Jehan, F.; Wion, D. Alteration in the levels of 1,25-(OH)2D3 and corticosterone found in experimental diabetes reduces nerve growth factor (NGF) gene expression in vitro. *Life Sci.* **1992**, *50*, 1769–1772. [CrossRef]
169. Lundqvist, J.; Norlin, M.; Wikvall, K. 1alpha,25-Dihydroxyvitamin D3 affects hormone production and expression of steroidogenic enzymes in human adrenocortical NCI-H295R cells. *Biochim. Biophys. Acta* **2010**, *1801*, 1056–1062. [CrossRef]
170. Obradovic, D.; Gronemeyer, H.; Lutz, B.; Rein, T. Cross-talk of vitamin D and glucocorticoids in hippocampal cells. *J. Neurochem.* **2006**, *96*, 500–509. [CrossRef]
171. Jiang, P.; Xue, Y.; Li, H.D.; Liu, Y.P.; Cai, H.L.; Tang, M.M.; Zhang, L.H. Dysregulation of vitamin D metabolism in the brain and myocardium of rats following prolonged exposure to dexamethasone. *Psychopharmacology* **2014**, *231*, 3345–3351. [CrossRef] [PubMed]
172. Camargo, A.; Dalmagro, A.P.; Platt, N.; Rosado, A.F.; Neis, V.B.; Zeni, A.L.B.; Kaster, M.P.; Rodrigues, A.L.S. Cholecalciferol abolishes depressive-like behavior and hippocampal glucocorticoid receptor impairment induced by chronic corticosterone administration in mice. *Pharmacol. Biochem. Behav.* **2020**, *196*, 172971. [CrossRef]
173. Camargo, A.; Dalmagro, A.P.; Rikel, L.; da Silva, E.B.; Simao da Silva, K.A.B.; Zeni, A.L.B. Cholecalciferol counteracts depressive-like behavior and oxidative stress induced by repeated corticosterone treatment in mice. *Eur. J. Pharmacol.* **2018**, *833*, 451–461. [CrossRef] [PubMed]
174. Koshkina, A.; Dudnichenko, T.; Baranenko, D.; Fedotova, J.; Drago, F. Effects of Vitamin D3 in Long-Term Ovariectomized Rats Subjected to Chronic Unpredictable Mild Stress: BDNF, NT-3, and NT-4 Implications. *Nutrients* **2019**, *11*, 1726. [CrossRef] [PubMed]
175. Sedaghat, K.; Yousefian, Z.; Vafaei, A.A.; Rashidy-Pour, A.; Parsaei, H.; Khaleghian, A.; Choobdar, S. Mesolimbic dopamine system and its modulation by vitamin D in a chronic mild stress model of depression in the rat. *Behav. Brain Res.* **2019**, *356*, 156–169. [CrossRef] [PubMed]
176. Sedaghat, K.; Naderian, R.; Pakdel, R.; Bandegi, A.R.; Ghods, Z. Regulatory effect of vitamin D on pro-inflammatory cytokines and anti-oxidative enzymes dysregulations due to chronic mild stress in the rat hippocampus and prefrontal cortical area. *Mol. Biol. Rep.* **2021**, *48*, 7865–7873. [CrossRef]
177. Neis, V.B.; Werle, I.; Moretti, M.; Rosa, P.B.; Camargo, A.; de, O. Dalsenter, Y.; Platt, N.; Rosado, A.F.; Engel, W.D.; de Almeida, G.R.L.; et al. Involvement of serotonergic neurotransmission in the antidepressant-like effect elicited by cholecalciferol in the chronic unpredictable stress model in mice. *Metab. Brain Dis.* **2022**, *37*, 1597–1608. [CrossRef]
178. Kaneko, I.; Sabir, M.S.; Dussik, C.M.; Whitfield, G.K.; Karrys, A.; Hsieh, J.C.; Haussler, M.R.; Meyer, M.B.; Pike, J.W.; Jurutka, P.W. 1,25-Dihydroxyvitamin D regulates expression of the tryptophan hydroxylase 2 and leptin genes: Implication for behavioral influences of vitamin D. *FASEB J.* **2015**, *29*, 4023–4035. [CrossRef]
179. Da Silva Souza, S.V.; da Rosa, P.B.; Neis, V.B.; Moreira, J.D.; Rodrigues, A.L.S.; Moretti, M. Effects of cholecalciferol on behavior and production of reactive oxygen species in female mice subjected to corticosterone-induced model of depression. *Naunyn-Schmiedeberg's Arch. Pharmacol.* **2020**, *393*, 111–120. [CrossRef]
180. Eyles, D.W.; Rogers, F.; Buller, K.; McGrath, J.J.; Ko, P.; French, K.; Burne, T.H. Developmental vitamin D (DVD) deficiency in the rat alters adult behaviour independently of HPA function. *Psychoneuroendocrinology* **2006**, *31*, 958–964. [CrossRef]

181. Tesic, D.; Hawes, J.E.; Zosky, G.R.; Wyrwoll, C.S. Vitamin D Deficiency in BALB/c Mouse Pregnancy Increases Placental Transfer of Glucocorticoids. *Endocrinology* **2015**, *156*, 3673–3679. [CrossRef]
182. Meaney, M.J. Maternal care, gene expression, and the transmission of individual differences in stress reactivity across generations. *Annu. Rev. Neurosci.* **2001**, *24*, 1161–1192. [CrossRef]
183. Rajabi-Naeeni, M.; Dolatian, M.; Qorbani, M.; Vaezi, A.A. Effect of omega-3 and vitamin D co-supplementation on psychological distress in reproductive-aged women with pre-diabetes and hypovitaminosis D: A randomized controlled trial. *Brain Behav.* **2021**, *11*, e2342. [CrossRef] [PubMed]
184. Tebben, P.J.; Singh, R.J.; Kumar, R. Vitamin D-Mediated Hypercalcemia: Mechanisms, Diagnosis, and Treatment. *Endocr. Rev.* **2016**, *37*, 521–547. [CrossRef] [PubMed]
185. Tuohimaa, P.; Keisala, T.; Minasyan, A.; Cachat, J.; Kalueff, A. Vitamin D, nervous system and aging. *Psychoneuroendocrinology* **2009**, *34*, S278–S286. [CrossRef] [PubMed]
186. Lima, G.O.; Menezes da Silva, A.L.; Azevedo, J.E.C.; Nascimento, C.P.; Vieira, L.R.; Hamoy, A.O.; Oliveira Ferreira, L.; Bahia, V.; Muto, N.A.; Lopes, D.C.F.; et al. 100 YEARS of VITAMIN D: Supraphysiological doses of vitamin D changes brainwave activity patterns in rats. *Endocr. Connect.* **2022**, *11*, e210457. [CrossRef] [PubMed]
187. Lam, V.; Takechi, R.; Mano, J. Vitamin D, Cerebrocapillary Integrity and Cognition in Murine Model of Accelerated Ageing. *Alzheimer's Dement.* **2017**, *13*, 1304. [CrossRef]
188. Razzaque, M.S.; Sitara, D.; Taguchi, T.; St-Arnaud, R.; Lanske, B. Premature aging-like phenotype in fibroblast growth factor 23 null mice is a vitamin D-mediated process. *FASEB J.* **2006**, *20*, 720–722. [CrossRef]
189. Momen, N.C.; Plana-Ripoll, O.; Agerbo, E.; Benros, M.E.; Borglum, A.D.; Christensen, M.K.; Dalsgaard, S.; Degenhardt, L.; de Jonge, P.; Debost, J.P.G.; et al. Association between Mental Disorders and Subsequent Medical Conditions. *N. Engl. J. Med.* **2020**, *382*, 1721–1731. [CrossRef]
190. Revez, J.A.; Lin, T.; Qiao, Z.; Xue, A.; Holtz, Y.; Zhu, Z.; Zeng, J.; Wang, H.; Sidorenko, J.; Kemper, K.E.; et al. Genome-wide association study identifies 143 loci associated with 25 hydroxyvitamin D concentration. *Nat. Commun.* **2020**, *11*, 1647. [CrossRef]
191. Wu, D.M.; Wen, X.; Han, X.R.; Wang, S.; Wang, Y.J.; Shen, M.; Fan, S.H.; Zhuang, J.; Li, M.Q.; Hu, B.; et al. Relationship between Neonatal Vitamin D at Birth and Risk of Autism Spectrum Disorders: The NBSIB Study. *J. Bone Miner. Res.* **2018**, *33*, 458–466. [CrossRef]
192. Windham, G.C.; Pearl, M.; Anderson, M.C.; Poon, V.; Eyles, D.; Jones, K.L.; Lyall, K.; Kharrazi, M.; Croen, L.A. Newborn vitamin D levels in relation to autism spectrum disorders and intellectual disability: A case-control study in california. *Autism. Res.* **2019**, *12*, 989–998. [CrossRef]
193. Saraf, R.; Morton, S.M.B.; Camargo, C.A.J.; Grant, C.C. Global summary of maternal and newborn vitamin D status—A systematic review. *Matern. Child Nutr.* **2016**, *12*, 647–668. [CrossRef] [PubMed]

 MDPI

Review

Comparing the Evidence from Observational Studies and Randomized Controlled Trials for Nonskeletal Health Effects of Vitamin D

William B. Grant [1,*], Barbara J. Boucher [2], Fatme Al Anouti [3] and Stefan Pilz [4]

1 Sunlight, Nutrition and Health Research Center, San Francisco, CA 94164-1603, USA
2 The London School of Medicine and Dentistry, The Blizard Institute, Barts, Queen Mary University of London, London E1 2AT, UK
3 Department of Health Sciences, College of Natural and Health Sciences, Zayed University, Abu Dhabi 144534, United Arab Emirates
4 Division of Endocrinology and Diabetology, Department of Internal Medicine, Medical University of Graz, 8036 Graz, Austria
* Correspondence: wbgrant@infionline.net

Abstract: Although observational studies of health outcomes generally suggest beneficial effects with, or following, higher serum 25-hydroxyvitamin D [25(OH)D] concentrations, randomized controlled trials (RCTs) have generally not supported those findings. Here we review results from observational studies and RCTs regarding how vitamin D status affects several nonskeletal health outcomes, including Alzheimer's disease and dementia, autoimmune diseases, cancers, cardiovascular disease, COVID-19, major depressive disorder, type 2 diabetes, arterial hypertension, all-cause mortality, respiratory tract infections, and pregnancy outcomes. We also consider relevant findings from ecological, Mendelian randomization, and mechanistic studies. Although clear discrepancies exist between findings of observational studies and RCTs on vitamin D and human health benefits these findings should be interpreted cautiously. Bias and confounding are seen in observational studies and vitamin D RCTs have several limitations, largely due to being designed like RCTs of therapeutic drugs, thereby neglecting vitamin D's being a nutrient with a unique metabolism that requires specific consideration in trial design. Thus, RCTs of vitamin D can fail for several reasons: few participants' having low baseline 25(OH)D concentrations, relatively small vitamin D doses, participants' having other sources of vitamin D, and results being analyzed without consideration of achieved 25(OH)D concentrations. Vitamin D status and its relevance for health outcomes can usefully be examined using Hill's criteria for causality in a biological system from results of observational and other types of studies before further RCTs are considered and those findings would be useful in developing medical and public health policy, as they were for nonsmoking policies. A promising approach for future RCT design is adjustable vitamin D supplementation based on interval serum 25(OH)D concentrations to achieve target 25(OH)D levels suggested by findings from observational studies.

Keywords: breast cancer; colorectal cancer; gestational diabetes; preeclampsia; preterm birth

Citation: Grant, W.B.; Boucher, B.J.; Al Anouti, F.; Pilz, S. Comparing the Evidence from Observational Studies and Randomized Controlled Trials for Nonskeletal Health Effects of Vitamin D. *Nutrients* **2022**, *14*, 3811. https://doi.org/10.3390/nu14183811

Academic Editor: Carsten Carlberg

Received: 11 August 2022
Accepted: 14 September 2022
Published: 15 September 2022

1. Introduction

The year 2022 marks a century since the discovery of vitamin D [1]. The National Library of Medicine's PubMed database currently includes 97,024 vitamin D–related publications, with 73,577 having "vitamin D" in the title or abstract [accessed August 5, 2022]. Most vitamin D research before 2000 was related to skeletal effects. Interest in nonskeletal effects had become the most important topic of vitamin D research by the year 2000 when there were 69,090 publications related to vitamin D, 59,104 having "vitamin D" in the title or abstract. Despite that vast body of literature, considerable confusion remains regarding vitamin D's role in determining health outcomes; this mainly results from disagreements between findings from observational studies of health outcome associations with serum 25-hydroxyvitamin D [25(OH)D] concentrations and those from randomized controlled trials (RCTs) of vitamin D supplementation. Findings in 2014 from the two approaches revealed widespread disagreement. Observational studies generally reported inverse correlations between serum 25(OH)D concentrations and health outcomes, whereas RCTs generally showed no significant health benefits of vitamin D supplementation [2]. Meta-analyses of prospective cohort studies of nonskeletal disorders had reported significantly reduced risk for highest versus lowest 25(OH)D concentrations for cardiovascular disease (CVD) incidence and mortality rates, diabetes incidence, colorectal cancer incidence, but not for several other cancers [2]. Because of these discrepancies, Autier and colleagues suggested that observational study findings could be the result of reverse bias—the lowering of 25(OH)D concentrations by illness—and of various biases and confounders inherent to observational study designs [2]. Some conditions do indeed reduce the level of 25(OH)D in serum. However, that conclusion ignores the fact that such reductions limit the supply of 25(OH)D to the tissues, thereby aggravating the effects of vitamin D inadequacy and might also reflect locally increased intracellular calcitriol formation following compensatory increases in intracellular 25(OH)D activating enzyme activity in some tissues [3].

By 2017, RCTs had failed to demonstrate significant nonskeletal benefits for vitamin D except for reductions in common upper respiratory tract infections (RTIs) and in asthma exacerbations [4–6]. By April 2019, however, RCTs had shown that vitamin D supplementation had beneficial effects for primary prevention of acute RTIs and reduced acute exacerbations of asthma and chronic obstructive pulmonary disease [7]. The situation deteriorated further in 2019, when results of two large vitamin D RCTs were reported, the VITamin D and OmegA-3 TriaL (*VITAL*) for cancer and CVD risks [8] and the Vitamin D and Type 2 *Diabetes* (*D2d*) trial [9]. Neither trial showed reductions of the disease of interest based on 'intention to treat' when comparing disease incidence between the treatment and placebo arms. Neither article's abstract mentioned the beneficial effects that secondary analyses later reported. Thus, the press reported that vitamin D did not prevent cancer, CVD, or diabetes and medical and public health bodies, media, and the public thought vitamin D's beneficial effects for nonskeletal health outcomes had been disproven.

Observational studies related to vitamin D are generally prospective cohort or nested case–control studies of health outcomes in relation to baseline serum 25(OH)D concentrations. Participants are enrolled in cohorts or studies and provide blood samples, other biological data, and information on lifestyle upon recruitment. Serum 25(OH)D concentrations are measured later in standardized assays using deep-frozen serum or plasma samples. Generally, no further information (beyond details on health status) is gathered from participants during the follow-up period, which can last from weeks or months to twenty years. At the end of follow-up, changes in health status are correlated with baseline serum 25(OH)D concentrations, with adjustments for data on other relevant factors at enrollment. In nested case–control studies, control subjects may or may not be carefully matched with cases. Case–control studies also can be done with 25(OH)D concentrations measured near the time of health events. However, such studies are unreliable for pre-disease serum 25(OH)D concentrations since disease can affect them; for example, severe infections can reduce 25(OH)D concentrations, an effect that is strongest for acute inflammatory diseases such as acute RTI [10] and is a concern with COVID-19 [10].

Several observational studies now exist based on vitamin D supplementation with measurements of 25(OH)D concentrations at baseline and at follow-up. Conditions studied include breast cancer [11], arterial hypertension [12], myocardial infarction (MI), all-cause mortality rate [13], and preterm birth [14]. According to a recent review [15], such observational studies require careful interpretation in light of their inherent limitations, as discussed later in this review.

Meta-analyses of prospective observational studies of cancer incidence showed significant inverse correlations between baseline serum 25(OH)D concentrations and incidence of 10 different types of cancers between 2016 and 2021 (including bladder, breast, colorectal, head and neck, liver, lung, ovarian, pancreatic, renal, and thyroid cancer) and a direct correlation for prostate cancer (Table 5 in [16]). However, without support from RCTs, such results from observational studies are generally overlooked or ignored.

Data from observational studies of vitamin D obtained from diet have been limited. The most important was that by Garland and colleagues, associating dietary vitamin D and calcium with reduced risk of colorectal cancer [17]. The main problem with using dietary supply values is that dietary sources generally account for only 10–20% of total vitamin D supply. Solar UVB exposure contributes most, as seen from the seasonal variation of serum 25(OH)D concentrations [18]. Meat is also an important source of vitamin D, due to its 25(OH)D content, [19] but 25(OH)D has only recently been included in food frequency tables.

Classically, RCTs are used to evaluate drugs for treatment of disease. People who might benefit from taking the drug of interest are invited to participate and, when enrolled, are randomly assigned to either the treatment or placebo arm. A basic assumption is that no participants obtain the drug outside the RCT. Drug efficacy and adverse effects are the investigated outcomes. Pharmacokinetics of new drugs are also characterized to identify optimum circulating levels of the drug for efficacy and for safety. Safety has often been studied for vitamin D, but the 25(OH) D concentrations needed for efficacy in different disorders have been largely ignored for vitamin D RCT designs until recently [20,21]. Moreover, even in the placebo group in vitamin D RCTs, there are no participants without any vitamin D intake, so that such trials can only compare groups with higher versus lower vitamin D supply.

Uncontrolled intervention studies of vitamin D to prevent dental caries were conducted on adolescents from 1924 to 1945 [22]. Vitamin D RCTs were first conducted in about 1973 for treatment of epileptic patients taking anticonvulsant drugs [23] and became more popular in the early 21st century for studies of disease prevention, being designed in accordance with the guidelines for trials of therapeutic agents, as discussed.

Most vitamin D RCTs for nonskeletal disorders report no benefits using intention-to-treat analyses [4]. However, problems with such RCTs have included that few participants had any degree of vitamin D deficiency [25(OH)D concentrations < 20 ng/mL] and that moderate, or even unspecified, vitamin D supplementation was permitted in the control or in both study arms. In addition, vitamin D doses were not optimized to achieve any specific target 25(OH)D concentration even though specific thresholds have been identified for many health benefits [15]. Thus, few nonskeletal benefits have been revealed to date from 'intention to treat' analyses. However, significant benefits identified have included reduced cancer mortality rates [24], acute RTI risks [25] and autoimmune disease risks [26], as mentioned, along with reductions in several adverse pregnancy outcomes, discussed later.

Heaney outlined guidelines for optimizing design and analysis of clinical studies of nutrients in 2014 [27]:

1. Basal nutrient status must be measured, used as an inclusion criterion for entry into study, and recorded in the report of the trial.
2. The intervention (change in nutrient exposure or intake) must be large enough to change nutrient status and must be quantified.
3. The change in nutrient status produced in trial participants must be measured by validated laboratory analyses and recorded in the report of the trial.
4. The hypothesis to be tested must be that a change in nutrient status (not just a change in diet) produces the sought effect.
5. Conutrient status must be optimized to ensure that the test nutrient is the only nutrition-related, limiting factor in the response.

A version of those guidelines specifically for vitamin D also has been published [28]. No large vitamin D RCTs reported to date for prevention of chronic and infectious diseases have followed these guidelines, partly because most of the trials were designed before 2014. Most RCTs still fail to use 25(OH)D concentrations as a criterion for participation. If they did, enrolling the desired number of participants would be hard. Many RCTs do not measure baseline or achieved 25(OH)D concentrations of any, let alone all, participants. Vitamin D_3 doses generally range from <1000 to 2000 IU/d and up to ~100,000 IU/month, with some trials using 4000 IU/d. No large vitamin D RCTs have yet optimized co-nutrient status. Magnesium concentration, for example, plays an important role in vitamin D metabolism and affects serum 25(OH)D concentrations upon vitamin D supplementation [29], and magnesium deficiency is common [30]. By contrast, some vitamin D RCTs gave the treatment arm calcium supplements which do not affect 25(OH)D concentrations but can affect health outcomes [31]. For example, calcium supplements but not high dietary calcium intakes may increase the risk of CVD [32].

Pilz and colleagues have also discussed secondary outcomes and subgroup analyses, noting that those analyses reported some beneficial effects of vitamin D supplementation but that they should be considered "explorative outcome" analyses [15]. The case could be made that even though the researchers did not propose looking for such outcomes in the trial design, if good mechanistic data exist to suggest that such results could be expected, they should be considered useful trial findings, especially since, given the large cost and effort needed to conduct large-scale vitamin D RCTs, it is doubtful that any further such trials will be carried out, as discussed later.

Mendelian randomization (MR) studies examine how genetic variation—typically single-nucleotide polymorphisms (SNPs)—affects health outcomes through the vitamin D pathways. Polymorphisms affecting serum 25(OH)D are used as proxies for long-term vitamin D provision [33]. SNPs are not thought to change in response to behavior or disease experiences, so that they should not be affected by confounding factors and are deemed suitable for assessing causality. However, whether epigenetic effects can modulate the effects of genetic variants throughout adulthood is the subject of ongoing research [34–36]. A total of 143 genetic variants associated with vitamin D have been identified in the UK Biobank dataset [37], and recent MR studies have used up to 77 SNPs [38]. Another recent study used 35 SNPs, accounting for 2.8% of the variation in 25(OH)D in the UK Biobank dataset [39]. MR analyses generally require many thousands, often ~100,000 participants, to obtain sufficient statistical power. Whereas linear MR analyses have indicated some significant effects of vitamin D, nonlinear analyses [for different levels of participant vitamin D status] may be more appropriate, since vitamin D effects are non-linear, particularly when J- or U-shaped curves are expected a priori [39,40].

Geographical and temporal ecological studies have been used to identify diseases affected by solar UVB exposure, an index for vitamin D production. Annual solar radiation was shown to be inversely correlated with U.S. colon cancer mortality rates by the brothers Garland in 1980 [41], who hypothesized vitamin D production as the mechanism. Many more ecological studies of cancer incidence and/or mortality rates with respect to indices of solar UVB doses have since been reported [16].

Ecological studies are often considered hypothesis generating and, indeed, have led to many further studies exploring vitamin D's role in reducing cancer risks [16]. Many other diseases also have incidence and/or mortality rates inversely correlated with solar UVB doses: anaphylaxis/food allergy, atopic dermatitis and eczema, attention deficit–hyperactivity disorder, autism, back pain, cancer, dental caries, type 1 diabetes, hypertension, inflammatory bowel disease, lupus, mononucleosis, multiple sclerosis (MS), Parkinson disease, pneumonia, rheumatoid arthritis, and sepsis. [42]; all those diseases have also been linked to low 25(OH)D concentrations. Unfortunately, geographical ecological studies are considered weak, not just being hypothesis generating, but because unmodeled confounders (residual confounding) might explain the findings. However, two ecological studies have looked at cancer mortality rates for whites in the United States with respect to summertime solar UVB doses. The findings with UVB alone [43] were later found to be virtually unchanged after including other cancer risk–modifying factors such as alcohol consumption, Hispanic heritage, lung cancer (as a proxy for smoking and diet), poverty, and degree of urbanization [44]. For diseases where solar UVB doses have been inversely correlated with risk, no non–vitamin D effects are apparent from UVB exposure. However, for diseases with pronounced seasonal variations, such as CVD, hypertension, and RTIs, it appears that UVA-induced increases in serum nitric oxide (NO) as well as changes in temperature could also affect risk [45], though those diseases show no significant inverse correlations with annual UVB doses.

Reconciling differences between observational studies and RCTs regarding vitamin D is important for several reasons. One is that observational studies are generally not considered able to establish causality because some degrees of bias and confounding can never be totally excluded. For example, 'associations' could be due to unmodeled factors such as non–vitamin D health benefits of UV exposure [45]. RCTs are generally considered able to show causality and are relied on in medicine and public health policies for guidance but major vitamin D RCTs for nonskeletal effects have not shown any significant effects in primary outcome analyses. By contrast, MR studies are now reporting significant inverse correlations between genetically determined 25(OH)D concentrations and the risk of diseases such as CVD [40]. If observational studies can be shown to determine 25(OH)D concentration–health outcome relationships reliably, at least for some important diseases or outcomes, those studies could provide a basis for public health policies instead of waiting for "proof" of causality from RCTs that may never be carried out, as was the case for lung cancer due to smoking [46].

This review aims to present findings from observational studies of 25(OH)D concentrations and from RCTs of vitamin D supplementation for major health outcomes and to evaluate each approach's strengths and weaknesses together with brief discussions of results from other approaches, such as MR studies, the mechanisms of vitamin D known to be relevant to each outcome, and ecological studies relevant to possible health effects of vitamin D. The primary health outcomes considered are Alzheimer's disease (AD) and dementia; autoimmune diseases; cancers; CVD; COVID-19; major depressive disorder (MDD), type 2 diabetes mellitus (T2DM); hypertension; mortality (all-cause); and RTIs, as well as pregnancy and birth outcomes. This review is a narrative rather than a systematic review. It aims to summarize the existing literature regarding some common health outcomes and focuses on comparing findings from different study designs and discussing their strengths and limitations.

2. Results

2.1. Diseases and Outcomes

2.1.1. Autoimmune Diseases

In autoimmune diseases, an aberrant immune response is directed against normal human proteins [47]. Calcitriol modulates the immune system through effects on B cells, $CD4^+$ and $CD8^+$ cells, dendritic cells, innate lymphoid cells, macrophages, and unconventional T cells [48].

By March 2019, observational studies reported an inverse association between vitamin D status and developing autoimmune diseases, such as systematic lupus erythematosus, thyrotoxicosis, type 1 diabetes, MS, iridocyclitis, Crohn's disease, ulcerative colitis, psoriasis vulgaris, seropositive rheumatoid arthritis, and polymyalgia rheumatica [47].

In the VITAL trial, vitamin D supplementation reduced the risk of incident autoimmune diseases [26]. In the vitamin D arm, 123 participants in the treatment group and 155 in the placebo group developed a confirmed autoimmune disease (hazard ratio [HR] = 0.78 [95% confidence interval (CI), 0.61–0.99; $p = 0.05$]). The incident autoimmune diseases that differ in the raw numbers in the vitamin D versus placebo group were unspecified autoimmune disease (40 vs. 56), polymyalgia rheumatica (31 vs. 43), autoimmune thyroid disease (21 vs. 11), rheumatoid arthritis (15 vs. 24), and psoriasis (15 vs. 23). In this context, viral or bacterial infections, through increasing inflammation, are risk factors for autoimmune thyroid disease [49], rheumatoid arthritis [50], and psoriasis [51]. The finding that vitamin D supplementation reduces autoimmune disease risks can be supported logically, by the ability of vitamin D to reduce the risk of many infections through inducing the secretion of cathelicidin (LL-37) [52] and by reducing inflammation per se [53].

2.1.2. Cancers

Cancers generally arise from genetic mutations of cells leading to tumor formation. The immune system maintains active cell surveillance to evaluate whether they belong in the organs or tissues where they are located. Vitamin D plays an important role in that process, regulating cellular differentiation, progression and apoptosis. Vitamin D also reduces angiogenesis around tumors and reduces the development of metastasis [16,54].

The first observational report of serum 25(OH)D and cancer outcome was a 1989 U.S. study [55] over 8 years in Maryland, involving 34 colon cancer cases diagnosed between August 1975 and January 1983 and 67 matched controls from a pool of 25,620 volunteers. The risk of colon cancer decreased with 25(OH)D concentrations above 20 ng/mL: for ranges of 4–19, 20–26, 27–32, 33–41, and 42–91 ng/mL, odds ratios (ORs) were 1.00, 0.48, 0.25, 0.21, and 0.73, respectively. The higher OR for 42–91 ng/mL was probably due to participants' taking vitamin D supplements, perhaps on medical advice to address concerns about osteoporosis [56]. A recent meta-analysis of colorectal cancer incidence with respect to serum 25(OH)D concentration in prospective studies found an insignificant increase in OR for 25(OH)D above 100 ng/mL compared to 87.5–100 ng/mL [57].

Many similar observational studies have been conducted, generally for colorectal cancer. A 2019 meta-analysis included 12 studies for men and 13 for women. Muñoz and Grant [16] point out that the authors of that meta-analysis did not consider or adjust the results for follow-up time. Thus, the reduced risk for higher 25(OH)D concentrations would have been underestimated, leading to the conclusion that men who developed colorectal cancer derived no benefit [57]. As shown in Figure 1 of [16], a linear regression fit to OR versus follow-up time had the equation OR = 0.74 + 0.031x years (r = 0.79) for men and OR = 0.77 + 0.081x years (r = 0.25) for women. Thus, adjusting the reported values for study duration could have yielded similar beneficial effects for men as for women because the apparent reductions decline with increasing follow-up time, probably due to changes in serum 25(OH)D concentrations [58–60]. The 2015 analysis showed that the regression fitted to the prospective observational studies of breast cancer incidence over time had an association for time zero, by extrapolation, that corresponded well with results from case–control studies in which 25(OH)D had been measured near the time of diagnosis [60].

An observational study of breast cancer incidence based on analysis of individual participant data from 3325 participants in two vitamin D RCTs [31,61] and 1713 participants in the GrassrootsHealth (GRH) prospective cohort study was reported in 2018 [11]. Serum 25(OH)D concentrations were measured at baseline and after a year of follow-up in the RCTs and every 6 months in the GRH cohort. Participants in the treatment arms of the RCTs took 1100 IU/d of vitamin D_3 plus 1450 mg/d of calcium in the first Lappe study and 2000 IU/d of vitamin D_3 plus 1500 mg/d of calcium in the second Lappe study while participants in the GRH study chose their own vitamin D supplemental intake. A total of 77 women developed breast cancer during the study periods, of whom 14 were from the GRH cohort. Women who achieved a 25(OH)D concentration >60 ng/mL compared with those achieving <20 ng/mL had a large risk reduction: HR = 0.20 (95% CI, 0.05–0.82, $p = 0.03$; $p_{\text{trend}} = 0.04$) after adjustment for age, body mass index (BMI), smoking status, and calcium supplement intake. That study's strengths included that many participants were from vitamin D RCTs, that all had 25(OH)D concentrations measured at baseline and follow-up, and that 25(OH)D concentrations ranged from 10 to 75 ng/mL.

Cancers—RCTs

The *VITAL* study was an ambitious project aiming to determine whether vitamin D and omega-3 supplementation reduced cancer rates, CVD incidence, and mortality rates [8]. VITAL enrolled 25,871 participants, including 18,046 white people, 5106 African Americans, and 2719 people of other or unknown race or ethnic group. Mean (SD) age was 67 ± 7 years, mean (SD) BMI was 28 ± 6 kg/m^2, and mean (SD) baseline 25(OH)D concentrations in the vitamin D treatment group were 28 ng/mL for 395 males and 32 ng/mL for 441 females as determined from using blood samples collected mainly in winter or spring. Participants were recruited between November 2011 and March 2014, and the intervention ended December 31, 2017, yielding a median follow-up period of 5.3 years. Participants in the vitamin D treatment arm took 2000 IU/d of vitamin D_3. However, all participants up to 70 years of age were permitted to take additional amounts of up to 600 IU/d of vitamin D and up to 800 IU/d if aged over 70 years.

The VITAL study did not show that supplementing with 2000 IU/d of vitamin D_3 reduced risk of incident cancer according to intention-to-treat analyses [8]. However, in secondary analyses, people with BMI < 25 kg/m^2 had a reduced risk (HR = 0.76 [95% CI, 0.63–0.90]). The baseline and achieved 25(OH)D concentrations for those participants with such data gave values of 33.3 and 45.9 ng/mL, respectively. The difference between baseline and achieved 25(OH)D concentration was 12 ng/mL for three BMI categories (healthy weight, <25; overweight, 25 to <30; and obese, ≥30 kg/m^2). Evidently, the vitamin D supplementation used did not counter obesity-related inflammation, because obesity increases vitamin D requirements [62]. In a 26-week RCT involving 52 participants aged 18–50 years given 7000 IU/d of vitamin D_3, mean serum 25(OH)D concentrations increased from 13 to 44 ng/mL, but inflammatory markers including high-sensitivity C-reactive protein (hsCRP) were unaffected [63]. Chronic inflammation is an important cancer risk factor and meta-analyses haves associated CRP with breast cancer (HR = 1.14 [95% CI, 1.01–1.28]; OR = 1.23 [95% CI, 1.05–1.43]); colorectal cancer (OR = 1.34 [95% CI, 1.11–1.59]); and lung cancer (HR = 2.03 [95% CI, 1.59–2.50]) [64].

Why have RCTs not shown that vitamin D supplementation reduces risk of cancer incidence? That could be explained in three ways:

1. With findings based on vitamin D dose rather than achieved 25(OH)D concentration, enrolled participants would include those with relatively high 25(OH)D concentrations, lowering the chances of detecting reduced cancer incidence.
2. Higher 25(OH)D concentration lower cancer mortality risks more strongly than it reduces cancer incidence rates [24].
3. Vitamin D simply has no significant effect on cancer incidence.

Table 1 shows that vitamin D's effect was always stronger for cancer mortality rate than incidence rate in studies comparing those outcomes.

One narrative review includes 25 papers (8 RCTs on cancer patients, 8 population RCTs, and 9 observational studies) found through March 2021, published between 2003 and 2020. That review revealed some evidence that vitamin D supplementation in cancer patients could improve cancer survival, but no significant effect was reported in RCTs [68]. Some observational studies reported evidence associating vitamin D supplementation with increased survival among cancer patients, and only one study indicated an opposite effect. Those findings, therefore, were inconclusive.

Prospective or retrospective cohort studies evaluating the association between blood 25(OH)D level and survival outcomes in women with breast cancer were included in another review [69]. Outcome measures included overall survival, breast cancer–specific survival, and disease-free survival. Twelve studies involving 8574 female breast cancer patients were identified and analyzed. In comparing the lowest with highest category of baseline 25(OH)D level, the pooled adjusted HR was 1.57 (95% CI, 1.35–1.83) for overall survival, 1.98 (95% CI, 1.55–2.53) for disease-free survival, and 1.44 (95% CI, 1.14–1.81) for breast cancer–specific survival.

Table 1. Findings regarding cancer incidence and mortality rates with respect to 25(OH)D concentrations or vitamin D supplementation.

Study	Change in 25(OH)D	Incidence, RR or HR (95% CI)	Mortality, RR or HR (95% CI)	Author
Observational Studies				
Harvard Health Professionals Follow-Up Study, all cancer	10 ng/mL	RR = 0.83 (0.74–0.92)	RR = 0.71 (0.60–0.83)	Giovannucci et al. [65]
Adult patients living in Olmsted County, Minnesota, all less skin cancer	<12 ng/mL	HR = 1.56 (1.03–2.36)	HR = 2.35 (1.01–5.48)	Johnson et al. [66]
Meta-analysis, breast cancer	High vs. low	RR = 0.92 (0.83–1.02)	RR = 0.58 (0.40–0.85)	Kim et al. [67]
VITAL, exclude first 2 years	Vitamin D treatment vs. placebo	HR = 0.94 (0.83–1.06)	HR = 0.75 (0.59–0.96)	Manson et al. [8]
RCTs				
RCTs, meta-analysis	All participants	(12 RCTs) SRR = 0.99 (0.04–1.03)	(6 RCT) SRR = 0.92 (0.82–1.03)	Keum et al. [24]
RCTs, meta-analysis	Normal-weight individuals	(1 RCT) SRR = 0.76 (0.60–0.94)		Keum et al. [24]

Abbreviations: 25(OH)D, 25-hydroxyvitamin D; 95% CI, 95% confidence interval; HR, hazard ratio; RCT, randomized controlled trial; RR, relative risk; SRR, summary RR; VITAL, VITamin D and OmegA-3 TriaL.

Cancers—Geographical Ecological Studies

Geographical ecological studies strongly support the UVB–vitamin D–cancer hypothesis as proposed in 1980 by the brothers Garland after comparing the U.S. colon cancer mortality rate map with the map of annual solar radiation [41]. In the latest review, four single-country or region ecological studies published between 2006 and 2012 reported inverse correlations between indices of solar UVB doses for incidence rates for 21 cancers. Furthermore, five single-country ecological studies published between 2006 and 2011 reported inverse correlations between indices of solar UVB doses and mortality rates for 24 different types of cancer [16].

The Centers for Disease Control and Prevention continues to post maps of cancer incidence rates averaged by state (https://gis.cdc.gov/Cancer/USCS/#/AtAGlance/, accessed 21 June 2022). The most recent maps are for 2019 and are similar to those in the *Atlas of Cancer Mortality Rates in the United States, 1950–94* [70]. A comparison of the lung cancer incidence maps with maps of U.S. obesity prevalence shows a high correlation between states with the highest obesity rates and states with the highest lung cancer rates. That finding is consistent with the finding that inflammation is an important risk factor for lung cancer [64] and that vitamin D does not reduce the raised CRP associated with obesity [63]. Similar findings are reported for colorectal cancer but not for female breast cancer [64]. Recent MR studies using nonlinear methodology do show significant effects of higher genetically determined vitamin D status in reducing CRP in deficient subjects [40].

Geographical ecological studies are based on indices of solar UVB doses. A concern is that solar UVB could be an index for more than one mechanism of solar UV, though that is considered unlikely. In support, a recent review examined the role of three UV mechanisms related to the seasonality of CVD, hypertension, and infectious diseases [45]. Reasonable evidence exists that solar UVB has non–vitamin D effects and that solar UVA increases serum NO concentrations, both of which can contribute to the seasonality of several diseases. However, cancers have little seasonal variation [71], and the mechanisms by which vitamin D reduces cancer risk are well known [16]. Thus, geographical ecological studies should be considered a strong support for the UVB–vitamin D–cancer hypothesis.

Such ecological studies are useful for several reasons:

- They are easy to conduct because they can be based on publicly available data.
- They include many participants.
- No participants are omitted.
- The analysis can include many other cancer risk–modifying factors averaged at the population level.
- They can be used to locate cancer hot spots globally.
- Analyses can be performed for different ethnicities and races and can be repeated for different periods.

Cancer—Mendelian Randomization Study

A 2021 MR study of genetically predicted cancer incidence used up to 77 independent SNPs for 25(OH)D, representing about 4% of the normal phenotypic variation in serum 25(OH)D, based on analysis of more than 400,000 UK Biobank participants [38]. Various cancer datasets with a genome-wide association study dataset were used. The total number of cancer cases included 122,977 for breast cancer, 25,509 for epithelial ovarian cancer, 12,906 for endometrial cancer, 79,148 for prostate cancer, 12,874 for melanoma, and 10,279 for esophageal cancer. The only cancer with a statistically significant result for increased genetically determined 25(OH)D was epithelial ovarian cancer, with an adjusted odds ratio (aOR) of 0.89 (95% CI, 0.82–0.96). Findings for endometrial, lung, mucinous, neuroblastoma, and pancreatic cancers showed aORs < 1.00 but were not statistically significant. The aORs for skin cancers and melanoma were in the expected direction: melanoma, 1.05 (95% CI, 0.90–1.23); squamous cell carcinoma of the skin, 1.02 (95% CI, 0.88–1.19); and basal cell carcinoma of the skin, 1.16 (95% CI, 1.04–1.28). The aOR for prostate cancer also was in the same direction, 1.11 (95% CI, 0.93 to 1.33). However, that MR study probably failed to detect significant associations between genetically determined 25(OH)D concentrations and cancer risk, having too few cancer cases, and since the genetic variant size effect on serum 25(OH)D values was no larger than the variance of the 25(OH)D assays used at time of measurement.

2.1.3. Cardiovascular Disease

Vitamin D could reduce CVD risk by reducing risks of vascular inflammation, endothelial dysfunction, proliferation of smooth muscle cells, hypertension, and secondary hyperparathyroidism [72]. Vitamin D also has several mechanisms to lower risk of metabolic syndrome [72], T2DM [73], seasonal influenza [74], and periodontal disease [75], that are all risk factors for CVD.

In the Framingham Offspring Study [76], serum 25(OH)D concentrations were measured in 1739 eligible participants between 1996 and 2001 followed up for a mean of 5.4 years. For the 688 hypertensive participants who developed a first CVD event, the fully adjusted HRs, compared with participants with 25(OH)D $\geq$ 15 ng/mL, were 2.07 (95% CI, 1.19–3.67) for 25(OH)D from 10 to <15 ng/mL and 2.43 (95% CI, 1.23–4.80) for those with 25(OH)D < 10 ng/mL; p_{trend} = 0.003. For those without hypertension, no significant differences were found.

A 2017 meta-analysis looked at total CVD events with respect to baseline 25(OH)D concentrations among 180,667 participants across 34 publications between 2008 and 2015 [77]. The RRs for an increase of 25(OH)D of 10 ng/mL varied as a function of follow-up time: for <5 years, RR = 0.84 (95% CI, 0.78–0.90); for 5 to <10 years, RR = 0.88 (95% CI, 0.77–1.01); and for $\geq$10 years, RR = 0.92 (95% CI, 0.89–0.96). The linear fit to the RR is RR = 0.81 + 0.0084x years. Again, the longer the follow-up, the lower the apparent beneficial effect of higher baseline 25(OH)D concentration. Figure 2 in that review showed that the RR varied from 1.00 at 0 ng/mL to 0.8 $\pm$ 0.02 at ~20 ng/mL.

Data from four prospective studies of CVD mortality rate versus baseline serum 25(OH)D concentrations were used to generate a graphical meta-analysis of HR versus 25(OH)D concentration for CVD mortality rate (Table 2 and Figure 1). The HR values were adjusted so that the aHR adjusted value at the highest 25(OH)D concentration for each study fell on the second-order fit to the data [78]. Thus, vitamin D is apparently more effective at reducing risks of CVD mortality than at reducing incidence rates, in general agreement with the finding for all-cancer incidence and mortality rates.

Table 2. Data on mean 25(OH)D concentration and aHR for CVD mortality extracted from four research articles as further adjusted as just discussed [78–81].

Mean 25(OH)D Concentration (ng/mL)	aHR	aHR Adjusted	aHR Adjusted, 95% CI Low	aHR Adjusted, 95% CI High	Author
12.5	1.20	1.39	1.01	1.90	Melamed et al. [79]
21.0	0.88	1.02	0.80	1.32	
28.3	0.83	0.96	0.75	1.24	
36.0	1.00	1.16	1.16	1.16	
7.00		2.36	1.17	4.75	Ginde et al. [78]
15.0		1.54	1.01	2.34	
25.0		1.26	0.85	1.88	
35.0		1.20	0.79	1.81	
45.0		1.00	1.00	1.00	
7.00	2.64	3.55	2.27	2.96	Semba et al. [80]
13.0	1.68	2.26	1.03	5.02	
21.0	2.19	2.95	1.42	6.20	
30.0	1.00	1.35	1.35	1.35	
15.0	1.52	1.79	1.52	2.11	Liu et al. [81]
25.0	1.09	1.29	1.11	1.51	
35.0	1.00	1.18	1.18	1.18	

Abbreviations: 95% CI, 95% confidence interval; 25(OH)D, 25-hydroxyvitamin D; aHR, adjusted hazard ratio.

A recent analysis of data from the UK Biobank dataset reported a linear inverse relationship between adjusted 25(OH)D concentration and odds of an incident CVD event [40]. The odds were 1.08 ± 0.03 at 4 ng/mL, 1.0 at 20 ng/mL, and 0.88 ± 0.03 at 52 ng/mL of 25(OH)D. Figure 3 in that article suggests that correcting 25(OH)D to above 20 and 40 ng/mL could reduce CVD incidence rates in the UK by 4% ± 2% and 6% ± 4%, respectively.

A long-term follow-up study of patients treated at the U.S. Veterans Health Administration from 1999 to 2018 showed reduced risk of MI for people supplementing with vitamin D [13]. The patients followed up were those with 25(OH)D concentration < 20 ng/mL at baseline. Many were counseled to take vitamin D supplements. For patients who achieved >30 ng/mL, the HR for MI was 0.65 (95% CI, 0.49–0.85) compared with those who achieved 20–30 ng/mL and 0.73 (95% CI, 0.55–0.96) as compared with those with 25(OH)D < 20 ng/mL. That effect is higher than that found in the UK Biobank study but only applied to MI. That study used propensity scores to correct for potential systematic differences between comparison groups. Included covariates were age, sex, BMI, hypertension, diabetes, coronary artery disease, congestive heart failure, peripheral arterial disease, chronic kidney disease, chronic obstructive pulmonary disease, smoking, concomitant therapies (aspirin, statin, and beta-blockers), and low-density lipoprotein cholesterol levels. The researchers also used propensity score–weighted, stabilized inverse probability of treatment weights to obtain unbiased estimates of treatment effects; hence strengthening the design and analysis. The propensity score is the probability of treatment assignment conditional on observed baseline characteristics and allows designing and analyzing an observational (nonrandomized) study so that it mimics some particular characteristics of an RCT [82].

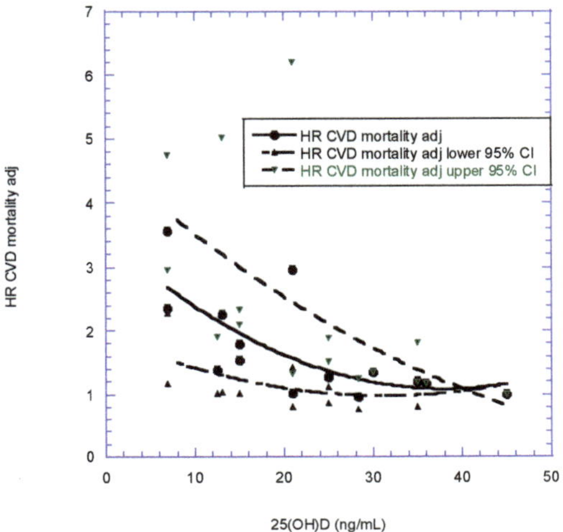

Figure 1. Hazard ratio (HR) and 95% confidence interval (CI) for CVD mortality according to mean serum 25-hydroxyvitamin D [25(OH)D] concentrations for data from four prospective observational studies.

Parathyroid hormone (PTH) plays an important role in CVD risk. Several observational studies report independent increases in risk for CVD for low 25(OH)D concentrations and high PTH concentrations [83–85]. Elevated PTH but not vitamin D deficiency has been associated with increased risk of heart failure [86]. An explanation for the difference in the relative contributions of 25(OH)D deficiency and higher PTH concentrations is that the PTH–25(OH)D relationship changes over time, as this ratio changes with age [87]. Figure 2 in that article shows median PTH values as a function of 25(OH)D concentration for ages < 20, 20–40, 40–60, and >60 years. When the PTH value of 65 pg/mL is used as the threshold concentration for significantly increased CVD events [84], the value is exceeded, with 25(OH)D concentrations of 7 ng/mL for those aged 40–60 years and of ~14 ng/mL for those > 60 years. This PTH effect may also help explain why CVD rates increase with age as well as why they are higher in winter than in summer [45].

A meta-analysis of observational studies of risk of atrial fibrillation with respect to serum 25(OH)D concentration as a continuous variable reported higher 25(OH)D concentrations associated with lower risk: OR = 0.96 (95% CI, 0.83–1.00; $p = 0.04$) from five cohort studies and 0.85 (95% CI, 0.79–0.92; $p < 0.0001$) from four case–control studies [88].

Cardiovascular Disease—RCTs

The VITAL study did not find that vitamin D supplementation reduced overall CVD risks [8], most likely because few participants had baseline 25(OH)D concentrations below 20 ng/mL (only 68/502 in the vitamin D treatment and placebo arms combined). A recent MR study however, clearly indicated that CVD risk decreases rapidly with increasing 25(OH)D concentration in those with baseline vitamin D inadequacy, [40] though the outcomes were not analyzed with respect to achieved 25(OH)D, PTH or season of the CVD event. However, half the participants had hypertension treated with medication and treating hypertension can significantly reduce CVD burden [89] and overall mortality rate [90].

Cardiovascular Disease—Mendelian Randomization

The MR study just mentioned showed that genetically predicted 25(OH)D concentrations correlated inversely with CVD risk [40], based on 44,510 CVD cases and 251,269 controls from the UK Biobank; it succeeded because the researchers calculated genetically instrumented 25(OH)D concentrations (as estimated using the 40 principal genetic factors affecting serum 25(OH)D) for each of 100 strata of baseline serum 25(OH)D concentration. Individual participants with 25(OH)D of 10 ng/mL had an OR of 1.11 (95% CI, 1.05–1.18) vs. 20 ng/mL. Individual participants with 25(OH)D of 30 ng/mL had an OR of 0.98 (95% CI, 0.97–0.99) vs. those at 20 ng/mL. The OR for the lowest genetically instrumented 25(OH)D concentration (4 ng/mL) was ~2.3 (95% CI, 1.4–3.5); their reductions in CVD risk reached 6% ± 3% as baseline 25(OH)D concentrations rose to 40 ng/mL.

The mechanisms by which vitamin D might reduce CVD risks are well understood. A recent meta-analysis reported inverse associations of vitamin D status with risks of metabolic syndrome and obesity, BMI, dyslipidemia, blood pressure (BP), insulin resistance, and dysglycemia. Meta-analysis of data from seven RCTs reported that supplementation reduced BP, abdominal obesity, and insulin resistance—all recognized CVD risk markers [63]. Mechanistic evidence shows that vitamin D reduces inflammation, an important factor in the progression of atherosclerosis. Vitamin D also lowers serum triglyceride levels and reduces the secretion of matrix metalloproteinase enzymes 2 and 9, which macrophages release when infiltrating arterial plaque, a process causing the plaque disruption that leads to acute arterial events through overlying clot formation [91]. Such findings are supported by the adverse effects of vitamin D deficiency seen experimentally on the vasculature [92].

2.1.4. COVID-19

Grant and colleagues suggested (April 2020) that higher 25(OH)D concentrations should reduce risk of COVID-19 incidence and death [93]. As of 16 June 2022, more than 2100 publications were found at pubmed.gov by searching for "vitamin D, COVID-19". Most observational reports were case–control studies with 25(OH)D measured around the time of a SARS-CoV-2–positive PCR test or COVID-19 symptoms.

Vitamin D's mechanisms to reduce risk of SARS-CoV-2 infection and COVID-19 incidence, severity, and death are now well known. These include reduced viral replication through inducing cathelicidin (LL-37) and reduced risk of the cytokine storm [93,94].

A large database of test results for SARS-CoV-2 positivity for patients who had serum 25(OH)D concentrations measured during the 12 months preceding the positive test by Quest Diagnostics between 9 March and 19 June 2020 was examined [95]. The 25(OH)D concentrations were seasonally adjusted, with a value of 32 ± 12 (mean ± SD) ng/mL. As analyzed by race/ethnicity, for white, non-Hispanic patients, SARS-CoV-2 positivity declined from 9% for 25(OH)D < 20 ng/mL to 5% for 60 ng/mL; for Hispanic patients, from 16% for <20 ng/mL to 8% at 50 ng/mL and from 19% for <20 ng/mL to 10% at 25(OH)D > 55 ng/mL in Black non-Hispanic subjects. However, such studies can be confounded by factors linked to both low circulating 25(OH)D concentrations and more severe COVID-19 [96].

More recently, two observational studies reported COVID-19 risk for people using vitamin D supplementation. One was from Barcelona, using data for vitamin D prescriptions and risk of SARS-CoV-2 or COVID-19 [97]. Most identifiable vitamin D use is by prescription rather than over-the-counter, though current usage is probably mainly over-the-counter. Patients on cholecalciferol treatment achieving 25(OH) D concentrations ≥30 ng/mL had lower risk of SARS-CoV2 infection, lower risk of severe COVID-19, and lower COVID-19 mortality than supplemented 25(OH)D–deficient patients (HR = 0.66 [95% CI, 0.46–0.93]; p = 0.02). Patients on calcifediol treatment achieving serum 25(OH)D concentrations ≥ 30 ng/mL had lower risks of SARS-CoV2 infection, of severe COVID-19, and of COVID-19 mortality than 25(OH)D–deficient patients not supplementing with vitamin D (HR = 0.56 [95% CI, 0.42–0.76]; p < 0.001).

The second study was based on 4599 veterans in U.S. Department of Veteran Affairs health care facilities. Participants received a positive SARS-CoV-2 test and a blood 25(OH)D test between 20 February 2020, and 8 November 2020, and were monitored for up to 60 days. After adjustment for all covariates, including race/ethnicity and poverty, a significant independent inverse dose–response relationship was evident between increasing 25(OH)D concentrations (from 15 to 60 ng/mL as a continuous variable) and decreasing probability of COVID-19–related hospitalization (from 24.1% to 18.7%; p = 0.009) and mortality (from 10.4% to 5.7%; p = 0.001) [98].

COVID-19 outcomes with respect to serum 25(OH)D concentrations from meta-analyses based on 72 observational studies published through 30 May 2021, were reported recently [99]; vitamin D deficiency/insufficiency increased odds of developing COVID-19, severe COVID-19, and death. Mean 25(OH)D concentrations were 4–5 ng/mL lower in people with COVID-19 than in controls for all outcomes. Associations between vitamin D deficiency/insufficiency and death were insignificant when studies with high risk of bias or reporting unadjusted effect estimates were excluded but bias and heterogeneity risks were high across all analyses. Discrepancies in timing of vitamin D testing, definitions of severe COVID-19, and of vitamin D deficiency/insufficiency contributed to that heterogeneity while serum 25(OH)D concentrations fall with severe COVID-19 illness [10], though whether that effect might increase the health benefits of adequate supplementation is a postulate urgently needing to be tested.

A meta-analysis of vitamin D supplementation on COVID-19–related outcomes from publications through January 2022 reported significantly reduced risk of admission to intensive care units (RR = 0.35 [95% CI, 0.20–0.62]) and reduced mortality (RR = 0.46 [95% CI, 0.30–0.70]) [100], though vitamin D status had no significant independent effect on COVID-19 incidence. Recently, positive effects of vitamin D supplementation on hospitalized COVID-19 patients were reported [101], adjuvant supplementation reducing hospital stay and duration of oxygen requirement.

Given that observational studies on the potential effectiveness of vitamin D supplementation for the prevention and treatment of COVID-9 are justifiably criticized for their limitations, inherent to observational study designs, readers should note that several public health recommendations during the COVID-19 pandemic, such as for the fourth SARS-CoV-2 vaccine dose, were also based solely on observational data and risk to benefit estimates but not on RCTs. Results from vitamin D RCTs are limited and inconsistent for COVID-19, so that no final conclusion can be drawn to date on the value of vitamin D supplementation regarding COVID-19.

2.1.5. Diabetes Mellitus Type 2

T2DM usually develops after a long period of increased insulin resistance (IR) where increased insulin concentrations become necessary to activate insulin effects in tissues such as liver and muscle. The demand for increased insulin secretion leads to islet beta cell damage and eventual inadequacy of insulin secretion with resultant hyperglycemia. Calcitriol is essential for normal insulin secretory responses to glucose and reduces the abnormal hepatic production of glucose and triglycerides seen in IR. Vitamin D effects also suppress inflammatory processes active in IR that contribute to the increased risks of both T2DM, and CVD seen with IR [102–105]. Vitamin D reduces inflammation and regulates intracellular Ca^{2+} level in many cell types, including islet beta cells and hepatocytes, contributing to reduced IR as reviewed by Szymczak-Pajor and colleagues [106].

Observational studies have long reported that 25(OH)D concentrations are inversely correlated with T2DM and with features of the metabolic syndrome [107]. A 2007 meta-analysis noted that for prevalence of T2DM in non-black people, the OR for highest versus lowest 25(OH)D concentration was 0.71 (95% CI, 0.57–0.89) [108]. A 2013 meta-analysis of 21 prospective studies involving 76,220 participants and 4996 incident T2DM cases showed that for each 4-ng/mL increase in 25(OH)D concentration, risk of T2DM decreased by 4% (95% CI, 3–6%), with the lowest risk near 60 ng/mL [109]. That analysis was updated by adding studies published up to 31 August 2016 [110]. With data from 31 (nested) case–control and cohort studies comparing participants with 25(OH)D values of approximately 20–30 ng/mL with those in the lowest category, the OR was 0.77 (95% CI, 0.72–0.82). With 23 (nested) case–control and cohort studies comparing participants with the highest 25(OH)D concentrations with those with the lowest 25(OH)D category, the OR was 0.66 (95% CI, 0.61–0.73). Again, reduced T2DM risks were seen up to 25(OH)D concentrations of ~50–60 ng/mL.

Diabetes Mellitus Type 2—RCTs

The *D2d* trial evaluated whether vitamin D supplementation could reduce risk of progressing from prediabetes to T2DM [9] and enrolled 2423 prediabetic participants, mean (SD) age 60 $\pm$ 10 years. Participants mean (SD) 25(OH)D concentration was 28 $\pm$ 10 ng/mL. Half were randomized to take 4000 IU/d of vitamin D_3 and half to take a placebo during a mean time of 2.5 years. On the basis of intention to treat, the HR for vitamin D compared with placebo was 0.88 (95% CI, 0.75–1.04; $p = 0.12$). In secondary analyses of this dataset, participants with BMI < 30 kg/m^2 had reduced T2DM risk, HR = 0.71 (95% CI, 0.53–0.95) as did those not given calcium supplements, HR = 0.81 (85% CI, 0.66–0.98).

A further secondary analysis of D2d data was then made, based on intratrial 25(OH)D concentrations [111]. For participants in the vitamin D treatment arm, each 10-ng/mL increase in 25(OH)D concentration above 20–30 ng/mL up to >50 ng/mL was associated with a significant HR of 0.75 (95% CI, 0.68–0.82) for progression to diabetes. No effect was seen in the placebo arm. Those findings strongly support the Heaney guidelines for basing analyses of vitamin D RCTs on achieved 25(OH)D concentrations, and show the importance of assessing vitamin D's effects on health by vitamin D status, not by dosages [27].

Diabetes Mellitus Type 2—Mendelian Randomization

Two MR analyses found that genetic variants of 25(OH)D were causally linked to risk of T2DM. The first one was reported in 2018 [112]. It used data from the China Kadoorie Biobank as well as other studies. A 10-ng/mL higher biochemically measured 25(OH)D was associated with a 9% (95% CI: 0–18%) lower risk of diabetes in the China Kadoorie Biobank. In a meta-analysis of all studies, a 10-ng/mL higher genetically instrumented 25(OH)D concentration was associated with a 14% (95% CI: 3–23%) lower risk of diabetes ($p = 0.01$) using two synthesis SNPs. An equivalent difference in 25(OH)D using a genetic score with 4 SNPs was not significantly associated with diabetes (odds ratio 8%, 95% CI: −1% to 16%, lower risk, $p = 0.07$), but had some evidence of pleiotropy.

A second MR analysis was performed using data on genetic variants for 25(OH)D from a genome-wide association study on UK Biobank subjects [n = 329,247] of European ancestry. A higher genetically instrumented 25(OH)D was causally linked to reduced risk of T2DM (OR per standard deviation increase in 25(OH)D = 0.95 [95% CI, 0.91–0.99]; $p = 0.01$) [113]. That study also confirmed vitamin D's causal role by studying two SNPs of the vitamin D–activating enzyme, where the HR for each 1-SD increase in 25(OH)D = 0.89 (95% CI, 0.82–0.98; $p = 0.02$).

2.1.6. Hypertension

Hypertension is generally defined by a systolic BP (SBP) >140 mmHg and a diastolic BP (DBP) > 90 mmHg. One risk causal factor for hypertension is endothelial dysfunction [114] that is related to lower availability of NO, an important vasodilator. Vitamin D can reduce endothelial dysfunction by suppressing renin production, which reduces activity of the renin–angiotensin–aldosterone system and increases expression of endothelial NO synthase (eNOS) [115]. Vitamin D also reduces production of reactive oxygen species and cyclooxygenase 1 (COX-1) mRNA and protein expression [116], thereby reducing production of endothelium-derived vasoconstrictive factors [117].

Researchers have found inversely correlated serum 25(OH)D concentrations with BP, starting with cross-sectional studies in 2005 [118] and nested case–control studies in 2007 [119]. Forman and colleagues included 613 men from the Health Professionals Follow-Up study and 1198 women from the Nurses' Health Study with measured baseline 25(OH)D concentrations who were followed up for 4–8 years. During the 4-year period of follow-up, 61 men and 129 women developed hypertension. The multivariable RRs comparing lowest with highest deciles of 25(OH)D were 2.31 (95% CI, 2.03–2.63) in men and 1.57 (95% CI, 1.44–1.72) in women.

A meta-analysis including 10 prospective studies from 2007 to 2015 reported that a linear increase of each 10 ng/mL in 25(OH)D resulted in an RR of 0.95 (95% CI, 0.90–1.00) [120].

Another meta-analysis of dose–response curves for hypertension versus 25(OH)D concentrations from 11 prospective cohort studies showed that the RR at 6 ng/mL versus 29 ng/mL = 1.37 (95% CI, 1.13–1.65). The RR decreased quasi-linearly up to 29 ng/mL before plateauing with 25(OH)Ds of ~50 ng/mL, where the RR was approximately 0.9 (95% CI, 0.8–1.0) [121]. The same meta-analysis found no effect for vitamin D supplementation on BP from the analysis of 27 RCTs.

A Canadian open-label vitamin D supplementation study looked at how raising serum 25(OH)D concentrations affected BP and hypertension [12]; the 8155 participants were given free vitamin D_3 and other supplements and counseled on how to achieve a 25(OH)D concentration > 40 ng/mL, with medically supervised dose adjustments. Of 592 participants hypertensive at enrollment, 71% were no longer hypertensive 12 ± 3 months later. Their mean (SD) SBP dropped by 18 ± 19 mmHg (95% CI, −24 to −12 mmHg) if not taking BP medications or by 14 ± 21 mmHg (95% CI, −18 to −9 mmHg) if taking BP medications after joining the program. Decreases in DBP were similar. However, for patients not hypertensive at enrollment, decreases in both SBP and DBP were insignificant. Though an observational study, the quality appears to be higher than that of many vitamin D RCTs. For example, many parameters were evaluated at baseline and tabulated by 25(OH)D concentration in 20-ng/mL increments. Values for most parameters, including SBP and DBP, did not vary significantly with baseline 25(OH)D concentration < 60 or 80 ng/mL. At the end of 1 year, mean (SD) serum 25(OH)D concentration had increased from 35 ± 15 ng/mL to 45 ± 16 ng/mL, but mean SBP, DBP, and pulse pressure were unchanged. After correction for possible confounding factors, only participants who were vitamin D insufficient (20–30 ng/mL) at baseline and achieved 25(OH)Ds > 40 ng/mL at follow-up had a lower risk of hypertension (OR = 0.10, 95% CI, 0.01–0.87; p = 0.03).

An unconsidered factor for BP risk is that solar UVA induces release of NO into the circulation, which can significantly lower BP as reported in 2014 by a team led by Weller [122]. The researchers showed that UVA irradiation of the skin lowers BP, that UVA increases nitrite production and reduces circulating nitrate, and that UVA increases blood flow in the forearm, the vasodilator NO being released by photolysis of nitrite to nitrate.

Hypertension—RCTs

Several vitamin D RCTs have looked at BP and hypertension. One studied black patients during the winters of 2008 and 2010 [123]. A total of 283 black people with median age 51 years were enrolled with a median 25(OH)D concentration of 16 ng/mL (interquartile range [IQR], 11–23 ng/mL), median SBP of 122 mmHg (IQR, 110–136 mmHg), and median DBP of 78 mmHg (IQR, 71–86 mmHg). Participants were assigned to four groups for 3 months of treatment in ~equal numbers for placebo, 1000, 2000, or 4000 IU/d of vitamin D_3 daily. The 3-month change in SBP per 1000 IU/d of vitamin D_3 given was −1.4 ± 0.7 mmHg (mean ± SE); p = 0.04, whereas that for DBP was −0.5 ± 0.5 mmHg.

Another vitamin D RCT regarding BP was conducted in Austria between June 2011 and August 2014 [124]. It enrolled 200 hypertensive participants with mean (SD) age 60 ± 11 years and 25(OH)D concentrations below 30 ng/mL. Half of the patients were assigned to the vitamin D treatment arm and received 2800 IU/d of vitamin D_3 for 8 weeks, whereas the other half received a placebo. The mean (SD) change in SBP was −0.4 (95% CI, −2.8 to 1.9) mmHg (p = 0.71). Data in this study [124] were reanalyzed for achieved 25(OH)D concentration [125], suggesting reduced SBP for high versus low

25(OH)D concentrations between 5 and 55 ng/mL (threshold of ~23 ng/mL). However, changes in this very short trial were not statistically significant.

Hypertension—Mendelian Randomization

An MR study regarding vitamin D and BP and hypertension was published in 2014 [126]. In a meta-analysis of data from 142,255 participants, the synthesis score for genetic increases in 25(OH)D was associated with a reduced risk of hypertension (OR per allele = 0.98 [95% CI, 0.96–0.99]; $p = 0.001$). In instrumental variable analyses, each 10% increase in genetically instrumented 25(OH)D concentration was associated with a change in SBP of −0.37 mmHg (95% CI, −0.73 to 0.003; $p = 0.052$), in DBP of −0.29 mmHg (95% CI, −0.52 to −0.07; $p = 0.01$), and an 8.1% decrease in hypertension (OR = 0.92 [95% CI, 0.87–0.97]; $p = 0.002$).

2.1.7. Mortality, All-Cause

All-cause mortality rate is mainly due to deaths from CVD, T2DM, respiratory disease, AD and other dementias, autoimmune diseases, and cancer, all of whose mortality rates are inversely correlated with serum 25(OH)D concentrations. Meta-analyses of prospective observational studies also report significant inverse correlations with all-cause mortality rates. One review involved 13 studies published between 2006 and 2010, including 5562 deaths among 62,548 participants [127]. That study showed a significant inverse relationship for 25(OH)D concentrations and mortality below 20 ng/mL. A second meta-analysis based on 32 studies published pre-2013 reported a similar cutoff value of 30 ng/mL [128]. A European individual participant data meta-analysis with 26,916 participants, of whom 6802 died during the median follow-up of 10.5 years, indicated a significant inverse correlation between baseline 25(OH)D concentration and all-cause mortality rates for values < 20 ng/mL [129]. The observational study based on veterans treated by the U.S. Veterans Administration Health System reported a HR of 0.61 (95% CI, 0.56–0.67) for all-cause mortality rate for participants who achieved 25(OH)Ds >30 ng/mL vs. those remaining at <20 ng/mL during 20 years of follow-up [13]. An MR study showed significant inverse correlations between genetically determined 25(OH)D concentrations and all-cause mortality rate [130], as a more recent study also reports [131].

2.1.8. Respiratory Tract Infections

Vitamin D reduces the risk of RTIs by increasing antimicrobial peptide gene expression, thereby increasing serum concentrations of human cathelicidin (LL-37) and defensins [52]; it also shifts the cytokine balance from proinflammatory T-helper (Th1) cell cytokine production to the production of anti-inflammatory Th2 cell cytokines [132].

Interest in vitamin D's role in RTIs began before the antibiotic era with a finding of reduced tuberculous illness with exposure to sunshine. That interest reignited with publication of the hypothesis that seasonal variations in solar UVB doses might explain seasonal variations in epidemic influenza, with peak rates in winter, from Cannell and colleagues [133]. That hypothesis was later shown to be only partially correct since temperature and absolute humidity also play important roles [134], while UV has a minor role [135]. However, that hypothesis did lead to further studies, e.g., showing that vitamin D supplementation reduced the risk of influenza A but not influenza B [136]. A meta-analysis of vitamin D RCTs for acute RTIs showed that daily or weekly supplementation with vitamin D significantly reduced risk of acute RTIs [6]. A later meta-analysis reported that maximal protection from vitamin D supplementation was achieved with daily doses of <400 IU for up to 12 months in younger participants (aged 1–16 years) with baseline 25(OH)D concentrations < 10 ng/mL [25].

2.1.9. Alzheimer's Disease and Other Dementias

The mechanisms by which vitamin D reduces risk of AD and dementia are reasonably well understood. Vitamin D triggers several neural pathways that may protect against those neurodegenerative mechanisms, including the deposition of amyloid plaques, inflammatory processes, neurofibrillary degeneration, glutamatergic excitotoxicity, excessive intraneuronal calcium influx, and oxidative stress [137].

Two recent meta-analyses looked at dementia and AD risks [138,139]. Jayedi and colleagues calculated the reduction in risk of dementia or AD for a 10-ng/mL increase in 25(OH)D for each study. Table 3 presents the data used in their meta-analysis. Values for mean 25(OH)D concentration were obtained from each article (except Littlejohns et al.). Mean 25(OH)D concentration was taken as the midpoint of 25(OH)D associated with a change in the HR in their Figure 2 and they found that higher 25(OH)D concentrations (up to 32 ng/mL) significantly reduced risk of dementia when Swedish study data [140] were omitted. Figure 2 shows the results for dementia (regression fit equation; HR = 3.3 − 0.16 × [25(OH)D] + 0.0027 × $[25(OH)D]^2$, r = 0.95). Figure 3 shows the same equation for AD; HR = 3.3 − 0.17 × [25(OH)D] + 0.0032 × $[25(OH)D]^2$, r = 0.75. This analysis highlights the value of considering different ways of doing meta-analyses of observational study data with vitamin D status. Graphs for follow-up time were also generated, but proved less informative than those based on 25(OH)D concentrations.

Table 3. Data associated with the observational studies in the meta-analysis by Jayedi and colleagues [138].

Country	Mean 25(OH)D (ng/mL)	Follow-Up (yrs)	Vascular Dementia, HR (95% CI) for 10 ng/mL Increase	Alzheimer's, HR (95% CI) for 10 ng/mL Increase	Author
US	12	5.6	0.57 (0.34–0.97)	0.61 (0.41–0.93)	Littlejohns et al. [141]
France	14	11.4	0.60 (0.47–0.78)	0.60 (0.47–0.78)	Feart et al. [142]
Finland	16	17.0	0.77 (0.62–0.92)		Knekt et al. [143]
Denmark	16	21.0		0.91 (0.82–1.02)	Afzal et al. [130]
Netherlands	20	13.3	0.77 (0.63–0.95)	0.73 (0.59–0.93)	Licher et al. [144]
US	22	16.6	0.93 (0.79–1.07)		Schneider et al. [145]
US	25	9.0	1.01 (0.88–1.14)	1.09 (0.95–1.12)	Karakis et al. [146]
Sweden	28	12.0	1.04 (0.93–1.17)	0.95 (0.81–1.12)	Olsson et al. [140]

Abbreviations: 95% CI, 95% confidence interval; 25(OH)D, 25-hydroxyvitamin D; HR, hazard ratio.

MR studies [147–149] have reported genetically predicted 25(OH)D concentrations correlating inversely with AD risk, as do vitamin D protein binding levels [150] with dementia risks [39].

2.1.10. Major Depressive Disorder

Neuroinflammation appears to be the key factor in onset and progression of MDD [151]. Mechanisms likely to underlie the beneficial effects reported for vitamin D in treating MDD [152] reviewed were, in particular, its antioxidant, anti-inflammatory, proneurogenic, and neuromodulatory properties. Vitamin D also modulates concentrations of gut microbiota, reducing bacterially induced activation of NF-κB in the intestine, thereby reducing remote inflammation further [153].

A meta-analysis of 29 RCTs with 4504 participants concluded that vitamin D supplementation reduced MDD incidence (standardized mean difference: −0.23) and improved responses to the treatment of depression (standardized mean difference: −0.92) [154]. The effects of 2800 IU/d vitamin D over >8 weeks were significant for both prevention and treatment. A meta-analysis of data from seven prospective observational studies of 16,287 older adults with 1157 cases of incident depression reported a pooled HR for depression per 10-ng/mL increase in 25(OH)D of 0.88 (95% CI, 0. 78–0.99) [155].

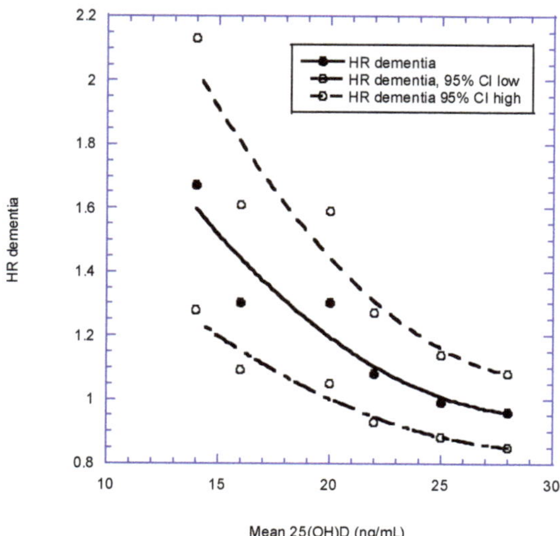

Figure 2. Hazard ratio (HR; mean and 95% confidence interval [CI]) for dementia according to mean 25-hydroxyvitamin D [25(OH)D] concentration in each of the six studies included in the meta-analysis [138].

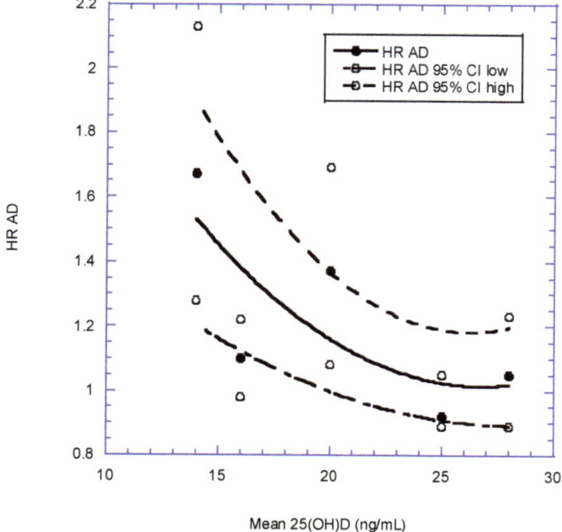

Figure 3. Hazard ratios (HR; mean and 95% confidence interval [CI]) for Alzheimer's disease (AD) according to mean 25-hydroxyvitamin D [25(OH)D] concentrations for each of the six studies included in the meta-analysis [138].

Another meta-analysis including 41 RCTs with 53,235 participants reported that vitamin D supplementation alleviated depressive symptoms (Hedges's $g = -0.32$ [95% CI, −0.41 to −0.23]; $p < 0.001$; $I^2 = 88\%$; grade, very low certainty) [156]. Vitamin D supplementation > 2000 IU/d reduced depressive symptoms ($g = -0.41$ [95% CI, −0.56 to −0.26]) much better than <2000 IU/d ($g = -0.18$ [95% CI, −0.29 to −0.08]). A ~linear decrease in

HR was seen with increases in serum 25(OH)D concentration, HR falling to 0.43 (95% CI, 0.20–0.92) at 65 ng/mL.

A nested case–control study conducted in Taiwan indicated that although moderate UVB exposure reduced risk of depression, high UVB exposure increased this risk [157]. The adjusted incidence rate ratio for moderate versus low UVB exposure was 0.89 (95% CI, 0.84–0.95); for high UVB, 1.12 (95% CI, 1.02–1.26); for very high UVB, 1.71 (95% CI, 1.51–1.95); and for extreme UVB, 2.79 (95% CI, 2.44–3.18). The authors proposed that high UVB exposure increased production of reactive oxygen species, thereby increasing inflammation. Those results also suggested that observational studies of risk of depression with changes in serum 25(OH)D concentration might be affected by whether 25(OH)D concentrations are raised by solar UVB exposure or by vitamin D supplementation in participants with higher rather than deficient baseline 25(OH)D concentrations.

A study based on the UK Biobank and two other databases showed that serum 25(OH)D concentrations were lower in participants with depression. However, MR analysis reported nonsignificant correlations between genetically determined 25(OH)D concentrations and risk of depression [158]. Thus, depression may lower 25(OH)D concentrations, perhaps by reducing sunlight exposure.

2.1.11. Pregnancy Disorders and Neonatal Outcomes

Vitamin D status during pregnancy is important for both mother and fetus. Wagner and Hollis published a comprehensive review on this topic in mid-2018 [159]. Observational studies reported significant inverse correlations between maternal 25(OH)D concentrations and risk of maternal problems, including preeclampsia [160], altered placental vascular pathology [161], cesarean delivery rates [162,163] gestational diabetes [164,165], and preterm birth rates [166,167]. Such studies also report similar associations for infant health outcomes, including brain dysfunction [168,169] and respiratory disorders [170]. Table 4 shows representative pregnancy outcomes with respect to maternal serum 25(OH)D concentrations from several observational studies [165–180].

Table 4. Representative pregnancy and infant outcomes with respect to maternal serum 25(OH)D concentration in observational studies.

Outcome	Setting	Outcome	Finding	Author
Birth weight				
Cesarean delivery, primary		Maternal 25(OH)D < 15 vs. >15 ng/mL	aOR = 3.8 (95% CI, 1.7–8.6)	Merewood et al. [162]
Cesarean delivery, primary		Maternal 25(OH)D < 15 vs. >15 ng/mL	aOR = 2.0 (95% CI, 1.2–3.3)	Scholl et al. [163]
Gestational diabetes	Meta-analysis, 29 studies	<20 vs. >20 ng/mL	OR = 1.39 (95% CI, 1.20–1.60)	Hu et al. [164]
Gestational diabetes	Meta-analysis, 27 studies	>20 vs. >30 ng/mL	OR = 1.26 (95% CI, 1.13–1.41)	Milajerdi et al. [165]
Preeclampsia	Hospital study	Early-onset severe preeclampsia, 10-ng/mL increase in 25(OH)D	aOR = 0.37 (95% CI, 0.22–0.62)	Robinson et al. [171]
Preeclampsia	Meta-analysis, 13 studies	Comparison of 25(OH)D	OR = 0.57 (95% CI, 0.51–0.65)	Serrano-Diaz et al. [172]
Preeclampsia	Meta-analysis, 11 studies	25(OH)D < 30 vs. > 30 ng/mL	OR = 1.44 (95% CI, 1.26–1.64)	Aguilar-Cordero et al. [160]
Preterm delivery	Hospital study	25(OH)D < 20 vs. > 40 ng/mL, <16 wks	OR = 3.8 (95% CI, 1.4–10.7)	Wagner et al. [166]

Table 4. *Cont.*

Outcome	Setting	Outcome	Finding	Author
Preterm delivery	Meta-analysis, 16 studies	25(OH)D < 20 vs. > 20 ng/mL	OR = 1.25 (95% CI, 1.13–1.38)	Zhou et al. [167]
Preterm delivery	Open-label vitamin D supplementation	25(OH)D > 40 vs. > 20 ng/mL	SES adjusted OR = 0.41 (95% CI, 0.24–0.72)	McDonnell et al. [14]
Infant outcomes				
Brain dysfunction	Language impairment in childhood vs. maternal 25(OH)D at 18 weeks pregnancy	6–18 vs. 29–62 ng/mL	aOR = 1.97 (95% CI, 1.00–3.93, $p < 0.05$)	Whitehouse et al. [168]
Brain dysfunction	Risk of ADHD, meta-analysis, 5 studies	High vs. low 25(OH)D	OR/RR = 0.72 (95% CI, 0.59–0.89)	Garcia-Serna et al. [169]
Respiratory dysfunction	Risk of asthma vs. maternal 25(OH)D, 11 studies	High vs. low 25(OH)D	OR = 0.78 (95% CI, 0.69–0.89)	Shi et al. [170]
Respiratory dysfunction	Risk of wheeze vs. maternal 25(OH)D, 14 studies	High vs. low 25(OH)D	OR = 0.65 (95% CI, 0.54–0.79)	Shi et al. [170]

Abbreviations: 25(OH)D, 25-hydroxyvitamin D; 95% CI, 95% confidence interval; ADHD, attention deficit–hyperactivity disorder; aOR, adjusted odds ratio; OR, odds ratio; SES socioeconomic status.

Pregnancy Outcomes in Interventional Studies

An open-label vitamin D supplementation observational study involved 1064 women delivering singleton births at the Medical University of South Carolina between September 2015 and December 2016 [14]. Women were given free bottles of 5000-IU vitamin D_3 and counseled on how to achieve a 25(OH)D concentration >40 ng/mL. Achieved 25(OH)D concentration was also measured. The goal was to see whether raising serum 25(OH)D concentration >40 ng/mL could reduce risk of preterm birth (<37 weeks of gestation). The socioeconomic status aOR was 0.41 (95% CI, 0.24–0.72). The gestation period increased from 36.8 $\pm$ 0.3 weeks with 25(OH)Ds~5 ng/mL to 38.3 $\pm$ 0.2 weeks at 25 ng/mL after supplementation

Meta-analyses of RCTs have largely, but not consistently, reported benefits of vitamin D supplementation in reducing risks of adverse pregnancy outcomes including low birth weight [173,174], cesarean delivery rates [175], gestational diabetes [174], preeclampsia [176], and preterm delivery [177] (see Table 5).

Table 5. Meta-analyses of vitamin D supplementation RCTs, comparing vitamin D supplementation with placebo.

Outcome	Setting	Finding	Author
Birth weight, low	Review of 5 RCTs	RR = 0.55 (95% CI, 0.35–0.87)	Palacios et al. [174]
Birth weight	Review of 11 RCTs	Increased weight, mean difference = 114 g (95% CI, 63–165 g)	Gallo et al. [173]
Cesarean delivery, primary	Review of 6 RCTs	OR = 0.9 (95% CI, 0.7–1.2)	Gallo et al. [173]
Cesarean delivery in Iran	Review of 5 RCTs	RR = 0.61 (95% CI, 0.44–0.83)	Saha and Saha, [175]
Gestational diabetes	Review of 4 RCTs	RR = 0.51 (95% CI, 0.27–0.97)	Palacios et al. [174]
Preeclampsia	Review of 27 RCTs	RR = 0.37 (95% CI, 0.26–0.52)	Fogacci et al. [176]
Preterm delivery	Review of 17 RCTs	RR = 0.70 (95% CI, 0.49–1.00)	Kinshella et al. [177]

Abbreviations: 95% CI, 95% confidence interval; OR, odds ratio; RCT, randomized controlled trial; RR, relative risk.

Maternal vitamin D deficiency leads to epigenetic changes in offspring and is associated with increased risks to bone health in childhood and to increased childhood obesity, those problems appearing to persist into later life. These changes serve as a further reason to ensure vitamin D adequacy in pregnancy [178,179].

3. Discussion

Why most vitamin D RCTs used small vitamin D doses—generally at or below 2000 IU/d—is puzzling. One reason may be the Institute of Medicine's 2011 report setting 4000 IU/d of vitamin D as the upper limit and recommending 600 IU/d for those aged up to 70 years and 800 IU/d for those aged above 70 years—intakes aiming to achieve 25(OH)D of at least 20 ng/mL [180]. The committee was concerned about reports of U-shaped 25(OH)D concentration–health outcome relationships at that time because of a National Cancer Institute review [181] of findings from prospective studies with respect to baseline 25(OH)D concentrations for breast, esophageal, pancreatic, and prostate cancer. The cited studies had drawbacks such as long follow-up times up to 15 years, which reduce the apparent benefit of higher baseline 25(OH)D concentration [16]; not evaluating whether participants changed vitamin D supplementation practices before or during the follow-up periods; and not evaluating vitamin D status during the study or at study completion. Some U-shaped 25(OH)D concentration–health outcome relationships seen have been proposed to be due to participants' starting vitamin D supplementation shortly before enrolling in prospective studies, for example, after recommendations by doctors over concerns for bone health [56]. Meta-analyses of prospective studies now report inverse relationships with 25(OH)D concentrations for breast cancer incidence from case–control and nested case–control studies, nonsignificant relationships for pancreatic cancer incidence, and direct relationships for prostate cancer incidence [16]. Prostate cancer is unique in that increased risk of mild prostate cancer is due to increased absorption of calcium and phosphorus [16]. However, a number of observational studies have reported U-shaped or reversed J-shaped relations for mortality rates with respect to serum 25(OH)D concentrations such as one that found a significantly increased risk of mortality rate for 25(OH)D concentrations > 120 ng/mL [182].

Several recent vitamin D supplementation observational studies showed that higher daily-dose vitamin D_3 supplementation was relatively safe and effective. One such study involving 19 lactating women reported that 6400 IU/d of vitamin D_3 supplementation safely raised 25(OH)D concentrations from 32 ± 4 to 59 ± 7 ng/mL [183]. That group of researchers provided concrete evidence about the safety of 6400 IU/d even for pregnant women to the NIH. A study in Canada had several thousand participants taking vitamin D_3 doses of their choice. A total of 2229 participants achieved a 25(OH)D concentration > 40 ng/mL by taking 2600 ± 2800 IU/d (for those with mean (SD) value of 47 ± 5 ng/mL) to 6300 ± 500 IU/d, for a mean (SD) value of 117 ± 15 ng/mL [12].

More recently, McCullough and colleagues [57] reported that supplementing over 400 hospital inpatients with 5000–50,000 IU/d of vitamin D_3 for 30 months was safe and significantly raised serum 25(OH)D concentrations and lowered PTH concentrations. Thus, future vitamin D RCTs should use higher doses of vitamin D_3 where necessary, for example, to achieve target status. However, although we have solid safety data on vitamin D supplementation, we should nevertheless be cautious, particularly with using high doses in older and/or ill people who may be more prone to vitamin D toxicity.

Hill's criteria for causality in a biological system.

Long before vitamin D RCTs were conducted, Sir Austin Bradford Hill outlined the criteria for causality in a biological system in an address to the Royal Society of Medicine in 1965 [184]. The criteria important for vitamin D include:

1. Strength of association
2. Consistency in findings
3. Temporality, that is, the risk factor must be experienced before the event
4. Biological gradient, that is, dose–response relationship.

5. Plausibility, for example, mechanisms that can explain the relationship
6. Coherence with known biological facts
7. Experiment, for example, RCT
8. Analogy with related associations

Added later [185]:

9. Accounting for confounding factors
10. Accounting for bias such as publication bias
11. Quality of study design

Hill stated that not all criteria need be satisfied to claim causality, but the more that are, the better.

Temporality seems to be interpreted as meaning that prospective studies after measurement of 25(OH)D should be used; however, as discussed, that approach can lead to underestimating the effect of higher 25(OH)D concentration. As long as the disease state does not affect serum 25(OH)D concentration, case–control studies with 25(OH)D measured near time of diagnosis should be acceptable.

Hill's criteria have been evaluated for vitamin D's role in several diseases, mainly using observational studies. Those studies reported beneficial effects of vitamin D on the basis of serum 25(OH)D concentrations, finding that nearly all criteria are satisfied except, in some cases, experimental verification for BP [12], cancer [186–189], CVD [190,191], COVID-19 [94,192], dementia [193], diabetes and pancreatic cancer [194], type 1 diabetes [195], MS [196,197], oral health [198], and periodontal disease [199].

4. Conclusions

Observational studies consistently report significant inverse correlations between serum 25(OH)D concentrations and health outcomes, but residual confounding cannot be completely excluded. Prospective studies are preferred over case–control studies on the basis of concern that the disease state could affect serum 25(OH)D concentrations. However, that concern appears to be mainly relevant for inflammatory and infectious diseases such as acute respiratory infections and not other outcomes. The same concern, however, exists when diseases are diagnosed at an advanced stage in their development, as with cancer. One major problem with prospective observational studies is that serum 25(OH)D concentrations change not only with season but also with respect to time, hence resulting in a potential underestimation of the effect of higher 25(OH)D concentrations as discussed earlier.

Another issue regarding observational studies is whether the cases and controls are well matched. Propensity score matching can help ensure good matches, given a sufficiently large pool of prospective controls. Another less-recognized limitation is that if cases and controls are not matched closely with respect to when blood was drawn to measure 25(OH)D concentration, additional bias may be introduced.

Another concern is whether all pertinent confounding factors were considered in the study's design and analysis. For causality, it is proposed that results of observational studies be included in the broader context of what is known about vitamin D in the health outcome of interest—for example, in relation to mechanistic data or on evaluation using Hill's criteria for causality [184].

RCTs have largely failed to support vitamin D's role in reducing risk of adverse health outcomes [15]. The main reasons appear to be that few vitamin D–deficient participants are enrolled, that low vitamin D doses are used, and that outcomes are evaluated by vitamin D dose rather than by achieved 25(OH)D concentrations. Moreover, secondary analyses not proposed in the trial protocols are generally ignored. Those analyses seem to be disregarded based on the concern that if many secondary analyses are conducted, some might accidentally find a significant result, in particular if multiple testing issues are not adequately considered. However, if the secondary outcome is one that researchers forgot to include in the protocol but which reasonably makes sense to include based on other evidence or on mechanistic data, such as the effect of race/ethnicity and BMI on cancer risk in the VITAL study [8]—then such evidence should rather be accepted. Given the time, effort and expense required for major vitamin D RCTs, it seems unlikely that many more will be conducted soon, or ever. Thus, it is imperative to learn to use findings from completed trials and from other research approaches as efficiently as possible.

Examining the role of genetic factors might also be helpful in understanding the inconsistency of results between observational studies and RCTs since supplements might interact differently according to specific genotypes and variants [200]. Unfortunately, such potential genotype/supplement interactions have rarely been examined in RCT settings.

Observational studies using variable vitamin D dose supplementation and frequent (annual or semiannual) measurement of 25(OH)D concentrations and other pertinent data have several apparent advantages over traditional observational studies and RCTs. Permitting variable vitamin D doses allows large 25(OH)D ranges to be covered, which should make any significant health benefits more likely to become apparent [11,12].

MR studies that establish vitamin D's causal role for several health outcomes are now being reported. By using genetically predicted 25(OH)D concentrations for many participants combined with reported health outcomes, MR studies of large cohorts average out lifestyle factors in ways that match the effects of randomization in RCTs. Recently, stratifying subjects into ~100 subgroups by baseline 25(OH)D concentrations for separate MR analyses (which allows analysis of nonlinear data) provides an excellent approach to examining the effects of genetic increases in serum 25(OH)D at low baseline 25(OH)D concentrations [40], especially when those effects cannot be adequately considered when all the data are included in a single analysis.

Ecological studies have historically been useful in highlighting that UVB radiation reduces risk of diseases such as MS and cancers [42]. The advantages of ecological studies include the fact that many participants are studied and that data for many population-level confounding factors can be used. Geographical ecological studies surpass temporal ecological studies because the multiple factors involved in seasonal variations are hard to untangle, including UVB production of vitamin D, UVB non–vitamin D mechanisms, UVA-induced increases in serum NO concentrations, and temperature [45].

Knowledge of the mechanisms whereby vitamin D reduces particular adverse health outcomes is fundamental to understanding the problems of deficiency and should be routinely considered in designing future trials. The mechanisms by which correcting deficiencies can reduce tissue dysfunction in each of the common disorders we have discussed are all reasonably well understood which supports the case for ensuring better vitamin D provision in populations commonly afflicted with deficiency.

Author Contributions: Conceptualization: W.B.G. and S.P.; Methodology—W.B.G. and S.P. Writing—Original Draft Preparation: W.B.G.; Writing—Review and Editing: W.B.G., B.J.B., F.A.A. and S.P.; Supervision—W.B.G. All authors have read and agreed to the published version of the manuscript.

Funding: This research received no external funding.

Data Availability Statement: Not applicable.

Conflicts of Interest: W.B.G. receives funding from Bio-Tech Pharmacal Inc. (Fayetteville, AR, USA). The other authors declare no conflict of interest.

References

1. Jones, G. 100 YEARS OF VITAMIN D: Historical aspects of vitamin D. *Endocr. Connect.* **2022**, *11*, e210594. [CrossRef] [PubMed]
2. Autier, P.; Boniol, M.; Pizot, C.; Mullie, P. Vitamin D status and ill health: A systematic review. *Lancet Diabetes Endocrinol.* **2014**, *2*, 76–89. [CrossRef]
3. Adams, J.S.; Rafison, B.; Witzel, S.; Reyes, R.E.; Shieh, A.; Chun, R.; Zavala, K.; Hewison, M.; Liu, P.T. Regulation of the extrarenal CYP27B1-hydroxylase. *J. Steroid Biochem. Mol. Biol.* **2014**, *144 Pt A*, 22–27. [CrossRef]
4. Autier, P.; Mullie, P.; Macacu, A.; Dragomir, M.; Boniol, M.; Coppens, K.; Pizot, C.; Boniol, M. Effect of vitamin D supplementation on non-skeletal disorders: A systematic review of meta-analyses and randomised trials. *Lancet Diabetes Endocrinol.* **2017**, *5*, 986–1004. [CrossRef]

5. Rejnmark, L.; Bislev, L.S.; Cashman, K.D.; Eiríksdottir, G.; Gaksch, M.; Grüebler, M.; Grimnes, G.; Gudnason, V.; Lips, P.; Pilz, S.; et al. Non-skeletal health effects of vitamin D supplementation: A systematic review on findings from meta-analyses summarizing trial data. *PLoS ONE* **2017**, *12*, e0180512. [CrossRef] [PubMed]
6. Martineau, A.R.; Jolliffe, D.A.; Hooper, R.L.; Greenberg, L.; Aloia, J.F.; Bergman, P.; Dubnov-Raz, G.; Esposito, S.; Ganmaa, D.; Ginde, A.A.; et al. Vitamin D supplementation to prevent acute respiratory tract infections: Systematic review and meta-analysis of individual participant data. *BMJ* **2017**, *356*, i6583. [CrossRef]
7. Maretzke, F.; Bechthold, A.; Egert, S.; Ernst, J.B.; Melo van Lent, D.; Pilz, S.; Reichrath, J.; Stangl, G.I.; Stehle, P.; Volkert, D.; et al. Role of Vitamin D in Preventing and Treating Selected Extraskeletal Diseases—An Umbrella Review. *Nutrients* **2020**, *12*, 969. [CrossRef]
8. Manson, J.E.; Cook, N.R.; Lee, I.M.; Christen, W.; Bassuk, S.S.; Mora, S.; Gibson, H.; Gordon, D.; Copeland, T.; D'Agostino, D.; et al. Vitamin D Supplements and Prevention of Cancer and Cardiovascular Disease. *N. Engl. J. Med.* **2019**, *380*, 33–44. [CrossRef]
9. Pittas, A.G.; Dawson-Hughes, B.; Sheehan, P.; Ware, J.H.; Knowler, W.C.; Aroda, V.R.; Brodsky, I.; Ceglia, L.; Chadha, C.; Chatterjee, R.; et al. Vitamin D Supplementation and Prevention of Type 2 Diabetes. *N. Engl. J. Med.* **2019**, *381*, 520–530. [CrossRef]
10. Smolders, J.; van den Ouweland, J.; Geven, C.; Pickkers, P.; Kox, M. Letter to the Editor: Vitamin D deficiency in COVID-19: Mixing up cause and consequence. *Metabolism* **2021**, *115*, 154434. [CrossRef]
11. McDonnell, S.L.; Baggerly, C.A.; French, C.B.; Baggerly, L.L.; Garland, C.F.; Gorham, E.D.; Hollis, B.W.; Trump, D.L.; Lappe, J.M. Breast cancer risk markedly lower with serum 25-hydroxyvitamin D concentrations ≥60 vs. <20 ng/mL (150 vs. 50 nmol/L): Pooled analysis of two randomized trials and a prospective cohort. *PLoS ONE* **2018**, *13*, e0199265. [CrossRef] [PubMed]
12. Mirhosseini, N.; Vatanparast, H.; Kimball, S.M. The Association between Serum 25(OH)D Status and Blood Pressure in Participants of a Community-Based Program Taking Vitamin D Supplements. *Nutrients* **2017**, *9*, 1244. [CrossRef] [PubMed]
13. Acharya, P.; Dalia, T.; Ranka, S.; Sethi, P.; Oni, O.A.; Safarova, M.S.; Parashara, D.; Gupta, K.; Barua, R.S. The Effects of Vitamin D Supplementation and 25-Hydroxyvitamin D Levels on the Risk of Myocardial Infarction and Mortality. *J. Endocr. Soc.* **2021**, *5*, bvab124. [CrossRef] [PubMed]
14. McDonnell, S.L.; Baggerly, K.A.; Baggerly, C.A.; Aliano, J.L.; French, C.B.; Baggerly, L.L.; Ebeling, M.D.; Rittenberg, C.S.; Goodier, C.G.; Mateus Nino, J.F.; et al. Maternal 25(OH)D concentrations ≥40 ng/mL associated with 60% lower preterm birth risk among general obstetrical patients at an urban medical center. *PLoS ONE* **2017**, *12*, e0180483. [CrossRef]
15. Pilz, S.; Trummer, C.; Theiler-Schwetz, V.; Grübler, M.R.; Verheyen, N.D.; Odler, B.; Karras, S.N.; Zittermann, A.; März, W. Critical Appraisal of Large Vitamin D Randomized Controlled Trials. *Nutrients* **2022**, *14*, 303. [CrossRef]
16. Muñoz, A.; Grant, W.B. Vitamin D and Cancer: An Historical Overview of the Epidemiology and Mechanisms. *Nutrients* **2022**, *14*, 1448. [CrossRef]
17. Garland, C.; Barrett-Connor, E.; Rossof, A.; Shekelle, R.; Criqui, M.; Paul, O. Dietary Vitamin D and Calcium and Risk of Colorectal Cancer: A 19-Year Prospective Study in Men. *Lancet* **1985**, *325*, 307–309. [CrossRef]
18. O'Neill, C.M.; Kazantzidis, A.; Kiely, M.; Cox, L.; Meadows, S.; Goldberg, G.; Prentice, A.; Kift, R.; Webb, A.R.; Cashman, K.D. A predictive model of serum 25-hydroxyvitamin D in UK white as well as black and Asian minority ethnic population groups for application in food fortification strategy development towards vitamin D deficiency prevention. *J. Steroid Biochem. Mol. Biol.* **2017**, *173*, 245–252. [CrossRef]
19. Crowe, F.L.; Steur, M.; Allen, N.E.; Appleby, P.N.; Travis, R.C.; Key, T.J. Plasma concentrations of 25-hydroxyvitamin D in meat eaters, fish eaters, vegetarians and vegans: Results from the EPIC–Oxford study. *Public Health Nutr.* **2011**, *14*, 340–346. [CrossRef]
20. Scragg, R. Limitations of vitamin D supplementation trials: Why observational studies will continue to help determine the role of vitamin D in health. *J. Steroid Biochem. Mol. Biol.* **2018**, *177*, 6–9. [CrossRef]
21. Boucher, B.J.; Grant, W.B. Re: Scragg–Emerging Evidence of Thresholds for Beneficial Effects from Vitamin D Supplementation. *Nutrients* **2019**, *11*, 1321. [CrossRef] [PubMed]
22. Hujoel, P.P. Vitamin D and dental caries in controlled clinical trials: Systematic review and meta-analysis. *Nutr. Rev.* **2013**, *71*, 88–97. [CrossRef] [PubMed]
23. Christiansen, C.; Rødbro, P.; Lund, M. Effect of Vitamin D on Bone Mineral Mass in Normal Subjects and in Epileptic Patients on Anticonvulsants: A Controlled Therapeutic Trial. *BMJ* **1973**, *2*, 208–209. [CrossRef] [PubMed]
24. Keum, N.; Chen, Q.-Y.; Lee, D.H.; Manson, J.E.; Giovannucci, E. Vitamin D supplementation and total cancer incidence and mortality by daily vs. infrequent large-bolus dosing strategies: A meta-analysis of randomised controlled trials. *Br. J. Cancer* **2022**, *127*, 872–878. [CrossRef] [PubMed]
25. Jolliffe, D.A.; Camargo, C.A.; Sluyter, J.D.; Aglipay, M.; Aloia, J.F.; Ganmaa, D.; Bergman, P.; Bischoff-Ferrari, H.A.; Borzutzky, A.; Damsgaard, C.T.; et al. Vitamin D supplementation to prevent acute respiratory infections: A systematic review and meta-analysis of aggregate data from randomised controlled trials. *Lancet Diabetes Endocrinol.* **2021**, *9*, 276–292. [CrossRef]
26. Hahn, J.; Cook, N.R.; Alexander, E.K.; Friedman, S.; Walter, J.; Bubes, V.; Kotler, G.; Lee, I.-M.; E Manson, J.; Costenbader, K.H. Vitamin D and marine omega 3 fatty acid supplementation and incident autoimmune disease: VITAL randomized controlled trial. *BMJ* **2022**, *376*, e066452. [CrossRef]
27. Heaney, R.P. Guidelines for optimizing design and analysis of clinical studies of nutrient effects. *Nutr. Rev.* **2014**, *72*, 48–54. [CrossRef]
28. Grant, W.B.; Boucher, B.J.; Bhattoa, H.P.; Lahore, H. Why vitamin D clinical trials should be based on 25-hydroxyvitamin D concentrations. *J. Steroid Biochem. Mol. Biol.* **2018**, *177*, 266–269. [CrossRef]

29. Lelieveld, J.; Klingmüller, K.; Pozzer, A.; Pöschl, U.; Fnais, M.; Daiber, A.; Münzel, T. Cardiovascular disease burden from ambient air pollution in Europe reassessed using novel hazard ratio functions. *Eur. Heart J.* **2019**, *40*, 1590–1596. [CrossRef]
30. DiNicolantonio, J.J.; O'Keefe, J.H.; Wilson, W. Subclinical magnesium deficiency: A principal driver of cardiovascular disease and a public health crisis. *Open Heart* **2018**, *5*, e000668. [CrossRef]
31. Lappe, J.M.; Travers-Gustafson, D.; Davies, K.M.; Recker, R.R.; Heaney, R.P. Vitamin D and calcium supplementation reduces cancer risk: Results of a randomized trial. *Am. J. Clin. Nutr.* **2007**, *85*, 1586–1591. [CrossRef] [PubMed]
32. Reid, I.R.; Birstow, S.M.; Bolland, M. Calcium and Cardiovascular Disease. *Endocrinol. Metab.* **2017**, *32*, 339–349. [CrossRef] [PubMed]
33. Wang, T.J.; Zhang, F.; Richards, J.B.; Kestenbaum, B.; van Meurs, J.B.; Berry, D.; Kiel, D.P.; Streeten, E.A.; Ohlsson, C.; Koller, D.L.; et al. Common genetic determinants of vitamin D insufficiency: A genome-wide association study. *Lancet* **2010**, *376*, 180–188. [CrossRef]
34. Hatchwell, E.; Greally, J.M. The potential role of epigenomic dysregulation in complex human disease. *Trends Genet.* **2007**, *23*, 588–595. [CrossRef] [PubMed]
35. Kilpinen, H.; Dermitzakis, E.T. Genetic and epigenetic contribution to complex traits. *Hum. Mol. Genet.* **2012**, *21*, R24–R28. [CrossRef]
36. Wong, A.K.; Sealfon, R.S.G.; Theesfeld, C.L.; Troyanskaya, O.G. Decoding disease: From genomes to networks to phenotypes. *Nat. Rev. Genet.* **2021**, *22*, 774–790. [CrossRef]
37. Revez, J.A.; Lin, T.; Qiao, Z.; Xue, A.; Holtz, Y.; Zhu, Z.; Zeng, J.; Wang, H.; Sidorenko, J.; Kemper, K.E.; et al. Genome-wide association study identifies 143 loci associated with 25 hydroxyvitamin D concentration. *Nat. Commun.* **2020**, *11*, 1647. [CrossRef]
38. Ong, J.S.; Dixon-Suen, S.C.; Han, X.; An, J.; Esophageal Cancer Consortium; 23 and Me Research Team; Liyanage, U.; Dusingize, J.C.; Schumacher, J.; Gockel, I.; et al. A comprehensive re-assessment of the association between vitamin D and cancer susceptibility using Mendelian randomization. *Nat. Commun.* **2021**, *12*, 246. [CrossRef]
39. Navale, S.S.; Mulugeta, A.; Zhou, A.; Llewellyn, D.J.; Hyppönen, E. Vitamin D and brain health: An observational and Mendelian randomization study. *Am. J. Clin. Nutr.* **2022**, *116*, 531–540. [CrossRef]
40. Zhou, A.; Selvanayagam, J.B.; Hyppönen, E. Non-linear Mendelian randomization analyses support a role for vitamin D deficiency in cardiovascular disease risk. *Eur. Heart J.* **2022**, *43*, 1731–1739. [CrossRef]
41. Garland, C.F.; Garland, F.C. Do sunlight and vitamin D reduce the likelihood of colon cancer? *Int. J. Epidemiol.* **1980**, *9*, 227–231. [CrossRef]
42. Grant, W.B. The role of geographical ecological studies in identifying diseases linked to UVB exposure and/or vitamin D. *Derm.-Endocrinol.* **2016**, *8*, e1137400. [CrossRef]
43. Grant, W.B. An estimate of premature cancer mortality in the U.S. due to inadequate doses of solar ultraviolet-B radiation. *Cancer* **2002**, *94*, 1867–1875. [CrossRef]
44. Grant, W.B.; Garland, C.F. The association of solar ultraviolet B (UVB) with reducing risk of cancer: Multifactorial ecologic analysis of geographic variation in age-adjusted cancer mortality rates. *Anticancer Res.* **2006**, *26*, 2687–2699.
45. Grant, W.B.; Boucher, B.J. An Exploration of How Solar Radiation Affects the Seasonal Variation of Human Mortality Rates and the Seasonal Variation in Some Other Common Disorders. *Nutrients* **2022**, *14*, 2519. [CrossRef]
46. Doll, R.; Peto, R.; Boreham, J.; Sutherland, I. Mortality from cancer in relation to smoking: 50 years observations on British doctors. *Br. J. Cancer* **2005**, *92*, 426–429. [CrossRef]
47. Murdaca, G.; Tonacci, A.; Negrini, S.; Greco, M.; Borro, M.; Puppo, F.; Gangemi, S. Emerging role of vitamin D in autoimmune diseases: An update on evidence and therapeutic implications. *Autoimmun. Rev.* **2019**, *18*, 102350. [CrossRef]
48. Dankers, W.; Colin, E.M.; van Hamburg, J.P.; Lubberts, E. Vitamin D in Autoimmunity: Molecular Mechanisms and Therapeutic Potential. *Front. Immunol.* **2017**, *7*, 697. [CrossRef]
49. Prummel, M.F.; Strieder, T.; Wiersinga, W.M. The environment and autoimmune thyroid diseases. *Eur. J. Endocrinol.* **2004**, *150*, 605–618. [CrossRef] [PubMed]
50. Oliver, J.E.; Silman, A.J. Risk factors for the development of rheumatoid arthritis. *Scand. J. Rheumatol.* **2006**, *35*, 169–174. [CrossRef]
51. Huerta, C.; Rivero, E.; Rodríguez, L.A.G. Incidence and Risk Factors for Psoriasis in the General Population. *Arch. Dermatol.* **2007**, *143*, 1559–1565. [CrossRef] [PubMed]
52. Gombart, A.F. The vitamin D–antimicrobial peptide pathway and its role in protection against infection. *Future Microbiol.* **2009**, *4*, 1151–1165. [CrossRef]
53. Guillot, X.; Semerano, L.; Saidenberg-Kermanac'h, N.; Falgarone, G.; Boissier, M.C. Vitamin D and inflammation. *Jt. Bone Spine* **2010**, *77*, 552–557. [CrossRef] [PubMed]
54. Carlberg, C.; Muñoz, A. An update on vitamin D signaling and cancer. *Semin. Cancer Biol.* **2022**, *79*, 217–230. [CrossRef] [PubMed]
55. Garland, C.; Garland, F.C.; Shaw, E.; Comstock, G.W.; Helsing, K.J.; Gorham, E.D. Serum 25-Hydroxyvitamin D and Colon Cancer: Eight-Year Prospective Study. *Lancet* **1989**, *334*, 1176–1178. [CrossRef]
56. Grant, W.B.; Karras, S.N.; Bischoff-Ferrari, H.A.; Annweiler, C.; Boucher, B.J.; Juzeniene, A.; Garland, C.F.; Holick, M.F. Do studies reporting 'U'-shaped serum 25-hydroxyvitamin D-health outcome relationships reflect adverse effects? *Derm.-Endocrinol.* **2016**, *8*, e1187349. [CrossRef] [PubMed]

57. McCullough, M.L.; Zoltick, E.S.; Weinstein, S.J.; Fedirko, V.; Wang, M.; Cook, N.R.; Eliassen, A.H.; Zeleniuch-Jacquotte, A.; Agnoli, C.; Albanes, D.; et al. Circulating Vitamin D and Colorectal Cancer Risk: An International Pooling Project of 17 Cohorts. *J. Natl. Cancer Inst.* **2019**, *111*, 158–169. [CrossRef]
58. Grant, W.B. Effect of interval between serum draw and follow-up period on relative risk of cancer incidence with respect to 25-hydroxyvitamin D level; implications for meta-analyses and setting vitamin D guidelines. *Derm.-Endocrinol.* **2011**, *3*, 199–204. [CrossRef]
59. Grant, W.B. Effect of follow-up time on the relation between prediagnostic serum 25-hydroxyvitamin D and all-cause mortality rate. *Derm.-Endocrinol.* **2012**, *4*, 198–202. [CrossRef]
60. Grant, W.B. 25-hydroxyvitamin D and breast cancer, colorectal cancer, and colorectal adenomas: Case-control versus nested case-control studies. *Anticancer Res.* **2015**, *35*, 1153–1160.
61. Lappe, J.; Watson, P.; Travers-Gustafson, D.; Recker, R.; Garland, C.; Gorham, E.; Baggerly, K.; McDonnell, S.L. Effect of Vitamin D and Calcium Supplementation on Cancer Incidence in Older Women: A Randomized Clinical Trial. *JAMA* **2017**, *317*, 1234–1243. [CrossRef]
62. Ekwaru, J.P.; Zwicker, J.D.; Holick, M.F.; Giovannucci, E.; Veugelers, P.J. The Importance of Body Weight for the Dose Response Relationship of Oral Vitamin D Supplementation and Serum 25-Hydroxyvitamin D in Healthy Volunteers. *PLoS ONE* **2014**, *9*, e111265. [CrossRef]
63. Wamberg, L.; Kampmann, U.; Stodkilde-Jorgensen, H.; Rejnmark, L.; Pedersen, S.B.; Richelsen, B. Effects of vitamin D supplementation on body fat accumulation, inflammation, and metabolic risk factors in obese adults with low vitamin D levels—Results from a randomized trial. *Eur. J. Intern. Med.* **2013**, *24*, 644–649. [CrossRef]
64. Michels, N.; van Aart, C.; Morisse, J.; Mullee, A.; Huybrechts, I. Chronic inflammation towards cancer incidence: A systematic review and meta-analysis of epidemiological studies. *Crit. Rev. Oncol.* **2021**, *157*, 103177. [CrossRef] [PubMed]
65. Giovannucci, E.; Liu, Y.; Rimm, E.B.; Hollis, B.W.; Fuchs, C.S.; Stampfer, M.J.; Willett, W.C. Prospective Study of Predictors of Vitamin D Status and Cancer Incidence and Mortality in Men. *J. Natl. Cancer Inst.* **2006**, *98*, 451–459. [CrossRef] [PubMed]
66. Johnson, C.R.; Dudenkov, D.V.; Mara, K.C.; Fischer, P.R.; Maxson, J.A.; Thacher, T.D. Serum 25-Hydroxyvitamin D and Subsequent Cancer Incidence and Mortality: A Population-Based Retrospective Cohort Study. *Mayo Clin. Proc.* **2021**, *96*, 2157–2167. [CrossRef] [PubMed]
67. Kim, Y.; Je, Y. Vitamin D intake, blood 25(OH)D levels, and breast cancer risk or mortality: A meta-analysis. *Br. J. Cancer* **2014**, *110*, 2772–2784. [CrossRef]
68. Gnagnarella, P.; Muzio, V.; Caini, S.; Raimondi, S.; Martinoli, C.; Chiocca, S.; Miccolo, C.; Bossi, P.; Cortinovis, D.; Chiaradonna, F.; et al. Vitamin D Supplementation and Cancer Mortality: Narrative Review of Observational Studies and Clinical Trials. *Nutrients* **2021**, *13*, 3285. [CrossRef]
69. Li, Z.; Wu, L.; Zhang, J.; Huang, X.; Thabane, L.; Li, G. Effect of Vitamin D Supplementation on Risk of Breast Cancer: A Systematic Review and Meta-Analysis of Randomized Controlled Trials. *Front. Nutr.* **2021**, *8*, 655727. [CrossRef]
70. Devesa, S.S.; Grauman, D.J.; Blot, W.J.; Pennelo, G.A.; Hoover, R.N.; Fraumeni, J.F., Jr. *Atlas of Cancer Mortality in the United States, 1950–1994*; National Institues of Health, National Cancer Institute: Bethesda, MD, USA, 1999; p. 360.
71. Marti-Soler, H.; Gonseth, S.; Gubelmann, C.; Stringhini, S.; Bovet, P.; Chen, P.-C.; Wojtyniak, B.; Paccaud, F.; Tsai, D.-H.; Zdrojewski, T.; et al. Seasonal Variation of Overall and Cardiovascular Mortality: A Study in 19 Countries from Different Geographic Locations. *PLoS ONE* **2014**, *9*, e113500. [CrossRef]
72. Mozos, I.; Marginean, O. Links between Vitamin D Deficiency and Cardiovascular Diseases. *BioMed Res. Int.* **2015**, *2015*, 109275. [CrossRef] [PubMed]
73. Glovaci, D.; Fan, W.; Wong, N.D. Epidemiology of Diabetes Mellitus and Cardiovascular Disease. *Curr. Cardiol. Rep.* **2019**, *21*, 21. [CrossRef] [PubMed]
74. Nguyen, J.L.; Yang, W.; Ito, K.; Matte, T.D.; Shaman, J.; Kinney, P.L. Seasonal Influenza Infections and Cardiovascular Disease Mortality. *JAMA Cardiol.* **2016**, *1*, 274–281. [CrossRef] [PubMed]
75. Liccardo, D.; Cannavo, A.; Spagnuolo, G.; Ferrara, N.; Cittadini, A.; Rengo, C.; Rengo, G. Periodontal Disease: A Risk Factor for Diabetes and Cardiovascular Disease. *Int. J. Mol. Sci.* **2019**, *20*, 1414. [CrossRef]
76. Wang, T.J.; Pencina, M.J.; Booth, S.L.; Jacques, P.F.; Ingelsson, E.; Lanier, K.; Benjamin, E.J.; D'Agostino, R.B.; Wolf, M.; Vasan, R.S. Vitamin D Deficiency and Risk of Cardiovascular Disease. *Circulation* **2008**, *117*, 503–511. [CrossRef]
77. Zhang, R.; Li, B.; Gao, X.; Tian, R.; Pan, Y.; Jiang, Y.; Gu, H.; Wang, Y.; Wang, Y.; Liu, G. Serum 25-hydroxyvitamin D and the risk of cardiovascular disease: Dose-response meta-analysis of prospective studies. *Am. J. Clin. Nutr.* **2017**, *105*, 810–819. [CrossRef]
78. Ginde, A.A.; Scragg, R.; Schwartz, R.S.; Camargo, C.A., Jr. Prospective Study of Serum 25-Hydroxyvitamin D Level, Cardiovascular Disease Mortality, and All-Cause Mortality in Older U.S. Adults. *J. Am. Geriatr. Soc.* **2009**, *57*, 1595–1603. [CrossRef]
79. Melamed, M.L.; Michos, E.D.; Post, W.; Astor, B. 25-Hydroxyvitamin D Levels and the Risk of Mortality in the General Population. *Arch. Intern. Med.* **2008**, *168*, 1629–1637. [CrossRef]
80. Semba, R.D.; Houston, D.; Bandinelli, S.; Sun, K.; Cherubini, A.; Cappola, A.R.; Guralnik, J.M.; Ferrucci, L. Relationship of 25-hydroxyvitamin D with all-cause and cardiovascular disease mortality in older community-dwelling adults. *Eur. J. Clin. Nutr.* **2010**, *64*, 203–209. [CrossRef]

81. Liu, L.; Chen, M.; Hankins, S.R.; Nùñez, A.E.; Watson, R.A.; Weinstock, P.J.; Newschaffer, C.J.; Eisen, H.J. Serum 25-Hydroxyvitamin D Concentration and Mortality from Heart Failure and Cardiovascular Disease, and Premature Mortality from All-Cause in United States Adults. *Am. J. Cardiol.* **2012**, *110*, 834–839. [CrossRef]
82. Austin, P.C. An Introduction to Propensity Score Methods for Reducing the Effects of Confounding in Observational Studies. *Multivar. Behav. Res.* **2011**, *46*, 399–424. [CrossRef] [PubMed]
83. Anderson, J.L.; Vanwoerkom, R.C.; Horne, B.D.; Bair, T.L.; May, H.T.; Lappe, D.L.; Muhlestein, J.B. Parathyroid hormone, vitamin D, renal dysfunction, and cardiovascular disease: Dependent or independent risk factors? *Am. Heart J.* **2011**, *162*, 331–339.e2. [CrossRef] [PubMed]
84. Kestenbaum, B.; Katz, R.; de Boer, I.; Hoofnagle, A.; Sarnak, M.J.; Shlipak, M.G.; Jenny, N.S.; Siscovick, D.S. Vitamin D, Parathyroid Hormone, and Cardiovascular Events Among Older Adults. *J. Am. Coll. Cardiol.* **2011**, *58*, 1433–1441. [CrossRef] [PubMed]
85. Chen, W.R.; Chen, Y.D.; Shi, Y.; Yin, D.W.; Wang, H.; Sha, Y. Vitamin D, parathyroid hormone and risk factors for coronary artery disease in an elderly Chinese population. *J. Cardiovasc. Med.* **2015**, *16*, 59–68. [CrossRef] [PubMed]
86. Wannamethee, S.G.; Welsh, P.; Papacosta, O.; Lennon, L.; Whincup, P.H.; Sattar, N. Elevated Parathyroid Hormone, But Not Vitamin D Deficiency, Is Associated with Increased Risk of Heart Failure in Older Men with and without Cardiovascular Disease. *Circ. Heart Fail.* **2014**, *7*, 732–739. [CrossRef]
87. Valcour, A.; Blocki, F.; Hawkins, D.M.; Rao, S.D. Effects of Age and Serum 25-OH-Vitamin D on Serum Parathyroid Hormone Levels. *J. Clin. Endocrinol. Metab.* **2012**, *97*, 3989–3995. [CrossRef]
88. Zhang, Z.; Yang, Y.; Ng, C.Y.; Wang, D.; Wang, J.; Li, G.; Liu, T. Meta-analysis of Vitamin D Deficiency and Risk of Atrial Fibrillation. *Clin. Cardiol.* **2016**, *39*, 537–543. [CrossRef]
89. Kahn, R.; Robertson, R.M.; Smith, R.; Eddy, D. The Impact of Prevention on Reducing the Burden of Cardiovascular Disease. *Diabetes Care* **2008**, *31*, 1686–1696. [CrossRef]
90. Farley, T.A.; Dalal, M.A.; Mostashari, F.; Frieden, T.R. Deaths Preventable in the U.S. by Improvements in Use of Clinical Preventive Services. *Am. J. Prev. Med.* **2010**, *38*, 600–609. [CrossRef]
91. Timms, P.M.; Mannan, N.; Hitman, G.A.; Noonan, K.; Mills, P.G.; Syndercombe-Court, D.; Aganna, E.; Price, C.P.; Boucher, B.J. Circulating MMP9, vitamin D and variation in the TIMP-1 response with VDR genotype: Mechanisms for inflammatory damage in chronic disorders? *QJM* **2002**, *95*, 787–796. [CrossRef]
92. Rimondi, E.; Marcuzzi, A.; Casciano, F.; Tornese, G.; Pellati, A.; Toffoli, B.; Secchiero, P.; Melloni, E. Role of vitamin D in the pathogenesis of atheromatosis. *Nutr. Metab. Cardiovasc. Dis.* **2021**, *31*, 344–353. [CrossRef]
93. Grant, W.B.; Lahore, H.; McDonnell, S.L.; Baggerly, C.A.; French, C.B.; Aliano, J.L.; Bhattoa, H.P. Evidence that Vitamin D Supplementation Could Reduce Risk of Influenza and COVID-19 Infections and Deaths. *Nutrients* **2020**, *12*, 988. [CrossRef] [PubMed]
94. Mercola, J.; Grant, W.B.; Wagner, C.L. Evidence Regarding Vitamin D and Risk of COVID-19 and Its Severity. *Nutrients* **2020**, *12*, 3361. [CrossRef] [PubMed]
95. Kaufman, H.W.; Niles, J.K.; Kroll, M.H.; Bi, C.; Holick, M.F. SARS-CoV-2 positivity rates associated with circulating 25-hydroxyvitamin D levels. *PLoS ONE* **2020**, *15*, e0239252.
96. Martineau, A.R.; Cantorna, M.T. Vitamin D for COVID-19: Where are we now? *Nat. Rev. Immunol.* **2022**, *22*, 529–530. [CrossRef] [PubMed]
97. Oristrell, J.; Oliva, J.C.; Casado, E.; Subirana, I.; Domínguez, D.; Toloba, A.; Balado, A.; Grau, M. Vitamin D supplementation and COVID-19 risk: A population-based, cohort study. *J. Endocrinol. Investig.* **2022**, *45*, 167–179. [CrossRef] [PubMed]
98. Seal, K.H.; Bertenthal, D.; Carey, E.; Grunfeld, C.; Bikle, D.D.; Lu, C.M. Association of Vitamin D Status and COVID-19-Related Hospitalization and Mortality. *J. Gen. Intern. Med.* **2022**, *37*, 853–861. [CrossRef]
99. Dissanayake, H.A.; de Silva, N.L.; Sumanatilleke, M.; de Silva, S.D.N.; Gamage, K.K.K.; Dematapitiya, C.; Kuruppu, D.C.; Ranasinghe, P.; Pathmanathan, S.; Katulanda, P. Prognostic and Therapeutic Role of Vitamin D in COVID-19: Systematic Review and Meta-analysis. *J. Clin. Endocrinol. Metab.* **2022**, *107*, 1484–1502. [CrossRef]
100. Hosseini, B.; El Abd, A.; Ducharme, F.M. Effects of Vitamin D Supplementation on COVID-19 Related Outcomes: A Systematic Review and Meta-Analysis. *Nutrients* **2022**, *14*, 2134. [CrossRef]
101. De Niet, S.; Trémège, M.; Coffiner, M.; Rousseau, A.-F.; Calmes, D.; Frix, A.-N.; Gester, F.; Delvaux, M.; Dive, A.-F.; Guglielmi, E.; et al. Positive Effects of Vitamin D Supplementation in Patients Hospitalized for COVID-19: A Randomized, Double-Blind, Placebo-Controlled Trial. *Nutrients* **2022**, *14*, 3048. [CrossRef]
102. Norman, A.W.; Frankel, B.J.; Heldt, A.M.; Grodsky, G.M. Vitamin D Deficiency Inhibits Pancreatic Secretion of Insulin. *Science* **1980**, *209*, 823–825. [CrossRef] [PubMed]
103. Hewison, M. Vitamin D and immune function: Autocrine, paracrine or endocrine? *Scand. J. Clin. Lab. Investig. Suppl.* **2012**, *243*, 92–102.
104. Cheng, Q.; Boucher, B.J.; Leung, P.S. Modulation of hypovitaminosis D-induced islet dysfunction and insulin resistance through direct suppression of the pancreatic islet renin–angiotensin system in mice. *Diabetologia* **2013**, *56*, 553–562. [CrossRef]
105. Leung, P.S. The Potential Protective Action of Vitamin D in Hepatic Insulin Resistance and Pancreatic Islet Dysfunction in Type 2 Diabetes Mellitus. *Nutrients* **2016**, *8*, 147. [CrossRef]
106. Szymczak-Pajor, I.; Drzewoski, J.; Śliwińska, A. The Molecular Mechanisms by Which Vitamin D Prevents Insulin Resistance and Associated Disorders. *Int. J. Mol. Sci.* **2020**, *21*, 6644. [CrossRef] [PubMed]

107. Boucher, B.J. Inadequate vitamin D status: Does it contribute to the disorders comprising syndrome 'X'? *Br. J. Nutr.* **1998**, *79*, 315–327. [CrossRef]
108. Pittas, A.G.; Lau, J.; Hu, F.B.; Dawson-Hughes, B. The Role of Vitamin D and Calcium in Type 2 Diabetes. A Systematic Review and Meta-Analysis. *J. Clin. Endocrinol. Metab.* **2007**, *92*, 2017–2029. [CrossRef] [PubMed]
109. Song, Y.; Wang, L.; Pittas, A.G.; Del Gobbo, L.C.; Zhang, C.; Manson, J.E.; Hu, F.B. Blood 25-hydroxy vitamin D levels and incident type 2 diabetes: A meta-analysis of prospective studies. *Diabetes Care* **2013**, *36*, 1422–1428. [CrossRef]
110. Ekmekcioglu, C.; Haluza, D.; Kundi, M. 25-Hydroxyvitamin D Status and Risk for Colorectal Cancer and Type 2 Diabetes Mellitus: A Systematic Review and Meta-Analysis of Epidemiological Studies. *Int. J. Environ. Res. Public Health* **2017**, *14*, 127. [CrossRef]
111. Dawson-Hughes, B.; Staten, M.A.; Knowler, W.C.; Nelson, J.; Vickery, E.M.; LeBlanc, E.S.; Neff, L.M.; Park, J.; Pittas, A.G. Intratrial Exposure to Vitamin D and New-Onset Diabetes Among Adults with Prediabetes: A Secondary Analysis from the Vitamin D and Type 2 Diabetes (D2d) Study. *Diabetes Care* **2020**, *43*, 2916–2922. [CrossRef]
112. Lu, L.; Bennett, D.A.; Millwood, I.Y.; Parish, S.; McCarthy, M.I.; Mahajan, A.; Lin, X.; Bragg, F.; Guo, Y.; Holmes, M.V.; et al. Association of vitamin D with risk of type 2 diabetes: A Mendelian randomisation study in European and Chinese adults. *PLOS Med.* **2018**, *15*, e1002566. [CrossRef] [PubMed]
113. Xu, Y.; Zhou, Y.; Liu, J.; Wang, C.; Qu, Z.; Wei, Z.; Zhou, D. Genetically increased circulating 25(OH)D level reduces the risk of type 2 diabetes in subjects with deficiency of vitamin D: A large-scale Mendelian randomization study. *Medicine* **2020**, *99*, e23672. [CrossRef] [PubMed]
114. Hadi, H.A.R.; Carr, C.S.; Al Suwaidi, J. Endothelial Dysfunction: Cardiovascular Risk Factors, Therapy, and Outcome. *Vasc. Health Risk Manag.* **2005**, *1*, 183–198. [PubMed]
115. Latic, N.; Erben, R.G. Vitamin D and Cardiovascular Disease, with Emphasis on Hypertension, Atherosclerosis, and Heart Failure. *Int. J. Mol. Sci.* **2020**, *21*, 6483. [CrossRef]
116. Wong, M.S.K.; Delansorne, R.; Man, R.Y.K.; Svenningsen, P.; Vanhoutte, P.M. Chronic treatment with vitamin D lowers arterial blood pressure and reduces endothelium-dependent contractions in the aorta of the spontaneously hypertensive rat. *Am. J. Physiol. Circ. Physiol.* **2010**, *299*, H1226–H1234. [CrossRef]
117. Kassi, E.; Adamopoulos, C.; Basdra, E.K.; Papavassiliou, A.G. Role of Vitamin D in Atherosclerosis. *Circulation* **2013**, *128*, 2517–2531. [CrossRef]
118. Ford, E.S.; Ajani, U.A.; McGuire, L.C.; Liu, S. Concentrations of Serum Vitamin D and the Metabolic Syndrome Among U.S. Adults. *Diabetes Care* **2005**, *28*, 1228–1230. [CrossRef]
119. Forman, J.P.; Giovannucci, E.; Holmes, M.D.; Bischoff-Ferrari, H.A.; Tworoger, S.S.; Willett, W.C.; Curhan, G.C. Plasma 25-Hydroxyvitamin D Levels and Risk of Incident Hypertension. *Hypertension* **2007**, *49*, 1063–1069. [CrossRef] [PubMed]
120. Mokhtari, E.; Hajhashemy, Z.; Saneei, P. Serum Vitamin D Levels in Relation to Hypertension and Pre-hypertension in Adults: A Systematic Review and Dose–Response Meta-Analysis of Epidemiologic Studies. *Front. Nutr.* **2022**, *9*, 829307. [CrossRef]
121. Zhang, D.; Cheng, C.; Wang, Y.; Sun, H.; Yu, S.; Xue, Y.; Liu, Y.; Li, W.; Li, X. Effect of Vitamin D on Blood Pressure and Hypertension in the General Population: An Update Meta-Analysis of Cohort Studies and Randomized Controlled Trials. *Prev. Chronic Dis.* **2020**, *17*, E03. [CrossRef]
122. Liu, D.; Fernandez, B.O.; Hamilton, A.; Lang, N.N.; Gallagher, J.M.; Newby, D.E.; Feelisch, M.; Weller, R.B. UVA Irradiation of Human Skin Vasodilates Arterial Vasculature and Lowers Blood Pressure Independently of Nitric Oxide Synthase. *J. Investig. Dermatol.* **2014**, *134*, 1839–1846. [CrossRef]
123. Forman, J.P.; Scott, J.B.; Ng, K.; Drake, B.; Suarez, E.G.; Hayden, D.L.; Bennett, G.G.; Chandler, P.; Hollis, B.W.; Emmons, K.M.; et al. Effect of Vitamin D Supplementation on Blood Pressure in Blacks. *Hypertension* **2013**, *61*, 779–785. [CrossRef] [PubMed]
124. Pilz, S.; Gaksch, M.; Kienreich, K.; Grübler, M.; Verheyen, N.; Fahrleitner-Pammer, A.; Treiber, G.; Drechsler, C.; ó Hartaigh, B.; Obermayer-Pietsch, B.; et al. Effects of vitamin D on blood pressure and cardiovascular risk factors: A randomized controlled trial. *Hypertension* **2015**, *65*, 1195–1201. [CrossRef] [PubMed]
125. Theiler-Schwetz, V.; Trummer, C.; Grübler, M.R.; Keppel, M.H.; Zittermann, A.; Tomaschitz, A.; Karras, S.N.; März, W.; Pilz, S.; Gängler, S. Effects of Vitamin D Supplementation on 24-Hour Blood Pressure in Patients with Low 25-Hydroxyvitamin D Levels: A Randomized Controlled Trial. *Nutrients* **2022**, *14*, 1360. [CrossRef]
126. Vimaleswaran, K.S.; Cavadino, A.; Berry, D.J.; Jorde, R.; Dieffenbach, A.K.; Lu, C.; Alves, A.C.; Heerspink, H.J.L.; Tikkanen, E.; Eriksson, J.; et al. Association of vitamin D status with arterial blood pressure and hypertension risk: A mendelian randomisation study. *Lancet Diabetes Endocrinol.* **2014**, *2*, 719–729. [CrossRef]
127. Zittermann, A.; Iodice, S.; Pilz, S.; Grant, W.; Bagnardi, V.; Gandini, S. Vitamin D deficiency and mortality risk in the general population: A meta-analysis of prospective cohort studies. *Am. J. Clin. Nutr.* **2012**, *95*, 91–100. [CrossRef] [PubMed]
128. Garland, C.F.; Kim, J.J.; Mohr, S.B.; Gorham, E.D.; Grant, W.B.; Giovannucci, E.L.; Baggerly, L.; Hofflich, H.; Ramsdell, J.W.; Zeng, K.; et al. Meta-analysis of All-Cause Mortality According to Serum 25-Hydroxyvitamin D. *Am. J. Public Health* **2014**, *104*, e43–e50. [CrossRef]
129. Gaksch, M.; Jorde, R.; Grimnes, G.; Joakimsen, R.; Schirmer, H.; Wilsgaard, T.; Mathiesen, E.B.; Njølstad, I.; Løchen, M.-L.; März, W.; et al. Vitamin D and mortality: Individual participant data meta-analysis of standardized 25-hydroxyvitamin D in 26916 individuals from a European consortium. *PLoS ONE* **2017**, *12*, e0170791. [CrossRef]
130. Afzal, S.; Brøndum-Jacobsen, P.; E Bojesen, S.; Nordestgaard, B.G. Genetically low vitamin D concentrations and increased mortality: Mendelian randomisation analysis in three large cohorts. *BMJ* **2014**, *349*, g6330. [CrossRef]

131. Sofianopoulou, E.; Kaptoge, S.K.; Afzal, S.; Jiang, T.; Gill, D.; Gundersen, T.E.; Bolton, T.R.; Allara, E.; Arnold, M.G.; Mason, A.M.; et al. Estimating dose-response relationships for vitamin D with coronary heart disease, stroke, and all-cause mortality: Observational and Mendelian randomisation analyses. *Lancet Diabetes Endocrinol.* **2021**, *9*, 837–846. [CrossRef]
132. Cantorna, M.T.; Snyder, L.; Lin, Y.-D.; Yang, L. Vitamin D and 1,25(OH)2D Regulation of T cells. *Nutrients* **2015**, *7*, 3011–3021. [CrossRef] [PubMed]
133. Cannell, J.J.; Vieth, R.; Umhau, J.C.; Holick, M.F.; Grant, W.B.; Madronich, S.; Garland, C.F.; Giovannucci, E. Epidemic influenza and vitamin D. *Epidemiol. Infect.* **2006**, *134*, 1129–1140. [CrossRef]
134. Shaman, J.; Kohn, M. Absolute humidity modulates influenza survival, transmission, and seasonality. *Proc. Natl. Acad. Sci. USA* **2009**, *106*, 3243–3248. [CrossRef]
135. Ianevski, A.; Zusinaite, E.; Shtaida, N.; Kallio-Kokko, H.; Valkonen, M.; Kantele, A.; Telling, K.; Lutsar, I.; Letjuka, P.; Metelitsa, N.; et al. Low Temperature and Low UV Indexes Correlated with Peaks of Influenza Virus Activity in Northern Europe during 2010–2018. *Viruses* **2019**, *11*, 207. [CrossRef]
136. Urashima, M.; Segawa, T.; Okazaki, M.; Kurihara, M.; Wada, Y.; Ida, H. Randomized trial of vitamin D supplementation to prevent seasonal influenza A in schoolchildren. *Am. J. Clin. Nutr.* **2010**, *91*, 1255–1260. [CrossRef]
137. Panza, F.; La Montagna, M.; Lampignano, L.; Zupo, R.; Bortone, I.; Castellana, F.; Sardone, R.; Borraccino, L.; Dibello, V.; Resta, E.; et al. Vitamin D in the development and progression of Alzheimer's disease: Implications for clinical management. *Expert Rev. Neurother.* **2021**, *21*, 287–301. [CrossRef]
138. Jayedi, A.; Rashidy-Pour, A.; Shab-Bidar, S. Vitamin D status and risk of dementia and Alzheimer's disease: A meta-analysis of dose-response. *Nutr. Neurosci.* **2019**, *22*, 750–759. [CrossRef]
139. Chai, B.; Gao, F.; Wu, R.; Dong, T.; Gu, C.; Lin, Q.; Zhang, Y. Vitamin D deficiency as a risk factor for dementia and Alzheimer's disease: An updated meta-analysis. *BMC Neurol.* **2019**, *19*, 284. [CrossRef]
140. Olsson, E.; Byberg, L.; Karlström, B.; Cederholm, T.; Melhus, H.; Sjögren, P.; Kilander, L. Vitamin D is not associated with incident dementia or cognitive impairment: An 18-y follow-up study in community-living old men. *Am. J. Clin. Nutr.* **2017**, *105*, 936–943. [CrossRef]
141. Littlejohns, T.J.; Kos, K.; Henley, W.E.; Kuzma, E.; Llewellyn, D.J. Vitamin D and Dementia. *J. Prev. Alzheimer's Dis.* **2016**, *3*, 43–52. [CrossRef]
142. Feart, C.; Helmer, C.; Merle, B.; Herrmann, F.R.; Annweiler, C.; Dartigues, J.; Delcourt, C.; Samieri, C. Associations of lower vitamin D concentrations with cognitive decline and long-term risk of dementia and Alzheimer's disease in older adults. *Alzheimer's Dement.* **2017**, *13*, 1207–1216. [CrossRef]
143. Knekt, P.; Sääksjärvi, K.; Järvinen, R.; Marniemi, J.; Männistö, S.; Kanerva, N.; Heliövaara, M. Serum 25-Hydroxyvitamin D Concentration and Risk of Dementia. *Epidemiology* **2014**, *25*, 799–804. [CrossRef] [PubMed]
144. Licher, S.; de Bruijn, R.F.; Wolters, F.J.; Zillikens, M.C.; Ikram, M.A. Vitamin D and the Risk of Dementia: The Rotterdam Study. *J. Alzheimer's Dis.* **2017**, *60*, 989–997. [CrossRef]
145. Schneider, A.L.; Lutsey, P.L.; Alonso, A.; Gottesman, R.F.; Sharrett, A.R.; Carson, K.A.; Gross, M.; Post, W.S.; Knopman, D.S.; Mosley, T.H.; et al. Vitamin D and cognitive function and dementia risk in a biracial cohort: The ARIC Brain MRI Study. *Eur. J. Neurol.* **2014**, *21*, 1211-e70. [CrossRef]
146. Karakis, I.; Pase, M.P.; Beiser, A.; Booth, S.L.; Jacques, P.F.; Rogers, G.; DeCarli, C.; Vasan, R.S.; Wang, T.J.; Himali, J.J.; et al. Association of Serum Vitamin D with the Risk of Incident Dementia and Subclinical Indices of Brain Aging: The Framingham Heart Study. *J. Alzheimer's Dis.* **2016**, *51*, 451–461. [CrossRef] [PubMed]
147. Mokry, L.E.; Ross, S.; Morris, J.A.; Manousaki, D.; Forgetta, V.; Richards, J.B. Genetically decreased vitamin D and risk of Alzheimer disease. *Neurology* **2016**, *87*, 2567–2574. [CrossRef]
148. Wang, L.; Qiao, Y.; Zhang, H.; Zhang, Y.; Hua, J.; Jin, S.; Liu, G. Circulating Vitamin D Levels and Alzheimer's Disease: A Mendelian Randomization Study in the IGAP and UK Biobank. *J. Alzheimer's Dis.* **2020**, *73*, 609–618. [CrossRef] [PubMed]
149. Meng, L.; Wang, Z.; Ming, Y.-C.; Shen, L.; Ji, H.-F. Are micronutrient levels and supplements causally associated with the risk of Alzheimer's disease? A two-sample Mendelian randomization analysis. *Food Funct.* **2022**, *13*, 6665–6673. [CrossRef]
150. Zhang, H.; Wang, T.; Han, Z.; Wang, L.; Zhang, Y.; Wang, L.; Liu, G. Impact of Vitamin D Binding Protein Levels on Alzheimer's Disease: A Mendelian Randomization Study. *J. Alzheimer's Dis.* **2020**, *74*, 991–998. [CrossRef]
151. De Haan, P.; Klein, H.C.; Hart, B.A. Autoimmune Aspects of Neurodegenerative and Psychiatric Diseases: A Template for Innovative Therapy. *Front. Psychiatry* **2017**, *8*, 46. [CrossRef]
152. Kouba, B.R.; Camargo, A.; Gil-Mohapel, J.; Rodrigues, A.L.S. Molecular Basis Underlying the Therapeutic Potential of Vitamin D for the Treatment of Depression and Anxiety. *Int. J. Mol. Sci.* **2022**, *23*, 7077. [CrossRef]
153. Luthold, R.V.; Fernandes, G.R.; Franco-De-Moraes, A.C.; Folchetti, L.G.; Ferreira, S.R.G. Gut microbiota interactions with the immunomodulatory role of vitamin D in normal individuals. *Metabolism* **2017**, *69*, 76–86. [CrossRef]
154. Xie, F.; Huang, T.; Lou, D.; Fu, R.; Ni, C.; Hong, J.; Ruan, L. Effect of vitamin D supplementation on the incidence and prognosis of depression: An updated meta-analysis based on randomized controlled trials. *Front. Public Health* **2022**, *10*, 903547. [CrossRef]
155. Li, H.; Sun, D.; Wang, A.; Pan, H.; Feng, W.; Ng, C.H.; Ungvari, G.S.; Tao, L.; Li, X.; Wang, W.; et al. Serum 25-Hydroxyvitamin D Levels and Depression in Older Adults: A Dose–Response Meta-Analysis of Prospective Cohort Studies. *Am. J. Geriatr. Psychiatry* **2019**, *27*, 1192–1202. [CrossRef]

156. Mikola, T.; Marx, W.; Lane, M.M.; Hockey, M.; Loughman, A.; Rajapolvi, S.; Rocks, T.; O'Neil, A.; Mischoulon, D.; Valkonen-Korhonen, M.; et al. The effect of vitamin D supplementation on depressive symptoms in adults: A systematic review and meta-analysis of randomized controlled trials. *Crit. Rev. Food Sci. Nutr.* **2022**, 1–18. [CrossRef]
157. Luo, C.-W.; Chen, S.-P.; Chiang, C.-Y.; Wu, W.-J.; Chen, C.-J.; Chen, W.-Y.; Kuan, Y.-H. Association between Ultraviolet B Exposure Levels and Depression in Taiwanese Adults: A Nested Case–Control Study. *Int. J. Environ. Res. Public Health* **2022**, *19*, 6846. [CrossRef]
158. Mulugeta, A.; Lumsden, A.; Hyppönen, E. Relationship between Serum 25(OH)D and Depression: Causal Evidence from a Bi-Directional Mendelian Randomization Study. *Nutrients* **2020**, *13*, 109. [CrossRef]
159. Wagner, C.L.; Hollis, B.W. The Implications of Vitamin D Status During Pregnancy on Mother and her Developing Child. *Front. Endocrinol.* **2018**, *9*, 500. [CrossRef]
160. Aguilar-Cordero, M.; Lasserrot-Cuadrado, A.; Mur-Villar, N.; León-Ríos, X.; Rivero-Blanco, T.; Pérez-Castillo, I. Vitamin D, preeclampsia and prematurity: A systematic review and meta-analysis of observational and interventional studies. *Midwifery* **2020**, *87*, 102707. [CrossRef]
161. Gernand, A.D.; Bodnar, L.M.; A Klebanoff, M.; Parks, W.T.; Simhan, H.N. Maternal serum 25-hydroxyvitamin D and placental vascular pathology in a multicenter US cohort. *Am. J. Clin. Nutr.* **2013**, *98*, 383–388. [CrossRef]
162. Merewood, A.; Mehta, S.D.; Chen, T.C.; Bauchner, H.; Holick, M.F. Association between Vitamin D Deficiency and Primary Cesarean Section. *J. Clin. Endocrinol. Metab.* **2009**, *94*, 940–945. [CrossRef] [PubMed]
163. Scholl, T.O.; Chen, X.; Stein, P. Maternal Vitamin D Status and Delivery by Cesarean. *Nutrients* **2012**, *4*, 319–330. [CrossRef]
164. Hu, L.; Zhang, Y.; Wang, X.; You, L.; Xu, P.; Cui, X.; Zhu, L.; Ji, C.; Guo, X.; Wen, J. Maternal Vitamin D Status and Risk of Gestational Diabetes: A Meta-Analysis. *Cell. Physiol. Biochem.* **2018**, *45*, 291–300. [CrossRef] [PubMed]
165. Milajerdi, A.; Abbasi, F.; Mousavi, S.M.; Esmaillzadeh, A. Maternal vitamin D status and risk of gestational diabetes mellitus: A systematic review and meta-analysis of prospective cohort studies. *Clin. Nutr.* **2021**, *40*, 2576–2586. [CrossRef] [PubMed]
166. Wagner, C.; Baggerly, C.; McDonnell, S.; Hamilton, S.; Winkler, J.; Warner, G.; Rodriguez, C.; Shary, J.; Smith, P.; Hollis, B. Post-hoc comparison of vitamin D status at three timepoints during pregnancy demonstrates lower risk of preterm birth with higher vitamin D closer to delivery. *J. Steroid Biochem. Mol. Biol.* **2015**, *148*, 256–260. [CrossRef]
167. Zhou, S.-S.; Tao, Y.-H.; Huang, K.; Zhu, B.-B.; Tao, F.-B. Vitamin D and risk of preterm birth: Up-to-date meta-analysis of randomized controlled trials and observational studies. *J. Obstet. Gynaecol. Res.* **2017**, *43*, 247–256. [CrossRef] [PubMed]
168. Whitehouse, A.J.O.; Holt, B.J.; Serralha, M.; Holt, P.G.; Kusel, M.M.H.; Hart, P.H. Maternal Serum Vitamin D Levels During Pregnancy and Offspring Neurocognitive Development. *Pediatrics* **2012**, *129*, 485–493. [CrossRef]
169. Garcia-Serna, A.M.; Morales, E. Neurodevelopmental effects of prenatal vitamin D in humans: Systematic review and meta-analysis. *Mol. Psychiatry* **2020**, *25*, 2468–2481. [CrossRef]
170. Shi, D.; Wang, D.; Meng, Y.; Chen, J.; Mu, G.; Chen, W. Maternal vitamin D intake during pregnancy and risk of asthma and wheeze in children: A systematic review and meta-analysis of observational studies. *J. Matern. Neonatal Med.* **2021**, *34*, 653–659. [CrossRef]
171. Robinson, C.J.; Alanis, M.C.; Wagner, C.L.; Hollis, B.W.; Johnson, D.D. Plasma 25-hydroxyvitamin D levels in early-onset severe preeclampsia. *Am. J. Obstet. Gynecol.* **2010**, *203*, 366.e1–366.e6. [CrossRef]
172. Serrano-Diaz, N.C.; Gamboa-Delgado, E.M.; Dominguez-Urrego, C.L.; Vesga-Varela, A.L.; Serrano-Gomez, S.E.; Quintero-Lesmes, D.C. Vitamin D and risk of preeclampsia: A systematic review and meta-analysis. *Biomedica* **2018**, *38* (Suppl. 1), 43–53. [CrossRef] [PubMed]
173. Gallo, S.; McDermid, J.M.; Al-Nimr, R.I.; Hakeem, R.; Moreschi, J.M.; Pari-Keener, M.; Stahnke, B.; Papoutsakis, C.; Handu, D.; Cheng, F.W. Vitamin D Supplementation during Pregnancy: An Evidence Analysis Center Systematic Review and Meta-Analysis. *J. Acad. Nutr. Diet.* **2020**, *120*, 898–924.e4. [CrossRef] [PubMed]
174. Palacios, C.; Kostiuk, L.K.; Pena-Rosas, J.P. Vitamin D supplementation for women during pregnancy. *Cochrane Database Syst. Rev.* **2019**, *7*, CD008873. [CrossRef] [PubMed]
175. Saha, S.; Saha, S. A comparison of the risk of cesarean section in gestational diabetes mellitus patients supplemented antenatally with vitamin D containing supplements versus placebo: A systematic review and meta-analysis of double-blinded randomized controlled trials. *J. Turk. Ger. Gynecol. Assoc.* **2020**, *21*, 201–212. [CrossRef]
176. Fogacci, S.; Fogacci, F.; Banach, M.; Michos, E.D.; Hernandez, A.V.; Lip, G.Y.; Blaha, M.J.; Toth, P.P.; Borghi, C.; Cicero, A.F.G. Vitamin D supplementation and incident preeclampsia: A systematic review and meta-analysis of randomized clinical trials. *Clin. Nutr.* **2020**, *39*, 1742–1752. [CrossRef]
177. Kinshella, M.-L.; Omar, S.; Scherbinsky, K.; Vidler, M.; Magee, L.; von Dadelszen, P.; Moore, S.; Elango, R. The PRECISE Conceptual Framework Working Group Effects of Maternal Nutritional Supplements and Dietary Interventions on Placental Complications: An Umbrella Review, Meta-Analysis and Evidence Map. *Nutrients* **2021**, *13*, 472. [CrossRef]
178. Godfrey, K.M.; Costello, P.M.; Lillycrop, K.A. Development, Epigenetics and Metabolic Programming. *Prev. Asp. Early Nutr.* **2016**, *85*, 71–80.
179. Moon, R.J.; Curtis, E.M.; Woolford, S.J.; Ashai, S.; Cooper, C.; Harvey, N.C. The importance of maternal pregnancy vitamin D for offspring bone health: Learnings from the MAVIDOS trial. *Ther. Adv. Musculoskelet. Dis.* **2021**, *13*, 1759720X211006979. [CrossRef]

180. Ross, A.C.; Manson, J.E.; Abrams, S.A.; Aloia, J.F.; Brannon, P.M.; Clinton, S.K.; Durazo-Arvizu, R.A.; Gallagher, J.C.; Gallo, R.L.; Jones, G.; et al. The 2011 Report on Dietary Reference Intakes for Calcium and Vitamin D from the Institute of Medicine: What Clinicians Need to Know. *J. Clin. Endocrinol. Metab.* **2011**, *96*, 53–58. [CrossRef]
181. Toner, C.D.; Davis, C.D.; Milner, J.A. The Vitamin D and Cancer Conundrum: Aiming at a Moving Target. *J. Am. Diet. Assoc.* **2010**, *110*, 1492–1500. [CrossRef]
182. Sempos, C.T.; Durazo-Arvizu, R.A.; Dawson-Hughes, B.; Yetley, E.A.; Looker, A.C.; Schleicher, R.L.; Cao, G.; Burt, V.; Kramer, H.; Bailey, R.L.; et al. Is There a Reverse J-Shaped Association Between 25-Hydroxyvitamin D and All-Cause Mortality? Results from the U.S. Nationally Representative NHANES. *J. Clin. Endocrinol. Metab.* **2013**, *98*, 3001–3009. [CrossRef]
183. Wagner, C.L.; Hulsey, T.C.; Fanning, D.; Ebeling, M.; Hollis, B.W. High-Dose Vitamin D_3 Supplementation in a Cohort of Breastfeeding Mothers and Their Infants: A 6-Month Follow-Up Pilot Study. *Breastfeed. Med.* **2006**, *1*, 59–70. [CrossRef]
184. Hill, A.B. The Environment and Disease: Association or Causation? *Proc. R. Soc. Med.* **1965**, *58*, 295–300. [CrossRef] [PubMed]
185. Potischman, N.; Weed, D.L. Causal criteria in nutritional epidemiology. *Am. J. Clin. Nutr.* **1999**, *69*, 1309S–1314S. [CrossRef] [PubMed]
186. Grant, W.B. How strong is the evidence that solar ultraviolet B and vitamin D reduce the risk of cancer? An examination using Hill's criteria for causality. *Derm.-Endocrinol.* **2009**, *1*, 17–24. [CrossRef] [PubMed]
187. Mohr, S.B.; Gorham, E.D.; Alcaraz, J.E.; Kane, C.I.; Macera, C.A.; Parsons, J.K.; Wingard, D.L.; Garland, C.F. Does the evidence for an inverse relationship between serum vitamin D status and breast cancer risk satisfy the Hill criteria? *Derm.-Endocrinol.* **2012**, *4*, 152–157. [CrossRef] [PubMed]
188. Robsahm, T.E.; Schwartz, G.G.; Tretli, S. The Inverse Relationship between 25-Hydroxyvitamin D and Cancer Survival: Discussion of Causation. *Cancers* **2013**, *5*, 1439–1455. [CrossRef] [PubMed]
189. Grant, W.B. Review of Recent Advances in Understanding the Role of Vitamin D in Reducing Cancer Risk: Breast, Colorectal, Prostate, and Overall Cancer. *Anticancer Res.* **2020**, *40*, 491–499. [CrossRef]
190. Weyland, P.G.; Grant, W.B.; Howie-Esquivel, J. Does Sufficient Evidence Exist to Support a Causal Association between Vitamin D Status and Cardiovascular Disease Risk? An Assessment Using Hill's Criteria for Causality. *Nutrients* **2014**, *6*, 3403–3430. [CrossRef]
191. Wimalawansa, S.J. Vitamin D and cardiovascular diseases: Causality. *J. Steroid Biochem. Mol. Biol.* **2018**, *175*, 29–43. [CrossRef] [PubMed]
192. Walsh, J.B.; McCartney, D.M.; Laird, É.; McCarroll, K.; Byrne, D.G.; Healy, M.; O'Shea, P.M.; Kenny, R.A.; Faul, J.L. Understanding a Low Vitamin D State in the Context of COVID-19. *Front. Pharmacol.* **2022**, *13*, 835480. [CrossRef] [PubMed]
193. Annweiler, C. Vitamin D in dementia prevention. *Ann. N. Y. Acad. Sci.* **2016**, *1367*, 57–63. [CrossRef] [PubMed]
194. Altieri, B.; Grant, W.B.; Della Casa, S.; Orio, F.; Pontecorvi, A.; Colao, A.; Sarno, G.; Muscogiuri, G. Vitamin D and pancreas: The role of sunshine vitamin in the pathogenesis of diabetes mellitus and pancreatic cancer. *Crit. Rev. Food Sci. Nutr.* **2017**, *57*, 3472–3488. [CrossRef]
195. Zipitis, C.S.; Akobeng, A.K. Vitamin D supplementation in early childhood and risk of type 1 diabetes: A systematic review and meta-analysis. *Arch. Dis. Child.* **2008**, *93*, 512–517. [CrossRef]
196. Giovannoni, G.; Ebers, G. Multiple sclerosis: The environment and causation. *Curr. Opin. Neurol.* **2007**, *20*, 261–268. [CrossRef]
197. Hanwell, H.E.; Banwell, B. Assessment of evidence for a protective role of vitamin D in multiple sclerosis. *Biochim. Biophys. Acta (BBA) Mol. Basis Dis.* **2011**, *1812*, 202–212. [CrossRef] [PubMed]
198. Uwitonze, A.M.; Murererehe, J.; Ineza, M.C.; Harelimana, E.I.; Nsabimana, U.; Uwambaye, P.; Gatarayiha, A.; Haq, A.; Razzaque, M.S. Effects of vitamin D status on oral health. *J. Steroid Biochem. Mol. Biol.* **2018**, *175*, 190–194. [CrossRef]
199. Grant, W.B.; Boucher, B.J. Are Hill's criteria for causality satisfied for vitamin D and periodontal disease? *Derm.-Endocrinol.* **2010**, *2*, 30–36. [CrossRef]
200. Carlberg, C. Nutrigenomics of Vitamin D. *Nutrients* **2019**, *11*, 676. [CrossRef]

Review

Genetic Determinants of 25-Hydroxyvitamin D Concentrations and Their Relevance to Public Health

Elina Hyppönen [1,2,*], Karani S. Vimaleswaran [3,4] and Ang Zhou [1,2]

1 Australian Centre for Precision Health, Clinical and Health Sciences, University of South Australia, Adelaide, SA 5001, Australia
2 South Australian Health and Medical Research Institute, Adelaide, SA 5001, Australia
3 Hugh Sinclair Unit of Human Nutrition, Department of Food and Nutritional Sciences, University of Reading, Reading RG6 6DZ, UK
4 The Institute for Food, Nutrition and Health (IFNH), University of Reading, Reading RG6 6DZ, UK
* Correspondence: elina.hypponen@unisa.edu.au

Abstract: Twin studies suggest a considerable genetic contribution to the variability in 25-hydroxyvitamin D (25(OH)D) concentrations, reporting heritability estimates up to 80% in some studies. While genome-wide association studies (GWAS) suggest notably lower rates (13–16%), they have identified many independent variants that associate with serum 25(OH)D concentrations. These discoveries have provided some novel insight into the metabolic pathway, and in this review we outline findings from GWAS studies to date with a particular focus on 35 variants which have provided replicating evidence for an association with 25(OH)D across independent large-scale analyses. Some of the 25(OH)D associating variants are linked directly to the vitamin D metabolic pathway, while others may reflect differences in storage capacity, lipid metabolism, and pathways reflecting skin properties. By constructing a genetic score including these 25(OH)D associated variants we show that genetic differences in 25(OH)D concentrations persist across the seasons, and the odds of having low concentrations (<50 nmol/L) are about halved for individuals in the highest 20% of vitamin D genetic score compared to the lowest quintile, an impact which may have notable influences on retaining adequate levels. We also discuss recent studies on personalized approaches to vitamin D supplementation and show how Mendelian randomization studies can help inform public health strategies to reduce adverse health impacts of vitamin D deficiency.

Keywords: 25-hydroxyvitamin D; vitamin D; genetic risk; heritability; personalized supplementation; genome-wide association study; Mendelian randomization

Citation: Hyppönen, E.; Vimaleswaran, K.S.; Zhou, A. Genetic Determinants of 25-Hydroxyvitamin D Concentrations and Their Relevance to Public Health. *Nutrients* **2022**, *14*, 4408. https://doi.org/10.3390/nu14204408

Academic Editor: Carsten Carlberg

Received: 30 September 2022
Accepted: 15 October 2022
Published: 20 October 2022

1. Introduction

Interest in the genetic architecture of 25-hydroxyvitamin D (25(OH)D) has been active during the past couple of decades, promoted by heritability estimates from twin studies which suggest that up to 80% of the variability in 25(OH)D concentrations might be explained by genetic variation in some populations [1–5]. Also an evolutionary perspective provides strong cues about the importance of genetic variation for vitamin D metabolism [6], as differences in skin colour are believed to have evolved at least in part as an adaptation to ultraviolet B radiation exposure during migration to more northern latitudes, where reduction in skin pigmentation became critical to vitamin D synthesis. Important insights into the genetic architecture of 25(OH)D concentrations have been obtained from genome-wide association studies (GWASs) which have used information across thousands of genomes to find polymorphisms which are statistically associated with 25(OH)D. For this paper, we systematically looked through the GWAS literature for genes that influence serum 25(OH)D levels. We describe key variants for which evidence has been provided from several independent studies and address the public health importance and some of the uses of this information.

2. Materials and Methods

For the systematic search of GWASs, we searched the MEDLINE, Embase, Cochrane, CINAHL and NHGRI-EBI GWAS catalogue [7] databases for original studies and meta-analyses of studies performed in humans and published in English from inception to February, 2022. The search terms used were ("vitamin D" OR "calcidiol" OR "25-hydroxyvitamin D" OR "25(OH)D") AND ("genome-wide association study" OR "genome-wide association scan" OR "genome-wide association analysis") along with the expanded MeSH search terms (in titles, abstracts, or keywords). The search identified 791 publications in total. After excluding duplicate entries, 522 publications remained, of which 29 were relevant. These were further scrutinized to identify sample overlap. We also scrutinized references within the selected articles, and from studies otherwise known to the authors, with evidence on gene function queried using the gene ontology (GO) resource (http://geneontology.org/, accessed on 27 September 2022), KEGG (https://www.genome.jp/kegg/, accessed on 27 September 2022), ConsensusPathDB (http://cpdb.molgen.mpg.de/, accessed on 27 September 2022) and other NCBI (https://www.ncbi.nlm.nih.gov/guide/all/, accessed on 27 September 2022) databases.

3. Results

3.1. Genome-Wide Association Studies on 25(OH)D

Our literature search identified 29 published GWASs that looked for SNPs associated with 25(OH)D [8–36], although many of these analyses were conducted using overlapping samples. Most of the studies only include adult participants of white European ancestry (n = 6722 to 443,374) [8,10,14,16,19,20,22,24,26,29,30,32–34,36]. There were seven studies including data from transethnic analyses [15] and studies with small to modest sample sizes (n = 697 to 9823) which had been conducted using data from African Americans (n = 697 in the discovery sample) [25], African descent in the UK (n = 9354) [35], Hispanic (n = 1190) [9] and Asian populations (n = 1387 to 9823) [13,23,28,35]. There were also five small GWASs on children/toddlers/new-borns [11,12,18,21,27].

Three loci, including *DHCR7*, *CYP2R1*, and *GC* were consistently reported across European [8,10,14,16,19,20,22,24,26,34] and several non-European GWASs [11,13,15,18,21, 23,25,27,28]. These loci were also confirmed in GWAS conducted in children/toddlers/new-borns [11,12,18,21,27]. These genes fit with existing evidence of the involvement of their corresponding proteins in the vitamin D metabolic pathway. The *DHCR7* gene encodes the 7-dehydrocholesterol reductase, which is an enzyme that converts dehydrocholesterol to cholesterol in the skin, and affects the substrate availability for vitamin D3 synthesis, which is a precursor of 25(OH)D [37] (Figure 1). The *CYP2R1* gene encodes the enzyme in the cytochrome P-450 family 2R1, and is the primary 25-hydroxylase in the liver, converting vitamin D to 25(OH)D. In the circulation, vitamin D metabolites including 25(OH)D are mainly found bound to vitamin D binding protein (encoded by *GC*), which in most of the GWASs to date, have come up with the strongest signal for 25(OH)D. GWAS on non-European cohorts have also reported some novel loci (e.g., *FOXA2/SSTR4* [13], *HSPG2* [25], *TINK* [25] and *KIF4B* [15]) which have not been identified in GWAS studies on white European ancestry. Independent replication is lacking with respect to most of these novel loci, and many have no clear link with the vitamin D metabolic pathway. One exception is *CYP2J2*, discovered in a multi-ethnic cohort of 942 pregnant women of Malay, Indian and Chinese ancestry [27]. *CYP2J2* encodes an enzyme (Cytochrome P-450 family 2, subfamily J, polypeptide 2) shown in vitro to act as a vitamin D hydroxylase [38]. There is thus a strong biological basis for this association. It is notable that variants coding the α1-hydroxylase (*CYP27B1*) or the vitamin D receptor (*VDR*) have not been identified by the GWASs on 25(OH)D conducted to date. Concentrations of active 1,25$(OH)_2$D in the circulation are ~1000 times lower than those of 25(OH)D. Therefore, it is possible that differences arising from the conversion of 25(OH)D to 1,25$(OH)_2$D, or those reflecting *VDR* related differences in the 'usage' of 1,25$(OH)_2$D, may simply be too small to detect.

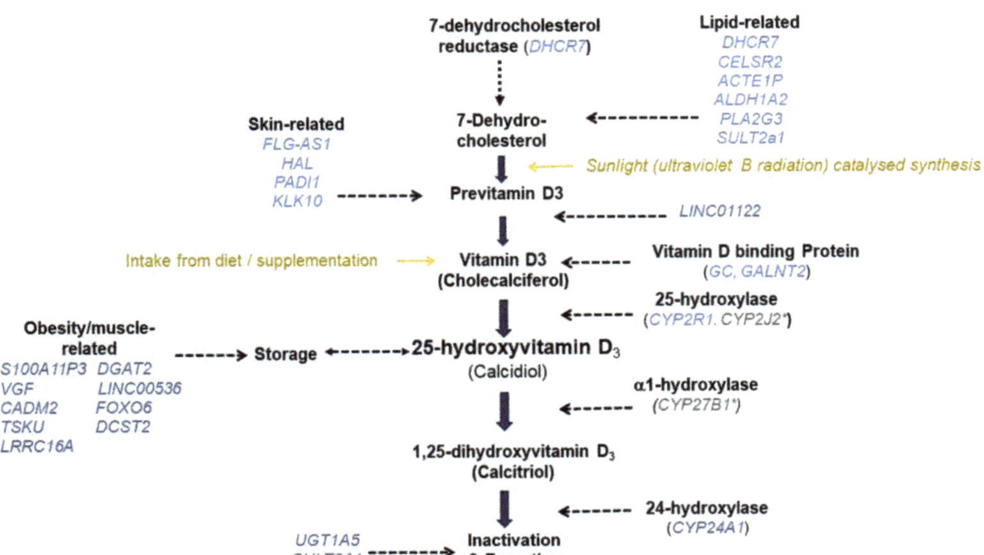

Figure 1. Possible role of selected replicated variants in the vitamin D pathway. * Indicates candidate genes with a confirmed role in vitamin D metabolism but which were not among the replicating variants. For the full list of single nucleotide polymorphism relating to each gene, please refer to Table 1.

Variants with Evidence for Replicated Association with 25(OH)D

In the largest GWAS published to date, identified variants for 25(OH)D concentrations were enriched in genes in the metabolic vitamin D pathway, lipid and lipoprotein pathways, and pathways related to skin properties. The liver, the brain and the skin were identified as the top three locations where 25(OH)D-associated loci may exert their actions [20]. In Table 1 we present more detailed information for the 35 common autosomal SNPs which were identified as top hits in the GWAS conducted in the UK Biobank [20], and which were associated in a consistent direction with serum 25(OH)D concentrations in the SUNLIGHT consortium meta-analyses [16,20,39]. In addition, there were 70 common variants that could not be replicated in the SUNLIGHT consortium meta-analyses. These included variants that are known to be pleiotropic (i.e., affect multiple traits) and/or affect cholesterol metabolism (e.g., *PCSK9, LIPC, ABCA1, CETP, APOE, APOB, APOC1, LIPG* and *LDLR*). It is possible that some of the differences between the UK Biobank findings and the SUNLIGHT consortium meta-analysis relates to the body mass index adjustment by the SUNLIGHT, which reduces the likelihood of adipose tissue related variants (reflecting differences in storage capacity) being detected. It should be noted, however, that replication does not imply causality, and the possible connection which we have identified with vitamin D metabolism/function is only based on literature available to date and may not fully describe the connection with 25(OH)D concentrations.

Table 1. Genetic variants with replicating evidence for an association with 25-hydroxyvitamin D concentrations.

Gene (SNP)	CHR	
PEX10 (rs6671730)	1	*PEX10* encodes a protein involved in import of peroxisomal matrix proteins. Mutations in *PEX10* gene have led to Zellweger syndrome [40] and osteopenia [41], for which vitamin D supplementation has been the treatment.
PADI1 (rs35408430)	1	*PADI1* encodes an enzyme, which catalyses the post-translational deimination of proteins by converting arginine residues into citrullines in the presence of calcium ions [42]. Deimination by PADIs occurs during epidermal differentiation [43], with possible influence on skin properties [20].
FOXO6 (rs7522116)	1	*FOXO6* encodes a protein that has been predicted to enable DNA-binding transcription factor activity, and RNA polymerase related DNA binding activity [44]. FoxO6 expression is downregulated in the brain of dietary obese mice [45].
CELSR2 (rs7528419)	1	*CELSR2* encodes the cadherin EGF LAG seven-pass G-type receptor 2 that is involved in contact-mediated communication, with cadherin domains acting as homophilic binding regions and the EGF-like domains involved in cell adhesion and receptor-ligand interactions [46].
FLG-AS1 (rs1933064)	1	FLG antisense RNA 1 (*FLG-AS1*) is an RNA Gene that is affiliated with the long non-coding RNA class. Skin pigmentation-related diseases such as Ichthyosis Vulgaris [47] and Peeling Skin Syndrome 6 [48] have been shown to be associated with FLG-AS1.
DCST2 (rs76798800)	1	*DCST2* gene encodes the DC-STAMP domain containing 2 protein that has been shown to be an important regulator of osteoclast cell-fusion in bone homeostasis [49]. *DCST2* gene is associated with early length and adult height [50].
GALNT2 (rs6672758)	1	*GALNT2* gene encodes the polypeptide N-acetylgalactosaminyltransferase 2 which is a member of the glycosyltransferase 2 protein family and which has been linked to post-translational modification of vitamin D-binding protein [51].
LINC01122 (rs727857)	2	*LINC01122* gene is an RNA gene that is affiliated with the lncRNA class [52]. *LINC01122* was one of the 989 differentially expressed genes which was significantly enriched in vitamin D3 biosynthesis [53].
CPS1 (rs1047891)	2	*CPS1* gene encodes the carbamoyl-phosphate synthase 1 which is a mitochondrial enzyme that catalyses synthesis of carbamoyl phosphate from ammonia and bicarbonate [54].
UGT1A5 * (rs2012736)	2	*UGT1A5* gene encodes the UDP glucuronosyltransferase family 1 member A5 which has been shown to transform small lipophilic molecules, into water-soluble, excretable metabolites [55]. Related isoenzymes have been identified as catalysts for 25(OH)D3 glucuronidation in the human liver [56].
CADM2 (rs6782190)	3	*CADM2* gene encodes the cell adhesion molecule 2 which is a member of the synaptic cell adhesion molecule 1 family [57]. In animal studies, *CADM2* is associated with metabolic traits [58], suggesting possible influence on vitamin D concentrations through its effect on obesity and storage capacity of 25(OH)D.
GC (rs705117, rs1352846)	4	*GC* gene encodes the vitamin D binding protein which binds to vitamin D and its plasma metabolites and transports them to target tissues [59].
CARMIL1/LRRC16A (rs78151190)	6	*CARMIL1* gene encodes the capping protein regulator and myosin 1 linker 1 with a role in actin filament network formation [60]. Approximately 10% of muscle tissue consists of actin, providing a possible link with 25(OH)D through storage capacity.
VGF (rs75741381)	7	*VGF* gene encodes a protein that is expressed in neuroendocrine cells and is upregulated by nerve growth factor [61]. *VGF* has been linked with appetite control [62], and diet-induced obesity [63], with a possible link through storage capacity.
LINC00536 (rs12056768)	8	*LINC00536* gene interacts with Wnt3a/β-Catenin signalling [64]. Wnt/β -Catenin signalling is an important signalling pathway in regulating adipose tissue lipogenesis with a possible link with 25(O)D through storage capacity.
GRID1 (rs77532868)	10	*GRID1* gene encodes the glutamate ionotropic receptor delta type subunit 1 which is a subunit of glutamate receptor channels that mediate the fast excitatory synaptic transmission in the central nervous system [65].
CYP2R1 (rs12794714)	11	*CYP2R1* gene encodes the cytochrome P450 family 2 subfamily R member 1 which acts as 25-hydroxylase of vitamin D [66].
TMEM151A (rs61891388)	11	*TMEM151A* has been predicted to be an integral component of membrane and *CD248* enables extracellular matrix binding activity and regulates endothelial cell apoptotic process.
AP002387.1/ACTE1P (rs1660839, rs12803256)	11	*ACTE1P* gene is an RNA gene. ACTE1P [67] and vitamin D [68] are both involved in adolescent idiopathic scoliosis (abnormal curvature of the spine), suggesting a possible role of ACTE1P in bone health.
S100A11P3 (rs12798050)	11	*S100A11P3* gene encodes the S100 calcium binding protein A11 pseudogene 3. It has multiple roles in buffering calcium ion concentration, participating in energy metabolism, regulating cell proliferation and differentiation [69].

Table 1. *Cont.*

Gene (SNP)	CHR	
DGAT2 (rs72997623)	11	*DGAT2* encodes the diacylglycerol O-acyltransferase 2, catalysing the synthesis of triglycerides [70]. Affects adipose tissue formation [71] with possible link to 25(OH)D storage.
GUCY2EP/TSKU (rs1149605)	11	*GUCY2EP* gene encodes guanylate cyclase 2E that is involved in chemosensation and *TSKU* gene encodes tsukushi, small leucine rich proteoglycan that has been predicted to act upstream/within several processes, including negative regulation of Wnt signaling pathway.
HAL (rs10859995)	12	*HAL* gene is upregulated during the differentiation of keratinocytes [72]. HAL deaminates L-histidine to trans-uronic acid [73], which in the stratum corneum absorbs UVB [74] and reduce the production 25(OH)D [75].
SEC23A (rs8018720)	14	*SEC23A* gene encodes the Sec23 homolog A, coat complex II component which plays a role in the ER-Golgi protein trafficking.
ALDH1A2 (rs261291)	15	*ALDH1A2* gene encodes aldehyde dehydrogenase 1 family member A2 which catalyses the synthesis of retinoic acid (RA) from retinaldehyde [76].
PDILT (rs77924615)	16	*PDILT/PDIA7* gene encodes the protein disulphide isomerase like, testis expressed which catalyses protein folding and thiol-disulphide interchange reactions [77].
SULT2A1 (rs212100)	19	*SULT2A1* gene encodes a liver- and intestine-expressed sulpho-conjugating enzyme that is responsible for the inactivation by sulphonation of 25(OH)D [78,79].
KLK10 (rs10426)	19	*KLK10* gene encodes the kallikrein related peptidase 10 that has been shown to play a role in dermal integrity [80].
CYP24A1 †	20	*CYP24A1* gene encodes cytochrome P450 family 24 subfamily A member 1 which is an important candidate for vitamin D metabolic pathway given that it initiates the degradation of 1,25-dihydroxyvitamin D3 by hydroxylation of the side chain [81]. In addition, this enzyme also plays a role in calcium homeostasis and vitamin D endocrine system [82].
PLA2G3 (rs2074735)	22	*PLA2G3* gene encodes the phospholipase A2 group III which functions in lipid metabolism and catalyses the calcium-dependent hydrolysis of the sn-2 acyl bond of phospholipids to release arachidonic acid and lysophospholipids [83].

* *UGT1A5, UGT1A6, UGT1A7, UGT1A8, UGT1A9, UGT1A10.* † rs6123359, rs17216707, rs2585442, rs2762943.

For many of the variants which have a replicated association with 25(OH)D we found some evidence that was compatible with a role in the vitamin D pathway (Figure 1, Table 1). Several of the variants were related to lipid levels, while others had been linked with skin integrity, suggesting a possible link with substrate availability for conversion to the circulating 25(OH)D metabolite. Despite body mass index adjustment in the SUNLIGHT consortium meta-analyses, for several of the replicating variants we observed suggested links with adiposity or muscle mass, which may be because these tissues serve as storage sites for 25(OH)D. In addition to confirming the role of 24-hydroxylation in the inactivation and removal of vitamin D metabolites, GWAS identified variants in *UGT1A5* and *SULT2A1* as potentially relevant. This suggests that sulphonation and glucuronidation pathways, which conjugate sulphur and glucuronide with vitamin D metabolites, are relevant for 25(OH)D excretion and/or recycling.

3.2. Heritability and the Genetic Contribution to the Prevalence of Deficiency

25(OH)D is a commonly used indicator of 'vitamin D status' with much of the concentrations determined based on availability of sunlight induced skin synthesis, with contributions from supplement intake and diet. According to twin- and family-based studies, there is great variability in the heritability of 25(OH)D, with overall estimates ranging from 28% to 80% [1–4]. As 'heritability' merely reflects the proportion of total variance that can be explained by genetic factors, it will be the higher with lower environmental contributions (total variance = genetic variance + environmental variance). Indeed, the highest heritability rates are seen in populations measured during winter, when contributions from sunlight synthesis (and hence, the environmental effects) are at their lowest. This may explain why a study on 510 middle-aged male twins found heritability estimates to be ~70% when assessed in winter compared to negligible (~0%) during summer [84]. Also a twin study of Hispanics and African Americans, reported heritabilities of 23%, 28% and 41% for 25(OH)D levels from data taken from California, Texas, and Colorado (n = 1530 individuals from 130 families), respectively [85]. These differences were broadly reflective of geographical location, such that the higher the latitude (and the less sunlight exposure) the higher the estimated heritability. However, seasonal differences in heritability have not been consistently observed, and for example in the recent 25(OH)D GWAS [20], heritability was higher in summer than in winter (0.19 vs. 0.10, respectively). Indeed, another way to estimate heritability is to use information from unrelated individuals using GWAS data (i.e., SNP heritability) and in the large UK Biobank GWAS on white Europeans, SNP heritability for serum 25(OH)D concentrations was estimated to be 13–16% [19,20]. Of practical relevance is the extent to which these genetic variants affect the circulating 25(OH)D levels and the prevalence of vitamin D deficiency. In Table 2, we show the distribution of 25(OH)D concentrations based on data from the 35 replicating variants in the UK Biobank. The odds of low concentrations are about halved for individuals in the highest 20% of vitamin D genetic risk score compared to the lowest quintile. Again, these genetic associations appear to be slightly stronger during summer compared to winter, possibly reflecting genetic differences in vitamin D skin synthesis. The overall difference in the mean 25(OH)D between individuals in the highest vs. lowest quartile of the GRS is about 9 nmol/L [40], which is similar to the association seen with self-reported use of vitamin D supplementation in the UK Biobank during winter (9.7 nmol/L) [84]. This suggests that if the supply from sunlight or diet is limited, differences in 25(OH)D concentrations by higher genetic burden may be of clinical relevance.

Table 2. Average 25-hydroxyvitamin D level and the odds of low concentrations by quintiles in vitamin D genetic risk score in the UK Biobank.

	Vitamin D Winter (n = 176,577)			Vitamin D Summer (n = 130,855)		
	25(OH)D Mean (SD)	<25 nmol/L OR (95% CI)	<50 nmol/L OR (95% CI)	25(OH)D Mean (SD)	<25 nmol/L OR (95% CI)	<50 nmol/L OR (95% CI)
Quintile 1 (Lowest 20%)	39.16 (17.41)	Reference	Reference	52.13 (17.37)	Reference	Reference
Quintile 2	41.84 (18.47)	0.79 (0.76–0.82)	0.75 (0.73–0.78)	56.16 (18.51)	0.72 (0.66–0.79)	0.68 (0.66–0.71)
Quintile 3	43.73 (19.36)	0.68 (0.66–0.71)	0.64 (0.61–0.66)	58.50 (19.20)	0.63 (0.57–0.69)	0.56 (0.54–0.58)
Quintile 4	45.40 (20.14)	0.60 (0.58–0.62)	0.54 (0.53–0.56)	60.65 (20.05)	0.53 (0.48–0.58)	0.49 (0.47–0.51)
Quintile 5	47.51 (21.25)	0.52 (0.50–0.54)	0.47 (0.45–0.48)	64.05 (21.17)	0.42 (0.38–0.47)	0.39 (0.37–0.40)

Genetic risk score calculated using 35 variants with replicated association with 25(OH)D concentrations. Adjusted for age, sex, month in which blood sample was taken, fasting time before blood sample was taken, sample aliquots for measurement, assessment centres, SNP array, and top 40 genetic principal components. Vitamin D winter classified as November to May and vitamin D summer as June to October, based on distribution of 25(OH)D concentrations in the UK biobank [86].

3.3. Genetic Differences in Response to Supplementation and the Need for Personalized Approaches

There were 25 independent loci which were suggestive of gene–environment (GxE) interaction in the recent GWAS [20], suggesting that the size of the genetic association with 25(OH)D can vary by environmental factors influencing serum 25(OH)D concentrations. For five loci, including *CYP2R1* and *SEC23A* there was a genome-wide significant interaction with season [19,20], where the carriers of 25(OH)D-lowering alleles appeared to be less responsive to season compared to non-carriers. This could suggest that some individuals may be more prone to low serum 25(OH)D levels regardless of the season of measurement [19]. In an earlier genome-wide GxE analysis, carriers of 25(OH)D-lowering allele at the *CYP2R1* locus were less responsive to dietary vitamin D intake [16]. A similar interaction with vitamin D lowering alleles has also been observed in the context of the *GC* locus and vitamin D supplementation [87], of vitamin D3-fortified bread and milk consumption [86,87] and UVB treatment [88,89].

There has been recent interest in genetic risk scores (GRS) that combine variants according to their vitamin D lowering alleles, and which look into whether individuals with genetically low 25(OH)D are less or more responsive to treatments for correcting low vitamin D status [87–89]. One study used a GRS combining variants in the *CYP2R1* and *GC* loci, and reported a somewhat more modest (~23%) increase in serum 25(OH)D concentrations in response to UVB treatment for individuals carrying four risk alleles compared to the 54% increase for those carrying no risk alleles [88]. They also found that individuals with four risk alleles benefitted the least from the consumption of vitamin D3–fortified bread and milk during this 6-month study [88]. GRS for 25(OH)D has been suggested to be useful for guiding the screening and treatment for vitamin D deficiency. This was tested in a recent study [26], where participants with serum 25(OH)D < 50 nmol/L were recommended to take vitamin D supplements, adjusting the dosage according to their genetic risk. Again, this study used a simple GRS (two SNPs only, taken from *GC* and *CYP2R1)*, and the individuals with three or four 25(OH)D-lowering alleles were instructed to take 50 µg (2000IU) per day, those with one to two risk alleles to take 20–30 µg/day and those with no risk alleles, 10–20 µg per day. In their study, recommendation to take 50 µg (2000IU) per day over 4 months was enough to reduce the gap between individuals carrying three or four risk alleles and those with no risk alleles both with respect to serum 25(OH)D concentration and the prevalence of 25(OH)D < 50 nmol/L. However, the prevalence of 25(OH)D < 50 nmol/L remained elevated for those with two risk alleles compared to no risk alleles. While these results are very interesting and even promising, they are tentative, as the higher vitamin D intakes were achieved by recommendations, and not by testing in a placebo controlled, and randomized context. This was also a relatively small study (n = 10 to

$n = 36$ per treatment group), so further trials with appropriate controls and a larger sample are required to examine possible benefits and effective approaches for personalized vitamin D supplementation. Given more profound genetic adaptations to differences in vitamin D intakes ('vitamin D scarcity') are possible, more research is also needed to establish target levels reflecting 'optimal' 25(OH)D concentrations and supplementation approaches for specific population groups, including indigenous Arctic and Tropical peoples [90].

3.4. Mendelian Randomization to Establish Evidence for Causal Effects of 25(OH)D

With the identification of genetic variants associated with serum 25(OH)D concentrations, it has become possible to use Mendelian randomization (MR) to examine evidence for the causal effect of vitamin D on other traits. This method is sometimes called the "natures controlled trial", as assuming random allocation of genetic variants during the gamete formation, individuals are randomized on different exposure groups based on the genetic variants they carry. Reliable causal inference based on MR studies is conditional to some key method assumptions, and where these hold, this method can help avoid bias due to confounding and reverse causation which more strongly affect other types of observational studies [91]. MR analyses on 25(OH)D have used several strategies, and many of the studies have restricted the variants used to those in the actual vitamin D pathway (including *DHCR7*, *CYP2R1*, *GC* and *CYP24A1*). With additional loci being discovered for serum 25(OH)D [19,20], MR studies now commonly incorporate these new loci into the analyses. While the inclusion of additional loci can improve statistical power, it is also important to keep in mind the potential for pleiotropic effects that these variants could bring into the models and which could bias the MR analysis [92].

MR studies on 25(OH)D have been conducted across a wide range of outcomes, with evidence supportive of a causal effect seen for multiple sclerosis [93], type 2 diabetes [94] and hypertension [95]. However, many of the newly discovered variants do not have clear or known function with respect to vitamin D metabolism, and some appear pleiotropic, with associations with other traits, such as BMI, and lipid measures. One approach to alleviate concerns relating to pleiotropy and residual genetic confounding affecting variant selection, is to restrict the analyses to variants which have consistent replicating association with 25(OH)D concentrations [39], as would be the case if we use the 35 SNPs described above. However, even there, pleiotropy is likely to remain a concern, and sensitivity analyses using different sets of variants and different analytical approaches will be required to help to assess the robustness of the findings. A multivariable MR approach, which directly accounts for pleiotropic effects by modelling the genetic effects on 25(OH)D simultaneously with pleiotropy related indicators, may also be helpful. However, to allow for the use of this approach, the relevant pleiotropic pathways will need to be hypothesized and relevant information must be available for the analyses. In the context of threshold effects, rigorously conducted MR studies that take into account non-linearity can increase the value of the genetic approach for vitamin D research, as evidence for an effect may only be seen at very low or high levels [96,97]. Recruiting people with vitamin D deficiency to supplementation trials is an important challenge, and often studies test the effects of supplementation in individuals who already have adequate concentrations, and who are typically allowed to take over-the-counter supplements [96,97]. Evidence for benefits with rectifying vitamin D deficiency with respect to outcomes such as mortality [98,99], cardiovascular disease [39] and dementia [100] has been obtained from recent studies using non-linear or stratified MR approaches. For dementia, evidence for a causal effect of vitamin D had already been provided by linear MR studies [101–103], while effects on mortality had been supported by RCT meta-analyses [104], but the non-linear studies in both contexts suggest that the benefits of increasing levels may be largely confined to the correction of clinical deficiency. These findings provide important insight into strategies that are likely to provide the greatest benefits, suggesting that large-dose supplementation is unlikely to be required, but population level strategies such as food fortification, which can ensure at least minimal intakes and eradicate severe deficiency across the range of population groups, is likely to work.

4. Conclusions

Vitamin D status is in part determined by genetic variation and GWAS studies have identified a large number of variants that are associated with circulating 25(OH)D concentrations. Some of them are linked to the actual vitamin D metabolic pathway, and others to lipid metabolism and skin properties. In terms of methodology, they may provide MR studies with the means to measure the various determinants of serum 25(OH)D concentrations. Further research is needed to understand how such genetic information may be used to personalize vitamin D supplementation and prevent vitamin D deficiency.

Author Contributions: E.H. conceptualized and wrote the paper. A.Z. conducted literature review, analysed data and wrote the paper. K.S.V. reviewed variants for gene function and prepared the figure with E.H. All authors approved the paper for submission. All authors have read and agreed to the published version of the manuscript.

Funding: This research was funded by National Health and Medical Research Council (Australia), GNT1123603.

Institutional Review Board Statement: The study was conducted in accordance with the Declaration of Helsinki. UK Biobank provided data for Table 2, with analyses conducted under project 20175. Ethical approval for the UK Biobank was granted by the National Information Governance Board for Health and Social Care and North West Multicentre Research Ethics Committee (11/NW/0382).

Informed Consent Statement: Informed consent was obtained from all subjects involved in the study.

Data Availability Statement: Original data for Table 2 is available from the UK Biobank upon application.

Conflicts of Interest: The funders had no role in the design of the study; in the collection, analyses, or interpretation of data; in the writing of the manuscript; or in the decision to publish the results.

References

1. Hunter, D.; De Lange, M.; Snieder, H.; MacGregor, A.J.; Swaminathan, R.; Thakker, R.V.; Spector, T.D. Genetic Contribution to Bone Metabolism, Calcium Excretion, and Vitamin D and Parathyroid Hormone Regulation. *J. Bone Miner. Res. Off. J. Am. Soc. Bone Miner. Res.* **2001**, *16*, 371–378. [CrossRef] [PubMed]
2. Orton, S.-M.; Morris, A.P.; Herrera, B.M.; Ramagopalan, S.V.; Lincoln, M.R.; Chao, M.J.; Vieth, R.; Sadovnick, A.D.; Ebers, G.C. Evidence for Genetic Regulation of Vitamin D Status in Twins with Multiple Sclerosis. *Am. J. Clin. Nutr.* **2008**, *88*, 441–447. [CrossRef] [PubMed]
3. Shea, M.K.; Benjamin, E.J.; Dupuis, J.; Massaro, J.M.; Jacques, P.F.; D'Agostino, R.B.; Ordovas, J.M.; O'Donnell, C.J.; Dawson-Hughes, B.; Vasan, R.S.; et al. Genetic and Non-Genetic Correlates of Vitamins K and D. *Eur. J. Clin. Nutr.* **2009**, *63*, 458–464. [CrossRef] [PubMed]
4. Wjst, M.; Altmüller, J.; Braig, C.; Bahnweg, M.; André, E. A Genome-Wide Linkage Scan for 25-OH-D(3) and 1,25-(OH)2-D3 Serum Levels in Asthma Families. *J. Steroid Biochem. Mol. Biol.* **2007**, *103*, 799–802. [CrossRef]
5. Berry, D.; Hypponen, E. Determinants of Vitamin D Status: Focus on Genetic Variations. *Curr. Opin. Nephrol. Hypertens.* **2011**, *20*, 331–336. [CrossRef]
6. Jablonski, N.G.; Chaplin, G. The Roles of Vitamin D and Cutaneous Vitamin D Production in Human Evolution and Health. *Int. J. Paleopathol.* **2018**, *23*, 54–59. [CrossRef]
7. Buniello, A.; MacArthur, J.A.L.; Cerezo, M.; Harris, L.W.; Hayhurst, J.; Malangone, C.; McMahon, A.; Morales, J.; Mountjoy, E.; Sollis, E.; et al. The NHGRI-EBI GWAS Catalog of Published Genome-Wide Association Studies, Targeted Arrays and Summary Statistics 2019. *Nucleic Acids Res.* **2019**, *47*, D1005–D1012. [CrossRef]
8. Ahn, J.; Yu, K.; Stolzenberg-Solomon, R.; Simon, K.C.; McCullough, M.L.; Gallicchio, L.; Jacobs, E.J.; Ascherio, A.; Helzlsouer, K.; Jacobs, K.B.; et al. Genome-Wide Association Study of Circulating Vitamin D Levels. *Hum. Mol. Genet.* **2010**, *19*, 2739–2745. [CrossRef]
9. Engelman, C.D.; Meyers, K.J.; Ziegler, J.T.; Taylor, K.D.; Palmer, N.D.; Haffner, S.M.; Fingerlin, T.E.; Wagenknecht, L.E.; Rotter, J.I.; Bowden, D.W.; et al. Genome-Wide Association Study of Vitamin D Concentrations in Hispanic Americans: The IRAS Family Study. *J. Steroid Biochem. Mol. Biol.* **2010**, *122*, 186–192. [CrossRef]

10. Wang, T.J.; Zhang, F.; Richards, J.B.; Kestenbaum, B.; Van Meurs, J.B.; Berry, D.; Kiel, D.P.; Streeten, E.A.; Ohlsson, C.; Koller, D.L.; et al. Common Genetic Determinants of Vitamin D Insufficiency: A Genome-Wide Association Study. *Lancet* **2010**, *376*, 180–188. [CrossRef]
11. Lasky-Su, J.; Lange, N.; Brehm, J.M.; Damask, A.; Soto-Quiros, M.; Avila, L.; Celedon, J.C.; Canino, G.; Cloutier, M.M.; Hollis, B.W.; et al. Genome-Wide Association Analysis of Circulating Vitamin D Levels in Children with Asthma. *Hum. Genet.* **2012**, *131*, 1495–1505. [CrossRef]
12. Anderson, D.; Holt, B.J.; Pennell, C.E.; Holt, P.G.; Hart, P.H.; Blackwell, J.M. Genome-Wide Association Study of Vitamin D Levels in Children: Replication in the Western Australian Pregnancy Cohort (Raine) Study. *Genes Immun.* **2014**, *15*, 578–583. [CrossRef]
13. Sapkota, B.R.; Hopkins, R.; Bjonnes, A.; Ralhan, S.; Wander, G.S.; Mehra, N.K.; Singh, J.R.; Blackett, P.R.; Saxena, R.; Sanghera, D.K. Genome-Wide Association Study of 25(OH) Vitamin D Concentrations in Punjabi Sikhs: Results of the Asian Indian Diabetic Heart Study. *J. Steroid Biochem. Mol. Biol.* **2016**, *158*, 149–156. [CrossRef]
14. Manousaki, D.; Dudding, T.; Haworth, S.; Hsu, Y.-H.; Liu, C.-T.; Medina-Gomez, C.; Voortman, T.; van der Velde, N.; Melhus, H.; Robinson-Cohen, C.; et al. Low-Frequency Synonymous Coding Variation in CYP2R1 Has Large Effects on Vitamin D Levels and Risk of Multiple Sclerosis. *Am. J. Hum. Genet.* **2017**, *101*, 227–238. [CrossRef]
15. Hong, J.; Hatchell, K.E.; Bradfield, J.P.; Bjonnes, A.; Chesi, A.; Lai, C.-Q.; Langefeld, C.D.; Lu, L.; Lu, Y.; Lutsey, P.L.; et al. Transethnic Evaluation Identifies Low-Frequency Loci Associated with 25-Hydroxyvitamin D Concentrations. *J. Clin. Endocrinol. Metab.* **2018**, *103*, 1380–1392. [CrossRef]
16. Jiang, X.; O'Reilly, P.F.; Aschard, H.; Hsu, Y.-H.; Richards, J.B.; Dupuis, J.; Ingelsson, E.; Karasik, D.; Pilz, S.; Berry, D.; et al. Genome-Wide Association Study in 79,366 European-Ancestry Individuals Informs the Genetic Architecture of 25-Hydroxyvitamin D Levels. *Nat. Commun.* **2018**, *9*, 260. [CrossRef]
17. O'Brien, K.M.; Shi, M.; Weinberg, C.R.; Sandler, D.P.; Harmon, Q.E.; Taylor, J.A. Genome-Wide Association Study of Serum 25-Hydroxyvitamin D in US Women. *Front. Genet.* **2018**, *9*, 67. [CrossRef]
18. Kampe, A.; Enlund-Cerullo, M.; Valkama, S.; Holmlund-Suila, E.; Rosendahl, J.; Hauta-Alus, H.; Pekkinen, M.; Andersson, S.; Makitie, O. Genetic Variation in GC and CYP2R1 Affects 25-Hydroxyvitamin D Concentration and Skeletal Parameters: A Genome-Wide Association Study in 24-Month-Old Finnish Children. *PLoS Genet.* **2019**, *15*, e1008530. [CrossRef]
19. Manousaki, D.; Mitchell, R.; Dudding, T.; Haworth, S.; Harroud, A.; Forgetta, V.; Shah, R.L.; Luan, J.; Langenberg, C.; Timpson, N.J.; et al. Genome-Wide Association Study for Vitamin D Levels Reveals 69 Independent Loci. *Am. J. Hum. Genet.* **2020**, *106*, 327–337. [CrossRef]
20. Revez, J.A.; Lin, T.; Qiao, Z.; Xue, A.; Holtz, Y.; Zhu, Z.; Zeng, J.; Wang, H.; Sidorenko, J.; Kemper, K.E.; et al. Genome-Wide Association Study Identifies 143 Loci Associated with 25 Hydroxyvitamin D Concentration. *Nat. Commun.* **2020**, *11*, 1647. [CrossRef]
21. Traglia, M.; Windham, G.C.; Pearl, M.; Poon, V.; Eyles, D.; Jones, K.L.; Lyall, K.; Kharrazi, M.; Croen, L.A.; Weiss, L.A. Genetic Contributions to Maternal and Neonatal Vitamin D Levels. *Genetics* **2020**, *214*, 1091–1102. [CrossRef]
22. Zheng, J.-S.; Luan, J.; Sofianopoulou, E.; Sharp, S.J.; Day, F.R.; Imamura, F.; Gundersen, T.E.; Lotta, L.A.; Sluijs, I.; Stewart, I.D.; et al. The Association between Circulating 25-Hydroxyvitamin D Metabolites and Type 2 Diabetes in European Populations: A Meta-Analysis and Mendelian Randomisation Analysis. *PLoS Med.* **2020**, *17*, e1003394. [CrossRef]
23. Kim, Y.A.; Yoon, J.W.; Lee, Y.; Choi, H.J.; Yun, J.W.; Bae, E.; Kwon, S.-H.; Ahn, S.E.; Do, A.-R.; Jin, H.; et al. Unveiling Genetic Variants Underlying Vitamin D Deficiency in Multiple Korean Cohorts by a Genome-Wide Association Study. *Endocrinol. Metab.* **2021**, *36*, 1189–1200. [CrossRef]
24. Ong, J.-S.; Dixon-Suen, S.C.; Han, X.; An, J.; Fitzgerald, R.; Buas, M.; Gammon, M.D.; Corley, D.A.; Shaheen, N.J.; Hardie, L.J.; et al. A Comprehensive Re-Assessment of the Association between Vitamin D and Cancer Susceptibility Using Mendelian Randomization. *Nat. Commun.* **2021**, *12*, 246. [CrossRef] [PubMed]
25. Palmer, N.D.; Lu, L.; Register, T.C.; Lenchik, L.; Carr, J.J.; Hicks, P.J.; Smith, S.C.; Xu, J.; Dimitrov, L.; Keaton, J.; et al. Genome-Wide Association Study of Vitamin D Concentrations and Bone Mineral Density in the African American-Diabetes Heart Study. *PLoS ONE* **2021**, *16*, e0251423. [CrossRef] [PubMed]
26. Sallinen, R.J.; Dethlefsen, O.; Ruotsalainen, S.; Mills, R.D.; Miettinen, T.A.; Jääskeläinen, T.E.; Lundqvist, A.; Kyllönen, E.; Kröger, H.; Karppinen, J.I.; et al. Genetic Risk Score for Serum 25-Hydroxyvitamin D Concentration Helps to Guide Personalized Vitamin D Supplementation in Healthy Finnish Adults. *J. Nutr.* **2021**, *151*, 281–292. [CrossRef] [PubMed]
27. Sampathkumar, A.; Tan, K.M.; Chen, L.; Chong, M.F.F.; Yap, F.; Godfrey, K.M.; Chong, Y.S.; Gluckman, P.D.; Ramasamy, A.; Karnani, N. Genetic Link Determining the Maternal-Fetal Circulation of Vitamin D. *Front. Genet.* **2021**, *12*, 721488. [CrossRef] [PubMed]
28. Zeng, H.; Ge, J.; Xu, W.; Ma, H.; Chen, L.; Xia, M.; Pan, B.; Lin, H.; Wang, S.; Gao, X. Type 2 Diabetes Is Causally Associated With Reduced Serum Osteocalcin: A Genomewide Association and Mendelian Randomization Study. *J. Bone Miner. Res.* **2021**, *36*, 1694–1707. [CrossRef]
29. Backman, J.D.; Li, A.H.; Marcketta, A.; Sun, D.; Mbatchou, J.; Kessler, M.D.; Benner, C.; Liu, D.; Locke, A.E.; Balasubramanian, S.; et al. Exome Sequencing and Analysis of 454,787 UK Biobank Participants. *Nature* **2021**, *599*, 628–634. [CrossRef]
30. Barton, A.R.; Sherman, M.A.; Mukamel, R.E.; Loh, P.-R. Whole-Exome Imputation within UK Biobank Powers Rare Coding Variant Association and Fine-Mapping Analyses. *Nat. Genet.* **2021**, *53*, 1260–1269. [CrossRef]

31. Benjamin, E.J.; Dupuis, J.; Larson, M.G.; Lunetta, K.L.; Booth, S.L.; Govindaraju, D.R.; Kathiresan, S.; Keaney, J.F.; Keyes, M.J.; Lin, J.-P.; et al. Genome-Wide Association with Select Biomarker Traits in the Framingham Heart Study. *BMC Med. Genet.* **2007**, *8* (Suppl. 1), S11. [CrossRef]
32. Locke, A.E.; Steinberg, K.M.; Chiang, C.W.K.; Service, S.K.; Havulinna, A.S.; Stell, L.; Pirinen, M.; Abel, H.J.; Chiang, C.C.; Fulton, R.S.; et al. Exome Sequencing of Finnish Isolates Enhances Rare-Variant Association Power. *Nature* **2019**, *572*, 323–328. [CrossRef]
33. Mbatchou, J.; Barnard, L.; Backman, J.; Marcketta, A.; Kosmicki, J.A.; Ziyatdinov, A.; Benner, C.; O'Dushlaine, C.; Barber, M.; Boutkov, B.; et al. Computationally Efficient Whole-Genome Regression for Quantitative and Binary Traits. *Nat. Genet.* **2021**, *53*, 1097–1103. [CrossRef]
34. Sinnott-Armstrong, N.; Tanigawa, Y.; Amar, D.; Mars, N.; Benner, C.; Aguirre, M.; Venkataraman, G.R.; Wainberg, M.; Ollila, H.M.; Kiiskinen, T.; et al. Genetics of 35 Blood and Urine Biomarkers in the UK Biobank. *Nat. Genet.* **2021**, *53*, 185–194. [CrossRef]
35. Sun, Q.; Graff, M.; Rowland, B.; Wen, J.; Huang, L.; Miller-Fleming, T.W.; Haessler, J.; Preuss, M.H.; Chai, J.-F.; Lee, M.P.; et al. Analyses of Biomarker Traits in Diverse UK Biobank Participants Identify Associations Missed by European-Centric Analysis Strategies. *J. Hum. Genet.* **2022**, *67*, 87–93. [CrossRef]
36. Richardson, T.G.; O'Nunain, K.; Relton, C.L.; Smith, G.D. Harnessing Whole Genome Polygenic Risk Scores to Stratify Individuals Based on Cardiometabolic Risk Factors and Biomarkers at Age 10 in the Lifecourse. *Arterioscler. Thromb. Vasc. Biol.* **2022**, *42*, ATVBAHA121316650. [CrossRef]
37. Christakos, S.; Dhawan, P.; Verstuyf, A.; Verlinden, L.; Carmeliet, G. Vitamin D: Metabolism, Molecular Mechanism of Action, and Pleiotropic Effects. *Physiol. Rev.* **2016**, *96*, 365–408. [CrossRef]
38. Aiba, I.; Yamasaki, T.; Shinki, T.; Izumi, S.; Yamamoto, K.; Yamada, S.; Terato, H.; Ide, H.; Ohyama, Y. Characterization of Rat and Human CYP2J Enzymes as Vitamin D 25-Hydroxylases. *Steroids* **2006**, *71*, 849–856. [CrossRef]
39. Zhou, A.; Selvanayagam, J.B.; Hyppönen, E. Non-Linear Mendelian Randomization Analyses Support a Role for Vitamin D Deficiency in Cardiovascular Disease Risk. *Eur. Heart J.* **2022**, *43*, 1731–1739. [CrossRef]
40. Collins, C.S.; Gould, S.J. Identification of a Common PEX1 Mutation in Zellweger Syndrome. *Hum. Mutat.* **1999**, *14*, 45–53. [CrossRef]
41. Rush, E.T.; Goodwin, J.L.; Braverman, N.E.; Rizzo, W.B. Low Bone Mineral Density Is a Common Feature of Zellweger Spectrum Disorders. *Mol. Genet. Metab.* **2016**, *117*, 33–37. [CrossRef]
42. Guerrin, M.; Ishigami, A.; Méchin, M.-C.; Nachat, R.; Valmary, S.; Sebbag, M.; Simon, M.; Senshu, T.; Serre, G. CDNA Cloning, Gene Organization and Expression Analysis of Human Peptidylarginine Deiminase Type I. *Biochem. J.* **2003**, *370*, 167–174. [CrossRef]
43. Méchin, M.C.; Enji, M.; Nachat, R.; Chavanas, S.; Charveron, M.; Ishida-Yamamoto, A.; Serre, G.; Takahara, H.; Simon, M. The Peptidylarginine Deiminases Expressed in Human Epidermis Differ in Their Substrate Specificities and Subcellular Locations. *Cell. Mol. Life Sci. CMLS* **2005**, *62*, 1984–1995. [CrossRef]
44. Jacobs, F.M.J.; van der Heide, L.P.; Wijchers, P.J.E.C.; Burbach, J.P.H.; Hoekman, M.F.M.; Smidt, M.P. FoxO6, a Novel Member of the FoxO Class of Transcription Factors with Distinct Shuttling Dynamics *. *J. Biol. Chem.* **2003**, *278*, 35959–35967. [CrossRef]
45. Zemva, J.; Schilbach, K.; Stöhr, O.; Moll, L.; Franko, A.; Krone, W.; Wiesner, R.J.; Schubert, M. Central FoxO3a and FoxO6 Expression Is Down-Regulated in Obesity Induced Diabetes but Not in Aging. *Exp. Clin. Endocrinol. Diabetes Off. J. Ger. Soc. Endocrinol. Ger. Diabetes Assoc.* **2012**, *120*, 340–350. [CrossRef]
46. Vincent, J.B.; Skaug, J.; Scherer, S.W. The Human Homologue of Flamingo, EGFL2, Encodes a Brain-Expressed Large Cadherin-like Protein with Epidermal Growth Factor-like Domains, and Maps to Chromosome 1p13.3-P21.1. *DNA Res.* **2000**, *7*, 233–235. [CrossRef]
47. Ichthyosis Vulgaris Disease: Malacards—Research Articles, Drugs, Genes, Clinical Trials. Available online: https://www.malacards.org/card/ichthyosis_vulgaris (accessed on 27 May 2022).
48. Peeling Skin Syndrome 6 Disease: Malacards—Research Articles, Drugs, Genes, Clinical Trials. Available online: https://www.malacards.org/card/peeling_skin_syndrome_6 (accessed on 27 May 2022).
49. Kukita, T.; Wada, N.; Kukita, A.; Kakimoto, T.; Sandra, F.; Toh, K.; Nagata, K.; Iijima, T.; Horiuchi, M.; Matsusaki, H.; et al. RANKL-Induced DC-STAMP Is Essential for Osteoclastogenesis. *J. Exp. Med.* **2004**, *200*, 941–946. [CrossRef]
50. Van der Valk, R.J.P.; Kreiner-Møller, E.; Kooijman, M.N.; Guxens, M.; Stergiakouli, E.; Sääf, A.; Bradfield, J.P.; Geller, F.; Hayes, M.G.; Cousminer, D.L.; et al. A Novel Common Variant in DCST2 Is Associated with Length in Early Life and Height in Adulthood. *Hum. Mol. Genet.* **2015**, *24*, 1155–1168. [CrossRef]
51. Borges, C.R.; Jarvis, J.W.; Oran, P.E.; Nelson, R.W. Population Studies of Vitamin D Binding Protein Microheterogeneity by Mass Spectrometry Lead to Characterization of Its Genotype-Dependent O-Glycosylation Patterns. *J. Proteome Res.* **2008**, *7*, 4143–4153. [CrossRef]
52. Hezroni, H.; Koppstein, D.; Schwartz, M.G.; Avrutin, A.; Bartel, D.P.; Ulitsky, I. Principles of Long Noncoding RNA Evolution Derived from Direct Comparison of Transcriptomes in 17 Species. *Cell Rep.* **2015**, *11*, 1110–1122. [CrossRef]
53. Ghatnatti, V.; Vastrad, B.; Patil, S.; Vastrad, C.; Kotturshetti, I. Identification of Potential and Novel Target Genes in Pituitary Prolactinoma by Bioinformatics Analysis. *AIMS Neurosci.* **2021**, *8*, 254–283. [CrossRef] [PubMed]
54. PubChem CPS1—Carbamoyl-Phosphate Synthase 1 (Human). Available online: https://pubchem.ncbi.nlm.nih.gov/gene/CPS1/human (accessed on 26 May 2022).

55. Meech, R.; Hu, D.G.; McKinnon, R.A.; Mubarokah, S.N.; Haines, A.Z.; Nair, P.C.; Rowland, A.; Mackenzie, P.I. The UDP-Glycosyltransferase (UGT) Superfamily: New Members, New Functions, and Novel Paradigms. *Physiol. Rev.* **2019**, *99*, 1153–1222. [CrossRef] [PubMed]
56. Wang, Z.; Wong, T.; Hashizume, T.; Dickmann, L.Z.; Scian, M.; Koszewski, N.J.; Goff, J.P.; Horst, R.L.; Chaudhry, A.S.; Schuetz, E.G.; et al. Human UGT1A4 and UGT1A3 Conjugate 25-Hydroxyvitamin D3: Metabolite Structure, Kinetics, Inducibility, and Interindividual Variability. *Endocrinology* **2014**, *155*, 2052–2063. [CrossRef]
57. Biederer, T. Bioinformatic Characterization of the SynCAM Family of Immunoglobulin-like Domain-Containing Adhesion Molecules. *Genomics* **2006**, *87*, 139–150. [CrossRef] [PubMed]
58. Yan, X.; Wang, Z.; Schmidt, V.; Gauert, A.; Willnow, T.E.; Heinig, M.; Poy, M.N. Cadm2 Regulates Body Weight and Energy Homeostasis in Mice. *Mol. Metab.* **2018**, *8*, 180–188. [CrossRef] [PubMed]
59. Chun, R.F. New Perspectives on the Vitamin D Binding Protein. *Cell Biochem. Funct.* **2012**, *30*, 445–456. [CrossRef] [PubMed]
60. Liang, Y.; Niederstrasser, H.; Edwards, M.; Jackson, C.E.; Cooper, J.A. Distinct Roles for CARMIL Isoforms in Cell Migration. *Mol. Biol. Cell* **2009**, *20*, 5290–5305. [CrossRef] [PubMed]
61. Canu, N.; Possenti, R.; Ricco, A.S.; Rocchi, M.; Levi, A. Cloning, Structural Organization Analysis, and Chromosomal Assignment of the Human Gene for the Neurosecretory Protein VGF. *Genomics* **1997**, *45*, 443–446. [CrossRef]
62. Benchoula, K.; Parhar, I.S.; Hwa, W.E. The Molecular Mechanism of Vgf in Appetite, Lipids, and Insulin Regulation. *Pharmacol. Res.* **2021**, *172*, 105855. [CrossRef]
63. Hahm, S.; Fekete, C.; Mizuno, T.M.; Windsor, J.; Yan, H.; Boozer, C.N.; Lee, C.; Elmquist, J.K.; Lechan, R.M.; Mobbs, C.V.; et al. VGF Is Required for Obesity Induced by Diet, Gold Thioglucose Treatment, and Agouti and Is Differentially Regulated in Pro-Opiomelanocortin- and Neuropeptide Y-Containing Arcuate Neurons in Response to Fasting. *J. Neurosci.* **2002**, *22*, 6929–6938. [CrossRef]
64. Li, R.; Zhang, L.; Qin, Z.; Wei, Y.; Deng, Z.; Zhu, C.; Tang, J.; Ma, L. High LINC00536 Expression Promotes Tumor Progression and Poor Prognosis in Bladder Cancer. *Exp. Cell Res.* **2019**, *378*, 32–40. [CrossRef] [PubMed]
65. Fossati, M.; Assendorp, N.; Gemin, O.; Colasse, S.; Dingli, F.; Arras, G.; Loew, D.; Charrier, C. Trans-Synaptic Signaling through the Glutamate Receptor Delta-1 Mediates Inhibitory Synapse Formation in Cortical Pyramidal Neurons. *Neuron* **2019**, *104*, 1081–1094.e7. [CrossRef] [PubMed]
66. Cheng, J.B.; Levine, M.A.; Bell, N.H.; Mangelsdorf, D.J.; Russell, D.W. Genetic Evidence That the Human CYP2R1 Enzyme Is a Key Vitamin D 25-Hydroxylase. *Proc. Natl. Acad. Sci. USA* **2004**, *101*, 7711–7715. [CrossRef]
67. Liu, J.; Zhou, Y.; Liu, S.; Song, X.; Yang, X.; Fan, Y.; Chen, W.; Akdemir, Z.C.; Yan, Z.; Zuo, Y.; et al. The Coexistence of Copy Number Variations (CNVs) and Single Nucleotide Polymorphisms (SNPs) at a Locus Can Result in Distorted Calculations of the Significance in Associating SNPs to Disease. *Hum. Genet.* **2018**, *137*, 553–567. [CrossRef] [PubMed]
68. Ng, S.-Y.; Bettany-Saltikov, J.; Cheung, I.Y.K.; Chan, K.K.Y. The Role of Vitamin D in the Pathogenesis of Adolescent Idiopathic Scoliosis. *Asian Spine J.* **2018**, *12*, 1127–1145. [CrossRef]
69. Donato, R.; Cannon, B.R.; Sorci, G.; Riuzzi, F.; Hsu, K.; Weber, D.J.; Geczy, C.L. Functions of S100 Proteins. *Curr. Mol. Med.* **2013**, *13*, 24–57. [CrossRef] [PubMed]
70. Cases, S.; Stone, S.J.; Zhou, P.; Yen, E.; Tow, B.; Lardizabal, K.D.; Voelker, T.; Farese, R.V. Cloning of DGAT2, a Second Mammalian Diacylglycerol Acyltransferase, and Related Family Members. *J. Biol. Chem.* **2001**, *276*, 38870–38876. [CrossRef]
71. Smith, S.J.; Cases, S.; Jensen, D.R.; Chen, H.C.; Sande, E.; Tow, B.; Sanan, D.A.; Raber, J.; Eckel, R.H.; Farese, R.V. Obesity Resistance and Multiple Mechanisms of Triglyceride Synthesis in Mice Lacking Dgat. *Nat. Genet.* **2000**, *25*, 87–90. [CrossRef]
72. Eckhart, L.; Schmidt, M.; Mildner, M.; Mlitz, V.; Abtin, A.; Ballaun, C.; Fischer, H.; Mrass, P.; Tschachler, E. Histidase Expression in Human Epidermal Keratinocytes: Regulation by Differentiation Status and All-Trans Retinoic Acid. *J. Dermatol. Sci.* **2008**, *50*, 209–215. [CrossRef]
73. Suchi, M.; Sano, H.; Mizuno, H.; Wada, Y. Molecular Cloning and Structural Characterization of the Human Histidase Gene (HAL). *Genomics* **1995**, *29*, 98–104. [CrossRef]
74. Welsh, M.M.; Karagas, M.R.; Applebaum, K.M.; Spencer, S.K.; Perry, A.E.; Nelson, H.H. A Role for Ultraviolet Radiation Immunosuppression in Non-Melanoma Skin Cancer as Evidenced by Gene-Environment Interactions. *Carcinogenesis* **2008**, *29*, 1950–1954. [CrossRef]
75. Landeck, L.; Jakasa, I.; Dapic, I.; Lutter, R.; Thyssen, J.P.; Skov, L.; Braun, A.; Schön, M.P.; John, S.M.; Kezic, S.; et al. The Effect of Epidermal Levels of Urocanic Acid on 25-Hydroxyvitamin D Synthesis and Inflammatory Mediators upon Narrowband UVB Irradiation. *Photodermatol. Photoimmunol. Photomed.* **2016**, *32*, 214–223. [CrossRef]
76. Simões-costa, M.S.; Azambuja, A.P.; Xavier-Neto, J. The search for non-chordate retinoic acid signaling: Lessons from chordates. *J. Exp. Zoolog. B Mol. Dev. Evol.* **2008**, *310B*, 54–72. [CrossRef]
77. Van Lith, M.; Hartigan, N.; Hatch, J.; Benham, A.M. PDILT, a Divergent Testis-Specific Protein Disulfide Isomerase with a Non-Classical SXXC Motif That Engages in Disulfide-Dependent Interactions in the Endoplasmic Reticulum. *J. Biol. Chem.* **2005**, *280*, 1376–1383. [CrossRef]
78. Kurogi, K.; Sakakibara, Y.; Suiko, M.; Liu, M.-C. Sulfation of Vitamin D3 -Related Compounds-Identification and Characterization of the Responsible Human Cytosolic Sulfotransferases. *FEBS Lett.* **2017**, *591*, 2417–2425. [CrossRef]

79. Wong, T.; Wang, Z.; Chapron, B.D.; Suzuki, M.; Claw, K.G.; Gao, C.; Foti, R.S.; Prasad, B.; Chapron, A.; Calamia, J.; et al. Polymorphic Human Sulfotransferase 2A1 Mediates the Formation of 25-Hydroxyvitamin D3-3-O-Sulfate, a Major Circulating Vitamin D Metabolite in Humans. *Drug Metab. Dispos.* **2018**, *46*, 367–379. [CrossRef]
80. Prassas, I.; Eissa, A.; Poda, G.; Diamandis, E.P. Unleashing the Therapeutic Potential of Human Kallikrein-Related Serine Proteases. *Nat. Rev. Drug Discov.* **2015**, *14*, 183–202. [CrossRef]
81. Jones, G.; Prosser, D.E.; Kaufmann, M. 25-Hydroxyvitamin D-24-Hydroxylase (CYP24A1): Its Important Role in the Degradation of Vitamin D. *Arch. Biochem. Biophys.* **2012**, *523*, 9–18. [CrossRef]
82. Shi, M.; Grabner, A.; Wolf, M. Importance of Extra-Renal CYP24A1 Expression for Maintaining Mineral Homeostasis. *J. Endocr. Soc.* **2021**, *5*, A234. [CrossRef]
83. Murase, R.; Taketomi, Y.; Miki, Y.; Nishito, Y.; Saito, M.; Fukami, K.; Yamamoto, K.; Murakami, M. Group III Phospholipase A2 Promotes Colitis and Colorectal Cancer. *Sci. Rep.* **2017**, *7*, 12261. [CrossRef]
84. Karohl, C.; Su, S.; Kumari, M.; Tangpricha, V.; Veledar, E.; Vaccarino, V.; Raggi, P. Heritability and Seasonal Variability of Vitamin D Concentrations in Male Twins. *Am. J. Clin. Nutr.* **2010**, *92*, 1393–1398. [CrossRef]
85. Engelman, C.D.; Fingerlin, T.E.; Langefeld, C.D.; Hicks, P.J.; Rich, S.S.; Wagenknecht, L.E.; Bowden, D.W.; Norris, J.M. Genetic and Environmental Determinants of 25-Hydroxyvitamin D and 1,25-Dihydroxyvitamin D Levels in Hispanic and African Americans. *J. Clin. Endocrinol. Metab.* **2008**, *93*, 3381–3388. [CrossRef] [PubMed]
86. Sutherland, J.P.; Zhou, A.; Leach, M.J.; Hyppönen, E. Differences and Determinants of Vitamin D Deficiency among UK Biobank Participants: A Cross-Ethnic and Socioeconomic Study. *Clin. Nutr.* **2021**, *40*, 3436–3447. [CrossRef] [PubMed]
87. Enlund-Cerullo, M.; Koljonen, L.; Holmlund-Suila, E.; Hauta-alus, H.; Rosendahl, J.; Valkama, S.; Helve, O.; Hytinantti, T.; Viljakainen, H.; Andersson, S.; et al. Genetic Variation of the Vitamin D Binding Protein Affects Vitamin D Status and Response to Supplementation in Infants. *J. Clin. Endocrinol. Metab.* **2019**, *104*, 5483–5498. [CrossRef] [PubMed]
88. Nissen, J.; Vogel, U.; Ravn-Haren, G.; Andersen, E.W.; Madsen, K.H.; Nexø, B.A.; Andersen, R.; Mejborn, H.; Bjerrum, P.J.; Rasmussen, L.B.; et al. Common Variants in CYP2R1 and GC Genes Are Both Determinants of Serum 25-Hydroxyvitamin D Concentrations after UVB Irradiation and after Consumption of Vitamin D_3-Fortified Bread and Milk during Winter in Denmark. *Am. J. Clin. Nutr.* **2015**, *101*, 218–227. [CrossRef] [PubMed]
89. Nissen, J.; Vogel, U.; Ravn-Haren, G.; Andersen, E.W.; Nexø, B.A.; Andersen, R.; Mejborn, H.; Madsen, K.H.; Rasmussen, L.B. Real-Life Use of Vitamin D3-Fortified Bread and Milk during a Winter Season: The Effects of CYP2R1 and GC Genes on 25-Hydroxyvitamin D Concentrations in Danish Families, the VitmaD Study. *Genes Nutr.* **2014**, *9*, 413. [CrossRef] [PubMed]
90. Frost, P. The Problem of Vitamin D Scarcity: Cultural and Genetic Solutions by Indigenous Arctic and Tropical Peoples. *Nutrients* **2022**, *14*, 4071. [CrossRef]
91. Davies, N.M.; Holmes, M.V.; Smith, G.D. Reading Mendelian Randomisation Studies: A Guide, Glossary, and Checklist for Clinicians. *BMJ* **2018**, *362*, k601. [CrossRef]
92. Burgess, S.; Davey Smith, G.; Davies, N.; Dudbridge, F.; Gill, D.; Glymour, M.; Hartwig, F.; Holmes, M.; Minelli, C.; Relton, C.; et al. Guidelines for Performing Mendelian Randomization Investigations [Version 2; Peer Review: 2 Approved]. *Wellcome Open Res.* **2020**, *4*, 186. [CrossRef]
93. Mokry, L.E.; Ross, S.; Ahmad, O.S.; Forgetta, V.; Smith, G.D.; Leong, A.; Greenwood, C.M.T.; Thanassoulis, G.; Richards, J.B. Vitamin D and Risk of Multiple Sclerosis: A Mendelian Randomization Study. *PLOS Med.* **2015**, *12*, e1001866. [CrossRef]
94. Lu, L.; Bennett, D.A.; Millwood, I.Y.; Parish, S.; McCarthy, M.I.; Mahajan, A.; Lin, X.; Bragg, F.; Guo, Y.; Holmes, M.V.; et al. Association of Vitamin D with Risk of Type 2 Diabetes: A Mendelian Randomisation Study in European and Chinese Adults. *PLOS Med.* **2018**, *15*, e1002566. [CrossRef]
95. Vimaleswaran, K.S.; Cavadino, A.; Berry, D.J.; Jorde, R.; Dieffenbach, A.K.; Lu, C.; Alves, A.C.; Heerspink, H.J.L.; Tikkanen, E.; Eriksson, J.; et al. Association of Vitamin D Status with Arterial Blood Pressure and Hypertension Risk: A Mendelian Randomisation Study. *Lancet Diabetes Endocrinol.* **2014**, *2*, 719–729. [CrossRef]
96. Morris, M.C.; Tangney, C.C. A Potential Design Flaw of Randomized Trials of Vitamin Supplements. *JAMA* **2011**, *305*, 1348–1349. [CrossRef]
97. Scragg, R. Limitations of Vitamin D Supplementation Trials: Why Observational Studies Will Continue to Help Determine the Role of Vitamin D in Health. *J. Steroid Biochem. Mol. Biol.* **2018**, *177*, 6–9. [CrossRef]
98. Sutherland, J.P.; Zhou, A.; Hyppönen, E. Vitamin D Deficiency Increases Mortality Risk in the UK Biobank: A Non-Linear Mendelian Randomization Study. *Ann. Intern. Med.* 2022, *in press*.
99. Sofianopoulou, E.; Kaptoge, S.K.; Afzal, S.; Jiang, T.; Gill, D.; Gundersen, T.E.; Bolton, T.R.; Allara, E.; Arnold, M.G.; Mason, A.M.; et al. Estimating Dose-Response Relationships for Vitamin D with Coronary Heart Disease, Stroke, and All-Cause Mortality: Observational and Mendelian Randomisation Analyses. *Lancet Diabetes Endocrinol.* **2021**, *9*, 837–846. [CrossRef]
100. Navale, S.S.; Mulugeta, A.; Zhou, A.; Llewellyn, D.J.; Hyppönen, E. Vitamin D and Brain Health: An Observational and Mendelian Randomization Study. *Am. J. Clin. Nutr.* **2022**, *116*, nqac107. [CrossRef]
101. Mokry, L.E.; Ross, S.; Forgetta, V.; Morris, J.A.; Manousaki, D.; Richards, J.B. Genetically Decreased Vitamin D and Risk of Alzheimer Disease. *Neurology* **2016**, *87*, 2567–2574. [CrossRef]
102. Wang, L.; Qiao, Y.; Zhang, H.; Zhang, Y.; Hua, J.; Jin, S.; Liu, G. Circulating Vitamin D Levels and Alzheimer's Disease: A Mendelian Randomization Study in the IGAP and UK Biobank. *J. Alzheimers Dis. JAD* **2020**, *73*, 609–618. [CrossRef]

103. Larsson, S.C.; Traylor, M.; Malik, R.; Dichgans, M.; Burgess, S.; Markus, H.S. Modifiable Pathways in Alzheimer's Disease: Mendelian Randomisation Analysis. *BMJ* **2017**, *359*, j5375. [CrossRef]
104. Bjelakovic, G.; Gluud, L.L.; Nikolova, D.; Whitfield, K.; Wetterslev, J.; Simonetti, R.G.; Bjelakovic, M.; Gluud, C. Vitamin D Supplementation for Prevention of Mortality in Adults. *Cochrane Database Syst. Rev.* **2014**, CD007470. [CrossRef]

MDPI
St. Alban-Anlage 66
4052 Basel
Switzerland
Tel. +41 61 683 77 34
Fax +41 61 302 89 18
www.mdpi.com

Nutrients Editorial Office
E-mail: nutrients@mdpi.com
www.mdpi.com/journal/nutrients

www.ingramcontent.com/pod-product-compliance
Lightning Source LLC
LaVergne TN
LVHW071614170726
843515LV00009B/2389
* 9 7 8 3 0 3 6 5 6 9 1 8 5 *